Complete Solutions Guide to Accompany

CALCULUS

Fifth Edition

Larson/Hostetler/Edwards

Dianna L. Zook

Indiana University
Purdue University at Fort Wayne, Indiana

Volume I

Chapters P–6

D. C. Heath and Company

Lexington, Massachusetts   Toronto

International Standard Book Number: 0-669-32712-3

10   9   8   7   6

TO THE STUDENT

The *Complete Solutions Guide for Calculus, Fifth Edition* is a supplement to the text by Roland E. Larson, Robert P. Hostetler, and Bruce H. Edwards. Solutions to every exercise in the text are given with all essential algebraic steps included. There are three volumes in the complete set of solutions guides. Volume I contains Chapters P–6, Volume II contains Chapters 7–12, and Volume III contains Chapters 13–16.

I have made every effort to see that the solutions are correct. However, I would appreciate hearing about any errors or other suggestions for improvement.

I would like to thank the staff at Larson Texts, Inc. for their help in the production of this guide. I would also like to thank my husband Edward L. Schlindwein for his support while I worked on this project.

Dianna L. Zook
Indiana University,
Purdue University
at Fort Wayne, Indiana 46805

CONTENTS

Chapter P **Prerequisites/The Cartesian Plane and Functions**

P.1 Real Numbers and the Real Line 1
P.2 The Cartesian Plane 8
P.3 Graphs of Equations 21
P.4 Lines in the Plane 30
P.5 Functions 41
P.6 Review of Trigonometric Functions 49
Review Exercises 62

Chapter 1 **Limits and Their Properties**

1.1 An Introduction to Limits 71
1.2 Properties of Limits 75
1.3 Techniques for Evaluating Limits 77
1.4 Continuity and One-Sided Limits 84
1.5 Infinite Limits 93
Review Exercises 98

Chapter 2 **Differentiation**

2.1 The Derivative and the Tangent Line Problem 103
2.2 Basic Differentiation Rules and Rates of Change 113
2.3 The Product and Quotient Rules and Higher-Order Derivatives 125
2.4 The Chain Rule 135
2.5 Implicit Differentiation 143
2.6 Related Rates 154
Review Exercises 167

Chapter 3 **Applications of Differentiation**

3.1 Extrema on an Interval 178
3.2 Rolle's Theorem and the Mean Value Theorem 184
3.3 Increasing and Decreasing Functions and the First Derivative Test 193
3.4 Concavity and the Second Derivative Test 207
3.5 Limits at Infinity 224
3.6 A Summary of Curve Sketching 236
3.7 Optimization Problems 252
3.8 Newton's Method 270
3.9 Differentials 283
3.10 Business and Economics Applications 290
Review Exercises 299

Chapter 4 Integration

4.1 Antiderivatives and Indefinite Integration 317
4.2 Area 325
4.3 Riemann Sums and Definite Integrals 336
4.4 The Fundamental Theorem of Calculus 343
4.5 Integration by Substitution 351
4.6 Numerical Integration 362
   Review Exercises 371

Chapter 5 Logarithmic, Exponential, and Other Transcendental Functions

5.1 The Natural Logarithmic Function and Differentiation 381
5.2 The Natural Logarithmic Function and Integration 391
5.3 Inverse Functions 396
5.4 Exponential Functions: Differentiation and Integration 407
5.5 Bases Other than *e* and Applications 419
5.6 Differential Equations: Growth and Decay 430
5.7 Inverse Trigonometric Functions and Differentiation 438
5.8 Inverse Trigonometric Functions: Integration and Completing the Square . 450
5.9 Hyperbolic Functions 457
   Review Exercises 467

Chapter 6 Applications of Integration

6.1 Area of a Region Between Two Curves 479
6.2 Volume: The Disc Method 491
6.3 Volume: The Shell Method 507
6.4 Arc Length and Surfaces of Revolution 518
6.5 Work 527
6.6 Fluid Pressure and Fluid Force 534
6.7 Moments, Centers of Mass, and Centroids 539
   Review Exercises 552

CHAPTER P
Prerequisites/The Cartesian Plane and Functions

Section P.1   Real Numbers and the Real Line

1. $0.7 = \frac{7}{10}$
   Rational

2. $-3678 = \frac{-3678}{1}$
   Rational

3. $\dfrac{3\pi}{2}$
   Irrational (since π is irrational)

4. $3\sqrt{2} - 1$
   Irrational (since $\sqrt{2}$ is irrational)

5. $4.345\overline{1451}$
   Rational

6. $\frac{22}{7}$
   Rational

7. $\sqrt[3]{64} = 4$
   Rational

8. 0.81778177
   Rational

9. $4\frac{5}{8} = \frac{45}{8}$
   Rational

10. $\left(\sqrt{2}\right)^3 = 2\sqrt{2}$
    Irrational

11. Let $x = 0.36\overline{36}$.

    $100x = 36.36\overline{36}$

    $\underline{-x = -0.36\overline{36}}$

    $99x = 36$

    $x = \frac{36}{99} = \frac{4}{11}$

12. Let $x = 0.318\overline{18}$.

    $1000x = 318.18\overline{18}$

    $\underline{-10x = -3.18\overline{18}}$

    $990x = 315$

    $x = \frac{315}{990} = \frac{7}{22}$

13. Let $x = 0.297\overline{297}$.

    $1000x = 297.297\overline{297}$

    $\underline{-x = -0.297\overline{297}}$

    $999x = 297$

    $x = \frac{297}{999} = \frac{11}{37}$

14. Let $x = 0.9900\overline{9900}$.

    $10,000x = 9900.9900\overline{9900}$

    $\underline{-x = \quad -0.9900\overline{9900}}$

    $9999x = 9900$

    $x = \frac{9900}{9999} = \frac{100}{101}$

15. Given $a < b$:

    a. $a + 2 < b + 2$;   True

    b. $5b < 5a$;   False

    c. $5 - a > 5 - b$;   True

    d. $\dfrac{1}{a} < \dfrac{1}{b}$;   False

    e. $(a - b)(b - a) > 0$;   False

    f. $a^2 < b^2$;   False

16.

Interval Notation	Set Notation	Graph
$[-2,\ 0)$	$\{x:\ -2 \leq x < 0\}$	
$(-\infty,\ -4]$	$\{x:\ x \leq -4\}$	
$\left[3,\ \frac{11}{2}\right]$	$\left\{x:\ 3 \leq x \leq \frac{11}{2}\right\}$	
$(-1,\ 7)$	$\{x:\ -1 < x < 7\}$	

17. $4x + 1 < 2x$

$\quad 4x < 2x - 1$

$\quad 2x < -1$

$\quad x < -\frac{1}{2}$

18. $2x + 7 < 3$

$\quad 2x < -4$

$\quad x < -2$

19. $2x - 1 \geq 0$

$\quad 2x \geq 1$

$\quad x \geq \frac{1}{2}$

20. $3x + 1 \geq 2x + 2$

$\quad 3x \geq 2x + 1$

$\quad x \geq 1$

21. $-4 < 2x - 3 < 4$

$\quad -1 < \quad 2x \quad < 7$

$\quad -\frac{1}{2} < \quad x \quad < \frac{7}{2}$

22. $\quad 0 \leq x + 3 < 5$

$\quad -3 \leq \quad x \quad < 2$

23. $\frac{3x}{4} > x + 1$

$\quad -\frac{x}{4} > 1$

$\quad x < -4$

24. $-1 < -\frac{x}{3} < \quad 1$

$\quad -3 < -x < \quad 3$

$\quad 3 > \quad x > -3$

$\quad -3 < \quad x < \quad 3$

25. $\dfrac{x}{2} + \dfrac{x}{3} > 5$

$3x + 2x > 30$

$5x > 30$

$x > 6$

26. $x > \dfrac{1}{x}$

$x - \dfrac{1}{x} > 0$

$\dfrac{x^2 - 1}{x} > 0$

$\dfrac{(x+1)(x-1)}{x} > 0$

Test intervals:

$(-\infty,\ -1),\ (-1,\ 0),\ (0,\ 1),\ (1,\ \infty)$

Solution: $-1 < x < 0$ or $x > 1$

27. $|x| < 1 \ \Rightarrow\ -1 < x < 1$

28. $\dfrac{x}{2} - \dfrac{x}{3} > 5$

$3x - 2x > 30$

$x > 30$

29. $\left|\dfrac{x-3}{2}\right| \geq 5$

$x - 3 \geq 10 \quad\text{or}\quad x - 3 \leq -10$

$x \geq 13 \qquad\qquad x \leq -7$

30. $\left|\dfrac{x}{2}\right| > 3 \ \Rightarrow\ x > 6$ or $x < -6$

31. $|x - a| < b$

$-b\ < x - a <\ \ b$

$a - b <\ \ x\ \ < a + b$

32. $|x + 2| < 5$

$-5 < x + 2 < 5$

$-7 <\ \ x\ \ < 3$

33. $|2x + 1| < 5$

$-5 < 2x + 1 < 5$

$-6 <\ \ 2x\ \ < 4$

$-3 <\ \ x\ \ < 2$

34. $|3x + 1| \geq 4$

$3x + 1 \geq 4 \quad\text{or}\quad 3x + 1 \leq -4$

$3x \geq 3 \qquad\qquad 3x \leq -5$

$x \geq 1 \qquad\qquad x \leq -\dfrac{5}{3}$

35. $\left|1 - \dfrac{2x}{3}\right| < 1$

$-1 < 1 - \dfrac{2x}{3} < 1$

$-2 < -\dfrac{2x}{3} < 0$

$3 > \quad x \quad > 0$

36. $|9 - 2x| < 1$

$-1 < 9 - 2x < \quad 1$

$-10 < \quad -2x \quad < -8$

$5 \quad > \quad x \quad > \quad 4$

37. $\qquad x^2 \le 3 - 2x$

$x^2 + 2x - 3 \le 0$

$(x + 3)(x - 1) \le 0$

Test intervals:

$\quad (-\infty, \ -3), \ (-3, \ 1), \ (1, \ \infty)$

Solution: $-3 \le x \le 1$

38. $\qquad x^4 - x \le 0$

$x(x^3 - 1) \le 0$

$x = 0$

$x = 1$

Test intervals:

$\quad (-\infty, \ 0), \ (0, \ 1), \ (1, \ \infty)$

Solution: $0 \le x \le 1$

39. $\qquad x^2 + x - 1 \le 5$

$x^2 + x - 6 \le 0$

$(x + 3)(x - 2) \le 0$

$x = -3$

$x = 2$

Test intervals:

$\quad (-\infty, \ -3), \ (-3, \ 2), \ (2, \ \infty)$

Solution: $-3 \le x \le 2$

40. $\qquad 2x^2 + 1 < 9x - 3$

$2x^2 - 9x + 4 < 0$

$(2x - 1)(x - 4) < 0$

$x = \tfrac{1}{2}$

$x = 4$

Test intervals:

$\quad \left(-\infty, \ \tfrac{1}{2}\right), \ \left(\tfrac{1}{2}, \ 4\right), \ (4, \ \infty)$

Solution: $\tfrac{1}{2} < x < 4$

41. $a = -1, \quad b = 3$

Directed distance from a to b: 4

Directed distance from b to a: -4

Distance between a and b: 4

42. $a = -\tfrac{5}{2}, \quad b = \tfrac{13}{4}$

Directed distance from a to b: $\tfrac{23}{4}$

Directed distance from b to a: $-\tfrac{23}{4}$

Distance between a and b: $\tfrac{23}{4}$

43. a. $a = 126, \quad b = 75$

Directed distance from a to b: -51

Directed distance from b to a: 51

Distance between a and b: 51

  b. $a = -126, \quad b = -75$

Directed distance from a to b: 51

Directed distance from b to a: -51

Distance between a and b: 51

44. a. $a = 9.34, \quad b = -5.65$

Directed distance from a to b: -14.99

Directed distance from b to a: 14.99

Distance between a and b: 14.99

  b. $a = \tfrac{16}{5}, \quad b = \tfrac{112}{75}$

Directed distance from a to b: $-\tfrac{128}{75}$

Directed distance from b to a: $\tfrac{128}{75}$

Distance between a and b: $\tfrac{128}{75}$

45. $a = -1, \quad b = 3$

Midpoint: $\dfrac{-1+3}{2} = 1$

46. $a = -5, \quad b = -\dfrac{3}{2}$

Midpoint: $\dfrac{-5+(-3/2)}{2} = -\dfrac{13}{4}$

47. a. $[7, \ 21]$

Midpoint: 14

b. $[8.6, \ 11.4]$

Midpoint: 10

48. a. $[-6.85, \ 9.35]$

Midpoint: 1.25

b. $[-4.6, \ -1.3]$

Midpoint: -2.95

49. $a = -2, \quad b = 2$

Midpoint: 0

Distance between midpoint and each endpoint: 2

$|x - 0| \leq 2$

$|x| \leq 2$

50. $a = -3, \quad b = 3$

Midpoint: 0

Distance between midpoint and each endpoint: 3

$|x - 0| \geq 3$

$|x| \geq 3$

51. $a = 0, \quad b = 4$

Midpoint: 2

Distance between midpoint and each endpoint: 2

$|x - 2| > 2$

52. $a = 20, \quad b = 24$

Midpoint: 22

Distance between midpoint and each endpoint: 2

$|x - 22| \geq 2$

53. a. All numbers that are at most 10 units from 12

$|x - 12| \leq 10$

b. All numbers that are at least 10 units from 12

$|x - 12| \geq 10$

54. a. y is at most 2 units from a: $\ |y - a| \leq 2$

b. y is less than δ units from c: $\ |y - c| < \delta$

55. $A = P + Prt$

$P = 1000, \quad A > 1250, \quad t = 2$

$1000 + 1000r(2) > 1250$

$2000r > 250$

$r > 0.1250$

$r > 12.5\%$

56. $R = 115.95x, \quad C = 95x + 750, \quad R > C$

$115.95x > 95x + 750$

$20.95x > 750$

$x > 35.7995$

$x \geq 36$ units

57. $C = 0.32m + 2300, \quad C < 10,000$

$0.32m + 2300 < 10,000$

$0.32m < 7700$

$m < 24,062.5$ miles

58. $\left| \dfrac{h - 68.5}{2.7} \right| \leq 1$

$-1 \ \leq \ \dfrac{h - 68.5}{2.7} \ \leq \ 1$

$-2.7 \ \leq \ h - 68.5 \ \leq \ 2.7$

$65.8 \text{ in.} \ \leq \quad h \quad \leq 71.2 \text{ in.}$

59. $\left| \dfrac{x - 50}{5} \right| \geq 1.645$

$\dfrac{x - 50}{5} \leq -1.645 \quad \text{or} \quad \dfrac{x - 50}{5} \geq 1.645$

$x - 50 \leq -8.225 \qquad\qquad x - 50 \geq 8.225$

$x \leq 41.775 \qquad\qquad\quad x \geq 58.225$

$x \leq 41 \qquad\qquad\qquad\quad x \geq 59$

60. $|p - 2,250,000| < 125,000$

$-125,000 < p - 2,250,000 < 125,000$

$2,125,000 < \quad p \quad < 2,375,000$

High = 2,375,000 barrels

Low = 2,125,000 barrels

61. a. $\pi \approx 3.1415926535$

$\frac{355}{113} = 3.141592920$

$\frac{355}{113} > \pi$

b. $\pi \approx 3.1415926535$

$\frac{22}{7} \approx 3.142857143$

$\frac{22}{7} > \pi$

62. a. $\frac{224}{151} \approx 1.483443709$

$\frac{144}{97} \approx 1.484536082$

$\frac{144}{97} > \frac{224}{151}$

b. $\frac{73}{81} \approx 0.901234568$

$\frac{6427}{7132} \approx 0.901149748$

$\frac{73}{81} > \frac{6427}{7132}$

63. Speed of light: 2.998×10^8 meters per second

Distance traveled in one year = rate $\times$ time

$$d = (2.998 \times 10^8) \times (365 \times \underset{\text{days}}{24} \times \underset{\text{hours}}{60} \times \underset{\text{minutes}}{60})$$
$$\qquad\qquad\qquad\quad \text{days} \;\times\; \text{hours} \times \text{minutes} \times \text{seconds}$$

$$= (2.998 \times 10^8) \times (3.1536 \times 10^7)$$

$$\approx\; 9.45 \times 10^{15}$$

This is best estimated by (b).

64. The significant digits of a number are the digits of the number beginning with the first nonzero digit to the left of the decimal point (or the first digit to the right of the decimal point if there isn't a nonzero digit to the left of the decimal point) and ending with the last digit to the right. The following examples all have three significant digits.

100, 307, 0.123, 0.012, 0.001, 1.23, 12.3, 0.120, 0.300

65. If $a \geq 0$ and $b \geq 0$, then $|ab| = ab = |a|\,|b|$.

If $a < 0$ and $b < 0$, then $|ab| = ab = (-a)(-b) = |a|\,|b|$.

If $a \geq 0$ and $b < 0$, then $|ab| = -ab = a(-b) = |a|\,|b|$.

If $a < 0$ and $b \geq 0$, then $|ab| = -ab = (-a)b = |a|\,|b|$.

66. $|a - b| = |(-1)(b - a)| = |-1|\,|b - a| = (1)|b - a| = |b - a|$

67. $\left|\dfrac{a}{b}\right| = \left|a\left(\dfrac{1}{b}\right)\right|$

$\qquad = |a|\left|\dfrac{1}{b}\right| = |a| \cdot \dfrac{1}{|b|} = \dfrac{|a|}{|b|}, \quad b \neq 0$

68. If $a \geq 0$, then $|a| = a = \sqrt{a^2}$.

If $a < 0$, then

$\qquad |a| = -a = \sqrt{(-a)^2} = \sqrt{a^2}$.

69. $n = 1, \quad |a| = |a|$

$n = 2, \quad |a^2| = |a \cdot a| = |a|\,|a| = |a|^2$

$n = 3, \quad |a^3| = |a^2 \cdot a| = |a^2|\,|a| = |a|^2|a| = |a|^3$

$\qquad\vdots$

$|a^n| = |a^{n-1}a| = |a^{n-1}|\,|a| = |a|^{n-1}|a| = |a|^n$

70. If $a \geq 0$, then $a = |a|$. Thus, $-|a| \leq a \leq |a|$.

If $a < 0$, then $a = -|a|$. Thus, $-|a| \leq a \leq |a|$.

71. $|a| \leq k \Leftrightarrow \sqrt{a^2} \leq k \Leftrightarrow a^2 \leq k^2 \Leftrightarrow a^2 - k^2 \leq 0 \Leftrightarrow (a+k)(a-k) \leq 0 \Leftrightarrow -k \leq a \leq k, \quad k > 0$

72. $k \leq |a| \Leftrightarrow k \leq \sqrt{a^2} \Leftrightarrow k^2 \leq a^2 \Leftrightarrow 0 \leq a^2 - k^2 \Leftrightarrow 0 \leq (a+k)(a-k) \Leftrightarrow k \leq a$ or $a \leq -k, \quad k > 0$

73. False; 2 is a nonzero integer and the reciprocal of 2 is $\frac{1}{2}$.

74. True; if x $(x \neq 0)$ is rational, then $x = p/q$ where p and q are nonzero integers. The reciprocal of x is $1/x = q/p$ which is also the ratio of two integers.

75. True

76. False; $|0| = 0$ which is not positive.

77. True; if $x < 0$, then $|x| = -x = \sqrt{x^2}$.

78. True; since a and b are **distinct**, $a \neq b$ and one of the numbers must be larger than the other one.

79.
$$\left.\begin{array}{l} |7 - 12| = |-5| = 5 \\ |7| - |12| = 7 - 12 = -5 \end{array}\right\} \quad |7 - 12| > |7| - |12|$$

$$\left.\begin{array}{l} |12 - 7| = |5| = 5 \\ |12| - |7| = 12 - 7 = 5 \end{array}\right\} \quad |12 - 7| = |12| - |7|$$

We know that $|a|\,|b| \geq ab$. Thus, $-2|a|\,|b| \leq -2ab$. Since $a^2 = |a|^2$ and $b^2 = |b|^2$, we have

$$|a|^2 + |b|^2 - 2|a|\,|b| \leq a^2 + b^2 - 2ab$$

$$0 \leq (|a| - |b|)^2 \leq (a - b)^2$$

$$\sqrt{(|a| - |b|)^2} \leq \sqrt{(a - b)^2}$$

$$\big||a| - |b|\big| \leq |a - b|.$$

Since $|a| - |b| \leq \big||a| - |b|\big|$, we have $|a| - |b| \leq |a - b|$. Thus, $|a - b| \geq |a| - |b|$.

80.
$$\frac{1}{2}(a + b + |a - b|) = \frac{1}{2}(a + b) + \frac{1}{2}|a - b|$$

$$= \frac{a + b}{2} + \frac{1}{2}|a - b|$$

$$= \text{Midpoint} + \tfrac{1}{2} \text{ the distance between } a \text{ and } b$$

$$= \max(a,\ b)$$

$$\min(a,\ b) = \text{Midpoint} - \tfrac{1}{2} \text{ the distance between } a \text{ and } b$$

$$= \frac{a + b}{2} - \frac{1}{2}|a - b|$$

$$= \frac{1}{2}(a + b - |a - b|)$$

Section P.2   The Cartesian Plane

1. $d = \sqrt{(4-2)^2 + (5-1)^2}$

$= \sqrt{4 + 16} = \sqrt{20} = 2\sqrt{5}$

Midpoint: $\left(\dfrac{4+2}{2}, \dfrac{5+1}{2}\right) = (3, \ 3)$

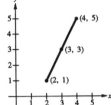

2. $d = \sqrt{(3+3)^2 + (-2-2)^2}$

$= \sqrt{36 + 16} = \sqrt{52} = 2\sqrt{13}$

Midpoint: $\left(\dfrac{-3+3}{2}, \dfrac{2+(-2)}{2}\right) = (0, \ 0)$

3. $d = \sqrt{\left(\dfrac{1}{2} + \dfrac{3}{2}\right)^2 + (1+5)^2}$

$= \sqrt{4 + 36} = \sqrt{40} = 2\sqrt{10}$

Midpoint:

$\left(\dfrac{(-3/2) + (1/2)}{2}, \dfrac{-5+1}{2}\right) = \left(-\dfrac{1}{2}, \ -2\right)$

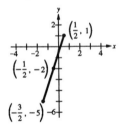

4. $d = \sqrt{\left(\dfrac{5}{6} - \dfrac{4}{6}\right)^2 + \left(\dfrac{3}{3} + \dfrac{1}{3}\right)^2}$

$= \sqrt{\dfrac{1}{36} + \dfrac{64}{36}} = \dfrac{\sqrt{65}}{6}$

Midpoint:

$\left(\dfrac{(2/3) + (5/6)}{2}, \dfrac{(-1/3) + 1}{2}\right) = \left(\dfrac{3}{4}, \ \dfrac{1}{3}\right)$

5. $d = \sqrt{(-1-1)^2 + (1-\sqrt{3})^2}$

$= \sqrt{4 + 1 - 2\sqrt{3} + 3} = \sqrt{8 - 2\sqrt{3}}$

Midpoint: $\left(\dfrac{-1+1}{2}, \dfrac{1+\sqrt{3}}{2}\right) = \left(0, \ \dfrac{1+\sqrt{3}}{2}\right)$

6. $d = \sqrt{(-2+0)^2 + (0-\sqrt{2})^2}$

$= \sqrt{4 + 2} = \sqrt{6}$

Midpoint: $\left(\dfrac{-2+0}{2}, \dfrac{0+\sqrt{2}}{2}\right) = \left(-1, \ \dfrac{\sqrt{2}}{2}\right)$

7. $d_1 = \sqrt{9 + 36} = \sqrt{45}$

$d_2 = \sqrt{4 + 1} = \sqrt{5}$

$d_3 = \sqrt{25 + 25} = \sqrt{50}$

$(d_1)^2 + (d_2)^2 = (d_3)^2$

Right triangle

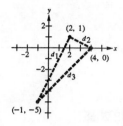

8. $d_1 = \sqrt{9 + 49} = \sqrt{58}$

$d_2 = \sqrt{25 + 4} = \sqrt{29}$

$d_3 = \sqrt{4 + 25} = \sqrt{29}$

$d_2 = d_3$

Isosceles triangle

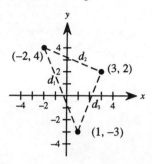

9. $d_1 = d_2 = d_3 = d_4 = \sqrt{5}$

Rhombus

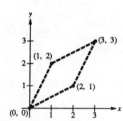

10. $d_1 = \sqrt{9 + 36} = \sqrt{45} = d_3$

$d_2 = \sqrt{1 + 9} = \sqrt{10} = d_4$

Parallelogram

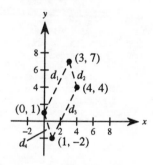

11. Let $x = 0$ correspond to 1950.

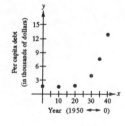

Given point	*Corresponding point*
(1950, $1688)	(0, 1.688)
(1960, $1572)	(10, 1.572)
(1970, $1807)	(20, 1.807)
(1980, $3981)	(30, 3.981)
(1985, $7614)	(35, 7.614)
(1990, $12,848)	(40, 12.848)

12. Let $x = 0$ correspond to 1950.

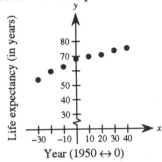

Life expectancy (in years)

Year (1950 ↔ 0)

Given point	*Corresponding point*
(1920, 54.1)	(−30, 54.1)
(1930, 59.7)	(−20, 59.7)
(1940, 62.9)	(−10, 62.9)
(1950, 68.2)	(0, 68.2)
(1960, 69.7)	(10, 69.7)
(1970, 70.8)	(20, 70.8)
(1980, 73.7)	(30, 73.7)
(1990, 75.2)	(40, 75.2)

13. $d_1 = \sqrt{4 + 16} = \sqrt{20} = 2\sqrt{5}$

$d_2 = \sqrt{1 + 4} = \sqrt{5}$

$d_3 = \sqrt{9 + 36} = 3\sqrt{5}$

$d_1 + d_2 = d_3$

Collinear

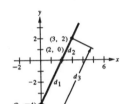

14. $d_1 = \sqrt{49 + 100} = \sqrt{149} \approx 12.2066$

$d_2 = \sqrt{25 + 49} = \sqrt{74} \approx 8.6023$

$d_3 = \sqrt{144 + 289} = \sqrt{433} \approx 20.8087$

$d_1 + d_2 \neq d_3$

Not collinear

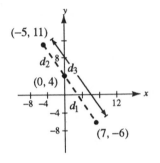

15. $d_1 = \sqrt{1 + 1} = \sqrt{2}$

$d_2 = \sqrt{9 + 4} = \sqrt{13}$

$d_3 = \sqrt{16 + 9} = 5$

$d_1 + d_2 \neq d_3$

Not collinear

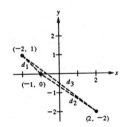

16. $d_1 = \sqrt{16 + 4} = \sqrt{20} = 2\sqrt{5}$

$d_2 = \sqrt{4 + 4} = \sqrt{8} = 2\sqrt{2}$

$d_3 = \sqrt{36 + 16} = \sqrt{52} = 2\sqrt{13}$

$d_1 + d_2 \neq d_3$

Not collinear

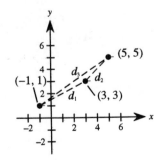

17. $5 = \sqrt{(x-0)^2 + (-4-0)^2}$

$\quad 5 = \sqrt{x^2 + 16}$

$\quad 25 = x^2 + 16$

$\quad 9 = x^2$

$\quad x = \pm 3$

18. $5 = \sqrt{(x-2)^2 + (2+1)^2}$

$\quad 5 = \sqrt{(x-2)^2 + 9}$

$\quad 25 = (x-2)^2 + 9$

$\quad 16 = (x-2)^2$

$\quad \pm 4 = x - 2$

$\quad x = 2 \pm 4 = -2, \ 6$

19. $8 = \sqrt{(3-0)^2 + (y-0)^2}$

$\quad 8 = \sqrt{9 + y^2}$

$\quad 64 = 9 + y^2$

$\quad 55 = y^2$

$\quad y = \pm\sqrt{55}$

20. $8 = \sqrt{(5-5)^2 + (y-1)^2}$

$\quad 8 = \sqrt{(y-1)^2}$

$\quad 8 = |y-1|$

$\quad y - 1 = 8 \quad$ or $\quad y - 1 = -8$

$\quad y = 9 \qquad\qquad y = -7$

21. $\sqrt{(x+2)^2 + (y-3)^2} = \sqrt{(x-4)^2 + (y+1)^2}$

$x^2 + 4x + 4 + y^2 - 6y + 9 = x^2 - 8x + 16 + y^2 + 2y + 1$

$\qquad 12x - 8y - 4 = 0$

$\qquad\quad 3x - 2y - 1 = 0$

22. $\sqrt{(x+7)^2 + (y+1)^2} = \sqrt{(x-3)^2 + \left(y - \dfrac{5}{2}\right)^2}$

$x^2 + 14x + 49 + y^2 + 2y + 1 = x^2 - 6x + 9 + y^2 - 5y + \dfrac{25}{4}$

$\qquad\quad 20x + 7y + 50 - \dfrac{61}{4} = 0$

$\qquad\quad 80x + 28y + 200 - 61 = 0$

$\qquad\qquad 80x + 28y + 139 = 0$

23. The midpoint of the given line segment is $\left(\dfrac{x_1 + x_2}{2}, \ \dfrac{y_1 + y_2}{2}\right)$.

The midpoint between $(x_1, \ y_1)$ and $\left(\dfrac{x_1 + x_2}{2}, \ \dfrac{y_1 + y_2}{2}\right)$ is

$\left(\dfrac{x_1 + (x_1 + x_2)/2}{2}, \ \dfrac{y_1 + (y_1 + y_2)/2}{2}\right) = \left(\dfrac{3x_1 + x_2}{4}, \ \dfrac{3y_1 + y_2}{4}\right)$.

The midpoint between $\left(\dfrac{x_1 + x_2}{2}, \ \dfrac{y_1 + y_2}{2}\right)$ and $(x_2, \ y_2)$ is

$\left(\dfrac{(x_1 + x_2)/2 + x_2}{2}, \ \dfrac{(y_1 + y_2)/2 + y_2}{2}\right) = \left(\dfrac{x_1 + 3x_2}{4}, \ \dfrac{y_1 + 3y_2}{4}\right)$.

Thus, the three points are

$\left(\dfrac{3x_1 + x_2}{4}, \ \dfrac{3y_1 + y_2}{4}\right), \ \left(\dfrac{x_1 + x_2}{2}, \ \dfrac{y_1 + y_2}{2}\right), \ \left(\dfrac{x_1 + 3x_2}{4}, \ \dfrac{y_1 + 3y_2}{4}\right)$.

24. a. $\left(\dfrac{3(1)+4}{4}, \dfrac{3(-2)+(-1)}{4}\right) = \left(\dfrac{7}{4}, -\dfrac{7}{4}\right)$

$\left(\dfrac{1+4}{2}, \dfrac{-2+(-1)}{2}\right) = \left(\dfrac{5}{2}, -\dfrac{3}{2}\right)$

$\left(\dfrac{1+3(4)}{4}, \dfrac{-2+3(-1)}{4}\right) = \left(\dfrac{13}{4}, -\dfrac{5}{4}\right)$

b. $\left(\dfrac{3(-2)+0}{4}, \dfrac{3(-3)+0}{4}\right) = \left(-\dfrac{3}{2}, -\dfrac{9}{4}\right)$

$\left(\dfrac{-2+0}{2}, \dfrac{-3+0}{2}\right) = \left(-1, -\dfrac{3}{2}\right)$

$\left(\dfrac{-2+3(0)}{4}, \dfrac{-3+3(0)}{4}\right) = \left(-\dfrac{1}{2}, -\dfrac{3}{4}\right)$

25. a. $x^2 + 5x = x^2 + 5x + \left(\dfrac{5}{2}\right)^2 - \left(\dfrac{5}{2}\right)^2$

$= x^2 + 5x + \dfrac{25}{4} - \dfrac{25}{4}$

$= \left(x + \dfrac{5}{2}\right)^2 - \dfrac{25}{4}$

b. $x^2 + 8x + 7 = x^2 + 8x + \left(\dfrac{8}{2}\right)^2 - \left(\dfrac{8}{2}\right)^2 + 7$

$= x^2 + 8x + 16 - 16 + 7$

$= (x+4)^2 - 9$

26. a. $4x^2 - 4x - 39 = 4\left[x^2 - x - \dfrac{39}{4}\right]$

$= 4\left[x^2 - x + \dfrac{1}{4} - \dfrac{1}{4} - \dfrac{39}{4}\right]$

$= 4\left[\left(x - \dfrac{1}{2}\right)^2 - 10\right]$

$= 4\left(x - \dfrac{1}{2}\right)^2 - 40$

b. $5x^2 + x = 5\left[x^2 + \dfrac{1}{5}x\right]$

$= 5\left[x^2 + \dfrac{1}{5}x + \dfrac{1}{100} - \dfrac{1}{100}\right]$

$= 5\left[\left(x + \dfrac{1}{10}\right)^2 - \dfrac{1}{100}\right]$

$= 5\left(x + \dfrac{1}{10}\right)^2 - \dfrac{1}{20}$

27. Center: $(0, 0)$
Radius: 1
Matches graph (c)

28. Center: $(1, 3)$
Radius: 2
Matches graph (b)

29. Center: $(1, 0)$
Radius: 0
Matches graph (a)

30. Center: $\left(-\dfrac{1}{2}, \dfrac{3}{4}\right)$
Radius: $\dfrac{1}{2}$
Matches graph (d)

31. $(x-0)^2 + (y-0)^2 = (3)^2$

$x^2 + y^2 - 9 = 0$

32. $(x-0)^2 + (y-0)^2 = (5)^2$

$x^2 + y^2 - 25 = 0$

33. $(x-2)^2 + (y+1)^2 = (4)^2$

$x^2 + y^2 - 4x + 2y - 11 = 0$

34. $(x+4)^2 + (y-3)^2 = \left(\dfrac{5}{8}\right)^2$

$64(x+4)^2 + 64(y-3)^2 = 25$

$64x^2 + 64y^2 + 512x - 384y + 1575 = 0$

35. Radius $= \sqrt{(-1-0)^2 + (2-0)^2} = \sqrt{5}$

$(x+1)^2 + (y-2)^2 = 5$

$x^2 + 2x + 1 + y^2 - 4y + 4 = 5$

$x^2 + y^2 + 2x - 4y = 0$

36. Radius $= \sqrt{[3-(-1)]^2 + (-2-1)^2} = 5$

$(x-3)^2 + (y+2)^2 = 25$

$x^2 - 6x + 9 + y^2 + 4y + 4 = 25$

$x^2 + y^2 - 6x + 4y - 12 = 0$

37. Center = Midpoint = $(3, \ 2)$

Radius = $\sqrt{10}$

$$(x - 3)^2 + (y - 2)^2 = (\sqrt{10})^2$$

$$x^2 - 6x + 9 + y^2 - 4y + 4 = 10$$

$$x^2 + y^2 - 6x - 4y + 3 = 0$$

38. Center = Midpoint = $(0, \ 0)$

Radius = $\sqrt{2}$

$$(x - 0)^2 + (y - 0)^2 = (\sqrt{2})^2$$

$$x^2 + y^2 - 2 = 0$$

39. $(0, 0), (0, 8), (6, 0)$

$$(x - h)^2 + (y - k)^2 = r^2$$

$(0 - h)^2 + (0 - k)^2 = r^2, \quad h^2 + k^2 = r^2$	Equation 1
$(0 - h)^2 + (8 - k)^2 = r^2, \quad h^2 + (8 - k)^2 = r^2$	Equation 2
$(6 - h)^2 + (0 - k)^2 = r^2, \quad (6 - h)^2 + k^2 = r^2$	Equation 3

$(8 - k)^2 - k^2 = 0, \quad 64 - 16k = 0, \quad k = 4$	Equation 2 − Equation 1
$(6 - h)^2 - h^2 = 0, \quad 36 - 12h = 0, \quad h = 3$	Equation 3 − Equation 1
$(3)^2 + (4)^2 = r^2, \quad 25 = r^2$	Equation 1

$$(x - 3)^2 + (y - 4)^2 = 25$$

$$x^2 - 6x + 9 + y^2 - 8y + 16 = 25$$

$$x^2 + y^2 - 6x - 8y = 0$$

Alternate Solution:

Note that the given points form the vertices of a right triangle. Thus, the hypotenuse of the triangle must lie on a diameter of the circle and the center must be the midpoint of the line segment joining $(0, 8)$ and $(6, 0)$.

$$(h, \ k) = (3, \ 4)$$

$$r = \sqrt{(3 - 6)^2 + (4 - 0)^2} = \sqrt{9 + 16} = 5$$

$$(x - 3)^2 + (y - 4)^2 = 25$$

$$x^2 - 6x + 9 + y^2 - 8y + 16 = 25$$

$$x^2 + y^2 - 6x - 8y = 0$$

40. $(1, \ -1), (2, \ -2), (0, \ -2)$

$$(x - h)^2 + (y - k)^2 = r^2$$

$(1 - h)^2 + (-1 - k)^2 = r^2$	Equation 1
$(2 - h)^2 + (-2 - k)^2 = r^2$	Equation 2
$h^2 + (-2 - k)^2 = r^2$	Equation 3

$(2 - h)^2 - h^2 = 0, \quad 4 - 4h = 0, \quad h = 1$	Equation 2 − Equation 3
$(-1 - k)^2 = r^2$	Equation 1
$(2 - 1)^2 + (-2 - k)^2 = r^2$	Equation 2

$$1 + (-2 - k)^2 = (-1 - k)^2$$

$$1 + 4 + 4k + k^2 = 1 + 2k + k^2, \quad 2k = -4, \quad k = -2$$

$$(1 - 1)^2 + (-1 + 2)^2 = r^2, \quad 1 = r^2$$

$$(x - 1)^2 + (y + 2)^2 = 1, \quad x^2 + y^2 - 2x + 4y + 4 = 0$$

41. Place the center of the earth at the origin. Then we have

$$x^2 + y^2 = (22,000 + 4000)^2$$

$$x^2 + y^2 = 26,000^2.$$

42. Let d be the diameter of the water pipe and z be the distance between the water pipe and the corner of the wall. If you let y equal the hypotenuse of the triangle whose one vertex is located at the center of the air duct, then $y = z + d + (D/2)$. The hypotenuse of the triangle whose one vertex is located at the center of the water pipe is $z + (d/2)$. Using the Pythagorean Theorem, you can find z as follows.

$$\left(z + \frac{d}{2}\right)^2 = \left(\frac{d}{2}\right)^2 + \left(\frac{d}{2}\right)^2$$

$$\left(z + \frac{d}{2}\right)^2 = \frac{d^2}{2}$$

$$z + \frac{d}{2} = \frac{d}{\sqrt{2}}$$

$$z = \frac{d}{\sqrt{2}} - \frac{d}{2}$$

Now solve for d, using the fact that these are similar triangles.

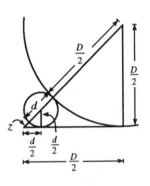

$$\frac{\dfrac{d}{2}}{z + \dfrac{d}{2}} = \frac{\dfrac{D}{2}}{y}$$

$$\frac{\dfrac{d}{2}}{\dfrac{d}{\sqrt{2}} - \dfrac{d}{2} + \dfrac{d}{2}} = \frac{\dfrac{D}{2}}{z + d + \dfrac{D}{2}}$$

$$\frac{\dfrac{d}{2}}{\dfrac{d}{\sqrt{2}}} = \frac{\dfrac{D}{2}}{\dfrac{d}{\sqrt{2}} - \dfrac{d}{2} + d + \dfrac{D}{2}}$$

$$\frac{d}{2}\left(\frac{d}{\sqrt{2}} + \frac{d}{2} + \frac{D}{2}\right) = \frac{d}{\sqrt{2}} \cdot \frac{D}{2}$$

$$d\left(\frac{1}{\sqrt{2}} + \frac{1}{2}\right) + \frac{D}{2} = \frac{D}{\sqrt{2}}$$

$$d\left(\frac{2 + \sqrt{2}}{2\sqrt{2}}\right) = \frac{D}{\sqrt{2}} - \frac{D}{2}$$

$$d\left(\frac{2 + \sqrt{2}}{2\sqrt{2}}\right) = D\left(\frac{2 - \sqrt{2}}{2\sqrt{2}}\right)$$

$$d = D\left(\frac{2 - \sqrt{2}}{2 + \sqrt{2}}\right)$$

The diameter of the largest water pipe which can be run in the right angle corner behind the air duct is

$$D\left(\frac{2 - \sqrt{2}}{2 + \sqrt{2}}\right).$$

43.
$$x^2 + y^2 - 2x + 6y + 6 = 0$$
$$(x^2 - 2x + 1) + (y^2 + 6y + 9) = -6 + 1 + 9$$
$$(x - 1)^2 + (y + 3)^2 = 4$$

Center: $(1, -3)$
Radius: 2

44.
$$x^2 + y^2 - 2x + 6y - 15 = 0$$
$$(x^2 - 2x + 1) + (y^2 + 6y + 9) = 15 + 1 + 9$$
$$(x - 1)^2 + (y + 3)^2 = 25$$

Center: $(1, -3)$
Radius: 5

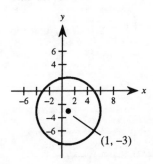

45.
$$x^2 + y^2 - 2x + 6y + 10 = 0$$
$$(x^2 - 2x + 1) + (y^2 + 6y + 9) = -10 + 1 + 9$$
$$(x - 1)^2 + (y + 3)^2 = 0$$

Only a point $(1, -3)$

46.
$$3x^2 + 3y^2 - 6y - 1 = 0$$
$$3x^2 + 3(y^2 - 2y + 1) = 1 + 3$$
$$x^2 + (y - 1)^2 = \frac{4}{3}$$

Center: $(0, 1)$

Radius: $\dfrac{2\sqrt{3}}{3}$

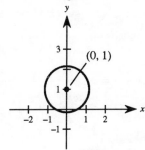

47.
$$2x^2 + 2y^2 - 2x - 2y - 3 = 0$$
$$2\left(x^2 - x + \frac{1}{4}\right) + 2\left(y^2 - y + \frac{1}{4}\right) = 3 + \frac{1}{2} + \frac{1}{2}$$
$$\left(x - \frac{1}{2}\right)^2 + \left(y - \frac{1}{2}\right)^2 = 2$$

Center: $\left(\dfrac{1}{2}, \dfrac{1}{2}\right)$

Radius: $\sqrt{2}$

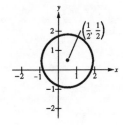

48. $$4x^2 + 4y^2 - 4x + 2y - 1 = 0$$

$$4\left(x^2 - x + \frac{1}{4}\right) + 4\left(y^2 + \frac{y}{2} + \frac{1}{16}\right) = 1 + 1 + \frac{1}{4}$$

$$4\left(x - \frac{1}{2}\right)^2 + 4\left(y + \frac{1}{4}\right)^2 = \frac{9}{4}$$

$$\left(x - \frac{1}{2}\right)^2 + \left(y + \frac{1}{4}\right)^2 = \frac{9}{16}$$

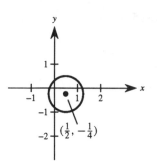

Center: $\left(\dfrac{1}{2}, \ -\dfrac{1}{4}\right)$

Radius: $\dfrac{3}{4}$

49. $$16x^2 + 16y^2 + 16x + 40y - 7 = 0$$

$$16\left(x^2 + x + \frac{1}{4}\right) + 16\left(y^2 + \frac{5y}{2} + \frac{25}{16}\right) = 7 + 4 + 25$$

$$16\left(x + \frac{1}{2}\right)^2 + 16\left(y + \frac{5}{4}\right)^2 = 36$$

$$\left(x + \frac{1}{2}\right)^2 + \left(y + \frac{5}{4}\right)^2 = \frac{9}{4}$$

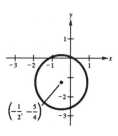

Center: $\left(-\dfrac{1}{2}, \ -\dfrac{5}{4}\right)$

Radius: $\dfrac{3}{2}$

50. $$x^2 + y^2 - 4x + 2y + 3 = 0$$

$$(x^2 - 4x + 4) + (y^2 + 2y + 1) = -3 + 4 + 1$$

$$(x - 2)^2 + (y + 1)^2 = 2$$

Center: $(2, \ -1)$

Radius: $\sqrt{2}$

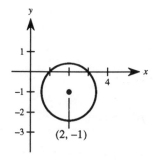

51. a.
$$4x^2 + 4y^2 - 4x + 24y - 63 = 0$$
$$x^2 + y^2 - x + 6y = \frac{63}{4}$$
$$\left(x^2 - x + \frac{1}{4}\right) + (y^2 + 6y + 9) = \frac{63}{4} + \frac{1}{4} + 9$$
$$\left(x - \frac{1}{2}\right)^2 + (y + 3)^2 = 25$$
$$(y + 3)^2 = 25 - \left(x - \frac{1}{2}\right)^2$$
$$y + 3 = \pm\sqrt{25 - \left(x - \frac{1}{2}\right)^2}$$
$$y = -3 \pm \sqrt{25 - \left(x - \frac{1}{2}\right)^2}$$
$$= \frac{-6 \pm \sqrt{99 + 4x - 4x^2}}{2}$$

b.

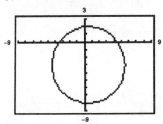

52. a.
$$x^2 + y^2 - 8x - 6y - 11 = 0$$
$$(x^2 - 8x + 16) + (y^2 - 6y + 9) = 11 + 16 + 9$$
$$(x - 4)^2 + (y - 3)^2 = 36$$
$$(y - 3)^2 = 36 - (x - 4)^2$$
$$y - 3 = \pm\sqrt{36 - (x - 4)^2}$$
$$y = 3 \pm \sqrt{20 + 8x - x^2}$$

b.

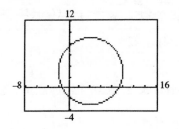

53.
$$x^2 + y^2 - 4x + 2y + 1 \le 0$$
$$(x^2 - 4x + 4) + (y^2 + 2y + 1) \le -1 + 4 + 1$$
$$(x - 2)^2 + (y + 1)^2 \le 4$$
Center: $(2, -1)$
Radius: 2

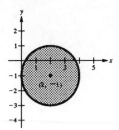

54. $(x - 1)^2 + \left(y - \frac{1}{2}\right)^2 > 1$

Center: $\left(1, \frac{1}{2}\right)$

Radius: 1

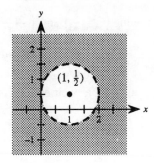

55. The distance between $(x_1, \ y_1)$ and $\left(\dfrac{2x_1 + x_2}{3}, \ \dfrac{2y_1 + y_2}{3} \right)$ is

$$d = \sqrt{\left(x_1 - \dfrac{2x_1 + x_2}{3} \right)^2 + \left(y_1 - \dfrac{2y_1 + y_2}{3} \right)^2}$$

$$= \sqrt{\left(\dfrac{x_1 - x_2}{3} \right)^2 + \left(\dfrac{y_1 - y_2}{3} \right)^2}$$

$$= \sqrt{\dfrac{1}{9}[(x_1 - x_2)^2 + (y_1 - y_2)^2]}$$

$$= \dfrac{1}{3}\sqrt{(x_1 - x_2)^2 + (y_1 - y_2)^2}$$

which is $\frac{1}{3}$ of the distance between $(x_1, \ y_1)$ and $(x_2, \ y_2)$.

$$\left(\dfrac{\left(\dfrac{2x_1 + x_2}{3} \right) + x_2}{2}, \ \dfrac{\left(\dfrac{2y_1 + y_2}{3} \right) + y_2}{2} \right) = \left(\dfrac{x_1 + 2x_2}{3}, \ \dfrac{y_1 + 2y_2}{3} \right)$$

is the second point of trisection.

56. a. $\left(\dfrac{2(1) + 4}{3}, \ \dfrac{2(-2) + 1}{3} \right) = (2, \ -1)$ **b.** $\left(\dfrac{2(-2) + 0}{3}, \ \dfrac{2(-3) + 0}{3} \right) = \left(-\dfrac{4}{3}, \ -2 \right)$

$\left(\dfrac{1 + 2(4)}{3}, \ \dfrac{-2 + 2(1)}{3} \right) = (3, \ 0)$ $\left(\dfrac{-2 + 2(0)}{3}, \ \dfrac{-3 + 2(0)}{3} \right) = \left(-\dfrac{2}{3}, \ -1 \right)$

57. Let one vertex be at $(0, 0)$ and another at $(a, \ 0)$.

Midpoint of $(0, 0)$ and $(d, \ e)$ is $\left(\dfrac{d}{2}, \ \dfrac{e}{2} \right)$.

Midpoint of $(b, \ c)$ and $(a, \ 0)$ is $\left(\dfrac{a + b}{2}, \ \dfrac{c}{2} \right)$.

Midpoint of $(0, 0)$ and $(a, \ 0)$ is $\left(\dfrac{a}{2}, \ 0 \right)$.

Midpoint of $(b, \ c)$ and $(d, \ e)$ is $\left(\dfrac{b + d}{2}, \ \dfrac{c + e}{2} \right)$.

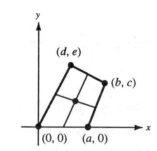

Midpoint of line segment joining $\left(\dfrac{d}{2}, \ \dfrac{e}{2} \right)$ and $\left(\dfrac{a + b}{2}, \ \dfrac{c}{2} \right)$ is $\left(\dfrac{a + b + d}{4}, \ \dfrac{c + e}{4} \right)$.

Midpoint of line segment joining $\left(\dfrac{a}{2}, \ 0 \right)$ and $\left(\dfrac{b + d}{2}, \ \dfrac{c + e}{2} \right)$ is $\left(\dfrac{a + b + d}{4}, \ \dfrac{c + e}{4} \right)$.

Therefore the line segments intersect at their midpoints.

58. Let the circle of radius r be centered at the origin. Let (a, b) and $(r, 0)$ be the endpoints of the chord. The midpoint M of the chord is $\left(\frac{a+r}{2}, \frac{b}{2}\right)$. We will show that OM is perpendicular to MR by verifying that $d_1{}^2 + d_2{}^2 = d_3{}^2$.

$$d_1{}^2 = \left(\frac{a+r}{2} - 0\right)^2 + \left(\frac{b}{2} - 0\right)^2 = \left(\frac{a+r}{2}\right)^2 + \left(\frac{b}{2}\right)^2$$

$$d_2{}^2 = \left(\frac{a+r}{2} - r\right)^2 + \left(\frac{b}{2} - 0\right)^2 = \left(\frac{a-r}{2}\right)^2 + \left(\frac{b}{2}\right)^2$$

$$d_1{}^2 + d_2{}^2 = \left(\frac{a^2 + 2ar + r^2}{4} + \frac{b^2}{4}\right) + \left(\frac{a^2 - 2ar + r^2}{4} + \frac{b^2}{4}\right)$$

$$= \frac{a^2}{2} + \frac{r^2}{2} + \frac{b^2}{2} = \frac{1}{2}(a^2 + b^2) + \frac{1}{2}r^2 = \frac{1}{2}r^2 + \frac{1}{2}r^2 = r^2 = d_3{}^2$$

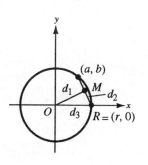

59. Let (a, b) be a point on the semicircle of radius r, centered at the origin. We will show that the angle at (a, b) is a right angle by verifying that $d_1{}^2 + d_2{}^2 = d_3{}^2$.

$$d_1{}^2 = (a + r)^2 + (b - 0)^2$$

$$d_2{}^2 = (a - r)^2 + (b - 0)^2$$

$$d_1{}^2 + d_2{}^2 = (a^2 + 2ar + r^2 + b^2) + (a^2 - 2ar + r^2 + b^2)$$

$$= 2a^2 + 2b^2 + 2r^2$$

$$= 2(a^2 + b^2) + 2r^2$$

$$= 2r^2 + 2r^2 = 4r^2 = (2r)^2 = d_3{}^2$$

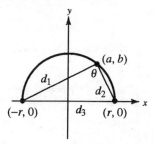

60. To show that $\left(\dfrac{x_1 + x_2}{2}, \dfrac{y_1 + y_2}{2}\right)$ is the midpoint of the line segment joining (x_1, y_1) and (x_2, y_2) we must show that $d_1 = d_2$ and $d_1 + d_2 = d_3$ (see graph).

$$d_1 = \sqrt{\left(\frac{x_1 + x_2}{2} - x_1\right)^2 + \left(\frac{y_1 + y_2}{2} - y_1\right)^2}$$

$$= \sqrt{\left(\frac{x_2 - x_1}{2}\right)^2 + \left(\frac{y_2 - y_1}{2}\right)^2}$$

$$= \frac{1}{2}\sqrt{(x_2 - x_1)^2 + (y_2 - y_1)^2}$$

$$d_2 = \sqrt{\left(x_2 - \frac{x_1 + x_2}{2}\right)^2 + \left(y_2 - \frac{y_1 - y_2}{2}\right)^2}$$

$$= \sqrt{\left(\frac{x_2 - x_1}{2}\right)^2 + \left(\frac{y_2 - y_1}{2}\right)^2}$$

$$= \frac{1}{2}\sqrt{(x_2 - x_1)^2 + (y_2 - y_1)^2}$$

$$d_3 = \sqrt{(x_2 - x_1)^2 + (y^2 - y_1)^2}$$

Therefore, $d_1 = d_2$ and $d_1 + d_2 = d_3$.

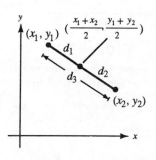

61. True; if $ab < 0$ then either a is positive and b is negative (Quadrant IV) or a is negative and b is positive (Quadrant II).

62. False

$$d = \sqrt{[(a+b)-(a-b)]^2 + (a-a)^2}$$

$$= \sqrt{(2b)^2 + 0^2}$$

$$= \sqrt{4b^2}$$

$$= 2|b|$$

63. True

64. True; if $ab = 0$ then $a = 0$ (y-axis) or $b = 0$ (x-axis).

65. Since $a > 0$ and $b > 0$, we have

$$(\sqrt{a} - \sqrt{b})^2 \geq 0$$

$$(\sqrt{a})^2 - 2\sqrt{a}\sqrt{b} + (\sqrt{b})^2 \geq 0$$

$$a - 2\sqrt{ab} + b \geq 0$$

$$a + b \geq 2\sqrt{ab}$$

$$\tfrac{1}{2}(a+b) \geq \sqrt{ab}.$$

66.

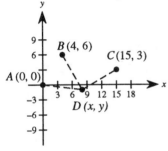

$$d(AD) = \sqrt{(x-0)^2 + (y-0)^2} = \sqrt{x^2 + y^2}$$

$$d(BD) = \sqrt{(x-4)^2 + (y-6)^2}$$

$$d(CD) = \sqrt{(x-15)^2 + (y-3)^2}$$

Since $d(AD) = d(BD) = d(CD)$, we have:

$$\sqrt{x^2 + y^2} = \sqrt{(x-4)^2 + (y-6)^2} \qquad \text{and} \qquad \sqrt{x^2 + y^2} = \sqrt{(x-15)^2 + (y-3)^2}$$

$$x^2 + y^2 = x^2 - 8x + 16 + y^2 - 12y + 36 \qquad\qquad x^2 + y^2 = x^2 - 30x + 225 + y^2 - 6y + 9$$

$$8x + 12y = 52 \qquad\qquad\qquad\qquad\qquad\qquad 30x + 6y = 234$$

Solving this system yields $x = 8$ and $y = -1$. The point is $(8, -1)$.

Section P.3   Graphs of Equations

1. $y = x - 2$
   x-intercept: $(2, 0)$
   y-intercept: $(0, -2)$
   Matches graph (c)

2. $y = -\frac{1}{2}x + 2$
   x-intercept: $(4, 0)$
   y-intercept: $(0, 2)$
   Matches graph (d)

3. $y = x^2 + 2x$
   x-intercepts: $(0, 0)$, $(-2, 0)$
   y-intercept: $(0, 0)$
   Matches graph (b)

4. $y = \sqrt{9 - x^2}$
   x-intercepts: $(-3, 0)$, $(3, 0)$
   y-intercept: $(0, 3)$
   Matches graph (f)

5. $y = 4 - x^2$
   x-intercepts: $(2, 0)$, $(-2, 0)$
   y-intercept: $(0, 4)$
   Matches graph (a)

6. $y = x^3 - x$
   x-intercepts: $(0, 0)$, $(-1, 0)$, $(1, 0)$
   y-intercept: $(0, 0)$
   Matches graph (e)

7. $y = 2x - 3$
   y-intercept: $y = 2(0) - 3 = -3$; $(0, -3)$
   x-intercept: $0 = 2x - 3$
   $$x = \tfrac{3}{2}; \ \left(\tfrac{3}{2}, 0\right)$$

8. $y = (x - 1)(x - 3)$
   y-intercept: $y = (0 - 1)(0 - 3)$
   $$y = 3; \ (0, 3)$$
   x-intercepts: $0 = (x - 1)(x - 3)$
   $$x = 1, 3; \ (1, 0), (3, 0)$$

9. $y = x^2 + x - 2$
   y-intercept: $y = 0^2 + 0 - 2$
   $$y = -2; \ (0, -2)$$
   x-intercepts: $0 = x^2 + x - 2$
   $$0 = (x + 2)(x - 1)$$
   $$x = -2, 1; \ (-2, 0), (1, 0)$$

10. $y^2 = x^3 - 4x$
   y-intercept: $y^2 = 0^3 - 4(0)$
   $$y = 0; \ (0, 0)$$
   x-intercepts: $0 = x^3 - 4x$
   $$0 = x(x - 2)(x + 2)$$
   $$x = 0, \pm 2; \ (0, 0), (\pm 2, 0)$$

11. $y = x^2\sqrt{9 - x^2}$
   y-intercept: $y = 0^2\sqrt{9 - 0^2}$
   $$y = 0; \ (0, 0)$$
   x-intercepts: $0 = x^2\sqrt{9 - x^2}$
   $$0 = x^2\sqrt{(3 - x)(3 + x)}$$
   $$x = 0, \pm 3; \ (0, 0), (\pm 3, 0)$$

12. $xy = 4$
   No intercepts

13. $y = \dfrac{x - 1}{x - 2}$

y-intercept: $y = \dfrac{0 - 1}{0 - 2}$

$$y = \frac{1}{2}; \ \left(0, \ \frac{1}{2}\right)$$

x-intercept: $0 = \dfrac{x - 1}{x - 2}$

$$x = 1; \ (1, \ 0)$$

14. $y = \dfrac{x^2 + 3x}{(3x + 1)^2}$

y-intercept: $y = \dfrac{0^2 + 3(0)}{[3(0) + 1]^2}$

$$y = 0; \ (0, \ 0)$$

x-intercepts: $0 = \dfrac{x^2 + 3x}{(3x + 1)^2}$

$$0 = \frac{x(x + 3)}{(3x + 1)^2}$$

$$x = 0, \ -3; \ (0, \ 0), \ (-3, \ 0)$$

15. $x^2 y - x^2 + 4y = 0$

y-intercept:

$$0^2(y) - 0^2 + 4y = 0$$

$$y = 0; \ (0, \ 0)$$

x-intercept:

$$x^2(0) - x^2 + 4(0) = 0$$

$$x = 0; \ (0, \ 0)$$

16. $y = 2x - \sqrt{x^2 + 1}$

y-intercept: $y = 2(0) - \sqrt{0^2 + 1}$

$$y = -1; \ (0, \ -1)$$

x-intercepts: $0 = 2x - \sqrt{x^2 + 1}$

$$2x = \sqrt{x^2 + 1}$$

$$4x^2 = x^2 + 1$$

$$3x^2 = 1$$

$$x^2 = \frac{1}{3}$$

$$x = \pm\frac{\sqrt{3}}{3}$$

$$x = \frac{\sqrt{3}}{3}; \ \left(\frac{\sqrt{3}}{3}, \ 0\right)$$

Note: $x = -\sqrt{3}/3$ is an extraneous solution.

17. Symmetric with respect to the y-axis since
$$y = (-x)^2 - 2 = x^2 - 2.$$

18. Symmetric with respect to the y-axis since
$$y = (-x)^4 - (-x)^2 + 3$$
$$= x^4 - x^2 + 3.$$

19. Symmetric with respect to the y-axis since
$$(-x)^2 y - (-x)^2 + 4y = 0$$
$$x^2 y - x^2 + 4y = 0.$$

20. Symmetric with respect to the origin since
$$(-x)(-y) - \sqrt{4 - (-x)^2} = 0$$
$$xy - \sqrt{4 - x^2} = 0.$$

21. Symmetric with respect to the x-axis since
$$(-y)^2 = y^2 = x^3 - 4x.$$

22. Symmetric with respect to the x-axis since
$$x(-y)^2 = xy^2 = -10.$$

23. Symmetric with respect to the origin since
$$(-y) = (-x)^3 + (-x)$$
$$-y = -x^3 - x$$
$$y = x^3 + x.$$

24. Symmetric with respect to the origin since
$$(-x)(-y) = xy = 1.$$

25. Symmetric with respect to the origin since

$$-y = \frac{-x}{(-x)^2 + 1}$$

$$y = \frac{x}{x^2 + 1}.$$

26. No symmetry with respect to either axis or the origin.

27. $y = 2x - 3$

(1, 2):	$2 \neq 2(1) - 3$	Not on the graph
(1, −1):	$-1 = 2(1) - 3$	On the graph
(4, 5):	$5 = 2(4) - 3$	On the graph

28. $x^2 + y^2 = 4$

$(1, -\sqrt{3})$:	$1^2 + (-\sqrt{3})^2 = 4$	On the graph
$\left(\frac{1}{2}, -1\right)$:	$\left(\frac{1}{2}\right)^2 + (-1)^2 \neq 4$	Not on the graph
$\left(\frac{3}{2}, \frac{7}{2}\right)$:	$\left(\frac{3}{2}\right)^2 + \left(\frac{7}{2}\right)^2 \neq 4$	Not on the graph

29. $x^2 y - x^2 + 4y = 0$

$$y = \frac{x^2}{x^2 + 4}$$

$\left(1, \frac{1}{5}\right)$:	$\frac{1}{5} = \frac{1}{1^2 + 4}$	On the graph
$\left(2, \frac{1}{2}\right)$:	$\frac{1}{2} = \frac{4}{2^2 + 4}$	On the graph
$(-1, -2)$:	$-2 \neq \frac{1}{(-1)^2 + 4}$	Not on the graph

30. $x^2 - xy + 4y = 3$

$$y = \frac{x^2 - 3}{x - 4}$$

$(0, 2)$:	$2 \neq \frac{0 - 3}{0 - 4}$	Not on the graph
$\left(-2, -\frac{1}{6}\right)$:	$-\frac{1}{6} = \frac{(-2)^2 - 3}{(-2) - 4}$	On the graph
$(3, -6)$:	$-6 = \frac{3^2 - 3}{3 - 4}$	On the graph

31. $y = x$

Intercept: (0, 0)

Symmetry: origin

32. $y = x - 2$

Intercepts:

(0, −2), (2, 0)

Symmetry: none

33. $y = x + 3$

Intercepts:

(−3, 0), (0, 3)

Symmetry: none

34. $y = 2x - 3$

Intercepts:

$\left(\frac{3}{2}, 0\right)$, $(0, -3)$

Symmetry: none

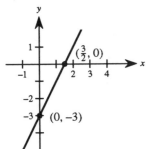

35. $y = -3x + 2$

Intercepts:

$\left(\frac{2}{3}, 0\right)$, $(0, 2)$

Symmetry: none

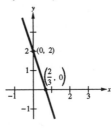

36. $y = -\frac{x}{2} + 2$

Intercepts:

$(4, 0)$, $(0, 2)$

Symmetry: none

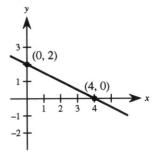

37. $y = \frac{x}{2} - 4$

Intercepts:

$(8, 0)$, $(0, -4)$

Symmetry: none

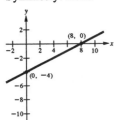

38. $y = x^2 + 3$

Intercept: $(0, 3)$

Symmetry: y-axis

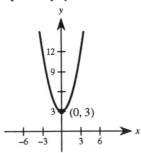

39. $y = 1 - x^2$

Intercepts:

$(1, 0)$, $(-1, 0)$, $(0, 1)$

Symmetry: y-axis

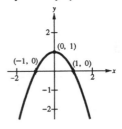

40. $y = 2x^2 + x = x(2x + 1)$

Intercepts:

$(0, 0)$, $\left(-\frac{1}{2}, 0\right)$

Symmetry: none

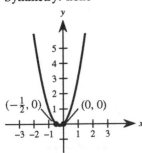

41. $y = x^3 + 2$

Intercepts:

$(-\sqrt[3]{2}, 0)$, $(0, 2)$

Symmetry: none

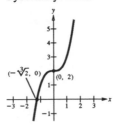

42. $y = \sqrt{9 - x^2}$

Intercepts:

$(-3, 0)$, $(3, 0)$, $(0, 3)$

Symmetry: y-axis

Domain: $[-3, 3]$

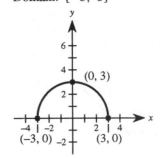

43. $y = (x + 2)^2$
Intercepts:
  $(-2, 0)$, $(0, 4)$
Symmetry: none

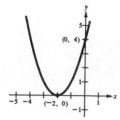

44. $x = y^2 - 4$
Intercepts:
  $(0, 2)$, $(0, -2)$, $(-4, 0)$
Symmetry: x-axis

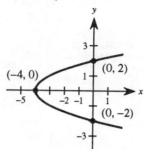

45. $y = \dfrac{1}{x}$
Intercepts: none
Symmetry: origin

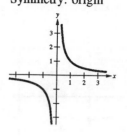

46. $y = 2x^4$
Intercept: $(0, 0)$
Symmetry: y-axis

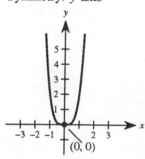

47. $y = -2x^2 + x + 1$
  $= (2x + 1)(-x + 1)$
Intercepts:
  $\left(-\frac{1}{2}, 0\right)$, $(1, 0)$, $(0, 1)$
Symmetry: none

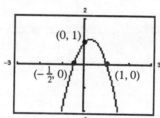

48. $y = x^3 - 1$
Intercepts: $(1, 0)$, $(0, -1)$
Symmetry: none

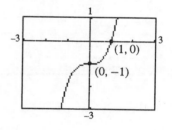

49. $y = x\sqrt{25 - x^2}$
Intercepts: $(0, 0)$, $(-5, 0)$, $(5, 0)$
Symmetry: origin
Domain: $[-5, 5]$

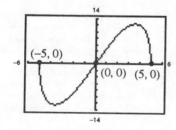

50. $y = \dfrac{5}{x^2 + 1} - 1$
Intercepts: $(0, 4)$, $(-2, 0)$, $(2, 0)$
Symmetry: y-axis

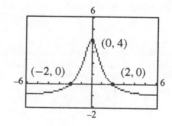

51. $x^2 + 4y^2 = 4 \Rightarrow y = \pm\dfrac{\sqrt{4 - x^2}}{2}$

Intercepts: $(-2, 0)$, $(2, 0)$, $(0, -1)$, $(0, 1)$

Symmetry: origin and both axes

Domain: $[-2, 2]$

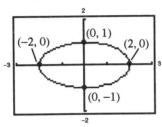

52. $9x^2 + y^2 = 9 \Rightarrow y = \pm 3\sqrt{1 - x^2}$

Intercepts: $(1, 0)$, $(-1, 0)$, $(0, 3)$, $(0, -3)$

Symmetry: origin and both axes

Domain: $[-1, 1]$

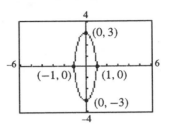

53. $y = (x + 2)(x - 4)(x - 6)$

54. $y = \left(x + \frac{5}{2}\right)(x - 2)\left(x - \frac{3}{2}\right)$

55. Some possible equations:

$$y = x$$
$$y = x^3$$
$$y = 3x^3 - x$$
$$y = \sqrt[3]{x}$$

56. Some possible equations:

$$x = y^2$$
$$x = |y|$$
$$x = y^4 + 1$$
$$x^2 + y^2 = 25$$

57. $x + y = 2 \Rightarrow y = 2 - x$

$2x - y = 1 \Rightarrow y = 2x - 1$

$$2 - x = 2x - 1$$
$$3 = 3x$$
$$1 = x$$

The corresponding y-value is $y = 1$.

Point of intersection: $(1, 1)$

58. $2x - 3y = 13 \Rightarrow y = \dfrac{2x - 13}{3}$

$5x + 3y = 1 \Rightarrow y = \dfrac{1 - 5x}{3}$

$$\dfrac{2x - 13}{3} = \dfrac{1 - 5x}{3}$$
$$2x - 13 = 1 - 5x$$
$$7x = 14$$
$$x = 2$$

The corresponding y-value is $y = -3$.

Point of intersection: $(2, -3)$

59. $x + y = 7 \Rightarrow y = 7 - x$

$3x - 2y = 11 \Rightarrow y = \dfrac{3x - 11}{2}$

$$7 - x = \dfrac{3x - 11}{2}$$
$$14 - 2x = 3x - 11$$
$$-5x = -25$$
$$x = 5$$

The corresponding y-value is $y = 2$.

Point of intersection: $(5, 2)$

60. $x^2 + y^2 = 25 \Rightarrow y^2 = 25 - x^2$

$2x + y = 10 \Rightarrow y = 10 - 2x$

$$25 - x^2 = (10 - 2x)^2$$
$$25 - x^2 = 100 - 40x + 4x^2$$
$$0 = 5x^2 - 40x + 75 = 5(x - 3)(x - 5)$$
$$x = 3 \text{ or } x = 5$$

The corresponding y-values are $y = 4$ and $y = 0$.

Points of intersection: $(3, 4)$, $(5, 0)$

61. $x^2 + y^2 = 5 \Rightarrow y^2 = 5 - x^2$

$\qquad x - y = 1 \Rightarrow y = x - 1$

$\qquad 5 - x^2 = (x-1)^2$

$\qquad 5 - x^2 = x^2 - 2x + 1$

$\qquad\quad 0 = 2x^2 - 2x - 4 = 2(x+1)(x-2)$

$\qquad\quad x = -1 \text{ or } x = 2$

The corresponding y-values are $y = -2$ and $y = 1$.

Points of intersection: $(-1, -2), (2, 1)$

62. $x^2 + y = 4 \Rightarrow y = 4 - x^2$

$\qquad 2x - y = 1 \Rightarrow y = 2x - 1$

$\qquad 4 - x^2 = 2x - 1$

$\qquad\quad 0 = x^2 + 2x - 5$

$\qquad\quad x = \dfrac{-2 \pm \sqrt{24}}{2} = -1 \pm 6$

The corresponding y-values are $y = -3 \pm 2\sqrt{6}$.

Points of intersection: $(-1 - \sqrt{6},\ -3 - 2\sqrt{6})$,

$\qquad\qquad\qquad\qquad\qquad (-1 + \sqrt{6},\ -3 + 2\sqrt{6})$

63. $\qquad y = x^3$

$\qquad\quad y = x$

$\qquad\quad x^3 = x$

$\qquad x^3 - x = 0$

$\qquad\quad x(x+1)(x-1) = 0$

$\qquad x = 0,\ x = -1,\ \text{or } x = 1$

The corresponding y-values are $y = 0,\ y = -1,$ and $y = 1$.

Points of intersection: $(0, 0),\ (-1, -1),\ (1, 1)$

64. $x = 3 - y^2 \Rightarrow y^2 = 3 - x$

$\qquad y = x - 1$

$\qquad 3 - x = (x-1)^2$

$\qquad 3 - x = x^2 - 2x + 1$

$\qquad\quad 0 = x^2 - x - 2 = (x+1)(x-2)$

$\qquad\quad x = -1 \text{ or } x = 2$

The corresponding y-values are $y = -2$ and $y = 1$.

Points of intersection: $(-1, -2),\ (2, 1)$

65. $\qquad\quad y = x^3 - 2x^2 + x - 1$

$\qquad\qquad y = -x^2 + 3x - 1$

$\quad x^3 - 2x^2 + x - 1 = -x^2 + 3x - 1$

$\qquad x^3 - x^2 - 2x = 0$

$\qquad x(x-2)(x+1) = 0$

$\qquad\qquad\qquad x = -1,\ 0,\ 2$

$(-1, -5),\ (0, -1),\ (2, 1)$

66. $\qquad\quad y = x^4 - 2x^2 + 1$

$\qquad\qquad y = 1 - x^2$

$\quad 1 - x^2 = x^4 - 2x^2 + 1$

$\qquad\quad 0 = x^4 - x^2$

$\qquad\quad 0 = x^2(x+1)(x-1)$

$\qquad\qquad x = -1,\ 0,\ 1$

$(-1, 0),\ (0, 1),\ (1, 0)$

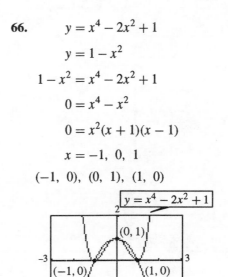

67. $8650x + 250{,}000 = 9950x$

$\qquad 250{,}000 = 1300x$

$\qquad\qquad\quad x \approx 192 \text{ units}$

68. $5.5\sqrt{x} + 10,000 = 3.29x$

$$(5.5\sqrt{x})^2 = (3.29x - 10,000)^2$$

$$30.25x = 10.8241x^2 - 65,800x + 100,000,000$$

$$0 = 10.8241x^2 - 65830.25x + 100,000,000 \quad \text{Use the Quadratic Formula}$$

$$x \approx 3133 \text{ units}$$

The other root, $x \approx 2949$, does not satisfy the equation $R = C$.

69. a. $\quad 4 = k(1)^3 \qquad k = 4$

   b. $\quad 1 = k(-2)^3 \qquad k = -\frac{1}{8}$

   c. $\quad 0 = k(0)^3 \qquad k = \text{any real number}$

   d. $\quad -1 = k(-1)^3 \qquad k = 1$

70. $y = kx + 5$ matches (b).

   Use $(1,\ 7)$: $\qquad\qquad 7 = k(1) + 5 \Rightarrow k = 2$, thus, $y = 2x + 5$.
   $y = x^2 + k$ matches (d).

   Use $(1,\ -9)$: $\qquad\qquad -9 = (1)^2 + k \Rightarrow k = -10$, thus, $y = x^2 - 10$.
   $y = kx^{3/2}$ matches (a).

   Use $(1,\ 3)$: $\qquad\qquad 3 = k(1)^{3/2} \Rightarrow k = 3$, thus, $y = 3x^{3/2}$.
   $xy = k$ matches (c).

   Use $(1,\ 36)$: $\qquad\qquad (1)(36) = k \Rightarrow k = 36$, thus, $xy = 36$.

71. a.

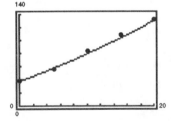

   b. $\dfrac{130.7}{38.8} \approx 3.369$

   c. $y = 0.05(30)^2 + 3.63(30) + 37.68$

   $\qquad \approx 191.58$ CPI for the year 2000

72. a.

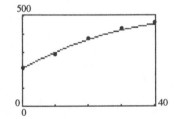

   b. $y = -0.09(50)^2 + 9.80(50) + 211.49$

   $\qquad \approx 476.49$ acres per farm in the year 2000

73. a.

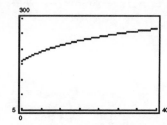

b. $y = 75.82 - 2.11(14.696) + 43.51\sqrt{14.696}$

  $\approx 211.61°$

c. When $x \approx 24.725$ lb/in.^{2}, $y \approx 240°$.

74. $y = \dfrac{10{,}770}{(2x)^2} - 0.37$

  $= \dfrac{1}{4}\left(\dfrac{10{,}770}{x^2}\right) - 0.37$

The resistance is changed by approximately a factor of $\frac{1}{4}$.

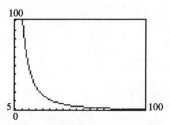

75. If $(x,\ y)$ is on the graph, then so is $(-x,\ y)$ by y-axis symmetry. Since $(-x,\ y)$ is on the graph, then so is $(-x,\ -y)$ by x-axis symmetry. Hence, the graph is symmetric with respect to the origin. The converse is not true. For example, $y = x^3$ has origin symmetry but is not symmetric with respect to either the x-axis or the y-axis.

76. Assume that the graph has x-axis and origin symmetry. If $(x,\ y)$ is on the graph, so is $(x,\ -y)$ by x-axis symmetry. Since $(x,\ -y)$ is on the graph, then so is $(-x,\ -(-y)) = (-x,\ y)$ by origin symmetry. Therefore, the graph is symmetric with respect to the y-axis. The argument is the same for y-axis and origin symmetry.

77. False; x-axis symmetry means that if $(1,\ -2)$ is on the graph, then $(1, 2)$ is also on the graph.

78. True

79. True; the x-intercepts are

  $\left(\dfrac{-b \pm \sqrt{b^2 - 4ac}}{2a},\ 0\right).$

80. True; the x-intercept is

  $\left(-\dfrac{b}{2a},\ 0\right).$

81. Distance to the origin $= K \times$ Distance to $(2, 0)$

$$\sqrt{x^2 + y^2} = K\sqrt{(x-2)^2 + y^2},\quad K \neq 1$$

$$x^2 + y^2 = K^2(x^2 - 4x + 4 + y^2)$$

$$(1 - K^2)x^2 + (1 - K^2)y^2 + 4K^2x - 4K^2 = 0$$

Note: This is the equation of a circle!

82. $x_n = \frac{1}{2}x_{n-1} + \frac{1}{x_{n-1}}, \quad n = 1, 2, \ldots, \text{ and } x_0 = 1$

$x_1 = \frac{1}{2}(1) + \frac{1}{1} = \frac{3}{2} = 1.5$

$x_2 = \frac{1}{2}\left(\frac{3}{2}\right) + \frac{1}{3/2} = \frac{17}{12} \approx 1.41667$

$x_3 = \frac{1}{2}\left(\frac{17}{12}\right) + \frac{1}{17/12} = \frac{577}{408} \approx 1.41422$

$x_4 = \frac{1}{2}\left(\frac{577}{408}\right) + \frac{1}{577/408} = \frac{665,857}{470,832} \approx 1.41421$

As n gets larger, $x_n \to \sqrt{2}$.

Section P.4   Lines in the Plane

1. $m = 1$ **2.** $m = 2$ **3.** $m = 0$ **4.** $m = -1$ **5.** $m = -12$ **6.** $m = \frac{40}{3}$

7.
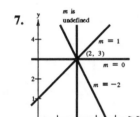

8.
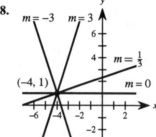

9. $m = \dfrac{2 - (-4)}{5 - 3}$

$= \dfrac{6}{2} = 3$

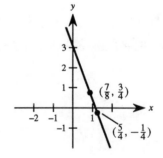

10. $m = \dfrac{5 - 1}{2 - 2}$

$= \dfrac{4}{0}$

undefined

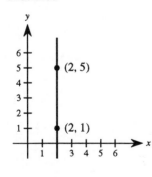

11. $m = \dfrac{4 - 2}{-2 - 1}$

$= -\dfrac{2}{3}$

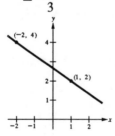

12. $m = \dfrac{(3/4) - (-1/4)}{(7/8) - (5/4)}$

$= \dfrac{1}{-3/8} = -\dfrac{8}{3}$

13. Since the slope is 0, the line is horizontal and its equation is $y = 1$. Therefore, three additional points are $(0, 1)$, $(1, 1)$, and $(3, 1)$.

14. Since the slope is undefined, the line is vertical and its equation is $x = -3$. Therefore, three additional points are $(-3, 2)$, $(-3, 3)$, and $(-3, 5)$.

15. The equation of this line is

$$y - 7 = -3(x - 1)$$

$$y = -3x + 10$$

Therefore, three additional points are $(0, 10)$, $(2, 4)$, and $(3, 1)$.

16. The equation of this line is

$$y + 2 = 2(x + 2)$$

$$y = 2x + 2$$

Therefore, three additional points are $(-3, -4)$, $(-1, 0)$, and $(0, 2)$.

17. a. Let $t = 0$ represent 1989.

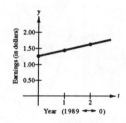

b. $(0, 1.26)$, $(2, 1.63)$

$$m = \frac{1.63 - 1.26}{2 - 0} = 0.185$$

For 1992, we have $y \approx 1.63 + 0.185 = \$1.82$.

18. a. $\dfrac{\Delta y}{\Delta x} = \dfrac{1}{3}$

b.

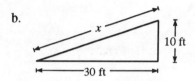

By the Pythagorean Theorem,

$$x^2 = 30^2 + 10^2 = 1000$$

$$x \approx 31.623 \text{ feet.}$$

19. $x + 5y = 20$

$$y = -\tfrac{1}{5}x + 4$$

Therefore, the slope is $m = -\tfrac{1}{5}$ and the y-intercept is $(0, 4)$.

20. $6x - 5y = 15$

$$y = \tfrac{6}{5}x - 3$$

Therefore, the slope is $m = \tfrac{6}{5}$ and the y-intercept is $(0, -3)$.

21. $x = 4$

The line is vertical. Therefore, the slope is undefined and there is no y-intercept.

22. $y = -1$

The line is horizontal. Therefore, the slope is $m = 0$ and the y-intercept is $(0, -1)$.

23. $m = \dfrac{1 - (-3)}{2 - 0} = 2$

$$y - 1 = 2(x - 2)$$

$$y - 1 = 2x - 4$$

$$0 = 2x - y - 3$$

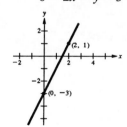

24. $m = \dfrac{4 - (-4)}{1 - (-3)} = \dfrac{8}{4} = 2$

$$y - 4 = 2(x - 1)$$

$$y - 4 = 2x - 2$$

$$0 = 2x - y + 2$$

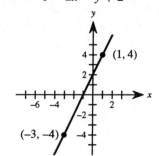

25. $m = \dfrac{3 - 0}{-1 - 0} = -3$

$$y - 0 = -3(x - 0)$$

$$y = -3x$$

$$3x + y = 0$$

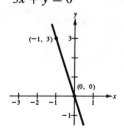

26. $m = \dfrac{6-2}{-3-1} = \dfrac{4}{-4} = -1$

$y - 2 = -1(x - 1)$

$y - 2 = -x + 1$

$x + y - 3 = 0$

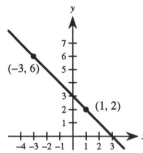

27. $m = 0$

$y = -2$

$y + 2 = 0$

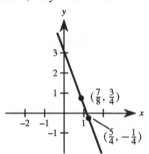

28. $m = \dfrac{(3/4) - (-1/4)}{(7/8) - (5/4)}$

$= \dfrac{1}{-3/8} = -\dfrac{8}{3}$

$y + \dfrac{1}{4} = \dfrac{-8}{3}\left(x - \dfrac{5}{4}\right)$

$12y + 3 = -32x + 40$

$32x + 12y - 37 = 0$

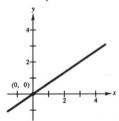

29. $y = \dfrac{3}{4}x + 3$

$4y = 3x + 12$

$0 = 3x - 4y + 12$

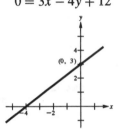

30. $x = -1$

$x + 1 = 0$

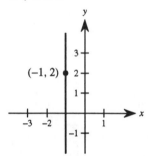

31. $y = \dfrac{2}{3}x$

$3y = 2x$

$2x - 3y = 0$

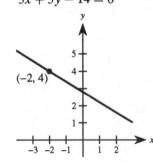

32. $y - 4 = -\dfrac{3}{5}(x + 2)$

$5y - 20 = -3x - 6$

$3x + 5y - 14 = 0$

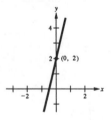

33. $y - 2 = 4(x - 0)$

$y = 4x + 2$

$0 = 4x - y + 2$

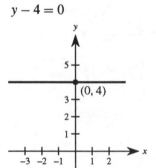

34. $y = 4$

$y - 4 = 0$

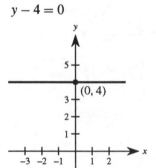

35. $x = 3$

$x - 3 = 0$

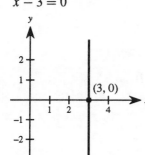

$(3, 0)$

36. $m = -\dfrac{b}{a}$

$y = \dfrac{-b}{a}x + b$

$\dfrac{b}{a}x + y = b$

$\dfrac{x}{a} + \dfrac{y}{b} = 1$

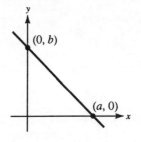

$(0, b)$

$(a, 0)$

37. $\dfrac{x}{2} + \dfrac{y}{3} = 1$

$3x + 2y - 6 = 0$

38. $\dfrac{x}{-2/3} + \dfrac{y}{-2} = 1$

$\dfrac{-3x}{2} - \dfrac{y}{2} = 1$

$3x + y = -2$

$3x + y + 2 = 0$

39. $\dfrac{x}{a} + \dfrac{y}{a} = 1$

$\dfrac{1}{a} + \dfrac{2}{a} = 1$

$\dfrac{3}{a} = 1$

$a = 3 \Rightarrow x + y = 3$

$x + y - 3 = 0$

40. $\dfrac{x}{a} + \dfrac{y}{a} = 1$

$\dfrac{-3}{a} + \dfrac{4}{a} = 1$

$\dfrac{1}{a} = 1$

$a = 1 \Rightarrow x + y = 1$

$x + y - 1 = 0$

41. $4x - 2y = 3$

$y = 2x - \tfrac{3}{2}$

$m = 2$

a. $y - 1 = 2(x - 2)$

$y - 1 = 2x - 4$

$2x - y - 3 = 0$

b. $y - 1 = -\tfrac{1}{2}(x - 2)$

$2y - 2 = -x + 2$

$x + 2y - 4 = 0$

42. $x + y = 7$

$y = -x + 7$

$m = -1$

a. $y - 2 = -1(x + 3)$

$y - 2 = -x - 3$

$x + y + 1 = 0$

b. $y - 2 = 1(x + 3)$

$y - 2 = x + 3$

$x - y + 5 = 0$

43. $5x + 3y = 0$

$$y = -\tfrac{5}{3}x$$

$$m = -\tfrac{5}{3}$$

a. $y - \tfrac{3}{4} = -\tfrac{5}{3}\left(x - \tfrac{7}{8}\right)$

$$24y - 18 = -40x + 35$$

$$40x + 24y - 53 = 0$$

b. $y - \tfrac{3}{4} = \tfrac{3}{5}\left(x - \tfrac{7}{8}\right)$

$$40y - 30 = 24x - 21$$

$$24x - 40y + 9 = 0$$

44. $3x + 4y = 7$

$$y = -\tfrac{3}{4}x + \tfrac{7}{4}$$

$$m = -\tfrac{3}{4}$$

a. $y - 4 = -\tfrac{3}{4}(x + 6)$

$$4y - 16 = -3x - 18$$

$$3x + 4y + 2 = 0$$

b. $y - 4 = \tfrac{4}{3}(x + 6)$

$$3y - 12 = 4x + 24$$

$$4x - 3y + 36 = 0$$

45. a. $x = 2 \Rightarrow x - 2 = 0$

b. $y = 5 \Rightarrow y - 5 = 0$

46. a. $y = 0$

b. $x = -1 \Rightarrow x + 1 = 0$

47. $y = -3$

$$y + 3 = 0$$

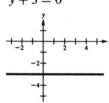

48. $x = 4$

$$x - 4 = 0$$

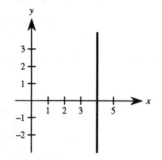

49. $2x - y - 3 = 0$

$$y = 2x - 3$$

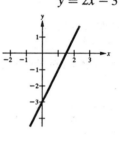

50. $x + 2y + 6 = 0$

$$y = -\tfrac{1}{2}x - 3$$

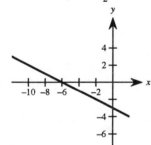

51. $y = -2x + 1$

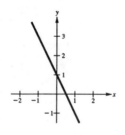

52. $y - 1 = 3(x + 4)$

$$y = 3x + 13$$

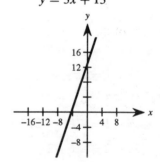

53.
$$x^2 = 4x - x^2$$
$$2x^2 - 4x = 0$$
$$2x(x - 2) = 0$$
$$x = 0 \qquad x = 2$$
$$y = 0 \qquad y = 4$$

Points of intersection: $(0, 0)$, $(2, 4)$
$$m = 2$$
$$y = 2x$$
$$0 = 2x - y$$

54.
$$x^2 - 4x + 3 = -x^2 + 2x + 3$$
$$2x^2 - 6x = 0$$
$$2x(x - 3) = 0$$
$$x = 0 \qquad x = 3$$
$$y = 3 \qquad y = 0$$

Points of intersection: $(0, 3)$, $(3, 0)$
$$m = -1$$
$$y - 3 = -(x - 0)$$
$$y = -x + 3$$
$$x + y - 3 = 0$$

55.
$$m_1 = \frac{1 - 0}{-2 - (-1)} = -1$$
$$m_2 = \frac{-2 - 0}{2 - (-1)} = -\frac{2}{3}$$
$$m_1 \neq m_2$$

The points are not collinear.

56.
$$m_1 = \frac{-6 - 4}{7 - 0} = -\frac{10}{7}$$
$$m_2 = \frac{11 - 4}{-5 - 0} = -\frac{7}{5}$$
$$m_1 \neq m_2$$

The points are not collinear.

57. Equations of perpendicular bisectors:
$$x = 0$$
$$y - \frac{c}{2} = \frac{a - b}{c}\left(x - \frac{a + b}{2}\right)$$
$$y - \frac{c}{2} = \frac{a + b}{-c}\left(x - \frac{b - a}{2}\right)$$

Solving simultaneously, the point of intersection is

$$\left(0, \frac{-a^2 + b^2 + c^2}{2c}\right).$$

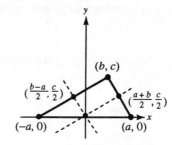

58. Equations of the medians:
$$y = \frac{c}{b}x$$
$$y = \frac{c}{3a + b}(x + a)$$
$$y = \frac{c}{-3a + b}(x - a)$$

Solving simultaneously, the point of intersection is

$$\left(\frac{b}{3}, \frac{c}{3}\right).$$

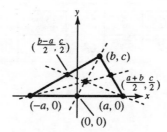

59. Equations of altitudes:

$$y = \frac{a-b}{c}(x+a)$$

$$x = b$$

$$y = -\frac{a+b}{c}(x-a)$$

Solving simultaneously, the point of intersection is

$$\left(b, \ \frac{a^2-b^2}{c}\right).$$

60. The slope of the line segment from $\left(\frac{b}{3}, \frac{c}{3}\right)$ to $\left(b, \ \frac{a^2-b^2}{c}\right)$ is:

$$m_1 = \frac{[(a^2-b^2)/c]-(c/3)}{b-(b/3)} = \frac{(3a^2-3b^2-c^2)/(3c)}{(2b)/3} = \frac{3a^2-3b^2-c^2}{2bc}.$$

The slope of the line segment from $\left(\frac{b}{3}, \frac{c}{3}\right)$ to $\left(0, \ \frac{-a^2+b^2+c^2}{2c}\right)$ is:

$$m_2 = \frac{[(-a^2+b^2+c^2)/(2c)]-(c/3)}{0-(b/3)} = \frac{(-3a^2+3b^2+3c^2-2c^2)/(6c)}{-b/3} = \frac{3a^2-3b^2-c^2}{2bc}$$

$$m_1 = m_2$$

Therefore, the points are collinear.

61. Find the equation of the line through the points (0, 32) and (100, 212).

$$m = \frac{180}{100} = \frac{9}{5}$$

$$F - 32 = \frac{9}{5}(C-0)$$

$$F = \frac{9}{5}C + 32$$

$$5F - 9C - 160 = 0$$

62.

C	−17.8°	−10°	10°	20°	32.2°	177°
F	0°	14°	50°	68°	90°	350.6°

63. $C = 0.27x + 125$

64. a. $W_1 = 9.50 + 0.75x$

$W_2 = 6.20 + 1.30x$

b. Point of intersection: (6, 14)

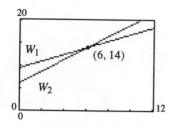

c. Both jobs pay \$14 per hour if 6 units are produced. For someone who can produce more than 6 units per hour, the second offer would pay more. For a worker who produces less than six units per hour, the first offer pays more.

65. a. Let $t = 0$ correspond to 1992.

(0, 28,500), (2, 32,900)

$$m = \frac{32,900 - 28,500}{2} = 2200$$

$S = 2200t + 28,500$

b. $S = 2200(5) + 28,500 = \$39,500$

66. a. Depreciation per year:

$\frac{875}{5} = \$175$

$y = 875 - 175x$

where $0 \le x \le 5$.

c. $y = 875 - 175(2) = \$525$

d. $200 = 875 - 175x$

$175x = 675$

$x \approx 3.86$ years

b.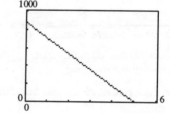

67. a. (50, 380), (47, 425)

$$m = \frac{425 - 380}{47 - 50} = -15$$

$p - 380 = -15(x - 50)$

$p = -15x + 1130$ or $x = \frac{1}{15}(1130 - p)$

b. $x = \frac{1}{15}(1130 - 455) = 45$ units

c. $x = \frac{1}{15}(1130 - 395) = 49$ units

68. a. Let $t = 0$ correspond to 1986.

(0, 37.5), (4, 54.0)

$$m = \frac{54.0 - 37.5}{4} = 4.125$$

$y = 4.125t + 37.5$

b. $y = 4.125(9) + 37.5$

$= 74.625$ million subscribers

c. $y = 4.125(2) + 37.5$

$= 45.75$ million subscribers

d. Since the slope is positive, the number of subscribers is increasing each year.

69. $4x + 3y - 10 = 0 \Rightarrow d = \dfrac{|4(0) + 3(0) - 10|}{\sqrt{4^2 + 3^2}} = \dfrac{10}{5} = 2$

70. $4x + 3y - 10 = 0 \Rightarrow d = \dfrac{|4(2) + 3(3) - 10|}{\sqrt{4^2 + 3^2}} = \dfrac{7}{5}$

71. $x - y - 2 = 0 \Rightarrow d = \dfrac{|1(-2) + (-1)(1) - 2|}{\sqrt{1^2 + 1^2}} = \dfrac{5}{\sqrt{2}} = \dfrac{5\sqrt{2}}{2}$

72. $x + 1 = 0 \Rightarrow d = \dfrac{|1(6) + (0)(2) + 1|}{\sqrt{1^2 + 0^2}} = 7$

73. A point on the line $x + y = 1$ is $(0, 1)$. The distance from the point $(0, 1)$ to $x + y - 5 = 0$ is

$$d = \frac{|1 - 5|}{\sqrt{2}} = \frac{4}{\sqrt{2}} = 2\sqrt{2}.$$

74. A point on the line $3x - 4y = 1$ is $(-1, -1)$. The distance from the point $(-1, -1)$ to $3x - 4y - 10 = 0$ is

$$d = \frac{|-3 + 4 - 10|}{5} = \frac{9}{5}.$$

75. For simplicity, let the vertices of the rhombus be $(0, 0)$, $(a, 0)$, (b, c), and $(a + b, c)$, as shown in the figure. The slopes of the diagonals are then

$$m_1 = \frac{c}{a + b} \quad \text{and} \quad m_2 = \frac{c}{b - a}.$$

Since the sides of the Rhombus are equal and $a^2 = b^2 + c^2$, we have

$$m_1 m_2 = \frac{c}{a + b} \cdot \frac{c}{b - a} = \frac{c^2}{b^2 - a^2} = \frac{c^2}{-c^2} = -1.$$

Therefore, the diagonals are perpendicular.

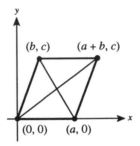

76. For simplicity, let the vertices of the quadrilateral be $(0, 0)$, $(a, 0)$, (b, c), and (d, e), as shown in the figure. The midpoints of the sides are

$$\left(\frac{a}{2}, 0\right), \quad \left(\frac{a + b}{c}, \frac{c}{2}\right), \quad \left(\frac{b + d}{2}, \frac{c + e}{2}\right), \quad \text{and} \quad \left(\frac{d}{2}, \frac{e}{2}\right).$$

The slopes of the opposite sides are equal.

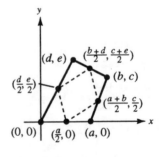

$$m_1 = \frac{\dfrac{c}{2} - 0}{\dfrac{a + b}{2} - \dfrac{a}{2}} = \frac{\dfrac{c + e}{2} - \dfrac{e}{2}}{\dfrac{b + d}{2} - \dfrac{d}{2}} = \frac{c}{b}$$

$$m_2 = \frac{0 - \dfrac{e}{2}}{\dfrac{a}{2} - \dfrac{d}{2}} = \frac{\dfrac{c}{2} - \dfrac{c + e}{2}}{\dfrac{a + b}{2} - \dfrac{b + d}{2}} = -\frac{e}{a - d}$$

Therefore, the figure is a parallelogram.

77. Since the triangles are similar, the result immediately follows.

$$\frac{y_2{}^* - y_1{}^*}{x_2{}^* - x_1{}^*} = \frac{y_2 - y_1}{x_2 - x_1}$$

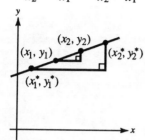

78. If $m_1 = -1/m_2$, then $m_1 m_2 = -1$. Let L_3 be a line with slope m_3 that is perpendicular to L_1. Then $m_1 m_3 = -1$. Hence, $m_2 = m_3 \Rightarrow L_2$ and L_3 are parallel. Therefore, L_2 and L_1 are also perpendicular.

79. True

$$ax + by = c_1 \Rightarrow y = -\frac{a}{b}x + \frac{c_1}{b} \Rightarrow m_1 = -\frac{a}{b}$$

$$bx - ay = c_2 \Rightarrow y = \frac{b}{a}x - \frac{c_2}{a} \Rightarrow m_2 = \frac{b}{a}$$

$$m_2 = -\frac{1}{m_1}$$

80. False; if m_1 is positive, then $m_2 = -1/m_1$ is negative.

81.
$$x^2 - 2x + y^2 = 0$$
$$x^2 - 2x + 1 + y^2 = 0 + 1$$
$$(x - 1)^2 + y^2 = 1$$

The line through (x, y) and $(4, 0)$ is perpendicular to the line through (x, y) and $(1, 0)$. Thus,

$$m_1 = \frac{y}{x - 4} \quad \text{and} \quad m_2 = \frac{y}{x - 1}.$$

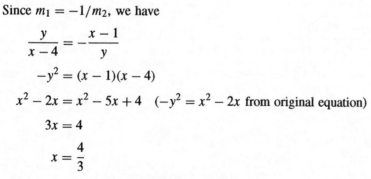

Since $m_1 = -1/m_2$, we have

$$\frac{y}{x - 4} = -\frac{x - 1}{y}$$

$$-y^2 = (x - 1)(x - 4)$$

$$x^2 - 2x = x^2 - 5x + 4 \quad (-y^2 = x^2 - 2x \text{ from original equation})$$

$$3x = 4$$

$$x = \frac{4}{3}$$

$$y = \pm\sqrt{2\left(\frac{4}{3}\right) - \left(\frac{4}{3}\right)^2} = \pm\frac{2\sqrt{2}}{3}$$

The two points of tangency are $(4/3, \pm 2\sqrt{2}/3)$. The slopes of the tangent lines are

$$m = \frac{(\pm 2\sqrt{2}/3) - 0}{(4/3) - 4} = \frac{\pm 2\sqrt{2}}{4 - 12} = \mp\frac{\sqrt{2}}{4}.$$

The equations of the tangent lines are

$$y = -\frac{\sqrt{2}}{4}x + \sqrt{2} \text{ and } y = \frac{\sqrt{2}}{4}x - \sqrt{2}.$$

82. The slope of the line $ax + by + c = 0$ is $-a/b$. The equation of the line through $(x_0,\ y_0)$ perpendicular to $ax + by + c = 0$ is:

$$y - y_0 = \frac{b}{a}(x - x_0)$$

$$ay - ay_0 = bx - bx_0$$

$$bx_0 - ay_0 = bx - ay$$

The point of intersection of these two lines is:

$$ax + by = -c \qquad \Rightarrow \quad a^2x + aby = -ac$$

$$\underline{bx - ay = bx_0 - ay_0 \ \Rightarrow \ b^2x - aby = b^2x_0 - aby_0}$$

$$(a^2 + b^2)x = -ac + b^2x_0 - aby_0$$

$$x = \frac{-ac + b^2x_0 - aby_0}{a^2 + b^2}$$

$$ax + by = -c \qquad \Rightarrow \quad abx + b^2y = -bc$$

$$\underline{bx - ay = bx_0 - ay_0 \ \Rightarrow \ -abx + a^2y = -abx_0 + a^2y_0}$$

$$(a^2 + b^2)y = -bc - abx_0 + a^2y_0$$

$$y = \frac{-bc - abx_0 + a^2y_0}{a^2 + b^2}$$

$$\left(\frac{-ac + b^2x_0 - aby_0}{a^2 + b^2},\ \frac{-bc - abx_0 + a^2y_0}{a^2 + b^2} \right)$$

The distance between $(x_0,\ y_0)$ and this point gives us the distance between $(x_0,\ y_0)$ and the line $ax + by + c = 0$.

$$d = \sqrt{\left[\frac{-ac + b^2x_0 - aby_0}{a^2 + b^2} - x_0 \right]^2 + \left[\frac{-bc - abx_0 + a^2y_0}{a^2 + b^2} - y_0 \right]^2}$$

$$= \sqrt{\left[\frac{-ac - aby_0 - a^2x_0}{a^2 + b^2} \right]^2 + \left[\frac{-bc - abx_0 - b^2y_0}{a^2 + b^2} \right]^2}$$

$$= \sqrt{\left[\frac{-a(c + by_0 + ax_0)}{a^2 + b^2} \right]^2 + \left[\frac{-b(c + ax_0 + by_0)}{a^2 + b^2} \right]^2}$$

$$= \sqrt{\frac{(a^2 + b^2)(c + ax_0 + by_0)^2}{(a^2 + b^2)^2}}$$

$$= \frac{|ax_0 + by_0 + c|}{\sqrt{a^2 + b^2}}$$

Section P.5 Functions

1. a. $f(0) = 2(0) - 3 = -3$
   b. $f(-3) = 2(-3) - 3 = -9$
   c. $f(b) = 2b - 3$
   d. $f(x - 1) = 2(x - 1) - 3 = 2x - 5$

2. a. $f\left(\dfrac{1}{2}\right) = \left(\dfrac{1}{2}\right)^2 - 2\left(\dfrac{1}{2}\right) + 2 = \dfrac{5}{4}$
   b. $f(-1) = (-1)^2 - 2(-1) + 2 = 5$
   c. $f(c) = c^2 - 2c + 2$
   d. $f(x + \Delta x) = (x + \Delta x)^2 - 2(x + \Delta x) + 2$
$$= x^2 + 2x\Delta x + (\Delta x)^2$$
$$- 2x - 2\Delta x + 2$$

3. a. $f(-2) = \sqrt{-2 + 3} = \sqrt{1} = 1$
   b. $f(6) = \sqrt{6 + 3} = \sqrt{9} = 3$
   c. $f(c) = \sqrt{c + 3}$
   d. $f(x + \Delta x) = \sqrt{x + \Delta x + 3}$

4. a. $f(2) = |2| + 4 = 2 + 4 = 6$
   b. $f(-2) = |-2| + 4 = 2 + 4 = 6$
   c. $f(x^2) = |x^2| + 4 = x^2 + 4$
   d. $f(x + \Delta x) - f(x) = |x + \Delta x| + 4 - (|x| + 4)$
$$= |x + \Delta x| - |x|$$

5. a. $f(-1) = 2(-1) + 1 = -1$
   b. $f(0) = 2(0) + 2 = 2$
   c. $f(2) = 2(2) + 2 = 6$
   d. $f(t^2 + 1) = 2(t^2 + 1) + 2 = 2t^2 + 4$
   (Note: $t^2 + 1 \geq 0$ for all real t.)

6. a. $f(-2) = (-2)^2 + 2 = 6$
   b. $f(0) = (0)^2 + 2 = 2$
   c. $f(1) = (1)^2 + 2 = 3$
   d. $f(s^2 + 2) = 2(s^2 + 2)^2 + 2 = 2s^2 + 8s + 10$
   (Note: $s^2 + 2 > 1$ for all real s.)

7. $\dfrac{f(2 + \Delta x) - f(2)}{\Delta x} = \dfrac{(2 + \Delta x)^2 - (2 + \Delta x) + 1 - [(2)^2 - 2 + 1]}{\Delta x}$

$$= \dfrac{4 + 4\Delta x + (\Delta x)^2 - 2 - \Delta x + 1 - 4 + 2 - 1}{\Delta x} = 4 - 1 + \Delta x = 3 + \Delta x, \quad \Delta x \neq 0$$

8. $\dfrac{f(1 + \Delta x) - f(1)}{\Delta x} = \dfrac{1/(1 + \Delta x) - 1}{\Delta x} = \dfrac{1 - (1 + \Delta x)}{\Delta x(1 + \Delta x)} = \dfrac{-1}{1 + \Delta x}, \quad \Delta x \neq 0$

9. $\dfrac{f(x + \Delta x) - f(x)}{\Delta x} = \dfrac{(x + \Delta x)^3 - x^3}{\Delta x} = \dfrac{x^3 + 3x^2\Delta x + 3x(\Delta x)^2 + (\Delta x)^3 - x^3}{\Delta x} = 3x^2 + 3x\Delta x + (\Delta x)^2, \quad \Delta x \neq 0$

10. $\dfrac{f(x) - f(1)}{x - 1} = \dfrac{3x - 1 - (3 - 1)}{x - 1} = \dfrac{3(x - 1)}{x - 1} = 3, \quad x \neq 1$

11. $\dfrac{f(x) - f(2)}{x - 2} = \dfrac{(1/\sqrt{x - 1} - 1)}{x - 2}$

$$= \dfrac{1 - \sqrt{x - 1}}{(x - 2)\sqrt{x - 1}} \cdot \dfrac{1 + \sqrt{x - 1}}{1 + \sqrt{x - 1}} = \dfrac{2 - x}{(x - 2)\sqrt{x - 1}(1 + \sqrt{x - 1})} = \dfrac{-1}{\sqrt{x - 1}(1 + \sqrt{x - 1})}, \quad x \neq 2$$

12. $\dfrac{f(x) - f(1)}{x - 1} = \dfrac{x^3 - x - 0}{x - 1} = \dfrac{x(x + 1)(x - 1)}{x - 1} = x(x + 1), \quad x \neq 1$

13. $f(x) = 4 - x$
Domain: $(-\infty, \infty)$
Range: $(-\infty, \infty)$

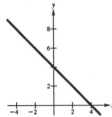

14. $g(x) = \dfrac{4}{x}$
Domain: $(-\infty, 0), (0, \infty)$
Range: $(-\infty, 0), (0, \infty)$

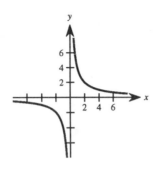

15. $h(x) = \sqrt{x - 1}$
Domain: $[1, \infty)$
Range: $[0, \infty)$

16. $f(x) = \frac{1}{2}x^3 + 2$
Domain: $(-\infty, \infty)$
Range: $(-\infty, \infty)$

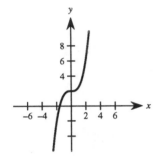

17. $f(x) = \sqrt{9 - x^2}$
Domain: $[-3, 3]$
Range: $[0, 3]$

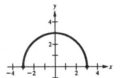

18. $h(x) = \sqrt{25 - x^2}$
Domain: $[-5, 5]$
Range: $[0, 5]$

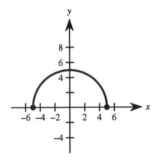

19. $f(x) = |x - 2|$
Domain: $(-\infty, \infty)$
Range: $[0, \infty)$

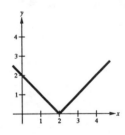

20. $f(x) = \dfrac{|x|}{x}$
Domain: $(-\infty, 0), (0, \infty)$
Range: $-1, 1$

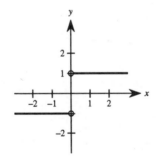

21. $f(x) = x + \sqrt{4 - x^2}$
Domain: $[-2, 2]$
Range:
$[-2, 2\sqrt{2}] \approx [-2, 2.83]$
y-intercept: $(0, 2)$
x-intercept: $(-\sqrt{2}, 0)$

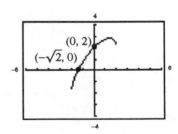

22. $f(x) = \dfrac{4}{\sqrt{x+3}+1}$

Domain: $[-3,\ \infty)$

Range: $(0,\ 4]$

y-intercept: $\left(0,\ \dfrac{4}{\sqrt{3}+1}\right) \approx (0,\ 1.46)$

x-intercept: none

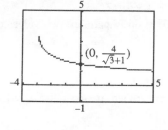

23. $p_1(x) = x^3 - x + 1$ has one zero. $p_2(x) = x^3 - x$ has three zeros. Every cubic polynomial has at least one zero. Given $p(x) = Ax^3 + Bx^2 + Cx + D$, we have $p \to -\infty$ as $x \to -\infty$ and $p \to \infty$ as $x \to \infty$ if $A > 0$ and $p \to \infty$ as $x \to -\infty$ and $p \to -\infty$ as $x \to \infty$ if $A < 0$. Since p is continuous, the graph must cross the x-axis at least one time.

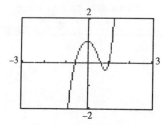

24. a. $f(x) = 3x^4 + 4x^3$

$f \to \infty$ as $x \to -\infty$

$f \to \infty$ as $x \to \infty$

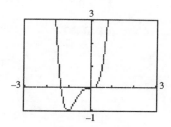

b. $f(x) = 2x^5 - 3x^2 + 1$

$f \to -\infty$ as $x \to -\infty$

$f \to \infty$ as $x \to \infty$

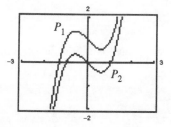

c. $f(x) = -2x^3 + 3x - 4$

$f \to \infty$ as $x \to -\infty$

$f \to -\infty$ as $x \to \infty$

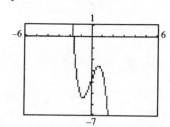

d. $f(x) = -10x^6 - 3x^5 + 2x$

$f \to -\infty$ as $x \to -\infty$

$f \to -\infty$ as $x \to \infty$

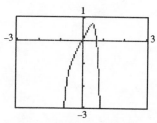

If the degree is even and the leading coefficient is positive, then the graph rises to the left and to the right. If the degree is even and the leading coefficient is negative, then the graph falls to the left and to the right. If the degree is odd and the leading coefficient is positive, then the graph rises to the right and falls to the left. If the degree is odd and the leading coefficient is negative, then the graph falls to the right and rises to the left.

25. $x - y^2 = 0 \Rightarrow y = \pm\sqrt{x}$

y is not a function of x.

26. $y = x^3 - 1$

y is a function of x.

27. $\sqrt{x^2 - 4} - y = 0 \Rightarrow y = \sqrt{x^2 - 4}$

y is a function of x.

28. $x^2 + y^2 = 9 \Rightarrow y = \pm\sqrt{9 - x^2}$

y is not a function of x.

29. $x^2 + y^2 = 4 \Rightarrow y = \pm\sqrt{4 - x^2}$

y is not a function of x since there are two values of y for some x.

30. $x = y^2 \Rightarrow y = \pm\sqrt{x}$

y is not a function of x since there are two values of y for some x.

31. $x^2 + y = 4 \Rightarrow y = 4 - x^2$

y is a function of x since there is one value of y for each x.

32. $x + y^2 = 4 \Rightarrow y = \pm\sqrt{4 - x}$

y is not a function of x since there are two values of y for some x.

33. $2x + 3y = 4 \Rightarrow y = \dfrac{4 - 2x}{3}$

y is a function of x since there is one value of y for each x.

34. $x^2 + y^2 - 4y = 0 \Rightarrow y^2 - 4y + 4 = -x^2 + 4$

$$(y - 2)^2 = 4 - x^2$$
$$y - 2 = \pm\sqrt{4 - x^2}$$
$$y = 2 \pm \sqrt{4 - x^2}$$

y is not a function of x since there are two values of y for some x.

35. $y^2 = x^2 - 1 \Rightarrow y = \pm\sqrt{x^2 - 1}$

y is not a function of x since there are two values of y for some x.

36. $x^2 y - x^2 + 4y = 0 \Rightarrow y = \dfrac{x^2}{x^2 + 4}$

y is a function of x since there is one value of y for each x.

37. a.

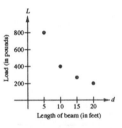

The value of L decreases as d increases.

b. L is a function of d.
Domain: $(0, \infty)$
Range: $(0, \infty)$

38.

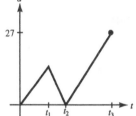

39. a. $y = \sqrt{x} + 2$

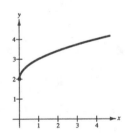

Vertical shift 2 units upward

b. $y = -\sqrt{x}$

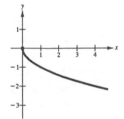

Reflection about the x-axis

39. –CONTINUED–

   c. $y = \sqrt{x - 2}$

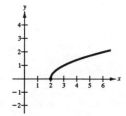

   Horizontal shift 2 units to the right

   d. $y = \sqrt{-x}$

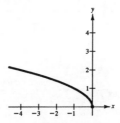

   Reflection about the y-axis

40. a. Horizontal shift 3 units to the left

     $y = (x + 3)^2$

  b. Vertical shift 3 units upward

     $y = x^2 + 3$

41. a. $g(x) = 4 - x^3 = -x^3 + 4$

    Reflection about the x-axis

    Vertical shift 4 units upward

  b. $g(x) = (x - 4)^3 + 2$

    Horizontal shift 4 units to the right

    Vertical shift 2 units upward

42. a. $h(x) = (x + 2)^3 + 1$

    Horizontal shift 2 units to the left

    Vertical shift 1 unit upward

  b. $h(x) = 5 - (x - 1)^3 = -(x - 1)^3 + 5$

    Reflection about the x-axis

    Horizontal shift 1 unit to the right

    Vertical shift 5 units upward

43. a. $f(g(1)) = f(0) = 0$

  b. $g(f(1)) = g(1) = 0$

  c. $g(f(0)) = g(0) = -1$

  d. $f(g(-4)) = f(15) = \sqrt{15}$

  e. $f(g(x)) = f(x^2 - 1) = \sqrt{x^2 - 1}$

  f. $g(f(x)) = g(\sqrt{x}) = (\sqrt{x})^2 - 1 = x - 1$

44. a. $f(g(2)) = f(3) = \dfrac{1}{3}$

  b. $g(f(2)) = g\left(\dfrac{1}{2}\right) = -\dfrac{3}{4}$

  c. $f\left(g\left(\dfrac{1}{\sqrt{2}}\right)\right) = f\left(-\dfrac{1}{2}\right) = -2$

  d. $g\left(f\left(\dfrac{1}{\sqrt{2}}\right)\right) = g(\sqrt{2}) = 1$

  e. $g(f(x)) = g\left(\dfrac{1}{x}\right) = \left(\dfrac{1}{x}\right)^2 - 1 = \dfrac{1 - x^2}{x^2}$

  f. $f(g(x)) = f(x^2 - 1) = \dfrac{1}{x^2 - 1}$

45. $f(x) = x^2,\ g(x) = \sqrt{x}$

  $(f \circ g)(x) = f(g(x)) = f(\sqrt{x}) = (\sqrt{x})^2 = x$

  Domain: $[0, \infty)$

  $(g \circ f)(x) = g(f(x)) = g(x^2) = \sqrt{x^2} = |x|$

  Domain: $(-\infty, \infty)$

  $(f \circ g) = (g \circ f)$ for $x \geq 0$.

46. $f(x) = x^3,\ g(x) = \sqrt[3]{x}$

  $(f \circ g)(x) = f(g(x)) = f(\sqrt[3]{x}) = (\sqrt[3]{x})^3 = x$

  Domain: $(-\infty, \infty)$

  $(g \circ f)(x) = g(f(x)) = g(x^3) = \sqrt[3]{x^3} = x$

  Domain: $(-\infty, \infty)$

  Yes, $(f \circ g) = (g \circ f)$.

47. $f(x) = x + 1$, $g(x) = \dfrac{1}{x}$

$(f \circ g)(x) = f(g(x)) = f\left(\dfrac{1}{x}\right) = \dfrac{1}{x} + 1 = \dfrac{1+x}{x}$

Domain: $(-\infty, 0)$, $(0, \infty)$

$(g \circ f)(x) = g(f(x)) = g(x + 1) = \dfrac{1}{x+1}$

Domain: $(-\infty, -1)$, $(-1, \infty)$

No, $(f \circ g) \neq (g \circ f)$.

48. $f(x) = x^2 - 1$, $g(x) = x$

$(f \circ g)(x) = f(g(x)) = f(x) = x^2 - 1$

Domain: $(-\infty, \infty)$

$(g \circ f)(x) = g(f(x)) = g(x^2 - 1) = x^2 - 1$

Domain: $(-\infty, \infty)$

Yes, $(f \circ g) = (g \circ f)$.

49. $(A \circ r)(t) = A(r(t)) = A(0.6t) = \pi(0.6t)^2 = 0.36\pi t^2$

$(A \circ r)(t)$ represents the area of the circle at time t.

50. a.

b. $H(1.6x) = 0.002(1.6x)^2 + 0.005(1.6x) - 0.029$

$= 0.00512x^2 + 0.008x - 0.029$, $\dfrac{10}{1.6} < x < \dfrac{100}{1.6} \Rightarrow 6.25 < x < 62.5$

51. $\dfrac{3}{x-1} + \dfrac{4}{x-2} = 0$

$\dfrac{3}{x-1} = -\dfrac{4}{x-2}$

$3(x - 2) = -4(x - 1)$

$3x - 6 = -4x + 4$

$7x = 10$

$x = \dfrac{10}{7}$

52. $x^3 - x = 0$

$x(x^2 - 1) = 0$

$x(x + 1)(x - 1) = 0$

$x = 0$, $x = -1$, $x = 1$

53. $f(-x) = 4 - (-x)^2 = 4 - x^2 = f(x)$

Even

54. $f(-x) = \sqrt[3]{-x} = -\sqrt[3]{x} = -f(x)$

Odd

55. $f(-x) = (-x)[4 - (-x)^2]$

$= -x(4 - x^2) = -f(x)$

Odd

56. $f(-x) = 4(-x) - (-x)^2 = -4x - x^2$

Neither odd nor even

57. $f(-x) = a_{2n+1}(-x)^{2n+1} + \cdots + a_3(-x)^3 + a_1(-x)$

$= -[a_{2n+1}x^{2n+1} + \cdots + a_3 x^3 + a_1 x]$

$= -f(x)$

Odd

58. $f(-x) = a_{2n}(-x)^{2n} + a_{2n-2}(-x)^{2n-2} + \cdots + a_2(-x)^2 + a_0$

$= a_{2n}x^{2n} + a_{2n-2}x^{2n-2} + \cdots + a_2x^2 + a_0$

$= f(x)$

Even

59. Let $F(x) = f(x)g(x)$ where f and g are even. Then

$F(-x) = f(-x)g(-x) = f(x)g(x) = F(x).$

Thus, $F(x)$ is even. Let $F(x) = f(x)g(x)$ where f and g are odd. Then

$F(-x) = f(-x)g(-x) = [-f(x)][-g(x)] = f(x)g(x) = F(x).$

Thus, $F(x)$ is even.

60. Let $F(x) = f(x)g(x)$ where f is even and g is odd. Then

$F(-x) = f(-x)g(-x) = f(x)[-g(x)] = -f(x)g(x) = -F(x).$

Thus, $F(x)$ is odd.

61. $f(x) = x^2 + 1$ and $g(x) = x^4$ are even. $f(x) = x^3 - x$ is odd and $g(x) = x^2$ is even.

$f(x)g(x) = (x^2+1)(x^4) = x^6 + x^4$ is even. $f(x)g(x) = (x^3-x)(x^2) = x^5 - x^3$ is odd.

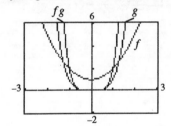   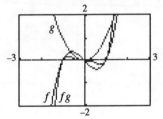

62. a. Let $F(x) = f(x) \pm g(x)$ where f and g are even. Then, $F(-x) = f(-x) \pm g(-x) = f(x) \pm g(x) = F(x)$.
Thus, $F(x)$ is even.

b. Let $F(x) = f(x) \pm g(x)$ where f and g are odd. Then, $F(-x) = f(-x) \pm g(-x) = -f(x) \mp g(x) = -F(x)$.
Thus, $F(x)$ is odd.

c. Let $F(x) = f(x) \pm g(x)$ where f is odd and g is even. Then, $F(-x) = f(-x) \pm g(-x) = -f(x) \pm g(x)$.
Thus, $F(x)$ is neither odd nor even.

63. $R = 4 - \dfrac{x^2}{2}$ **64.** $R = x^2$ **65.** $h = x^2$ **66.** $h = \sqrt{4 - x^2}$

$r = 2$ $r = x^3$ $p = x$ $p = x$

67. $2x + 2y = 100 \Rightarrow A = xy = \dfrac{x(100 - 2x)}{2} = x(50 - x)$

68. $4x + 3y = 200 \Rightarrow A = 2xy = 2x\left[\dfrac{200 - 4x}{3}\right] = \dfrac{8}{3}(50x - x^2)$

69. $V = lwh = (12 - 2x)^2 x$ **70.** $y = \sqrt{25 - x^2}$

$= 4x(6 - x)^2$ $A = 2xy = 2x\sqrt{25 - x^2}$

71. $4x + y = 108$

$V = x^2 y = x^2(108 - 4x) = 108x^2 - 4x^3$

72. $2x^2 + 4xy = 100$

$$V = x^2 y = x^2 \left(\frac{100 - 2x^2}{4x} \right) = 25x - \frac{x^3}{2}$$

73. $T = \dfrac{D}{R}$

$T = T_{\text{row}} + T_{\text{walk}}$

$$= \frac{\sqrt{x^2 + 4}}{2} + \frac{\sqrt{1 + (3 - x)^2}}{4}$$

$$= \frac{\sqrt{x^2 + 4}}{2} + \frac{\sqrt{x^2 - 6x + 10}}{4}$$

74. $A = \pi r^2 = \pi y^2 = \pi (\sqrt{x})^2 = \pi x$

75. False; let $f(x) = x^2$.

Then $f(-3) = f(3) = 9$, but $-3 \neq 3$.

76. True

77. True

78. False; let $f(x) = x^2$. Then $f(3x) = (3x)^2 = 9x^2$ and $3f(x) = 3x^2$. Thus, $3f(x) \neq f(3x)$.

79. Let

$$f(x) = \frac{f(x) + f(-x) - f(-x) + f(x)}{2} = \left[\frac{f(x) + f(-x)}{2} \right] + \left[\frac{f(x) - f(-x)}{2} \right] = f_1(x) + f_2(x).$$

Then,

$$f_1(-x) = \frac{f(-x) + f(x)}{2} = f_1(x) \quad \text{and} \quad f_2(-x) = \frac{f(-x) - f(x)}{2} = -f_2(x).$$

Thus, $f_1(x)$ is even and $f_2(x)$ is odd.

80. $g(x) = \dfrac{1}{1 - x}$

Domain: $(-\infty, \ 1) \cup (1, \ \infty)$

Range: $(-\infty, \ 0) \cup (0, \ \infty)$

$$(g \circ g \circ g)(x) = g(g(g(x))) = g\left(g\left(\frac{1}{1 - x} \right) \right) = g\left(\frac{1}{1 - \dfrac{1}{1 - x}} \right) = g\left(\frac{1 - x}{-x} \right) = \frac{1}{1 - \left(\dfrac{1 - x}{-x} \right)} = x$$

Domain: All real x except $x = 0, \ 1$

Range: All real x except $x = 0, \ 1$

81. a. $f(x) = |x - 3| - 2$

If $x \geq 3$, then $|x - 3| = x - 3$ and $f(x) = (x - 3) - 2 = x - 5$.

If $x < 3$, then $|x - 3| = -(x - 3)$ and $f(x) = -(x - 3) - 2 = -x + 1$.

Thus,

$$f(x) = \begin{cases} 1 - x, & x < 3 \\ x - 5, & x \geq 3. \end{cases}$$

b. $f(x) = |x| + |x - 2|$

If $x < 0$, then $f(x) = -x - (x - 2) = -2x + 2 = 2(1 - x)$.

If $0 \leq x < 2$, then $f(x) = x - (x - 2) = 2$.

If $x \geq 2$, then $f(x) = x + (x - 2) = 2x - 2 = 2(x - 1)$.

Thus,

$$f(x) = \begin{cases} 2(1 - x), & x < 0 \\ 2, & 0 \leq x < 2 \\ 2(x - 1), & x \geq 2. \end{cases}$$

Section P.6   Review of Trigonometric Functions

1. a. $396°$, $-324°$

   b. $240°$, $-480°$

2. a. $660°$, $-60°$

   b. $300°$, $-60°$

3. a. $\dfrac{19\pi}{9}$, $-\dfrac{17\pi}{9}$

   b. $\dfrac{10\pi}{3}$, $-\dfrac{2\pi}{3}$

4. a. $\dfrac{7\pi}{4}$, $-\dfrac{\pi}{4}$

   b. $\dfrac{26\pi}{9}$, $-\dfrac{10\pi}{9}$

5. a. $30\left(\dfrac{\pi}{180}\right) = \dfrac{\pi}{6} \approx 0.524$

   b. $150\left(\dfrac{\pi}{180}\right) = \dfrac{5\pi}{6} \approx 2.618$

   c. $315\left(\dfrac{\pi}{180}\right) = \dfrac{7\pi}{4} \approx 5.498$

   d. $120\left(\dfrac{\pi}{180}\right) = \dfrac{2\pi}{3} \approx 2.094$

6. a. $-20\left(\dfrac{\pi}{180}\right) = -\dfrac{\pi}{9} \approx -0.349$

   b. $-240\left(\dfrac{\pi}{180}\right) = -\dfrac{4\pi}{3} \approx -4.189$

   c. $-270\left(\dfrac{\pi}{180}\right) = -\dfrac{3\pi}{2} \approx -4.712$

   d. $144\left(\dfrac{\pi}{180}\right) = \dfrac{4\pi}{5} \approx 2.513$

7. a. $\dfrac{3\pi}{2}\left(\dfrac{180}{\pi}\right) = 270°$

   b. $\dfrac{7\pi}{6}\left(\dfrac{180}{\pi}\right) = 210°$

   c. $-\dfrac{7\pi}{12}\left(\dfrac{180}{\pi}\right) = -105°$

   d. $-2.367\left(\dfrac{180}{\pi}\right) \approx -135.6°$

8. a. $\dfrac{7\pi}{3}\left(\dfrac{180}{\pi}\right) = 420°$

   b. $-\dfrac{11\pi}{30}\left(\dfrac{180}{\pi}\right) = -66°$

   c. $\dfrac{11\pi}{6}\left(\dfrac{180}{\pi}\right) = 330°$

   d. $0.438\left(\dfrac{180}{\pi}\right) \approx 25.1°$

9.

r	8 ft	15 in.	85 cm	24 in.	$\dfrac{12963}{\pi}$ mi
s	12 ft	24 in.	63.75π cm	96 in.	8642 mi
θ	1.5	1.6	$\dfrac{3\pi}{4}$	4	$\dfrac{2\pi}{3}$

10.　　$s = r\theta$

$\frac{1}{2}$ in. $= 2$ in. θ

$\theta = \frac{1}{4}$ radian $\approx 14.32°$

11.　　$s = r\theta$

12 in. $= 4$ in. θ

$\theta = 3$ radians $\approx 171.89°$

12. a. 50 mph $= \dfrac{50(5280)}{60} = 4400$ ft/min

Circumference of tire: $C = 2.5\pi$ feet

Revolutions per minute: $\dfrac{4400}{2.5\pi} \approx 560.2$

b. $\theta = \dfrac{4400}{2.5\pi}(2\pi) = 3520$ radians

Angular speed:

$\dfrac{\theta}{t} = \dfrac{3520 \text{ radians}}{1 \text{ minute}} = 3520$ rad/min

13. a. $x = 3, \quad y = 4, \quad r = 5$

$\sin\theta = \frac{4}{5}$　　$\csc\theta = \frac{5}{4}$

$\cos\theta = \frac{3}{5}$　　$\sec\theta = \frac{5}{3}$

$\tan\theta = \frac{4}{3}$　　$\cot\theta = \frac{3}{4}$

b. $x = -12, \quad y = -5, \quad r = 13$

$\sin\theta = -\frac{5}{13}$　　$\csc\theta = -\frac{13}{5}$

$\cos\theta = -\frac{12}{13}$　　$\sec\theta = -\frac{13}{12}$

$\tan\theta = \frac{5}{12}$　　$\cot\theta = \frac{12}{5}$

14. a. $x = 8, \quad y = -15, \quad r = 17$

$\sin\theta = -\dfrac{15}{17}$　　$\csc\theta = -\dfrac{17}{15}$

$\cos\theta = \dfrac{8}{17}$　　$\sec\theta = \dfrac{17}{8}$

$\tan\theta = -\dfrac{15}{8}$　　$\cot\theta = -\dfrac{8}{15}$

b. $x = 1, \quad y = -1, \quad r = \sqrt{2}$

$\sin\theta = -\dfrac{\sqrt{2}}{2}$　$\csc\theta = -\sqrt{2}$

$0\cos\theta = \dfrac{\sqrt{2}}{2}$　$\sec\theta = \sqrt{2}$

$\tan\theta = -1$　　$\cot\theta = -1$

15. a. $\sin\theta < 0 \Rightarrow \theta$ is in Quadrant III or IV.

$\cos\theta < 0 \Rightarrow \theta$ is in Quadrant II or III.

$\sin\theta < 0$ **and** $\cos\theta < 0 \Rightarrow \theta$ is in Quadrant III.

b. $\sec\theta > 0 \Rightarrow \theta$ is in Quadrant I or IV.

$\cot\theta < 0 \Rightarrow \theta$ is in Quadrant II or IV.

$\sec\theta > 0$ **and** $\cot\theta < 0 \Rightarrow \theta$ is in Quadrant IV.

16. a. $\sin\theta > 0 \Rightarrow \theta$ is in Quadrant I or II.

$\cos\theta < 0 \Rightarrow \theta$ is in Quadrant II or III.

$\sin\theta > 0$ **and** $\cos\theta < 0 \Rightarrow \theta$ is in Quadrant II.

b. $\csc\theta < 0 \Rightarrow \theta$ is in Quadrant III or IV.

$\tan\theta > 0 \Rightarrow \theta$ is in Quadrant I or III.

$\csc\theta < 0$ **and** $\tan\theta > 0 \Rightarrow \theta$ is in Quadrant III.

17. $x^2 + 1^2 = 2^2 \Rightarrow x = \sqrt{3}$

$$\cos\theta = \frac{x}{2} = \frac{\sqrt{3}}{2}$$

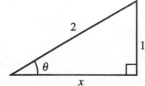

18. $x^2 + 1^2 = 3^2 \Rightarrow x = \sqrt{8} = 2\sqrt{2}$

$$\tan\theta = \frac{1}{x} = \frac{1}{2\sqrt{2}} = \frac{\sqrt{2}}{4}$$

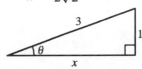

19. $4^2 + y^2 = 5^2 \Rightarrow y = 3$

$$\cot\theta = \frac{4}{y} = \frac{4}{3}$$

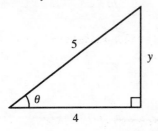

20. $5^2 + y^2 = 13^2 \Rightarrow y = 12$

$$\csc\theta = \frac{13}{y} = \frac{13}{12}$$

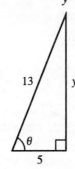

21. $15^2 + 8^2 = r^2 \Rightarrow r = 17$

$$\sec\theta = \frac{r}{15} = \frac{17}{15}$$

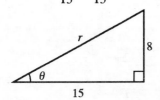

22. $2^2 + 1^2 = r^2 \Rightarrow r = \sqrt{5}$

$$\sin\theta = \frac{1}{r} = \frac{1}{\sqrt{5}} = \frac{\sqrt{5}}{5}$$

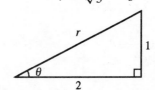

23. a. $\sin 60° = \dfrac{\sqrt{3}}{2}$

$\cos 60° = \dfrac{1}{2}$

$\tan 60° = \sqrt{3}$

c. $\sin \dfrac{\pi}{4} = \dfrac{\sqrt{2}}{2}$

$\cos \dfrac{\pi}{4} = \dfrac{\sqrt{2}}{2}$

$\tan \dfrac{\pi}{4} = 1$

b. $\sin 120° = \sin 60° = \dfrac{\sqrt{3}}{2}$

$\cos 120° = -\cos 60° = -\dfrac{1}{2}$

$\tan 120° = -\tan 60° = -\sqrt{3}$

d. $\sin \dfrac{5\pi}{4} = -\sin \dfrac{\pi}{4} = -\dfrac{\sqrt{2}}{2}$

$\cos \dfrac{5\pi}{4} = -\cos \dfrac{\pi}{4} = -\dfrac{\sqrt{2}}{2}$

$\tan \dfrac{5\pi}{4} = \tan \dfrac{\pi}{4} = 1$

24. a. $\sin(-30°) = -\sin 30° = -\dfrac{1}{2}$

$\cos(-30°) = \cos 30° = \dfrac{\sqrt{3}}{2}$

$\tan(-30°) = -\tan 30° = -\dfrac{\sqrt{3}}{3}$

c. $\sin\left(-\dfrac{\pi}{6}\right) = -\sin\dfrac{\pi}{6} = -\dfrac{1}{2}$

$\cos\left(-\dfrac{\pi}{6}\right) = \cos\dfrac{\pi}{6} = \dfrac{\sqrt{3}}{2}$

$\tan\left(-\dfrac{\pi}{6}\right) = -\tan\dfrac{\pi}{6} = -\dfrac{\sqrt{3}}{3}$

b. $\sin 150° = \sin 30° = \dfrac{1}{2}$

$\cos 150° = -\cos 30° = -\dfrac{\sqrt{3}}{2}$

$\tan 150° = -\tan 30° = -\dfrac{\sqrt{3}}{3}$

d. $\sin\dfrac{\pi}{2} = 1$

$\cos\dfrac{\pi}{2} = 0$

$\tan\dfrac{\pi}{2}$ is undefined.

25. a. $\sin 225° = -\sin 45° = -\dfrac{\sqrt{2}}{2}$

$\cos 225° = -\cos 45° = -\dfrac{\sqrt{2}}{2}$

$\tan 225° = \tan 45° = 1$

c. $\sin\dfrac{5\pi}{3} = -\sin\dfrac{\pi}{3} = -\dfrac{\sqrt{3}}{2}$

$\cos\dfrac{5\pi}{3} = \cos\dfrac{\pi}{3} = \dfrac{1}{2}$

$\tan\dfrac{5\pi}{3} = -\tan\dfrac{\pi}{3} = -\sqrt{3}$

b. $\sin(-225°) = \sin 45° = \dfrac{\sqrt{2}}{2}$

$\cos(-225°) = -\cos 45° = -\dfrac{\sqrt{2}}{2}$

$\tan(-225°) = -\tan 45° = -1$

d. $\sin\dfrac{11\pi}{6} = -\sin\dfrac{\pi}{6} = -\dfrac{1}{2}$

$\cos\dfrac{11\pi}{6} = \cos\dfrac{\pi}{6} = \dfrac{\sqrt{3}}{2}$

$\tan\dfrac{11\pi}{6} = -\tan\dfrac{\pi}{6} = -\dfrac{\sqrt{3}}{3}$

26. a. $\sin 750° = \sin 30° = \dfrac{1}{2}$

$\cos 750° = \cos 30° = \dfrac{\sqrt{3}}{2}$

$\tan 750° = \tan 30° = \dfrac{\sqrt{3}}{3}$

c. $\sin\dfrac{10\pi}{3} = -\sin\dfrac{\pi}{3} = -\dfrac{\sqrt{3}}{2}$

$\cos\dfrac{10\pi}{3} = -\cos\dfrac{\pi}{3} = -\dfrac{1}{2}$

$\tan\dfrac{10\pi}{3} = \tan\dfrac{\pi}{3} = \sqrt{3}$

b. $\sin 510° = \sin 30° = \dfrac{1}{2}$

$\cos 510° = -\cos 30° = -\dfrac{\sqrt{3}}{2}$

$\tan 510° = -\tan 30° = -\dfrac{\sqrt{3}}{3}$

d. $\sin\dfrac{17\pi}{3} = -\sin\dfrac{\pi}{3} = -\dfrac{\sqrt{3}}{2}$

$\cos\dfrac{17\pi}{3} = \cos\dfrac{\pi}{3} = \dfrac{1}{2}$

$\tan\dfrac{17\pi}{3} = -\tan\dfrac{\pi}{3} = -\sqrt{3}$

27. a. $\sin 10° \approx 0.1736$
b. $\csc 10° \approx 5.759$

28. a. $\sec 225° \approx -1.414$
b. $\sec 135° \approx -1.414$

29. a. $\tan\dfrac{\pi}{9} \approx 0.3640$

b. $\tan\dfrac{10\pi}{9} \approx 0.3640$

30. a. $\cot 1.35 \approx 0.2245$
b. $\tan 1.35 \approx 4.455$

31. a. $\cos \theta = \dfrac{\sqrt{2}}{2}$

$\theta = \dfrac{\pi}{4}, \ \dfrac{7\pi}{4}$

b. $\cos \theta = -\dfrac{\sqrt{2}}{2}$

$\theta = \dfrac{3\pi}{4}, \ \dfrac{5\pi}{4}$

33. a. $\tan \theta = 1$

$\theta = \dfrac{\pi}{4}, \ \dfrac{5\pi}{4}$

b. $\cot \theta = -\sqrt{3}$

$\theta = \dfrac{5\pi}{6}, \ \dfrac{11\pi}{6}$

35. $2 \sin^2 \theta = 1$

$\sin \theta = \pm \dfrac{\sqrt{2}}{2}$

$\theta = \dfrac{\pi}{4}, \ \dfrac{3\pi}{4}, \ \dfrac{5\pi}{4}, \ \dfrac{7\pi}{4}$

37. $\tan^2 \theta - \tan \theta = 0$

$\tan \theta (\tan \theta - 1) = 0$

$\tan \ \theta = 0 \qquad\qquad \tan \ \theta = 1$

$\theta = 0, \ \pi \qquad\qquad \theta = \dfrac{\pi}{4}, \ \dfrac{5\pi}{4}$

39. $\sec \theta \csc \theta - 2 \csc \theta = 0$

$\csc \theta (\sec \theta - 2) = 0$

$(\csc \theta \neq 0 \text{ for any value of } \theta)$

$\sec \theta = 2$

$\theta = \dfrac{\pi}{3}, \ \dfrac{5\pi}{3}$

41. $\cos^2 \theta + \sin \theta = 1$

$1 - \sin^2 \theta + \sin \theta = 1$

$\sin^2 \theta - \sin \theta = 0$

$\sin \theta (\sin \theta - 1) = 0$

$\sin \theta = 0 \qquad\qquad \sin \theta = 1$

$\theta = 0, \ \pi \qquad\qquad \theta = \pi/2$

32. a. $\sec \theta = 2$

$\theta = \dfrac{\pi}{3}, \ \dfrac{5\pi}{3}$

b. $\sec \theta = -2$

$\theta = \dfrac{2\pi}{3}, \ \dfrac{4\pi}{3}$

34. a. $\sin \theta = \dfrac{\sqrt{3}}{2}$

$\theta = \dfrac{\pi}{3}, \ \dfrac{2\pi}{3}$

b. $\sin \theta = -\dfrac{\sqrt{3}}{2}$

$\theta = \dfrac{4\pi}{3}, \ \dfrac{5\pi}{3}$

36. $\tan^2 \theta = 3$

$\tan \theta = \pm \sqrt{3}$

$\theta = \dfrac{\pi}{3}, \ \dfrac{2\pi}{3}, \ \dfrac{4\pi}{3}, \ \dfrac{5\pi}{3}$

38. $2 \cos^2 \theta - \cos \theta - 1 = 0$

$(2 \cos \theta + 1)(\cos \theta - 1) = 0$

$\cos \ \theta = -\dfrac{1}{2} \qquad\qquad \cos \ \theta = 1$

$\theta = \dfrac{2\pi}{3}, \ \dfrac{4\pi}{3} \qquad\qquad \theta = 0$

40. $\sin \theta = \cos \theta$

$\tan \theta = 1$

$\theta = \dfrac{\pi}{4}, \ \dfrac{5\pi}{4}$

42. $\cos(\theta/2) - \cos \theta = 1$

$\cos(\theta/2) = \cos \theta + 1$

$\sqrt{(1/2)(1 + \cos \theta)} = \cos \theta + 1$

$(1/2)(1 + \cos \theta) = \cos^2 \theta + 2 \cos \theta + 1$

$0 = \cos^2 \theta + (3/2) \cos \theta + (1/2)$

$0 = (1/2)(2 \cos^2 \theta + 3 \cos \theta + 1)$

$0 = (1/2)(2 \cos \theta + 1)(\cos \theta + 1)$

$\cos \theta = -1/2 \qquad \cos \theta = -1$

$\theta = 2\pi/3 \qquad\qquad \theta = \pi$

$(\theta = 4\pi/3 \text{ is extraneous.})$

43. $\tan 30° = \dfrac{y}{100}$

$$y = 100\left(\dfrac{1}{\sqrt{3}}\right) = \dfrac{100\sqrt{3}}{3}$$

44. $\cos 60° = \dfrac{x}{10}$

$$x = 10\left(\dfrac{1}{2}\right) = 5$$

45. $\tan 60° = \dfrac{25}{x}$

$$x = \dfrac{25}{\sqrt{3}} = \dfrac{25\sqrt{3}}{3}$$

46. $\sin 45° = \dfrac{30}{r}$

$$r = \dfrac{30}{1/\sqrt{2}} = 30\sqrt{2}$$

47. $(275\ \text{ft/sec})(60\ \text{sec}) = 16{,}500$ feet

$$\sin 18° = \dfrac{a}{16{,}500}$$

$$a = 16{,}500\sin 18° \approx 5099 \text{ feet}$$

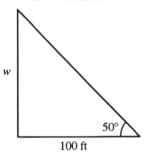

48. $\tan 50° = \dfrac{w}{100}$

$$w = 100\tan 50° \approx 119.175 \text{ feet}$$

49.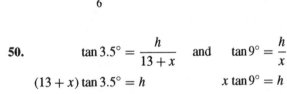

Diameter $= 2 + 2x$ where $\tan 3° = x/6$.

$x = 6\tan 3°$

Thus, $d = 2 + 2(6\tan 3°) \approx 2.63$ inches.

50. $\tan 3.5° = \dfrac{h}{13 + x}$ and $\tan 9° = \dfrac{h}{x}$

$(13 + x)\tan 3.5° = h$ $x\tan 9° = h$

$13\tan 3.5° + x\tan 3.5° = x\tan 9°$

$13\tan 3.5° = x(\tan 9° - \tan 3.5°)$

$\dfrac{13\tan 3.5°}{\tan 9° - \tan 3.5°} = x$

$h = x\tan 9° = \dfrac{13\tan 3.5° \tan 9°}{\tan 9° - \tan 3.5°} \approx 1.295$ miles or 6839.307 feet

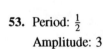

51. a. Period: π

Amplitude: 2

b. Period: 2

Amplitude: $\frac{1}{2}$

52. a. Period: 4π

Amplitude: $\frac{3}{2}$

b. Period: 6π

Amplitude: 2

53. Period: $\frac{1}{2}$

Amplitude: 3

54. Period: 20
Amplitude: $\frac{2}{3}$

55. Period: $\frac{\pi}{2}$

56. Period: $\frac{1}{2}$

57. Period: $\frac{2\pi}{5}$

58. Period: $\frac{\pi}{2}$

59. a. $f(x) = c \sin x$; changing c changes the amplitude.
When $c = -2$: $f(x) = -2 \sin x$.
When $c = -1$: $f(x) = -\sin x$.
When $c = 1$: $f(x) = \sin x$.
When $c = 2$: $f(x) = 2 \sin x$.

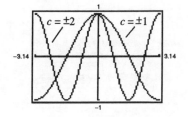

b. $f(x) = \cos(cx)$; changing c changes the period.
When $c = -2$: $f(x) = \cos(-2x) = \cos 2x$.
When $c = -1$: $f(x) = \cos(-x) = \cos x$.
When $c = 1$: $f(x) = \cos x$.
When $c = 2$: $f(x) = \cos 2x$.

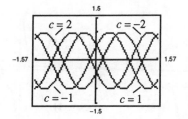

c. $f(x) = \cos(\pi x - c)$; changing c causes a horizontal shift.
When $c = -2$: $f(x) = \cos(\pi x + 2)$.
When $c = -1$: $f(x) = \cos(\pi x + 1)$.
When $c = 1$: $f(x) = \cos(\pi x - 1)$.
When $c = 2$: $f(x) = \cos(\pi x - 2)$.

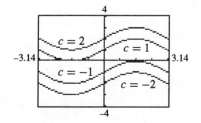

60. a. $f(x) = \sin x + c$; changing c causes a vertical shift.
When $c = -2$: $f(x) = \sin x - 2$.
When $c = -1$: $f(x) = \sin x - 1$.
When $c = 1$: $f(x) = \sin x + 1$.
When $c = 2$: $f(x) = \sin x + 2$.

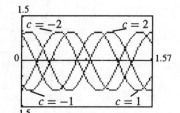

b. $f(x) = -\sin(2\pi x - c)$; changing c causes a horizontal shift.
When $c = -2$: $f(x) = -\sin(2\pi x + 2)$.
When $c = -1$: $f(x) = -\sin(2\pi x + 1)$.
When $c = 1$: $f(x) = -\sin(2\pi x - 1)$.
When $c = 2$: $f(x) = -\sin(2\pi x - 2)$.

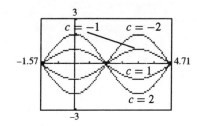

c. $f(x) = c \cos x$; changing c changes the amplitude.
When $c = -2$: $f(x) = -2 \cos x$.
When $c = -1$: $f(x) = -\cos x$.
When $c = 1$: $f(x) = \cos x$.
When $c = 2$: $f(x) = 2 \cos x$.

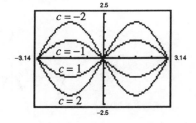

61. $y = \sin \dfrac{x}{2}$
Period: 4π
Amplitude: 1

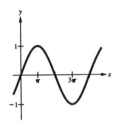

62. $y = 2\cos 2x$
Period: π
Amplitude: 2

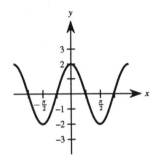

63. $y = -\sin \dfrac{2\pi x}{3}$
Period: 3
Amplitude: 1

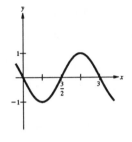

64. $y = 2\tan x$
Period: π

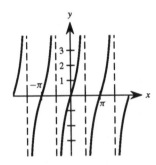

65. $y = \csc \dfrac{x}{2}$
Period: 4π

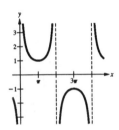

66. $y = \tan 2x$
Period: $\dfrac{\pi}{2}$

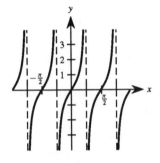

67. $y = 2\sec 2x$
Period: π

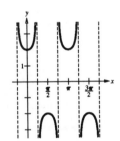

68. $y = \csc 2\pi x$
Period: 1

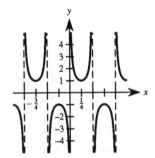

69. $y = \sin(x + \pi)$
Period: 2π
Amplitude: 1

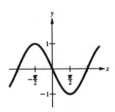

70. $y = \cos\left(x - \dfrac{\pi}{3}\right)$

Period: 2π

Amplitude: 1

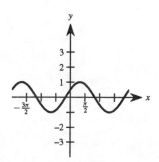

71. $y = 1 + \cos\left(x - \dfrac{\pi}{2}\right)$

Period: 2π

Amplitude: 1

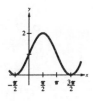

72. $y = 1 + \sin\left(x + \dfrac{\pi}{2}\right)$

Period: 2π

Amplitude: 1

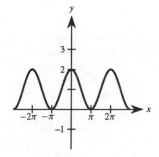

73. a. $f(x) = 2\cos 2x + 3\sin 3x$ has a period of 2π.

$g(x) = 2\cos 2x + 3\sin 4x$ has a period of π.

b. $h(x) = A\cos\alpha x + B\sin\beta x$ is periodic. The period of $a\cos\alpha x$ is $2\pi/\alpha$. The period of $B\sin\beta x$ is $2\pi/\beta$. The period of h is the least common multiple of these two periods.

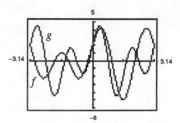

74. a. $5.35 - 2 = 3.35$

$5.35 + 2 = 7.35$

b. $5.35 - 2(3) = -0.65$

c. $13.35 = 5.35 + 2(4)$

$-4.65 = 5.35 - 2(5)$

True; since f and g have periods of 2 and intersect at $x = 5.35$, $f(13.35) = g(-4.65)$.

75. $y = a\cos(bx - c)$

From the graph, we see that the amplitude is 3, the period is 4π and the horizontal shift is π. Thus,

$a = 3$

$\dfrac{2\pi}{b} = 4\pi \;\Rightarrow\; b = \dfrac{1}{2}$

$\dfrac{c}{b} = \pi \;\Rightarrow\; c = \dfrac{\pi}{2}.$

Therefore, $y = 3\cos[(1/2)x - (\pi/2)]$.

76. $y = a\sin(bx - c)$

From the graph, we see that the amplitude is $\frac{1}{2}$, the period is π and the horizontal shift is 0. Also, the graph is reflected about the x-axis. Thus,

$a = -\dfrac{1}{2}$

$\dfrac{2\pi}{b} = \pi \;\Rightarrow\; b = 2$

$\dfrac{c}{b} = 0 \;\Rightarrow\; c = 0.$

Therefore, $y = -\frac{1}{2}\sin 2x$.

77. a.

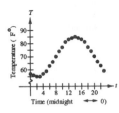

c.

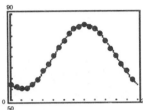

b. Amplitude:
$$\frac{1}{2}(85 - 55) = 15 \;\Rightarrow\; b = 15$$
Vertical shift:
$$85 - 15 = 55 + 15 = 70 \;\Rightarrow\; a = 70$$
Length of $\frac{1}{2}$ cycle:
$$20 - 8 = 12$$
Period:
$$2(12) = 24 \;\Rightarrow\; \frac{2\pi}{\omega} = 24 \;\Rightarrow\; \omega = \frac{\pi}{12}$$
Shift:
$$\frac{\delta}{\omega} = 8 \;\Rightarrow\; \frac{\delta}{\pi/12} = 8 \;\Rightarrow\; \delta = \frac{2\pi}{3}$$
Thus, $T(t) = 70 + 15\sin\left(\dfrac{\pi}{12}t - \dfrac{2\pi}{3}\right).$

78. If $h = 50 + 50\sin\left(8\pi t - \dfrac{\pi}{2}\right)$, then $h = 0$ when $t = 0$.

79. $y = 0.001\sin 880\pi t$

a. $p = \dfrac{2\pi}{880\pi} = \dfrac{1}{440}$

b. $f = \dfrac{1}{p} = 440$

c.

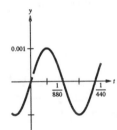

80. $P = 100 - 20\cos\dfrac{5\pi t}{3}$

a. Period: $\dfrac{2\pi}{5\pi/3} = \dfrac{6}{5}$ sec

b. $60 \cdot \dfrac{5}{6} = 50$ heartbeats per min

c.

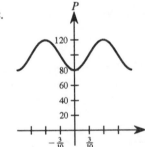

81.

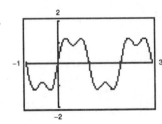

$$f(x) = \frac{4}{\pi}\left(\sin \pi x + \frac{1}{3}\sin 3\pi x\right)$$

$$g(x) = \frac{4}{\pi}\left(\sin \pi x + \frac{1}{3}\sin 3\pi x + \frac{1}{5}\sin 5\pi x\right)$$

Pattern:
$$f(x) = \frac{4}{\pi}\left(\sin \pi x + \frac{1}{3}\sin 3\pi x + \frac{1}{5}\sin 5\pi x + \cdots + \frac{1}{2n-1}\sin(2n-1)\pi x\right), \quad n = 1,\ 2,\ 3,\ \ldots$$

82.

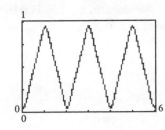

$$f(x) = \frac{1}{2} - \frac{4}{\pi^2}\left(\cos \pi x + \frac{1}{9}\cos 3\pi x\right)$$

$$g(x) = \frac{1}{2} - \frac{4}{\pi^2}\left(\cos \pi x + \frac{1}{9}\cos 3\pi x + \frac{1}{25}\cos 5\pi x\right)$$

Pattern:

$$f(x) = \frac{1}{2} - \frac{4}{\pi^2}\left(\cos \pi x + \frac{1}{9}\cos 3\pi x + \frac{1}{25}\cos 5\pi x + \cdots + \frac{1}{(2n-1)^2}\cos(2n-1)\pi x\right), \quad n = 1, 2, 3, \ldots$$

83.

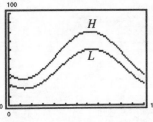

$$H(t) = 54.33 - 20.38\cos\frac{\pi t}{6} - 15.69\sin\frac{\pi t}{6}$$

$$L(t) = 39.36 - 15.70\cos\frac{\pi t}{6} - 14.16\sin\frac{\pi t}{6}$$

a. The difference is greatest during June–August. The difference is smallest during December–February.

b. The warmest temperature occurs when $t \approx 7.2$ which approximately corresponds to July 6. The difference between June 21 and July 6 is 15 days.

84. $S = 58.3 + 32.5\cos\dfrac{\pi t}{6}$

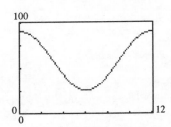

Sales exceed 75,000 during the months of January, November, and December.

85. Let m be the slope of the line and let θ be the angle the line makes with the x-axis. The slope of the line is related to θ by the equation $m = \tan\theta$. If $m = 0$, then the line is horizontal and $\theta = 0$ ($m = 0 = \tan 0$). If m is undefined, then the line is vertical and $\theta = \pi/2$ (m is undefined and $\tan(\pi/2)$ is undefined). If the line has a positive slope, then it will intersect the x-axis. Let this point be $(x_1, 0)$ (see figure). If (x_2, y_2) is a second point on the line, then the slope is given by

$$m = \frac{y_2 - 0}{x_2 - x_1} = \frac{y_2}{x_2 - x_1} = \tan\theta.$$

If the line has a negative slope, then it will intersect the x-axis. Let this point be $(x_3, 0)$ (see figure). If (x_4, y_4) is a second point on the line, then the slope is given by

$$m = \frac{0 - y_4}{x_3 - x_4} = -\frac{y_4}{x_3 - x_4} = \tan\theta.$$

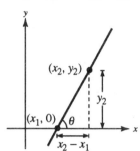   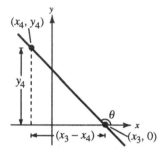

86.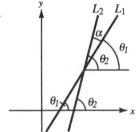

Let m_1 = slope of L_1 and m_2 = slope of L_2. From Exercise 85, we have $m_1 = \tan\theta_1$ and $m_2 = \tan\theta_2$. Let $\alpha = \theta_2 - \theta_1$. Then

$$\tan\alpha = \tan(\theta_2 - \theta_1) = \frac{\tan\theta_2 - \tan\theta_1}{1 + \tan\theta_2 \tan\theta_1} = \frac{m_2 - m_1}{1 + m_2 m_1}.$$

Since we want α to be the acute angle between L_1 and L_2, we take the absolute value of $(m_2 - m_1)/(1 + m_2 m_1)$. Thus,

$$\tan\alpha = \left| \frac{m_2 - m_1}{1 + m_2 m_1} \right| \quad \text{if } m_1 m_2 \neq -1.$$

87. $f(x) = \sin x$

$g(x) = |\sin x|$

$h(x) = \sin |x|$

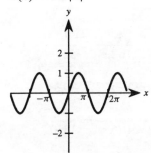 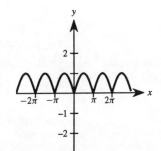 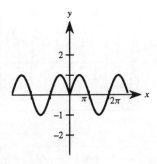

The graph of $|f(x)|$ will reflect any parts of the graph below the x-axis about the y-axis.

The graph of $f(|x|)$ will reflect the part of the graph to the right of the y-axis about the y-axis.

88.

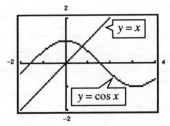

Since $y = x$ and $y = \cos x$ intersect in the interval $[0, 1]$, there exists an x between 0 and 1 for which $x = \cos x$.

$x_0 = 1$

$x_1 = \cos 1 \approx 0.5403$

$x_2 = \cos 0.5403 \approx 0.8576$

$x_3 = \cos 0.8576 \approx 0.6543$

Graphically, this sequence starts at $(1, 1)$ on the line and moves to $(1, \cos 1)$ on the cosine curve. Then you move to $(\cos 1, \cos 1)$ on the line and move on to $(\cos 1, \cos(\cos 1))$ on the cosine curve. You continue this process until you zero onto the point of intersection. Thus, $x \approx 0.739$.

89. $p(t) = \dfrac{1}{4\pi}[p_1(t) + 30p_2(t) + p_3(t) + p_5(t) + 30p_6(t)]$ where $p_n(t) = \dfrac{1}{n}\sin(524n\pi t)$.

a. $p(0.0025) \approx 0.8845$

b.
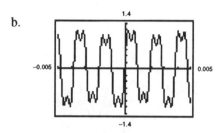

d. Frequency = 1/period = 262

e. The sound whose wave is given by p sounds like "high C" because $30p_2(t)$ has a frequency of 524 and it is the loudest component of the sound.

c. $p(t)$ is periodic.

The period of $p_1(t)$ is $\dfrac{2\pi}{524\pi} = \dfrac{1}{262}$.

The period of $p_2(t)$ is $\dfrac{2\pi}{(524)(2\pi)} = \dfrac{1}{524}$.

The period of $p_3(t)$ is $\dfrac{2\pi}{(524)(3\pi)} = \dfrac{1}{786}$.

The period of $p_5(t)$ is $\dfrac{2\pi}{(524)(5\pi)} = \dfrac{1}{1310}$.

The period of $p_6(t)$ is $\dfrac{2\pi}{(524)(6\pi)} = \dfrac{1}{1572}$.

The least common multiple of these periods is $\frac{1}{262}$. The period of $p(t)$ is $\frac{1}{262}$.

Chapter P   Review Exercises

1. $|x - 2| \le 3$

$-3 \le x - 2 \le 3$

$-1 \le x \le 5$

2. $|3x - 2| \le 0 \Rightarrow 3x - 2 = 0$

$x = \frac{2}{3}$

3. $4 < (x + 3)^2$

$4 < x^2 + 6x + 9$

$0 < x^2 + 6x + 5$

$0 < (x + 1)(x + 5)$

$x < -5$ or $x > -1$

4. $\dfrac{1}{|x|} < 1$

$|x| > 1$

$x > 1$ or $x < -1$

5. $\dfrac{(7/8) + (10/4)}{2} = \dfrac{27/8}{2} = \dfrac{27}{16}$

6. $\dfrac{-1 + (3/2)}{2} = \dfrac{1}{4}$

7. $\left(\dfrac{1 - 3}{2}, \dfrac{4 + 2}{2}\right) = (-1, \ 3)$

$\left(\dfrac{1 + 5}{2}, \dfrac{4 + 0}{2}\right) = (3, \ 2)$

$\left(\dfrac{5 - 3}{2}, \dfrac{0 + 2}{2}\right) = (1, \ 1)$

8. Let the coordinates of the vertices be (x_1, y_1), (x_2, y_2), and (x_3, y_3).

$$\left(\frac{x_1 + x_2}{2}, \frac{y_1 + y_2}{2}\right) = (0, 2) \quad \Rightarrow \quad \frac{x_1 + x_2}{2} = 0, \quad \frac{y_1 + y_2}{2} = 2$$

$$\left(\frac{x_1 + x_3}{2}, \frac{y_1 + y_3}{2}\right) = (2, 1) \quad \Rightarrow \quad \frac{x_1 + x_3}{2} = 2, \quad \frac{y_1 + y_3}{2} = 1$$

$$\left(\frac{x_2 + x_3}{2}, \frac{y_2 + y_3}{2}\right) = (1, -1) \quad \Rightarrow \quad \frac{x_2 + x_3}{2} = 1, \quad \frac{y_2 + y_3}{2} = -1$$

From the first set of equations we have $x_2 = -x_1$, $y_2 = 4 - y_1$.

From the second set of equations we have $x_3 = 4 - x_1$, $y_3 = 2 - y_1$.

From the third set of equations we have

$$\frac{-x_1 + (4 - x_1)}{2} = 1 \Rightarrow x_1 = 1$$

$$\frac{(4 - y_1) + (2 - y_1)}{2} = -1 \Rightarrow y_1 = 4$$

Using these values we have $x_2 = -1$, $y_2 = 0$, $x_3 = 3$, and $y_3 = -2$. Therefore, the vertices are $(1, 4)$, $(-1, 0)$, and $(3, -2)$.

9. $(x^2 + 6x + 9) + (y^2 - 2y + 1) = -1 + 9 + 1$

$$(x + 3)^2 + (y - 1)^2 = 9$$

Center: $(-3, 1)$

Radius: 3

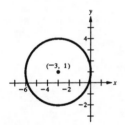

10. $4\left(x^2 - x + \frac{1}{4}\right) + 4(y^2 + 2y + 1) = 11 + 1 + 4$

$$\left(x - \tfrac{1}{2}\right)^2 + (y + 1)^2 = 4$$

Center: $\left(\frac{1}{2}, -1\right)$

Radius: 2

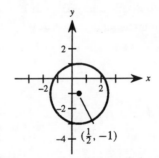

11. $(x^2 + 6x + 9) + (y^2 - 2y + 1) = -10 + 9 + 1$

$$(x + 3)^2 + (y - 1)^2 = 0$$

Point: $(-3, 1)$

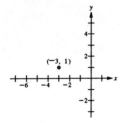

12. $(x^2 - 6x + 9) + (y^2 + 8y + 16) = 9 + 16$

$$(x - 3)^2 + (y + 4)^2 = 25$$

Center: $(3, -4)$

Radius: 5

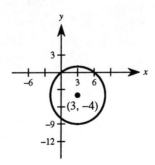

13. $(x^2 - 6x + 9) + (y^2 + 8y + 16) = c + 9 + 16$

$$(x - 3)^2 + (y + 4)^2 = c + 25$$

If the radius is 2, then

$$c + 25 = 4$$

$$c = -21.$$

14. $\sqrt{(x + 2)^2 + (y - 0)^2} = 2\sqrt{(x - 3)^2 + (y - 1)^2}$

$$(x + 2)^2 + y^2 = 4[(x - 3)^2 + (y - 1)^2]$$

$$x^2 + 4x + 4 + y^2 = 4x^2 - 24x + 4y^2 - 8y + 40$$

$$0 = 3x^2 + 3y^2 - 28x - 8y + 36$$

Completing the square, you obtain: $\left(x - \dfrac{14}{3}\right)^2 + \left(y - \dfrac{4}{3}\right)^2 = \dfrac{104}{9}$

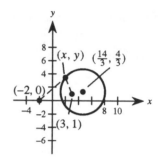

15. $(x - 1)^2 + (y - 2)^2 = 9$

$$x^2 - 2x + 1 + y^2 - 4y + 4 = 9$$

$$x^2 + y^2 - 2x - 4y = 4$$

a. (1, 5)   On the circle since $1^2 + 5^2 - 2(1) - 4(5) = 4$
b. (0, 0)   Inside the circle since $0^2 + 0^2 - 2(0) - 4(0) < 4$
c. (−2, 1)   Outside the circle since $(-2)^2 + 1^2 - 2(-2) - 4(1) > 4$
d. (0, 4)   Inside the circle since $0^2 + 4^2 - 2(0) - 4(4) < 4$

16. $(x - 2)^2 + (y - 1)^2 = 4$

$$x^2 + y^2 - 4x - 2y = -1$$

a. (1, 1)   Inside the circle since $(1)^2 + (1)^2 - 4(1) - 2(1) < -1$
b. (4, 2)   Outside the circle since $(4)^2 + (2)^2 - 4(4) - 2(2) > -1$
c. (0, 1)   On the circle since $(0)^2 + (1)^2 - 4(0) - 2(1) = -1$
d. (3, 1)   Inside the circle since $(3)^2 + (1)^2 - 4(3) - 2(1) < -1$

17. $\dfrac{1 - t}{1 - 0} = \dfrac{1 - 5}{1 - (-2)}$

$$1 - t = -\dfrac{4}{3}$$

$$t = \dfrac{7}{3}$$

18. $\dfrac{3 - (-1)}{-3 - t} = \dfrac{3 - 6}{-3 - 8}$

$$\dfrac{4}{-3 - t} = \dfrac{-3}{-11}$$

$$-44 = 9 + 3t$$

$$-53 = 3t$$

$$t = -\dfrac{53}{3}$$

19. $y = -\frac{1}{2}x + \frac{3}{2}$

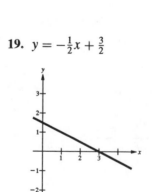

20. $4x - 2y = 6$

$y = 2x - 3$

Slope: 2

y-intercept: -3

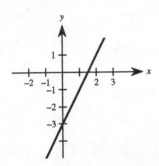

21. $-\frac{1}{3}x + \frac{5}{6}y = 1$

$-\frac{2}{5}x + y = \frac{6}{5}$

$y = \frac{2}{5}x + \frac{6}{5}$

Slope: $\frac{2}{5}$

y-intercept: $\frac{6}{5}$

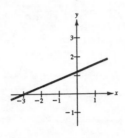

22. $0.02x + 0.15y = 0.25$

$2x + 15y = 25$

$y = -\frac{2}{15}x + \frac{5}{3}$

Slope: $-\frac{2}{15}$

y-intercept: $\frac{5}{3}$

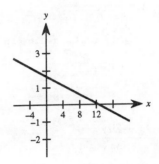

23. $y = 7 - 6x - x^2$

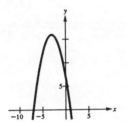

24. $y = x(6 - x)$

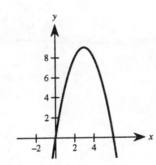

25. $y = \sqrt{5 - x}$

Domain: $(-\infty, \ 5]$

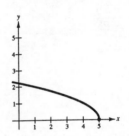

26. $y = |x - 4| - 4$

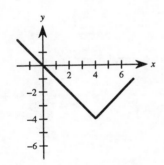

27. a. $y - 4 = \dfrac{7}{16}(x + 2)$

$16y - 64 = 7x + 14$

$0 = 7x - 16y + 78$

b. $y - 4 = \dfrac{5}{3}(x + 2)$

$3y - 12 = 5x + 10$

$0 = 5x - 3y + 22$

c. $m = \dfrac{4 - 0}{-2 - 0} = -2$

$y = -2x$

$2x + y = 0$

d. $x = -2$

$x + 2 = 0$

28. a. $y - 3 = -\dfrac{2}{3}(x - 1)$

$3y - 9 = -2x + 2$

$2x + 3y - 11 = 0$

b. $y - 3 = 1(x - 1)$

$y = x + 2$

$0 = x - y + 2$

c. $m = \dfrac{4 - 3}{2 - 1} = 1$

$y - 3 = 1(x - 1)$

$y = x + 2$

$0 = x - y + 2$

d. $y = 3$

$y - 3 = 0$

29. $\left(\dfrac{x+2}{2},\ \dfrac{y+3}{2}\right) = (-1,\ 4)$

$$x = -4$$
$$y = 5$$

Other endpoint: $(-4,\ 5)$

30. $\sqrt{(x-0)^2 + (y-0)^2} = \sqrt{(x-2)^2 + (y-3)^2} = \sqrt{(x-3)^2 + (y+2)^2}$

$x^2 + y^2 = x^2 + y^2 - 4x - 6y + 13 = x^2 + y^2 - 6x + 4y + 13$

$$4x + 6y = 13 \qquad\qquad 2x - 10y = 0$$

$$\begin{array}{l} 4x + 6y = 13 \\ \underline{-4x + 20y = 0} \\ 26y = 13 \\ y = \tfrac{1}{2} \end{array}$$

$2x - 10(\tfrac{1}{2}) = 0$

$x = \tfrac{5}{2}$

Point: $\left(\tfrac{5}{2},\ \tfrac{1}{2}\right)$

31. $3x - 4y = 8$

$$\begin{array}{l} \underline{4x + 4y = 20} \\ 7x = 28 \end{array}$$

$$x = 4$$
$$y = 1$$

Point: $(4,\ 1)$

32. $y = x + 1$

$(x + 1) - x^2 = 7$

$0 = x^2 - x + 6$

No real solution

No points of intersection

33. The speed of the plane is 560 miles per hour.

34. a. $C = 9.25t + 13.50t + 36{,}500$

$ = 22.75t + 36{,}500$

b. $R = 30t$

c. $30t = 22.75t + 36{,}500$

$7.25t = 36{,}500$

$t \approx 5034.48$ hours to break even

35.

$2x + 2y = 24$

$y = 12 - x$

$A = xy = x(12 - x) = 12x - x^2$

Domain: $\{x:\ 0 < x < 12\}$

36.

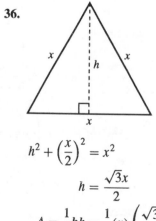

$h^2 + \left(\dfrac{x}{2}\right)^2 = x^2$

$h = \dfrac{\sqrt{3}x}{2}$

$A = \dfrac{1}{2}bh = \dfrac{1}{2}(x)\left(\dfrac{\sqrt{3}}{2}x\right) = \dfrac{\sqrt{3}}{4}x^2$

Domain: $\{x:\ x > 0\}$

37. $d = 45t$

Domain: $\{t: \ t \geq 0\}$

38. $v = 1250a + 500{,}000$

Domain: $\{a: \ a \geq 0\}$

39. $x - y^2 = 0$

$$y = \pm\sqrt{x}$$

Not a function of x since there are two values of y for some x.

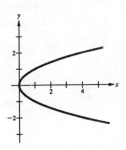

40. $x^2 - y = 0$

Function of x since there is one value for y for each x.

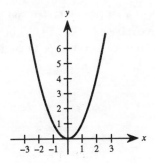

41. $y = x^2 - 2x$

Function of x since there is one value of y for each x.

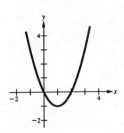

42. $x = 9 - y^2$

Not a function of x since there are two values of y for some x.

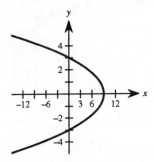

43. a. $f(x) = x^3 + c, \ c = -2, \ 0, \ 2$

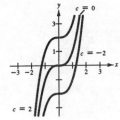

b. $f(x) = (x - c)^3, \ c = -2, \ 0, \ 2$

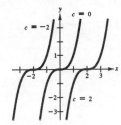

c. $f(x) = (x - 2)^3 + c, \ c = -2, \ 0, \ 2$

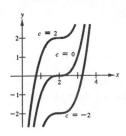

d. $f(x) = cx^3, \ c = -2, \ 0, \ 2$

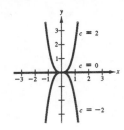

44. a.

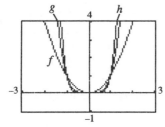

The graphs of f, g, and h all rise to the left and to the right. As the degree increases, the "width" of the graph decreases. All three graphs pass through $(0, 0)$, $(1, 1)$, and $(-1, 1)$.

b.

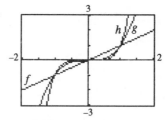

The graphs of f, g, and h all rise to the right and fall to the left. As the degree increases, the graph rises and falls more steeply. All three graphs pass through $(0, 0)$, $(1, 1)$, and $(-1, -1)$.

45. a.

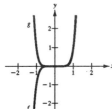

b.

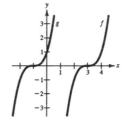

46. a. $f(x) = x^2(x-6)^2$ **b.** $g(x) = x^3(x-6)^2$ **c.** $h(x) = x^3(x-6)^3$

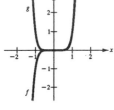 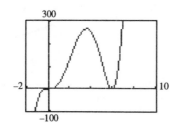 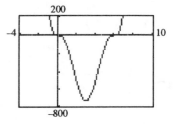

47. $f(x) = 1 - x^2$ and $g(x) = 2x + 1$

    a. $f(x) - g(x) = (1 - x^2) - (2x + 1) = -x^2 - 2x$

    b. $f(x)g(x) = (1-x^2)(2x+1) = -2x^3 - x^2 + 2x + 1$

    c. $g(f(x)) = g(1 - x^2) = 2(1 - x^2) + 1 = 3 - 2x^2$

48. $f(x) = 2x - 3$ and $g(x) = \sqrt{x+1}$

    a. $f(x) + g(x) = 2x - 3 + \sqrt{x+1}$

    b. $\dfrac{f(x)}{g(x)} = \dfrac{2x-3}{\sqrt{x+1}}, \quad x > -1$

    c. $f(g(x)) = f(\sqrt{x+1}) = 2\sqrt{x+1} - 3$

49. a.

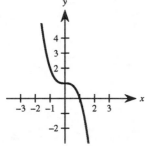

f must be at least a third degree polynomial. Since f rises to the left and falls to the right, the leading coefficient must be negative.

b.

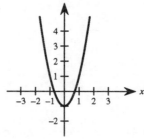

f must be at least a second degree polynomial. Since f rises to the left and to the right, the leading coefficient must be positive.

50. a.

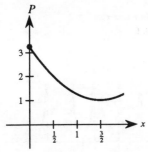

The profit for this product is decreasing for about the first year and a half; then it makes a recovery and begins to increase.

b.

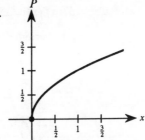

The profit for this product is increasing over the two year period.

51. $\tan 32° = \dfrac{35,000}{y} \Rightarrow y = \dfrac{35,000}{\tan 32°}$

$\tan 76° = \dfrac{35,000}{x} \Rightarrow x = \dfrac{35,000}{\tan 76°}$

Distance between towns:

$y - x = \dfrac{35,000}{\tan 32°} - \dfrac{35,000}{\tan 76°}$

$\approx 47,285$ feet

≈ 9 miles

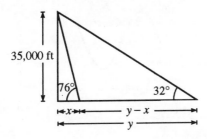

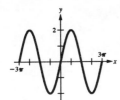

52. $\sin 30° = \dfrac{y_1}{20} = \dfrac{1}{2} \Rightarrow y_1 = 10$

$\cos 30° = \dfrac{x_1}{20} = \dfrac{\sqrt{3}}{2} \Rightarrow x_1 = 10\sqrt{3}$

$\sin 60° = \dfrac{y_2}{20} = \dfrac{\sqrt{3}}{2} \Rightarrow y_2 = 10\sqrt{3}$

$\cos 60° = \dfrac{x_2}{20} = \dfrac{1}{2} \Rightarrow x_2 = 10$

The coordinates of the center of each hole are $(10\sqrt{3},\ 10)$ and $(10,\ 10\sqrt{3})$.

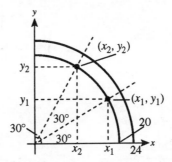

53. $f(x) = 2\sin\left(\dfrac{2x}{3}\right)$

Amplitude: 2
Period: 3π

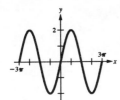

54. $f(x) = \dfrac{1}{2}\cos\left(\dfrac{x}{3}\right)$

Amplitude: $\dfrac{1}{2}$
Period: 6π

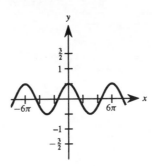

55. $f(x) = \cos\left(2x - \dfrac{\pi}{3}\right)$

Amplitude: 1
Period: π
Shift: $\dfrac{\pi}{6}$

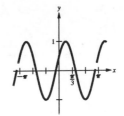

56. $f(x) = -\sin\left(2x + \dfrac{\pi}{2}\right)$

Amplitude: 1
Period: π
Shift: $-\dfrac{\pi}{4}$

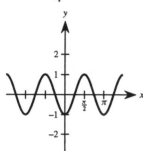

57. $f(x) = \tan\left(\dfrac{x}{2}\right)$

Period: 2π

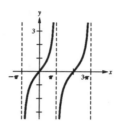

58. $f(x) = \csc 2x$

Period: π

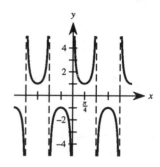

59. $f(x) = \sec\left(x - \dfrac{\pi}{4}\right)$

Period: 2π
Shift: $\dfrac{\pi}{4}$

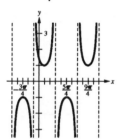

60. $f(x) = \cot 3x$

Period: $\dfrac{\pi}{3}$

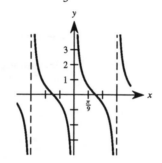

61. $S = 74.50 + 43.75 \sin \dfrac{\pi t}{6}$

Amplitude: 43.75
Period: 12
Vertical shift: 74.50 units upward

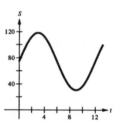

62.

$a = b = 10$

$\tan 35° = \dfrac{c}{10} \;\Rightarrow\; c = d = e = 10\tan 35° \approx 7.002$

$f = g = h = i = \sqrt{10^2 + 7.002^2} \approx 12.208$

CHAPTER 1
Limits and Their Properties

Section 1.1 An Introduction to Limits

1.

x	1.9	1.99	1.999	2.001	2.01	2.1
$f(x)$	0.3448	0.3344	0.3334	0.3332	0.3322	0.3226

$$\lim_{x \to 2} \frac{x-2}{x^2 - x - 2} \approx 0.3333 \quad \text{(Actual limit is } \tfrac{1}{3}.\text{)}$$

2.

x	1.9	1.99	1.999	2.001	2.01	2.1
$f(x)$	0.2564	0.2506	0.2501	0.2499	0.2494	0.2439

$$\lim_{x \to 2} \frac{x-2}{x^2 - 4} \approx 0.25 \quad \text{(Actual limit is } \tfrac{1}{4}.\text{)}$$

3.

x	-0.1	-0.01	-0.001	0.001	0.01	0.1
$f(x)$	0.2911	0.2889	0.2887	0.2887	0.2884	0.2863

$$\lim_{x \to 0} \frac{\sqrt{x+3} - \sqrt{3}}{x} \approx 0.2887. \quad \text{(Actual limit is } 1/(2\sqrt{3}).\text{)}$$

4.

x	-3.1	-3.01	-3.001	-2.999	-2.99	-2.9
$f(x)$	-0.2485	-0.2498	-0.2500	-0.2500	-0.2502	-0.2516

$$\lim_{x \to -3} \frac{\sqrt{1-x} - 2}{x+3} \approx -0.25 \quad \text{(Actual limit is } -\tfrac{1}{4}.\text{)}$$

5.

x	2.9	2.99	2.999	3.001	3.01	3.1
$f(x)$	-0.0641	-0.0627	-0.0625	-0.0625	-0.0623	-0.0610

$$\lim_{x \to 3} \frac{[1/(x+1)] - (1/4)}{x-3} \approx -0.0625 \quad \text{(Actual limit is } -\tfrac{1}{16}.\text{)}$$

6.

x	3.9	3.99	3.999	4.001	4.01	4.1
$f(x)$	0.0408	0.0401	0.0400	0.0400	0.0399	0.0392

$$\lim_{x \to 4} \frac{[x/(x+1)] - (4/5)}{x-4} \approx 0.04 \quad \text{(Actual limit is } \tfrac{1}{25}.\text{)}$$

7.

x	-0.1	-0.01	-0.001	0.001	0.01	0.1
$f(x)$	0.9983	0.99998	1.0000	1.0000	0.99998	0.9983

$$\lim_{x \to 0} \frac{\sin x}{x} \approx 1.0000 \quad \text{(Actual limit is 1.)}$$

8.

x	-0.1	-0.01	-0.001	0.001	0.01	0.1
$f(x)$	0.0500	0.0050	0.0005	-0.0005	-0.0050	-0.0500

$$\lim_{x \to 0} \frac{\cos x - 1}{x} \approx 0.0000 \quad \text{(Actual limit is 0.)}$$

9. $\lim\limits_{x \to 3} (4 - x) = 1$

10. $\lim\limits_{x \to 1} (x^2 + 2) = 3$

11. $\lim\limits_{x \to 2} f(x) = \lim\limits_{x \to 2} (4 - x) = 2$

12. $\lim\limits_{x \to 1} f(x) = \lim\limits_{x \to 1} (x^2 + 2) = 3$

13. $\lim\limits_{x \to 5} \dfrac{|x - 5|}{x - 5}$ does not exist since

$\lim\limits_{x \to 5^-} \dfrac{|x - 5|}{x - 5} = -1$ and $\lim\limits_{x \to 5^+} \dfrac{|x - 5|}{x - 5} = 1$.

14. $\lim\limits_{x \to 3} \dfrac{1}{x - 3}$ does not exist since the function increases and decreases without bound as x approaches 3.

15. $\lim\limits_{x \to \pi/2} \tan x$ does not exist since the function increases and decreases without bound as x approaches $\pi/2$.

16. $\lim\limits_{x \to 0} \sec x = 1$

17. $\lim\limits_{x \to 0} \cos(1/x)$ does not exist since the function oscillates between -1 and 1 as x approaches 0.

18. $\lim\limits_{x \to 1} \sin(\pi x) = 0$

19. $\lim\limits_{x \to 2} (3x + 2) = 8$

$|(3x + 2) - 8| < 0.01$

$|3x - 6| < 0.01$

$3|x - 2| < 0.01$

$0 < |x - 2| < \dfrac{0.01}{3} \approx 0.0033 = \delta$

20. $\lim\limits_{x \to 4} \left(4 - \dfrac{x}{2}\right) = 2$

$\left|\left(4 - \dfrac{x}{2}\right) - 2\right| < 0.01$

$\left|2 - \dfrac{x}{2}\right| < 0.01$

$\left|-\dfrac{1}{2}(x - 4)\right| < 0.01$

$0 < |x - 4| < 0.02 = \delta$

21. $\lim\limits_{x \to 2} (x^2 - 3) = 1$

$$|(x^2 - 3) - 1| < 0.01$$

$$|x^2 - 4| < 0.01$$

$$|(x + 2)(x - 2)| < 0.01$$

$$|x + 2|\,|x - 2| < 0.01$$

$$|x - 2| < \frac{0.01}{|x + 2|}$$

If we assume $1 < x < 3$, then $\delta = 0.01/5 = 0.002$.

22. $\lim\limits_{x \to 5} (x^2 + 4) = 29$

$$|(x^2 + 4) - 29| < 0.01$$

$$|x^2 - 25| < 0.01$$

$$|(x + 5)(x - 5)| < 0.01$$

$$|x - 5| < \frac{0.01}{|x + 5|}$$

If we assume $4 < x < 6$, then $\delta = 0.01/11 \approx 0.0009$.

23. $\lim\limits_{x \to 2} (x + 3) = 5$

Given $\epsilon > 0$:

$$|(x + 3) - 5| < \epsilon$$

$$|x - 2| < \epsilon = \delta$$

Hence, let $\delta = \epsilon$.

24. $\lim\limits_{x \to -3} (2x + 5) = -1$

Given $\epsilon > 0$:

$$|(2x + 5) - (-1)| < \epsilon$$

$$|2x + 6| < \epsilon$$

$$2|x + 3| < \epsilon$$

$$|x + 3| < \frac{\epsilon}{2} = \delta$$

Hence, let $\delta = \epsilon/2$.

25. $\lim\limits_{x \to 6} 3 = 3$

Given $\epsilon > 0$:

$$|3 - 3| < \epsilon$$

$$0 < \epsilon$$

Hence, any δ will work.

26. $\lim\limits_{x \to 2} (-1) = -1$

Given $\epsilon > 0$:

$$|-1 - (-1)| < \epsilon$$

$$0 < \epsilon$$

Hence, any δ will work.

27. $\lim\limits_{x \to 0} \sqrt[3]{x} = 0$

Given $\epsilon > 0$:

$$|\sqrt[3]{x} - 0| < \epsilon$$

$$|\sqrt[3]{x}| < \epsilon$$

$$|x| < \epsilon^3 = \delta$$

Hence, let $\delta = \epsilon^3$.

28. $\lim\limits_{x \to 3} |x - 3| = 0$

Given $\epsilon > 0$:

$$|(x - 3) - 0| < \epsilon$$

$$|x - 3| < \epsilon = \delta$$

Hence, let $\delta = \epsilon$.

29. $\lim\limits_{x \to 1} (x^2 + 1) = 2$

Given $\epsilon > 0$:

$$|(x^2 + 1) - 2| < \epsilon$$

$$|x^2 - 1| < \epsilon$$

$$|(x + 1)(x - 1)| < \epsilon$$

$$|x - 1| < \frac{\epsilon}{|x + 1|}$$

If we assume $0 < x < 2$, then $\delta = \epsilon/3$.

30. $\lim\limits_{x \to -3} (x^2 + 3x) = 0$

Given $\epsilon > 0$:

$$|(x^2 + 3x) - 0| < \epsilon$$

$$|x(x + 3)| < \epsilon$$

$$|x + 3| < \frac{\epsilon}{|x|}$$

If we assume $-4 < x < -2$, then $\delta = \epsilon/4$.

31. False; $f(x) = (\sin x)/x$ is undefined when $x = 0$. From Exercise 7, we have

$$\lim\limits_{x \to 0} \frac{\sin x}{x} = 1.$$

32. True

33. False; let

$$f(x) = \begin{cases} x^2 - 4x, & x \neq 4 \\ 10, & x = 4. \end{cases}$$

$f(4) = 10$

$\lim_{x \to 4} f(x) = \lim_{x \to 4} (x^2 - 4x) = 0 \neq 10$

34. False; let

$$f(x) = \begin{cases} x^2 - 4x, & x \neq 4 \\ 10, & x = 4. \end{cases}$$

$\lim_{x \to 4} f(x) = \lim_{x \to 4} (x^2 - 4x) = 0$ and $f(4) = 10 \neq 0$

35. $f(x) = \dfrac{\sqrt{x+5} - 3}{x - 4}$

$\lim_{x \to 4} f(x) = \dfrac{1}{6}$

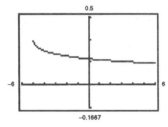

The graphing utility does not show the hole at $\left(4, \frac{1}{6}\right)$.

36. $f(x) = \dfrac{x - 4}{x^2 - 5x + 4}$

$\lim_{x \to 4} f(x) = \dfrac{1}{3}$

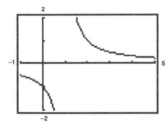

The graphing utility does not show the hole at $\left(4, \frac{1}{3}\right)$.

37. $\lim_{x \to 4} \dfrac{x^2 - x - 12}{x - 4} = 7$

n	$4 + [0.1]^n$	$f(4 + [0.1]^n)$
1	4.1	7.1
2	4.01	7.01
3	4.001	7.001
4	4.0001	7.0001

n	$4 - [0.1]^n$	$f(4 - [0.1]^n)$
1	3.9	6.9
2	3.99	6.99
3	3.999	6.999
4	3.9999	6.9999

38. If $\lim_{x \to c} f(x) = L_1$ and $\lim_{x \to c} f(x) = L_2$, then for every $\epsilon > 0$, there exists $\delta_1 > 0$ and $\delta_2 > 0$ such that $|x - c| < \delta_1 \Rightarrow |f(x) - L_1| < \epsilon$ and $|x - c| < \delta_2 \Rightarrow |f(x) - L_2| < \epsilon$. Let δ equal the smaller of δ_1 and δ_2. Then for $|x - c| < \delta$, we have

$$|L_1 - L_2| = |L_1 - f(x) + f(x) - L_2| \leq |L_1 - f(x)| + |f(x) - L_2| < \epsilon + \epsilon.$$

Therefore, $|L_1 - L_2| < 2\epsilon$. Since $\epsilon > 0$ is arbitrary, it follows that $L_1 = L_2$.

39. $f(x) = mx + b, \quad m \neq 0$

$\lim_{x \to c} (mx + b) = mc + b$

$$|(mx + b) - (mc + b)| < \epsilon$$

$$|m(x - c)| < \epsilon$$

$$|x - c| < \dfrac{\epsilon}{|m|} = \delta$$

Hence for $\epsilon > 0$, let $\delta = \epsilon/|m|$.

40. $\lim_{x \to c} [f(x) - L] = 0$ means that for every $\epsilon > 0$ there exists $\delta > 0$ such that if $0 < |x - c| < \delta$, then $|(f(x) - L) - 0| < \epsilon$. This is the same as $|f(x) - L| < \epsilon$ when $0 < |x - c| < \delta$. Thus, $\lim_{x \to c} f(x) = L$.

41. $f(x) = (1+x)^{1/x}$

$$\lim_{x \to 0} (1+x)^{1/x} = e \approx 2.71828$$

x	$f(x)$
-0.1	2.867972
-0.01	2.731999
-0.001	2.719642
-0.0001	2.718418
-0.00001	2.718295
-0.000001	2.718283

x	$f(x)$
0.1	2.593742
0.01	2.704814
0.001	2.716924
0.0001	2.718146
0.00001	2.718268
0.000001	2.718280

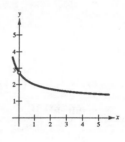

Section 1.2 Properties of Limits

1. a. $\lim_{x \to 0} f(x) = 1$

  b. $\lim_{x \to -1} f(x) = 3$

2. a. $\lim_{x \to 1} f(x) = -2$

  b. $\lim_{x \to 3} f(x) = 0$

3. a. $\lim_{x \to \pi/3} g(x) = \dfrac{\sqrt{3}}{2}$

  b. $\lim_{x \to \pi/2} g(x) = 1$

4. a. $\lim_{x \to 0} h(x) = 0$

  b. $\lim_{x \to 3\pi/4} h(x) = -1$

5. $\lim_{x \to 4} x^2 = 4^2 = 16$

6. $\lim_{x \to -3} (3x + 2) = 3(-3) + 2 = -7$

7. $\lim_{x \to 0} (2x - 1) = 2(0) - 1 = -1$

8. $\lim_{x \to 1} (-x^2 + 1) = -(1)^2 + 1 = 0$

9. $\lim_{x \to 2} (-x^2 + x - 2) = -(2)^2 + (2) - 2 = -4$

10. $\lim_{x \to 1} (3x^3 - 2x^2 + 4) = 3(1)^3 - 2(1)^2 + 4 = 5$

11. $\lim_{x \to 3} \sqrt{x + 1} = \sqrt{3 + 1} = 2$

12. $\lim_{x \to 4} \sqrt[3]{x + 4} = \sqrt[3]{4 + 4} = 2$

13. $\lim_{x \to -4} (x + 3)^2 = (-4 + 3)^2 = 1$

14. $\lim_{x \to 0} (2x - 1)^3 = [2(0) - 1]^3 = -1$

15. $\lim_{x \to 2} \dfrac{1}{x} = \dfrac{1}{2}$

16. $\lim_{x \to -3} \dfrac{2}{x + 2} = \dfrac{2}{-3 + 2} = -2$

17. $\lim_{x \to -1} \dfrac{x^2 + 1}{x} = \dfrac{(-1)^2 + 1}{-1} = -2$

18. $\lim_{x \to 3} \dfrac{\sqrt{x + 1}}{x - 4} = \dfrac{\sqrt{3 + 1}}{3 - 4} = -2$

19. $\lim_{x \to \pi/2} \sin x = \sin \dfrac{\pi}{2} = 1$

20. $\lim_{x \to \pi} \tan x = \tan \pi = 0$

21. $\lim_{x \to 1} \cos \pi x = \cos \pi = -1$

22. $\lim_{x \to 1} \sin \dfrac{\pi x}{2} = \sin \dfrac{\pi}{2} = 1$

23. $\lim_{x \to 0} \sec 2x = \sec 0 = 1$

24. $\lim_{x \to \pi} \cos 3x = \cos 3\pi = -1$

25. $\lim_{x \to 5\pi/6} \sin x = \sin \dfrac{5\pi}{6} = \dfrac{1}{2}$

26. $\lim_{x \to 5\pi/3} \cos x = \cos \dfrac{5\pi}{3} = \dfrac{1}{2}$

27. $\lim\limits_{x\to 3} \tan\left(\dfrac{\pi x}{4}\right) = \tan\dfrac{3\pi}{4} = -1$

28. $\lim\limits_{x\to 7} \sec\left(\dfrac{\pi x}{6}\right) = \sec\dfrac{7\pi}{6} = \dfrac{-2\sqrt{3}}{3}$

29. a. $\lim\limits_{x\to c} [5g(x)] = 5\lim\limits_{x\to c} g(x) = 5(3) = 15$

  b. $\lim\limits_{x\to c} [f(x) + g(x)] = \lim\limits_{x\to c} f(x) + \lim\limits_{x\to c} g(x) = 2 + 3 = 5$

  c. $\lim\limits_{x\to c} [f(x)g(x)] = \left[\lim\limits_{x\to c} f(x)\right]\left[\lim\limits_{x\to c} g(x)\right] = (2)(3) = 6$

  d. $\lim\limits_{x\to c} \dfrac{f(x)}{g(x)} = \dfrac{\lim\limits_{x\to c} f(x)}{\lim\limits_{x\to c} g(x)} = \dfrac{2}{3}$

30. a. $\lim\limits_{x\to c} [4f(x)] = 4\lim\limits_{x\to c} f(x) = 4\left(\dfrac{3}{2}\right) = 6$

  b. $\lim\limits_{x\to c} [f(x) + g(x)] = \lim\limits_{x\to c} f(x) + \lim\limits_{x\to c} g(x) = \dfrac{3}{2} + \dfrac{1}{2} = 2$

  c. $\lim\limits_{x\to c} [f(x)g(x)] = \left[\lim\limits_{x\to c} f(x)\right]\left[\lim\limits_{x\to c} g(x)\right] = \left(\dfrac{3}{2}\right)\left(\dfrac{1}{2}\right) = \dfrac{3}{4}$

  d. $\lim\limits_{x\to c} \dfrac{f(x)}{g(x)} = \dfrac{\lim\limits_{x\to c} f(x)}{\lim\limits_{x\to c} g(x)} = \dfrac{3/2}{1/2} = 3$

31. a. $\lim\limits_{x\to c} [f(x)]^3 = \left[\lim\limits_{x\to c} f(x)\right]^3 = (4)^3 = 64$

  b. $\lim\limits_{x\to c} \sqrt{f(x)} = \sqrt{\lim\limits_{x\to c} f(x)} = \sqrt{4} = 2$

  c. $\lim\limits_{x\to c} [3f(x)] = 3\lim\limits_{x\to c} f(x) = 3(4) = 12$

  d. $\lim\limits_{x\to c} [f(x)]^{3/2} = \left[\lim\limits_{x\to c} f(x)\right]^{3/2} = (4)^{3/2} = 8$

32. a. $\lim\limits_{x\to c} \sqrt[3]{f(x)} = \sqrt[3]{\lim\limits_{x\to c} f(x)} = \sqrt[3]{27} = 3$

  b. $\lim\limits_{x\to c} \dfrac{f(x)}{18} = \dfrac{\lim\limits_{x\to c} f(x)}{\lim\limits_{x\to c} 18} = \dfrac{27}{18} = \dfrac{3}{2}$

  c. $\lim\limits_{x\to c} [f(x)]^2 = \left[\lim\limits_{x\to c} f(x)\right]^2 = (27)^2 = 729$

  d. $\lim\limits_{x\to c} [f(x)]^{2/3} = \left[\lim\limits_{x\to c} f(x)\right]^{2/3} = (27)^{2/3} = 9$

33. $f(x) = \dfrac{x - 9}{\sqrt{x} - 3}$

Domain: $[0,\ 9) \cup (9,\ \infty)$

$\lim\limits_{x\to 9} \dfrac{x - 9}{\sqrt{x} - 3} = 6$

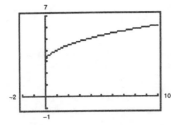

The graphing utility does not show the hole at $(9,\ 6)$.

34. $f(x) = \dfrac{x - 3}{x^2 - 9}$

Domain: $(-\infty,\ -3) \cup (-3,\ 3) \cup (3,\ \infty)$

$\lim\limits_{x\to 3} \dfrac{x - 3}{x^2 - 9} = \dfrac{1}{6}$

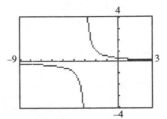

The graphing utility does not show the hole at $\left(3,\ \tfrac{1}{6}\right)$.

35. Let $f(x) = 1/x$ and $g(x) = -1/x$. $\lim\limits_{x\to 0} f(x)$ and $\lim\limits_{x\to 0} g(x)$ do not exist.

$\lim\limits_{x\to 0} [f(x) + g(x)] = \lim\limits_{x\to 0} \left[\dfrac{1}{x} + \left(-\dfrac{1}{x}\right)\right] = \lim\limits_{x\to 0} [0] = 0$

36. Suppose, on the contrary, that $\lim\limits_{x \to c} g(x)$ exists. Then, since $\lim\limits_{x \to c} f(x)$ exists, so would $\lim\limits_{x \to c} [f(x) + g(x)]$, which is a contradiction. Hence, $\lim\limits_{x \to c} g(x)$ does not exist.

37. Given $f(x) = b$, show that for every $\epsilon > 0$ there exists a $\delta > 0$ such that $|f(x) - b| < \epsilon$ whenever $|x - c| < \delta$. Since $|f(x) - b| = |b - b| = 0 < \epsilon$ for any $\epsilon > 0$, then any value of δ will work.

38. Given $f(x) = x^n$, n a positive integer, then

$$\lim_{x \to c} x^n = \lim_{x \to c} (xx^{n-1}) = \left[\lim_{x \to c} x \right] \left[\lim_{x \to c} x^{n-1} \right]$$

$$= c \left[\lim_{x \to c} (xx^{n-2}) \right] = c \left[\lim_{x \to c} x \right] \left[\lim_{x \to c} x^{n-2} \right] = c(c) \lim_{x \to c} (xx^{n-3}) = \cdots = c^n.$$

39. If $b = 0$, then the property is true because both sides are equal to 0. If $b \neq 0$, let $\epsilon > 0$ be given. Since $\lim\limits_{x \to c} f(x) = L$, there exists $\delta > 0$ such that $|f(x) - L| < \epsilon/|b|$ whenever $0 < |x - c| < \delta$. Hence, wherever $0 < |x - c| < \delta$, we have

$$|b| \, |f(x) - L| < \epsilon \quad \text{or} \quad |fb(x) - bL| < \epsilon$$

which implies that $\lim\limits_{x \to c} [bf(x)] = bL$.

40. False. Let $f(x) = \frac{1}{2}x^2$ and $g(x) = x^2$. Then $f(x) < g(x)$ for all $x \neq 0$. But $\lim\limits_{x \to 0} f(x) = \lim\limits_{x \to 0} g(x) = 0$.

Section 1.3   Techniques for Evaluating Limits

1. $f(x) = -2x + 1$ and $g(x) = \dfrac{-2x^2 + x}{x}$ agree except at $x = 0$.

  a. $\lim\limits_{x \to 0} g(x) = \lim\limits_{x \to 0} f(x) = 1$

  b. $\lim\limits_{x \to -1} g(x) = \lim\limits_{x \to -1} f(x) = 3$

2. $f(x) = x - 3$ and $h(x) = \dfrac{x^2 - 3x}{x}$ agree except at $x = 0$.

  a. $\lim\limits_{x \to -2} h(x) = \lim\limits_{x \to -2} f(x) = -5$

  b. $\lim\limits_{x \to 0} h(x) = \lim\limits_{x \to 0} f(x) = -3$

3. $f(x) = x(x + 1)$ and $g(x) = \dfrac{x^3 - x}{x - 1}$ agree except at $x = 1$.

  a. $\lim\limits_{x \to 1} g(x) = \lim\limits_{x \to 1} f(x) = 2$

  b. $\lim\limits_{x \to -1} g(x) = \lim\limits_{x \to -1} f(x) = 0$

4. $f(x) = \dfrac{1}{x - 1}$ and $g(x) = \dfrac{x + 1}{x^2 - 1}$ agree except at $x = -1$.

  a. $\lim\limits_{x \to 1} f(x)$ does not exist.

  b. $\lim\limits_{x \to 2} f(x) = 1$

5. $f(x) = \dfrac{x^2 - 1}{x + 1}$ and $g(x) = x - 1$

agree except at $x = -1$.

$$\lim_{x \to -1} f(x) = \lim_{x \to -1} g(x) = -2$$

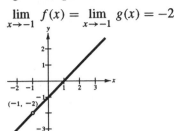

$(-1, -2)$

6. $f(x) = \dfrac{2x^2 - x - 3}{x + 1}$ and $g(x) = 2x - 3$

agree except at $x = -1$.

$$\lim_{x \to -1} f(x) = \lim_{x \to -1} g(x) = -5$$

$(-1, -5)$

7. $f(x) = \dfrac{x^3 + 8}{x + 2}$ and $g(x) = x^2 - 2x + 4$

agree except at $x = -2$.

$$\lim_{x \to -2} f(x) = \lim_{x \to -2} g(x) = 12$$

$(-2, 12)$

8. $f(x) = \dfrac{x^3 + 1}{x + 1}$ and $g(x) = x^2 - x + 1$

agree except at $x = -1$.

$$\lim_{x \to -1} f(x) = \lim_{x \to -1} g(x) = 3$$

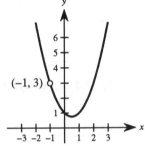

$(-1, 3)$

9. $\displaystyle \lim_{x \to 5} \frac{x - 5}{x^2 - 25} = \lim_{x \to 5} \frac{x - 5}{(x + 5)(x - 5)}$

$$= \lim_{x \to 5} \frac{1}{x + 5} = \frac{1}{10}$$

10. $\displaystyle \lim_{x \to 2} \frac{2 - x}{x^2 - 4} = \lim_{x \to 2} \frac{-(x - 2)}{(x - 2)(x + 2)}$

$$= \lim_{x \to 2} \frac{-1}{x + 2} = -\frac{1}{4}$$

11. $\displaystyle \lim_{x \to 1} \frac{x^2 + x - 2}{x^2 - 1} = \lim_{x \to 1} \frac{(x + 2)(x - 1)}{(x + 1)(x - 1)} = \lim_{x \to 1} \frac{x + 2}{x + 1} = \frac{3}{2}$

12. $\displaystyle \lim_{x \to 0} \frac{\sqrt{2 + x} - \sqrt{2}}{x} = \lim_{x \to 0} \frac{\sqrt{2 + x} - \sqrt{2}}{x} \cdot \frac{\sqrt{2 + x} + \sqrt{2}}{\sqrt{2 + x} + \sqrt{2}}$

$$= \lim_{x \to 0} \frac{2 + x - 2}{(\sqrt{2 + x} + \sqrt{2})x} = \lim_{x \to 0} \frac{1}{\sqrt{2 + x} + \sqrt{2}} = \frac{1}{2\sqrt{2}} = \frac{\sqrt{2}}{4}$$

13. $\displaystyle \lim_{x \to 0} \frac{\sqrt{3 + x} - \sqrt{3}}{x} = \lim_{x \to 0} \frac{\sqrt{3 + x} - \sqrt{3}}{x} \cdot \frac{\sqrt{3 + x} + \sqrt{3}}{\sqrt{3 + x} + \sqrt{3}}$

$$= \lim_{x \to 0} \frac{3 + x - 3}{x(\sqrt{3 + x} + \sqrt{3})} = \lim_{x \to 0} \frac{1}{\sqrt{3 + x} + \sqrt{3}} = \frac{1}{2\sqrt{3}} = \frac{\sqrt{3}}{6}$$

14. $\displaystyle \lim_{x \to 0} \frac{\dfrac{1}{x + 4} - \dfrac{1}{4}}{x} = \lim_{x \to 0} \frac{\dfrac{4 - (x + 4)}{4(x + 4)}}{x} = \lim_{x \to 0} \frac{-1}{4(x + 4)} = -\frac{1}{16}$

15. $\lim\limits_{x\to 0} \dfrac{\dfrac{1}{2+x} - \dfrac{1}{2}}{x} = \lim\limits_{x\to 0} \dfrac{\dfrac{2-(2+x)}{2(2+x)}}{x} = \lim\limits_{x\to 0} \dfrac{-1}{2(2+x)} = -\dfrac{1}{4}$

16. $\lim\limits_{x\to 3} \dfrac{\sqrt{x+1}-2}{x-3} = \lim\limits_{x\to 3} \dfrac{\sqrt{x+1}-2}{x-3} \cdot \dfrac{\sqrt{x+1}+2}{\sqrt{x+1}+2} = \lim\limits_{x\to 3} \dfrac{x-3}{(x-3)[\sqrt{x+1}+2]} = \lim\limits_{x\to 3} \dfrac{1}{\sqrt{x+1}+2} = \dfrac{1}{4}$

17. $\lim\limits_{\Delta x\to 0} \dfrac{2(x+\Delta x) - 2x}{\Delta x} = \lim\limits_{\Delta x\to 0} \dfrac{2x + 2\Delta x - 2x}{\Delta x} = \lim\limits_{\Delta x\to 0} 2 = 2$

18. $\lim\limits_{\Delta x\to 0} \dfrac{(x+\Delta x)^2 - x^2}{\Delta x} = \lim\limits_{\Delta x\to 0} \dfrac{x^2 + 2x\Delta x + (\Delta x)^2 - x^2}{\Delta x} = \lim\limits_{\Delta x\to 0} \dfrac{\Delta x(2x + \Delta x)}{\Delta x} = 2x$

19. $\lim\limits_{\Delta x\to 0} \dfrac{(x+\Delta x)^2 - 2(x+\Delta x) + 1 - (x^2 - 2x + 1)}{\Delta x}$

$= \lim\limits_{\Delta x\to 0} \dfrac{x^2 + 2x\Delta x + (\Delta x)^2 - 2x - 2\Delta x + 1 - x^2 + 2x - 1}{\Delta x} = \lim\limits_{\Delta x\to 0} (2x + \Delta x - 2) = 2x - 2$

20. $\lim\limits_{\Delta x\to 0} \dfrac{(1+\Delta x)^3 - 1}{\Delta x} = \lim\limits_{\Delta x\to 0} \dfrac{1 + 3\Delta x + 3(\Delta x)^2 + (\Delta x)^3 - 1}{\Delta x} = \lim\limits_{\Delta x\to 0} [3 + 3\Delta x + (\Delta x)^2] = 3$

21.

x	-0.1	-0.01	-0.001	0	0.001	0.01	0.1
$f(x)$	0.358	0.354	0.354	?	0.354	0.353	0.349

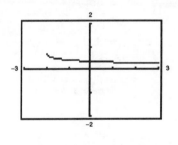

$\lim\limits_{x\to 0} \dfrac{\sqrt{x+2}-\sqrt{2}}{x} \approx 0.354$

$\lim\limits_{x\to 0} \dfrac{\sqrt{x+2}-\sqrt{2}}{x} = \lim\limits_{x\to 0} \dfrac{\sqrt{x+2}-\sqrt{2}}{x} \cdot \dfrac{\sqrt{x+2}+\sqrt{2}}{\sqrt{x+2}+\sqrt{2}}$

$= \lim\limits_{x\to 0} \dfrac{x+2-2}{x(\sqrt{x+2}+\sqrt{2})}$

$= \lim\limits_{x\to 0} \dfrac{1}{\sqrt{x+2}+\sqrt{2}} = \dfrac{1}{2\sqrt{2}} = \dfrac{\sqrt{2}}{4}$

22.

x	0.9	0.99	0.999	1	1.001	1.01	1.1
$f(x)$	-0.202	-0.200	-0.200	?	-0.200	-0.200	-0.198

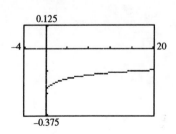

$\lim\limits_{x\to 1} \dfrac{4-\sqrt{x}}{x-16} = -0.20$

$\lim\limits_{x\to 1} \dfrac{4-\sqrt{x}}{x-16} = \dfrac{4-\sqrt{1}}{1-16} = \dfrac{3}{-15} = -\dfrac{1}{5}$

23.

x	1.9	1.99	1.999	1.9999	2.0	2.0001	2.001	2.01	2.1
$f(x)$	72.39	79.20	79.92	79.99	?	80.01	80.08	80.80	88.41

$$\lim_{x \to 2} \frac{x^5 - 32}{x - 2} = 80$$

$$\lim_{x \to 2} \frac{x^5 - 32}{x - 2} = \lim_{x \to 2} \frac{(x - 2)(x^4 + 2x^3 + 4x^2 + 8x + 16)}{x - 2}$$

$$= \lim_{x \to 2} (x^4 + 2x^3 + 4x^2 + 8x + 16)$$

$$= 80$$

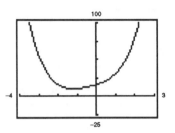

(*Hint:*   Use long division to factor $x^5 - 32$.)

24.

x	-0.1	-0.01	-0.001	0	0.001	0.01	0.1
$f(x)$	-0.263	-0.251	-0.250	?	-0.250	-0.249	-0.238

$$\lim_{x \to 0} \frac{\dfrac{1}{2 + x} - \dfrac{1}{2}}{x} = -\frac{1}{4}$$

$$\lim_{x \to 0} \frac{\dfrac{1}{2 + x} - \dfrac{1}{2}}{x} = \lim_{x \to 0} \frac{2 - (2 + x)}{2(2 + x)} \cdot \frac{1}{x}$$

$$= \lim_{x \to 0} \frac{-x}{2(2 + x)} \cdot \frac{1}{x}$$

$$= \lim_{x \to 0} \frac{-1}{2(2 + x)} = -\frac{1}{4}$$

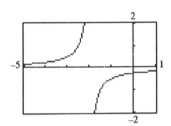

25. $\displaystyle\lim_{x \to 0} \frac{\sin x}{5x} = \lim_{x \to 0} \left[\left(\frac{\sin x}{x}\right)\left(\frac{1}{5}\right) \right]$

$$= (1)\left(\frac{1}{5}\right) = \frac{1}{5}$$

26. $\displaystyle\lim_{x \to 0} \frac{3(1 - \cos x)}{x} = \lim_{x \to 0} \left[3\left(\frac{1 - \cos x}{x}\right) \right]$

$$= (3)(0) = 0$$

27. $\displaystyle\lim_{\theta \to 0} \frac{\sec \theta - 1}{\theta \sec \theta} = \lim_{\theta \to 0} \frac{1 - \cos \theta}{\theta} = 0$

28. $\displaystyle\lim_{\theta \to 0} \frac{\cos \theta \tan \theta}{\theta} = \lim_{\theta \to 0} \frac{\sin \theta}{\theta} = 1$

29. $\displaystyle\lim_{\theta \to 0} \frac{\sin^2 x}{x} = \lim_{\theta \to 0} \left[\frac{\sin x}{x} \sin x \right]$

$$= (1) \sin 0 = 0$$

30. $\displaystyle\lim_{x \to 0} \frac{\tan^2 x}{x} = \lim_{x \to 0} \frac{\sin^2 x}{x \cos^2 x}$

$$= \lim_{x \to 0} \left[\frac{\sin x}{x} \cdot \frac{\sin x}{\cos^2 x} \right]$$

$$= (1)(0) = 0$$

31. $\displaystyle\lim_{h \to 0} \frac{(1 - \cos h)^2}{h} = \lim_{h \to 0} \left[\frac{1 - \cos h}{h}(1 - \cos h) \right]$

$$= (0)(0) = 0$$

32. $\displaystyle\lim_{\phi \to \pi} \phi \sec \phi = -\pi$

33. $\lim\limits_{x\to\pi/2} \dfrac{\cos x}{\cot x} = \lim\limits_{x\to\pi/2} \sin x = 1$

34. $\lim\limits_{x\to\pi/4} \dfrac{1-\tan x}{\sin x - \cos x} = \lim\limits_{x\to\pi/4} \dfrac{\cos x - \sin x}{\sin x \cos x - \cos^2 x}$

$$= \lim\limits_{x\to\pi/4} \dfrac{-(\sin x - \cos x)}{\cos x(\sin x - \cos x)}$$

$$= \lim\limits_{x\to\pi/4} \dfrac{-1}{\cos x}$$

$$= \lim\limits_{x\to\pi/4} (-\sec x)$$

$$= -\sqrt{2}$$

35. $\lim\limits_{t\to0} \dfrac{\sin^2 t}{t^2} = \lim\limits_{t\to0} \left(\dfrac{\sin t}{t}\right)^2 = (1)^2 = 1$

36. $\lim\limits_{x\to0} \dfrac{\sin 2x}{\sin 3x} = \lim\limits_{x\to0} \left[2\left(\dfrac{\sin 2x}{2x}\right)\left(\dfrac{1}{3}\right)\left(\dfrac{3x}{\sin 3x}\right)\right]$

$$= 2(1)\left(\dfrac{1}{3}\right)(1) = \dfrac{2}{3}$$

37. $\lim\limits_{t\to0} \dfrac{\sin 3t}{t} = \lim\limits_{t\to0} \left[3\left(\dfrac{\sin 3t}{3t}\right)\right] = 3(1) = 3$

38. $\lim\limits_{h\to0} (1 + \cos 2h) = 1 + 1 = 2$

39. $\lim\limits_{x\to0} \dfrac{\sin x^2}{x} = \lim\limits_{x\to0} \left[x\left(\dfrac{\sin x^2}{x^2}\right)\right] = (0)(1) = 0$

40. $\lim\limits_{x\to0} \dfrac{\sin x}{\sqrt[3]{x}} = \lim\limits_{x\to0} \left[\sqrt[3]{x^2}\left(\dfrac{\sin x}{x}\right)\right] = (0)(1) = 0$

41. $\lim\limits_{x\to0} (4 - x^2) \le \lim\limits_{x\to0} f(x) \le \lim\limits_{x\to0} (4 + x^2)$

$$4 \le \lim\limits_{x\to0} f(x) \le 4$$

Therefore, $\lim\limits_{x\to0} f(x) = 4$.

42. $\lim\limits_{x\to a} [b - |x - a|] \le \lim\limits_{x\to a} f(x) \le \lim\limits_{x\to a} [b + |x - a|]$

$$b \le \lim\limits_{x\to a} f(x) \le b$$

Therefore, $\lim\limits_{x\to a} f(x) = b$.

43. $f(x) = x \cos x$

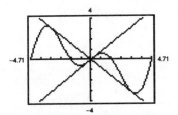

$$\lim\limits_{x\to0} (x \cos x) = 0$$

44. $f(x) = |x \sin x|$

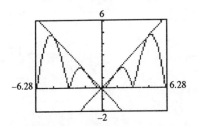

$$\lim\limits_{x\to0} |x \sin x| = 0$$

45. $f(x) = |x| \sin x$

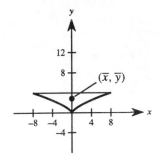

$$\lim\limits_{x\to0} |x| \sin x = 0$$

46. $f(x) = |x| \cos x$

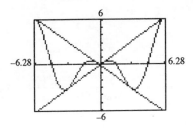

$$\lim\limits_{x\to0} |x| \cos x = 0$$

47. $f(x) = x \sin \dfrac{1}{x}$

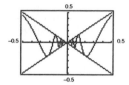

$$\lim_{x \to 0} \left(x \sin \frac{1}{x} \right) = 0$$

48. $f(x) = x \cos \dfrac{1}{x}$

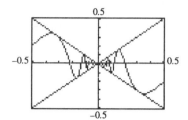

$$\lim_{x \to 0} \left(x \cos \frac{1}{x} \right) = 0$$

49. $f(x) = x, \ g(x) = \sin x, \ h(x) = \dfrac{\sin x}{x}$

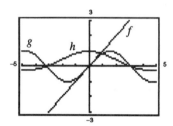

When you are "close to" 0 the magnitude of f is approximately equal to the magnitude of g. Thus, $|g|/|f| \approx 1$ when x is "close to" 0.

50. $f(x) = x, \ g(x) = \sin^2 x, \ h(x) = \dfrac{\sin^2 x}{x}$

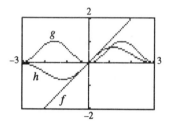

When you are "close to" 0 the magnitude of g is "smaller" than the magnitude of f and the magnitude of g is approaching zero "faster" than the magnitude of f. Thus, $|g|/|f| \approx 0$ when x is "close to" 0.

51. $s(t) = -16t^2 + 1000$

$$\lim_{t \to 5} \frac{s(5) - s(t)}{5 - t} = \lim_{t \to 5} \frac{600 - (-16t^2 + 1000)}{5 - t} = \lim_{t \to 5} \frac{16(t + 5)(t - 5)}{-(t - 5)} = \lim_{t \to 5} -16(t + 5) = -160 \text{ ft/sec}$$

52. $s(t) = -16t^2 + 1000 = 0$ when $t = \sqrt{\dfrac{1000}{16}} = \dfrac{5\sqrt{10}}{2}$ seconds

$$\lim_{t \to 5\sqrt{10}/2} \frac{s\left(\dfrac{5\sqrt{10}}{2} \right) - s(t)}{\dfrac{5\sqrt{10}}{2} - t} = \lim_{t \to 5\sqrt{10}/2} \frac{0 - (-16t^2 + 1000)}{\dfrac{5\sqrt{10}}{2} - t}$$

$$= \lim_{t \to 5\sqrt{10}/2} \frac{16\left(t^2 - \dfrac{125}{2} \right)}{\dfrac{5\sqrt{10}}{2} - t} = \lim_{t \to 5\sqrt{10}/2} \frac{16\left(t + \dfrac{5\sqrt{10}}{2} \right)\left(t - \dfrac{5\sqrt{10}}{2} \right)}{-\left(t - \dfrac{5\sqrt{10}}{2} \right)}$$

$$= \lim_{t \to 5\sqrt{10}/2} -16\left(t + \dfrac{5\sqrt{10}}{2} \right) = -80\sqrt{10} \text{ ft/sec}$$

53. Given $\lim\limits_{x \to c} f(x) = 0$:

For every $\epsilon > 0$, there exists $\delta > 0$ such that $|f(x) - 0| < \epsilon$ whenever $0 < |x - c| < \delta$.
Now $|f(x) - 0| = |f(x)| = \big||f(x)| - 0\big| < \epsilon$ for $|x - c| < \delta$. Therefore, $\lim\limits_{x \to c} |f(x)| = 0$.

54. $-M|f(x)| \le f(x)g(x) \le M|f(x)|$

$$\lim_{x \to c} (-M|f(x)|) \le \lim_{x \to c} f(x)g(x) \le \lim_{x \to c} (M|f(x)|)$$

$$-M(0) \le \lim_{x \to c} f(x)g(x) \le M(0)$$

$$0 \le \lim_{x \to c} f(x)g(x) \le 0$$

Therefore, $\lim_{x \to c} f(x)g(x) = 0$.

55. If $\lim_{x \to c} |f(x)| = 0$, then $\lim_{x \to c} [-|f(x)|] = 0$.

$$-|f(x)| \le f(x) \le |f(x)|$$

$$\lim_{x \to c} [-|f(x)|] \le \lim_{x \to c} f(x) \le \lim_{x \to c} |f(x)|$$

$$0 \le \lim_{x \to c} f(x) \le 0$$

Therefore, $\lim_{x \to c} f(x) = 0$.

56. Given $\lim_{x \to c} f(x) = L$:

For every $\epsilon > 0$, there exists $\delta > 0$ such that $|f(x) - L| < \epsilon$ whenever $0 < |x - c| < \delta$.
Since $\big||f(x)| - |L|\big| \le |f(x) - L| < \epsilon$ for $|x - c| < \delta$, then $\lim_{x \to c} |f(x)| = |L|$.

57. Let
$$f(x) = \begin{cases} 4, & \text{if } x \ge 0 \\ -4, & \text{if } x < 0 \end{cases}$$

$$\lim_{x \to 0} |f(x)| = \lim_{x \to 0} 4 = 4.$$

$\lim_{x \to 0} f(x)$ does not exist since $\lim_{x \to 0^-} f(x) = -4$ and $\lim_{x \to 0^+} f(x) = 4$.

58. $\lim_{x \to 0} \dfrac{1 - \cos x}{x} = \lim_{x \to 0} \dfrac{1 - \cos x}{x} \cdot \dfrac{1 + \cos x}{1 + \cos x} = \lim_{x \to 0} \dfrac{1 - \cos^2 x}{x(1 + \cos x)} = \lim_{x \to 0} \dfrac{\sin^2 x}{x(1 + \cos x)}$

$$= \lim_{x \to 0} \frac{\sin x}{x} \cdot \frac{\sin x}{1 + \cos x}$$

$$= \left[\lim_{x \to 0} \frac{\sin x}{x}\right]\left[\lim_{x \to 0} \frac{\sin x}{1 + \cos x}\right]$$

$$= (1)(0) = 0$$

59. $f(x) = \begin{cases} 0, & \text{if } x \text{ is rational} \\ 1, & \text{if } x \text{ is irrational} \end{cases}$

$g(x) = \begin{cases} 0, & \text{if } x \text{ is rational} \\ x, & \text{if } x \text{ is irrational} \end{cases}$

$\lim_{x \to 0} f(x)$ does not exist.

No matter how "close to" 0 x is, there are still an infinite number of rational and irrational numbers so that $\lim_{x \to 0} f(x)$ does not exist.

$\lim_{x \to 0} g(x) = 0$.

When x is "close to" 0, both parts of the function are "close to" 0.

60. $\displaystyle\lim_{x\to 0} \frac{\sec x - 1}{x^2} = \lim_{x\to 0} \frac{\sec x - 1}{x^2} \cdot \frac{\sec x + 1}{\sec x + 1}$

$\displaystyle\qquad = \lim_{x\to 0} \frac{\tan^2 x}{x^2(\sec x + 1)} = \lim_{x\to 0} \left(\frac{1}{\cos^2 x}\right)\left(\frac{\sin^2 x}{x^2}\right)\left(\frac{1}{\sec x + 1}\right) = (1)(1)\left(\frac{1}{2}\right) = \frac{1}{2}$

x	-0.1	-0.01	-0.001	0	0.001	0.01	0.1
$\dfrac{\sec x - 1}{x^2}$	0.502	0.500	0.500	?	0.500	0.500	0.502

61. $\displaystyle\lim_{x\to 0} \frac{1 - \cos x}{x^2} = \lim_{x\to 0} \frac{\sec x - 1}{x^2 \sec x} = \lim_{x\to 0} \left(\frac{\sec x - 1}{x^2}\right)\left(\frac{1}{\sec x}\right) = \left(\frac{1}{2}\right)(1) = \frac{1}{2}$

(See Exercise 60.) For x "close to" 0, we have

$$\frac{1 - \cos x}{x^2} \approx \frac{1}{2}.$$

Thus,

$$1 - \cos x \approx \frac{1}{2}x^2 \Rightarrow \cos x \approx 1 - \frac{1}{2}x^2$$

$$\cos(0.1) \approx 1 - \frac{1}{2}(0.1)^2 = 0.995.$$

Section 1.4   Continuity and One-Sided Limits

1. a. $\displaystyle\lim_{x\to 3^+} f(x) = 1$

   b. $\displaystyle\lim_{x\to 3^-} f(x) = 1$

   c. $\displaystyle\lim_{x\to 3} f(x) = 1$

2. a. $\displaystyle\lim_{x\to -2^+} f(x) = -2$

   b. $\displaystyle\lim_{x\to -2^-} f(x) = -2$

   c. $\displaystyle\lim_{x\to -2} f(x) = -2$

3. a. $\displaystyle\lim_{x\to 3^+} f(x) = 0$

   b. $\displaystyle\lim_{x\to 3^-} f(x) = 0$

   c. $\displaystyle\lim_{x\to 3} f(x) = 0$

4. a. $\displaystyle\lim_{x\to -2^+} f(x) = 2$

   b. $\displaystyle\lim_{x\to -2^-} f(x) = 2$

   c. $\displaystyle\lim_{x\to -2} f(x) = 2$

5. a. $\displaystyle\lim_{x\to 3^+} f(x) = 3$

   b. $\displaystyle\lim_{x\to 3^-} f(x) = -3$

   c. $\displaystyle\lim_{x\to 3} f(x)$ does not exist.

6. a. $\displaystyle\lim_{x\to -1^+} f(x) = 0$

   b. $\displaystyle\lim_{x\to -1^-} f(x) = 2$

   c. $\displaystyle\lim_{x\to -1} f(x)$ does not exist.

7. $\displaystyle\lim_{x\to 5^+} \frac{x-5}{x^2-25} = \lim_{x\to 5^+} \frac{1}{x+5} = \frac{1}{10}$

8. $\displaystyle\lim_{x\to 2^+} \frac{2-x}{x^2-4} = \lim_{x\to 2^+} -\frac{1}{x+2} = -\frac{1}{4}$

9. $\displaystyle\lim_{x\to 2^+} \frac{x}{\sqrt{x^2-4}}$ does not exist since $\dfrac{x}{\sqrt{x^2-4}} \to \infty$ as $x \to 2^+$.

10. $\displaystyle\lim_{x\to 4^-} \frac{\sqrt{x}-2}{x-4} = \lim_{x\to 4^-} \frac{\sqrt{x}-2}{x-4} \cdot \frac{\sqrt{x}+2}{\sqrt{x}+2} = \lim_{x\to 4^-} \frac{x-4}{(x-4)(\sqrt{x}+2)} = \lim_{x\to 4^-} \frac{1}{\sqrt{x}+2} = \frac{1}{4}$

11. $\displaystyle\lim_{x\to 0} \frac{|x|}{x}$ does not exist since

$$\lim_{x\to 0^+} \frac{|x|}{x} = 1 \text{ and } \lim_{x\to 0^-} \frac{|x|}{x} = -1.$$

12. $\displaystyle\lim_{x\to 2^+} \frac{|x-2|}{x-2} = \lim_{x\to 2^+} \frac{x-2}{x-2} = 1$

$$\lim_{x\to 2^-} \frac{|x-2|}{x-2} = \lim_{x\to 2^-} \frac{-(x-2)}{x-2} = -1$$

$$\lim_{x\to 2^+} \frac{|x-2|}{x-2} \neq \lim_{x\to 2^-} \frac{|x-2|}{x-2}$$

Thus, the limit does not exist.

13. $\displaystyle\lim_{\Delta x\to 0^-} \frac{\dfrac{1}{x+\Delta x} - \dfrac{1}{x}}{\Delta x} = \lim_{\Delta x\to 0^-} \frac{x-(x+\Delta x)}{x(x+\Delta x)} \cdot \frac{1}{\Delta x} = \lim_{\Delta x\to 0^-} \frac{-\Delta x}{x(x+\Delta x)} \cdot \frac{1}{\Delta x}$

$$= \lim_{\Delta x\to 0^-} \frac{-1}{x(x+\Delta x)}$$

$$= \frac{-1}{x(x+0)}$$

$$= -\frac{1}{x^2}$$

14. $\displaystyle\lim_{\Delta x\to 0^+} \frac{(x+\Delta x)^2 + (x+\Delta x) - (x^2+x)}{\Delta x} = \lim_{\Delta x\to 0^+} \frac{x^2 + 2x(\Delta x) + (\Delta x)^2 + x + \Delta x - x^2 - x}{\Delta x}$

$$= \lim_{\Delta x\to 0^+} \frac{2x(\Delta x) + (\Delta x)^2 + \Delta x}{\Delta x}$$

$$= \lim_{\Delta x\to 0^+} (2x + \Delta x + 1)$$

$$= 2x + 0 + 1$$

$$= 2x + 1$$

15. $\displaystyle\lim_{x\to 3^+} f(x) = \lim_{x\to 3^+} \frac{12-2x}{3} = 2$

$$\lim_{x\to 3^-} f(x) = \lim_{x\to 3^-} \frac{x+2}{2} = \frac{5}{2}$$

$\displaystyle\lim_{x\to 3} f(x)$ does not exist.

16. $\displaystyle\lim_{x\to 2^+} f(x) = \lim_{x\to 2^+} (-x^2 + 4x - 2) = 2$

$$\lim_{x\to 2^-} f(x) = \lim_{x\to 2^-} (x^2 - 4x + 6) = 2$$

$$\lim_{x\to 2} f(x) = 2$$

17. $\displaystyle\lim_{x\to 1^+} f(x) = \lim_{x\to 1^+} (x+1) = 2$

$$\lim_{x\to 1^-} f(x) = \lim_{x\to 1^-} (x^3+1) = 2$$

$$\lim_{x\to 1} f(x) = 2$$

18. $\displaystyle\lim_{x\to 1^+} f(x) = \lim_{x\to 1^+} (1-x) = 0$

$$\lim_{x\to 1^-} f(x) = \lim_{x\to 1^-} (x) = 1$$

$\displaystyle\lim_{x\to 1} f(x)$ does not exist.

19. $\displaystyle\lim_{x\to \pi} \cot x$ does not exist since

$$\lim_{x\to \pi^+} \cot x = \infty \text{ and } \lim_{x\to \pi^-} \cot x = -\infty.$$

20. $\displaystyle\lim_{x\to \pi/2} \sec x$ does not exist since

$$\lim_{x\to (\pi/2)^+} \sec x = -\infty \text{ and } \lim_{x\to (\pi/2)^-} \sec x = \infty.$$

21. $\displaystyle\lim_{x\to 3^-} (2[\![x]\!] - 1) = 2(2) - 1 = 3$

22. $\displaystyle\lim_{x\to 2^+} (2x - [\![x]\!]) = 2(2) - 2 = 2$

23. $f(x) = \dfrac{1}{x^2 - 4}$

has discontinuities at $x = -2$ and $x = 2$ since $f(-2)$ and $f(2)$ are not defined.

24. $f(x) = \dfrac{x^2 - 1}{x + 1}$

has a discontinuity at $x = -1$ since $f(-1)$ is not defined.

25. $f(x) = \dfrac{[\![x]\!]}{2} + x$ has discontinuities at each integer k since $\lim\limits_{x \to k^-} f(x) \neq \lim\limits_{x \to k^+} f(x)$.

26. $f(x) = \begin{cases} x, & x < 1 \\ 2, & x = 1 \\ 2x - 1, & x > 1 \end{cases}$ has a discontinuity at $x = 1$ since $f(1) \neq \lim\limits_{x \to 1} f(x)$.

27. $f(x) = x^2 - 2x + 1$ is continuous for all real x.

28. $f(x) = \dfrac{1}{x^2 + 1}$ is continuous for all real x.

29. $f(x) = x + \sin x$ is continuous for all real x.

30. $f(x) = \cos \dfrac{\pi x}{2}$ is continuous for all real x.

31. $f(x) = \dfrac{1}{x - 1}$ has a nonremovable discontinuity at $x = 1$ since $\lim\limits_{x \to 1} f(x)$ does not exist.

32. $f(x) = \dfrac{x}{x^2 - 1}$ has nonremovable discontinuities at $x = 1$ and $x = -1$ since $\lim\limits_{x \to 1} f(x)$ and $\lim\limits_{x \to -1} f(x)$ do not exist.

33. $f(x) = \dfrac{x}{x^2 + 1}$

is continuous for all real x.

34. $f(x) = \dfrac{x - 3}{x^2 - 9}$

has a nonremovable discontinuity at $x = -3$ since $\lim\limits_{x \to -3} f(x)$ does not exist, and has a removable discontinuity at $x = 3$ since

$$\lim\limits_{x \to 3} f(x) = \lim\limits_{x \to 3} \dfrac{1}{x + 3} = \dfrac{1}{6}.$$

35. $f(x) = \dfrac{x - 1}{(x + 2)(x - 1)}$

has a nonremovable discontinuity at $x = -2$ since $\lim\limits_{x \to -2} f(x)$ does not exist, and has a removable discontinuity at $x = 1$ since

$$\lim\limits_{x \to 1} f(x) = \lim\limits_{x \to 1} \dfrac{1}{x + 2} = \dfrac{1}{3}.$$

36. $f(x) = \dfrac{x + 2}{(x + 2)(x - 5)}$

has a nonremovable discontinuity at $x = 5$ since $\lim\limits_{x \to 5} f(x)$ does not exist, and has a removable discontinuity at $x = -2$ since

$$\lim\limits_{x \to -2} f(x) = \lim\limits_{x \to -2} \dfrac{1}{x - 5} = -\dfrac{1}{7}.$$

37. $f(x) = \dfrac{|x + 2|}{x + 2}$

has a nonremovable discontinuity at $x = -2$ since $\lim\limits_{x \to -2} f(x)$ does not exist.

38. $f(x) = \dfrac{|x - 3|}{x - 3}$

has a nonremovable discontinuity at $x = 3$ since $\lim\limits_{x \to 3} f(x)$ does not exist.

39. $f(x) = \begin{cases} x, & x \le 1 \\ x^2, & x > 1 \end{cases}$ has a **possible** discontinuity at $x = 1$.

1. $f(1) = 1$

2. $\left. \begin{array}{l} \lim\limits_{x \to 1^-} f(x) = \lim\limits_{x \to 1^-} x = 1 \\ \lim\limits_{x \to 1^+} f(x) = \lim\limits_{x \to 1^+} x^2 = 1 \end{array} \right\}$ $\lim\limits_{x \to 1} f(x) = 1$

3. $f(1) = \lim\limits_{x \to 1} f(x)$

f is continuous at $x = 1$, therefore, f is continuous for all real x.

40. $f(x) = \begin{cases} -2x + 3, & x < 1 \\ x^2, & x \ge 1 \end{cases}$ has a **possible** discontinuity at $x = 1$.

1. $f(1) = 1^2 = 1$

2. $\left. \begin{array}{l} \lim\limits_{x \to 1^-} f(x) = \lim\limits_{x \to 1^-} (-2x + 3) = 1 \\ \lim\limits_{x \to 1^+} f(x) = \lim\limits_{x \to 1^+} x^2 = 1 \end{array} \right\}$ $\lim\limits_{x \to 1} f(x) = 1$

3. $f(1) = \lim\limits_{x \to 1} f(x)$

f is continuous at $x = 1$, therefore, f is continuous for all real x.

41. $f(x) = \begin{cases} \dfrac{x}{2} + 1, & x \le 2 \\ 3 - x, & x > 2 \end{cases}$ has a **possible** discontinuity at $x = 2$.

1. $f(2) = \dfrac{2}{2} + 1 = 2$

2. $\left. \begin{array}{l} \lim\limits_{x \to 2^-} f(x) = \lim\limits_{x \to 2^-} \left(\dfrac{x}{2} + 1 \right) = 2 \\ \lim\limits_{x \to 2^+} f(x) = \lim\limits_{x \to 2^+} (3 - x) = 1 \end{array} \right\}$ $\lim\limits_{x \to 2} f(x)$ does not exist.

Therefore, f has a nonremovable discontinuity at $x = 2$.

42. $f(x) = \begin{cases} -2x, & x \le 2 \\ x^2 - 4x + 1, & x > 2 \end{cases}$ has a **possible** discontinuity at $x = 2$.

1. $f(2) = -2(2) = -4$

2. $\left. \begin{array}{l} \lim\limits_{x \to 2^-} f(x) = \lim\limits_{x \to 2^-} (-2x) = -4 \\ \lim\limits_{x \to 2^+} f(x) = \lim\limits_{x \to 2^+} (x^2 - 4x + 1) = -3 \end{array} \right\}$ $\lim\limits_{x \to 2} f(x)$ does not exist.

Therefore, f has a nonremovable discontinuity at $x = 2$.

43. $f(x) = \begin{cases} \csc \dfrac{\pi x}{6}, & |x - 3| \le 2 \\ 2, & |x - 3| > 2 \end{cases} = \begin{cases} \csc \dfrac{\pi x}{6}, & 1 \le x \le 5 \\ 2, & x < 1 \text{ or } x > 5 \end{cases}$ has **possible** discontinuities at $x = 1$, $x = 5$.

1. $f(1) = \csc \dfrac{\pi}{6} = 2$ $f(5) = \csc \dfrac{5\pi}{6} = 2$

2. $\lim\limits_{x \to 1} f(x) = 2$ $\lim\limits_{x \to 5} f(x) = 2$

3. $f(1) = \lim\limits_{x \to 1} f(x)$ $f(5) = \lim\limits_{x \to 5} f(x)$

f is continuous at $x = 1$ and $x = 5$, therefore, f is continuous for all real x.

44. $f(x) = \begin{cases} \tan\dfrac{\pi x}{4}, & |x| < 1 \\ x, & |x| \geq 1 \end{cases} = \begin{cases} \tan\dfrac{\pi x}{4}, & -1 < x < 1 \\ x, & x \leq -1 \text{ or } x \geq 1 \end{cases}$ has **possible** discontinuities at $x = -1$ and $x = 1$.

1. $f(-1) = -1$ $f(1) = 1$

2. $\displaystyle\lim_{x\to-1} f(x) = -1$ $\displaystyle\lim_{x\to 1} f(x) = 1$

3. $f(-1) = \displaystyle\lim_{x\to-1} f(x)$ $f(1) = \displaystyle\lim_{x\to 1} f(x)$

f is continuous at $x = \pm 1$, therefore, f is continuous for all real x.

45. $f(x) = \csc 2x$ had nonremovable discontinuities at integer multiples of $\pi/2$.

46. $f(x) = \tan\dfrac{\pi x}{2}$ has nonremovable discontinuities at each $2k + 1$, k is an integer.

47. $f(x) = [\![x - 1]\!]$ has nonremovable discontinuities at each integer k.

48. $f(x) = x - [\![x]\!]$ has nonremovable discontinuities at each integer k.

49. $f(2) = 8$

Find a so that $\displaystyle\lim_{x\to 2^+} ax^2 = 8 \;\Rightarrow\; a = \dfrac{8}{2^2} = 2.$

50. Find a and b such that $\displaystyle\lim_{x\to-1^+} (ax + b) = -a + b = 2$ and $\displaystyle\lim_{x\to 3^-} (ax + b) = 3a + b = -2.$

$$\begin{array}{rl} a - b = -2 & \\ \underline{(+)\ 3a + b = -2} & \\ 4a\qquad = -4 & \\ a = -1 & \\ b = \ 2 + (-1) = 1 & \end{array} \qquad f(x) = \begin{cases} 2, & x \leq -1 \\ -x + 1, & -1 < x < 3 \\ -2, & x \geq 3 \end{cases}$$

51. $\displaystyle\lim_{x\to 0^-} g(x) = \lim_{x\to 0^-} \dfrac{4\sin x}{x} = 4$

$\displaystyle\lim_{x\to 0^+} g(x) = \lim_{x\to 0^+} (a - 2x) = a$

Let $a = 4$.

52. $\displaystyle\lim_{x\to a} g(x) = \lim_{x\to a} \dfrac{x^2 - a^2}{x - a}$

$= \displaystyle\lim_{x\to a} (x + a) = 2a$

Find a such that $2a = 8 \Rightarrow a = 4$.

53. $f(g(x)) = (x - 1)^2$

Continuous for all real x

54. $f(g(x)) = \dfrac{1}{\sqrt{x - 1}}$

Nonremovable discontinuity at $x = 1$

Continuous for all $x > 1$

55. $f(g(x)) = \dfrac{1}{(x^2 + 5) - 6} = \dfrac{1}{x^2 - 1}$

Nonremovable discontinuities at $x = \pm 1$

56. $f(g(x)) = \sin x^2$

Continuous for all real x

57. $f(x) = \dfrac{x}{x^2 + 1}$

Continuous on $(-\infty, \infty)$

58. $f(x) = x\sqrt{x + 3}$

Continuous on $[-3, \infty)$

59. $f(x) = \csc \dfrac{x}{2}$

Continuous on:

$\ldots, \ (-2\pi, \ 0), \ (0, \ 2\pi), \ (2\pi, \ 4\pi), \ \ldots$

60. $f(x) = \dfrac{x + 1}{\sqrt{x}}$

Continuous on $(0, \infty)$

61. $y = [\![x]\!] - x$

Nonremovable discontinuity at each integer

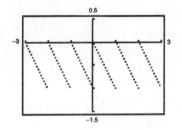

62. $h(x) = \dfrac{1}{(x + 1)(x - 2)}$

Nonremovable discontinuities at $x = -1$ and $x = 2$

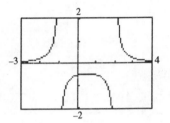

63. $f(x) = \begin{cases} 2x - 4, & x \le 3 \\ x^2 - 2x, & x > 3 \end{cases}$

Nonremovable discontinuity at $x = 3$

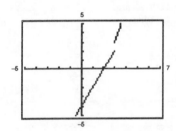

64. $f(x) = \begin{cases} \dfrac{\cos x - 1}{x}, & x < 0 \\ 5x, & x \ge 0 \end{cases}$

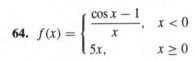

$f(0) = 5(0) = 0$

$\displaystyle \lim_{x \to 0^-} f(x) = \lim_{x \to 0^-} \frac{(\cos x - 1)}{x} = 0$

$\displaystyle \lim_{x \to 0^+} f(x) = \lim_{x \to 0^+} (5x) = 0$

Therefore, $\displaystyle \lim_{x \to 0} f(x) = 0 = f(0)$ and f is continuous on the entire real line. ($x = 0$ was the only possible discontinuity.)

65. $f(x) = \dfrac{\sin x}{x}$

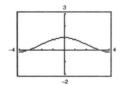

The graph **appears** to be continuous on the interval $[-4, 4]$. Since $f(0)$ is not defined, we know that f has a discontinuity at $x = 0$. This discontinuity is removable so it does not show up on the graph.

66. $f(x) = \dfrac{x^3 - 8}{x - 2}$

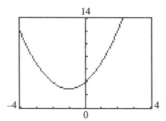

The graph **appears** to be continuous on the interval $[-4, 4]$. Since $f(2)$ is not defined, we know that f has a discontinuity at $x = 2$. This discontinuity is removable so it does not show up on the graph.

67. $f(x) = x^2 - 4x + 3$

$f(x)$ is continuous on $[2, 4]$.

$f(2) = -1$ and $f(4) = 3$

By the Intermediate Value Theorem, $f(x) = 0$ for at least one value c between 2 and 4.

68. $f(x) = x^3 + 3x - 2$

$f(x)$ is continuous on $[0, 1]$.

$f(0) = -2$ and $f(1) = 2$

By the Intermediate Value Theorem, $f(x) = 0$ for at least one value c between 0 and 1.

69. $f(x) = x^3 + x - 1$

$f(x)$ is continuous on $[0, 1]$.

$f(0) = -1$ and $f(1) = 1$

By the Intermediate Value Theorem, $f(x) = 0$ for at least one value c between 0 and 1. Using a graphing utility, we find that $x \approx 0.68$.

70. $f(x) = x^3 + 3x - 2$

$f(x)$ is continuous on $[0, 1]$.

$f(0) = -2$ and $f(1) = 2$

By the Intermediate Value Theorem, $f(x) = 0$ for at least one value c between 0 and 1. Using a graphing utility, we find that $x \approx 0.60$.

71. $f(x) = x^2 + x - 1$

f is continuous on $[0, 5]$.

$f(0) = -1$ and $f(5) = 29$

$-1 < 11 < 29$

The Intermediate Value Theorem applies.

$$x^2 + x - 1 = 11$$
$$x^2 + x - 12 = 0$$
$$(x + 4)(x - 3) = 0$$
$$x = -4 \text{ or } x = 3$$
$$c = 3 \quad (x = -4 \text{ is not in the interval.})$$

Thus, $f(3) = 11$.

72. $f(x) = x^2 - 6x + 8$

f is continuous on $[0, 3]$.

$f(0) = 8$ and $f(3) = -1$

$-1 < 0 < 8$

The Intermediate Value Theorem applies.

$$x^2 - 6x + 8 = 0$$
$$(x - 2)(x - 4) = 0$$
$$x = 2 \text{ or } x = 4$$
$$c = 2 \quad (x = 4 \text{ is not in the interval.})$$

Thus, $f(2) = 0$.

73. $f(x) = x^3 - x^2 + x - 2$

f is continuous on $[0, 3]$.

$f(0) = -2$ and $f(3) = 19$

$-2 < 4 < 19$

The Intermediate Value Theorem applies.

$$x^3 - x^2 + x - 2 = 4$$

$$x^3 - x^2 + x - 6 = 0$$

$$(x - 2)(x^2 + x + 3) = 0$$

$$x = 2$$

$(x^2 + x + 3$ has no real solution.)

$$c = 2$$

Thus, $f(2) = 4$.

74. $f(x) = \dfrac{x^2 + x}{x - 1}$

f is continuous on $\left[\frac{5}{2}, 4\right]$. The nonremovable discontinuity, $x = 1$, lies outside the interval.

$$f\left(\frac{5}{2}\right) = \frac{35}{6} \quad \text{and} \quad f(4) = \frac{20}{3}$$

$$\frac{35}{6} < 6 < \frac{20}{3}$$

The Intermediate Value Theorem applies.

$$\frac{x^2 + x}{x - 1} = 6$$

$$x^2 + x = 6x - 6$$

$$x^2 - 5x + 6 = 0$$

$$(x - 2)(x - 3) = 0$$

$$x = 2 \text{ or } x = 3$$

$c = 3$ ($x = 2$ is not in the interval.)

Thus, $f(3) = 6$.

75. $S(t) = 28,500(1.09)^{[\![t]\!]}$

Discontinuous at every positive integer

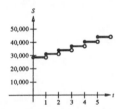

76. $C = \begin{cases} 1.04, & 0 < t \le 2 \\ 1.04 + 0.36[\![t - 1]\!], & t > 2, \quad t \text{ is not an integer} \\ 1.04 + 0.36(t - 2), & t > 2, \quad t \text{ is an integer} \end{cases}$

Nonremovable discontinuity at each integer greater than 2.

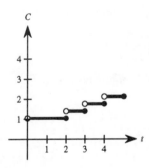

77. $N(t) = 25\left(2\left[\!\left[\dfrac{t + 2}{2}\right]\!\right] - t\right)$

t	0	1	1.8	2	3	3.8
$N(t)$	50	25	5	50	25	5

Discontinuous at every positive even integer. The company replenishes its inventory every two months.

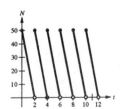

78. Let $s(t)$ be the position function for the run up to the campsite. $s(0) = 0$ ($t = 0$ corresponds to 8:00 A.M., $s(20) = k$ (distance to campsite). Let $r(t)$ be the position function for the run back down the mountain: $r(0) = k$, $r(10) = 0$. Let $f(t) = s(t) - r(t)$.

When $t = 0$ (8:00 A.M.), $f(0) = s(0) - r(0) = 0 - k < 0$

When $t = 10$ (8:10 A.M.), $f(10) = s(10) - r(10) > 0$

Since $f(0) < 0$ and $f(10) > 0$, then there must be a value t in the interval $(0, 10)$ such that $f(t) = 0$. If $f(t) = 0$, then $s(t) - r(t) = 0$, which gives us $s(t) = r(t)$. Therefore, at some time t, where $0 < t < 10$, the position functions for the run up and down are equal.

79. Suppose there exists x_1 in $[a, b]$ such that $f(x_1) > 0$ and there exists x_2 in $[a, b]$ such that $f(x_2) < 0$. Then by the Intermediate Value Theorem, $f(x)$ must equal zero for some value of x in $[x_1, x_2]$ (or $[x_2, x_1]$ if $x_2 < x_1$). Thus, f would have a zero in $[a, b]$, which is a contradiction. Therefore, $f(x) > 0$ for all x in $[a, b]$ or $f(x) < 0$ for all x in $[a, b]$.

80. Let c be any real number. Then, $\lim\limits_{x \to c} f(x)$ does not exist since there are both rational and irrational numbers arbitrarily close to c. Therefore, f is not continuous at c.

81. True

1. $f(c) = L$ is defined.

2. $\lim\limits_{x \to c} = L$ exists.

3. $f(c) = \lim\limits_{x \to c} f(x)$

All of the conditions for continuity are met.

82. True; if $f(x) = g(x)$, $x \neq c$, then $\lim\limits_{x \to c} f(x) = \lim\limits_{x \to c} g(x)$ and at least one of these limits (if they exist) does not equal the corresponding function at $x = c$.

83. False; a rational function can be written as $P(x)/Q(x)$ where P and Q are polynomials of degree m and n, respectively. It can have, at most, n discontinuities.

84. False; $f(1)$ is not defined and $\lim\limits_{x \to 1} f(x)$ does not exist.

85. 1. $f(x_0)$ is defined.

2. $\lim\limits_{x \to x_0} f(x) = \lim\limits_{\Delta x \to 0} f(x_0 + \Delta x) = f(x_0)$

(Let $x = x_0 + \Delta x$. As $x \to x_0$, $\Delta x \to 0$.)

3. $f(x_0) = \lim\limits_{x \to x_0} f(x)$

Therefore, f is continuous at x_0.

86. $\text{sgn}(x) = \begin{cases} -1, & \text{if } x < 0 \\ 0, & \text{if } x = 0 \\ 1, & \text{if } x > 0 \end{cases}$

a. $\lim\limits_{x \to 0^-} \text{sgn}(x) = -1$

b. $\lim\limits_{x \to 0^+} \text{sgn}(x) = 1$

c. $\lim\limits_{x \to 0} \text{sgn}(x)$ does not exist.

d. $f(x) = |\text{sgn}(x)|$ is not continuous on any interval containing $x = 0$ since $f(0) = 0$ and $\lim\limits_{x \to 0} |\text{sgn}(x)| = 1$.

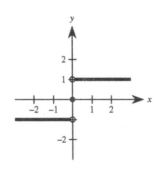

87. Let y be a real number. If $y = 0$, then $x = 0$. If $y > 0$, then let $0 < x_0 < \pi/2$ such that $M = \tan x_0 > y$ (this is possible since the tangent function increases without bound on $[0, \pi/2)$). By the Intermediate Value Theorem, $f(x) = \tan x$ is continuous on $[0, x_0]$ and $0 < y < M$, which implies that there exists x between 0 and x_0 such that $\tan x = y$. The argument is similar if $y < 0$.

88. $f(x) = \dfrac{\sqrt{x + c^2} - c}{x}, \quad c > 0$

Domain: $x + c^2 \geq 0 \Rightarrow x \geq -c^2$ and $x \neq 0, \quad [-c^2, 0) \cup (0, \infty)$

$$\lim_{x \to 0} \frac{\sqrt{x + c^2} - c}{x} = \lim_{x \to 0} \frac{\sqrt{x + c^2} - c}{x} \cdot \frac{\sqrt{x + c^2} + c}{\sqrt{x + c^2} + c}$$

$$= \lim_{x \to 0} \frac{(x + c^2) - c^2}{x[\sqrt{x + c^2} + c]} = \lim_{x \to 0} \frac{1}{\sqrt{x + c^2} + c} = \frac{1}{2c}$$

Define $f(0) = \dfrac{1}{2c}$ to make f continuous at $x = 0$.

89. $h(x) = x[\![x]\!]$

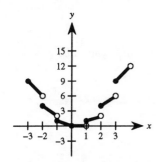

h has nonremovable discontinuities at $x = \pm 1, \pm 2, \pm 3, \ldots$.

90. Define $f(x) = f_2(x) - f_1(x)$. Since f_1 and f_2 are continuous on $[a, b]$, so is f.

$$f(a) = f_2(a) - f_1(a) > 0 \quad \text{and}$$
$$f(b) = f_2(b) - f_1(b) < 0.$$

By the Intermediate Value Theorem, there exists c in $[a, b]$ such that $f(c) = 0$.

$$f(c) = f_2(c) - f_1(c) = 0 \Rightarrow f_1(c) = f_2(c)$$

Section 1.5   Infinite Limits

1. $\displaystyle\lim_{x \to -2^+} \frac{1}{(x + 2)^2} = \infty$

$\displaystyle\lim_{x \to -2^-} \frac{1}{(x + 2)^2} = \infty$

2. $\displaystyle\lim_{x \to -2^+} \frac{1}{x + 2} = \infty$

$\displaystyle\lim_{x \to -2^-} \frac{1}{x + 2} = -\infty$

3. $\displaystyle\lim_{x \to -2^+} \tan \frac{\pi x}{4} = -\infty$

$\displaystyle\lim_{x \to -2^-} \tan \frac{\pi x}{4} = \infty$

4. $\displaystyle\lim_{x \to -2^+} \sec \frac{\pi x}{4} = \infty$

$\displaystyle\lim_{x \to -2^-} \sec \frac{\pi x}{4} = -\infty$

5. $\displaystyle\lim_{x \to -3^+} \frac{1}{x^2 - 9} = -\infty$

$\displaystyle\lim_{x \to -3^-} \frac{1}{x^2 - 9} = \infty$

6. $\displaystyle\lim_{x \to -3^+} \frac{x}{x^2 - 9} = \infty$

$\displaystyle\lim_{x \to -3^-} \frac{x}{x^2 - 9} = -\infty$

7. $\displaystyle\lim_{x \to -3^+} \frac{x^2}{x^2 - 9} = -\infty$

$\displaystyle\lim_{x \to -3^-} \frac{x^2}{x^2 - 9} = \infty$

8. $\displaystyle\lim_{x \to -3^+} \sec \frac{\pi x}{6} = \infty$

$\displaystyle\lim_{x \to -3^-} \sec \frac{\pi x}{6} = -\infty$

9. $\lim\limits_{x \to 0^+} \dfrac{1}{x^2} = \infty = \lim\limits_{x \to 0^-} \dfrac{1}{x^2}$

Therefore, $x = 0$ is a vertical asymptote.

10. $\lim\limits_{x \to 2^+} \dfrac{4}{(x-2)^3} = \infty$

$\lim\limits_{x \to 2^-} \dfrac{4}{(x-2)^3} = -\infty$

Therefore, $x = 2$ is a vertical asymptote.

11. $\lim\limits_{x \to 2^+} \dfrac{x^2 - 2}{(x-2)(x+1)} = \infty$

$\lim\limits_{x \to 2^-} \dfrac{x^2 - 2}{(x-2)(x+1)} = -\infty$

Therefore, $x = 2$ is a vertical asymptote.

$\lim\limits_{x \to -1^+} \dfrac{x^2 - 2}{(x-2)(x+1)} = \infty$

$\lim\limits_{x \to -1^-} \dfrac{x^2 - 2}{(x-2)(x+1)} = -\infty$

Therefore, $x = -1$ is a vertical asymptote.

12. $\lim\limits_{x \to 1^+} \dfrac{2+x}{1-x} = -\infty$

$\lim\limits_{x \to 1^-} \dfrac{2+x}{1-x} = \infty$

Therefore, $x = 1$ is a vertical asymptote.

13. $\lim\limits_{x \to -1^+} \dfrac{x^3}{x^2 - 1} = \infty$

$\lim\limits_{x \to -1^-} \dfrac{x^3}{x^2 - 1} = -\infty$

Therefore, $x = -1$ is a vertical asymptote.

$\lim\limits_{x \to 1^+} \dfrac{x^3}{x^2 - 1} = \infty$

$\lim\limits_{x \to 1^-} \dfrac{x^3}{x^2 - 1} = -\infty$

Therefore, $x = 1$ is a vertical asymptote.

14. No vertical asymptote since the denominator is never zero.

15. $\lim\limits_{x \to 0^+} \left(1 - \dfrac{4}{x^2}\right) = -\infty = \lim\limits_{x \to 0^-} \left(1 - \dfrac{4}{x^2}\right)$

Therefore, $x = 0$ is a vertical asymptote.

16. $\lim\limits_{x \to 2^+} \dfrac{-2}{(x-2)^2} = -\infty = \lim\limits_{x \to 2^-} \dfrac{-2}{(x-2)^2}$

Therefore, $x = 2$ is a vertical asymptote.

17. $\lim\limits_{x \to -2^+} \dfrac{x}{(x+2)(x-1)} = \infty$

$\lim\limits_{x \to -2^-} \dfrac{x}{(x+2)(x-1)} = -\infty$

Therefore, $x = -2$ is a vertical asymptote.

$\lim\limits_{x \to 1^+} \dfrac{x}{(x+2)(x-1)} = \infty$

$\lim\limits_{x \to 1^-} \dfrac{x}{(x+2)(x-1)} = -\infty$

Therefore, $x = 1$ is a vertical asymptote.

18. $\lim\limits_{x \to -3^+} \dfrac{1}{(x+3)^4} = \infty = \lim\limits_{x \to -3^-} \dfrac{1}{(x+3)^4}$

Therefore, $x = -3$ is a vertical asymptote.

19. $f(x) = \tan 2x = \dfrac{\sin 2x}{\cos 2x}$

has vertical asymptotes at

$x = \dfrac{(2n+1)\pi}{4} = \dfrac{\pi}{4} + \dfrac{n\pi}{2}$, n any integer.

20. $f(x) = \sec \pi x = \dfrac{1}{\cos \pi x}$

has vertical asymptotes at

$x = \dfrac{2n+1}{2}$, n any integer.

21. $f(x) = \dfrac{x^3+1}{x+1} = \dfrac{(x+1)(x^2-x+1)}{x+1}$

has no vertical asymptotes since

$\displaystyle\lim_{x\to -1} f(x) = \lim_{x\to -1}(x^2 - x + 1) = 3$, not infinity.

22. $f(x) = \dfrac{x^2-4}{x^3+2x^2+x+2} = \dfrac{(x+2)(x-2)}{(x+2)(x^2+1)}$

has no vertical asymptotes since

$\displaystyle\lim_{x\to -2} f(x) = \lim_{x\to -2}\dfrac{x-2}{x^2+1} = -\dfrac{4}{5}$, not infinity.

23. $f(x) = \dfrac{x}{\sin x}$ has vertical asymptotes at $x = n\pi$, n a nonzero integer. There is no vertical asymptote at $x = 0$ since

$\displaystyle\lim_{x\to 0}\dfrac{x}{\sin x} = 1.$

24. $f(x) = \dfrac{\tan x}{x} = \dfrac{\sin x}{x\cos x}$ has vertical asymptotes at

$x = \dfrac{(2n+1)\pi}{2} = \dfrac{\pi}{2} + n\pi$, n any integer.

There is no vertical asymptote at $x = 0$ since

$\displaystyle\lim_{x\to 0}\dfrac{\tan x}{x} = 1.$

25. $\displaystyle\lim_{x\to -1}\dfrac{x^2-1}{x+1} = \lim_{x\to -1}(x-1) = -2$

Removable discontinuity at $x = -1$

26. $\displaystyle\lim_{x\to -1}\dfrac{x^2-6x-7}{x+1} = \lim_{x\to -1}(x-7) = -8$

Removable discontinuity at $x = -1$

27. $\displaystyle\lim_{x\to -1^+}\dfrac{x^2+1}{x+1} = \infty$

$\displaystyle\lim_{x\to -1^-}\dfrac{x^2+1}{x+1} = -\infty$

Vertical asymptote at $x = -1$

28. $\displaystyle\lim_{x\to -1}\dfrac{\sin(x+1)}{x+1} = 1$

Removable discontinuity at $x = -1$

29. $\displaystyle\lim_{x\to 2^+}\dfrac{x-3}{x-2} = -\infty$

30. $\displaystyle\lim_{x\to 1^+}\dfrac{2+x}{1-x} = -\infty$

31. $\displaystyle\lim_{x\to 4^+}\dfrac{x^2}{x^2-16} = \infty$

32. $\displaystyle\lim_{x\to 4}\dfrac{x^2}{x^2+16} = \dfrac{1}{2}$

33. $\displaystyle\lim_{x\to -3^-}\dfrac{x^2+2x-3}{x^2+x-6} = \lim_{x\to -3^-}\dfrac{x-1}{x-2} = \dfrac{4}{5}$

34. $\displaystyle\lim_{x\to -(1/2)^+}\dfrac{6x^2+x-1}{4x^2-4x-3} = \lim_{x\to -(1/2)^+}\dfrac{3x-1}{2x-3} = \dfrac{5}{8}$

35. $\displaystyle\lim_{x\to 0^-}\left(1 + \dfrac{1}{x}\right) = -\infty$

36. $\displaystyle\lim_{x\to 0^-}\left(x^2 - \dfrac{1}{x}\right) = \infty$

37. $\displaystyle\lim_{x\to 0^+}\dfrac{2}{\sin x} = \infty$

38. $\displaystyle\lim_{x\to (\pi/2)^+}\dfrac{-2}{\cos x} = \infty$

39. $\displaystyle\lim_{x\to 1}\dfrac{x^2-x}{(x^2+1)(x-1)} = \lim_{x\to 1}\dfrac{x}{x^2+1} = \dfrac{1}{2}$

40. $\displaystyle\lim_{x\to 3}\dfrac{x-2}{x^2} = \dfrac{1}{9}$

41. $\displaystyle\lim_{x\to 1^+}\dfrac{x^2+x+1}{x^3-1} = \lim_{x\to 1^+}\dfrac{1}{x-1} = \infty$

42. $\displaystyle\lim_{x\to 1}\dfrac{x^3-1}{x^2+x+1} = \lim_{x\to 1}(x-1) = 0$

43. $f(x) = \dfrac{1}{x^2 - 25}$

$\lim\limits_{x \to 5^-} f(x) = -\infty$

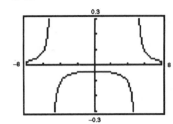

44. $f(x) = \sec \dfrac{\pi x}{6}$

$\lim\limits_{x \to 3^+} f(x) = -\infty$

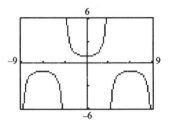

45. $Q = \dfrac{k}{\sqrt{t - 4}}$

$\lim\limits_{t \to 4^+} \dfrac{k}{\sqrt{t - 4}} = k(\infty) = \pm\infty$

depending upon the sign of k.

46. $P = \dfrac{k}{V}$

$\lim\limits_{V \to 0^+} \dfrac{k}{V} = k(\infty) = \infty$

In this case, we know that $k > 0$.

47. a. $r = \dfrac{2(7)}{\sqrt{625 - 49}} = \dfrac{7}{12}$ ft/sec

b. $r = \dfrac{2(15)}{\sqrt{625 - 225}} = \dfrac{3}{2}$ ft/sec

c. $\lim\limits_{x \to 25^-} \dfrac{2x}{\sqrt{625 - x^2}} = \infty$

48. a. $r = 50\pi \sec^2 \dfrac{\pi}{6} = \dfrac{200\pi}{3}$ ft/sec

b. $r = 50\pi \sec^2 \dfrac{\pi}{3} = 200\pi$ ft/sec

c. $\lim\limits_{\theta \to (\pi/2)^-} [50\pi \sec^2 \theta] = \infty$

49. $C = \dfrac{528x}{100 - x}, \quad 0 \le x < 100$

a. $C(25) = \$176$ million

c. $C(75) = \$1584$ million

b. $C(50) = \$528$ million

d. $\lim\limits_{x \to 100^-} \dfrac{528x}{100 - x} = \infty$

Thus, it is not possible.

50. a. Average speed $= \dfrac{\text{Total distance}}{\text{Total time}}$

$50 = \dfrac{2d}{(d/x) + (d/y)}$

$50 = \dfrac{2xy}{y + x}$

$50y + 50x = 2xy$

$50x = 2xy - 50y$

$50x = 2y(x - 25)$

$\dfrac{25x}{x - 25} = y$

Domain: $x > 25$

b.

x	30	40	50	60
y	150	66.667	50	42.857

c. $\lim\limits_{x \to 25^+} \dfrac{25x}{x - 25} = \infty$

51.

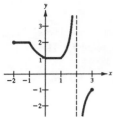

52. Let $f(x) = \dfrac{1}{x^2}$ and $g(x) = \dfrac{1}{x^4}$.

$\displaystyle\lim_{x\to 0} \frac{1}{x^2} = \infty$ and $\displaystyle\lim_{x\to 0} \frac{1}{x^4} = \infty$, but

$$\lim_{x\to 0}\left(\frac{1}{x^2} - \frac{1}{x^4}\right) = \lim_{x\to 0}\left(\frac{x^2-1}{x^4}\right) = -\infty \neq 0.$$

53. False; let

$$f(x) = \frac{x^2-1}{x-1}.$$

The graph of f has a hole at $(1,\ 2)$, not a vertical asymptote.

54. False; see Exercise 52.

55. False; let

$$F(x) = \frac{1}{x^2+4}.$$

This rational function has no vertical asymptotes.

56. True

57. False; let

$$F(x) = \begin{cases} \dfrac{1}{x}, & x \neq 0 \\ 3, & x = 0. \end{cases}$$

The graph of F has a vertical asymptote at $x = 0$, but $F(0) = 3$.

58. True

59. Given $\displaystyle\lim_{x\to c} f(x) = \infty$ and $\displaystyle\lim_{x\to c} g(x) = L$:

   (2) Product:

   If $L > 0$, then for $\epsilon = L/2 > 0$ there exists $\delta_1 > 0$ such that $|g(x) - L| < L/2$ whenever $0 < |x - c| < \delta_1$. Thus, $L/2 < g(x) < 3L/2$. Since $\displaystyle\lim_{x\to c} f(x) = \infty$ then for $M > 0$, there exists $\delta_2 > 0$ such that $f(x) > M(2/L)$ whenever $|x - c| < \delta_2$. Let δ be the smaller of δ_1 and δ_2, then for $0 < |x - c| < \delta$, we have $f(x)g(x) > M(2/L)(L/2) = M$. Therefore, $\displaystyle\lim_{x\to c} f(x)g(x) = \infty$. The proof is similar for $L < 0$.

   (3) Quotient: Let $\epsilon > 0$ be given.

   There exists $\delta_1 > 0$ such that $f(x) > 3L/2\epsilon$ whenever $0 < |x - c| < \delta_1$ and there exists $\delta_2 > 0$ such that $|g(x) - L| < L/2$ whenever $0 < |x - c| < \delta_2$. This inequality gives us $L/2 < g(x) < 3L/2$. Let δ be the smaller of δ_1 and δ_2, then for $0 < |x - c| < \delta$, we have

   $$\left|\frac{g(x)}{f(x)}\right| < \frac{3L/2}{3L/2\epsilon} = \epsilon.$$

   Therefore, $\displaystyle\lim_{x\to c} \frac{g(x)}{f(x)} = 0.$

60. Given $\lim\limits_{x \to c} f(x) = \infty$, let $g(x) = 1$.

Then $\lim\limits_{x \to c} \dfrac{g(x)}{f(x)} = \lim\limits_{x \to c} \dfrac{1}{f(x)} = 0$

by Theorem 1.15.

61. Given $\lim\limits_{x \to c} \dfrac{1}{f(x)} = 0$:

Suppose $\lim\limits_{x \to c} f(x)$ exists and equals L. Then,

$$\lim_{x \to c} \frac{1}{f(x)} = \frac{\lim\limits_{x \to c} 1}{\lim\limits_{x \to c} f(x)} = \frac{1}{L} = 0.$$

This is not possible. Thus, $\lim\limits_{x \to c} f(x)$ does not exist.

Chapter 1   Review Exercises

1. $\lim\limits_{x \to 2} (5x - 3) = 5(2) - 3 = 7$

2. $\lim\limits_{x \to 2} (3x + 5) = 3(2) + 5 = 11$

3. $\lim\limits_{x \to 2} (5x - 3)(3x + 5) = [5(2) - 3][3(2) + 5]$

$$= 7 \cdot 11 = 77$$

4. $\lim\limits_{x \to 2} \left(\dfrac{3x + 5}{5x - 3} \right) = \dfrac{3(2) + 5}{5(2) - 3} = \dfrac{11}{7}$

5. $\lim\limits_{x \to 3} \dfrac{t^2 + 1}{t} = \dfrac{3^2 + 1}{3} = \dfrac{10}{3}$

6. $\lim\limits_{t \to 3} \dfrac{t^2 - 9}{t - 3} = \lim\limits_{t \to 3} (t + 3) = 6$

7. $\lim\limits_{t \to -2} \dfrac{t + 2}{t^2 - 4} = \lim\limits_{t \to -2} \dfrac{1}{t - 2} = -\dfrac{1}{4}$

8. $\lim\limits_{x \to 0} \dfrac{\sqrt{4 + x} - 2}{x} = \lim\limits_{x \to 0} \dfrac{\sqrt{4 + x} - 2}{x} \cdot \dfrac{\sqrt{4 + x} + 2}{\sqrt{4 + x} + 2}$

$$= \lim_{x \to 0} \frac{1}{\sqrt{4 + x} + 2} = \frac{1}{4}$$

9. $\lim\limits_{x \to 0} \dfrac{[1/(x + 1)] - 1}{x} = \lim\limits_{x \to 0} \dfrac{-1}{x + 1} = -1$

10. $\lim\limits_{s \to 0} \dfrac{(1/\sqrt{1 + s}) - 1}{s} = \lim\limits_{s \to 0} \left[\dfrac{(1/\sqrt{1 + s}) - 1}{s} \cdot \dfrac{(1/\sqrt{1 + s}) + 1}{(1/\sqrt{1 + s}) + 1} \right]$

$$= \lim_{s \to 0} \frac{[1/(1 + s)] - 1}{s[(1/\sqrt{1 + s}) + 1]} = \lim_{s \to 0} \frac{-1}{(1 + s)[(1/\sqrt{1 + s}) + 1]} = -\frac{1}{2}$$

11. $\lim\limits_{x \to -5} \dfrac{x^3 + 125}{x + 5} = \lim\limits_{x \to -5} \dfrac{(x + 5)(x^2 - 5x + 25)}{x + 5}$

$$= \lim_{x \to -5} (x^2 - 5x + 25)$$

$$= 75$$

12. $\lim\limits_{x \to -2} \dfrac{x^2 - 4}{x^3 + 8} = \lim\limits_{x \to -2} \dfrac{(x + 2)(x - 2)}{(x + 2)(x^2 - 2x + 4)}$

$$= \lim_{x \to -2} \frac{x - 2}{x^2 - 2x + 4}$$

$$= -\frac{4}{12} = -\frac{1}{3}$$

13. $\lim\limits_{x \to 0^+} \left(x - \dfrac{1}{x^3} \right) = -\infty$

14. $\lim\limits_{x \to 2^+} \dfrac{1}{\sqrt[3]{x^2 - 4}} = \infty$

$$\lim_{x \to 2^-} \frac{1}{\sqrt[3]{x^2 - 4}} = -\infty$$

Thus, $\lim\limits_{x \to 2} \dfrac{1}{\sqrt[3]{x^2 - 4}}$ does not exist.

15. $\displaystyle\lim_{\Delta x \to 0} \frac{\sin[(\pi/6) + \Delta x] - (1/2)}{\Delta x} = \lim_{\Delta x \to 0} \frac{\sin(\pi/6)\cos\Delta x + \cos(\pi/6)\sin\Delta x - (1/2)}{\Delta x}$

$$= \lim_{\Delta x \to 0} \frac{1}{2} \cdot \frac{(\cos\Delta x - 1)}{\Delta x} + \lim_{\Delta x \to 0} \frac{\sqrt{3}}{2} \cdot \frac{\sin\Delta x}{\Delta x}$$

$$= 0 + \frac{\sqrt{3}}{2}(1)$$

$$= \frac{\sqrt{3}}{2}$$

16. $\displaystyle\lim_{\Delta x \to 0} \frac{\cos(\pi + \Delta x) + 1}{\Delta x} = \lim_{\Delta x \to 0} \frac{\cos\pi\cos\Delta x - \sin\pi\sin\Delta x + 1}{\Delta x}$

$$= \lim_{\Delta x \to 0} \left[-\frac{(\cos\Delta x - 1)}{\Delta x}\right] - \lim_{\Delta x \to 0} \left[\sin\pi\frac{\sin\Delta x}{\Delta x}\right]$$

$$= -0 - (0)(1)$$

$$= 0$$

17. $\displaystyle\lim_{x \to -2^-} \frac{2x^2 + x + 1}{x + 2} = -\infty$

18. $\displaystyle\lim_{x \to (1/2)^+} \frac{x}{2x - 1} = \infty$

19. $\displaystyle\lim_{x \to -1^+} \frac{x + 1}{x^3 + 1} = \lim_{x \to -1^+} \frac{1}{x^2 - x + 1} = \frac{1}{3}$

20. $\displaystyle\lim_{x \to -1^-} \frac{x + 1}{x^4 - 1} = \lim_{x \to -1^-} \frac{1}{(x^2 + 1)(x - 1)} = -\frac{1}{4}$

21. $\displaystyle\lim_{x \to 1^-} \frac{x^2 + 2x + 1}{x - 1} = -\infty$

22. $\displaystyle\lim_{x \to -1^+} \frac{x^2 - 2x + 1}{x + 1} = \infty$

23. $\displaystyle\lim_{x \to 0^+} \frac{\sin 4x}{5x} = \frac{4}{5}$

24. $\displaystyle\lim_{x \to 0^+} \frac{\sec x}{x} = \infty$

25. $\displaystyle\lim_{x \to 0^+} \frac{\csc 2x}{x} = \infty$

26. $\displaystyle\lim_{x \to 0^-} \frac{\cos^2 x}{x} = -\infty$

27. $f(x) = \dfrac{\sqrt{2x + 1} - \sqrt{3}}{x - 1}$

a.

x	1.1	1.01	1.001	1.0001
$f(x)$	0.5680	0.5764	0.5772	0.5773

$$\lim_{x \to 1^+} \frac{\sqrt{2x + 1} - \sqrt{3}}{x - 1} \approx 0.577 \quad \text{(Actual limit is } \sqrt{3}/3.)$$

b. $\displaystyle\lim_{x \to 1^+} \frac{\sqrt{2x + 1} - \sqrt{3}}{x - 1} = \lim_{x \to 1^+} \frac{\sqrt{2x + 1} - \sqrt{3}}{x - 1} \cdot \frac{\sqrt{2x + 1} + \sqrt{3}}{\sqrt{2x + 1} + \sqrt{3}} = \lim_{x \to 1^+} \frac{(2x + 1) - 3}{(x - 1)(\sqrt{2x + 1} + \sqrt{3})}$

$$= \lim_{x \to 1^+} \frac{2}{\sqrt{2x + 1} + \sqrt{3}}$$

$$= \frac{2}{2\sqrt{3}} = \frac{1}{\sqrt{3}} = \frac{\sqrt{3}}{3}$$

28. $f(x) = \dfrac{1 - \sqrt[3]{x}}{x - 1}$

a.

x	1.1	1.01	1.001	1.0001
$f(x)$	-0.3228	-0.3322	-0.3332	-0.3333

$$\lim_{x \to 1^+} \frac{1 - \sqrt[3]{x}}{x - 1} \approx -0.333 \quad \text{(Actual limit is } -\tfrac{1}{3}.\text{)}$$

b. $\displaystyle\lim_{x \to 1^+} \frac{1 - \sqrt[3]{x}}{x - 1} = \lim_{x \to 1^+} \frac{1 - \sqrt[3]{x}}{x - 1} \cdot \frac{1 + \sqrt[3]{x} + (\sqrt[3]{x})^2}{1 + \sqrt[3]{x} + (\sqrt[3]{x})^2} = \lim_{x \to 1^+} \frac{1 - x}{(x - 1)[1 + \sqrt[3]{x} + (\sqrt[3]{x})^2]}$

$$= \lim_{x \to 1^+} \frac{-1}{1 + \sqrt[3]{x} + (\sqrt[3]{x})^2}$$

$$= -\frac{1}{3}$$

29. $f(x) = [\![x + 3]\!]$

$\displaystyle\lim_{x \to k^+} [\![x + 3]\!] = k + 3$ where k is an integer.

$\displaystyle\lim_{x \to k^-} [\![x + 3]\!] = k + 2$ where k is an integer.

Nonremovable discontinuity at each integer k

Continuous on $(k, \ k + 1)$ for all integers k

30. $f(x) = \dfrac{3x^2 - x - 2}{x - 1} = \dfrac{(3x + 2)(x - 1)}{x - 1}$

$\displaystyle\lim_{x \to 1} f(x) = \lim_{x \to 1} (3x + 2) = 5$

Removable discontinuity at $x = 1$

Continuous on $(-\infty, \ 1) \cup (1, \ \infty)$

31. $f(x) = \begin{cases} \dfrac{3x^2 - x - 2}{x - 1}, & x \neq 1 \\ 0, & x = 1 \end{cases}$

$\displaystyle\lim_{x \to 1} f(x) = \lim_{x \to 1} \frac{3x^2 - x - 2}{x - 1}$

$\displaystyle\qquad = \lim_{x \to 1} (3x + 2) = 5 \neq 0$

Removable discontinuity at $x = 1$

Continuous on $(-\infty, \ 1) \cup (1, \ \infty)$

32. $f(x) = \begin{cases} 5 - x, & x \leq 2 \\ 2x - 3, & x > 2 \end{cases}$

$\displaystyle\lim_{x \to 2^-} (5 - x) = 3$

$\displaystyle\lim_{x \to 2^+} (2x - 3) = 1$

Nonremovable discontinuity at $x = 2$

Continuous on $(-\infty, \ 2) \cup (2, \ \infty)$

33. $f(x) = \dfrac{1}{(x - 2)^2}$

$\displaystyle\lim_{x \to 2} \frac{1}{(x - 2)^2} = \infty$

Nonremovable discontinuity at $x = 2$

Continuous on $(-\infty, \ 2) \cup (2, \ \infty)$

34. $f(x) = \sqrt{\dfrac{x + 1}{x}} = \sqrt{1 + \dfrac{1}{x}}$

$\displaystyle\lim_{x \to 0^+} \sqrt{1 + \frac{1}{x}} = \infty$

Nonremovable discontinuity at $x = 0$

Continuous on $(-\infty, \ -1] \cup (0, \ \infty)$

35. $f(x) = \dfrac{3}{x + 1}$

$\displaystyle\lim_{x \to 1^-} f(x) = -\infty$

$\displaystyle\lim_{x \to 1^+} f(x) = \infty$

Nonremovable discontinuity at $x = -1$

Continuous on $(-\infty, \ -1) \cup (-1, \ \infty)$

36. $f(x) = \dfrac{x + 1}{2x + 2}$

$\displaystyle\lim_{x \to -1} \frac{x + 1}{2(x + 1)} = \frac{1}{2}$

Removable discontinuity at $x = -1$

Continuous on $(-\infty, \ -1) \cup (-1, \ \infty)$

37. $f(x) = \csc \dfrac{\pi x}{2}$

Nonremovable discontinuities at each even integer
Continuous on $(2k,\ 2k+2)$ for all integers k

38. $f(x) = \tan 2x$

Nonremovable discontinuities when

$$x = \frac{(2n+1)\pi}{4}$$

Continuous on $\left(\dfrac{(2n-1)\pi}{4},\ \dfrac{(2n+1)\pi}{4}\right)$ for all integers n

39. $f(2) = 5$

Find c so that $\displaystyle\lim_{x \to 2^+} (cx + 6) = 5$.

$$c(2) + 6 = 5$$
$$2c = -1$$
$$c = -\tfrac{1}{2}$$

40. $\displaystyle\lim_{x \to 1^+} (x+1) = 2$

$\displaystyle\lim_{x \to 3^-} (x+1) = 4$

Find b and c so that $\displaystyle\lim_{x \to 1^-} (x^2 + bx + c) = 2$ and $\displaystyle\lim_{x \to 3^+} (x^2 + bx + c) = 4$.

Consequently we get $1 + b + c = 2$ and $9 + 3b + c = 4$.

Solving simultaneously, $b = -3$ and $c = 4$.

41. $A = 5000(1.06)^{[\![2t]\!]}$

Nonremovable discontinuity every 6 months

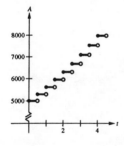

42. $C = 9.80 + 2.50[-[\![-x]\!] - 1],\quad x > 0$

$\qquad = 9.80 - 2.50[[\![-x]\!] + 1]$

C has a nonremovable discontinuity at each integer.

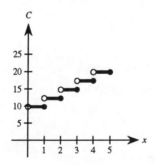

43. $g(x) = 1 + \dfrac{2}{x}$

Vertical asymptote at $x = 0$

44. $h(x) = \dfrac{4x}{4 - x^2}$

Vertical asymptotes at $x = 2$ and $x = -2$

45. $f(x) = \dfrac{8}{(x-10)^2}$

Vertical asymptote at $x = 10$

46. $f(x) = \csc \pi x$

Vertical asymptote at every integer k

47. $C = \dfrac{80,000p}{100 - p}, \quad 0 \le p < 100$

   a. $C(15) \approx \$14,117.65$

   b. $C(50) = \$80,000$

   c. $C(90) = \$720,000$

   d. $\displaystyle\lim_{p \to 100^-} \dfrac{80,000p}{100 - p} = \infty$

48. $f(x) = \dfrac{\tan 2x}{x}$

   a.

x	-0.1	-0.01	-0.001	0.001	0.01	0.1
$f(x)$	2.0271	2.0003	2.0000	2.0000	2.0003	2.0271

$$\lim_{x \to 0} \frac{\tan 2x}{x} = 2$$

   b. Yes, define

$$f(x) = \begin{cases} \dfrac{\tan 2x}{x}, & x \ne 0 \\ 2, & x = 0 \end{cases}$$

      Now $f(x)$ is continuous at $x = 0$.

49. $\displaystyle\lim_{x \to 0} \dfrac{|x|}{x} = 1$ is false, $\displaystyle\lim_{x \to 0} \dfrac{|x|}{x}$ does not exist,

   since $\displaystyle\lim_{x \to 0^-} \dfrac{|x|}{x} = -1$ and $\displaystyle\lim_{x \to 0^+} \dfrac{|x|}{x} = 1$.

50. $\displaystyle\lim_{x \to 0} x^3 = 0$ is true.

51. True; see Theorem 1.7.

52. False; let

$$f(x) = \frac{x - 1}{x^2 - 1}.$$

   Then,

$$\lim_{x \to 1} f(x) = \lim_{x \to 1} \frac{1}{x + 1} = \frac{1}{2}$$

   but $f(1)$ is not defined.

53. True

54. $\displaystyle\lim_{x \to 2} f(x) = 3$ is false since

$$\lim_{x \to 2^-} f(x) = 3 \text{ and } \lim_{x \to 2^+} f(x) = 0.$$

55. $\displaystyle\lim_{x \to 3} f(x) = 1$ is true since

$$\lim_{x \to 3^-} (x - 2) = 1 \text{ and } \lim_{x \to 3^+} (-x^2 + 8x - 14) = 1.$$

56. $f(x) = \dfrac{x^2 - 4}{|x - 2|} = (x + 2)\left[\dfrac{x - 2}{|x - 2|}\right]$

   $\displaystyle\lim_{x \to 2^-} f(x) = -4$

   $\displaystyle\lim_{x \to 2^+} f(x) = 4$

   $\displaystyle\lim_{x \to 2} f(x)$ does not exist.

57. $f(x) = \sqrt{(x - 1)x}$

   Domain: $(-\infty, \ 0] \cup [1, \ \infty)$

   $\displaystyle\lim_{x \to 0^-} f(x) = 0$

   $\displaystyle\lim_{x \to 1^+} f(x) = 0$

CHAPTER 2
Differentiation

Section 2.1 The Derivative and the Tangent Line Problem

1. a. $m = 0$

  b. $m = -3$

2. a. $m = \frac{1}{4}$

  b. $m = 1$

3.

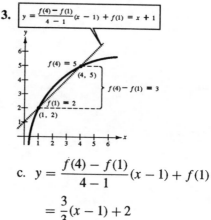
$$y = \frac{f(4) - f(1)}{4 - 1}(x - 1) + f(1) = x + 1$$

  c. $y = \dfrac{f(4) - f(1)}{4 - 1}(x - 1) + f(1)$

  $= \dfrac{3}{3}(x - 1) + 2$

  $= 1(x - 1) + 2$

  $= x + 1$

4. a. $\dfrac{f(4) - f(1)}{4 - 1} = \dfrac{5 - 2}{3} = 1$

  $\dfrac{f(4) - f(3)}{4 - 3} \approx \dfrac{5 - 4.75}{1} = 0.25$

  Thus, $\dfrac{f(4) - f(1)}{4 - 1} > \dfrac{f(4) - f(3)}{4 - 3}$.

  b. The slope of the tangent line at $(1, \ 2)$ equals $f'(1)$. This slope is steeper than the slope of the line through $(1, \ 2)$ and $(4, \ 5)$. Thus,

  $$\dfrac{f(4) - f(1)}{4 - 1} < f'(1).$$

5. $f(x) = 3$

$f'(x) = \lim\limits_{\Delta x \to 0} \dfrac{f(x + \Delta x) - f(x)}{\Delta x}$

$= \lim\limits_{\Delta x \to 0} \dfrac{3 - 3}{\Delta x}$

$= \lim\limits_{\Delta x \to 0} 0 = 0$

6. $f(x) = 3x + 2$

$f'(x) = \lim\limits_{\Delta x \to 0} \dfrac{f(x + \Delta x) - f(x)}{\Delta x}$

$= \lim\limits_{\Delta x \to 0} \dfrac{[3(x + \Delta x) + 2] - [3x + 2]}{\Delta x}$

$= \lim\limits_{\Delta x \to 0} \dfrac{3\Delta x}{\Delta x}$

$= \lim\limits_{\Delta x \to 0} 3 = 3$

7. $f(x) = -5x$

$f'(x) = \lim\limits_{\Delta x \to 0} \dfrac{f(x + \Delta x) - f(x)}{\Delta x}$

$= \lim\limits_{\Delta x \to 0} \dfrac{-5(x + \Delta x) - (-5x)}{\Delta x}$

$= \lim\limits_{\Delta x \to 0} -5 = -5$

8. $g(x) = 9 - \dfrac{1}{2}x$

$g'(x) = \lim\limits_{\Delta x \to 0} \dfrac{g(x + \Delta x) - g(x)}{\Delta x}$

$= \lim\limits_{\Delta x \to 0} \dfrac{[9 - (1/2)(x + \Delta x)] - [9 - (1/2)x]}{\Delta x}$

$= \lim\limits_{\Delta x \to 0} \left(-\dfrac{1}{2}\right) = -\dfrac{1}{2}$

9. $f(x) = 2x^2 + x - 1$

$$f'(x) = \lim_{\Delta x \to 0} \frac{f(x + \Delta x) - f(x)}{\Delta x}$$

$$= \lim_{\Delta x \to 0} \frac{[2(x + \Delta x)^2 + (x + \Delta x) - 1] - [2x^2 + x - 1]}{\Delta x}$$

$$= \lim_{\Delta x \to 0} \frac{(2x^2 + 4x\Delta x + 2(\Delta x)^2 + x + \Delta x - 1) - (2x^2 + x - 1)}{\Delta x}$$

$$= \lim_{\Delta x \to 0} \frac{4x\Delta x + 2(\Delta x)^2 + \Delta x}{\Delta x} = \lim_{\Delta x \to 0} (4x + 2\Delta x + 1) = 4x + 1$$

10. $f(x) = 1 - x^2$

$$f'(x) = \lim_{\Delta x \to 0} \frac{f(x + \Delta x) - f(x)}{\Delta x}$$

$$= \lim_{\Delta x \to 0} \frac{[1 - (x + \Delta x)^2] - [1 - x^2]}{\Delta x}$$

$$= \lim_{\Delta x \to 0} \frac{1 - x^2 - 2x\Delta x - (\Delta x)^2 - 1 + x^2}{\Delta x}$$

$$= \lim_{\Delta x \to 0} \frac{-2x\Delta x - (\Delta x)^2}{\Delta x} = \lim_{\Delta x \to 0} (-2x - \Delta x) = -2x$$

11. $f(x) = x^3 - 12x$

$$f'(x) = \lim_{\Delta x \to 0} \frac{f(x + \Delta x) - f(x)}{\Delta x}$$

$$= \lim_{\Delta x \to 0} \frac{[(x + \Delta x)^3 - 12(x + \Delta x)] - [x^3 - 12x]}{\Delta x}$$

$$= \lim_{\Delta x \to 0} \frac{x^3 + 3x^2\Delta x + 3x(\Delta x)^2 + (\Delta x)^3 - 12x - 12\Delta x - x^3 + 12x}{\Delta x}$$

$$= \lim_{\Delta x \to 0} \frac{3x^2\Delta x + 3x(\Delta x)^2 + (\Delta x)^3 - 12\Delta x}{\Delta x}$$

$$= \lim_{\Delta x \to 0} (3x^2 + 3x\Delta x + (\Delta x)^2 - 12) = 3x^2 - 12$$

12. $f(x) = x^3 + x^2$

$$f'(x) = \lim_{\Delta x \to 0} \frac{f(x + \Delta x) - f(x)}{\Delta x}$$

$$= \lim_{\Delta x \to 0} \frac{[(x + \Delta x)^3 + (x + \Delta x)^2] - [x^3 + x^2]}{\Delta x}$$

$$= \lim_{\Delta x \to 0} \frac{x^3 + 3x^2\Delta x + 3x(\Delta x)^2 + (\Delta x)^3 + x^2 + 2x\Delta x + (\Delta x)^2 - x^3 - x^2}{\Delta x}$$

$$= \lim_{\Delta x \to 0} \frac{3x^2\Delta x + 3x(\Delta x)^2 + (\Delta x)^3 + 2x\Delta x + (\Delta x)^2}{\Delta x}$$

$$= \lim_{\Delta x \to 0} (3x^2 + 3x\Delta x + (\Delta x)^2 + 2x + (\Delta x)) = 3x^2 + 2x$$

13. $f(x) = \dfrac{1}{x-1}$

$$f'(x) = \lim_{\Delta x \to 0} \frac{f(x+\Delta x) - f(x)}{\Delta x}$$

$$= \lim_{\Delta x \to 0} \frac{\dfrac{1}{x+\Delta x-1} - \dfrac{1}{x-1}}{\Delta x}$$

$$= \lim_{\Delta x \to 0} \frac{(x-1)-(x+\Delta x-1)}{\Delta x(x+\Delta x-1)(x-1)}$$

$$= \lim_{\Delta x \to 0} \frac{-\Delta x}{\Delta x(x+\Delta x-1)(x-1)}$$

$$= \lim_{\Delta x \to 0} \frac{-1}{(x+\Delta x-1)(x-1)}$$

$$= -\frac{1}{(x-1)^2}$$

14. $f(x) = \dfrac{1}{x^2}$

$$f'(x) = \lim_{\Delta x \to 0} \frac{f(x+\Delta x) - f(x)}{\Delta x}$$

$$= \lim_{\Delta x \to 0} \frac{\dfrac{1}{(x+\Delta x)^2} - \dfrac{1}{x^2}}{\Delta x}$$

$$= \lim_{\Delta x \to 0} \frac{x^2 - (x+\Delta x)^2}{\Delta x(x+\Delta x)^2 x^2}$$

$$= \lim_{\Delta x \to 0} \frac{-2x\Delta x - (\Delta x)^2}{\Delta x(x+\Delta x)^2 x^2}$$

$$= \lim_{\Delta x \to 0} \frac{-2x - \Delta x}{(x+\Delta x)^2 x^2}$$

$$= \frac{-2x}{x^4}$$

$$= -\frac{2}{x^3}$$

15. $f(x) = \sqrt{x-4}$

$$f'(x) = \lim_{\Delta x \to 0} \frac{f(x+\Delta x) - f(x)}{\Delta x} = \lim_{\Delta x \to 0} \frac{\sqrt{x+\Delta x-4} - \sqrt{x-4}}{\Delta x} \cdot \frac{\sqrt{x+\Delta x-4} + \sqrt{x-4}}{\sqrt{x+\Delta x-4} + \sqrt{x-4}}$$

$$= \lim_{\Delta x \to 0} \frac{(x+\Delta x-4)-(x-4)}{\Delta x[\sqrt{x+\Delta x-4} + \sqrt{x-4}]} = \lim_{\Delta x \to 0} \frac{1}{\sqrt{x+\Delta x-4} + \sqrt{x-4}} = \frac{1}{2\sqrt{x-4}}$$

16. $f(x) = \dfrac{1}{\sqrt{x}}$

$$f'(x) = \lim_{\Delta x \to 0} \frac{f(t+\Delta x) - f(x)}{\Delta x} = \lim_{\Delta x \to 0} \frac{\dfrac{1}{\sqrt{x+\Delta x}} - \dfrac{1}{\sqrt{x}}}{\Delta x}$$

$$= \lim_{\Delta x \to 0} \frac{\sqrt{x} - \sqrt{x+\Delta x}}{\Delta x\sqrt{x+\Delta x}\sqrt{x}} \cdot \frac{\sqrt{x} + \sqrt{x+\Delta x}}{\sqrt{x} + \sqrt{x+\Delta x}} = \lim_{\Delta x \to 0} \frac{x-(x+\Delta x)}{\Delta x\sqrt{x+\Delta x}\sqrt{x}[\sqrt{x} + \sqrt{x+\Delta x}]}$$

$$= \lim_{\Delta x \to 0} \frac{-1}{\sqrt{x+\Delta x}\sqrt{x}[\sqrt{x} + \sqrt{x+\Delta x}]} = -\frac{1}{\sqrt{x}\sqrt{x}(\sqrt{x} + \sqrt{x})} = -\frac{1}{2x\sqrt{x}}$$

17. $f(x) = x^2 + 1$

$$f'(x) = \lim_{\Delta x \to 0} \frac{f(x+\Delta x) - f(x)}{\Delta x}$$

$$= \lim_{\Delta x \to 0} \frac{[(x+\Delta x)^2 + 1] - [x^2 + 1]}{\Delta x}$$

$$= \lim_{\Delta x \to 0} \frac{2x\Delta x + (\Delta x)^2}{\Delta x}$$

$$= \lim_{\Delta x \to 0} (2x + \Delta x) = 2x$$

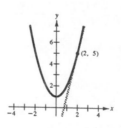

At $(2, 5)$, the slope of the tangent line is
$m = 2(2) = 4$. The equation of the tangent line is

$$y - 5 = 4(x - 2)$$

$$y - 5 = 4x - 8$$

$$y = 4x - 3.$$

18. $f(x) = x^2 + 2x + 1$

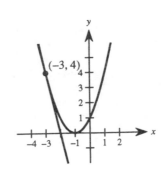

$$f'(x) = \lim_{\Delta x \to 0} \frac{f(x + \Delta x) - f(x)}{\Delta x}$$

$$= \lim_{\Delta x \to 0} \frac{[(x + \Delta x)^2 + 2(x + \Delta x) + 1] - [x^2 + 2x + 1]}{\Delta x}$$

$$= \lim_{\Delta x \to 0} \frac{2x\Delta x + (\Delta x)^2 + 2\Delta x}{\Delta x}$$

$$= \lim_{\Delta x \to 0} (2x + \Delta x + 2) = 2x + 2$$

At $(-3, 4)$, the slope of the tangent line is $m = 2(-3) + 2 = -4$. The equation of the tangent line is

$$y - 4 = -4(x + 3)$$

$$y = -4x - 8.$$

19. $f(x) = x^3$

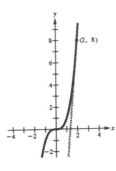

$$f'(x) = \lim_{\Delta x \to 0} \frac{f(x + \Delta x) - f(x)}{\Delta x}$$

$$= \lim_{\Delta x \to 0} \frac{(x + \Delta x)^3 - x^3}{\Delta x}$$

$$= \lim_{\Delta x \to 0} \frac{3x^2\Delta x + 3x(\Delta x)^2 + (\Delta x)^3}{\Delta x}$$

$$= \lim_{\Delta x \to 0} (3x^2 + 3x\Delta x + (\Delta x)^2) = 3x^2$$

At $(2, 8)$, the slope of the tangent line is
$m = 3(2)^2 = 12$. The equation of the tangent line is

$$y - 8 = 12(x - 2)$$

$$y = 12x - 16.$$

20. $f(x) = \sqrt{x}$

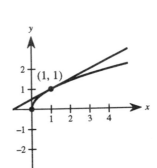

$$f'(x) = \lim_{\Delta x \to 0} \frac{f(x + \Delta x) - f(x)}{\Delta x}$$

$$= \lim_{\Delta x \to 0} \frac{\sqrt{x + \Delta x} - \sqrt{x}}{\Delta x} \cdot \frac{\sqrt{x + \Delta x} + \sqrt{x}}{\sqrt{x + \Delta x} + \sqrt{x}}$$

$$= \lim_{\Delta x \to 0} \frac{(x + \Delta x) - x}{\Delta x(\sqrt{x + \Delta x} + \sqrt{x})}$$

$$= \lim_{\Delta x \to 0} \frac{1}{\sqrt{x + \Delta x} + \sqrt{x}} = \frac{1}{2\sqrt{x}}$$

At $(1, 1)$, the slope of the tangent line is

$$m = \frac{1}{2\sqrt{1}} = \frac{1}{2}.$$

The equation of the tangent line is

$$y - 1 = \frac{1}{2}(x - 1)$$

$$y = \frac{1}{2}x + \frac{1}{2}.$$

21. $f(x) = x + \dfrac{1}{x}$

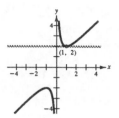

$$f'(x) = \lim_{\Delta x \to 0} \frac{f(x + \Delta x) - f(x)}{\Delta x}$$

$$= \lim_{\Delta x \to 0} \frac{\left[(x + \Delta x) + \dfrac{1}{x + \Delta x} \right] - \left[x + \dfrac{1}{x} \right]}{\Delta x}$$

$$= \lim_{\Delta x \to 0} \frac{\Delta x + \dfrac{x - (x + \Delta x)}{(x + \Delta x)x}}{\Delta x}$$

$$= \lim_{\Delta x \to 0} \left[1 + \frac{-\Delta x}{\Delta x(x + \Delta x)x} \right]$$

$$= \lim_{\Delta x \to 0} \left[1 - \frac{1}{(x + \Delta x)x} \right]$$

$$= 1 - \frac{1}{x^2}$$

At $(1, \; 2)$, the slope of the tangent line is

$$m = 1 - \frac{1}{1^2} = 0.$$

The equation of the tangent line is

$$y - 2 = 0(x - 1)$$

$$y = 2.$$

22. $f(x) = \dfrac{1}{x + 1}$

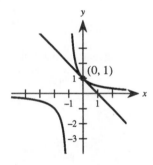

$$f'(x) = \lim_{\Delta x \to 0} \frac{f(x + \Delta x) - f(x)}{\Delta x}$$

$$= \lim_{\Delta x \to 0} \frac{\dfrac{1}{x + \Delta x + 1} - \dfrac{1}{x + 1}}{\Delta x}$$

$$= \lim_{\Delta x \to 0} \frac{(x + 1) - (x + \Delta x + 1)}{\Delta x(x + \Delta x + 1)(x + 1)}$$

$$= \lim_{\Delta x \to 0} -\frac{1}{(x + \Delta x + 1)(x + 1)}$$

$$= -\frac{1}{(x + 1)^2}$$

At $(0, \; 1)$, the slope of the tangent line is

$$m = \frac{-1}{(0 + 1)^2} = -1.$$

The equation of the tangent line is $y = -x + 1$.

23. From Exercise 19 we know that $f'(x) = 3x^2$. Since the slope of the given line is 3, we have

$$3x^2 = 3$$

$$x = \pm 1.$$

Therefore, at the points $(1, 1)$ and $(-1, -1)$ the tangent lines are parallel to $3x - y + 1 = 0$. These lines have equations

$$y - 1 = 3(x - 1) \quad \text{and} \quad y + 1 = 3(x + 1)$$

$$y = 3x - 2 \qquad\qquad\qquad y = 3x + 2.$$

24. From Exercise 16 we know that

$$f'(x) = \frac{-1}{2x\sqrt{x}}.$$

Since the slope of the given line is $-\frac{1}{2}$, we have

$$-\frac{1}{2x\sqrt{x}} = -\frac{1}{2}$$

$$x = 1.$$

Therefore, at the point $(1, 1)$ the tangent line is parallel to $x + 2y - 6 = 0$. The equation of this line is

$$y - 1 = -\frac{1}{2}(x - 1)$$

$$y - 1 = -\frac{1}{2}x + \frac{1}{2}$$

$$y = -\frac{1}{2}x + \frac{3}{2}.$$

25. Given the equation $y = 4x - x^2$ and the point $(2, 5)$, by the limit definition of the derivative, we have $dy/dx = 4 - 2x$. The equation of the tangent line is $y - 5 = (4 - 2x)(x - 2)$. Substituting the given equation for y, we have

$$(4x - x^2) - 5 = (4 - 2x)(x - 2)$$

$$4x - x^2 - 5 = -2x^2 + 8x - 8$$

$$x^2 - 4x + 3 = 0$$

$$(x - 1)(x - 3) = 0.$$

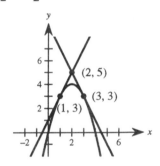

Therefore, the tangent lines intersect the parabola at $(1, 3)$ and $(3, 3)$ and their equations are

$$y - 5 = 2(x - 2) \quad \text{and} \quad y - 5 = -2(x - 2)$$

$$y = 2x + 1 \qquad\qquad\qquad y = -2x + 9.$$

26. Given the equation $y = x^2$ and the point $(1, -3)$, by the limit definition of the derivative $dy/dx = 2x$. The equation of the tangent line is $y + 3 = 2x(x - 1)$. Substituting the given equation for y we have

$$x^2 + 3 = 2x(x - 1)$$

$$x^2 - 2x - 3 = 0$$

$$(x + 1)(x - 3) = 0$$

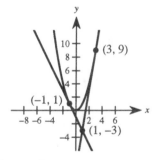

Therefore, the tangent lines intersect the curve at $(-1, 1)$ and $(3, 9)$ and their equations are

$$y - 1 = -2(x + 1) \quad \text{and} \quad y - 9 = 6(x - 3)$$

$$y - 1 = -2x - 2 \qquad\qquad\qquad y - 9 = 6x - 18$$

$$y = -2x - 1 \qquad\qquad\qquad\qquad y = 6x - 9.$$

27. $f(x) = x$

$f'(x) = 1$

Matches graph (b).

28. $f(x) = x^2$

$f'(x) = 2x$

Matches graph (e).

29. $f(x) = \sqrt{x}$

$f'(x) = \dfrac{1}{2\sqrt{x}}$

Matches graph (c).

30. $f(x) = \dfrac{1}{x}$

$f'(x) = -\dfrac{1}{x^2}$

Matches graph (a).

31. $f(x) = |x| = \begin{cases} x, & \text{if } x \geq 0 \\ -x, & \text{if } x < 0 \end{cases}$

$f'(x) = \begin{cases} 1, & \text{if } x > 0 \\ -1, & \text{if } x < 0 \end{cases}$

Matches graph (f).

32. $f(x) = \sin x$

The slope, $f'(x)$, is periodic.

Matches graph (d).

33. $f(x) = \tfrac{1}{4}x^3$

By the limit definition of the derivative we have $f'(x) = \tfrac{3}{4}x^2$.

x	-2	-1.5	-1	-0.5	0	0.5	1	1.5	2
$f(x)$	-2	$-\frac{27}{32}$	$-\frac{1}{4}$	$-\frac{1}{32}$	0	$\frac{1}{32}$	$\frac{1}{4}$	$\frac{27}{32}$	2
$f'(x)$	3	$\frac{27}{16}$	$\frac{3}{4}$	$\frac{3}{16}$	0	$\frac{3}{16}$	$\frac{3}{4}$	$\frac{27}{16}$	3

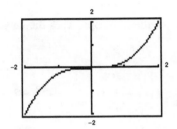

34. $f(x) = \tfrac{1}{2}x^2$

By the limit definition of the derivative we have $f'(x) = x$.

x	-2	-1.5	-1	-0.5	0	0.5	1	1.5	2
$f(x)$	2	1.125	0.5	0.125	0	0.125	0.5	1.125	2
$f'(x)$	-2	-1.5	-1	-0.5	0	0.5	1	1.5	2

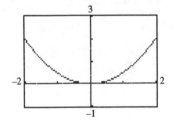

35. $f(x) = 4 - (x-3)^2$

$$S_{\Delta x}(x) = \frac{f(2 + \Delta x) - f(2)}{\Delta x}(x - 2) + f(2)$$

$$= \frac{4 - (2 + \Delta x - 3)^2 - 3}{\Delta x}(x - 2) + 3 = \frac{1 - (\Delta x - 1)^2}{\Delta x}(x - 2) + 3 = (-\Delta x + 2)(x - 2) + 3$$

a. $\Delta x = 1$: $S_{\Delta x} = (x - 2) + 3 = x + 1$

$\Delta x = 0.5$: $S_{\Delta x} = \left(\dfrac{3}{2}\right)(x - 2) + 3 = \dfrac{3}{2}x$

$\Delta x = 0.1$: $S_{\Delta x} = \left(\dfrac{19}{10}\right)(x - 2) + 3 = \dfrac{19}{10}x - \dfrac{4}{5}$

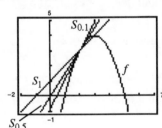

b. As $\Delta x \to 0$, the line approaches the tangent line to f at $(2, 3)$.

36. $f(x) = x + \dfrac{1}{x}$

$$S_{\Delta x}(x) = \frac{f(2 + \Delta x) - f(2)}{\Delta x}(x - 2) + f(2) = \frac{(2 + \Delta x) + \dfrac{1}{2 + \Delta x} - \dfrac{5}{2}}{\Delta x}(x - 2) + \frac{5}{2}$$

$$= \frac{2(2 + \Delta x)^2 + 2 - 5(2 + \Delta x)}{2(2 + \Delta x)\Delta x}(x - 2) + \frac{5}{2} = \frac{(2\Delta x + 3)}{2(2 + \Delta x)}(x - 2) + \frac{5}{2}$$

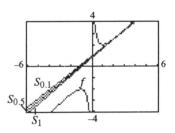

a. $\Delta x = 1$: $S_{\Delta x} = \dfrac{5}{6}(x - 2) + \dfrac{5}{2} = \dfrac{5}{6}x + \dfrac{5}{6}$

   $\Delta x = 0.5$: $S_{\Delta x} = \dfrac{4}{5}(x - 2) + \dfrac{5}{2} = \dfrac{4}{5}x + \dfrac{9}{10}$

   $\Delta x = 0.1$: $S_{\Delta x} = \dfrac{16}{21}(x - 2) + \dfrac{5}{2} = \dfrac{16}{21}x + \dfrac{41}{42}$

b. As $\Delta x \to 0$, the line approaches the tangent line to f at $\left(2, \frac{5}{2}\right)$.

37. $f(x) = x^2 - 1$, $c = 2$

$$f'(2) = \lim_{x \to 2} \frac{f(x) - f(2)}{x - 2} = \lim_{x \to 2} \frac{(x^2 - 1) - 3}{x - 2} = \lim_{x \to 2} (x + 2) = 4$$

38. $f(x) = x^3 + 2x$, $c = 1$

$$f'(1) = \lim_{x \to 1} \frac{f(x) - f(1)}{x - 1} = \lim_{x \to 1} \frac{x^3 + 2x - 3}{x - 1} = \lim_{x \to 1} (x^2 + x + 3) = 5$$

39. $f(x) = x^3 + 2x^2 + 1$, $c = -2$

$$f'(-2) = \lim_{x \to -2} \frac{f(x) - f(-2)}{x + 2} = \lim_{x \to -2} \frac{(x^3 + 2x^2 + 1) - 1}{x + 2} = \lim_{x \to -2} x^2 = 4$$

40. $f(x) = \dfrac{1}{x}$, $c = 3$

$$f'(3) = \lim_{x \to 3} \frac{f(x) - f(3)}{x - 3} = \lim_{x \to 3} \frac{(1/x) - (1/3)}{x - 3} = \lim_{x \to 3} \frac{3 - x}{3x} \cdot \frac{1}{x - 3} = \lim_{x \to 3} \left(-\frac{1}{3x}\right) = -\frac{1}{9}$$

41. $f(x) = (x - 1)^{2/3}$, $c = 1$

$$f'(1) = \lim_{x \to 1} \frac{f(x) - f(1)}{x - 1} = \lim_{x \to 1} \frac{(x - 1)^{2/3} - 0}{x - 1} = \lim_{x \to 1} \frac{1}{(x - 1)^{1/3}}$$

The limit does not exist. Thus, f is not differentiable at $x = 1$.

42. $f(x) = |x - 2|$, $c = 2$

$$f'(2) = \lim_{x \to 2} \frac{f(x) - f(2)}{x - 2} = \lim_{x \to 2} \frac{|x - 2|}{x - 2}$$

The limit does not exist. Thus, f is not differentiable at $x = 2$.

43. $f(x)$ is differentiable everywhere except at $x = -3$. (Sharp turn in the graph)

44. $f(x)$ is differentiable everywhere except at $x = \pm 3$. (Sharp turns in the graph)

45. $f(x)$ is differentiable everywhere except at $x = -1$. (Discontinuity)

46. $f(x)$ is differentiable everywhere except at $x = 1$. (Discontinuity)

47. $f(x)$ is differentiable everywhere except at $x = 3$. (Vertical tangent)

48. $f(x)$ is differentiable everywhere except $x = 0$. (Vertical tangent)

49. $f(x)$ is differentiable on the interval $(1, \infty)$. (At $x = 1$ the tangent line is vertical.)

50. $f(x)$ is differentiable everywhere except at $x = \pm 2$. (Discontinuities)

51. $f(x)$ is differentiable everywhere except at $x = 0$. (Discontinuity)

52. $f(x)$ is differentiable everywhere except at $x = 1$. (Discontinuity)

53. $f(x) = |x - 1|$

The derivative from the left is $\displaystyle\lim_{x \to 1^-} \frac{f(x) - f(1)}{x - 1} = \lim_{x \to 1^-} \frac{|x-1| - 0}{x - 1} = -1.$

The derivative from the right is $\displaystyle\lim_{x \to 1^+} \frac{f(x) - f(1)}{x - 1} = \lim_{x \to 1^+} \frac{|x-1| - 0}{x - 1} = 1.$

The one-sided limits are not equal. Therefore, f is not differentiable at $x = 1$.

54. $f(x) = \sqrt{1 - x^2}$

The derivative from the left is

$$\lim_{x \to 1^-} \frac{f(x) - f(1)}{x - 1} = \lim_{x \to 1^-} \frac{\sqrt{1-x^2} - 0}{x - 1} = \lim_{x \to 1^-} \frac{\sqrt{1-x^2}}{x - 1} \cdot \frac{\sqrt{1-x^2}}{\sqrt{1-x^2}} = \lim_{x \to 1^-} -\frac{1+x}{\sqrt{1-x^2}} = -\infty.$$

The limit from the right does not exist since f is undefined for $x > 1$. Therefore, f is not differentiable at $x = 1$.

55. $f(x) = \begin{cases} (x-1)^3, & x \le 1 \\ (x-1)^2, & x > 1 \end{cases}$

The derivative from the left is

$$\lim_{x \to 1^-} \frac{f(x) - f(1)}{x - 1} = \lim_{x \to 1^-} \frac{(x-1)^3 - 0}{x - 1} = \lim_{x \to 1^-} (x-1)^2 = 0.$$

The derivative from the right is

$$\lim_{x \to 1^+} \frac{f(x) - f(1)}{x - 1} = \lim_{x \to 1^+} \frac{(x-1)^2 - 0}{x - 1} = \lim_{x \to 1^+} (x-1) = 0.$$

These one-sided limits are equal. Therefore, f is differentiable at $x = 1$ ($f'(1) = 0$).

56. $f(x) = \begin{cases} x, & x \le 1 \\ x^2, & x > 1 \end{cases}$

The derivative from the left is

$$\lim_{x \to 1^-} \frac{f(x) - f(1)}{x - 1} = \lim_{x \to 1^-} \frac{x - 1}{x - 1} = \lim_{x \to 1^-} 1 = 1.$$

The derivative from the right is

$$\lim_{x \to 1^+} \frac{f(x) - f(1)}{x - 1} = \lim_{x \to 1^+} \frac{x^2 - 1}{x - 1} = \lim_{x \to 1^+} (x + 1) = 2.$$

These one-sided limits are not equal. Therefore, f is not differentiable at $x = 1$.

57. $f(x) = \begin{cases} x^2 + 1, & x \le 2 \\ 4x - 3, & x > 2 \end{cases}$

The derivative from the left is

$$\lim_{x \to 2^-} \frac{f(x) - f(2)}{x - 2} = \lim_{x \to 2^-} \frac{(x^2 + 1) - 5}{x - 2} = \lim_{x \to 2^-} (x + 2) = 4.$$

The derivative from the right is

$$\lim_{x \to 2^+} \frac{f(x) - f(2)}{x - 2} = \lim_{x \to 2^+} \frac{(4x - 3) - 5}{x - 2} = \lim_{x \to 2^+} 4 = 4.$$

The one-sided limits are equal. Therefore, f is differentiable at $x = 2$. $(f'(2) = 4)$

58. $f(x) = \begin{cases} \frac{1}{2}x + 1, & x < 2 \\ \sqrt{2x}, & x \ge 2 \end{cases}$

The derivative from the left is

$$\lim_{x \to 2^-} \frac{f(x) - f(2)}{x - 2} = \lim_{x \to 2^-} \frac{\left(\frac{1}{2}x + 1\right) - 2}{x - 2} = \lim_{x \to 2^-} \frac{\frac{1}{2}(x - 2)}{x - 2} = \frac{1}{2}.$$

The derivative from the right is

$$\lim_{x \to 2^+} \frac{f(x) - f(2)}{x - 2} = \lim_{x \to 2^+} \frac{\sqrt{2x} - 2}{x - 2} \cdot \frac{\sqrt{2x} + 2}{\sqrt{2x} + 2}$$

$$= \lim_{x \to 2^+} \frac{2x - 4}{(x - 2)(\sqrt{2x} + 2)} = \lim_{x \to 2^+} \frac{2(x - 2)}{(x - 2)(\sqrt{2x} + 2)} = \lim_{x \to 2^+} \frac{2}{\sqrt{2x} + 2} = \frac{1}{2}.$$

The one-sided limits are equal. Therefore, f is differentiable at $x = 2$. $\left(f'(2) = \frac{1}{2}\right)$

59. a. If $f(x)$ is odd, then the graph of f is symmetric about the origin.

So, if $f'(c) = 3$, then $f'(-c) = 3$.

b. If $f(x)$ is even, then the graph of f is symmetric about the y-axis.

So, if $f'(c) = 3$, then $f'(-c) = -3$.

60. a. $f(x) = x^2, \quad f'(x) = 2x$

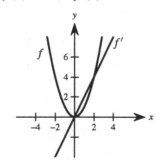

b. $f(x) = x^3, \quad f'(x) = 3x^2$

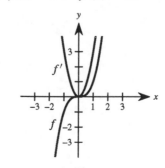

61. $y = x^2$

Slope: $m = y' = 2x$

True, the slope is different at every point on the curve.

62. False. $y = |x - 2|$ is continuous at $x = 2$, but is not differentiable at $x = 2$. (Sharp turn in the graph)

63. True. Differentiability implies continuity.

64. False. If the derivative from the left of a point does not equal the derivative from the right of a point, then the derivative does not exist at that point. For example, if $f(x) = |x|$, then the derivative from the left of $x = 0$ is -1 and the derivative from the right of $x = 0$ is 1. At $x = 0$, the derivative does not exist.

65. $f(x) = \begin{cases} x \sin \dfrac{1}{x}, & x \neq 0 \\ 0, & x = 0 \end{cases}$

Using the Squeeze Theorem, we have $-|x| \leq x \sin(1/x) \leq |x|$, $x \neq 0$. Thus, $\lim\limits_{x \to 0} x \sin(1/x) = 0 = f(0)$ and f is continuous at $x = 0$. Using the alternative form of the derivative we have

$$\lim_{x \to 0} \frac{f(x) - f(0)}{x - 0} = \lim_{x \to 0} \frac{x \sin \dfrac{1}{x} - 0}{x - 0} = \lim_{x \to 0} \left(\sin \frac{1}{x} \right).$$

Since this limit does not exist (it oscillates between -1 and 1), the function is not differentiable at $x = 0$.

$$g(x) = \begin{cases} x^2 \sin \dfrac{1}{x}, & x \neq 0 \\ 0, & x = 0 \end{cases}$$

Using the Squeeze Theorem again we have $-x^2 \leq x^2 \sin(1/x) \leq x^2$, $x \neq 0$. Thus, $\lim\limits_{x \to 0} x^2 \sin(1/x) = 0 = f(0)$ and f is continuous at $x = 0$. Using the alternative form of the derivative again we have

$$\lim_{x \to 0} \frac{f(x) - f(0)}{x - 0} = \lim_{x \to 0} \frac{x^2 \sin \dfrac{1}{x} - 0}{x - 0} = \lim_{x \to 0} x \sin \frac{1}{x} = 0.$$

Therefore, g is differentiable at $x = 0$, $g'(0) = 0$.

Section 2.2   Basic Differentiation Rules and Rates of Change

1. a. $y = x^{1/2}$

$y' = \frac{1}{2}x^{-1/2}$

$y'(1) = \frac{1}{2}$

b. $y = x^{3/2}$

$y' = \frac{3}{2}x^{1/2}$

$y'(1) = \frac{3}{2}$

c. $y = x^2$

$y' = 2x$

$y'(1) = 2$

d. $y = x^3$

$y' = 3x^2$

$y'(1) = 3$

2. a. $y = x^{-1/2}$

$y' = -\frac{1}{2}x^{-3/2}$

$y(1) = -\frac{1}{2}$

b. $y = x^{-1}$

$y' = -x^{-2}$

$y'(1) = -1$

c. $y = x^{-3/2}$

$y' = -\frac{3}{2}x^{-5/2}$

$y'(1) = -\frac{3}{2}$

d. $y = x^{-2}$

$y' = -2x^{-3}$

$y'(1) = -2$

3. $y = 3$

$y' = 0$

4. $f(x) = -2$

$f'(x) = 0$

5. $f(x) = x + 1$

$f'(x) = 1$

6. $g(x) = 3x - 1$

$g'(x) = 3$

7. $g(x) = x^2 + 4$

$g'(x) = 2x$

8. $y = t^2 + 2t - 3$

$y' = 2t + 2$

9. $f(t) = -2t^2 + 3t - 6$

$f'(t) = -4t + 3$

10. $y = x^3 - 9$

$y' = 3x^2$

11. $s(t) = t^3 - 2t + 4$

$s'(t) = 3t^2 - 2$

12. $f(x) = 2x^3 - x^2 + 3x$

$f'(x) = 6x^2 - 2x + 3$

13. $y = x^2 - \frac{1}{2}\cos x$

$y' = 2x + \frac{1}{2}\sin x$

14. $y = 5 + \sin x$

$y' = \cos x$

15. $y = \frac{1}{x} - 3\sin x$

$y' = -\frac{1}{x^2} - 3\cos x$

16. $g(t) = \pi \cos t$

$g'(t) = -\pi \sin t$

	Function	_Rewrite_	_Derivative_	_Simplify_
17.	$y = \dfrac{1}{3x^3}$	$y = \dfrac{1}{3}x^{-3}$	$y' = -x^{-4}$	$y' = -\dfrac{1}{x^4}$
18.	$y = \dfrac{2}{3x^2}$	$y = \dfrac{2}{3}x^{-2}$	$y' = -\dfrac{4}{3}x^{-3}$	$y' = -\dfrac{4}{3x^3}$
19.	$y = \dfrac{1}{(3x)^3}$	$y = \dfrac{1}{27}x^{-3}$	$y' = -\dfrac{1}{9}x^{-4}$	$y' = -\dfrac{1}{9x^4}$
20.	$y = \dfrac{\pi}{(3x)^2}$	$y = \dfrac{\pi}{9}x^{-2}$	$y' = -\dfrac{2\pi}{9}x^{-3}$	$y' = -\dfrac{2\pi}{9x^3}$
21.	$y = \dfrac{\sqrt{x}}{x}$	$y = x^{-1/2}$	$y' = -\dfrac{1}{2}x^{-3/2}$	$y' = -\dfrac{1}{2x^{3/2}}$
22.	$y = \dfrac{4}{x^{-3}}$	$y = 4x^3$	$y' = 12x^2$	$y' = 12x^2$

23. $f(x) = \dfrac{1}{x}$, $(1, 1)$

$f'(x) = -\dfrac{1}{x^2}$

$f'(1) = -1$

24. $f(t) = 3 - \dfrac{3}{5t}$, $\left(\dfrac{3}{5}, 2\right)$

$f'(t) = \dfrac{3}{5t^2}$

$f'\left(\dfrac{3}{5}\right) = \dfrac{5}{3}$

25. $f(x) = -\dfrac{1}{2} + \dfrac{7}{5}x^3$, $\left(0, -\dfrac{1}{2}\right)$

$f'(x) = \dfrac{21}{5}x^2$

$f'(0) = 0$

26. $y = 3x\left(x^2 - \dfrac{2}{x}\right)$, $(2, 18)$

$= 3x^3 - 6$

$y' = 9x^2$

$y'(2) = 36$

27. $y = (2x + 1)^2$, $(0, 1)$

$= 4x^2 + 4x + 1$

$y' = 8x + 4$

$y'(0) = 4$

28. $f(x) = 3(5 - x)^2$, $(5, 0)$

$= 3x^2 - 30x + 75$

$f'(x) = 6x - 30$

$f'(5) = 0$

29. $f(\theta) = 4\sin\theta - \theta$, $(0, 0)$

$f'(\theta) = 4\cos\theta - 1$

$f'(0) = 4(1) - 1 = 3$

30. $g(t) = 2 + 3\cos t$, $(\pi, -1)$

$g'(t) = -3\sin t$

$g'(\pi) = 0$

31. $f(x) = x^3 - 3x - 2x^{-4}$

$\quad f'(x) = 3x^2 - 3 + 8x^{-5}$

$\qquad = 3x^2 - 3 + \dfrac{8}{x^5}$

32. $f(x) = x^2 - 3x - 3x^{-2}$

$\quad f'(x) = 2x - 3 + 6x^{-3}$

$\qquad = 2x - 3 + \dfrac{6}{x^3}$

33. $g(t) = t^2 - 4t^{-1}$

$\quad g'(t) = 2t + 4t^{-2}$

$\qquad = 2t + \dfrac{4}{t^2}$

34. $f(x) = x + x^{-2}$

$\quad f'(x) = 1 - 2x^{-3}$

$\qquad = 1 - \dfrac{2}{x^3}$

35. $f(x) = \dfrac{x^3 - 3x^2 + 4}{x^2} = x - 3 + 4x^{-2}$

$\quad f'(x) = 1 - \dfrac{8}{x^3} = \dfrac{x^3 - 8}{x^3}$

36. $h(x) = \dfrac{2x^2 - 3x + 1}{x} = 2x - 3 + x^{-1}$

$\quad h'(x) = 2 - \dfrac{1}{x^2} = \dfrac{2x^2 - 1}{x^2}$

37. $y = x(x^2 + 1) = x^3 + x$

$\quad y' = 3x^2 + 1$

38. $f(x) = \sqrt[3]{x} + \sqrt[5]{x} = x^{1/3} + x^{1/5}$

$\quad f'(x) = \dfrac{1}{3}x^{-2/3} + \dfrac{1}{5}x^{-4/5} = \dfrac{1}{3x^{2/3}} + \dfrac{1}{5x^{4/5}}$

39. $h(s) = s^{4/5}$

$\quad h'(s) = \dfrac{4}{5}s^{-1/5} = \dfrac{4}{5s^{1/5}}$

40. $f(t) = t^{1/3} - 1$

$\quad f'(t) = \dfrac{1}{3}t^{-2/3} = \dfrac{1}{3t^{2/3}}$

41. $f(x) = 4\sqrt{x} + 3\cos x$

$\quad f'(x) = 2x^{-1/2} - 3\sin x$

$\qquad = \dfrac{2}{\sqrt{x}} - 3\sin x$

42. $f(x) = 2\sin x + 3\cos x$

$\quad f'(x) = 2\cos x - 3\sin x$

43. $y = x^4 - 3x^2 + 2$

$\quad y' = 4x^3 - 6x$

At $(1, 0)$: $y' = 4(1)^3 - 6(1) = -2$

Tangent line:

$\qquad y - 0 = -2(x - 1)$

$\qquad 2x + y - 2 = 0$

44. $y = x^3 + x$

$\quad y' = 3x^2 + 1$

At $(-1, -2)$: $y' = 3(-1)^2 + 1 = 4$

Tangent line:

$\qquad y + 2 = 4(x + 1)$

$\qquad 4x - y + 2 = 0$

45. $f(x) = \dfrac{1}{\sqrt[3]{x^2}} = x^{-2/3}$

$\quad f'(x) = -\dfrac{2}{3}x^{-5/3} = -\dfrac{2}{3\sqrt[3]{x^5}}$

At $\left(8, \dfrac{1}{4}\right)$: $y' = -\dfrac{2}{3(\sqrt[3]{8})^5} = -\dfrac{1}{48}$

Tangent line:

$\qquad y - \dfrac{1}{4} = -\dfrac{1}{48}(x - 8)$

$\qquad -48y + 12 = x - 8$

$\qquad 0 = x + 48y - 20$

46. $y = (x^2 + 2x)(x + 1)$

$\qquad = x^3 + 3x^2 + 2x$

$\quad y' = 3x^2 + 6x + 2$

At $(1, 6)$: $y' = 3(1)^2 + 6(1) + 2 = 11$

Tangent line:

$\qquad y - 6 = 11(x - 1)$

$\qquad 0 = 11x - y - 5$

47. $y = x^4 - 3x^2 + 2$

$y' = 4x^3 - 6x$

$\qquad = 2x(2x^2 - 3) = 0 \Rightarrow x = 0 \text{ or } x = \pm\sqrt{\frac{3}{2}}$

At $x = 0$, $y = 2$, and at $x = \pm\sqrt{\frac{3}{2}}$, $y = -\frac{1}{4}$.

Horizontal tangents: $(0, 2)$, $\left(\pm\sqrt{\frac{3}{2}}, -\frac{1}{4}\right)$

48. $y = x^3 + x$

$y' = 3x^2 + 1 > 0$ for all x.

Therefore, there are no horizontal tangents.

49. $y = \dfrac{1}{x^2} = x^{-2}$

$y' = -2x^{-3} = \dfrac{-2}{x^3}$ cannot equal zero.

Therefore, there are no horizontal tangents.

50. $y = x^2 + 1$

$y' = 2x = 0 \Rightarrow x = 0$

At $x = 0$, $y = 1$.

Horizontal tangent: $(0, 1)$

51. $y = x + \sin x, \quad 0 \le x < 2\pi$

$y' = 1 + \cos x = 0$

$\cos x = -1 \Rightarrow x = \pi$

At $x = \pi$, $y = \pi$.

Horizontal tangent: (π, π)

52. $y = \sqrt{3}\,x + 2\cos x, \quad 0 \le x < 2\pi$

$y' = \sqrt{3} - 2\sin x = 0$

$\sin x = \dfrac{\sqrt{3}}{2} \Rightarrow x = \dfrac{\pi}{3} \text{ or } \dfrac{2\pi}{3}$

At $x = \dfrac{\pi}{3}$, $y = \dfrac{\sqrt{3}\,\pi + 3}{3}$.

At $x = \dfrac{2\pi}{3}$, $y = \dfrac{2\sqrt{3}\,\pi - 3}{3}$.

Horizontal tangents:

$$\left(\frac{\pi}{3}, \frac{\sqrt{3}\,\pi + 3}{3}\right), \; \left(\frac{2\pi}{3}, \frac{2\sqrt{3}\,\pi - 3}{3}\right)$$

53.

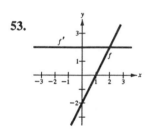

If f is linear then its derivative is a constant function.

$f(x) = ax + b$

$f'(x) = a$

54.

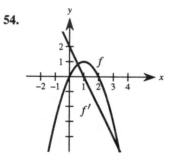

If f is quadratic its derivative is a linear function.

$f(x) = ax^2 + bx + c$

$f'(x) = 2ax + b$

55. Let (x_1, y_1) and (x_2, y_2) be the points of tangency on $y = x^2$ and $y = -x^2 + 6x - 5$, respectively. The derivatives of these functions are

$$y' = 2x \implies m = 2x_1 \text{ and } y' = -2x + 6 \implies m = -2x_2 + 6.$$

$$m = 2x_1 = -2x_2 + 6$$

$$x_1 = -x_2 + 3$$

Since $y_1 = x_1^2$ and $y_2 = -x_2^2 + 6x_2 - 5$,

$$m = \frac{y_2 - y_1}{x_2 - x_1} = \frac{(-x_2^2 + 6x_2 - 5) - (x_1^2)}{x_2 - x_1} = -2x_2 + 6.$$

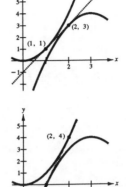

$$\frac{(-x_2^2 + 6x_2 - 5) - (-x_2 + 3)^2}{x_2 - (-x_2 + 3)} = -2x_2 + 6$$

$$(-x_2^2 + 6x_2 - 5) - (x_2^2 - 6x_2 + 9) = (-2x_2 + 6)(2x_2 - 3)$$

$$-2x_2^2 + 12x_2 - 14 = -4x_2^2 + 18x_2 - 18$$

$$2x_2^2 - 6x_2 + 4 = 0$$

$$2(x_2 - 2)(x_2 - 1) = 0$$

$$x_2 = 1 \text{ or } 2$$

$$x_2 = 1 \implies y_2 = 0, \quad x_1 = 2 \text{ and } y_1 = 4$$

Thus, the tangent line through $(1, 0)$ and $(2, 4)$ is

$$y - 0 = \left(\frac{4 - 0}{2 - 1}\right)(x - 1) \implies y = 4x - 4.$$

$$x_2 = 2 \implies y_2 = 3, \quad x_1 = 1 \text{ and } y_1 = 1$$

Thus, the tangent line through $(2, 3)$ and $(1, 1)$ is

$$y - 1 = \left(\frac{3 - 1}{2 - 1}\right)(x - 1) \implies y = 2x - 1.$$

56. m_1 is the slope of the line tangent to $y = x$. m_2 is the slope of the line tangent to $y = 1/x$. Since

$$y = x \implies y' = 1 \implies m_1 = 1 \text{ and } y = \frac{1}{x} \implies y' = \frac{-1}{x^2} \implies m_2 = \frac{-1}{x^2}.$$

The points of intersection of $y = x$ and $y = 1/x$ are

$$x = \frac{1}{x} \implies x^2 = 1 \implies x = \pm 1.$$

At $x = \pm 1$, $m_2 = -1$. Since $m_2 = -1/m_1$, these tangent lines are perpendicular at the points of intersection.

57. $f(x) = \sqrt{x}, \quad (-4, 0)$

$$f'(x) = \frac{1}{2}x^{-1/2} = \frac{1}{2\sqrt{x}}$$

$$\frac{1}{2\sqrt{x}} = \frac{0-y}{-4-x}$$

$$4 + x = 2\sqrt{x}\,y$$

$$4 + x = 2\sqrt{x}\sqrt{x}$$

$$4 + x = 2x$$

$$x = 4, \quad y = 2$$

The point $(4, 2)$ is on the graph of f.
Tangent line:

$$y - 2 = \frac{0-2}{-4-4}(x-4)$$

$$4y - 8 = x - 4$$

$$0 = x - 4y + 4$$

58. $f(x) = \dfrac{2}{x}, \quad (5, 0)$

$$f'(x) = -\frac{2}{x^2}$$

$$-\frac{2}{x^2} = \frac{0-y}{5-x}$$

$$-10 + 2x = -x^2 y$$

$$-10 + 2x = -x^2\left(\frac{2}{x}\right)$$

$$-10 + 2x = -2x$$

$$4x = 10$$

$$x = \frac{5}{2}, \quad y = \frac{4}{5}$$

The point $\left(\frac{5}{2}, \frac{4}{5}\right)$ is on the graph of f. The slope of
the tangent line is $f'\left(\frac{5}{2}\right) = -\frac{8}{25}$.
Tangent line:

$$y - \frac{4}{5} = -\frac{8}{25}\left(x - \frac{5}{2}\right)$$

$$25y - 20 = -8x + 20$$

$$8x + 25y - 40 = 0$$

59. $f(x) = x^{3/2}$

a.
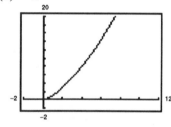

b. $f'(x) = \frac{3}{2}x^{1/2} = \frac{3}{2}\sqrt{x}$

$\quad T(x) = f'(4)(x-4) + f(4)$

$\qquad = 3(x-4) + 8$

$\qquad = 3x - 4$

c. As you move away from $(4, 8)$ the approximation
becomes less and less accurate.

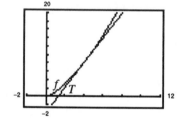

d.

Δx	-3	-2	-1	-0.5	-0.1	0
$f(4+\Delta x)$	1	2.8284	5.1962	6.5479	7.7019	8
$T(4+\Delta x)$	-1	2	5	6.5	7.7	8

Δx	0.1	0.5	1	2	3
$f(4+\Delta x)$	8.3019	9.5459	11.1803	14.6969	18.5203
$T(4+\Delta x)$	8.3	9.5	11	14	17

60. $f(x) = x^3$, (1, 1)

a.

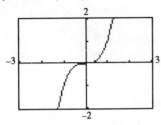

b. $f'(x) = 3x^2$

$$T(x) = f'(1)(x - 1) + f(1)$$
$$= 3(x - 1) + 1$$
$$= 3x - 2$$

c. As you move away from (1, 1), the accuracy of the approximation decreases. In Exercise 59, the graph of f was almost linear near (4, 8). Here the graph is not as "linear" near (1, 1). $T(x)$ is only a good approximation for $f(x)$ for values very close to $x = 1$.

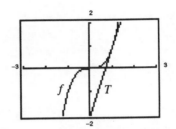

d.

Δx	-3	-2	-1	-0.5	-0.1	0	0.1	0.5	1	2	3
$f(1 + \Delta x)$	-8	-1	0	0.125	0.729	1	1.331	3.375	8	27	64
$T(1 + \Delta x)$	-8	-5	-2	-0.500	0.700	1	1.300	2.500	4	7	10

61. False. Let $f(x) = x^2$ and $g(x) = x^2 + 4$. Then $f'(x) = g'(x) = 2x$, but $f(x) \neq g(x)$.

62. True. If $f(x) = g(x) + c$, then $f'(x) = g'(x) + 0 = g'(x)$.

63. False. If $y = \pi^2$, then $dy/dx = 0$. (π^2 is a constant.)

64. True. If $y = x/\pi = (1/\pi) \cdot x$, then $dy/dx = 1/\pi(1) = 1/\pi$.

65.

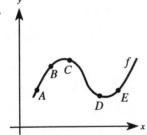

c.

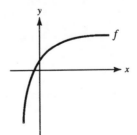

d. The average rates of change are approximately equal between B and C and D and E.

a. The slope appears to be steepest between A and B.

b. The average rate of change between A and B is **greater** than the instantaneous rate of change at B.

66. The graph of a function f such that $f' > 0$ for all x and the rate of change the function is decreasing would, in general, look like the graph at the right. One example would be $f(x) = \sqrt{x + 2} - 4$.

67. $f(t) = 2t + 7$, [1, 2]

$f'(t) = 2$

Instantaneous rate of change is the constant 2.

Average rate of change:

$$\frac{f(2) - f(1)}{2 - 1} = \frac{[2(2) + 7] - [2(1) + 7]}{1} = 2$$

68. $f(t) = t^2 - 3$, [2, 2.1]

$f'(t) = 2t$

Instantaneous rates of change:

$$(2, \ 1) \ \Rightarrow \ f'(2) = 2(2) = 4$$

$$(2.1, \ 1.41) \ \Rightarrow \ f'(2.1) = 4.2$$

Average rate of change:

$$\frac{f(2.1) - f(2)}{2.1 - 2} = \frac{1.41 - 1}{0.1} = 4.1$$

69. $f(x) = -\dfrac{1}{x}$, [1, 2]

$f'(x) = \dfrac{1}{x^2}$

Instantaneous rate of change:

$$(1, \ -1) \ \Rightarrow \ f'(1) = 1$$

$$\left(2, \ -\frac{1}{2}\right) \ \Rightarrow \ f'(2) = \frac{1}{4}$$

Average rate of change:

$$\frac{f(2) - f(1)}{2 - 1} = \frac{(-1/2) - (-1)}{2 - 1} = \frac{1}{2}$$

70. $f(x) = \sin x$, $\left[0, \dfrac{\pi}{6}\right]$

$f'(x) = \cos x$

Instantaneous rate of change:

$$(0, \ 0) \ \Rightarrow \ f'(0) = 1$$

$$\left(\frac{\pi}{6}, \frac{1}{2}\right) \ \Rightarrow \ f'\left(\frac{\pi}{6}\right) = \frac{\sqrt{3}}{2}$$

Average rate of change:

$$\frac{f(\pi/6) - f(0)}{(\pi/6) - 0} = \frac{(1/2) - 0}{(\pi/6) - 0} = \frac{3}{\pi}$$

71. a. $s(t) = -16t^2 + 1362$

$v(t) = -32t$

b. $\dfrac{s(2) - s(1)}{2 - 1} = 1298 - 1346 = -48$ ft/sec

c. $v(t) = s'(t) = -32t$

When $t = 1$: $v(1) = -32$ ft/sec.

When $t = 2$: $v(2) = -64$ ft/sec.

d. $-16t^2 + 1362 = 0$

$$t^2 = \frac{1362}{16} \ \Rightarrow \ t = \frac{\sqrt{1362}}{4} \approx 9.226 \text{ sec}$$

e. $v\left(\dfrac{\sqrt{1362}}{4}\right) = -32\left(\dfrac{\sqrt{1362}}{4}\right)$

$$= -8\sqrt{1362} \approx -295.242 \text{ ft/sec}$$

72.

$$s(t) = -16t^2 - 22t + 220$$

$$v(t) = -32t - 22$$

$$v(3) = -118 \text{ ft/sec}$$

$$s(t) = -16t^2 - 22t + 220$$

$$= 112 \quad \text{(height after falling 108 ft)}$$

$$-16t^2 - 22t + 108 = 0$$

$$-2(t - 2)(8t + 27) = 0$$

$$t = 2$$

$$v(2) = -32(2) - 22$$

$$= -86 \text{ ft/sec}$$

73. $s(t) = -16t^2 + 384t$

$v(t) = -32t + 384$

$v(5) = 224$ ft/sec

$v(10) = 64$ ft/sec

74. $s(t) = -16t^2 + s_0 = 0$ when $t = 6.8$

$s_0 = 16(6.8)^2 \approx 740$ ft

75.

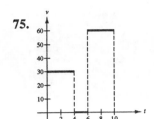

76.

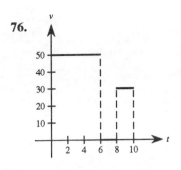

77. $v = 40$ mph $= \frac{2}{3}$ mi/min

$\left(\frac{2}{3} \text{ mi/min}\right)(6 \text{ min}) = 4$ mi

$v = 60$ mph $= 1$ mi/min

$(1 \text{ mi/min})(2 \text{ min}) = 2$ mi

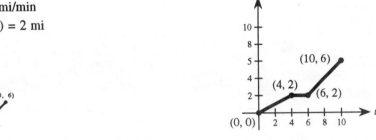

78. This graph corresponds with Exercise 75.

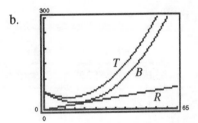

79. a. $T(x) = R(x) + B(x)$

$= 1.1x + (0.14x^2 - 4.43x + 58.40)$

$= 0.14x^2 - 3.33x + 58.40$

c. $T'(x) = 0.28x - 3.33$

$T'(30) = 5.07$ ft/mph

$T'(40) = 7.87$ ft/mph

$T'(55) = 12.07$ ft/mph

b.

80. $s(t) = -\frac{1}{2}at^2 + c$

Average velocity: $\dfrac{s(t_0 + \Delta t) - s(t_0 - \Delta t)}{(t_0 + \Delta t) - (t_0 - \Delta t)} = \dfrac{\left[-\frac{1}{2}a(t_0 + \Delta t)^2 + c\right] - \left[-\frac{1}{2}a(t_0 - \Delta t)^2 + c\right]}{2\Delta t}$

$= \dfrac{-\frac{1}{2}a(t_0^2 + 2t_0\Delta t + (\Delta t)^2) + \frac{1}{2}a(t_0^2 - 2t_0\Delta t + (\Delta t)^2)}{2\Delta t}$

$= \dfrac{-2at_0\Delta t}{2\Delta t}$

$= -at_0$

$= s'(t_0)$ Instantaneous velocity at $t = t_0$

81. $A = s^2$, $\dfrac{dA}{ds} = 2s$

When $s = 4$ m, $\dfrac{dA}{ds} = 8$ m^2.

82. $V = s^3$, $\dfrac{dV}{ds} = 3s^2$

When $s = 4$ cm, $\dfrac{dV}{ds} = 48$ cm^3.

83. $E = \dfrac{1}{27}(9t + 3t^2 - t^3)$

$\dfrac{dE}{dt} = \dfrac{1}{9}(3 + 2t - t^2)$

   a. When $t = 1$, $\dfrac{dE}{dt} = \dfrac{4}{9}$.

   b. When $t = 2$, $\dfrac{dE}{dt} = \dfrac{1}{3}$.

   c. When $t = 3$, $\dfrac{dE}{dt} = 0$.

   d. When $t = 4$, $\dfrac{dE}{dt} = -\dfrac{5}{9}$.

84. $P = 50\sqrt{x} - 0.5x - 500$

$\dfrac{dP}{dx} = \dfrac{25}{\sqrt{x}} - 0.5$

   a. When $x = 900$, $\dfrac{dP}{dx} = \dfrac{1}{3}$.

   b. When $x = 1600$, $\dfrac{dP}{dx} = \dfrac{1}{8}$.

   c. When $x = 2500$, $\dfrac{dP}{dx} = 0$.

   d. When $x = 3600$, $\dfrac{dP}{dx} = -\dfrac{1}{12}$.

85. $$C = \dfrac{1{,}008{,}000}{Q} + 6.3Q$$

$$\dfrac{dC}{dQ} = -\dfrac{1{,}008{,}000}{Q^2} + 6.3$$

$C(351) - C(350) \approx 5083.095 - 5085 \approx -\$1.91$

When $Q = 350$, $\dfrac{dC}{dQ} \approx -\$1.93$.

86. $C = $ (gallons of fuel used)(cost per gallon) $= (15{,}000/x)(1.10)$. Thus, $C = 16{,}500/x$.

$\dfrac{dC}{dx} = -\dfrac{16{,}500}{x^2}$

x	10	15	20	25	30	35	40
C	\$1650	\$1100	\$825	\$660	\$550	\$471.43	\$412.50
$\dfrac{dC}{dx}$	-165	-73.33	-41.25	-26.40	-18.33	-13.47	-10.31

The driver who gets 15 miles per gallon would benefit more from a 1 mile per gallon increase in fuel efficiency. The rate of change is much larger when $x = 15$.

87. $F = K\dfrac{m_1 m_2}{r^2} = K m_1 m_2 r^{-2}, \quad r > 0$

$\dfrac{dF}{dr} = K m_1 m_2 (-2r^{-3}) = -2K\dfrac{m_1 m_2}{r^3}$

dF/dr is negative since the force decreases as the distance between particles increases.

88. $\dfrac{dT}{dt} = K(T - T_a)$

89. $y = ax^2 + bx + c$

Since the parabola passes through (0, 1) and (1, 0), we have

$(0, 1): \quad 1 = a(0)^2 + b(0) + c \implies c = 1$

$(1, 0): \quad 0 = a(1)^2 + b(1) + 1 \implies b = -a - 1$

Thus, $y = ax^2 + (-a-1)x + 1$. From the tangent line $x = y + 1$ we know that the derivative is 1 at the point (1, 0).

$y' = 2ax + (-a-1)$

$1 = 2a(1) + (-a-1)$

$1 = a - 1$

$a = 2$

$b = -a - 1 = -3$

Therefore, $y = 2x^2 - 3x + 1$.

90.

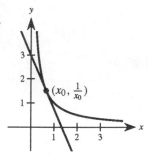

$y = \dfrac{1}{x}, \quad x > 0$

$y' = -\dfrac{1}{x^2}$

At (a, b), the equation of the tangent line is

$y - \dfrac{1}{a} = -\dfrac{1}{a^2}(x - a)$

or

$y = -\dfrac{x}{a^2} + \dfrac{2}{a}.$

The x-intercept is $(2a, 0)$.

The y-intercept is $(0, 2/a)$.

The area of the triangle is

$A = \dfrac{1}{2}bh = \dfrac{1}{2}(2x_0)\left(\dfrac{2}{x_0}\right) = 2.$

91. $y = x^3 - 9x$

$y' = 3x^2 - 9$

Tangent lines through (1, −9):

$y + 9 = (3x^2 - 9)(x - 1)$

$(x^3 - 9x) + 9 = 3x^3 - 3x^2 - 9x + 9$

$0 = 2x^3 - 3x^2 = x^2(2x - 3)$

$x = 0 \quad \text{or} \quad x = \dfrac{3}{2}$

The points of tangency are $(0, 0)$ and $\left(\dfrac{3}{2}, -\dfrac{81}{8}\right)$. At $(0, 0)$ the slope is $y'(0) = -9$. At $\left(\dfrac{3}{2}, -\dfrac{81}{8}\right)$ the slope is $y'\left(\dfrac{3}{2}\right) = -\dfrac{9}{4}$.

Tangent lines:

$y - 0 = -9(x - 0) \qquad \text{and} \qquad y + \dfrac{81}{8} = -\dfrac{9}{4}\left(x - \dfrac{3}{2}\right)$

$\qquad y = -9x \qquad\qquad\qquad\qquad y = -\dfrac{9}{4}x - \dfrac{27}{4}$

$9x + y = 0 \qquad\qquad\qquad 9x + 4y + 27 = 0$

92. $y = x^2$

$y' = 2x$

a. Tangent lines through $(0, \ a)$:

$$y - a = 2x(x - 0)$$

$$x^2 - a = 2x^2$$

$$-a = x^2$$

$$\pm\sqrt{-a} = x$$

The points of tangency are $(\pm\sqrt{-a}, \ -a)$. At $(\sqrt{-a}, \ -a)$ the slope is $y'(\sqrt{-a}) = 2\sqrt{-a}$. At $(-\sqrt{-a}, \ -a)$ the slope is $y'(-\sqrt{-a}) = -2\sqrt{-a}$.

Tangent lines:

$$y + a = 2\sqrt{-a}(x - \sqrt{-a}) \quad \text{and} \quad y + a = -2\sqrt{-a}(x + \sqrt{-a})$$

$$y = 2\sqrt{-a}\,x + a \qquad\qquad\qquad y = -2\sqrt{-a}\,x + a$$

Restriction: a must be negative.

b. Tangent lines through $(a, \ 0)$:

$$y - 0 = 2x(x - a)$$

$$x^2 = 2x^2 - 2ax$$

$$0 = x^2 - 2ax = x(x - 2a)$$

The points of tangency are $(0, 0)$ and $(2a, \ 4a^2)$. At $(0, 0)$ the slope is $y'(0) = 0$. At $(2a, \ 4a^2)$ the slope is $y'(2a) = 4a$.

Tangent lines:

$$y - 0 = 0(x - 0) \quad \text{and} \quad y - 4a^2 = 4a(x - 2a)$$

$$y = 0 \qquad\qquad\qquad\qquad y = 4ax - 4a^2$$

Restriction: None, a can be any real number.

93. $f(x) = \begin{cases} ax^3, & x \le 2 \\ x^2 + b, & x > 2 \end{cases}$

f must be continuous at $x = 2$ to be differentiable at $x = 2$.

$$\left. \begin{array}{l} \lim\limits_{x \to 2^-} f(x) = \lim\limits_{x \to 2^-} ax^3 = 8a \\[2mm] \lim\limits_{x \to 2^+} f(x) = \lim\limits_{x \to 2^+} (x^2 + b) = 4 + b \end{array} \right\} \quad \begin{array}{l} 8a = 4 + b \\[2mm] 8a - 4 = b \end{array}$$

$$f'(x) = \begin{cases} 3ax^2, & x < 2 \\ 2x, & x > 2 \end{cases}$$

For f to be differentiable at $x = 2$, the left derivative must equal the right derivative.

$$3a(2)^2 = 2(2)$$

$$12a = 4$$

$$a = \tfrac{1}{3}$$

$$b = 8a - 4 = -\tfrac{4}{3}$$

94. $f_1(x) = |\sin x|$

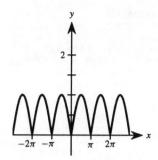

f_1 is not differentiable at $x = k\pi$ where k is any integer.

$f_2(x) = \sin |x|$

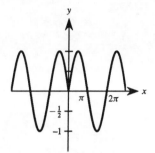

f_2 is not differentiable at $x = 0$.

95. Let $f(x) = \cos x$. Then,

$$f'(x) = \lim_{\Delta x \to 0} \frac{f(x + \Delta x) - f(x)}{\Delta x} = \lim_{\Delta x \to 0} \frac{\cos(x + \Delta x) - \cos x}{\Delta x}$$

$$= \lim_{\Delta x \to 0} \frac{\cos x \cos \Delta x - \sin x \sin \Delta x - \cos x}{\Delta x}$$

$$= \lim_{\Delta x \to 0} \left[\frac{\cos x (\cos \Delta x - 1)}{\Delta x} - \frac{\sin x \sin \Delta x}{\Delta x} \right]$$

$$= \lim_{\Delta x \to 0} \left[(\cos x) \left(\frac{\cos \Delta x - 1}{\Delta x} \right) - (\sin x) \left(\frac{\sin \Delta x}{\Delta x} \right) \right]$$

$$= (\cos x)(0) - (\sin x)(1)$$

$$= -\sin x.$$

Section 2.3   The Product and Quotient Rules and Higher-Order Derivatives

1. $f(x) = \frac{1}{3}(2x^3 - 4)$

$f'(x) = \frac{1}{3}(6x^2) = 2x^2$

$f'(0) = 0$

2. $f(x) = (x^2 - 2x + 1)(x^3 - 1)$

$f'(x) = (x^2 - 2x + 1)(3x^2) + (x^3 - 1)(2x - 2)$

$\qquad = 3x^2(x - 1)^2 + 2(x - 1)^2(x^2 + x + 1)$

$\qquad = (x - 1)^2(5x^2 + 2x + 2)$

$f'(1) = 0$

3. $f(x) = (x^3 - 3x)(2x^2 + 3x + 5)$

$f'(x) = (x^3 - 3x)(4x + 3) + (2x^2 + 3x + 5)(3x^2 - 3)$

$\qquad = 10x^4 + 12x^3 - 3x^2 - 18x - 15$

$f'(0) = -15$

4. $f(x) = \dfrac{x + 1}{x - 1}$

$f'(x) = \dfrac{(x - 1)(1) - (x + 1)(1)}{(x - 1)^2}$

$\qquad = \dfrac{x - 1 - x - 1}{(x - 1)^2}$

$\qquad = -\dfrac{2}{(x - 1)^2}$

$f'(2) = -\dfrac{2}{(2 - 1)^2} = -2$

5. $f(x) = x \cos x$

$f'(x) = (x)(-\sin x) + (\cos x)(1)$

$\qquad = \cos x - x \sin x$

$f'\left(\dfrac{\pi}{4}\right) = \dfrac{\sqrt{2}}{2} - \dfrac{\pi}{4}\left(\dfrac{\sqrt{2}}{2}\right)$

$\qquad = \dfrac{\sqrt{2}}{8}(4 - \pi)$

6. $f(x) = \dfrac{\sin x}{x}$

$f'(x) = \dfrac{(x)(\cos x) - (\sin x)(1)}{x^2}$

$\qquad = \dfrac{x \cos x - \sin x}{x^2}$

$f'\left(\dfrac{\pi}{6}\right) = \dfrac{(\pi/6)(\sqrt{3}/2) - (1/2)}{\pi^2/36}$

$\qquad = \dfrac{3\sqrt{3}\pi - 18}{\pi^2}$

$\qquad = \dfrac{3(\sqrt{3}\pi - 6)}{\pi^2}$

	Function	*Rewrite*	*Derivative*	*Simplify*
7.	$y = \dfrac{x^2 + 2x}{x}$	$y = x + 2$	$y' = 1$	$y' = 1$
8.	$y = \dfrac{4x^{3/2}}{x}$	$y = 4\sqrt{x}$	$y' = 2x^{-1/2}$	$y' = \dfrac{2}{\sqrt{x}}$
9.	$y = \dfrac{7}{3x^3}$	$y = \dfrac{7}{3}x^{-3}$	$y' = -7x^{-4}$	$y' = -\dfrac{7}{x^4}$
10.	$y = \dfrac{4}{5x^2}$	$y = \dfrac{4}{5}x^{-2}$	$y' = -\dfrac{8}{5}x^{-3}$	$y' = -\dfrac{8}{5x^3}$
11.	$y = \dfrac{3x^2 - 5}{7}$	$y = \dfrac{1}{7}(3x^2 - 5)$	$y' = \dfrac{1}{7}(6x)$	$y' = \dfrac{6x}{7}$
12.	$y = \dfrac{x^2 - 4}{x + 2}$	$y = x - 2$	$y' = 1$	$y' = 1$

13. $f(x) = \dfrac{3x - 2}{2x - 3}$

$f'(x) = \dfrac{(2x - 3)(3) - (3x - 2)(2)}{(2x - 3)^2}$

$\qquad = \dfrac{-5}{(2x - 3)^2}$

14. $f(x) = \dfrac{x^3 + 3x + 2}{x^2 - 1}$

$f'(x) = \dfrac{(x^2 - 1)(3x^2 + 3) - (x^3 + 3x + 2)(2x)}{(x^2 - 1)^2}$

$\qquad = \dfrac{x^4 - 6x^2 - 4x - 3}{(x^2 - 1)^2}$

15. $f(x) = \dfrac{3 - 2x - x^2}{x^2 - 1}$

$f'(x) = \dfrac{(x^2 - 1)(-2 - 2x) - (3 - 2x - x^2)(2x)}{(x^2 - 1)^2}$

$\qquad = \dfrac{2x^2 - 4x + 2}{(x^2 - 1)^2} = \dfrac{2(x - 1)^2}{(x^2 - 1)^2}$

$\qquad = \dfrac{2}{(x + 1)^2}, \quad x \ne 1$

16. $f(x) = x^4\left[1 - \dfrac{2}{x + 1}\right] = x^4\left[\dfrac{x - 1}{x + 1}\right]$

$f'(x) = x^4\left[\dfrac{(x + 1) - (x - 1)}{(x + 1)^2}\right] + \left[\dfrac{x - 1}{x + 1}\right](4x^3)$

$\qquad = 2x^3\left[\dfrac{2x^2 + x - 2}{(x + 1)^2}\right]$

17. $f(x) = \dfrac{x+1}{\sqrt{x}}$

$f'(x) = \dfrac{\sqrt{x}(1) - (x+1)[1/(2\sqrt{x})]}{x}$

$= \dfrac{x-1}{2x^{3/2}}$

Alternate solution:

$f(x) = \dfrac{x+1}{\sqrt{x}} = x^{1/2} + x^{-1/2}$

$f'(x) = \dfrac{1}{2}x^{-1/2} - \dfrac{1}{2}x^{-3/2}$

$= \dfrac{1}{2x^{1/2}} - \dfrac{1}{2x^{3/2}}$

18. $f(x) = \sqrt[3]{x}(\sqrt{x} + 3) = x^{1/3}(x^{1/2} + 3)$

$f'(x) = x^{1/3}\left(\dfrac{1}{2}x^{-1/2}\right) + (x^{1/2} + 3)\left(\dfrac{1}{3}x^{-2/3}\right)$

$= \dfrac{5}{6}x^{-1/6} + x^{-2/3}$

$= \dfrac{5}{6x^{1/6}} + \dfrac{1}{x^{2/3}}$

Alternate solution:

$f(x) = \sqrt[3]{x}(\sqrt{x} + 3)$

$= x^{5/6} + 3x^{1/3}$

$f'(x) = \dfrac{5}{6}x^{-1/6} + x^{-2/3}$

$= \dfrac{5}{6x^{1/6}} + \dfrac{1}{x^{2/3}}$

19. $h(s) = (s^3 - 2)^2 = s^6 - 4s^3 + 4$

$h'(s) = 6s^5 - 12s^2 = 6s^2(s^3 - 2)$

20. $h(x) = (x^2 - 1)^2 = x^4 - 2x^2 + 1$

$h'(x) = 4x^3 - 4x = 4x(x^2 - 1)$

21. $h(t) = \dfrac{t+1}{t^2 + 2t + 2}$

$h'(t) = \dfrac{(t^2 + 2t + 2)(1) - (t + 1)(2t + 2)}{(t^2 + 2t + 2)^2}$

$= \dfrac{-t^2 - 2t}{(t^2 + 2t + 2)^2}$

22. $f(x) = \dfrac{x(x^2 - 1)}{x + 3} = \dfrac{x^3 - x}{x + 3}$

$f'(x) = \dfrac{(x + 3)(3x^2 - 1) - (x^3 - x)}{(x + 3)^2}$

$= \dfrac{2x^3 + 9x^2 - 3}{(x + 3)^2}$

23. $f(x) = (3x^3 + 4x)(x - 5)(x + 1)$

$f'(x) = (9x^2 + 4)(x - 5)(x + 1) + (3x^3 + 4x)(1)(x + 1) + (3x^3 + 4x)(x - 5)(1)$

$= (9x^2 + 4)(x^2 - 4x - 5) + 3x^4 + 3x^3 + 4x^2 + 4x + 3x^4 - 15x^3 + 4x^2 - 20x$

$= 9x^4 - 36x^3 - 41x^2 - 16x - 20 + 6x^4 - 12x^3 + 8x^2 - 16x$

$= 15x^4 - 48x^3 - 33x^2 - 32x - 20$

24. $f(x) = (x^2 - x)(x^2 + 1)(x^2 + x + 1)$

$f'(x) = (2x - 1)(x^2 + 1)(x^2 + x + 1) + (x^2 - x)(2x)(x^2 + x + 1) + (x^2 - x)(x^2 + 1)(2x + 1)$

$= (2x - 1)(x^4 + x^3 + 2x^2 + x + 1) + (x^2 - x)(2x^3 + 2x^2 + 2x) + (x^2 - x)(2x^3 + x^2 + 2x + 1)$

$= 2x^5 + x^4 + 3x^3 + x - 1 + 2x^5 - 2x^2 + 2x^5 - x^4 + x^3 - x^2 - x$

$= 6x^5 + 4x^3 - 3x^2 - 1$

25. $g(x) = \left[\dfrac{x+1}{x+2}\right](2x - 5)$

$g'(x) = \left[\dfrac{x+1}{x+2}\right](2) + (2x - 5)\left[\dfrac{(x+2)(1) - (x+1)(1)}{(x+2)^2}\right] = \dfrac{2x^2 + 6x + 4 + 2x - 5}{(x+2)^2} = \dfrac{2x^2 + 8x - 1}{(x+2)^2}$

26. $f(x) = \left(\dfrac{x^2 - x - 3}{x^2 + 1} \right)(x^2 + x + 1)$

$f'(x) = \left(\dfrac{x^2 - x - 3}{x^2 + 1} \right)(2x + 1) + (x^2 + x + 1)\left[\dfrac{(x^2 + 1)(2x - 1) - (x^2 - x - 3)(2x)}{(x^2 + 1)^2} \right]$

$\quad = \dfrac{2x^3 - x^2 - 7x - 3}{x^2 + 1} + \dfrac{x^4 + 9x^3 + 8x^2 + 7x - 1}{(x^2 + 1)^2}$

$\quad = \dfrac{2x^5 + 4x^3 + 4x^2 - 4}{(x^2 + 1)^2}$

27. $f(x) = \dfrac{x^2 + c^2}{x^2 - c^2}$

$f'(x) = \dfrac{(x^2 - c^2)(2x) - (x^2 + c^2)(2x)}{(x^2 - c^2)^2}$

$\quad = \dfrac{-4xc^2}{(x^2 - c^2)^2}$

28. $f(x) = \dfrac{c^2 - x^2}{c^2 + x^2}$

$f'(x) = \dfrac{(c^2 + x^2)(-2x) - (c^2 - x^2)(2x)}{(c^2 + x^2)^2}$

$\quad = \dfrac{-4xc^2}{(c^2 + x^2)^2}$

29. $f(t) = t^2 \sin t$

$f'(t) = t^2 \cos t + 2t \sin t$

$\quad = t(t \cos t + 2 \sin t)$

30. $f(\theta) = (\theta + 1) \cos \theta$

$f'(\theta) = (\theta + 1)(-\sin \theta) + (\cos \theta)(1)$

$\quad = \cos \theta - (\theta + 1) \sin \theta$

31. $f(t) = \dfrac{\cos t}{t}$

$f'(t) = \dfrac{-t \sin t - \cos t}{t^2} = -\dfrac{t \sin t + \cos t}{t^2}$

32. $f(x) = \dfrac{\sin x}{x}$

$f'(x) = \dfrac{x \cos x - \sin x}{x^2}$

33. $f(x) = -x + \tan x$

$f'(x) = -1 + \sec^2 x = \tan^2 x$

34. $y = x + \cot x$

$y' = 1 - \csc^2 x = -\cot^2 x$

35. $g(t) = \sqrt{t} + 4 \sec t$

$g'(t) = \dfrac{1}{2} t^{-1/2} + 4 \sec t \tan t$

$\quad = \dfrac{1}{2\sqrt{t}} + 4 \sec t \tan t$

36. $h(s) = \dfrac{1}{s} - 10 \csc s$

$h'(s) = -\dfrac{1}{s^2} + 10 \csc s \cot s$

37. $y = 5x \csc x$

$y' = -5x \csc x \cot x + 5 \csc x$

$\quad = 5 \csc x(-x \cot x + 1)$

$\quad = 5 \csc x(1 - x \cot x)$

38. $y = \dfrac{\sec x}{x}$

$y' = \dfrac{x \sec x \tan x - \sec x}{x^2}$

$\quad = \dfrac{\sec x(x \tan x - 1)}{x^2}$

39. $y = -\csc x - \sin x$

$y' = \csc x \cot x - \cos x$

$\quad = \dfrac{\cos x}{\sin^2 x} - \cos x$

$\quad = \cos x(\csc^2 x - 1)$

$\quad = \cos x \cot^2 x$

40. $y = x \sin x + \cos x$

$y' = x \cos x + \sin x - \sin x = x \cos x$

41. $y = x^2 \sin x + 2x \cos x$

$y' = x^2 \cos x + 2x \sin x - 2x \sin x + 2 \cos x$

$\quad = x^2 \cos x + 2 \cos x$

42. $f(x) = \sin x \cos x$

$f'(x) = \sin x(-\sin x) + \cos x(\cos x)$

$\quad = \cos 2x$

43. $f(x) = x^2 \tan x$

$f'(x) = x^2 \sec^2 x + 2x \tan x$

$\quad = x(x \sec^2 x + 2 \tan x)$

44. $h(\theta) = 5 \sec \theta + \tan \theta$

$h'(\theta) = 5 \sec \theta \tan \theta + \sec^2 \theta$

$\quad = \sec \theta(5 \tan \theta + \sec \theta)$

45. $g(\theta) = \dfrac{\theta}{1 - \sin \theta}$

$g'(\theta) = \dfrac{(1 - \sin \theta)(1) - \theta(-\cos \theta)}{(1 - \sin \theta)^2}$

$\quad = \dfrac{1 - \sin \theta + \theta \cos \theta}{(1 - \sin \theta)^2}$

46. $f(\theta) = \dfrac{\sin \theta}{1 - \cos \theta}$

$f'(\theta) = \dfrac{(1 - \cos \theta) \cos \theta - \sin \theta(\sin \theta)}{(1 - \cos \theta)^2}$

$\quad = \dfrac{\cos \theta - 1}{(1 - \cos \theta)^2} = \dfrac{1}{\cos \theta - 1}$

47. $\quad y = \dfrac{1 + \csc x}{1 - \csc x}$

$y' = \dfrac{(1 - \csc x)(-\csc x \cot x) - (1 + \csc x)(\csc x \cot x)}{(1 - \csc x)^2} = \dfrac{-2 \csc x \cot x}{(1 - \csc x)^2}$

$y'\left(\dfrac{\pi}{6}\right) = \dfrac{-2(2)(\sqrt{3})}{(1 - 2)^2} = -4\sqrt{3}$

48. $f(x) = \tan x \cot x = 1$

$f'(x) = 0$

$f'(1) = 0$

49. $h(t) = \dfrac{\sec t}{t}$

$h'(t) = \dfrac{t(\sec t \tan t) - (\sec t)(1)}{t^2}$

$\quad = \dfrac{\sec t(t \tan t - 1)}{t^2}$

$h'(\pi) = \dfrac{\sec \pi(\pi \tan \pi - 1)}{\pi^2} = \dfrac{1}{\pi^2}$

50. $\quad f(x) = \sin x(\sin x + \cos x)$

$f'(x) = \sin x(\cos x - \sin x) + (\sin x + \cos x) \cos x$

$\quad = \sin x \cos x - \sin^2 x + \sin x \cos x + \cos^2 x$

$\quad = \sin 2x + \cos 2x$

$f'\left(\dfrac{\pi}{4}\right) = \sin \dfrac{\pi}{2} + \cos \dfrac{\pi}{2} = 1$

51. $f(x) = \dfrac{x}{x - 1}, \quad (2, 2)$

$f'(x) = \dfrac{(x - 1)(1) - x(1)}{(x - 1)^2} = \dfrac{-1}{(x - 1)^2}$

$f'(2) = \dfrac{-1}{(2 - 1)^2} = -1 = \text{ slope at } (2, 2)$

Tangent line:

$\quad y - 2 = -1(x - 2) \ \Rightarrow \ y = -x + 4$

52. $f(x) = (x - 1)(x^2 - 2), \quad (0, 2)$

$f'(x) = (x - 1)(2x) + (x^2 - 2)(1) = 3x^2 - 2x - 2$

$f'(0) = -2 = \text{ slope at } (0, 2)$

Tangent line:

$\quad y - 2 = -2x \ \Rightarrow \ y = -2x + 2$

53. $f(x) = (x^3 - 3x + 1)(x + 2)$, $(1, -3)$

$f'(x) = (x^3 - 3x + 1)(1) + (x + 2)(3x^2 - 3)$

$\qquad = 4x^3 + 6x^2 - 6x - 5$

$f'(1) = -1 = $ slope at $(1, -3)$

Tangent line:

$\qquad y + 3 = -1(x - 1) \implies y = -x - 2$

54. $f(x) = \dfrac{x - 1}{x + 1}$, $\left(2, \dfrac{1}{3}\right)$

$f'(x) = \dfrac{(x + 1)(1) - (x - 1)(1)}{(x + 1)^2} = \dfrac{2}{(x + 1)^2}$

$f'(2) = \dfrac{2}{9} = $ slope at $\left(2, \dfrac{1}{3}\right)$

Tangent line:

$\qquad y - \dfrac{1}{3} = \dfrac{2}{9}(x - 2) \implies y = \dfrac{2}{9}x - \dfrac{1}{9}$

55. $f(x) = \tan x$, $\left(\dfrac{\pi}{4}, 1\right)$

$f'(x) = \sec^2 x$

$f'\left(\dfrac{\pi}{4}\right) = 2 = $ slope at $\left(\dfrac{\pi}{4}, 1\right)$

Tangent line:

$\qquad y - 1 = 2\left(x - \dfrac{\pi}{4}\right)$

$\qquad y - 1 = 2x - \dfrac{\pi}{2}$

$\qquad 4x - 2y - \pi + 2 = 0$

56. $f(x) = \sec x$, $\left(\dfrac{\pi}{3}, 2\right)$

$f'(x) = \sec x \tan x$

$f'\left(\dfrac{\pi}{3}\right) = 2\sqrt{3} = $ slope at $\left(\dfrac{\pi}{3}, 2\right)$

Tangent line:

$\qquad y - 2 = 2\sqrt{3}\left(x - \dfrac{\pi}{3}\right)$

$\qquad 6\sqrt{3}x - 3y + 6 - 2\sqrt{3}\pi = 0$

57. $f(x) = \dfrac{x^2}{x - 1}$

$f'(x) = \dfrac{(x - 1)(2x) - x^2(1)}{(x - 1)^2}$

$\qquad = \dfrac{x^2 - 2x}{(x - 1)^2} = \dfrac{x(x - 2)}{(x - 1)^2}$

$f'(x) = 0$ when $x = 0$ or $x = 2$

Horizontal tangents are at $(0, 0)$ and $(2, 4)$.

58. $f(x) = \dfrac{x^2}{x^2 + 1}$

$f'(x) = \dfrac{(x^2 + 1)(2x) - (x^2)(2x)}{(x^2 + 1)^2} = \dfrac{2x}{(x^2 + 1)^2}$

$f'(x) = 0$ when $x = 0$

Horizontal tangent is at $(0, 0)$.

59. $f(x) = 2g(x) + h(x)$

$f'(x) = 2g'(x) + h'(x)$

$f'(2) = 2g'(2) + h'(2)$

$\qquad = 2(-2) + 4$

$\qquad = 0$

60. $f(x) = 4 - h(x)$

$f'(x) = -h'(x)$

$f'(2) = -h'(2) = -4$

61. $f(x) = \dfrac{g(x)}{h(x)}$

$f'(x) = \dfrac{h(x)g'(x) - g(x)h'(x)}{[h(x)]^2}$

$f'(2) = \dfrac{h(2)g'(2) - g(2)h'(2)}{[h(2)]^2}$

$\qquad = \dfrac{(-1)(-2) - (3)(4)}{(-1)^2}$

$\qquad = -10$

62. $f(x) = g(x)h(x)$

$f'(x) = g(x)h'(x) + h(x)g'(x)$

$f'(2) = g(2)h'(2) + h(2)g'(2)$

$\qquad = (3)(4) + (-1)(-2)$

$\qquad = 14$

63. $f(x) = x^n \sin x$

$f'(x) = x^n \cos x + nx^{n-1} \sin x$

$\qquad = x^{n-1}(x \cos x + n \sin x)$

When $n = 1$: $f'(x) = x \cos x + \sin x$.
When $n = 2$: $f'(x) = x(x \cos x + 2 \sin x)$.
When $n = 3$: $f'(x) = x^2(x \cos x + 3 \sin x)$.
When $n = 4$: $f'(x) = x^3(x \cos x + 4 \sin x)$.

64. $f(x) = \dfrac{\cos x}{x^n} = x^{-n} \cos x$

$f'(x) = -x^{-n} \sin x - nx^{-n-1} \cos x$

$\qquad = -x^{-n-1}(x \sin x + n \cos x)$

$\qquad = -\dfrac{x \sin x + n \cos x}{x^{n+1}}$

When $n = 1$: $f'(x) = -\dfrac{x \sin x + \cos x}{x^2}$.

When $n = 2$: $f'(x) = -\dfrac{x \sin x + 2 \cos x}{x^3}$.

When $n = 3$: $f'(x) = -\dfrac{x \sin x + 3 \cos x}{x^4}$.

When $n = 4$: $f'(x) = -\dfrac{x \sin x + 4 \cos x}{x^5}$.

65. $C = 100 \left(\dfrac{200}{x^2} + \dfrac{x}{x+30} \right), \quad 1 \le x$

$\dfrac{dC}{dx} = 100 \left(-\dfrac{400}{x^3} + \dfrac{30}{(x+30)^2} \right)$

a. When $x = 10$: $\dfrac{dC}{dx} = -\$38.13$.

b. When $x = 15$: $\dfrac{dC}{dx} = -\$10.37$.

c. When $x = 20$: $\dfrac{dC}{dx} = -\$3.80$.

As the order size increases, the cost decreases.

66. $P = \dfrac{k}{V}$

$\dfrac{dP}{dV} = -\dfrac{k}{V^2}$

67. $P(t) = 500 \left[1 + \dfrac{4t}{50 + t^2} \right]$

$P'(t) = 500 \left[\dfrac{(50 + t^2)(4) - (4t)(2t)}{(50 + t^2)^2} \right] = 500 \left[\dfrac{200 - 4t^2}{(50 + t^2)^2} \right] = 2000 \left[\dfrac{50 - t^2}{(50 + t^2)^2} \right]$

$P'(2) \approx 31.55$

68. $f(x) = \sec x$

$g(x) = \csc x, \quad [0, \, 2\pi)$

$f'(x) = g'(x)$

$\sec x \tan x = -\csc x \cot x \ \Rightarrow \ \dfrac{\sec x \tan x}{\csc x \cot x} = -1 \ \Rightarrow \ \dfrac{\dfrac{1}{\cos x} \cdot \dfrac{\sin x}{\cos x}}{\dfrac{1}{\sin x} \cdot \dfrac{\cos x}{\sin x}} = -1 \ \Rightarrow$

$\dfrac{\sin^3 x}{\cos^3 x} = -1 \ \Rightarrow \ \tan^3 x = -1 \ \Rightarrow \ \tan x = -1$

$x = \dfrac{3\pi}{4}, \ \dfrac{7\pi}{4}$

69. a. $\sec x = \dfrac{1}{\cos x}$

$$\frac{d}{dx}[\sec x] = \frac{d}{dx}\left[\frac{1}{\cos x}\right] = \frac{(\cos x)(0) - (1)(-\sin x)}{(\cos x)^2} = \frac{\sin x}{\cos x \cos x} = \frac{1}{\cos x} \cdot \frac{\sin x}{\cos x} = \sec x \tan x$$

b. $\csc x = \dfrac{1}{\sin x}$

$$\frac{d}{dx}[\csc x] = \frac{d}{dx}\left[\frac{1}{\sin x}\right] = \frac{(\sin x)(0) - (1)(\cos x)}{(\sin x)^2} = -\frac{\cos x}{\sin x \sin x} = -\frac{1}{\sin x} \cdot \frac{\cos x}{\sin x} = -\csc x \cot x$$

c. $\cot x = \dfrac{\cos x}{\sin x}$

$$\frac{d}{dx}[\cot x] = \frac{d}{dx}\left[\frac{\cos x}{\sin x}\right] = \frac{\sin x(-\sin x) - (\cos x)(\cos x)}{(\sin x)^2} = -\frac{\sin^2 x + \cos^2 x}{\sin^2 x} = -\frac{1}{\sin^2 x} = -\csc^2 x$$

70. The graph of a differentiable function f such that $f > 0$ and $f' < 0$ for all real numbers x would in general look like the graph at the right. An example for this case is

$$f(x) = \frac{1}{\sqrt{x+4}}.$$

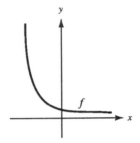

71. $f(x) = 4x^{3/2}$

$f'(x) = 6x^{1/2}$

$f''(x) = 3x^{-1/2} = \dfrac{3}{\sqrt{x}}$

72. $f(x) = \dfrac{x^2 + 2x - 1}{x} = x + 2 - \dfrac{1}{x}$

$f'(x) = 1 + \dfrac{1}{x^2}$

$f''(x) = -\dfrac{2}{x^3}$

73. $f(x) = \dfrac{x}{x-1}$

$f'(x) = \dfrac{(x-1)(1) - x(1)}{(x-1)^2} = \dfrac{-1}{(x-1)^2}$

$f''(x) = \dfrac{2}{(x-1)^3}$

74. $f(x) = x + \dfrac{32}{x^2}$

$f'(x) = 1 - \dfrac{64}{x^3}$

$f''(x) = \dfrac{192}{x^4}$

75. $f(x) = 3 \sin x$

$f'(x) = 3 \cos x$

$f''(x) = -3 \sin x$

76. $f(x) = \sec x$

$f'(x) = \sec x \tan x$

$f''(x) = \sec x(\sec^2 x) + \tan x(\sec x \tan x)$

$ = \sec x(\sec^2 x + \tan^2 x)$

77. $f'(x) = x^2$

$f''(x) = 2x$

78. $f''(x) = 2 - 2x^{-1}$

$f'''(x) = 2x^{-2} = \dfrac{2}{x^2}$

79. $f'''(x) = 2\sqrt{x}$

$f^{(4)}(x) = \dfrac{1}{2}(2)x^{-1/2} = \dfrac{1}{\sqrt{x}}$

80. $f^{(4)}(x) = 2x + 1$

$f^{(5)}(x) = 2$

$f^{(6)}(x) = 0$

81. $f(x) = g(x)h(x)$

a. $f'(x) = g(x)h'(x) + h(x)g'(x)$

$$f''(x) = g(x)h''(x) + g'(x)h'(x) + h(x)g''(x) + h'(x)g'(x)$$

$$= g(x)h''(x) + 2g'(x)h'(x) + h(x)g''(x)$$

$$f'''(x) = g(x)h'''(x) + g'(x)h''(x) + 2g'(x)h''(x) + 2g''(x)h'(x) + h(x)g'''(x) + h'(x)g''(x)$$

$$= g(x)h'''(x) + 3g'(x)h''(x) + 3g''(x)h'(x) + g'''(x)h(x)$$

$$f^{(4)}(x) = g(x)h^{(4)}(x) + g'(x)h'''(x) + 3g'(x)h'''(x) + 3g''(x)h''(x) + 3g''(x)h''(x) + 3g'''(x)h'(x)$$

$$+ g'''(x)h'(x) + g^{(4)}(x)h(x)$$

$$= g(x)h^{(4)}(x) + 4g'(x)h'''(x) + 6g''(x)h''(x) + 4g'''(x)h'(x) + g^{(4)}(x)h(x)$$

b. $f^{(n)}(x) = g(x)h^{(n)}(x) + \dfrac{n(n-1)(n-2)\cdots(2)(1)}{1[(n-1)(n-2)\cdots(2)(1)]} g'(x)h^{(n-1)}(x)$

$$+ \dfrac{n(n-1)(n-2)\cdots(2)(1)}{(2)(1)[(n-2)(n-3)\cdots(2)(1)]} g''(x)h^{(n-2)}(x)$$

$$+ \dfrac{n(n-1)(n-2)\cdots(2)(1)}{(3)(2)(1)[(n-3)(n-4)\cdots(2)(1)]} g'''(x)h^{(n-3)}(x) + \cdots$$

$$+ \dfrac{n(n-1)(n-2)\cdots(2)(1)}{[(n-1)(n-2)\cdots(2)(1)](1)} g^{(n-1)}(x)h'(x) + g^{(n)}(x)h(x)$$

$$= g(x)h^{(n)}(x) + \dfrac{n!}{1!(n-1)!} g'(x)h^{(n-1)}(x) + \dfrac{n!}{2!(n-2)!} g''(x)h^{(n-2)}(x) + \cdots$$

$$+ \dfrac{n!}{(n-1)!1!} g^{(n-1)}(x)h'(x) + g^{(n)}(x)h(x)$$

Note: For a definition of $n!$ (read "n factorial"), see Section 8.1 of the text.

82. a. $f(x) = x^n$

$$f^n(x) = n(n-1)(n-2)\cdots(2)(1) = n!$$

b. $f(x) = \dfrac{1}{x}$

$$f^{(n)}(x) = \dfrac{(-1)^n (n)(n-1)(n-2)\cdots(2)(1)}{x^{n+1}}$$

$$= \dfrac{(-1)^n n!}{x^{n+1}}$$

Note: For a definition of $n!$ (read "n factorial"), see Section 8.1 of the text.

83. $v(t) = 36 - t^2, \quad 0 \le t \le 6$

$a(t) = -2t$

$v(3) = 27$ m/sec

$a(3) = -6$ m/sec^2

The speed of the object is decreasing, but the rate of that decrease is increasing.

84. $v(t) = \dfrac{100t}{2t + 15}$

$$a(t) = \dfrac{(2t + 15)(100) - (100t)(2)}{(2t + 15)^2}$$

$$= \dfrac{1500}{(2t + 15)^2}$$

a. $a(5) = \dfrac{1500}{[2(5) + 15]^2} = 2.4$ ft/sec^2

b. $a(10) = \dfrac{1500}{[2(10) + 15]^2} \approx 1.2$ ft/sec^2

c. $a(20) = \dfrac{1500}{[2(20) + 15]^2} \approx 0.5$ ft/sec^2

85. $s(t) = -8.25t^2 + 66t$

$v(t) = -16.50t + 66$

$a(t) = -16.50$

t (sec)	0	1	2	3	4
$s(t)$ (ft)	0	57.75	99	123.75	132
$v(t) = s'(t)$ (ft/sec)	66	49.5	33	16.5	0
$a(t) = v'(t)$ (ft/sec^2)	-16.5	-16.5	-16.5	-16.5	-16.5

Average velocity on:

[0, 1] is $\dfrac{57.75 - 0}{1 - 0} = 57.75.$

[1, 2] is $\dfrac{99 - 57.75}{2 - 1} = 41.25.$

[2, 3] is $\dfrac{123.75 - 99}{3 - 2} = 24.75.$

[3, 4] is $\dfrac{132 - 123.75}{4 - 3} = 8.25.$

86. $s = -\frac{27}{10}t^2 + 27t + 6$

a. $v(t) = -\frac{27}{5}t + 27$

$a(t) = -\frac{27}{5} = -5.4$ ft/sec^2

b. $-\frac{27}{5}t + 27 = 0$ when $t = 5$ seconds.

$s(5) = 73.5$ feet

c. On earth, $a = -32$ ft/sec^2.

87. False. If $y = f(x)g(x)$, then

$\dfrac{dy}{dx} = f(x)g'(x) + g(x)f'(x).$

88. True. y is a fourth-degree polynomial.

$\dfrac{d^n y}{dx^n} = 0$ when $n > 4.$

89. True

$h'(c) = f(c)g'(c) + g(c)f'(c)$

$\quad = f(c)(0) + g(c)(0)$

$\quad = 0$

90. True

91. True

92. True. If $v(t) = c$ then $a(t) = v'(t) = 0.$

93. $f(x) = x|x| = \begin{cases} x^2, & \text{if } x \geq 0 \\ -x^2, & \text{if } x < 0 \end{cases}$

$f'(x) = \begin{cases} 2x, & \text{if } x \geq 0 \\ -2x, & \text{if } x < 0 \end{cases}$

$f''(x) = \begin{cases} 2, & \text{if } x > 0 \\ -2, & \text{if } x < 0 \end{cases}$

$f''(0)$ does not exist since the left and right derivatives are not equal.

94. a. $(fg' - f'g)' = fg'' + f'g' - f'g' - f''g$

$\qquad\qquad\qquad = fg'' - f''g \qquad$ True

b. $(fg)'' = (fg' + f'g)'$

$\qquad\quad = fg'' + f'g' + f'g' + f''g$

$\qquad\quad = fg'' + 2f'g' + f''g$

$\qquad\quad \neq fg'' + f''g \qquad$ False

95.

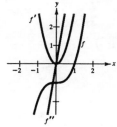

It appears that f is cubic; so f' would be quadratic and f'' would be linear.

96.

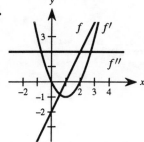

It appears that f is quadratic; so f' would be linear and f'' would be constant.

Section 2.4 The Chain Rule

$y = f(g(x))$	$u = g(x)$	$y = f(u)$
1. $y = (6x - 5)^4$	$u = 6x - 5$	$y = u^4$
2. $y = \dfrac{1}{\sqrt{x+1}}$	$u = x + 1$	$y = u^{-1/2}$
3. $y = \sqrt{x^2 - 1}$	$u = x^2 - 1$	$y = \sqrt{u}$
4. $y = \tan(\pi x + 1)$	$u = \pi x + 1$	$y = \tan u$
5. $y = \csc^3 x$	$u = \csc x$	$y = u^3$
6. $y = \cos \dfrac{3x}{2}$	$u = \dfrac{3x}{2}$	$y = \cos u$

7. $y = (2x - 7)^3$

$y' = 3(2x - 7)^2(2) = 6(2x - 7)^2$

8. $y = (3x^2 + 1)^4$

$y' = 4(3x^2 + 1)^3(6x) = 24x(3x^2 + 1)^3$

9. $g(x) = 3(4 - 9x)^4$

$g'(x) = 12(4 - 9x)^3(-9) = -108(4 - 9x)^3$

10. $f(x) = 2(1 - x^2)^3$

$f'(x) = 6(1 - x^2)^2(-2x) = -12x(1 - x^2)^2$

11. $f(x) = (9 - x^2)^{2/3}$

$f'(x) = \dfrac{2}{3}(9 - x^2)^{-1/3}(-2x) = -\dfrac{4x}{3(9 - x^2)^{1/3}}$

12. $f(t) = (9t + 2)^{2/3}$

$f'(t) = \dfrac{2}{3}(9t + 2)^{-1/3}(9) = \dfrac{6}{\sqrt[3]{9t + 2}}$

13. $f(t) = (1 - t)^{1/2}$

$f'(t) = \dfrac{1}{2}(1 - t)^{-1/2}(-1) = -\dfrac{1}{2\sqrt{1 - t}}$

14. $g(x) = (3 - 2x)^{1/2}$

$g'(x) = \dfrac{1}{2}(3 - 2x)^{-1/2}(-2) = -\dfrac{1}{\sqrt{3 - 2x}}$

15. $y = (9x^2 + 4)^{1/3}$

$y' = \dfrac{1}{3}(9x^2 + 4)^{-2/3}(18x) = \dfrac{6x}{(9x^2 + 4)^{2/3}}$

16. $g(x) = \sqrt{x^2 - 2x + 1} = \sqrt{(x - 1)^2} = |x - 1|$

$g'(x) = \begin{cases} 1, & x > 1 \\ -1, & x < 1 \end{cases}$

17. $y = 2(4 - x^2)^{1/2}$

$y' = (4 - x^2)^{-1/2}(-2x) = -\dfrac{2x}{\sqrt{4 - x^2}}$

18. $f(x) = -3(2 - 9x)^{1/4}$

$f'(x) = -\dfrac{3}{4}(2 - 9x)^{-3/4}(-9) = \dfrac{27}{4(2 - 9x)^{3/4}}$

19. $y = (x - 2)^{-1}$

$$y' = -1(x-2)^{-2}(1) = \frac{-1}{(x-2)^2}$$

20. $s(t) = (t^2 + 3t - 1)^{-1}$

$$s'(t) = -1(t^2 + 3t - 1)^{-2}(2t + 3)$$
$$= \frac{-(2t+3)}{(t^2 + 3t - 1)^2}$$

21. $f(t) = (t - 3)^{-2}$

$$f'(t) = -2(t-3)^{-3} = \frac{-2}{(t-3)^3}$$

22. $y = -4(t + 2)^{-2}$

$$y' = 8(t+2)^{-3} = \frac{8}{(t+2)^3}$$

23. $y = (x + 2)^{-1/2}$

$$\frac{dy}{dx} = -\frac{1}{2}(x+2)^{-3/2} = -\frac{1}{2(x+2)^{3/2}}$$

24. $g(t) = (t^2 - 2)^{-1/2}$

$$g'(t) = -\frac{1}{2}(t^2 - 2)^{-3/2}(2t) = -\frac{t}{(t^2 - 2)^{3/2}}$$

25. $f(x) = x^2(x - 2)^4$

$$f'(x) = x^2[4(x-2)^3(1)] + (x-2)^4(2x)$$
$$= 2x(x-2)^3[2x + (x-2)]$$
$$= 2x(x-2)^3(3x - 2)$$

26. $f(x) = x(3x - 9)^3$

$$f'(x) = x[3(3x-9)^2(3)] + (3x-9)^3(1)$$
$$= (3x-9)^2[9x + 3x - 9]$$
$$= 27(x-3)^2(4x - 3)$$

27. $y = x\sqrt{1 - x^2} = x(1 - x^2)^{1/2}$

$$y' = x\left[\frac{1}{2}(1-x^2)^{-1/2}(-2x)\right] + (1-x^2)^{1/2}(1)$$
$$= -x^2(1-x^2)^{-1/2} + (1-x^2)^{1/2}$$
$$= (1-x^2)^{-1/2}[-x^2 + (1-x^2)]$$
$$= \frac{1 - 2x^2}{\sqrt{1-x^2}}$$

28. $y = x^2\sqrt{9 - x^2} = x^2(9 - x^2)^{1/2}$

$$y' = x^2\left[\frac{1}{2}(9-x^2)^{-1/2}(-2x)\right] + (9-x^2)^{1/2}(2x)$$
$$= -x^3(9-x^2)^{-1/2} + 2x(9-x^2)^{1/2}$$
$$= x(9-x^2)^{-1/2}[-x^2 + 2(9-x^2)]$$
$$= \frac{x(18 - 3x^2)}{\sqrt{9-x^2}}$$
$$= \frac{3x(6 - x^2)}{\sqrt{9-x^2}}$$

29. $y = \dfrac{x}{\sqrt{x^2 + 1}} = x(x^2 + 1)^{-1/2}$

$$y' = x\left[-\frac{1}{2}(x^2+1)^{-3/2}(2x)\right] + (x^2+1)^{-1/2}(1)$$
$$= -x^2(x^2+1)^{-3/2} + (x^2+1)^{-1/2}$$
$$= (x^2+1)^{-3/2}[-x^2 + (x^2 + 1)]$$
$$= \frac{1}{(x^2+1)^{3/2}}$$

30. $y = \sqrt{x + \sqrt{x}} = (x + x^{1/2})^{1/2}$

$$y' = \frac{1}{2}(x + x^{1/2})^{-1/2}\left(1 + \frac{1}{2}x^{-1/2}\right)$$
$$= \frac{1 + [1/(2\sqrt{x})]}{2\sqrt{x + \sqrt{x}}}$$
$$= \frac{2\sqrt{x} + 1}{4\sqrt{x}\sqrt{x + \sqrt{x}}}$$

31. $y = \dfrac{\sqrt{x}+1}{x^2+1}$

$y' = \dfrac{1-3x^2-4x^{3/2}}{2\sqrt{x}(x^2+1)^2}$

The zero of y' corresponds to the point on the graph of y where the tangent line is horizontal.

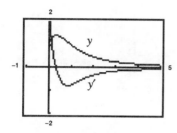

32. $y = \sqrt{\dfrac{2x}{x+1}}$

$y' = \dfrac{1}{\sqrt{2x}(x+1)^{3/2}}$

y' has no zeros.

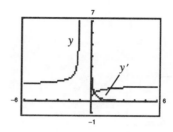

33. $g(t) = \dfrac{3t^2}{\sqrt{t^2+2t-1}}$

$g'(t) = \dfrac{3t(t^2+3t-2)}{(t^2+2t-1)^{3/2}}$

The zeros of g' correspond to the points on the graph of g where the tangent lines are horizontal.

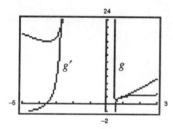

34. $f(x) = \sqrt{x}(2-x)^2$

$f'(x) = \dfrac{(x-2)(5x-2)}{2\sqrt{x}}$

The zeros of f' correspond to the points on the graph of f where the tangent lines are horizontal.

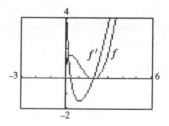

35. $y = \sqrt{\dfrac{x+1}{x}}$

$y' = -\dfrac{\sqrt{(x+1)/x}}{2x(x+1)}$

y' has no zeros.

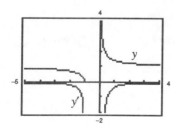

36. $y = (t^2-9)\sqrt{t+2}$

$y' = \dfrac{5t^2+8t-9}{2\sqrt{t+2}}$

The zero of y' corresponds to the point on the graph of y where the tangent line is horizontal.

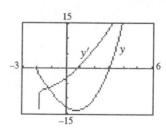

37. $s(t) = \dfrac{-2(2-t)\sqrt{1+t}}{3}$

$s'(t) = \dfrac{t}{\sqrt{1+t}}$

The zero of $s'(t)$ corresponds to the point on the graph of $s(t)$ where the tangent line is horizontal.

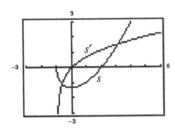

38. $g(x) = \sqrt{x-1} + \sqrt{x+1}$

$g'(x) = \dfrac{1}{2\sqrt{x-1}} + \dfrac{1}{2\sqrt{x+1}}$

g' has no zeros.

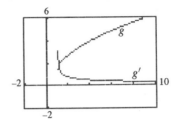

39. a. $y = \sin x$

$y' = \cos x$

$y'(0) = 1$

1 cycle in $[0,\ 2\pi]$

b. $y = \sin 2x$

$y' = 2\cos 2x$

$y'(0) = 2$

2 cycles in $[0,\ 2\pi]$

40. a. $y = \sin 3x$

$y' = 3\cos 3x$

$y'(0) = 3$

3 cycles in $[0,\ 2\pi]$

b. $y = \sin\left(\dfrac{x}{2}\right)$

$y' = \left(\dfrac{1}{2}\right)\cos\left(\dfrac{x}{2}\right)$

$y'(0) = \dfrac{1}{2}$

Half cycle in $[0,\ 2\pi]$

41. $y = \cos 3x$

$\dfrac{dy}{dx} = -3\sin 3x$

42. $y = \sin \pi x$

$\dfrac{dy}{dx} = \pi \cos \pi x$

43. $g(x) = 3\tan 4x$

$g'(x) = 12\sec^2 4x$

44. $h(x) = \sec(x^2)$

$h'(x) = 2x\sec(x^2)\tan(x^2)$

45. $f(\theta) = \tfrac{1}{4}\sin^2 2\theta = \tfrac{1}{4}(\sin 2\theta)^2$

$f'(\theta) = 2\left(\tfrac{1}{4}\right)(\sin 2\theta)\cos 2\theta(2)$

$= \sin 2\theta \cos 2\theta = \tfrac{1}{2}\sin 4\theta$

46. $g(t) = 5\cos^2 \pi t = 5(\cos \pi t)^2$

$g'(t) = 10\cos \pi t(-\sin \pi t)(\pi)$

$= -10\pi(\sin \pi t)(\cos \pi t) = -5\pi \sin 2\pi t$

47. $y = \sqrt{x} + \dfrac{1}{4}\sin(2x)^2$

$= \sqrt{x} + \dfrac{1}{4}\sin(4x^2)$

$\dfrac{dy}{dx} = \dfrac{1}{2}x^{-1/2} + \dfrac{1}{4}\cos(4x^2)(8x)$

$= \dfrac{1}{2\sqrt{x}} + 2x\cos(2x)^2$

48. $y = 3x - 5\cos(\pi x)^2$

$= 3x - 5\cos(\pi^2 x^2)$

$\dfrac{dy}{dx} = 3 + 5\sin(\pi^2 x^2)(2\pi^2 x)$

$= 3 + 10\pi^2 x \sin(\pi x)^2$

49. $y = \sin(\cos x)$

$\dfrac{dy}{dx} = \cos(\cos x) \cdot (-\sin x)$

$\qquad = -\sin x \cos(\cos x)$

50. $y = \sin \sqrt{x} + \sqrt{\sin x}$

$\dfrac{dy}{dx} = \cos \sqrt{x} \cdot \dfrac{1}{2} x^{-1/2} + \dfrac{1}{2} (\sin x)^{-1/2} \cdot \cos x$

$\qquad = \dfrac{\cos \sqrt{x}}{2\sqrt{x}} + \dfrac{\cos x}{2\sqrt{\sin x}}$

51. $y = \dfrac{\cos \pi x + 1}{x}$

$\dfrac{dy}{dx} = \dfrac{-\pi x \sin \pi x - \cos \pi x - 1}{x^2}$

$\qquad = -\dfrac{\pi x \sin \pi x + \cos \pi x + 1}{x^2}$

The zeros of y' correspond to the points on the graph of y where the tangent lines are horizontal.

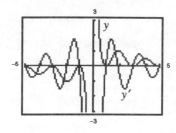

52. $y = x^2 \tan \dfrac{1}{x}$

$\dfrac{dy}{dx} = 2x \tan \dfrac{1}{x} - \sec^2 \dfrac{1}{x}$

The zeros of y' correspond to the points on the graph of y where the tangent lines are horizontal.

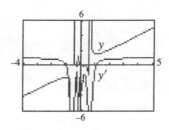

53. $s(t) = (t^2 + 2t + 8)^{1/2}, \quad (2,\ 4)$

$s'(t) = \dfrac{1}{2}(t^2 + 2t + 8)^{-1/2}(2t + 2)$

$\qquad = \dfrac{t + 1}{\sqrt{t^2 + 2t + 8}}$

$s'(2) = \dfrac{3}{4}$

54. $y = (3x^3 + 4x)^{1/5}, \quad (2,\ 2)$

$y' = \dfrac{1}{5}(3x^3 + 4x)^{-4/5}(9x^2 + 4)$

$\quad = \dfrac{9x^2 + 4}{5(3x^3 + 4x)^{4/5}}$

$y'(2) = \dfrac{1}{2}$

55. $f(x) = \dfrac{3}{x^3 - 4} = 3(x^3 - 4)^{-1}, \quad \left(-1,\ -\dfrac{3}{5}\right)$

$f'(x) = -3(x^3 - 4)^{-2}(3x^2) = -\dfrac{9x^2}{(x^3 - 4)^2}$

$f'(-1) = -\dfrac{9}{25}$

56. $f(x) = \dfrac{1}{(x^2 - 3x)^2} = (x^2 - 3x)^{-2}, \quad \left(4,\ \dfrac{1}{16}\right)$

$f'(x) = -2(x^2 - 3x)^{-3}(2x - 3) = \dfrac{-2(2x - 3)}{(x^2 - 3x)^3}$

$f'(4) = -\dfrac{5}{32}$

57. $f(t) = \dfrac{3t + 2}{t - 1}, \quad (0,\ -2)$

$f'(t) = \dfrac{(t - 1)(3) - (3t + 2)(1)}{(t - 1)^2} = \dfrac{-5}{(t - 1)^2}$

$f'(0) = -5$

58. $f(x) = \dfrac{x + 1}{2x - 3}, \quad (2,\ 3)$

$f'(x) = \dfrac{(2x - 3)(1) - (x + 1)(2)}{(2x - 3)^2} = \dfrac{-5}{(2x - 3)^2}$

$f'(2) = -5$

59. $y = 37 - \sec^3(2x),$ $(0,\ 36)$

$y' = -3\sec^2(2x)[2\sec(2x)\tan(2x)]$

$\quad = -6\sec^3(2x)\tan(2x)$

$y'(0) = 0$

60. $y = \dfrac{1}{x} + \sqrt{\cos x},$ $\left(\dfrac{\pi}{2},\ \dfrac{2}{\pi}\right)$

$y' = -\dfrac{1}{x^2} - \dfrac{\sin x}{2\sqrt{\cos x}}$

$y'\left(\dfrac{\pi}{2}\right)$ is undefined.

61. $f(x) = \sqrt{3x^2 - 2},$ $(3,\ 5)$

$f'(x) = \dfrac{1}{2}(3x^2 - 2)^{-1/2}(6x)$

$\quad = \dfrac{3x}{\sqrt{3x^2 - 2}}$

$f'(3) = \dfrac{9}{5}$

Tangent line:

$y - 5 = \dfrac{9}{5}(x - 3) \Rightarrow 9x - 5y - 2 = 0$

62. $f(x) = x\sqrt{x^2 + 5},$ $(2,\ 6)$

$f'(x) = x\left[\dfrac{1}{2}(x^2 + 5)^{-1/2}(2x)\right] + (x^2 + 5)^{1/2}$

$\quad = \dfrac{x^2}{\sqrt{x^2 + 5}} + \sqrt{x^2 + 5} = \dfrac{2x^2 + 5}{\sqrt{x^2 + 5}}$

$f'(2) = \dfrac{13}{3}$

Tangent line:

$y - 6 = \dfrac{13}{3}(x - 2) \Rightarrow 13x - 3y - 8 = 0$

63. $f(x) = \sin 2x,$ $(\pi,\ 0)$

$f'(x) = 2\cos 2x$

$f'(\pi) = 2$

Tangent line:

$y = 2(x - \pi) \Rightarrow 2x - y - 2\pi = 0$

64. $f(x) = \tan^2 x,$ $\left(\dfrac{\pi}{4},\ 1\right)$

$f'(x) = 2\tan x \sec^2 x$

$f'\left(\dfrac{\pi}{4}\right) = 2(1)(2) = 4$

Tangent line:

$y - 1 = 4\left(x - \dfrac{\pi}{4}\right) \Rightarrow 4x - y + (1 - \pi) = 0$

65.
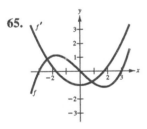

The zeros of f' correspond to the points where the graph of f has horizontal tangents.

66.
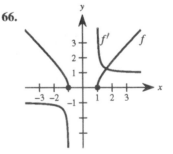

f is decreasing on $(-\infty,\ -1)$ so f' must be negative there. f is increasing on $(1,\ \infty)$ so f' must be positive there.

67.

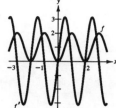

The zeros of f' correspond to the points where the graph of f has horizontal tangents.

68.

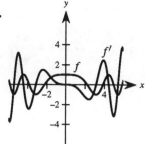

The zeros of f' correspond to the points where the graph of f has horizontal tangents.

69. $f(x) = 2(x^2 - 1)^3$

$f'(x) = 6(x^2 - 1)^2(2x)$

$\quad = 12x(x^4 - 2x^2 + 1)$

$\quad = 12x^5 - 24x^3 + 12x$

$f''(x) = 60x^4 - 72x^2 + 12$

$\quad = 12(5x^2 - 1)(x^2 - 1)$

70. $f(x) = (x - 2)^{-1}$

$f'(x) = -(x - 2)^{-2} = \dfrac{-1}{(x - 2)^2}$

$f''(x) = 2(x - 2)^{-3} = \dfrac{2}{(x - 2)^3}$

71. $f(x) = \sin x^2$

$f'(x) = 2x \cos x^2$

$f''(x) = 2x[2x(-\sin x^2)] + 2\cos x^2 = 2[\cos x^2 - 2x^2 \sin x^2]$

72. $f(x) = \sec^2 \pi x$

$f'(x) = 2 \sec \pi x (\pi \sec \pi x \tan \pi x)$

$\quad = 2\pi \sec^2 \pi x \tan \pi x$

$f''(x) = 2\pi \sec^2 \pi x (\sec^2 \pi x)(\pi) + 2\pi \tan \pi x (2\pi \sec^2 \pi x \tan \pi x)$

$\quad = 2\pi^2 \sec^4 \pi x + 4\pi^2 \sec^2 \pi x \tan^2 \pi x$

$\quad = 2\pi^2 \sec^2 \pi x (\sec^2 \pi x + 2 \tan^2 \pi x)$

$\quad = 2\pi^2 \sec^2 \pi x (3 \sec^2 \pi x - 2)$

73. a. $f(x) = g(x)h(x)$

$f'(x) = g(x)h'(x) + g'(x)h(x)$

$f'(5) = (-3)(-2) + (6)(3) = 24$

b. $f(x) = \dfrac{g(x)}{h(x)}$

$f'(x) = \dfrac{h(x)g'(x) - g(x)h'(x)}{[h(x)]^2}$

$f'(5) = \dfrac{(3)(6) - (-3)(-2)}{(3)^2} = \dfrac{12}{9} = \dfrac{4}{3}$

c. $f(x) = g(h(x))$

$f'(x) = g'(h(x))h'(x)$

$f'(5) = g'(3)(-2) = -2g'(3)$

Need $g'(3)$ to find $f'(5)$.

d. $f(x) = [g(x)]^3$

$f'(x) = 3[g(x)]^2 g'(x)$

$f'(5) = 3(-3)^2(6) = 162$

74. $|u| = \sqrt{u^2}$

$$\frac{d}{dx}[|u|] = \frac{d}{dx}\left[\sqrt{u^2}\right] = \frac{1}{2}(u^2)^{-1/2}(2uu') = \frac{uu'}{\sqrt{u^2}} = u'\frac{u}{|u|}, \quad u \neq 0$$

75. $g(x) = |2x - 3|$

$$g'(x) = 2\left(\frac{2x - 3}{|2x - 3|}\right)$$

76. $f(x) = |x^2 - 4|$

$$f'(x) = 2x\left(\frac{x^2 - 4}{|x^2 - 4|}\right)$$

77. $h(x) = |x| \cos x$

$$h'(x) = -|x| \sin x + \frac{x}{|x|} \cos x$$

78. $f(x) = |\sin x|$

$$f'(x) = \cos x \left(\frac{\sin x}{|\sin x|}\right)$$

79. a. $f = 132{,}400(331 - v)^{-1}$

$$f' = (-1)(132{,}400)(331 - v)^{-2}(-1)$$

$$= \frac{132{,}400}{(331 - v)^2}$$

When $v = 30$, $f' \approx 1.461$.

   b. $f = 132{,}400(331 + v)^{-1}$

$$f' = (-1)(132{,}400)(331 + v)^{-2}(1)$$

$$= \frac{-132{,}400}{(331 + v)^2}$$

When $v = 30$, $f' \approx -1.016$.

80. $y = \frac{1}{3} \cos 12t - \frac{1}{4} \sin 12t$

$$v = y' = \frac{1}{3}[-12 \sin 12t] - \frac{1}{4}[12 \cos 12t]$$

$$= -4 \sin 12t - 3 \cos 12t$$

When $t = \pi/8$, $y = 0.25$ feet and $v = 4$ feet per second.

81. $\theta = 0.2 \cos 8t$

The maximum angular displacement is $\theta = 0.2$ (since $-1 \leq \cos 8t \leq 1$).

$$\frac{d\theta}{dt} = 0.2[-8 \sin 8t] = -1.6 \sin 8t$$

When $t = 3$, $d\theta/dt = -1.6 \sin 24 \approx 1.4489$ radians per second.

82. $y = A \cos \omega t$

a. Amplitude: $3.5 \Rightarrow A = \dfrac{3.5}{2} = 1.75$

$$y = 1.75 \cos \omega t$$

Period: $10 \Rightarrow \omega = \dfrac{2\pi}{10} = \dfrac{\pi}{5}$

$$y = 1.75 \cos \frac{\pi t}{5}$$

b. $v = y' = 1.75\left[-\dfrac{\pi}{5} \sin \dfrac{\pi t}{5}\right]$

$$= -0.35\pi \sin \frac{\pi t}{5}$$

83. $S = C(R^2 - r^2)$

$$\frac{dS}{dt} = C\left(2R\frac{dR}{dt} - 2r\frac{dr}{dt}\right)$$

Since r is constant, we have $dr/dt = 0$ and

$$\frac{dS}{dt} = (1.76 \times 10^5)(2)(1.2 \times 10^{-2})(10^{-5})$$

$$= 4.224 \times 10^{-2} = 0.04224.$$

84. $f(x) = \sec^2 x$

$$f'(x) = 2 \sec x \sec x \tan x = 2 \sec^2 x \tan x$$

$$g(x) = \tan^2 x$$

$$g'(x) = 2 \tan x \sec^2 x$$

Thus, $f'(x) = g'(x)$.

85. $f(x) = \sin \beta x$

    a. $f'(x) = \beta \cos \beta x$

        $f''(x) = -\beta^2 \sin \beta x$

        $f'''(x) = -\beta^3 \cos \beta x$

        $f^{(4)}(x) = \beta^4 \sin \beta x$

    b. $f''(x) + \beta^2 f(x) = -\beta^2 \sin \beta x + \beta^2(\sin \beta x) = 0$

    c. $f^{(2k)}(x) = (-1)^k \beta^{2k} \sin \beta x$

        $f^{(2k-1)}(x) = (-1)^{k+1} \beta^{2k-1} \cos \beta x$

86. $g(x) = \sin^2 x + \cos^2 x = 1$

     $g'(x) = 0$

87. False. If $y = (1-x)^{1/2}$, then

    $y' = \frac{1}{2}(1-x)^{-1/2}(-1)$.

88. False. If $f(x) = \sin^2 2x$, then

    $f'(x) = 2(\sin 2x)(2 \cos 2x)$.

89. True

90. If $f(-x) = -f(x)$, then

$$\frac{d}{dx}[f(-x)] = \frac{d}{dx}[-f(x)]$$

$$f'(-x)(-1) = -f'(x)$$

$$f'(-x) = f'(x).$$

Thus, $f'(x)$ is even.

91. $g = \sqrt{x(x+n)}$

$$= \sqrt{x^2 + nx}$$

$$\frac{dg}{dx} = \frac{1}{2}(x^2+nx)^{-1/2}(2x+n)$$

$$= \frac{2x+n}{2\sqrt{x^2+nx}}$$

$$= \frac{(2x+n)/2}{\sqrt{x(x+n)}}$$

$$= \frac{[x+(x+n)]/2}{\sqrt{x(x+n)}}$$

$$= \frac{a}{g}$$

Section 2.5   Implicit Differentiation

1. $x^2 + y^2 = 16$

    $2x + 2yy' = 0$

       $y' = \dfrac{-x}{y}$

2. $x^2 - y^2 = 16$

    $2x - 2yy' = 0$

       $y' = \dfrac{x}{y}$

3. $x^{1/2} + y^{1/2} = 9$

    $\dfrac{1}{2}x^{-1/2} + \dfrac{1}{2}y^{-1/2}y' = 0$

       $y' = -\dfrac{x^{-1/2}}{y^{-1/2}} = -\sqrt{\dfrac{y}{x}}$

4. $x^3 + y^3 = 8$

    $3x^2 + 3y^2y' = 0$

       $y' = -\dfrac{x^2}{y^2}$

5. $x^3 - xy + y^2 = 4$

    $3x^2 - xy' - y + 2yy' = 0$

       $(2y - x)y' = y - 3x^2$

           $y' = \dfrac{y - 3x^2}{2y - x}$

6. $x^2y + y^2x = -2$

    $x^2y' + 2xy + y^2 + 2yxy' = 0$

       $(x^2 + 2xy)y' = -(y^2 + 2xy)$

           $y' = \dfrac{-y(y + 2x)}{x(x + 2y)}$

7.
$$x^3y^3 - y - x = 0$$
$$3x^3y^2y' + 3x^2y^3 - y' - 1 = 0$$
$$(3x^3y^2 - 1)y' = 1 - 3x^2y^3$$
$$y' = \frac{1 - 3x^2y^3}{3x^3y^2 - 1}$$

8.
$$(xy)^{1/2} - x + 2y = 0$$
$$\frac{1}{2}(xy)^{-1/2}(xy' + y) - 1 + 2y' = 0$$
$$\frac{x}{2\sqrt{xy}}y' + \frac{y}{2\sqrt{xy}} - 1 + 2y' = 0$$
$$xy' + y - 2\sqrt{xy} + 4\sqrt{xy}\,y' = 0$$
$$y' = \frac{2\sqrt{xy} - y}{4\sqrt{xy} + x}$$

9.
$$x^3 - 2x^2y + 3xy^2 = 38$$
$$3x^2 - 2x^2y' - 4xy + 6xyy' + 3y^2 = 0$$
$$2x(3y - x)y' = 4xy - 3x^2 - 3y^2$$
$$y' = \frac{4xy - 3x^2 - 3y^2}{2x(3y - x)}$$

10.
$$2\sin x \cos y = 1$$
$$2[\sin x(-\sin y)y' + \cos y(\cos x)] = 0$$
$$y' = \frac{\cos x \cos y}{\sin x \sin y}$$
$$= \cot x \cot y$$

11.
$$\sin x + 2\cos 2y = 1$$
$$\cos x - 4(\sin 2y)y' = 0$$
$$y' = \frac{\cos x}{4\sin 2y}$$

12.
$$(\sin \pi x + \cos \pi y)^2 = 2$$
$$2(\sin \pi x + \cos \pi y)[\pi \cos \pi x - \pi(\sin \pi y)y'] = 0$$
$$\pi \cos \pi x - \pi(\sin \pi y)y' = 0$$
$$y' = \frac{\cos \pi x}{\sin \pi y}$$

13.
$$\sin x = x(1 + \tan y)$$
$$\cos x = x(\sec^2 y)y' + (1 + \tan y)(1)$$
$$y' = \frac{\cos x - \tan y - 1}{x\sec^2 y}$$

14.
$$\cot y = x - y$$
$$(-\csc^2 y)y' = 1 - y'$$
$$y' = \frac{1}{1 - \csc^2 y}$$
$$= \frac{1}{-\cot^2 y} = -\tan^2 y$$

15.
$$y = \sin(xy)$$
$$y' = [xy' + y]\cos(xy)$$
$$y' = \frac{y\cos(xy)}{1 - x\cos(xy)}$$

16.
$$x = \sec \frac{1}{y}$$
$$1 = -\frac{y'}{y^2}\sec \frac{1}{y}\tan \frac{1}{y}$$
$$y' = \frac{-y^2}{\sec(1/y)\tan(1/y)}$$

17.
$$xy = 4$$
$$xy' + y(1) = 0$$
$$xy' = -y$$
$$y' = \frac{-y}{x}$$
At $(-4, -1)$: $y' = -\frac{1}{4}$

18.
$$x^2 - y^3 = 0$$
$$2x - 3y^2y' = 0$$
$$y' = \frac{2x}{3y^2}$$
At $(1, 1)$: $y' = \frac{2}{3}$

19. $y^2 = \dfrac{x^2 - 9}{x^2 + 9}$

$2yy' = \dfrac{(x^2 + 9)(2x) - (x^2 - 9)2x}{(x^2 + 9)^2}$

$y' = \dfrac{18x}{(x^2 + 9)^2 y}$

At $(3, \ 0)$: y' is undefined.

20. $(x + y)^3 = x^3 + y^3$

$x^3 + 3x^2 y + 3xy^2 + y^3 = x^3 + y^3$

$3x^2 y + 3xy^2 = 0$

$x^2 y + xy^2 = 0$

$x^2 y' + 2xy + 2xyy' + y^2 = 0$

$(x^2 + 2xy)y' = -(y^2 + 2xy)$

$y' = -\dfrac{y(y + 2x)}{x(x + 2y)}$

At $(-1, \ 1)$: $y' = -1$

21. $x^{2/3} + y^{2/3} = 5$

$\dfrac{2}{3}x^{-1/3} + \dfrac{2}{3}y^{-1/3}y' = 0$

$y' = \dfrac{-x^{-1/3}}{y^{-1/3}} = -\sqrt[3]{\dfrac{y}{x}}$

At $(8, \ 1)$: $y' = -\dfrac{1}{2}$

22. $x^3 + y^3 - 2xy = 0$

$3x^2 + 3y^2 y' - 2xy' - 2y = 0$

$(3y^2 - 2x)y' = 2y - 3x^2$

$y' = \dfrac{2y - 3x^2}{3y^2 - 2x}$

At $(1, \ 1)$: $y' = -1$

23. $\tan(x + y) = x$

$(1 + y')\sec^2(x + y) = 1$

$y' = \dfrac{1 - \sec^2(x + y)}{\sec^2(x + y)}$

$= \dfrac{-\tan^2(x + y)}{\tan^2(x + y) + 1}$

$= -\dfrac{x^2}{x^2 + 1}$

At $(0, 0)$: $y' = 0$

24. $x \cos y = 1$

$x[-y' \sin y] + \cos y = 0$

$y' = \dfrac{\cos y}{x \sin y}$

$= \dfrac{1}{x}\cot y = \dfrac{\cot y}{x}$

At $\left(2, \ \dfrac{\pi}{3}\right)$: $y' = \dfrac{1}{2\sqrt{3}}$

25. $(x^2 + 4)y = 8$

$(x^2 + 4)y' + y(2x) = 0$

$y' = \dfrac{-2xy}{x^2 + 4}$

$= \dfrac{-2x[8/(x^2 + 4)]}{x^2 + 4}$

$= \dfrac{-16x}{(x^2 + 4)^2}$

At $(2, \ 1)$: $y' = \dfrac{-32}{64} = -\dfrac{1}{2}$

26. $(4 - x)y^2 = x^3$

$(4 - x)(2yy') + y^2(-1) = 3x^2$

$y' = \dfrac{3x^2 + y^2}{2y(4 - x)}$

At $(2, \ 2)$: $y' = 2$

27.
$$(x^2 + y^2)^2 = 4x^2 y$$

$$2(x^2 + y^2)(2x + 2yy') = 4x^2 y' + y(8x)$$

$$4x^3 + 4x^2 yy' + 4xy^2 + 4y^3 y' = 4x^2 y' + 8xy$$

$$4x^2 yy' + 4y^3 y' - 4x^2 y' = 8xy - 4x^3 - 4xy^2$$

$$4y'(x^2 y + y^3 - x^2) = 4(2xy - x^3 - xy^2)$$

$$y' = \frac{2xy - x^3 - xy^2}{x^2 y + y^3 - x^2}$$

At $(1, \ 1)$: $y' = 0$

28.
$$x^3 + y^3 - 6xy = 0$$

$$3x^2 + 3y^2 y' - 6xy' - 6y = 0$$

$$y'(3y^2 - 6x) = 6y - 3x^2$$

$$y' = \frac{6y - 3x^2}{3y^2 - 6x} = \frac{2y - x^2}{y^2 - 2x}$$

At $\left(\dfrac{4}{3}, \dfrac{8}{3}\right)$: $y' = \dfrac{(16/3) - (16/9)}{(64/9) - (8/3)} = \dfrac{32}{40} = \dfrac{4}{5}$

29.
$$\sqrt{x} + \sqrt{y} = 3$$

$$\frac{1}{2}x^{-1/2} + \frac{1}{2}y^{-1/2}y' = 0$$

$$y' = -\frac{(1/2)x^{-1/2}}{(1/2)y^{-1/2}} = -\frac{\sqrt{y}}{\sqrt{x}}$$

At $(4, 1)$: $y' = -\dfrac{1}{2}$.

Tangent line:

$$y - 1 = -\frac{1}{2}(x - 4)$$

$$y = -\frac{1}{2}x + 3$$

$$x + 2y - 6 = 0$$

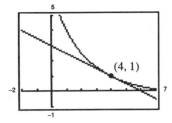

30.
$$y^2 = \frac{x - 1}{x^2 + 1}$$

$$2yy' = \frac{(x^2 + 1)(1) - (x - 1)(2x)}{(x^2 + 1)^2}$$

$$= \frac{x^2 + 1 - 2x^2 + 2x}{(x^2 + 1)^2}$$

$$y' = \frac{1 + 2x - x^2}{2y(x^2 + 1)^2}$$

At $\left(2, \dfrac{\sqrt{5}}{5}\right)$: $y' = \dfrac{1 + 4 - 4}{[(2\sqrt{5})/5](4 + 1)^2} = \dfrac{1}{10\sqrt{5}}$

Tangent line:

$$y - \frac{\sqrt{5}}{5} = \frac{1}{10\sqrt{5}}(x - 2)$$

$$10\sqrt{5}\,y - 10 = x - 2$$

$$x - 10\sqrt{5}\,y + 8 = 0$$

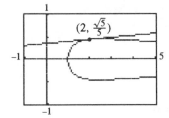

31. a. $x^2 + y^2 = 16$

$$y^2 = 16 - x^2$$

$$y = \pm\sqrt{16 - x^2}$$

c. Explicitly:

$$\frac{dy}{dx} = \pm\frac{1}{2}(16 - x^2)^{-1/2}(-2x)$$

$$= \frac{\mp x}{\sqrt{16 - x^2}} = -\frac{x}{y}$$

b.

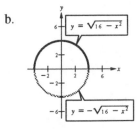

d. Implicitly:

$$2x + 2yy' = 0$$

$$y' = -\frac{x}{y}$$

32. a. $(x^2 - 4x + 4) + (y^2 + 6y + 9) = -9 + 4 + 9$

$\qquad (x - 2)^2 + (y + 3)^2 = 4$

$\qquad\qquad (y + 3)^2 = 4 - (x - 2)^2$

$\qquad\qquad\qquad y = -3 \pm \sqrt{4 - (x - 2)^2}$

c. Explicitly:

$$\frac{dy}{dx} = \pm\frac{1}{2}[4 - (x - 2)^2]^{-1/2}(-2)(x - 2)$$

$$= \frac{\mp(x - 2)}{(\sqrt{4 - (x - 2)^2} - 3) + 3} = \frac{-(x - 2)}{y + 3}$$

b.

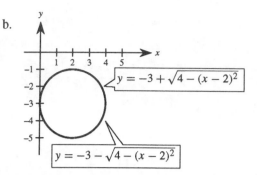

$y = -3 + \sqrt{4 - (x - 2)^2}$

$y = -3 - \sqrt{4 - (x - 2)^2}$

d. Implicitly:

$$2x + 2yy' - 4 + 6y' = 0$$

$$(2y + 6)y' = -2(x - 2)$$

$$y' = \frac{-(x - 2)}{y + 3}$$

33. a. $16y^2 = 144 - 9x^2$

$$y^2 = \frac{1}{16}(144 - 9x^2) = \frac{9}{16}(16 - x^2)$$

$$y = \pm\frac{3}{4}\sqrt{16 - x^2}$$

c. Explicitly:

$$\frac{dy}{dx} = \pm\frac{3}{8}(16 - x^2)^{-1/2}(-2x)$$

$$= \mp\frac{3x}{4\sqrt{16 - x^2}} = \frac{-3x}{4(4/3)y} = \frac{-9x}{16y}$$

b.

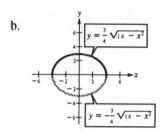

$y = \frac{3}{4}\sqrt{16 - x^2}$

$y = -\frac{3}{4}\sqrt{16 - x^2}$

d. Implicitly:

$$18x + 32yy' = 0$$

$$y' = \frac{-9x}{16y}$$

34. a. $y^2 = 1 + \dfrac{x^2}{4} = \dfrac{x^2 + 4}{4}$

$$y = \pm\frac{1}{2}\sqrt{x^2 + 4}$$

c. Explicitly:

$$\frac{dy}{dx} = \pm\frac{1}{4}(x^2 + 4)^{-1/2}(2x)$$

$$= \frac{\pm x}{2\sqrt{x^2 + 4}}$$

$$= \frac{\pm x}{4(1/2)\sqrt{x^2 + 4}} = \frac{x}{4y}$$

b.

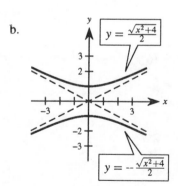

$y = \dfrac{\sqrt{x^2 + 4}}{2}$

$y = -\dfrac{\sqrt{x^2 + 4}}{2}$

d. Implicitly:

$$2yy' - \frac{1}{2}x = 0$$

$$y' = \frac{x}{4y}$$

35.
$$x^2 + xy = 5$$
$$2x + xy' + y = 0$$
$$y' = \frac{-(2x+y)}{x}$$
$$2 + xy'' + y' + y' = 0$$
$$xy'' = -2(1+y')$$
$$y'' = \frac{-2[1-(2x+y)/x]}{x} = \frac{2(x+y)}{x^2}$$
$$y'' = \frac{2(x+y)}{x^2} \cdot \frac{x}{x} = \frac{10}{x^3}$$

(**Note:** You could write $y = (5-x^2)/x$ and calculate y' and y'' directly.)

36.
$$x^2y^2 - 2x = 3$$
$$2x^2yy' + 2xy^2 - 2 = 0$$
$$x^2yy' + xy^2 - 1 = 0$$
$$y' = \frac{1-xy^2}{x^2y}$$
$$2xyy' + x^2(y')^2 + x^2yy'' + 2xyy' + y^2 = 0$$
$$4xyy' + x^2(y')^2 + x^2yy'' + y^2 = 0$$
$$\frac{4-4xy^2}{x} + \frac{(1-xy^2)^2}{x^2y^2} + x^2yy'' + y^2 = 0$$
$$4xy^2 - 4x^2y^4 + 1 - 2xy^2 + x^2y^4 + x^4y^3y'' + x^2y^4 = 0$$
$$x^4y^3y'' = 2x^2y^4 - 2xy^2 - 1$$
$$y'' = \frac{2x^2y^4 - 2xy^2 - 1}{x^4y^3}$$

37.
$$x^2 - y^2 = 16$$
$$2x - 2yy' = 0$$
$$y' = \frac{x}{y}$$
$$x - yy' = 0$$
$$1 - yy'' - (y')^2 = 0$$
$$1 - yy'' - \left(\frac{x}{y}\right)^2 = 0$$
$$y'' = \frac{y^2 - x^2}{y^3} = \frac{-16}{y^3}$$

38. $1 - xy = x - y$
$$y - xy = x - 1$$
$$y = \frac{x-1}{1-x} = -1$$
$$y' = 0$$
$$y'' = 0$$

39. $y^2 = x^3$

$2yy' = 3x^2$

$y' = \dfrac{3x^2}{2y} = \dfrac{3x^2}{2y} \cdot \dfrac{xy}{xy} = \dfrac{3y}{2x} \cdot \dfrac{x^3}{y^2} = \dfrac{3y}{2x}$

$y'' = \dfrac{2x(3y') - 3y(2)}{4x^2}$

$\quad = \dfrac{2x[3 \cdot (3y/2x)] - 6y}{4x^2}$

$\quad = \dfrac{3y}{4x^2} = \dfrac{3x}{4y}$

40. $y^2 = 4x$

$2yy' = 4$

$y' = \dfrac{2}{y}$

$y'' = -2y^{-2}y' = \left[\dfrac{-2}{y^2}\right] \cdot \dfrac{2}{y} = \dfrac{-4}{y^3}$

41. $x^2 + y^2 = 25$

$y' = \dfrac{-x}{y}$

At $(4, 3)$:

Tangent line: $y - 3 = \dfrac{-4}{3}(x - 4) \implies 4x + 3y - 25 = 0$

Normal line: $y - 3 = \dfrac{3}{4}(x - 4) \implies 3x - 4y = 0$

At $(-3, 4)$:

Tangent line: $y - 4 = \dfrac{3}{4}(x + 3) \implies 3x - 4y + 25 = 0$

Normal line: $y - 4 = \dfrac{-4}{3}(x + 3) \implies 4x + 3y = 0$

42. $x^2 + y^2 = 9$

$y' = \dfrac{-x}{y}$

At $(0, 3)$:

Tangent line: $y = 3$

Normal line: $x = 0$

At $(2, \sqrt{5})$:

Tangent line: $y - \sqrt{5} = \dfrac{-2}{\sqrt{5}}(x - 2) \implies 2x + \sqrt{5}\,y - 9 = 0$

Normal line: $y - \sqrt{5} = \dfrac{\sqrt{5}}{2}(x - 2) \implies \sqrt{5}\,x - 2y = 0$

43. $x^2 + y^2 = r^2$

$2x + 2yy' = 0$

$$y' = \frac{-x}{y} = \text{ slope of tangent line}$$

$$\frac{y}{x} = \text{ slope of normal line}$$

Let $(x_0, \ y_0)$ be a point on the circle. If $x_0 = 0$, then the tangent line is horizontal, the normal line is vertical and, hence, passes through the origin. If $x_0 \neq 0$, then the equation of the normal line is

$$y - y_0 = \frac{y_0}{x_0}(x - x_0)$$

$$y = \frac{y_0}{x_0}x$$

which passes through the origin.

44. $y^2 = 4x$

$2yy' = 4$

$$y' = \frac{2}{y} = 1 \text{ at } (1, \ 2)$$

Equation of normal at $(1, 2)$ is $y - 2 = -1(x - 1)$, $\ y = 3 - x$. The centers of the circles must be on the normal and at a distance of 4 units from $(1, 2)$. Therefore,

$$(x - 1)^2 + [(3 - x) - 2]^2 = 16$$

$$2(x - 1)^2 = 16$$

$$x = 1 \pm 2\sqrt{2}.$$

Centers of the circles: $\left(1 + 2\sqrt{2}, \ 2 - 2\sqrt{2}\right)$ and $\left(1 - 2\sqrt{2}, \ 2 + 2\sqrt{2}\right)$

Equations: $\left(x - 1 - 2\sqrt{2}\right)^2 + \left(y - 2 + 2\sqrt{2}\right)^2 = 16$

$$\left(x - 1 + 2\sqrt{2}\right)^2 + \left(y - 2 - 2\sqrt{2}\right)^2 = 16$$

45. $25x^2 + 16y^2 + 200x - 160y + 400 = 0$

$$50x + 32yy' + 200 - 160y' = 0$$

$$y' = \frac{200 + 50x}{160 - 32y}$$

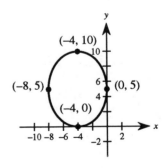

Horizontal tangents occur when $x = -4$:

$$25(16) + 16y^2 + 200(-4) - 160y + 400 = 0$$

$$y(y - 10) = 0 \ \Rightarrow \ y = 0, \ 10$$

Horizontal tangents: $(-4, \ 0)$, $(-4, \ 10)$

Vertical tangents occur when $y = 5$:

$$25x^2 + 400 + 200x - 800 + 400 = 0$$

$$25x(x + 8) = 0 \ \Rightarrow \ x = 0, \ -8$$

Vertical tangents: $(0, \ 5)$, $(-8, \ 5)$

46. $4x^2 + y^2 - 8x + 4y + 4 = 0$

$8x + 2yy' - 8 + 4y' = 0$

$$y' = \frac{8 - 8x}{2y + 4} = \frac{4 - 4x}{y + 2}$$

Horizontal tangents occur when $x = 1$:

$4(1)^2 + y^2 - 8(1) + 4y + 4 = 0$

$y^2 + 4y = y(y + 4) = 0 \implies y = 0, \ -4$

Horizontal tangents: $(1, \ 0), \ (1, \ -4)$

Vertical tangents occur when $y = -2$:

$4x^2 + (-2)^2 - 8x + 4(-2) + 4 = 0$

$4x^2 - 8x = 4x(x - 2) = 0 \implies x = 0, \ 2$

Vertical tangents: $(0, \ -2), \ (2, \ -2)$

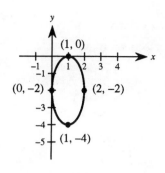

47. Find the points of intersection by letting $y^2 = 4x$ in the equation $2x^2 + y^2 = 6$.

$2x^2 + 4x = 6 \quad$ and $\quad (x + 3)(x - 1) = 0$

The curves intersect at $(1, \ \pm 2)$.

Ellipse: *Parabola:*

$4x + 2yy' = 0 \quad 2yy' = 4$

$y' = -\dfrac{2x}{y} \qquad y' = \dfrac{2}{y}$

At $(1, 2)$, the slopes are:

   $y' = -1 \qquad\qquad y' = 1$

At $(1, \ -2)$, the slopes are:

   $y' = 1 \qquad\qquad y' = -1$

Tangents are perpendicular.

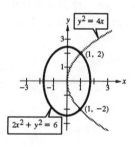

48. Find the points of intersection by letting $y^2 = x^3$ in the equation $2x^2 + 3y^2 = 5$.

$2x^2 + 3x^3 = 5 \quad$ and $\quad 3x^3 + 2x^2 - 5 = 0$

Intersect when $x = 1$

Points of intersection: $(1, \ \pm 1)$

$\underline{y^2 = x^3:} \qquad \underline{2x^2 + 3y^2 = 5:}$

$2yy' = 3x^2 \qquad 4x + 6yy' = 0$

$y' = \dfrac{3x^2}{2y} \qquad y' = -\dfrac{2x}{3y}$

At $(1, 1)$, the slopes are:

$y' = \dfrac{3}{2} \qquad\qquad y' = -\dfrac{2}{3}$

At $(1, \ -1)$, the slopes are:

$y' = -\dfrac{3}{2} \qquad\qquad y' = \dfrac{2}{3}$

Tangents are perpendicular.

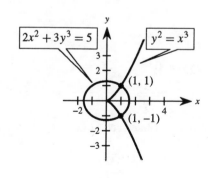

49. $y = -x$ and $x = \sin y$

Point of intersection: $(0, 0)$

$y = -x$:	$x = \sin y$:
$y' = -1$	$1 = y' \cos y$
	$y' = \sec y$

At $(0, 0)$, the slopes are:

$$y' = -1 \qquad y' = 1$$

Tangents are perpendicular.

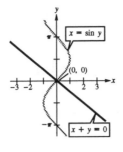

50. Rewriting each equation and differentiating,

$$x^3 = 3(y - 1) \qquad x(3y - 29) = 3$$

$$y = \frac{x^3}{3} + 1 \qquad y = \frac{1}{3}\left(\frac{3}{x} + 29\right)$$

$$y' = x^2 \qquad y' = -\frac{1}{x^2}$$

For each value of x, the derivatives are negative reciprocals of each other. Thus, the tangent lines are orthogonal at both points of intersection.

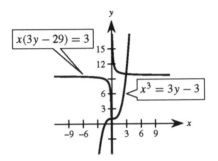

51.

$xy = C$	$x^2 - y^2 = K$
$xy' + y = 0$	$2x - 2yy' = 0$
$y' = -\dfrac{y}{x}$	$y' = \dfrac{x}{y}$

At any point of intersection (x, y) the product of the slopes is $(-y/x)(x/y) = -1$. The curves are orthogonal.

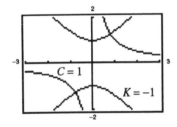
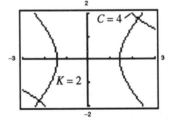

52.

$x^2 + y^2 = C^2$	$y = Kx$
$2x + 2yy' = 0$	$y' = K$
$y' = -\dfrac{x}{y}$	

At the point of intersection (x, y) the product of the slopes is $(-x/y)(K) = (-x/Kx)(K) = -1$. The curves are orthogonal.

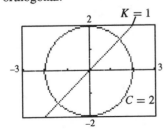
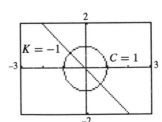

53. $\tan y = x$

$y' \sec^2 y = 1$

$$y' = \frac{1}{\sec^2 y}, \quad -\frac{\pi}{2} < y < \frac{\pi}{2}$$

$$\sec^2 y = 1 + \tan^2 y = 1 + x^2$$

$$y' = \frac{1}{1 + x^2}$$

54. $(x - h)^2 + (y - k)^2 = r^2$

$2(x - h) + 2(y - k)y' = 0$

$$y' = \frac{-2(x - h)}{2(y - k)}$$

$$= -\frac{x - h}{y - k}$$

55. Let $f(x) = x^n = x^{p/q}$, where p and q are nonzero integers and $q > 0$. First consider the case where $p = 1$. The derivative of $f(x) = x^{1/q}$ is given by

$$\frac{d}{dx}[x^{1/q}] = \lim_{\Delta x \to 0} \frac{f(x + \Delta x) - f(x)}{\Delta x} = \lim_{t \to x} \frac{f(t) - f(x)}{t - x}$$

where $t = x + \Delta x$. Observe that

$$\frac{f(t) - f(x)}{t - x} = \frac{t^{1/q} - x^{1/q}}{t - x} = \frac{t^{1/q} - x^{1/q}}{(t^{1/q})^q - (x^{1/q})^q}$$

$$= \frac{t^{1/q} - x^{1/q}}{(t^{1/q} - x^{1/q})(t^{1-(1/q)} + t^{1-(2/q)}x^{1/q} + \cdots + t^{1/q}x^{1-(2/q)} + x^{1-(1/q)})}$$

$$= \frac{1}{(t^{1-(1/q)} + t^{1-(2/q)}x^{1/q} + \cdots + t^{1/q}x^{1-(2/q)} + x^{1-(1/q)})}.$$

As $t \to x$, the denominator approaches $qx^{1-(1/q)}$. That is,

$$\frac{d}{dx}[x^{1/q}] = \frac{1}{qx^{1-(1/q)}} = \frac{1}{q}x^{(1/q)-1}.$$

Now consider $f(x) = x^{p/q} = (x^p)^{1/q}$. From the Chain Rule,

$$f'(x) = \frac{1}{q}(x^p)^{(1/q)-1}\frac{d}{dx}[x^p]$$

$$= \frac{1}{q}(x^p)^{(1/q)-1}px^{p-1} = \frac{p}{q}x^{[(p/q)-p]+(p-1)} = \frac{p}{q}x^{(p/q)-1} = nx^{n-1} \qquad \left(n = \frac{p}{q}\right)$$

56. $\sqrt{x} + \sqrt{y} = \sqrt{c}$

$$\frac{1}{2\sqrt{x}} + \frac{1}{2\sqrt{y}}\frac{dy}{dx} = 0$$

$$\frac{dy}{dx} = -\frac{\sqrt{y}}{\sqrt{x}}$$

Tangent line at (x_0, y_0):

$$y - y_0 = -\frac{\sqrt{y_0}}{\sqrt{x_0}}(x - x_0)$$

x-intercept: $(x_0 + \sqrt{x_0}\sqrt{y_0}, \ 0)$

y-intercept: $(0, \ y_0 + \sqrt{x_0}\sqrt{y_0})$

Sum of intercepts:

$$(x_0 + \sqrt{x_0}\sqrt{y_0}) + (y_0 + \sqrt{x_0}\sqrt{y_0}) = x_0 + 2\sqrt{x_0}\sqrt{y_0} + y_0 = (\sqrt{x_0} + \sqrt{y_0})^2 = (\sqrt{c})^2 = c$$

57. $x^2 + y^2 = 25$

$$2x + 2y\frac{dy}{dx} = 0$$

$$\frac{dy}{dx} = -\frac{x}{y}$$

When $x = 3$ and $y = 4$, $dy/dx = -3/4$.

58. $xy = 4$

$$x\frac{dy}{dx} + y(1) = 0$$

$$\frac{dy}{dx} = -\frac{y}{x}$$

When $x = 1$ and $y = 4$, $dy/dx = -4$.

59. $\sqrt{3}x = y$

$$\sqrt{3} = \frac{dy}{dx}$$

When $x = \sqrt{3}$ and $y = 3$, $dy/dx = \sqrt{3}$.

60. $y = \frac{2}{3}\cos x$

$$\frac{dy}{dx} = -\frac{2}{3}\sin x$$

When $x = \pi/3$ and $y = 1/3$,

$$\frac{dy}{dx} = -\frac{2}{3}\left(\frac{\sqrt{3}}{2}\right) = -\frac{\sqrt{3}}{3}.$$

Section 2.6   Related Rates

1. $y = \sqrt{x}$

$$\frac{dy}{dt} = \left(\frac{1}{2\sqrt{x}}\right)\frac{dx}{dt}$$

$$\frac{dx}{dt} = 2\sqrt{x}\frac{dy}{dt}$$

a. When $x = 4$ and $dx/dt = 3$,

$$\frac{dy}{dt} = \frac{1}{2\sqrt{4}}(3) = \frac{3}{4}.$$

b. When $x = 25$ and $dy/dt = 2$,

$$\frac{dx}{dt} = 2\sqrt{25}(2) = 20.$$

2. $y = x^2 - 3x$

$$\frac{dy}{dt} = (2x - 3)\frac{dx}{dt}$$

$$\frac{dx}{dt} = \left(\frac{1}{2x - 3}\right)\frac{dy}{dt}$$

a. When $x = 3$ and $dx/dt = 2$,

$$\frac{dy}{dt} = [2(3) - 3](2) = 6.$$

b. When $x = 1$ and $dy/dt = 5$,

$$\frac{dx}{dt} = \frac{1}{2(1) - 3}(5) = -5.$$

3. $xy = 4$

$$x\frac{dy}{dt} + y\frac{dx}{dt} = 0$$

$$\frac{dy}{dt} = \left(-\frac{y}{x}\right)\frac{dx}{dt}$$

$$\frac{dx}{dt} = \left(-\frac{x}{y}\right)\frac{dy}{dt}$$

a. When $x = 8$, $y = 1/2$, and $dx/dt = 10$,

$$\frac{dy}{dt} = -\frac{1/2}{8}(10) = -\frac{5}{8}.$$

b. When $x = 1$, $y = 4$, and $dy/dt = -6$,

$$\frac{dx}{dt} = -\frac{1}{4}(-6) = \frac{3}{2}.$$

4. $x^2 + y^2 = 25$

$$2x\frac{dx}{dt} + 2y\frac{dy}{dt} = 0$$

$$\frac{dy}{dt} = \left(-\frac{x}{y}\right)\frac{dx}{dt}$$

$$\frac{dx}{dt} = \left(-\frac{y}{x}\right)\frac{dy}{dt}$$

a. When $x = 3$, $y = 4$, and $dx/dt = 8$,

$$\frac{dy}{dt} = -\frac{3}{4}(8) = -6.$$

b. When $x = 4$, $y = 3$, and $dy/dt = -2$,

$$\frac{dx}{dt} = -\frac{3}{4}(-2) = \frac{3}{2}.$$

5. $y = x^2 + 1$

$\dfrac{dx}{dt} = 2$

$\dfrac{dy}{dt} = 2x\dfrac{dx}{dt}$

a. When $x = -1$,

$\dfrac{dy}{dt} = 2(-1)(2) = -4$ cm/sec.

b. When $x = 0$,

$\dfrac{dy}{dt} = 2(0)(2) = 0$ cm/sec.

c. When $x = 1$,

$\dfrac{dy}{dt} = 2(1)(2) = 4$ cm/sec.

d. When $x = 3$,

$\dfrac{dy}{dt} = 2(3)(2) = 12$ cm/sec.

7. $y = \tan x$

$\dfrac{dx}{dt} = 2$

$\dfrac{dy}{dt} = \sec^2 x\dfrac{dx}{dt}$

a. When $x = -\pi/3$,

$\dfrac{dy}{dt} = (2)^2(2) = 8$ cm/sec.

b. When $x = -\pi/4$,

$\dfrac{dy}{dt} = (\sqrt{2})^2(2) = 4$ cm/sec.

c. When $x = 0$,

$\dfrac{dy}{dt} = (1)^2(2) = 2$ cm/sec.

d. When $x = 1$,

$\dfrac{dy}{dt} = (\sec 1)^2 2 \approx 6.8510$ cm/sec.

6. $y = \dfrac{1}{1 + x^2}$

$\dfrac{dx}{dt} = 2$

$\dfrac{dy}{dt} = \left[\dfrac{-2x}{(1 + x^2)^2}\right]\dfrac{dx}{dt}$

a. When $x = -2$,

$\dfrac{dy}{dt} = \dfrac{-2(-2)(2)}{25} = \dfrac{8}{25}$ cm/sec.

b. When $x = 0$,

$\dfrac{dy}{dt} = 0$ cm/sec.

c. When $x = 2$,

$\dfrac{dy}{dt} = \dfrac{-2(2)(2)}{25} = \dfrac{-8}{25}$ cm/sec.

d. When $x = 10$,

$\dfrac{dy}{dt} = \dfrac{-2(10)(2)}{(101)^2} = \dfrac{-40}{10,201} \approx -0.0039$ cm/sec.

8. $y = \sin x$

$\dfrac{dx}{dt} = 2$

$\dfrac{dy}{dt} = \cos x\dfrac{dx}{dt}$

a. When $x = \pi/6$,

$\dfrac{dy}{dt} = \left(\cos\dfrac{\pi}{6}\right)(2) = \sqrt{3}$ cm/sec.

b. When $x = \pi/4$,

$\dfrac{dy}{dt} = \left(\cos\dfrac{\pi}{4}\right)(2) = \sqrt{2}$ cm/sec.

c. When $x = \pi/3$,

$\dfrac{dy}{dt} = \left(\cos\dfrac{\pi}{3}\right)(2) = 1$ cm/sec.

d. When $x = \pi/2$,

$\dfrac{dy}{dt} = \left(\cos\dfrac{\pi}{2}\right)(2) = 0$ cm/sec.

9. $D = \sqrt{x^2 + y^2} = \sqrt{x^2 + (x^2+1)^2} = \sqrt{x^4 + 3x^2 + 1}$

$\dfrac{dx}{dt} = 2$

$\dfrac{dD}{dt} = \dfrac{1}{2}(x^4 + 3x^2 + 1)^{-1/2}(4x^3 + 6x)\dfrac{dx}{dt} = \dfrac{2x^3 + 3x}{\sqrt{x^4 + 3x^2 + 1}}\dfrac{dx}{dt}$

a. When $x = -1$, $\dfrac{dD}{dt} = \dfrac{-5}{\sqrt{5}}(2) = -2\sqrt{5} \approx -4.472$ cm/sec.

b. When $x = 0$, $\dfrac{dD}{dt} = \dfrac{0}{1}(2) = 0$ cm/sec.

c. When $x = 1$, $\dfrac{dD}{dt} = \dfrac{5}{\sqrt{5}}(2) = 2\sqrt{5} \approx 4.472$ cm/sec.

d. When $x = 3$, $\dfrac{dD}{dt} = \dfrac{63}{\sqrt{109}}(2) = \dfrac{126\sqrt{109}}{109} \approx 12.069$ cm/sec.

10. $D = \sqrt{x^2 + y^2} = \sqrt{x^2 + \sin^2 x}$

$\dfrac{dx}{dt} = 2$

$\dfrac{dD}{dt} = \dfrac{1}{2}(x^2 + \sin^2 x)^{-1/2}(2x + 2\sin x \cos x)\dfrac{dx}{dt} = \dfrac{x + \sin x \cos x}{\sqrt{x^2 + \sin^2 x}}\dfrac{dx}{dt}$

a. When $x = \dfrac{\pi}{6}$, $\dfrac{dD}{dt} = \dfrac{(\pi/6) + (1/2)(\sqrt{3}/2)}{\sqrt{(\pi^2/36) + (1/4)}}(2) = \dfrac{2\pi + 3\sqrt{3}}{\sqrt{\pi^2 + 9}}$ cm/sec.

b. When $x = \dfrac{\pi}{4}$, $\dfrac{dD}{dt} = \dfrac{(\pi/4) + (\sqrt{2}/2)(\sqrt{2}/2)}{\sqrt{(\pi^2/16) + (1/2)}}(2) = \dfrac{2(\pi + 2)}{\sqrt{\pi^2 + 8}}$ cm/sec.

c. When $x = \dfrac{\pi}{3}$, $\dfrac{dD}{dt} = \dfrac{(\pi/3) + (\sqrt{3}/2)(1/2)}{\sqrt{(\pi^2/9) + (3/4)}}(2) = \dfrac{4\pi + 3\sqrt{3}}{\sqrt{4\pi^2 + 27}}$ cm/sec.

d. When $x = \dfrac{\pi}{2}$, $\dfrac{dD}{dt} = \dfrac{(\pi/2) + (1)(0)}{\sqrt{(\pi^2/4) + 1}}(2) = \dfrac{2\pi}{\sqrt{\pi^2 + 4}}$ cm/sec.

11. $A = \pi r^2$

$\dfrac{dr}{dt} = 2$

$\dfrac{dA}{dt} = 2\pi r \dfrac{dr}{dt}$

(a) When $r = 6$,

$\dfrac{dA}{dt} = 2\pi(6)(2) = 24\pi$ in²/min.

(b) When $r = 24$,

$\dfrac{dA}{dt} = 2\pi(24)(2) = 96\pi$ in²/min.

12. $A = \pi r^2$

$\dfrac{dA}{dt} = 2\pi r \dfrac{dr}{dt}$

If dr/dt is constant, dA/dt is not constant.

13. a. $\sin\dfrac{\theta}{2} = \dfrac{(1/2)b}{s} \Rightarrow b = 2s\sin\dfrac{\theta}{2}$

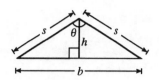

$\cos\dfrac{\theta}{2} = \dfrac{h}{s} \Rightarrow h = s\cos\dfrac{\theta}{2}$

$A = \dfrac{1}{2}bh = \dfrac{1}{2}\left(2s\sin\dfrac{\theta}{2}\right)\left(s\cos\dfrac{\theta}{2}\right)$

$ = \dfrac{s^2}{2}\left(2\sin\dfrac{\theta}{2}\cos\dfrac{\theta}{2}\right) = \dfrac{s^2}{2}\sin\theta$

b. $\dfrac{dA}{dt} = \dfrac{s^2}{2}\cos\theta\,\dfrac{d\theta}{dt}$ where $\dfrac{d\theta}{dt} = \dfrac{1}{2}$ rad/min.

When $\theta = \dfrac{\pi}{6}$, $\dfrac{dA}{dt} = \dfrac{s^2}{2}\left(\dfrac{\sqrt{3}}{2}\right)\left(\dfrac{1}{2}\right) = \dfrac{\sqrt{3}\,s^2}{8}$ rad/min.

When $\theta = \dfrac{\pi}{3}$, $\dfrac{dA}{dt} = \dfrac{s^2}{2}\left(\dfrac{1}{2}\right)\left(\dfrac{1}{2}\right) = \dfrac{s^2}{8}$ rad/min.

c. If $d\theta/dt$ is constant, dA/dt is proportional to $\cos\theta$. If $d\theta/dt$ is constant, dA/dt is proportional to $\cos\theta$.

14. $V = \dfrac{4}{3}\pi r^3$

$\dfrac{dr}{dt} = 2$

$\dfrac{dV}{dt} = 4\pi r^2\dfrac{dr}{dt}$

a. When $r = 6$, $\dfrac{dV}{dt} = 4\pi(6)^2(2) = 288\pi$ in^3/min.

When $r = 24$, $\dfrac{dV}{dt} = 4\pi(24)^2(2) = 4608\pi$ in^3/min.

b. If dr/dt is constant, dV/dt is proportional to r^2.

15. $V = \dfrac{4}{3}\pi r^3$

$\dfrac{dV}{dt} = 20$

$\dfrac{dV}{dt} = 4\pi r^2\dfrac{dr}{dt}$

$\dfrac{dr}{dt} = \left(\dfrac{1}{4\pi r^2}\right)\dfrac{dV}{dt}$

a. When $r = 1$,

$\dfrac{dr}{dt} = \dfrac{1}{4\pi(1)^2}(20) = \dfrac{5}{\pi}$ ft/min.

b. When $r = 2$,

$\dfrac{dr}{dt} = \dfrac{1}{4\pi(2)^2}(20) = \dfrac{5}{4\pi}$ ft/min.

16. $V = x^3$

$\dfrac{dx}{dt} = 3$

$\dfrac{dV}{dt} = 3x^2\dfrac{dx}{dt}$

a. When $x = 1$,

$\dfrac{dV}{dt} = 3(1)^2(3) = 9$ cm^3/sec.

b. When $x = 10$,

$\dfrac{dV}{dt} = 3(10)^2(3) = 900$ cm^3/sec.

17. $s = 6x^2$

$$\frac{dx}{dt} = 3$$

$$\frac{ds}{dt} = 12x\frac{dx}{dt}$$

a. When $x = 1$,

$$\frac{ds}{dt} = 12(1)(3) = 36 \text{ cm}^2/\text{sec}.$$

b. When $x = 10$,

$$\frac{ds}{dt} = 12(10)(3) = 360 \text{ cm}^2/\text{sec}.$$

18. $V = \frac{1}{3}\pi r^2 h = \frac{1}{3}\pi r^2(3r) = \pi r^3$

$$\frac{dr}{dt} = 2$$

$$\frac{dV}{dt} = 3\pi r^2\frac{dr}{dt}$$

a. When $r = 6$,

$$\frac{dV}{dt} = 3\pi(6)^2(2) = 216\pi \text{ in}^3/\text{min}.$$

b. When $r = 24$,

$$\frac{dV}{dt} = 3\pi(24)^2(2) = 3456\pi \text{ in}^3/\text{min}.$$

19. $V = \frac{1}{3}\pi r^2 h = \frac{1}{3}\pi\left(\frac{9}{4}h^2\right)h$ [since $2r = 3h$]

$$= \frac{3\pi}{4}h^3$$

$$\frac{dV}{dt} = 10$$

$$\frac{dV}{dt} = \frac{9\pi}{4}h^2\frac{dh}{dt} \Rightarrow \frac{dh}{dt} = \frac{4(dV/dt)}{9\pi h^2}$$

When $h = 15$, $\frac{dh}{dt} = \frac{4(10)}{9\pi(15)^2} = \frac{8}{405\pi}$ ft/min.

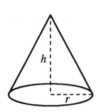

20. $V = \frac{1}{3}\pi r^2 h = \frac{1}{3}\pi\frac{25}{144}h^3 = \frac{25\pi}{3(144)}h^3$

$$r = \frac{5}{12}h$$

$$\frac{dV}{dt} = 10$$

$$\frac{dV}{dt} = \frac{25\pi}{144}h^2\frac{dh}{dt} \Rightarrow \frac{dh}{dt} = \left(\frac{144}{25\pi h^2}\right)\frac{dV}{dt}$$

When $h = 8$, $\frac{dh}{dt} = \frac{144}{25\pi(64)}(10) = \frac{9}{10\pi}$ ft/min.

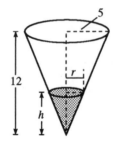

21. a. Total volume $= \left(\frac{1}{2}\right)(40)(5)(20) + (40)(4)(20) = 5200 \text{ ft}^3$

Volume of 4 ft. of water $= \left(\frac{1}{2}\right)(32)(4)(20) = 1280 \text{ ft}^3$

% of pool filled $= \frac{1280}{5200}(100\%) \approx 24.6\%$

b. $(b = 8h)$, $0 \le h \le 5$

$$V = \left(\frac{1}{2}\right)bh(20) = 10bh = 80h^2$$

$$\frac{dV}{dt} = 160h\frac{dh}{dt} \Rightarrow \frac{dh}{dt} = \left(\frac{1}{160h}\right)\frac{dV}{dt}$$

When $h = 4$ and $\frac{dV}{dt} = 10$, $\frac{dh}{dt} = \frac{1}{160(4)}(10) = \frac{1}{64}$ ft/min.

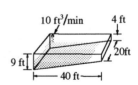

22. $V = \left(\dfrac{1}{2}\right)bh(12) = 6bh = 6h^2$ (since $b = h$)

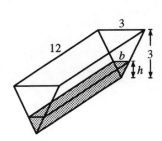

$\dfrac{dV}{dt} = 12h\dfrac{dh}{dt} \Rightarrow \dfrac{dh}{dt} = \left(\dfrac{1}{12h}\right)\dfrac{dV}{dt}$

When $h = 1$ and $\dfrac{dV}{dt} = 2$,

$\dfrac{dh}{dt} = \dfrac{1}{12}(2) = \dfrac{1}{6}$ ft/min = 2 in/min.

23. $\qquad x^2 + y^2 = 25^2$

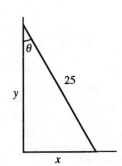

$2x\dfrac{dx}{dt} + 2y\dfrac{dy}{dt} = 0$

$\qquad \dfrac{dy}{dt} = \dfrac{-x}{y} \cdot \dfrac{dx}{dt} = \dfrac{-2x}{y}$ since $\dfrac{dx}{dt} = 2$.

a. When $x = 7$, $y = \sqrt{576} = 24$, $\dfrac{dy}{dt} = \dfrac{-2(7)}{24} = \dfrac{-7}{12}$ ft/sec.

$\qquad$ When $x = 15$, $y = \sqrt{400} = 20$, $\dfrac{dy}{dt} = \dfrac{-2(15)}{20} = \dfrac{-3}{2}$ ft/sec.

$\qquad$ When $x = 24$, $y = 7$, $\dfrac{dy}{dt} = \dfrac{-2(24)}{7} = \dfrac{-48}{7}$ ft/sec.

b. $\quad A = \dfrac{1}{2}xy$

$\dfrac{dA}{dt} = \dfrac{1}{2}\left(x\dfrac{dy}{dt} + y\dfrac{dx}{dt}\right)$

$\qquad$ From part a we have $x = 7$, $y = 24$, $\dfrac{dx}{dt} = 2$, and $\dfrac{dy}{dt} = -\dfrac{7}{12}$.

$\qquad$ Thus, $\dfrac{dA}{dt} = \dfrac{1}{2}\left[7\left(-\dfrac{7}{12}\right) + 24(2)\right] = \dfrac{527}{24} \approx 21.96$ ft²/sec.

c. $\qquad \tan\theta = \dfrac{x}{y}$

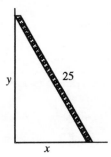

$\sec^2\theta\dfrac{d\theta}{dt} = \dfrac{1}{y} \cdot \dfrac{dx}{dt} - \dfrac{x}{y^2} \cdot \dfrac{dy}{dt}$

$\qquad \dfrac{d\theta}{dt} = \cos^2\theta\left[\dfrac{1}{y} \cdot \dfrac{dx}{dt} - \dfrac{x}{y^2} \cdot \dfrac{dy}{dt}\right]$

$\qquad$ Using $x = 7$, $y = 24$, $\dfrac{dx}{dt} = 2$, $\dfrac{dy}{dt} = -\dfrac{7}{12}$ and $\cos\theta = \dfrac{24}{25}$, we have

$\qquad \dfrac{d\theta}{dt} = \left(\dfrac{24}{25}\right)^2\left[\dfrac{1}{24}(2) - \dfrac{7}{(24)^2}\left(-\dfrac{7}{12}\right)\right] = \dfrac{1}{12}$ rad/sec.

24. $x^2 + y^2 = 16^2$

$2x\dfrac{dx}{dt} + 2y\dfrac{dy}{dt} = 0$

$\quad\quad\quad \dfrac{dx}{dt} = -\dfrac{y}{x} \cdot \dfrac{dy}{dt} = -\dfrac{0.5y}{x}$ since $\dfrac{dy}{dt} = 0.5$

When $x = 8$,

$\quad y = \sqrt{192} = 8\sqrt{3}, \ \ \dfrac{dx}{dt} = -\dfrac{0.5(8\sqrt{3})}{8} = -\dfrac{\sqrt{3}}{2}$ ft/sec.

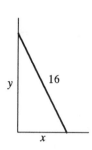

25. When $y = 20$, $\ x = \sqrt{40^2 - 20^2} = 20\sqrt{3}$ and $s = 40$.

$\quad\quad\quad x^2 + (40 - y)^2 = s^2$

$\quad 2x\dfrac{dx}{dt} + 2(40 - y)(-1)\dfrac{dy}{dt} = 2s\dfrac{ds}{dt}$

$\quad\quad x\dfrac{dx}{dt} + (y - 40)\dfrac{dy}{dt} = s\dfrac{ds}{dt}$

Also,

$\quad\quad x^2 + y^2 = 40^2$

$\quad 2x\dfrac{dx}{dt} + 2y\dfrac{dy}{dt} = 0 \ \Rightarrow \ \dfrac{dy}{dt} = -\dfrac{x}{y}\dfrac{dx}{dt}.$

Thus,

$\quad x\dfrac{dx}{dt} + (y - 40)\left(-\dfrac{x}{y}\dfrac{dx}{dt}\right) = s\dfrac{ds}{dt}$

$\quad\quad \dfrac{dx}{dt}\left[x - x + \dfrac{40x}{y}\right] = s\dfrac{ds}{dt}$

$\quad\quad\quad \dfrac{dx}{dt} = \dfrac{sy}{40x}\dfrac{ds}{dt} = \dfrac{(40)(20)}{(40)(20\sqrt{3})}(-0.5) = -\dfrac{\sqrt{3}}{6}$ ft/sec

$\quad\quad\quad \dfrac{dy}{dt} = -\dfrac{x}{y}\dfrac{dx}{dt} = -\dfrac{20\sqrt{3}}{20}\left(-\dfrac{\sqrt{3}}{6}\right) = \dfrac{1}{2}$ ft/sec.

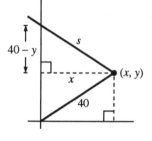

26. $L^2 = 144 + x^2$

$2L\dfrac{dL}{dt} = 2x\dfrac{dx}{dt}$

$\quad \dfrac{dx}{dt} = \dfrac{L}{x} \cdot \dfrac{dL}{dt} = -\dfrac{4L}{x}$ since $\dfrac{dL}{dt} = -4$ ft/sec

When $L = 13$,

$\quad x = \sqrt{L^2 - 144} = \sqrt{169 - 144} = 5$

$\dfrac{dx}{dt} = -\dfrac{4(13)}{5} = -\dfrac{52}{5} = -10.4$ ft/sec.

Speed of the boat increases as it approaches the dock.

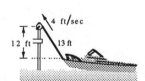

27. a. $s^2 = x^2 + y^2$

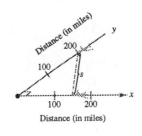

$$\frac{dx}{dt} = -450$$

$$\frac{dy}{dt} = -600$$

$$2s\frac{ds}{dt} = 2x\frac{dx}{dt} + 2y\frac{dy}{dt}$$

$$\frac{ds}{dt} = \frac{x(dx/dt) + y(dy/dt)}{s}$$

When $x = 150$ and $y = 200$, $s = 250$ and

$$\frac{ds}{dt} = \frac{150(-450) + 200(-600)}{250} = -750 \text{ mph.}$$

b. $t = \dfrac{250}{750} = \dfrac{1}{3}$ hr $= 20$ min

28. $x^2 + y^2 = s^2$

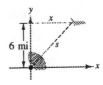

$$x = \sqrt{s^2 - y^2}$$

$$\frac{ds}{dt} = -240 \text{ mph}$$

$$y = 6 \text{ mi}$$

$$x = \sqrt{s^2 - 36}$$

$$\frac{dx}{dt} = \left(\frac{1}{2}\right)(s^2 - 36)^{-1/2}\left(2s\frac{ds}{dt}\right) = \frac{s}{\sqrt{s^2 - 36}} \cdot \frac{ds}{dt}$$

When $s = 10$, $\dfrac{dx}{dt} = \dfrac{10}{\sqrt{10^2 - 36}}(-240)$

$$= \frac{10}{8}(-240) = -300 \text{ mph.}$$

The speed of the plane is 300 mph.

29. $s^2 = 90^2 + x^2$

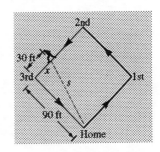

$$x = 30$$

$$\frac{dx}{dt} = -28$$

$$2s\frac{ds}{dt} = 2x\frac{dx}{dt} \implies \frac{ds}{dt} = \frac{x}{s} \cdot \frac{dx}{dt}$$

When $x = 30$,

$$s = \sqrt{90^2 + 30^2} = 30\sqrt{10}$$

$$\frac{ds}{dt} = \frac{30}{30\sqrt{10}}(-28) = \frac{-28}{\sqrt{10}} \approx -8.85 \text{ ft/sec.}$$

30. $s^2 = 90^2 + x^2$

$x = 60$

$\dfrac{dx}{dt} = 28$

$\dfrac{ds}{dt} = \dfrac{x}{s} \cdot \dfrac{dx}{dt}$

When $x = 60$,

$$s = \sqrt{90^2 + 60^2} = 30\sqrt{13}$$

$$\dfrac{ds}{dt} = \dfrac{60}{30\sqrt{13}}(28) = \dfrac{56}{\sqrt{13}} \approx 15.53 \text{ ft/sec.}$$

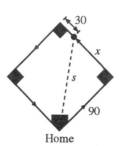

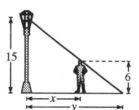

Home

31. a. $\dfrac{15}{6} = \dfrac{y}{y-x} \Rightarrow 15y - 15x = 6y$

$y = \dfrac{5}{3}x$

$\dfrac{dx}{dt} = 5$

$\dfrac{dy}{dt} = \dfrac{5}{3} \cdot \dfrac{dx}{dt} = \dfrac{5}{3}(5) = \dfrac{25}{3}$ ft/sec

b. $\dfrac{d(y-x)}{dt} = \dfrac{dy}{dt} - \dfrac{dx}{dt} = \dfrac{25}{3} - 5 = \dfrac{10}{3}$ ft/sec

32. $s = \sqrt{(2-0)^2 + (0-y)^2}$ Distance Formula

$s = \sqrt{4 + y^2} = (4 + y^2)^{1/2}$

$\dfrac{ds}{dt} = \dfrac{1}{2}(4 + y^2)^{-1/2}\left(2y\dfrac{dy}{dt}\right)$

$\dfrac{ds}{dt} = \dfrac{y}{\sqrt{4 + y^2}}\dfrac{dy}{dt}$

$y = 2\sin\dfrac{\pi t}{2}$

$\dfrac{dy}{dt} = 2\cos\dfrac{\pi t}{2}\left(\dfrac{\pi}{2}\right) = \pi\cos\dfrac{\pi t}{2}$

When $t = \dfrac{3}{2}$, $y = 2\sin\dfrac{\pi(3/2)}{2} = 2\left(\dfrac{\sqrt{2}}{2}\right) = \sqrt{2}$ and $\dfrac{dy}{dt} = \pi\cos\dfrac{\pi(3/2)}{2} = -\dfrac{\pi\sqrt{2}}{2}$.

The rate of change of the length of the mechanical arm when $t = \dfrac{3}{2}$ is $\dfrac{ds}{dt} = \dfrac{\sqrt{2}}{\sqrt{4+2}} \cdot \dfrac{-\pi\sqrt{2}}{2} = -\dfrac{\pi}{\sqrt{6}}$.

When $t = 3$, $y = 2\sin\dfrac{\pi(3)}{2} = 2(-1) = -2$ and $\dfrac{dy}{dt} = \pi\cos\dfrac{\pi(3)}{2} = \pi(0) = 0$.

The rate of change of the length of the mechanical arm when $t = 3$ is

$$\dfrac{ds}{dt} = \dfrac{-2}{\sqrt{4+4}}(0) = 0.$$

33. $x(t) = 2\sin\dfrac{\pi t}{6}$

a. Period: $\dfrac{2\pi}{\pi/6} = 12$ sec

b. When $x = \pm 2$, $y = \sqrt{3^2 - 2^2} = \sqrt{5}$.
Lowest point: $(0, \sqrt{5})$

c. When $x = 1$, $y = \sqrt{3^2 - 1^2} = 2\sqrt{2}$ and $t = 1$:

$$\frac{dx}{dt} = 2\left(\frac{\pi}{6}\right)\cos\frac{\pi t}{6} = \frac{\pi}{3}\cos\frac{\pi t}{6}$$

$$x^2 + y^2 = 3^2$$

$$2x\frac{dx}{dt} + 2y\frac{dy}{dt} = 0$$

$$\frac{dy}{dt} = -\frac{x}{y}\frac{dx}{dt}$$

When $x = 1$:

$$\frac{dy}{dt} = -\frac{1}{2\sqrt{2}}\left(\frac{\pi}{3}\cos\frac{\pi}{6}\right)$$

$$= -\frac{\pi}{6\sqrt{2}}\cdot\frac{\sqrt{3}}{2} = -\frac{\sqrt{6}\pi}{24}$$

$$\text{Speed} = \left|-\frac{\sqrt{6}\pi}{24}\right| \approx 0.32 \text{ ft/sec}$$

34. $x(t) = 2\sin\pi t$

a. Period: $\dfrac{2\pi}{\pi} = 2$ sec

b. When $x = \pm 2$, $y = \sqrt{5}$.
Lowest point: $(0, \sqrt{5})$

c. When $x = 1$, $y = 2\sqrt{2}$ and $t = \frac{1}{6}$:

$$\frac{dx}{dt} = 2\pi\cos\pi t$$

$$x^2 + y^2 = 3^2$$

$$2x\frac{dx}{dt} + 2y\frac{dy}{dt} = 0$$

$$\frac{dy}{dt} = -\frac{x}{y}\frac{dx}{dt}$$

When $x = 1$:

$$\frac{dy}{dt} = -\frac{1}{2\sqrt{2}}\left(2\pi\cos\frac{\pi}{6}\right) = -\frac{\sqrt{6}\pi}{4}$$

$$\text{Speed} = \left|-\frac{\sqrt{6}\pi}{4}\right| \approx 1.92 \text{ ft/sec.}$$

35. Since the evaporation rate is proportional to the surface area, $dV/dt = k(4\pi r^2)$. However, since $V = (4/3)\pi r^3$, we have

$$\frac{dV}{dt} = 4\pi r^2\frac{dr}{dt}.$$

Therefore,

$$k(4\pi r^2) = 4\pi r^2\frac{dr}{dt} \implies k = \frac{dr}{dt}.$$

36.

$$\frac{1}{R} = \frac{1}{R_1} + \frac{1}{R_2}$$

$$\frac{dR_1}{dt} = 1$$

$$\frac{dR_2}{dt} = 1.5$$

$$\frac{1}{R^2}\cdot\frac{dR}{dt} = \frac{1}{R_1^2}\cdot\frac{dR_1}{dt} + \frac{1}{R_2^2}\cdot\frac{dR_2}{dt}$$

When $R_1 = 50$ and $R_2 = 75$,

$$R = 30$$

$$\frac{dR}{dt} = (30)^2\left[\frac{1}{(50)^2}(1) + \frac{1}{(75)^2}(1.5)\right]$$

$$= 0.6 \text{ ohms/sec.}$$

37.
$$pv^{1.3} = k$$

$$1.3pv^{0.3}\frac{dv}{dt} + v^{1.3}\frac{dp}{dt} = 0$$

$$v^{0.3}\left(1.3p\frac{dv}{dt} + v\frac{dp}{dt}\right) = 0$$

38.
$$rg\tan\theta = v^2$$

$$32r\tan\theta = v^2, \quad r \text{ is a constant.}$$

$$32r\sec^2\theta\frac{d\theta}{dt} = 2v\frac{dv}{dt}$$

$$\frac{dv}{dt} = \frac{16r}{v}\sec^2\theta\frac{d\theta}{dt}$$

Likewise, $\dfrac{d\theta}{dt} = \dfrac{v}{16r}\cos^2\theta\dfrac{dv}{dt}.$

39. $\tan \theta = \dfrac{y}{100}$

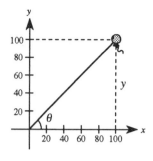

$\dfrac{dy}{dt} = 10 \text{ ft/sec}$

$(\sec^2 \theta)\dfrac{d\theta}{dt} = \dfrac{1}{100} \cdot \dfrac{dy}{dt}$

$\dfrac{d\theta}{dt} = \dfrac{1}{100} \cos^2 \theta \dfrac{dy}{dt}$

When $y = 100$,

$\theta = \dfrac{\pi}{4}$

$\dfrac{d\theta}{dt} = \dfrac{1}{100}\left(\dfrac{1}{2}\right)(10) = \dfrac{1}{20} \text{ rad/sec.}$

40. $\sin \theta = \dfrac{15}{x}$

$\dfrac{dx}{dt} = -1 \text{ ft/sec}$

$\cos \theta \left(\dfrac{d\theta}{dt}\right) = \dfrac{-15}{x^2} \cdot \dfrac{dx}{dt}$

$\dfrac{d\theta}{dt} = \dfrac{-15}{x^2}(\sec \theta)\dfrac{dx}{dt}$

$= \dfrac{-15}{625}\left(\dfrac{25}{20}\right)(-1) = \dfrac{3}{100} \text{ rad/sec}$

41. $\tan \theta = \dfrac{y}{x}, \quad y = 5$

$\dfrac{dx}{dt} = -600 \text{ mi/hr}$

$(\sec^2 \theta)\dfrac{d\theta}{dt} = -\dfrac{5}{x^2} \cdot \dfrac{dx}{dt}$

$\dfrac{d\theta}{dt} = \cos^2 \theta \left(-\dfrac{5}{x^2}\right)\dfrac{dx}{dt} = \dfrac{x^2}{L^2}\left(-\dfrac{5}{x^2}\right)\dfrac{dx}{dt}$

$= \left(-\dfrac{5^2}{L^2}\right)\left(\dfrac{1}{5}\right)\dfrac{dx}{dt} = (-\sin^2 \theta)\left(\dfrac{1}{5}\right)(-600) = 120 \sin^2 \theta$

a. When $\theta = 30°$, $\dfrac{d\theta}{dt} = \dfrac{120}{4} = 30 \text{ rad/hr} = \dfrac{1}{2} \text{ rad/min.}$

b. When $\theta = 60°$, $\dfrac{d\theta}{dt} = 120\left(\dfrac{3}{4}\right) = 90 \text{ rad/hr} = \dfrac{3}{2} \text{ rad/min.}$

c. When $\theta = 75°$, $\dfrac{d\theta}{dt} = 120 \sin^2 75° \approx 111.96 \text{ rad/hr} \approx 1.87 \text{ rad/min.}$

42. $\tan \theta = \dfrac{x}{50}$

$$\frac{d\theta}{dt} = 30(2\pi) = 60\pi \text{ rad/min} = \pi \text{ rad/sec}$$

$$\sec^2 \theta \left(\frac{d\theta}{dt} \right) = \frac{1}{50} \left(\frac{dx}{dt} \right)$$

$$\frac{dx}{dt} = 50 \sec^2 \theta \left(\frac{d\theta}{dt} \right)$$

a. When $\theta = 30°$, $\dfrac{dx}{dt} = \dfrac{200\pi}{3}$ ft/sec.

b. When $\theta = 60°$, $\dfrac{dx}{dt} = 200\pi$ ft/sec.

c. When $\theta = 70°$, $\dfrac{dx}{dt} \approx 427.43\pi$ ft/sec.

43. $x = -\cos \theta$

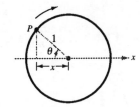

$$\frac{d\theta}{dt} = 10(2\pi) = 20\pi \text{ rad/sec}$$

$$\frac{dx}{dt} = \sin \theta \frac{d\theta}{dt} = 20\pi \sin \theta$$

a. When $\theta = 0°$, $\dfrac{dx}{dt} = 0$ ft/sec.

b. When $\theta = 30°$, $\dfrac{dx}{dt} = 10\pi$ ft/sec.

c. When $\theta = 60°$, $\dfrac{dx}{dt} = 10\sqrt{3}\,\pi$ ft/sec.

44. $\sin 22° = \dfrac{x}{y}$

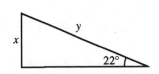

$$0 = -\frac{x}{y^2} \cdot \frac{dy}{dt} + \frac{1}{y} \cdot \frac{dx}{dt}$$

$$\frac{dx}{dt} = \frac{x}{y} \cdot \frac{dy}{dt} = (\sin 22°)(240) \approx 89.9056 \text{ mi/hr}$$

45. $x^2 + y^2 = 25$; acceleration of the top of the ladder $= \dfrac{d^2 y}{dt^2}$

First derivative: $2x\dfrac{dx}{dt} + 2y\dfrac{dy}{dt} = 0$

$$x\dfrac{dx}{dt} + y\dfrac{dy}{dt} = 0$$

Second derivative: $x\dfrac{d^2 x}{dt^2} + \dfrac{dx}{dt} \cdot \dfrac{dx}{dt} + y\dfrac{d^2 y}{dt^2} + \dfrac{dy}{dt} \cdot \dfrac{dy}{dt} = 0$

$$\dfrac{d^2 y}{dt^2} = \left(\dfrac{1}{y}\right)\left[-x\dfrac{d^2 x}{dt^2} - \left(\dfrac{dx}{dt}\right)^2 - \left(\dfrac{dy}{dt}\right)^2\right]$$

When $x = 7$, $y = 24$, $\dfrac{dy}{dt} = -\dfrac{7}{12}$, and $\dfrac{dx}{dt} = 2$ (see Exercise 23). Since $\dfrac{dx}{dt}$ is constant, $\dfrac{d^2 x}{dt^2} = 0$.

$$\dfrac{d^2 y}{dt^2} = \dfrac{1}{24}\left[-7(0) - (2)^2 - \left(-\dfrac{7}{12}\right)^2\right] = \dfrac{1}{24}\left[-4 - \dfrac{49}{144}\right] = \dfrac{1}{24}\left[-\dfrac{625}{144}\right] \approx -0.1808 \text{ ft/sec}^2$$

46. $L^2 = 144 + x^2$; acceleration of the boat $= \dfrac{d^2 x}{dt^2}$

First derivative: $2L\dfrac{dL}{dt} = 2x\dfrac{dx}{dt}$

$$L\dfrac{dL}{dt} = x\dfrac{dx}{dt}$$

Second derivative: $L\dfrac{d^2 L}{dt^2} + \dfrac{dL}{dt} \cdot \dfrac{dL}{dt} = x\dfrac{d^2 x}{dt^2} + \dfrac{dx}{dt} \cdot \dfrac{dx}{dt}$

$$\dfrac{d^2 x}{dt^2} = \left(\dfrac{1}{x}\right)\left[L\dfrac{d^2 L}{dt^2} + \left(\dfrac{dL}{dt}\right)^2 - \left(\dfrac{dx}{dt}\right)^2\right]$$

When $L = 13$, $x = 5$, $\dfrac{dx}{dt} = -10.4$, and $\dfrac{dL}{dt} = -4$ (see Exercise 26). Since $\dfrac{dL}{dt}$ is constant, $\dfrac{d^2 L}{dt^2} = 0$.

$$\dfrac{d^2 x}{dt^2} = \dfrac{1}{5}[13(0) + (-4)^2 - (-10.4)^2]$$

$$= \dfrac{1}{5}[16 - 108.16] = \dfrac{1}{5}[-92.16] = -18.432 \text{ ft/sec}^2$$

Chapter 2   Review Exercises

1. $f(x) = x^2 - 2x + 3$

$$f'(x) = \lim_{\Delta x \to 0} \frac{f(x + \Delta x) - f(x)}{\Delta x}$$

$$= \lim_{\Delta x \to 0} \frac{[(x + \Delta x)^2 - 2(x + \Delta x) + 3] - [x^2 - 2x + 3]}{\Delta x}$$

$$= \lim_{\Delta x \to 0} \frac{(x^2 + 2x(\Delta x) + (\Delta x)^2 - 2x - 2(\Delta x) + 3) - (x^2 - 2x + 3)}{\Delta x}$$

$$= \lim_{\Delta x \to 0} \frac{2x(\Delta x) + (\Delta x)^2 - 2(\Delta x)}{\Delta x} = \lim_{\Delta x \to 0} (2x + \Delta x - 2) = 2x - 2$$

2. $f(x) = \dfrac{x + 1}{x - 1}$

$$f'(x) = \lim_{\Delta x \to 0} \frac{f(x + \Delta x) - f(x)}{\Delta x} = \lim_{\Delta x \to 0} \frac{\dfrac{x + \Delta x + 1}{x + \Delta x - 1} - \dfrac{x + 1}{x - 1}}{\Delta x}$$

$$= \lim_{\Delta x \to 0} \frac{(x + \Delta x + 1)(x - 1) - (x + \Delta x - 1)(x + 1)}{\Delta x(x + \Delta x - 1)(x - 1)}$$

$$= \lim_{\Delta x \to 0} \frac{(x^2 + x\Delta x + x - x - \Delta x - 1) - (x^2 + x\Delta x - x + x + \Delta x - 1)}{\Delta x(x + \Delta x - 1)(x - 1)}$$

$$= \lim_{\Delta x \to 0} \frac{-2\Delta x}{\Delta x(x + \Delta x - 1)(x - 1)} = \lim_{\Delta x \to 0} \frac{-2}{(x + \Delta x - 1)(x - 1)} = \frac{-2}{(x - 1)^2}$$

3. $f(x) = x\sqrt{4 - x}$

a.

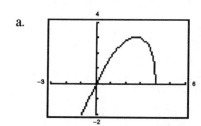

c. $x = 0$

Δx	$f(x + \Delta x)$	$f(x)$	$\dfrac{f(x + \Delta x) - f(x)}{\Delta x}$
2	2.8284	0	1.4142
1	1.7321	0	1.7321
0.5	0.9354	0	1.8708
0.1	0.1975	0	1.9748

b. $f'(x) = x\left(\dfrac{1}{2}\right)(4 - x)^{-1/2}(-1) + (4 - x)^{1/2}$

$$= \frac{1}{2}(4 - x)^{-1/2}[-x + 2(4 - x)]$$

$$= \frac{8 - 3x}{2\sqrt{4 - x}}$$

When $x = 0$, $f'(0) = 2$.

Tangent line: $y = 2x$

d. $\dfrac{f(x + \Delta x) - f(x)}{\Delta x}$ represents the slope of the line through $(x, \ f(x))$ and $(x + \Delta x, \ f(x + \Delta x))$. As $\Delta x \to 0$, this quantity approaches $f'(x)$. That is, the numbers in the last column approach $f'(0) = 2$.

4. a.

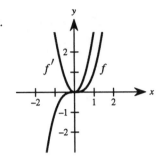

b.

5. $f(x) = x^3 - 3x^2$

$f'(x) = 3x^2 - 6x = 3x(x-2)$

6. $f(x) = x^{1/2} - x^{-1/2}$

$f'(x) = \frac{1}{2}x^{-1/2} + \frac{1}{2}x^{-3/2} = \frac{x+1}{2x^{3/2}}$

7. $f(x) = 2x - x^{-2}$

$f'(x) = 2 + 2x^{-3} = 2\left(1 + \frac{1}{x^3}\right)$

$= \frac{2(x^3+1)}{x^3}$

8. $f(x) = \frac{x+1}{x-1}$

$f'(x) = \frac{(x-1)(1) - (x+1)(1)}{(x-1)^2}$

$= \frac{-2}{(x-1)^2}$

9. $g(t) = \frac{2}{3}t^{-2}$

$g'(t) = \frac{-4}{3}t^{-3} = \frac{-4}{3t^3}$

10. $h(x) = \frac{2}{9}x^{-2}$

$h'(x) = \frac{-4}{9}x^{-3} = \frac{-4}{9x^3}$

11. $f(x) = (1-x^3)^{1/2}$

$f'(x) = \frac{1}{2}(1-x^3)^{-1/2}(-3x^2)$

$= -\frac{3x^2}{2\sqrt{1-x^3}}$

12. $f(x) = (x^2-1)^{1/3}$

$f'(x) = \frac{1}{3}(x^2-1)^{-2/3}(2x)$

$= \frac{2x}{3(x^2-1)^{2/3}}$

13. $f(x) = (3x^2+7)(x^2-2x+3)$

$f'(x) = (3x^2+7)(2x-2) + (x^2-2x+3)(6x)$

$= 2(6x^3 - 9x^2 + 16x - 7)$

14. $f(x) = \left(x^2 + \frac{1}{x}\right)^5$

$f'(x) = 5\left(x^2 + \frac{1}{x}\right)^4\left(2x - \frac{1}{x^2}\right)$

15. $f(s) = (s^2-1)^{5/2}(s^3+5)$

$f'(s) = (s^2-1)^{5/2}(3s^2) + (s^3+5)\left(\frac{5}{2}\right)(s^2-1)^{3/2}(2s)$

$= s(s^2-1)^{3/2}[3s(s^2-1) + 5(s^3+5)]$

$= s(s^2-1)^{3/2}(8s^3 - 3s + 25)$

16. $h(\theta) = \frac{\theta}{(1-\theta)^3}$

$h'(\theta) = \frac{(1-\theta)^3 - \theta[3(1-\theta)^2(-1)]}{(1-\theta)^6}$

$= \frac{(1-\theta)^2(1-\theta+3\theta)}{(1-\theta)^6} = \frac{2\theta+1}{(1-\theta)^4}$

17. $f(x) = \frac{x^2+x-1}{x^2-1}$

$f'(x) = \frac{(x^2-1)(2x+1) - (x^2+x-1)(2x)}{(x^2-1)^2}$

$= \frac{-(x^2+1)}{(x^2-1)^2}$

18. $f(x) = \frac{6x-5}{x^2+1}$

$f'(x) = \frac{(x^2+1)(6) - (6x-5)(2x)}{(x^2+1)^2}$

$= \frac{2(3 + 5x - 3x^2)}{(x^2+1)^2}$

19. $f(x) = (4 - 3x^2)^{-1}$

$f'(x) = -(4 - 3x^2)^{-2}(-6x) = \dfrac{6x}{(4 - 3x^2)^2}$

20. $f(x) = 9(3x^2 - 2x)^{-1}$

$f'(x) = -9(3x^2 - 2x)^{-2}(6x - 2) = \dfrac{18(1 - 3x)}{(3x^2 - 2x)^2}$

21. $y = 3\cos(3x + 1)$

$y' = -9\sin(3x + 1)$

22. $y = 1 - \cos 2x + 2\cos^2 x$

$y' = 2\sin 2x - 4\cos x \sin x$

$\quad = 2[2\sin x \cos x] - 4\sin x \cos x$

$\quad = 0$

23. $y = \frac{1}{2}\csc 2x$

$y' = \frac{1}{2}(-\csc 2x \cot 2x)(2) = -\csc 2x \cot 2x$

24. $y = \csc 3x + \cot 3x$

$y' = -3\csc 3x \cot 3x - 3\csc^2 3x$

$\quad = -3\csc 3x(\cot 3x + \csc 3x)$

25. $y = \dfrac{x}{2} - \dfrac{\sin 2x}{4}$

$y' = \dfrac{1}{2} - \dfrac{1}{4}\cos 2x(2)$

$\quad = \dfrac{1}{2}(1 - \cos 2x) = \sin^2 x$

26. $y = \dfrac{1 + \sin x}{1 - \sin x}$

$y' = \dfrac{(1 - \sin x)\cos x - (1 + \sin x)(-\cos x)}{(1 - \sin x)^2}$

$\quad = \dfrac{2\cos x}{(1 - \sin x)^2}$

27. $y = \dfrac{2}{3}\sin^{3/2} x - \dfrac{2}{7}\sin^{7/2} x$

$y' = \sin^{1/2} x \cos x - \sin^{5/2} x \cos x$

$\quad = (\cos x)\sqrt{\sin x}(1 - \sin^2 x)$

$\quad = (\cos^3 x)\sqrt{\sin x}$

28. $y = \dfrac{\sec^7 x}{7} - \dfrac{\sec^5 x}{5}$

$y' = \sec^6 x(\sec x \tan x) - \sec^4 x(\sec x \tan x)$

$\quad = \sec^5 x \tan x(\sec^2 x - 1)$

$\quad = \sec^5 x \tan^3 x$

29. $y = -x\tan x$

$y' = -x\sec^2 x - \tan x$

30. $y = x\cos x - \sin x$

$y' = -x\sin x + \cos x - \cos x = -x\sin x$

31. $y = \dfrac{\sin x}{x^2}$

$y' = \dfrac{(x^2)\cos x - (\sin x)(2x)}{x^4} = \dfrac{x\cos x - 2\sin x}{x^3}$

32. $y = \dfrac{\cos(x - 1)}{x - 1}$

$y' = \dfrac{-(x - 1)\sin(x - 1) - \cos(x - 1)(1)}{(x - 1)^2}$

$\quad = -\dfrac{1}{(x - 1)^2}[(x - 1)\sin(x - 1) + \cos(x - 1)]$

33. $f(t) = t^2(t-1)^5$

$f'(t) = t(t-1)^4(7t-2)$

The zeros of f' correspond to the points on the graph of f where the tangent line is horizontal.

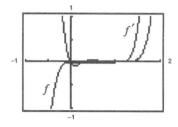

34. $f(x) = (x^2 + 2x - 8)^2$

$f'(x) = 4(x^3 + 3x^2 - 6x - 8)$

$\quad\quad = 4(x-2)(x+1)(x+4)$

The zeros of f' correspond to the points on the graph of f where the tangent line is horizontal.

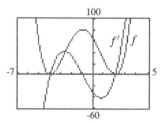

35. $g(x) = 2x(x+1)^{-1/2}$

$g'(x) = \dfrac{x+2}{(x+1)^{3/2}}$

g' does not equal zero for any value of x in the domain. The graph of g has no horizontal tangent lines.

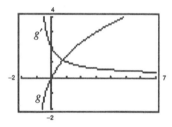

36. $g(x) = x(x^2+1)^{1/2}$

$g'(x) = \dfrac{2x^2+1}{\sqrt{x^2+1}}$

g' does not equal zero for any value of x. The graph of g has no horizontal tangent lines.

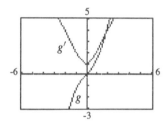

37. $f(t) = (t+1)^{1/2}(t+1)^{1/3} = (t+1)^{5/6}$

$f'(t) = \dfrac{5}{6(t+1)^{1/6}}$

f' does not equal zero for any x in the domain. The graph of f has no horizontal tangent lines.

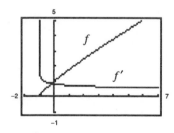

38. $y = \sqrt{3x}(x+2)^3$

$y' = \dfrac{3(x+2)^2(7x+2)}{2\sqrt{3x}}$

y' does not equal zero for any x in the domain. The graph has no horizontal tangent lines.

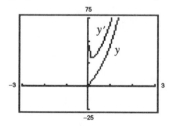

39. $y = \tan\sqrt{1-x}$

$$y' = -\frac{\sec^2\sqrt{1-x}}{2\sqrt{1-x}}$$

y' does not equal zero for any x in the domain. The graph has no horizontal tangent lines.

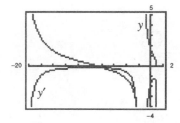

40. $y = 2\csc^3\sqrt{x}$

$$y' = -\frac{3}{\sqrt{x}}\csc^3\sqrt{x}\cot\sqrt{x}$$

The zero of y' corresponds to the point on the graph of y where the tangent line is horizontal.

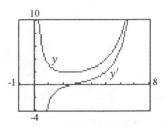

41. $y = 2x^2 + \sin 2x$

$y' = 4x + 2\cos 2x$

$y'' = 4 - 4\sin 2x$

42. $y = x^{-1} + \tan x$

$y' = -x^{-2} + \sec^2 x$

$y'' = 2x^{-3} + 2\sec x(\sec x\tan x)$

$\quad = \dfrac{2}{x^3} + 2\sec^2 x\tan x$

43. $f(x) = \cot x$

$f'(x) = -\csc^2 x$

$f''(x) = 2\csc^2 x\cot x$

44. $y = \sin^2 x$

$y' = 2\sin x\cos x = \sin 2x$

$y'' = 2\cos 2x$

45. $f(t) = \dfrac{t}{(1-t)^2}$

$f'(t) = \dfrac{t+1}{(1-t)^3}$

$f''(t) = \dfrac{2(t+2)}{(1-t)^4}$

46. $g(x) = \dfrac{6x-5}{x^2+1}$

$g'(x) = \dfrac{2(-3x^2+5x+3)}{(x^2+1)^2}$

$g''(x) = \dfrac{2(6x^3-15x^2-18x+5)}{(x^2+1)^3}$

47. $g(x) = x\tan x$

$g'(x) = x\sec^2 x + \tan x$

$g''(x) = 2\sec^2 x(x\tan x + 1)$

48. $h(x) = x\sqrt{x^2-1}$

$h'(x) = \dfrac{2x^2-1}{\sqrt{x^2-1}}$

$h''(x) = \dfrac{x(2x^2-3)}{(x^2-1)^{3/2}}$

49. $\qquad x^2 + 3xy + y^3 = 10$

$2x + 3xy' + 3y + 3y^2y' = 0$

$\qquad 3(x+y^2)y' = -(2x+3y)$

$\qquad\qquad y' = \dfrac{-(2x+3y)}{3(x+y^2)}$

50. $x^2 + 9y^2 - 4x + 3y - 7 = 0$

$2x + 18yy' - 4 + 3y' = 0$

$\qquad 3(6y+1)y' = 4 - 2x$

$\qquad\qquad y' = \dfrac{4-2x}{3(6y+1)}$

51.
$$y\sqrt{x} - x\sqrt{y} = 16$$

$$y\left(\frac{1}{2}x^{-1/2}\right) + x^{1/2}y' - x\left(\frac{1}{2}y^{-1/2}y'\right) - y^{1/2} = 0$$

$$\left(\sqrt{x} - \frac{x}{2\sqrt{y}}\right)y' = \sqrt{y} - \frac{y}{2\sqrt{x}}$$

$$\frac{2\sqrt{xy} - x}{2\sqrt{y}}y' = \frac{2\sqrt{xy} - y}{2\sqrt{x}}$$

$$y' = \frac{2\sqrt{xy} - y}{2\sqrt{x}} \cdot \frac{2\sqrt{y}}{2\sqrt{xy} - x} = \frac{2y\sqrt{x} - y\sqrt{y}}{2x\sqrt{y} - x\sqrt{x}}$$

52.
$$y^2 = x^3 - x^2y + xy - y^2$$

$$0 = x^3 - x^2y + xy - 2y^2$$

$$0 = 3x^2 - x^2y' - 2xy + xy' + y - 4yy'$$

$$(x^2 - x + 4y)y' = 3x^2 - 2xy + y$$

$$y' = \frac{3x^2 - 2xy + y}{x^2 - x + 4y}$$

53.
$$x \sin y = y \cos x$$

$$(x \cos y)y' + \sin y = -y \sin x + y' \cos x$$

$$y'(x \cos y - \cos x) = -y \sin x - \sin y$$

$$y' = \frac{y \sin x + \sin y}{\cos x - x \cos y}$$

54.
$$\cos(x + y) = x$$

$$-(1 + y')\sin(x + y) = 1$$

$$-y' \sin(x + y) = 1 + \sin(x + y)$$

$$y' = -\frac{1 + \sin(x + y)}{\sin(x + y)}$$

$$= -\csc(x + 1) - 1$$

55. $y = (x + 3)^3$

$y' = 3(x + 3)^2$

At $(-2, 1)$: $y' = 3$

Tangent line: $y - 1 = 3(x + 2)$

$$3x - y + 7 = 0$$

Normal line: $y - 1 = -\frac{1}{3}(x + 2)$

$$x + 3y - 1 = 0$$

56. $y = (x - 2)^2$

$y' = 2(x - 2)$

At $(2, 0)$: $y' = 0$

Tangent line: $y = 0$

Normal line: $x = 2$

57. $x^2 + y^2 = 20$

$2x + 2yy' = 0$

$$y' = -\frac{x}{y}$$

At $(2, 4)$: $y' = -\frac{1}{2}$

Tangent line: $y - 4 = -\frac{1}{2}(x - 2)$

$$x + 2y - 10 = 0$$

Normal line: $y - 4 = 2(x - 2)$

$$2x - y = 0$$

58. $x^2 - y^2 = 16$

$2x - 2yy' = 0$

$$y' = \frac{x}{y}$$

At $(5, 3)$: $y' = \frac{5}{3}$

Tangent line: $y - 3 = \frac{5}{3}(x - 5)$

$$5x - 3y - 16 = 0$$

Normal line: $y - 3 = -\frac{3}{5}(x - 5)$

$$3x + 5y - 30 = 0$$

59. $y = (x-2)^{2/3}$

$y' = \dfrac{2}{3}(x-2)^{-1/3} = \dfrac{2}{3\sqrt[3]{x-2}}$

At $(3, 1)$: $y' = \dfrac{2}{3}$

Tangent line: $y - 1 = \dfrac{2}{3}(x-3)$

$2x - 3y - 3 = 0$

Normal line: $y - 1 = -\dfrac{3}{2}(x-3)$

$3x + 2y - 11 = 0$

60. $y = \dfrac{2x}{1-x^2}$

$y' = \dfrac{2(1-x^2) - 2x(-2x)}{(1-x^2)^2} = \dfrac{2(x^2+1)}{(1-x^2)^2}$

At $(0, 0)$: $y' = 2$

Tangent line: $y = 2x$

Normal line: $y = -\dfrac{1}{2}x$

61. $f(x) = \dfrac{1}{3}x^3 + x^2 - x - 1$

$f'(x) = x^2 + 2x - 1$

a. $x^2 + 2x - 1 = -1$

$x(x+2) = 0$

$(0, -1), \left(-2, \dfrac{7}{3}\right)$

b. $x^2 + 2x - 1 = 2$

$(x+3)(x-1) = 0$

$(-3, 2), \left(1, -\dfrac{2}{3}\right)$

c. $x^2 + 2x - 1 = 0$

$(x+1)^2 = 2$

$x = -1 \pm \sqrt{2}$

$\left(-1 + \sqrt{2}, \dfrac{2(1 - 2\sqrt{2})}{3}\right),$

$\left(-1 - \sqrt{2}, \dfrac{2(1 + 2\sqrt{2})}{3}\right)$

62. $f(x) = x^2 + 1$

$f'(x) = 2x$

a. $2x = -1$

$x = -\dfrac{1}{2}$

$\left(-\dfrac{1}{2}, \dfrac{5}{4}\right)$

b. $2x = 0$

$x = 0$

$(0, 1)$

c. $2x = 1$

$x = \dfrac{1}{2}$

$\left(\dfrac{1}{2}, \dfrac{5}{4}\right)$

63. $f(x) = 4 - |x - 2|$

a. Continuous at $x = 2$

b. Not differentiable at $x = 2$ because of the sharp turn in the graph

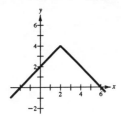

64. $f(x) = \begin{cases} x^2 + 4x + 2, & \text{if } x < -2 \\ 1 - 4x - x^2, & \text{if } x \ge -2 \end{cases}$

a. Nonremovable discontinuity at $x = -2$

b. Not differentiable at $x = -2$ because the function is discontinuous there.

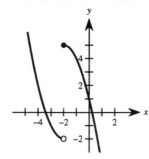

65.
$$y = 2\sin x + 3\cos x$$
$$y' = 2\cos x - 3\sin x$$
$$y'' = -2\sin x - 3\cos x$$
$$y'' + y = -(2\sin x + 3\cos x) + (2\sin x + 3\cos x)$$
$$= 0$$

66.
$$y = \frac{(10 - \cos x)}{x}$$
$$xy + \cos x = 10$$
$$xy' + y - \sin x = 0$$
$$xy' = \sin x - y$$
$$xy' + y = (\sin x - y) + y = \sin x$$

67. $T = 700(t^2 + 4t + 10)^{-1}$
$$T' = \frac{-1400(t + 2)}{(t^2 + 4t + 10)^2}$$

a. When $t = 1$,
$$T' = \frac{-1400(1 + 2)}{(1 + 4 + 10)^2} \approx -18.667 \text{ deg/hr.}$$

b. When $t = 3$,
$$T' = \frac{-1400(3 + 2)}{(9 + 12 + 10)^2} \approx -7.284 \text{ deg/hr.}$$

c. When $t = 5$,
$$T' = \frac{-1400(5 + 2)}{(25 + 20 + 10)^2} \approx -3.240 \text{ deg/hr.}$$

d. When $t = 10$,
$$T' = \frac{-1400(10 + 2)}{(100 + 40 + 10)^2} \approx -0.747 \text{ deg/hr.}$$

68. $v = \sqrt{2gh} = \sqrt{2(32)h} = 8\sqrt{h}$
$$\frac{dv}{dh} = \frac{4}{\sqrt{h}}$$

a. When $h = 9$, $\dfrac{dv}{dh} = \dfrac{4}{3}$ ft/sec.

b. When $h = 4$, $\dfrac{dv}{dh} = 2$ ft/sec.

69. $F = 200\sqrt{T}$
$$F'(t) = \frac{100}{\sqrt{T}}$$

a. When $T = 4$, $F'(4) = 50$ vibrations/sec/lb.

b. When $T = 9$, $F'(9) = 33\frac{1}{3}$ vibrations/sec/lb.

70. $s = -16t^2 + s_0$
First ball:
$$-16t^2 + 100 = 0$$
$$t = \sqrt{\frac{100}{16}} = \frac{10}{4} = 2.5 \text{ seconds to hit ground}$$
Second ball:
$$-16t^2 + 75 = 0$$
$$t^2 = \sqrt{\frac{75}{16}} = \frac{5\sqrt{3}}{4} \approx 2.165 \text{ seconds to hit ground}$$
Since the second ball was released one second after the first ball, the first ball will hit the ground first. The second ball will hit the ground $3.165 - 2.5 = 0.665$ second later.

71. Assume that the stone is thrown from an initial height of $s_0 = 0$. Thus, the position equation is $s = -16t^2 + v_0t$. The maximum value of s occurs when $ds/dt = 0$ and thus

$$\frac{ds}{dt} = -32t + v_0 = 0$$

$$-32t = -v_0$$

$$t = \frac{v_0}{32}.$$

This means that the maximum height is

$$s = -16\left(\frac{v_0}{32}\right)^2 + v_0\left(\frac{v_0}{32}\right) = \frac{v_0^2}{64}.$$

If s is to attain a value of 49, then

$$\frac{v_0^2}{64} = 49$$

$$v_0^2 = 3136$$

$$v_0 = 56 \text{ ft/sec.}$$

72. $s(t) = -16t^2 + 14,400 = 0$

$16t^2 = 14,400$

$t = 30 \text{ sec}$

Since 600 mph $= \frac{1}{6}$ mi/sec, in 30 seconds the bomb will move horizontally $\left(\frac{1}{6}\right)(30) = 5$ miles.

73. a.

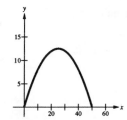

   Total horizontal distance: 50
   b. $0 = x - 0.02x^2$

   $0 = x\left(1 - \frac{x}{50}\right)$ implies $x = 50$.

   c. Ball reaches maximum height when $x = 25$.
   d. $y = x - 0.02x^2$

   $y' = 1 - 0.04x$

   $y'(0) = 1$

   $y'(10) = 0.6$

   $y'(25) = 0$

   $y'(30) = -0.2$

   $y'(50) = -1$

   e. $y'(25) = 0$

74. a.

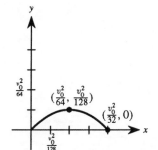

   b. $y = x - \dfrac{32}{v_0^2}x^2 = x\left(1 - \dfrac{32}{v_0^2}x\right)$

   $= 0$ if $x = 0$ or $x = \dfrac{v_0^2}{32}$.

   Projectile strikes the ground when $x = \dfrac{v_0^2}{32}$.

   Projectile reaches its maximum height at

   $x = \dfrac{v_0^2}{64}.$

   c. $y' = 1 - \dfrac{64}{v_0^2}x$

   When $x = \dfrac{v_0^2}{64}$, $\quad y' = 1 - \dfrac{64}{v_0^2}\left(\dfrac{v_0^2}{64}\right) = 0.$

75. $y = x - \dfrac{32}{v_0{}^2}x^2 = x\left(1 - \dfrac{32}{v_0{}^2}x\right) = 0$

when $x = 0$ and $x = x_0{}^2/32$. Therefore, the range is $x = v_0{}^2/32$. When the initial velocity is doubled the range is

$$x = \frac{(2v_0)^2}{32} = \frac{4v_0{}^2}{32}$$

or four times the initial range. From Exercise 74, the maximum height occurs when $x = v_0{}^2/64$. The maximum height is

$$y\left(\frac{v_0{}^2}{64}\right) = \frac{v_0{}^2}{64} - \frac{32}{v_0{}^2}\left(\frac{v_0{}^2}{64}\right)^2 = \frac{v_0{}^2}{64} - \frac{v_0{}^2}{128} = \frac{v_0{}^2}{128}.$$

If the initial velocity is doubled, the maximum height is

$$y\left[\frac{(2v_0)^2}{64}\right] = \frac{(2v_0)^2}{128} = 4\left(\frac{v_0{}^2}{128}\right)$$

or four times the original maximum height.

76. $v_0 = 70$ ft/sec

Range: $\;x = \dfrac{v_0{}^2}{32} = \dfrac{(70)^2}{32} = 153.125$ ft

Maximum height: $\;y = \dfrac{v_0{}^2}{128} = \dfrac{(70)^2}{128} \approx 38.28$ ft

77. $y = \sqrt{x}$

$\dfrac{dy}{dt} = 2$ units/sec

$\dfrac{dy}{dt} = \dfrac{1}{2\sqrt{x}}\dfrac{dx}{dt} \Rightarrow \dfrac{dx}{dt} = 2\sqrt{x}\dfrac{dy}{dt} = 4\sqrt{x}$

a. When $x = \dfrac{1}{2}$, $\;\dfrac{dx}{dt} = 2\sqrt{2}$ units/sec.

b. When $x = 1$, $\;\dfrac{dx}{dt} = 4$ units/sec.

c. When $x = 4$, $\;\dfrac{dx}{dt} = 8$ units/sec.

78. $y = \sqrt{x}$

$L^2 = x^2 + y^2$

$\dfrac{dy}{dt} = 2$ units/sec

$L^2 = y^4 + y^2$

$2L\dfrac{dL}{dt} = (4y^3 + 2y)\dfrac{dy}{dt}$

$\dfrac{dL}{dt} = \dfrac{4y^3 + 2y}{2L}\dfrac{dy}{dt} = \dfrac{4y^3 + 2y}{L} = \dfrac{(4x + 2)\sqrt{x}}{L}$

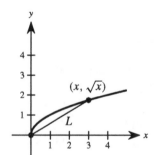

a. When $x = \dfrac{1}{2}$, $\;L = \sqrt{\left(\dfrac{1}{2}\right)^2 + \left(\dfrac{1}{\sqrt{2}}\right)^2} = \dfrac{\sqrt{3}}{2}$ and $\dfrac{dL}{dt} = \dfrac{(2 + 2)(1/\sqrt{2})}{\sqrt{3}/2} = \dfrac{8}{\sqrt{6}}$ units/sec.

b. When $x = 1$, $\;L = \sqrt{(1)^2 + (1)^2} = \sqrt{2}$ and $\dfrac{dL}{dt} = \dfrac{(4 + 2)(1)}{\sqrt{2}} = 3\sqrt{2}$ units/sec.

c. When $x = 4$, $\;L = \sqrt{(4)^2 + (2)^2} = 2\sqrt{5}$ and $\dfrac{dL}{dt} = \dfrac{(16 + 2)(2)}{2\sqrt{5}} = \dfrac{18}{\sqrt{5}}$ units/sec.

79. $\dfrac{s}{h} = \dfrac{1/2}{2}$

$s = \dfrac{1}{4}h$

$\dfrac{dV}{dt} = 1$

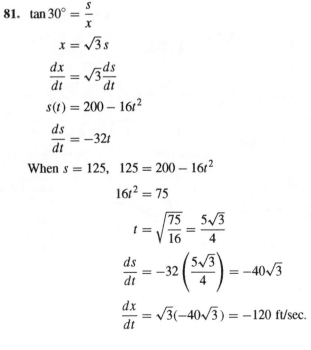

Width of water at depth h: $w = 2 + 2\left(\dfrac{1}{4}h\right) = \dfrac{4+h}{2}$

$V = \dfrac{5}{2}\left(2 + \dfrac{4+h}{2}\right)h = \dfrac{5}{4}(8+h)h$

$\dfrac{dV}{dt} = \dfrac{5}{2}(4+h)\dfrac{dh}{dt}$

$\dfrac{dh}{dt} = \dfrac{2(dV/dt)}{5(4+h)}$

When $h = 1$, $\dfrac{dh}{dt} = \dfrac{2}{25}$ ft/min.

80. $\tan\theta = x$

$\dfrac{d\theta}{dt} = 3(2\pi)$ rad/min

$\sec^2\theta\left(\dfrac{d\theta}{dt}\right) = \dfrac{dx}{dt}$

$\dfrac{dx}{dt} = (\tan^2\theta + 1)(6\pi) = 6\pi(x^2 + 1)$

When $x = \dfrac{1}{2}$, $\dfrac{dx}{dt} = 6\pi\left(\dfrac{1}{4} + 1\right) = \dfrac{15\pi}{2}$ mi/min $= 450\pi$ mph.

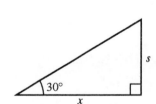

81. $\tan 30° = \dfrac{s}{x}$

$x = \sqrt{3}\,s$

$\dfrac{dx}{dt} = \sqrt{3}\dfrac{ds}{dt}$

$s(t) = 200 - 16t^2$

$\dfrac{ds}{dt} = -32t$

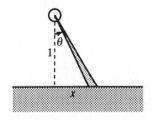

When $s = 125$, $125 = 200 - 16t^2$

$16t^2 = 75$

$t = \sqrt{\dfrac{75}{16}} = \dfrac{5\sqrt{3}}{4}$

$\dfrac{ds}{dt} = -32\left(\dfrac{5\sqrt{3}}{4}\right) = -40\sqrt{3}$

$\dfrac{dx}{dt} = \sqrt{3}(-40\sqrt{3}) = -120$ ft/sec.

CHAPTER 3
Applications of Differentiation

Section 3.1   Extrema on an Interval

1. $f(x) = \dfrac{x^2}{x^2+4}$

$f'(x) = \dfrac{(x^2+4)(2x) - (x^2)(2x)}{(x^2+4)^2} = \dfrac{8x}{(x^2+4)^2}$

$f'(0) = 0$

2. $f(x) = \cos\dfrac{\pi x}{2}$

$f'(x) = -\dfrac{\pi}{2}\sin\dfrac{\pi x}{2}$

$f'(0) = 0$

$f'(2) = 0$

3. $f(x) = x + \dfrac{32}{x^2}$

$f'(x) = 1 - \dfrac{64}{x^3}$

$f'(4) = 0$

4. $f(x) = -3x\sqrt{x+1}$

$f'(x) = -3x\left[\tfrac{1}{2}(x+1)^{-1/2}\right] + \sqrt{x+1}\,(-3)$

$\quad = -\tfrac{3}{2}(x+1)^{-1/2}[x + 2(x+1)]$

$\quad = -\tfrac{3}{2}(x+1)^{-1/2}(3x+2)$

$f'\left(-\tfrac{2}{3}\right) = 0$

5. $f(x) = (x+2)^{2/3}$

$f'(x) = \dfrac{2}{3}(x+2)^{-1/3}$

$f'(-2)$ is undefined.

6. Using the limit definition of the derivative,

$\displaystyle\lim_{x\to 0^-} \dfrac{f(x) - f(0)}{x - 0} = \lim_{x\to 0^-}\dfrac{(4 - |x|) - 4}{x} = 1$

$\displaystyle\lim_{x\to 0^+} \dfrac{f(x) - f(0)}{x - 0} = \lim_{x\to 0^+}\dfrac{(4 - |x|) - 4}{x - 0} = -1$

$f'(0)$ does not exist, since the one-sided derivatives are not equal.

7. $f(x) = x^2(x-3) = x^3 - 3x^2$

$f'(x) = 3x^2 - 6x = 3x(x-2)$

Critical numbers: $x = 0,\ x = 2$

8. $g(x) = x^2(x^2 - 4) = x^4 - 4x^2$

$g'(x) = 4x^3 - 8x = 4x(x^2 - 2)$

Critical numbers: $x = 0,\ x = \pm\sqrt{2}$

9. $g(t) = t\sqrt{4-t}$

$g'(t) = t\left[\dfrac{1}{2}(4-t)^{-1/2}(-1)\right] + (4-t)^{1/2}$

$\quad = \dfrac{1}{2}(4-t)^{-1/2}[-t + 2(4-t)]$

$\quad = \dfrac{8 - 3t}{2\sqrt{4-t}}$

Critical numbers: $t = 4,\ t = \tfrac{8}{3}$

10. $f(x) = \dfrac{4x}{x^2+1}$

$f'(x) = \dfrac{(x^2+1)(4) - (4x)(2x)}{(x^2+1)^2} = \dfrac{4(1-x^2)}{(x^2+1)^2}$

Critical numbers: $x = \pm 1$

11. $h(x) = \sin^2 x + \cos x, \quad 0 \le x < 2\pi$

$h'(x) = 2 \sin x \cos x - \sin x = \sin x (2 \cos x - 1)$

Critical numbers: $x = 0, \; x = \dfrac{\pi}{3}, \; x = \pi, \; x = \dfrac{5\pi}{3}$

12. $f(\theta) = 2 \sec \theta + \tan \theta, \quad 0 \le \theta < 2\pi$

$f'(\theta) = 2 \sec \theta \tan \theta + \sec^2 \theta$

$\qquad = \sec \theta (2 \tan \theta + \sec \theta)$

$\qquad = \sec \theta \left[2 \left(\dfrac{\sin \theta}{\cos \theta} \right) + \dfrac{1}{\cos \theta} \right]$

$\qquad = \sec^2 \theta (2 \sin \theta + 1)$

Critical numbers: $\theta = \dfrac{7\pi}{6}, \; \theta = \dfrac{11\pi}{6}$

13. $f(x) = 2(3 - x), \quad [-1, \; 2]$

$f'(x) = -2 \; \Rightarrow \;$ No critical numbers

Left endpoint: $(-1, \; 8)$ Maximum
Right endpoint: $(2, \; 2)$ Minimum

14. $f(x) = \dfrac{2x + 5}{3}, \quad [0, \; 5]$

$f'(x) = \dfrac{2}{3} \; \Rightarrow \;$ No critical numbers

Left endpoint: $\left(0, \; \dfrac{5}{3} \right)$ Minimum

Right endpoint: $(5, \; 5)$ Maximum

15. $f(x) = -x^2 + 3x, \quad [0, \; 3]$

$f'(x) = -2x + 3$

Left endpoint: $(0, \; 0)$ Minimum
Critical number: $\left(\dfrac{3}{2}, \; \dfrac{9}{4} \right)$ Maximum
Right endpoint: $(3, \; 0)$ Minimum

16. $f(x) = x^2 + 2x - 4, \quad [-1, \; 1]$

$f'(x) = 2x + 2 = 2(x + 1)$

Left endpoint: $(-1, \; -5)$ Minimum
Right endpoint: $(1, \; -1)$ Maximum

17. $f(x) = x^3 - 3x^2, \quad [-1, \; 3]$

$f'(x) = 3x^2 - 6x = 3x(x - 2)$

Left endpoint: $(-1, \; -4)$ Minimum
Critical number: $(0, \; 0)$ Maximum
Critical number: $(2, \; -4)$ Minimum
Right endpoint: $(3, \; 0)$ Maximum

18. $f(x) = x^3 - 12x, \quad [0, \; 4]$

$f'(x) = 3x^2 - 12 = 3(x^2 - 4)$

Left endpoint: $(0, \; 0)$
Critical number: $(2, \; -16)$ Minimum
Right endpoint: $(4, \; 16)$ Maximum
Note: $x = -2$ is not in the interval.

19. $f(x) = 3x^{2/3} - 2x, \quad [-1, \; 1]$

$f'(x) = 2x^{-1/3} - 2 = \dfrac{2(1 - \sqrt[3]{x})}{\sqrt[3]{x}}$

Left endpoint: $(-1, \; 5)$ Maximum
Critical number: $(0, \; 0)$ Minimum
Right endpoint: $(1, \; 1)$

20. $g(x) = \sqrt[3]{x}, \quad [-1, \; 1]$

$g'(x) = \dfrac{1}{3x^{2/3}}$

Left endpoint: $(-1, \; -1)$ Minimum
Critical number: $(0, \; 0)$
Right endpoint: $(1, \; 1)$ Maximum

21. $h(t) = 4 - |t - 4|$, [1, 6]

From the graph of the function on the interval [1, 6] you can determine the following.

Left endpoint: (1, 1)   Minimum

Critical number: (4, 4)   Maximum

Right endpoint: (6, 2)

22. $g(t) = \dfrac{t^2}{t^2 + 3}$, [-1, 1]

$g'(t) = \dfrac{6t}{(t^2 + 3)^2}$

Left endpoint: $\left(-1, \dfrac{1}{4}\right)$ Maximum

Critical number: (0, 0)   Minimum

Right endpoint: $\left(1, \dfrac{1}{4}\right)$ Maximum

23. $h(s) = \dfrac{1}{s - 2}$, [0, 1]

$h'(s) = \dfrac{-1}{(s - 2)^2}$

Left endpoint: $\left(0, -\dfrac{1}{2}\right)$ Maximum

Right endpoint: (1, −1)   Minimum

24. $h(t) = \dfrac{t}{t - 2}$, [3, 5]

$h'(t) = \dfrac{-2}{(t - 2)^2}$

Left endpoint: (3, 3)   Maximum

Right endpoint: $\left(5, \dfrac{5}{3}\right)$ Minimum

25. $f(x) = \cos \pi x$, $\left[0, \dfrac{1}{6}\right]$

$f'(x) = -\pi \sin \pi x$

Left endpoint: (0, 1)   Maximum

Right endpoint: $\left(\dfrac{1}{6}, \dfrac{\sqrt{3}}{2}\right)$ Minimum

26. $g(x) = \csc x$, $\left[\dfrac{\pi}{6}, \dfrac{\pi}{3}\right]$

$g'(x) = -\csc x \cot x$

Left endpoint: $\left(\dfrac{\pi}{6}, 2\right)$ Maximum

Right endpoint: $\left(\dfrac{\pi}{3}, \dfrac{2\sqrt{3}}{3}\right)$ Minimum

27. $f(x) = \tan x$

f is continuous on $[0, \pi/4]$ but not on $[0, \pi]$.

28. Let $f(x) = 1/x$. f is continuous on (0, 1) but does not have a maximum. f is also continuous on (−1, 0) but does not have a minimum. This can occur if one of the endpoints is an infinite discontinuity.

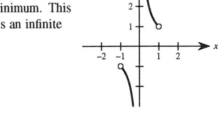

29. a. Yes
b. No

30. a. Yes
b. No

31. a. No
b. Yes

32. a. No
b. Yes

33. a. Minimum: (0, −3)
   Maximum: (2, 1)
b. Minimum: (0, −3)
c. Maximum: (2, 1)
d. No extrema

34. a. Minimum: (4, 1)
   Maximum: (1, 4)
b. Maximum: (1, 4)
c. Minimum: (4, 1)
d. No extrema

35. $f(x) = \begin{cases} 2x + 2, & 0 \le x \le 1 \\ 4x^2, & 1 < x \le 3 \end{cases}$

Left endpoint: (0, 2)   Minimum

Right endpoint: (3, 36)   Maximum

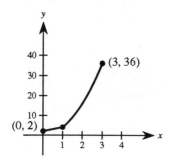

36. $f(x) = \begin{cases} 2 - x^2, & 1 \le x < 3 \\ 2 - 3x, & 3 \le x \le 5 \end{cases}$

Left endpoint: (1, 1)   Maximum

Right endpoint: (5, −13)   Minimum

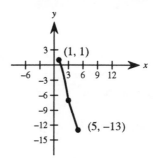

37. $f(x) = \dfrac{3}{x - 1}$,   (1, 4]

Right endpoint: (4, 1)   Minimum

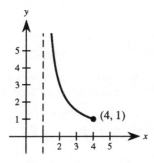

38. $f(x) = \dfrac{2}{2 - x}$,   [0, 2)

Left endpoint: (0, 1)   Minimum

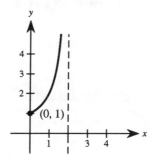

39. $f(x) = (1 + x^3)^{1/2}$,   [0, 2]

$f'(x) = \dfrac{3}{2}x^2(1 + x^3)^{-1/2}$

$f''(x) = \dfrac{3}{4}(x^4 + 4x)(1 + x^3)^{-3/2}$

$f'''(x) = -\dfrac{3}{8}(x^6 + 20x^3 - 8)(1 + x^3)^{-5/2}$

Setting $f''' = 0$, we have $x^6 + 20x^3 - 8 = 0$.

$x^3 = \dfrac{-20 \pm \sqrt{400 - 4(1)(-8)}}{2}$

$x = \sqrt[3]{-10 \pm \sqrt{108}}$

In the interval [0, 2], choose $x = \sqrt[3]{-10 + \sqrt{108}} \approx$ 0.732.

$\left| f''\left(\sqrt[3]{-10 + \sqrt{108}}\right) \right| \approx 1.47$ is the maximum value.

40. $f(x) = \dfrac{1}{x^2 + 1}$, $\left[\dfrac{1}{2},\ 3\right]$

$f'(x) = \dfrac{-2x}{(x^2 + 1)^2}$

$f''(x) = \dfrac{-2(1 - 3x^2)}{(x^2 + 1)^3}$

$f'''(x) = \dfrac{24x - 24x^3}{(x^2 + 1)^4}$

Setting $f''' = 0$, we have $x = 0,\ \pm 1$.

$|f''(1)| = \dfrac{1}{2}$ is the maximum value.

41. $f(x) = (x+1)^{2/3}, \quad [0, 2]$

$f'(x) = \frac{2}{3}(x+1)^{-1/3}$

$f''(x) = -\frac{2}{9}(x+1)^{-4/3}$

$f'''(x) = \frac{8}{27}(x+1)^{-7/3}$

$f^{(4)}(x) = -\frac{56}{81}(x+1)^{-10/3}$

$f^{(5)}(x) = \frac{560}{243}(x+1)^{-13/3}$

$|f^{(4)}(0)| = \frac{56}{81}$ is the maximum value.

42. $f(x) = \frac{1}{x^2+1}, \quad [-1, 1]$

$f'''(x) = \frac{24x - 24x^3}{(x^2+1)^4}$ (See Exercise 40.)

$f^{(4)}(x) = \frac{24(5x^4 - 10x^2 + 1)}{(x^2+1)^5}$

$f^{(5)}(x) = \frac{-240x(3x^4 - 10x^2 + 3)}{(x^2+1)^6}$

$|f^{(4)}(0)| = 24$ is the maximum value.

43. $P = VI - RI^2 = 12I - 0.5I^2, \quad 0 \le I \le 15$

$P = 0$ when $I = 0$.

$P = 67.5$ when $I = 15$.

$P' = 12 - I = 0$

Critical number: $I = 12$ amps

When $I = 12$ amps, $P = 72$, the maximum output.

44. $C = 2x + \frac{300,000}{x}, \quad 1 \le x \le 300$

$C(0)$ is undefined.

$C(300) = 1600$

$C' = 2 - \frac{300,000}{x^2} = 0$

$2x^2 = 300,000$

$x^2 = 150,000$

$x = 100\sqrt{15} \approx 387 > 300$

C is minimized when $x = 300$ units.

45. $x = \frac{v^2 \sin 2\theta}{32}, \quad \frac{\pi}{4} \le \theta \le \frac{3\pi}{4}$

$\frac{d\theta}{dt}$ is constant.

$\frac{dx}{dt} = \frac{dx}{d\theta}\frac{d\theta}{dt}$ (by the Chain Rule)

$= \frac{v^2 \cos 2\theta}{16}\frac{d\theta}{dt}$

In the interval $[\pi/4, 3\pi/4]$, $\theta = \pi/4, 3\pi/4$ indicate minimums for dx/dt and $\theta = \pi/2$ indicates a maximum for dx/dt. This implies that the sprinkler waters longest when $\theta = \pi/4$ and $3\pi/4$. Thus, the lawn farthest from the sprinkler gets the most water.

46. $S = 6hs + \frac{3s^2}{2}\left(\frac{\sqrt{3} - \cos\theta}{\sin\theta}\right), \quad \frac{\pi}{6} \le \theta \le \frac{\pi}{2}$

$\frac{dS}{d\theta} = \frac{3s^2}{2}(-\sqrt{3}\csc\theta\cot\theta + \csc^2\theta)$

$= \frac{3s^2}{2}\csc\theta(-\sqrt{3}\cot\theta + \csc\theta) = 0$

$\csc\theta = \sqrt{3}\cot\theta$

$\sec\theta = \sqrt{3}$

$\theta = \text{arcsec}\sqrt{3} \approx 0.9553$ radians

$S\left(\frac{\pi}{6}\right) = 6hs + \frac{3s^2}{2}(\sqrt{3})$

$S\left(\frac{\pi}{2}\right) = 6hs + \frac{3s^2}{2}(\sqrt{3})$

$S(\text{arcsec}\sqrt{3}) = 6hs + \frac{3s^2}{2}(\sqrt{2})$

S is minimum when $\theta = \text{arcsec}\sqrt{3} \approx 0.9553$ radians.

47. a.

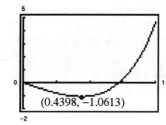

(0.4398, −1.0613)

b.

$$f(x) = 3.2x^5 + 5x^3 - 3.5x, \quad [0, \ 1]$$

$$f'(x) = 16x^4 + 15x^2 - 3.5$$

$$16x^4 + 15x^2 - 3.5 = 0$$

$$x^2 = \frac{-15 \pm \sqrt{(15)^2 - 4(16)(-3.5)}}{2(16)}$$

$$= \frac{-15 \pm \sqrt{449}}{32}$$

$$x = \sqrt{\frac{-15 + \sqrt{449}}{32}} \approx 0.4398$$

$$f(0) = 0$$

$$f(1) = 4.7$$

$$f\left(\sqrt{\frac{-15 + \sqrt{449}}{32}}\right) \approx -1.0613$$

Minimum: (0.4398, −1.0613)

48. a.

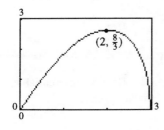

$(2, \frac{8}{3})$

b. $f(x) = \frac{4}{3}x\sqrt{3 - x}, \quad [0, \ 3]$

$$f'(x) = \frac{4}{3}\left[x\left(\frac{1}{2}\right)(3 - x)^{-1/2}(-1) + (3 - x)^{1/2}(1)\right]$$

$$= \frac{4}{3}(3 - x)^{-1/2}\left(\frac{1}{2}\right)[-x + 2(3 - x)]$$

$$= \frac{2(6 - 3x)}{3\sqrt{3 - x}} = \frac{6(2 - x)}{3\sqrt{3 - x}} = \frac{2(2 - x)}{\sqrt{3 - x}}$$

Critical number: $x = 2$

$$f(0) = 0$$

$$f(3) = 0$$

$$f(2) = \frac{8}{3}$$

Maximum: $\left(2, \ \frac{8}{3}\right)$

49. True. See Exercise 17.

50. True. This is stated in the Extreme Value Theorem.

51. True

52. False. Let $f(x) = x^2$. $x = 0$ is a critical number of f.

$$g(x) = f(x - k)$$
$$= (x - k)^2$$

$x = k$ is a critical number of g.

53. $f(x) = [\![x]\!]$

The derivative of f is undefined at every integer and is zero at any noninteger real number. All real numbers are critical numbers.

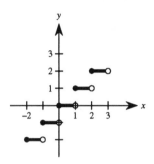

Section 3.2   Rolle's Theorem and the Mean Value Theorem

1. Rolle's Theorem does not apply to $f(x) = 1 - |x - 1|$ over $[0, 2]$ since f is not differentiable at $x = 1$.

2. Rolle's Theorem does not apply to $f(x) = \cot(x/2)$ over $[\pi, \ 3\pi]$ since f is not continuous at $x = 2\pi$.

3. $f(x) = x^2 - 2x, \quad [0, \ 2]$

$f(0) = f(2) = 0$

f is continuous on $[0, \ 2]$. f is differentiable on $(0, \ 2)$. Rolle's Theorem applies.

$f'(x) = 2x - 2$

$2x - 2 = 0 \implies x = 1$

c value: 1

4. $f(x) = x^2 - 3x + 2, \quad [1, \ 2]$

$f(1) = f(2) = 0$

f is continuous on $[1, \ 2]$. f is differentiable on $(1, \ 2)$. Rolle's Theorem applies.

$f'(x) = 2x - 3$

$2x - 3 = 0 \implies x = \frac{3}{2} = 1.5$

c value: $\frac{3}{2} = 1.5$

5. $f(x) = (x - 1)(x - 2)(x - 3), \quad [1, \ 3]$

$f(1) = f(3) = 0$

f is continuous on $[1, \ 3]$. f is differentiable on $(1, \ 3)$. Rolle's Theorem applies.

$$f(x) = x^3 - 6x^2 + 11x - 6$$
$$f'(x) = 3x^2 - 12x + 11$$

$$3x^2 - 12x + 11 = 0 \implies x = \frac{6 \pm \sqrt{3}}{3}$$

$$c = \frac{6 - \sqrt{3}}{3}, \ c = \frac{6 + \sqrt{3}}{3}$$

6. $f(x) = (x - 3)(x + 1)^2, \quad [-1, \ 3]$

$f(-1) = f(3) = 0$

f is continuous on $[-1, \ 3]$. f is differentiable on $(-1, \ 3)$. Rolle's Theorem applies.

$$f'(x) = (x - 3)(2)(x + 1) + (x + 1)^2$$
$$= (x + 1)[2x - 6 + x + 1]$$
$$= (x + 1)(3x - 5)$$

c value: $\frac{5}{3}$

7. $f(x) = |x| - 1, \quad [-1, \ 1]$

$f(-1) = f(1) = 0$

f is continuous on $[-1, \ 1]$. f is not differentiable on $(-1, \ 1)$ since $f'(0)$ does not exist. Rolle's Theorem does not apply.

8. $f(x) = 3 - |x - 3|, \quad [0, \ 6]$

$f(0) = f(6) = 0$

f is continuous on $[0, 6]$. f is not differentiable on $(0, 6)$ since $f'(3)$ does not exist. Rolle's Theorem does not apply.

9. $f(x) = x^{2/3} - 1,$ $[-8, 8]$

$f(-8) = f(8) = 3$

f is continuous on $[-8, 8]$. f is not differentiable on $(-8, 8)$ since $f'(0)$ does not exist. Rolle's Theorem does not apply.

10. $f(x) = x - x^{1/3},$ $[0, 1]$

$f(0) = f(1) = 0$

f is continuous on $[0, 1]$. f is differentiable on $(0, 1)$. (**Note:** f is not differentiable at $x = 0$.) Rolle's Theorem applies.

$$f'(x) = 1 - \frac{1}{3\sqrt[3]{x^2}} = 0$$

$$1 = \frac{1}{3\sqrt[3]{x^2}}$$

$$\sqrt[3]{x^2} = \frac{1}{3}$$

$$x^2 = \frac{1}{27}$$

$$x = \sqrt{\frac{1}{27}} = \frac{\sqrt{3}}{9}$$

c value: $\dfrac{\sqrt{3}}{9} \approx 0.1925$

11. $f(x) = \dfrac{x^2 - 2x - 3}{x + 2},$ $[-1, 3]$

$f(-1) = f(3) = 0$

f is continuous on $[-1, 3]$. (**Note:**   The discontinuity, $x = -2$, is not in the interval.) f is differentiable on $(-1, 3)$. Rolle's Theorem applies.

$$f'(x) = \frac{(x + 2)(2x - 2) - (x^2 - 2x - 3)(1)}{(x + 2)^2} = 0$$

$$\frac{x^2 + 4x - 1}{(x + 2)^2} = 0$$

$$x = \frac{-4 \pm 2\sqrt{5}}{2} = -2 \pm \sqrt{5}$$

c value: $-2 + \sqrt{5}$

12. $f(x) = \dfrac{x^2 - 1}{x},$ $[-1, 1]$

$f(-1) = f(1) = 0$

f is not continuous on $[-1, 1]$ since $f(0)$ does not exist. Rolle's Theorem does not apply.

13. $f(x) = \sin x,$ $[0, 2\pi]$

$f(0) = f(2\pi) = 0$

f is continuous on $[0, 2\pi]$. f is differentiable on $(0, 2\pi)$. Rolle's Theorem applies.

$$f'(x) = \cos x$$

c values: $\dfrac{\pi}{2}, \dfrac{3\pi}{2}$

14. $f(x) = \cos x,$ $[0, 2\pi]$

$f(0) = f(2\pi) = 1$

f is continuous on $[0, 2\pi]$. f is differentiable on $(0, 2\pi)$. Rolle's Theorem applies.

$$f'(x) = -\sin x$$

c value: π

15. $f(x) = \sin 2x, \quad \left[\dfrac{\pi}{6}, \dfrac{\pi}{3}\right]$

$f\left(\dfrac{\pi}{6}\right) = f\left(\dfrac{\pi}{3}\right) = \dfrac{\sqrt{3}}{2}$

f is continuous on $[\pi/6,\ \pi/3]$. f is differentiable on $(\pi/6,\ \pi/3)$. Rolle's Theorem applies.

$f'(x) = 2\cos 2x$

$2\cos 2x = 0$

$x = \dfrac{\pi}{4}$

c value: $\dfrac{\pi}{4}$

16. $f(x) = 4x - \tan \pi x, \quad \left[-\dfrac{1}{4}, \dfrac{1}{4}\right]$

$f\left(-\dfrac{1}{4}\right) = f\left(\dfrac{1}{4}\right) = 0$

f is continuous on $[-1/4,\ 1/4]$. f is differentiable on $(-1/4,\ 1/4)$. Rolle's Theorem applies.

$f'(x) = 4 - \pi \sec^2 \pi x = 0$

$\sec^2 \pi x = \dfrac{4}{\pi}$

$\sec \pi x = \pm \dfrac{2}{\sqrt{\pi}}$

$x = \pm\dfrac{1}{\pi}\operatorname{arcsec}\dfrac{2}{\sqrt{\pi}} = \pm\dfrac{1}{\pi}\arccos\dfrac{\sqrt{\pi}}{2}$

$\approx \pm 0.1533$ radian

c values: ± 0.1533 radian

17. $f(x) = \dfrac{x}{2} - \sin\dfrac{\pi x}{6}, \quad [-1,\ 0]$

$f(-1) = f(0) = 0$

f is continuous on $[-1,\ 0]$. f is differentiable on $(-1,\ 0)$. Rolle's Theorem applies.

$f'(x) = \dfrac{1}{2} - \dfrac{\pi}{6}\cos\dfrac{\pi x}{6} = 0$

$\cos\dfrac{\pi x}{6} = \dfrac{3}{\pi}$

$x = -\dfrac{6}{\pi}\arccos\dfrac{3}{\pi}$ [Value needed in $(-1,\ 0)$]

≈ -0.5756 radian

c value: -0.5756

18. $f(x) = \dfrac{6x}{\pi} - 4\sin^2 x, \quad \left[0,\ \dfrac{\pi}{6}\right]$

$f(0) = f\left(\dfrac{\pi}{6}\right) = 0$

f is continuous on $[0,\ \pi/6]$. f is differentiable on $(0,\ \pi/6)$. Rolle's Theorem applies.

$f'(x) = \dfrac{6}{\pi} - 8\sin x \cos x = 0$

$\dfrac{6}{\pi} = 8\sin x \cos x$

$\dfrac{3}{4\pi} = \dfrac{1}{2}\sin 2x$

$\dfrac{3}{2\pi} = \sin 2x$

$\dfrac{1}{2}\arcsin\left(\dfrac{3}{2\pi}\right) = x$

$x \approx 0.2489$

c value: 0.2489

19. $f(x) = \tan x, \quad [0,\ \pi]$

$f(0) = f(\pi) = 0$

f is not continuous on $[0,\ \pi]$ since $f(\pi/2)$ does not exist. Rolle's Theorem does not apply.

20. $f(x) = \sec x, \quad \left[-\dfrac{\pi}{4}, \dfrac{\pi}{4}\right]$

$f\left(-\dfrac{\pi}{4}\right) = f\left(\dfrac{\pi}{4}\right) = \sqrt{2}$

f is continuous on $[-\pi/4,\ \pi/4]$. f is differentiable on $(-\pi/4,\ \pi/4)$. Rolle's Theorem applies.

$f'(x) = \sec x \tan x$

$\sec x \tan x = 0$

$x = 0$

c value: 0

21. $f(t) = -16t^2 + 48t + 32$

   a. $f(1) = f(2) = 64$

   b. $v = f'(t)$ must be 0 at some time in $[1, 2]$.

$$f'(t) = -32t + 48 = 0$$

$$t = \tfrac{3}{2} \text{ seconds}$$

22. $C(x) = 10\left(\dfrac{1}{x} + \dfrac{x}{x+3}\right)$

   a. $C(3) = C(6) = \dfrac{25}{3}$

   b.
$$C'(x) = 10\left(-\dfrac{1}{x^2} + \dfrac{3}{(x+3)^2}\right) = 0$$

$$\dfrac{3}{x^2 + 6x + 9} = \dfrac{1}{x^2}$$

$$2x^2 - 6x - 9 = 0$$

$$x = \dfrac{6 \pm \sqrt{108}}{4}$$

$$= \dfrac{6 \pm 6\sqrt{3}}{4} = \dfrac{3 \pm 3\sqrt{3}}{2}$$

In the interval $[3, 6]$: $c = \dfrac{3 + 3\sqrt{3}}{2} \approx 4.098$

23. No. Let $f(x) = x^2$ on $[-1, \ 2]$.

$$f'(x) = 2x$$

$f'(0) = 0$ and zero is in the interval $(-1, \ 2)$ but $f(-1) \neq f(2)$.

24. $f(a) = f(b)$ and $f'(c) = 0$ where c is in the interval $(a, \ b)$.

   a. $g(x) = f(x) + k$

$$g(a) = g(b) = f(a) + k$$

$$g'(x) = f'(x) \ \Rightarrow \ g'(c) = 0$$

Interval: $[a, \ b]$

   b. $g(x) = f(x - k)$

$$g(a + k) = g(b + k) = f(a)$$

$$g'(x) = f'(x - k)$$

$$g'(c + k) = f'(c) = 0$$

Interval: $[a + k, \ b + k]$

   c. $g(x) = f(kx)$

$$g\left(\dfrac{a}{k}\right) = g\left(\dfrac{b}{k}\right) = f(a)$$

$$g'(x) = kf'(kx)$$

$$g'\left(\dfrac{c}{k}\right) = kf'(c) = 0$$

Interval: $\left[\dfrac{a}{k}, \ \dfrac{b}{k}\right]$

25. $f(x) = x^2$ is continuous on $[-2, \ 1]$ and differentiable on $(-2, \ 1)$.

$$\dfrac{f(1) - f(-2)}{1 - (-2)} = \dfrac{1 - 4}{3} = -1$$

$f'(x) = 2x = -1$ when $x = -\tfrac{1}{2}$. Therefore, $c = -\tfrac{1}{2}$.

26. $f(x) = x(x^2 - x - 2)$ is continuous on $[-1, \ 1]$ and differentiable on $(-1, \ 1)$.

$$\dfrac{f(1) - f(-1)}{1 - (-1)} = -1$$

$$f'(x) = 3x^2 - 2x - 2 = -1$$

$$(3x + 1)(x - 1) = 0$$

$$c = -\dfrac{1}{3}$$

27. $f(x) = x^{2/3}$ is continuous on $[0, \ 1]$ and differentiable on $(0, \ 1)$.

$$\frac{f(1) - f(0)}{1 - 0} = 1$$

$$f'(x) = \frac{2}{3}x^{-1/3} = 1$$

$$x = \left(\frac{2}{3}\right)^3 = \frac{8}{27}$$

$$c = \frac{8}{27}$$

28. $f(x) = (x + 1)/x$ is continuous on $[1/2, 2]$ and differentiable on $(1/2, \ 2)$.

$$\frac{f(2) - f(1/2)}{2 - (1/2)} = \frac{(3/2) - 3}{3/2} = -1$$

$$f'(x) = \frac{-1}{x^2} = -1$$

$$x^2 = 1$$

$$c = 1$$

29. $f(x) = x/(x + 1)$ is continuous on $[-1/2, \ 2]$ and differentiable on $(-1/2, \ 2)$.

$$\frac{f(2) - f(-1/2)}{2 - (-1/2)} = \frac{(2/3) - (-1)}{5/2} = \frac{2}{3}$$

$$f'(x) = \frac{1}{(x + 1)^2} = \frac{2}{3}$$

$$(x + 1)^2 = \frac{3}{2}$$

$$x = -1 \pm \frac{\sqrt{6}}{2} = \frac{-2 \pm \sqrt{6}}{2}$$

In the interval $\left(-\frac{1}{2}, \ 2\right)$: $c = \frac{-2 + \sqrt{6}}{2}$

30. $f(x) = \sqrt{x - 2}$ is continuous on $[2, \ 6]$ and differentiable on $(2, \ 6)$.

$$\frac{f(6) - f(2)}{6 - 2} = \frac{2 - 0}{4} = \frac{1}{2}$$

$$f'(x) = \frac{1}{2\sqrt{x - 2}} = \frac{1}{2}$$

$$\sqrt{x - 2} = 1$$

$$c = 3$$

31. $f(x) = x^3$ is continuous on $[0, 1]$ and differentiable on $(0, 1)$.

$$\frac{f(1) - f(0)}{1 - 0} = \frac{1 - 0}{1} = 1$$

$$f'(x) = 3x^2 = 1$$

$$x = \pm\frac{\sqrt{3}}{3}$$

In the interval $(0, 1)$: $c = \frac{\sqrt{3}}{3}$

32. $f(x) = \sin x$ is continuous on $[0, \ \pi]$ and differentiable on $(0, \ \pi)$.

$$\frac{f(\pi) - f(0)}{\pi - 0} = \frac{0 - 0}{\pi} = 0$$

$$f'(x) = \cos x = 0$$

$$c = \frac{\pi}{2}$$

33. $f(x) = x - 2\sin x$ is continuous on $[-\pi, \ \pi]$ and differentiable on $(-\pi, \ \pi)$.

$$\frac{f(\pi) - f(-\pi)}{\pi - (-\pi)} = \frac{\pi - (-\pi)}{2\pi} = 1$$

$$f'(x) = 1 - 2\cos x = 1$$

$$\cos x = 0$$

$$c = \pm \frac{\pi}{2}$$

34. $f(x) = 2\sin x + \sin 2x$ is continuous on $[0, \ \pi]$ and differentiable on $(0, \ \pi)$.

$$\frac{f(\pi) - f(0)}{\pi - 0} = \frac{0 - 0}{\pi} = 0$$

$$f'(x) = 2\cos x + 2\cos 2x = 0$$

$$2[\cos x + 2\cos^2 x - 1] = 0$$

$$2(2\cos x - 1)(\cos x + 1) = 0$$

$$\cos x = \frac{1}{2}$$

$$\cos x = -1$$

$$x = \frac{\pi}{3}, \ \frac{3\pi}{2}, \ \frac{5\pi}{3}$$

In the interval $(0, \ \pi)$: $c = \dfrac{\pi}{3}$

35. $f(x) = \dfrac{1}{x - 3}, \quad [0, \ 6]$

f has a discontinuity at $x = 3$.

36. $f(x) = |x - 3|, \quad [0, \ 6]$

f is not differentiable at $x = 3$.

37. f is continuous on $[-5, \ 5]$ and does not satisfy the conditions of the Mean Value Theorem.

$\Rightarrow \ f$ is not differentiable on $(-5, \ 5)$.

Example: $f(x) = |x|$

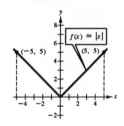

38. f is not continous on $[-5, \ 5]$.

Example: $f(x) = \dfrac{1}{x}$

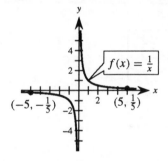

39. $s(t) = -16t^2 + 500$

a. $V_{\text{avg}} = \dfrac{s(3) - s(0)}{3 - 0} = -\dfrac{144}{3} = -48$ ft/sec

b. $s(t)$ is continuous on $[0, 3]$ and differentiable on $(0, 3)$. Therefore, the Mean Value Theorem applies.

$$v(t) = s'(t) = -32t = -48$$

$$t = \frac{3}{2} \text{ seconds}$$

In the interval $[0, 3]$: $c = \dfrac{3}{2}$

40. $S(t) = 200\left(5 - \dfrac{9}{2+t}\right)$

a. $\dfrac{S(12) - S(0)}{12 - 0} = \dfrac{200[5 - (9/14)] - 200[5 - (9/2)]}{12} = \dfrac{450}{7}$

b. $S'(t) = 200\left(\dfrac{9}{(2+t)^2}\right) = \dfrac{450}{7}$

   $\dfrac{1}{(2+t)^2} = \dfrac{1}{28}$

   $2 + t = 2\sqrt{7}$

   $t = 2\sqrt{7} - 2 \approx 3.2915 \text{ months}$

   $S'(t)$ is equal to the average value in April.

41. $f(x) = \sqrt{x}, \quad [1, \ 9]$

$(1, \ 1), \quad (9, \ 3)$

$m = \dfrac{3 - 1}{9 - 1} = \dfrac{1}{4}$

Secant line:

   $y - 1 = \dfrac{1}{4}(x - 1)$

   $y = \dfrac{1}{4}x + \dfrac{3}{4}$

   $0 = x - 4y + 3$

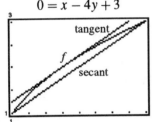

$f'(x) = \dfrac{1}{2\sqrt{x}}$

$\dfrac{f(9) - f(1)}{9 - 1} = \dfrac{1}{4}$

   $\dfrac{1}{2\sqrt{c}} = \dfrac{1}{4}$

   $\sqrt{c} = 2$

   $c = 4$

$(c, \ f(c)) = (4, \ 2)$

   $m = f'(4) = \dfrac{1}{4}$

Tangent line:

   $y - 2 = \dfrac{1}{4}(x - 4)$

   $y = \dfrac{1}{4}x + 1$

   $0 = x - 4y + 4$

42. $f(x) = -x^4 + 4x^3 + 8x^2 + 5$, $(0, 5)$, $(5, 80)$

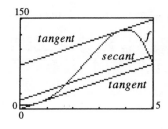

$m = \dfrac{80 - 5}{5 - 0} = 15$

Secant line: $y - 5 = 15(x - 0)$

$0 = 15x - y + 5$

$f'(x) = -4x^3 + 12x^2 + 16x$

$\dfrac{f(5) - f(1)}{5 - 1} = 15$

$-4c^3 + 12c^2 + 16c = 15$

$0 = 4c^3 - 12c^2 - 16c + 15$

$c \approx 0.67$ or $c \approx 3.79$

First tangent line: $y - f(c) = m(x - c)$ Second tangent line: $y - f(c) = m(x - c)$

$y - 9.59 = 15(x - 0.67)$ $y - 131.35 = 15(x - 3.79)$

$0 = 15x - y - 0.46$ $0 = 15x - y + 74.5$

43. False. $f(x) = 1/x$ has a discontinuity at $x = 0$.

44. False. f must also be continuous *and* differentiable on each interval. Let

$$f(x) = \frac{x^3 - 4x}{x^2 - 1}.$$

45. True. A polynomial is continuous and differentiable everywhere.

46. True

47. True. Both equal the slope of the line.

48. True

49. Suppose that $p(x) = x^{2n+1} + ax + b$ has two real roots x_1 and x_2. Then by Rolle's Theorem, since $p(x_1) = p(x_2) = 0$, there exists c in (x_1, x_2) such that $p'(c) = 0$. But $p'(x) = (2n + 1)x^{2n} + a \neq 0$, since $n > 0$, $a > 0$. Therefore, $p(x)$ cannot have two real roots.

50. a. Consider two consecutive zeros of $p'(x)$ and suppose that between them there are two zeros of $p(x)$. Then by Rolle's Theorem, there would be another zero of $p'(x)$ between the zeros of $p(x)$. This contradicts the fact that the zeros of $p'(x)$ are consecutive.

   b. Let x_1, x_2, and x_3 be the zeros of $p(x)$ where $x_1 < x_2 < x_3$. Then by Rolle's Theorem, there exists c_1 and c_2, where $x_1 < c_1 < x_2$ and $x_2 < c_2 < x_3$, such that $p'(c_1) = p'(c_2) = 0$. Then, by Rolle's Theorem again, there exists d, where $c_1 < d < c_2$ such that $p''(d) = 0$.

51. Suppose $f(x)$ is not constant on $[a, b]$. Then there exists x_1 and x_2 in $[a, b]$ such that $f(x_1) \neq f(x_2)$. Then by the Mean Value Theorem, there exists c in (a, b) such that

$$f'(c) = \frac{f(x_2) - f(x_1)}{x_2 - x_1} \neq 0.$$

This contradicts the fact that $f'(x) = 0$ for all x in (a, b).

52. If $p(x) = Ax^2 + Bx + C$, then

$$p'(x) = 2Ax + B = \frac{f(b) - f(a)}{b - a} = \frac{(Ab^2 + Bb + C) - (Aa^2 + Ba + C)}{b - a}$$

$$= \frac{A(b^2 - a^2) + B(b - a)}{b - a}$$

$$= \frac{(b - a)[A(b + a) + B]}{b - a}$$

$$= A(b + a) + B.$$

Thus, $2Ax = A(b + a)$ and $x = (b + a)/2$ which is the midpoint of $[a, b]$.

53. Let $h(x) = f(x) - g(x)$. Then $h'(x) = f'(x) - g'(x) = 0$. By Exercise 51, $h(x)$ is constant. Therefore, $h(x) = f(x) - g(x) = C$ and $f(x) = g(x) + C$.

54. Suppose $f(x)$ has two fixed points c_1 and c_2. Then, by the Mean Value Theorem, there exists c such that

$$f'(c) = \frac{f(c_2) - f(c_1)}{c_2 - c_1} = \frac{c_2 - c_1}{c_2 - c_1} = 1.$$

This contradicts the fact that $f'(x) < 1$ for all x.

55. $f(x) = \frac{1}{2}\cos x$ is differentiable on $(-\infty, \infty)$.

$$f'(x) = -\frac{1}{2}\sin x$$

$$-\frac{1}{2} \le f'(x) \le \frac{1}{2} \implies f'(x) < 1 \text{ for all real numbers.}$$

Thus, from Exercise 54, f has, at most, one fixed point.

56. Let $f(x) = \cos x$. f is continuous and differentiable for all real numbers. By the Mean Value Theorem, for any interval $[a, b]$, there exists c in (a, b) such that

$$\frac{f(b) - f(a)}{b - a} = f'(c)$$

$$\frac{\cos b - \cos a}{b - a} = -\sin c$$

$$\cos b - \cos a = (-\sin c)(b - a)$$

$$|\cos b - \cos a| = |-\sin c| |b - a|$$

$$|\cos b - \cos a| \le |b - a| \text{ since } |-\sin c| \le 1.$$

57. Given $f(x) = g'(x)$ and $g(x) = -f'(x)$. We have

$$\frac{d}{dx}[f(x)^2 + g(x)^2] = 2f(x)f'(x) + 2g(x)g'(x)$$

$$= 2g'(x)f'(x) + 2(-f'(x))g'(x)$$

$$= 0.$$

Thus, $f(x)^2 + g(x)^2$ is a constant. Example: $f(x) = \cos x$, $g(x) = \sin x$

Section 3.3   Increasing and Decreasing Functions and the First Derivative Test

1. $f(x) = x^2 - 6x + 8$

Increasing on: $(3, \infty)$

Decreasing on: $(-\infty, 3)$

2. $y = -(x + 1)^2$

Increasing on: $(-\infty, -1)$

Decreasing on: $(-1, \infty)$

3. $y = \dfrac{x^3}{4} - 3x$

Increasing on: $(-\infty, -2), (2, \infty)$

Decreasing on: $(-2, 2)$

4. $f(x) = x^4 - 2x^2$

Increasing on: $(-1, 0), (1, \infty)$

Decreasing on: $(-\infty, -1), (0, 1)$

5. $f(x) = \dfrac{1}{x^2}$

Increasing on: $(-\infty, 0)$

Decreasing on: $(0, \infty)$

6. $y = \dfrac{x^2}{x + 1}$

Increasing on: $(-\infty, -2), (0, \infty)$

Decreasing on: $(-2, -1), (-1, 0)$

7. $f(x) = -2x^2 + 4x + 3$

$f'(x) = -4x + 4 = 0$

Critical number: $x = 1$

Test intervals:	$-\infty < x < 1$	$1 < x < \infty$
Sign of $f'(x)$:	$f' > 0$	$f' < 0$
Conclusion:	Increasing	Decreasing

Increasing on: $(-\infty, 1)$

Decreasing on: $(1, \infty)$

Relative maximum: $(1, 5)$

8. $f(x) = x^2 + 8x + 10$

$f'(x) = 2x + 8 = 0$

Critical number: $x = -4$

Test intervals:	$-\infty < x < -4$	$-4 < x < \infty$
Sign of $f'(x)$:	$f' < 0$	$f' > 0$
Conclusion:	Decreasing	Increasing

Increasing on: $(-4, \infty)$

Decreasing on: $(-\infty, -4)$

Relative minimum: $(-4, -6)$

9. $f(x) = x^2 - 6x$

$f'(x) = 2x - 6 = 0$

Critical number: $x = 3$

Test intervals:	$-\infty < x < 3$	$3 < x < \infty$
Sign of $f'(x)$:	$f' < 0$	$f' > 0$
Conclusion:	Decreasing	Increasing

Increasing on: $(3, \infty)$

Decreasing on: $(-\infty, 3)$

Relative minimum: $(3, -9)$

10. $f(x) = (x-1)^2(x+2)$

$f'(x) = (x-1)^2(1) + (x+2)(2)(x-1) = (x-1)[(x-1) + 2(x+2)] = 3(x-1)(x+1) = 0$

Critical numbers: $x = -1, \ 1$

Test intervals:	$-\infty < x < -1$	$-1 < x < 1$	$1 < x < \infty$
Sign of $f'(x)$:	$f' > 0$	$f' < 0$	$f' > 0$
Conclusion:	Increasing	Decreasing	Increasing

Increasing on: $(-\infty, \ -1), \ (1, \ \infty)$
Decreasing on: $(-1, \ 1)$
Relative maximum: $(-1, \ 4)$
Relative minimum: $(1, 0)$

11. $f(x) = 2x^3 + 3x^2 - 12x$

$f'(x) = 6x^2 + 6x - 12 = 6(x+2)(x-1) = 0$

Critical numbers: $x = -2, \ 1$

Test intervals:	$-\infty < x < -2$	$-2 < x < 1$	$1 < x < \infty$
Sign of $f'(x)$:	$f' > 0$	$f' < 0$	$f' > 0$
Conclusion:	Increasing	Decreasing	Increasing

Increasing on: $(-\infty, \ -2), \ (1, \ \infty)$
Decreasing on: $(-2, \ 1)$
Relative maximum: $(-2, \ 20)$
Relative minimum: $(1, \ -7)$

12. $f(x) = (x-3)^3$

$f'(x) = 3(x-3)^2 = 0$

Critical number: $x = 3$

Test intervals:	$-\infty < x < 3$	$3 < x < \infty$
Sign of $f'(x)$:	$f' > 0$	$f' > 0$
Conclusion:	Increasing	Increasing

Increasing on: $(-\infty, \ \infty)$
No relative extrema

13. $f(x) = \dfrac{x^5 - 5x}{5}$

$f'(x) = x^4 - 1$

Critical numbers: $x = -1,\ 1$

Test intervals:	$-\infty < x < -1$	$-1 < x < 1$	$1 < x < \infty$
Sign of $f'(x)$:	$f' > 0$	$f' < 0$	$f' > 0$
Conclusion:	Increasing	Decreasing	Increasing

Increasing on: $(-\infty,\ -1),\ (1,\ \infty)$
Decreasing on: $(-1,\ 1)$
Relative maximum: $\left(-1,\ \frac{4}{5}\right)$
Relative minimum: $\left(1,\ -\frac{4}{5}\right)$

14. $f(x) = x^4 - 32x + 4$

$f'(x) = 4x^3 - 32 = 4(x^3 - 8)$

Critical number: $x = 2$

Test intervals:	$-\infty < x < 2$	$2 < x < \infty$
Sign of $f'(x)$:	$f' < 0$	$f' > 0$
Conclusion:	Decreasing	Increasing

Increasing on: $(2,\ \infty)$
Decreasing on: $(-\infty,\ 2)$
Relative minimum: $(2,\ -44)$

15. $f(x) = x^{1/3} + 1$

$f'(x) = \dfrac{1}{3}x^{-2/3} = \dfrac{1}{3x^{2/3}}$

Critical number: $x = 0$

Test intervals:	$-\infty < x < 0$	$0 < x < \infty$
Sign of $f'(x)$:	$f' > 0$	$f' > 0$
Conclusion:	Increasing	Increasing

Increasing on: $(-\infty,\ \infty)$
No relative extrema

16. $f(x) = x^{2/3}(x - 5) = x^{5/3} - 5x^{2/3}$

$f'(x) = \frac{5}{3}x^{2/3} - \frac{10}{3}x^{-1/3} = \frac{5}{3}x^{-1/3}(x - 2)$

Critical numbers: $x = 0, \ 2$

Test intervals:	$-\infty < x < 0$	$0 < x < 2$	$2 < x < \infty$
Sign of $f'(x)$:	$f' > 0$	$f' < 0$	$f' > 0$
Conclusion:	Increasing	Decreasing	Increasing

Increasing on: $(-\infty, \ 0), \ (2, \ \infty)$
Decreasing on: $(0, 2)$
Relative maximum: $(0, 0)$
Relative minimum: $(2, \ -3\sqrt[3]{4})$

17. $f(x) = 5 - |x - 5|$

$f'(x) = -\dfrac{x - 5}{|x - 5|}$

Critical number: $x = 5$

Test intervals:	$-\infty < x < 5$	$5 < x < \infty$
Sign of $f'(x)$:	$f' > 0$	$f' < 0$
Conclusion:	Increasing	Decreasing

Increasing on: $(-\infty, \ 5)$
Decreasing on: $(5, \ \infty)$
Relative maximum: $(5, 5)$

18. $f(x) = |x + 3| - 1$

$f'(x) = \dfrac{x + 3}{|x + 3|}$

Critical number: $x = -3$

Test intervals:	$-\infty < x < -3$	$-3 < x < \infty$
Sign of $f'(x)$:	$f' < 0$	$f' > 0$
Conclusion:	Decreasing	Increasing

Increasing on: $(-3, \ \infty)$
Decreasing on: $(-\infty, \ -3)$
Relative minimum: $(-3, \ -1)$

19. $f(x) = \dfrac{x^2}{x^2 - 9}$

$f'(x) = \dfrac{(x^2 - 9)(2x) - (x^2)(2x)}{(x^2 - 9)^2} = \dfrac{-18x}{(x^2 - 9)^2}$

Critical number: $x = 0$
Discontinuities: $x = -3, \ 3$

Test intervals:	$-\infty < x < -3$	$-3 < x < 0$	$0 < x < 3$	$3 < x < \infty$
Sign of $f'(x)$:	$f' > 0$	$f' > 0$	$f' < 0$	$f' < 0$
Conclusion:	Increasing	Increasing	Decreasing	Decreasing

Increasing on: $(-\infty, \ -3), \ (-3, \ 0)$
Decreasing on: $(0, \ 3), \ (3, \ \infty)$
Relative maximum: $(0, 0)$

20. $f(x) = \dfrac{x+3}{x^2} = \dfrac{1}{x} + \dfrac{3}{x^2}$

$\qquad f'(x) = -\dfrac{1}{x^2} - \dfrac{6}{x^3} = \dfrac{-(x+6)}{x^3}$

Critical number: $x = -6$

Discontinuity: $x = 0$

Test intervals:	$-\infty < x < -6$	$-6 < x < 0$	$0 < x < \infty$
Sign of $f'(x)$:	$f' < 0$	$f' > 0$	$f' < 0$
Conclusion:	Decreasing	Increasing	Decreasing

Increasing on : $(-6,\ 0)$

Decreasing on: $(-\infty,\ -6),\ (0,\ \infty)$

Relative minimum: $\left(-6,\ -\frac{1}{12}\right)$

21. $f(x) = x^3 - 6x^2 + 15$

$\qquad f'(x) = 3x^2 - 12x = 3x(x-4) = 0$

Critical numbers: $x = 0,\ 4$

Test intervals:	$-\infty < x < 0$	$0 < x < 4$	$4 < x < \infty$
Sign of $f'(x)$:	$f' > 0$	$f' < 0$	$f' > 0$
Conclusion:	Increasing	Decreasing	Increasing

Increasing on: $(-\infty,\ 0),\ (4,\ \infty)$

Decreasing on: $(0, 4)$

Relative maximum: $(0, 15)$

Relative minimum: $(4,\ -17)$

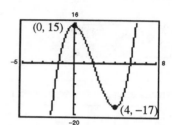

22. $f(x) = x^4 - 2x^3$

$\qquad f'(x) = 4x^3 - 6x^2 = 2x^2(2x - 3) = 0$

Critical numbers: $x = 0,\ \frac{3}{2}$

Test intervals:	$-\infty < x < 0$	$0 < x < \frac{3}{2}$	$\frac{3}{2} < x < \infty$
Sign of $f'(x)$:	$f' < 0$	$f' < 0$	$f' > 0$
Conclusion:	Decreasing	Decreasing	Increasing

Decreasing on: $\left(-\infty,\ \frac{3}{2}\right)$

Increasing on: $\left(\frac{3}{2},\ \infty\right)$

Relative minimum: $\left(\frac{3}{2},\ -\frac{27}{16}\right)$

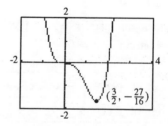

23. $f(x) = (x - 1)^{2/3}$

$f'(x) = \dfrac{2}{3(x-1)^{1/3}}$

Critical number: $x = 1$

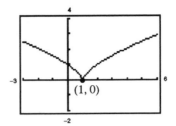

Test intervals:	$-\infty < x < 1$	$1 < x < \infty$
Sign of $f'(x)$:	$f' < 0$	$f' > 0$
Conclusion:	Decreasing	Increasing

Increasing on: $(1, \infty)$
Decreasing on: $(-\infty, 1)$
Relative minimum: $(1, 0)$

24. $f(x) = (x - 1)^{1/3}$

$f'(x) = \dfrac{1}{3(x-1)^{2/3}}$

Critical number: $x = 1$

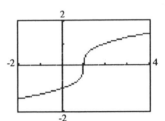

Test intervals:	$-\infty < x < 1$	$1 < x < \infty$
Sign of $f'(x)$:	$f' > 0$	$f' > 0$
Conclusion:	Increasing	Increasing

Increasing on: $(-\infty, \infty)$
No relative extrema

25. $f(x) = x + \dfrac{1}{x}$

$f'(x) = 1 - \dfrac{1}{x^2} = \dfrac{x^2 - 1}{x^2}$

Critical numbers: $x = -1, \ 1$
Discontinuity: $x = 0$

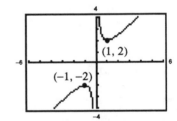

Test intervals:	$-\infty < x < -1$	$-1 < x < 0$	$0 < x < 1$	$1 < x < \infty$
Sign of $f'(x)$:	$f' > 0$	$f' < 0$	$f' < 0$	$f' > 0$
Conclusion:	Increasing	Decreasing	Decreasing	Increasing

Increasing on: $(-\infty, -1), \ (1, \infty)$
Decreasing on: $(-1, 0), \ (0, 1)$
Relative maximum: $(-1, -2)$
Relative minimum: $(1, 2)$

26. $f(x) = \dfrac{x}{x+1}$

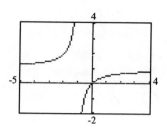

$f'(x) = \dfrac{(x+1)(1) - (x)(1)}{(x+1)^2} = \dfrac{1}{(x+1)^2}$

Discontinuity: $x = -1$

Test intervals:	$-\infty < x < -1$	$-1 < x < \infty$
Sign of $f'(x)$:	$f' > 0$	$f' > 0$
Conclusion:	Increasing	Increasing

Increasing on: $(-\infty, -1)$, $(-1, \infty)$
No relative extrema

27. $f(x) = \dfrac{x^2 - 2x + 1}{x+1}$

$f'(x) = \dfrac{(x+1)(2x-2) - (x^2 - 2x + 1)(1)}{(x+1)^2} = \dfrac{x^2 + 2x - 3}{(x+1)^2} = \dfrac{(x+3)(x-1)}{(x+1)^2}$

Critical numbers: $x = -3, \ 1$
Discontinuity: $x = -1$

Test intervals:	$-\infty < x < -3$	$-3 < x < -1$	$-1 < x < 1$	$1 < x < \infty$
Sign of $f'(x)$:	$f' > 0$	$f' < 0$	$f' < 0$	$f' > 0$
Conclusion:	Increasing	Decreasing	Decreasing	Increasing

Increasing on: $(-\infty, -3)$, $(1, \infty)$
Decreasing on: $(-3, -1)$, $(-1, 1)$
Relative maximum: $(-3, -8)$
Relative minimum: $(1, 0)$

28. $f(x) = \dfrac{x^2 - 3x - 4}{x - 2}$

$f'(x) = \dfrac{(x-2)(2x-3) - (x^2 - 3x - 4)(1)}{(x-2)^2} = \dfrac{x^2 - 4x + 10}{(x-2)^2}$

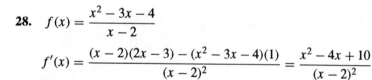

Discontinuity: $x = 2$

Test intervals:	$-\infty < x < 2$	$2 < x < \infty$
Sign of $f'(x)$:	$f' > 0$	$f' > 0$
Conclusion:	Increasing	Increasing

Increasing on: $(-\infty, 2)$, $(2, \infty)$
No relative extrema

29. $f(x) = \dfrac{x}{2} + \cos x, \quad 0 < x < 2\pi$

$f'(x) = \dfrac{1}{2} - \sin x = 0$

Critical numbers: $x = \dfrac{\pi}{6}, \dfrac{5\pi}{6}$

Test intervals:	$0 < x < \dfrac{\pi}{6}$	$\dfrac{\pi}{6} < x < \dfrac{5\pi}{6}$	$\dfrac{5\pi}{6} < x < 2\pi$
Sign of $f'(x)$:	$f' > 0$	$f' < 0$	$f' > 0$
Conclusion:	Increasing	Decreasing	Increasing

Increasing on: $\left(0, \dfrac{\pi}{6}\right), \left(\dfrac{5\pi}{6}, 2\pi\right)$

Decreasing on: $\left(\dfrac{\pi}{6}, \dfrac{5\pi}{6}\right)$

Relative maximum: $\left(\dfrac{\pi}{6}, \dfrac{\pi + 6\sqrt{3}}{12}\right)$

Relative minimum: $\left(\dfrac{5\pi}{6}, \dfrac{5\pi - 6\sqrt{3}}{12}\right)$

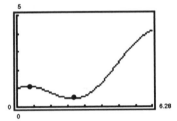

30. $f(x) = \sin x \cos x = \dfrac{1}{2}\sin 2x, \quad 0 < x < 2\pi$

$f'(x) = \cos 2x = 0$

Critical numbers: $x = \dfrac{\pi}{4}, \dfrac{3\pi}{4}, \dfrac{5\pi}{4}, \dfrac{7\pi}{4}$

Test intervals:	$0 < x < \dfrac{\pi}{4}$	$\dfrac{\pi}{4} < x < \dfrac{3\pi}{4}$	$\dfrac{3\pi}{4} < x < \dfrac{5\pi}{4}$	$\dfrac{5\pi}{4} < x < \dfrac{7\pi}{4}$	$\dfrac{7\pi}{4} < x < 2\pi$
Sign of $f'(x)$:	$f' > 0$	$f' < 0$	$f' > 0$	$f' < 0$	$f' > 0$
Conclusion:	Increasing	Decreasing	Increasing	Decreasing	Increasing

Increasing on: $\left(0, \dfrac{\pi}{4}\right), \left(\dfrac{3\pi}{4}, \dfrac{5\pi}{4}\right), \left(\dfrac{7\pi}{4}, 2\pi\right)$

Decreasing on: $\left(\dfrac{\pi}{4}, \dfrac{3\pi}{4}\right), \left(\dfrac{5\pi}{4}, \dfrac{7\pi}{4}\right)$

Relative maxima: $\left(\dfrac{\pi}{4}, \dfrac{1}{2}\right), \left(\dfrac{5\pi}{4}, \dfrac{1}{2}\right)$

Relative minima: $\left(\dfrac{3\pi}{4}, -\dfrac{1}{2}\right), \left(\dfrac{7\pi}{4}, -\dfrac{1}{2}\right)$

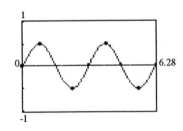

31. $f(x) = \sin^2 x + \sin x, \quad 0 < x < 2\pi$

$f'(x) = 2\sin x \cos x + \cos x = \cos x(2\sin x + 1) = 0$

Critical numbers: $x = \dfrac{\pi}{2}, \ \dfrac{7\pi}{6}, \ \dfrac{3\pi}{2}, \ \dfrac{11\pi}{6}$

Test intervals:	$0 < x < \dfrac{\pi}{2}$	$\dfrac{\pi}{2} < x < \dfrac{7\pi}{6}$	$\dfrac{7\pi}{6} < x < \dfrac{3\pi}{2}$	$\dfrac{3\pi}{2} < x < \dfrac{11\pi}{6}$	$\dfrac{11\pi}{6} < x < 2\pi$
Sign of $f'(x)$:	$f' > 0$	$f' < 0$	$f' > 0$	$f' < 0$	$f' > 0$
Conclusion:	Increasing	Decreasing	Increasing	Decreasing	Increasing

Increasing on: $\left(0, \dfrac{\pi}{2}\right), \ \left(\dfrac{7\pi}{6}, \dfrac{3\pi}{2}\right), \ \left(\dfrac{11\pi}{6}, 2\pi\right)$

Decreasing on: $\left(\dfrac{\pi}{2}, \dfrac{7\pi}{6}\right), \ \left(\dfrac{3\pi}{2}, \dfrac{11\pi}{6}\right)$

Relative minima: $\left(\dfrac{7\pi}{6}, -\dfrac{1}{4}\right), \ \left(\dfrac{11\pi}{6}, -\dfrac{1}{4}\right)$

Relative maxima: $\left(\dfrac{\pi}{2}, 2\right), \ \left(\dfrac{3\pi}{2}, 0\right)$

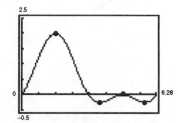

32. $f(x) = \dfrac{\cos x}{1 + \sin^2 x}, \quad 0 < x < 2\pi$

$f'(x) = \dfrac{(1 + \sin^2 x)(-\sin x) - \cos x(2\sin x \cos x)}{(1 + \sin^2 x)^2} = \dfrac{-\sin x(2 + \cos^2 x)}{(1 + \sin^2 x)^2} = 0$

Critical number: $x = \pi$

Test intervals:	$0 < x < \pi$	$\pi < x < 2\pi$
Sign of $f'(x)$:	$f' < 0$	$f' > 0$
Conclusion:	Decreasing	Increasing

Increasing on: $(\pi, \ 2\pi)$
Decreasing on: $(0, \ \pi)$
Relative minimum: $(\pi, \ -1)$

33. $f(x) = 2x\sqrt{9 - x^2}, \quad [-3, \ 3]$

a. $f'(x) = \dfrac{2(9 - 2x^2)}{\sqrt{9 - x^2}}$

b.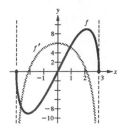

c. $\dfrac{2(9 - 2x^2)}{\sqrt{9 - x^2}} = 0$

Critical numbers: $x = \pm\dfrac{3}{\sqrt{2}} = \pm\dfrac{3\sqrt{2}}{2}$

d. Intervals:

$\left(-3, \ -\dfrac{3\sqrt{2}}{2}\right)$	$\left(-\dfrac{3\sqrt{2}}{2}, \ \dfrac{3\sqrt{2}}{2}\right)$	$\left(\dfrac{3\sqrt{2}}{2}, \ 3\right)$
$f'(x) < 0$	$f'(x) > 0$	$f'(x) < 0$
Decreasing	Increasing	Decreasing

f is increasing when f' is positive and decreasing when f' is negative.

34. $f(x) = 10(5 - \sqrt{x^2 - 3x + 16})$, $[0,\ 5]$

   a. $f'(x) = -\dfrac{5(2x-3)}{\sqrt{x^2 - 3x + 16}}$

   b.
   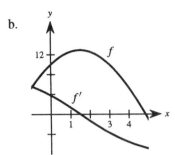

   c. $-\dfrac{5(2x-3)}{\sqrt{x^2 - 3x + 16}} = 0$

   Critical number: $x = \dfrac{3}{2}$

   d. Intervals:

   $\left(0,\ \dfrac{3}{2}\right)$ $\left(\dfrac{3}{2},\ 5\right)$

   $f'(x) > 0$ $f'(x) < 0$

   Increasing   Decreasing

   f is increasing when f' is positive and decreasing when f' is negative.

35. $f(t) = t^2 \sin t$, $[0,\ 2\pi]$

   a. $f'(t) = t^2 \cos t + 2t \sin t$

   $= t(t \cos t + 2 \sin t)$

   b.
   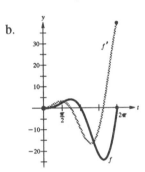

   c. $t(t \cos t + 2 \sin t) = 0$

   $t = 0$ or $t = -2 \tan t$

   $t \cot t = -2$

   $t \approx 2.2889,\ \ 5.0870$

   Critical numbers: $t = 2.2889,\ t = 5.0870$

   d. Intervals:

   $(0,\ 2.2889)$ $(2.2889,\ 5.0870)$ $(5.0870,\ 2\pi)$

   $f'(t) > 0$ $\quad f'(t) < 0$ $\quad\quad f'(t) > 0$

   Increasing $\quad$ Decreasing $\quad\quad$ Increasing

   f is increasing when f' is positive and decreasing when f' is negative.

36. $f(x) = \dfrac{x}{2} + \cos \dfrac{x}{2}$, $[0,\ 4\pi]$

   a. $f'(x) = \dfrac{1}{2} - \dfrac{1}{2} \sin \dfrac{x}{2}$

   b.
   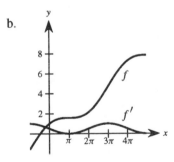

   c. $\dfrac{1}{2} - \dfrac{1}{2} \sin \dfrac{x}{2} = 0$

   $\sin \dfrac{x}{2} = 1$

   $\dfrac{x}{2} = \dfrac{\pi}{2}$

   Critical number: $x = \pi$

   d. Intervals:

   $(0,\ \pi)$ $\quad (\pi,\ 4\pi)$

   $f'(x) > 0$ $\quad f'(x) > 0$

   Increasing   Increasing

   f is increasing when f' is positive.

In Exercises 37–42, $f'(x) > 0$ on $(-\infty,\ -4)$, $f'(x) < 0$ on $(-4,\ 6)$ and $f'(x) > 0$ on $(6,\ \infty)$.

37. $g(x) = f(x) + 5$

   $g'(x) = f'(x)$

   $g'(0) = f'(0) < 0$

38. $g(x) = 3f(x) - 3$

   $g'(x) = 3f'(x)$

   $g'(-5) = 3f'(-5) > 0$

39. $g(x) = -f(x)$

$g'(x) = -f'(x)$

$g'(-6) = -f'(-6) < 0$

40. $g(x) = -f(x)$

$g'(x) = -f'(x)$

$g'(0) = -f'(0) > 0$

41. $g(x) = f(x - 10)$

$g'(x) = f'(x - 10)$

$g'(0) = f'(-10) > 0$

42. $g(x) = f(x - 10)$

$g'(x) = f'(x - 10)$

$g'(8) = f'(-2) < 0$

43. $f'(x) = \begin{cases} > 0, & x < 4 \;\Rightarrow\; f \text{ is increasing on } (-\infty,\ 4) \\ \text{undefined}, & x = 4 \\ < 0, & x > 4 \;\Rightarrow\; f \text{ is decreasing on } (4,\ \infty) \end{cases}$

Two possibilities for $f(x)$ are given below.

a.

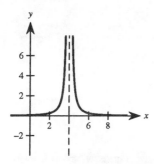

b.

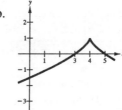

44. Critical number: $x = 5$

$f'(4) = -2.5 \;\Rightarrow\; f$ is decreasing at $x = 4$

$f'(6) = 3 \;\Rightarrow\; f$ is increasing at $x = 6$

$(5,\ f(5))$ is a relative minimum.

45. The critical numbers are in the intervals
$(-0.50,\ -0.25)$ and $(0.25,\ 0.50)$ since the sign
of f' changes in these intervals. f is decreasing
on approximately $(-1,\ -0.40)$, $(0.48,\ 1)$, and
increasing on $(-0.40,\ 0.48)$.

Relative minimum when $x \approx -0.40$.
Relative maximum when $x \approx 0.48$.

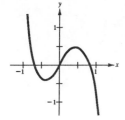

46. a. $f(x) = x$

$g(x) = \sin x$ on $(0,\ \pi)$

Let $h(x) = f(x) - g(x) = x - \sin x$

$h'(x) = 1 - \cos x > 0$ on $(0,\ \pi)$.

Therefore, $h(x)$ is increasing on $(0,\ \pi)$.

Since $h(0) = 0 \Rightarrow h(x) > 0$ on $(0,\ \pi)$.

$x - \sin x > 0 \Rightarrow x > \sin x \Rightarrow f(x) > g(x)$

b.

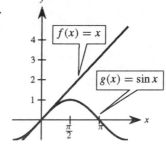

47. $a(t) = -16(\sin\theta)t^2$

$v(t) = -16(\sin\theta)\dfrac{t^3}{3}$

The velocity *can* be changed by changing θ. The velocity will be maximum at a given time t when $\theta = \pi/2$ since $\sin\pi/2 = 1$.

48. $C = \dfrac{3t}{27 + t^3}$

$C' = \dfrac{(27 + t^3)(3) - (3t)(3t^2)}{(27 + t^3)^2}$

$= \dfrac{3(27 - 2t^3)}{(27 + t^3)^2} = 0$

$t = \dfrac{3}{\sqrt[3]{2}} \approx 2.38 \text{ hours}$

49. $v = k(R - r)r^2 = k(Rr^2 - r^3)$

$v' = k(2Rr - 3r^2)$

$= kr(2R - 3r) = 0$

$r = 0 \text{ or } \tfrac{2}{3}R$

Maximum when $r = \tfrac{2}{3}R$

50. $P = 2.44x - \dfrac{x^2}{20{,}000} - 5000, \quad 0 \le x \le 35{,}000$

$P' = 2.44 - \dfrac{x}{10{,}000} = 0$

$x = 24{,}400$

Test intervals:	$0 < x < 24{,}400$	$24{,}400 < x < 35{,}000$
Sign of P':	$P' > 0$	$P' < 0$

Increasing when $0 < x < 24{,}400$ hamburgers
Decreasing when $24{,}400 < x < 35{,}000$ hamburgers

51. $W = 0.033t^2 - 0.3974t + 7.3032, \quad 0 \le t \le 14$

$W' = 0.066t - 0.3974 = 0$

$t \approx 6.02$

Test intervals:	$0 < t < 6.02$	$6.02 < t < 14$
Sign of W':	$W' < 0$	$W' > 0$

Increasing when $6.02 < t < 14$ days
Decreasing when $0 < t < 6.02$ days

52. $P = \dfrac{vR_1R_2}{(R_1 + R_2)^2}, \quad v \text{ and } R_1 \text{ are constant}$

$\dfrac{dP}{dR_2} = \dfrac{(R_1 + R_2)^2(vR_1) - vR_1R_2[2(R_1 + R_2)(1)]}{(R_1 + R_2)^4}$

$= \dfrac{vR_1(R_1 - R_2)}{(R_1 + R_2)^3} = 0 \;\Rightarrow\; R_2 = R_1$

Maximum when $R_1 = R_2$

53. $R = \sqrt{0.001T^4 - 4T + 100}$

a. $R' = \dfrac{0.004T^3 - 4}{2\sqrt{0.001T^4 - 4T + 100}} = 0$

$T = 10°, \quad R \approx 8.3666\Omega$

b.
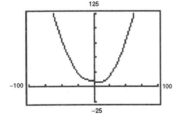

54. $T = 45 - 23\cos\left[\dfrac{2\pi(t - 32)}{365}\right]$

a.
$T' = 23\left(\dfrac{2\pi}{365}\right)\sin\left[\dfrac{2\pi(t - 32)}{365}\right] = 0$

$\dfrac{2\pi(t - 32)}{365} = 0 \quad \text{or} \quad \dfrac{2\pi(t - 32)}{365} = \pi$

$t = 32 \text{ (February 1)}$

$t = 214.5 \text{ (August 2 and 3)}$

Warmest days: August 2 and 3
Coldest day: February 1

b.
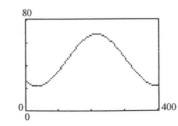

55. $f(x) = ax^3 + bx^2 + cx + d$

Relative minimum: $(0, 0)$

Relative maximum: $(2, 2)$

Since $f(0) = 0 \Rightarrow d = 0$

$$f'(x) = 3ax^2 + 2bx + c$$

Since $(0, 0)$ is a relative minimum $\Rightarrow f'(0) = 0 \Rightarrow c = 0$. Since $(2, 2)$ is a relative maximum
$f'(2) = 0 \Rightarrow 12a + 4b = 0$.

$$f(2) = 2 \Rightarrow 8a + 4b = 2$$

Solving this system of equations simultaneously, we have $a = -\frac{1}{2}$ and $b = \frac{3}{2}$. Therefore, $f(x) = -\frac{1}{2}x^3 + \frac{3}{2}x^2$.

56. $f(x) = ax^3 + bx^2 + cx + d$

Relative minimum: $(0, 0)$

Relative maximum: $(4, 1000)$

Since $f(0) = 0 \Rightarrow d = 0$

$$f'(x) = 3ax^2 + 2bx + c$$

Since $(0, 0)$ is a relative minimum $\Rightarrow f'(0) = 0 \Rightarrow c = 0$. Since $(4, 1000)$ is a relative maximum
$\Rightarrow f'(4) = 48a + 8b = 0$.

$$f(4) = 1000 \Rightarrow 64a + 16b = 1000$$

Solving this system of equations simultaneously, we have $a = -\frac{125}{4}$ and $b = \frac{375}{2}$. Therefore, $f(x) = -\frac{125}{4}x^3 + \frac{375}{2}x^2$.

57. True

Let $h(x) = f(x) + g(x)$ where f and g are increasing. Then $h'(x) = f'(x) + g'(x) > 0$ since $f'(x) > 0$ and $g'(x) > 0$.

58. False

Let $h(x) = f(x)g(x)$ where $f(x) = g(x) = x$. Then $h(x) = x^2$ is decreasing on $(-\infty, 0)$.

59. True

$$f(x) = ax^3 + b \Rightarrow f'(x) = 3ax^2$$

If f is increasing on $(-1, 1)$, then $3ax^2 > 0$ on $(-1, 1)$. Thus, $a > 0$ on $(-1, 1)$.

60. False

Let $f(x) = x^3$, then $f'(x) = 3x^2$ and f only has one critical number. Or, let $f(x) = x^3 + 3x + 1$, then $f'(x) = 3(x^2 + 1)$ has no critical numbers.

61. True

If $f(x)$ is an nth-degree polynomial, then the degree of $f'(x)$ is $n - 1$.

62. False

Let $f(x) = x^3$. Then $f'(x) = 3x^2$ and $c = 0$ is a critical number. This function has no relative extrema; it is increasing on $(-\infty, \infty)$.

63. Assume that $f'(x) < 0$ for all x in the interval (a, b) and let $x_1 < x_2$ be any two points in the interval. By the Mean Value Theorem, we know there exists a number c such that $x_1 < c < x_2$, and

$$f'(c) = \frac{f(x_2) - f(x_1)}{x_2 - x_1}.$$

Since $f'(c) < 0$ and $x_2 - x_1 > 0$, then $f(x_2) - f(x_1) < 0$, which implies that $f(x_2) < f(x_1)$. Thus, f is decreasing on the interval.

64. Suppose $f'(x)$ changes from positive to negative at c. Then there exists a and b in I such that $f'(x) > 0$ for all x in (a, c) and $f'(x) < 0$ for all x in (c, b). By Theorem 3.5, f is increasing on (a, c) and decreasing on (c, b). Therefore, $f(c)$ is a maximum of f on (a, b) and thus, a relative maximum of f.

65. Let $f(x) = (1+x)^n - nx - 1$. Then

$$f'(x) = n(1+x)^{n-1} - n$$

$$= n[(1+x)^{n-1} - 1] > 0 \text{ since } x > 0 \text{ and } n > 1.$$

Thus, $f(x)$ is increasing on $(0, \infty)$. Since $f(0) = 0 \implies f(x) > 0$ on $(0, \infty)$

$$(1+x)^n - nx - 1 > 0 \implies (1+x)^n > 1 + nx.$$

66. a. $D = \pi + 2\alpha - 4\sin^{-1}\left(\dfrac{1}{k}\sin\alpha\right)$ where $k \approx 1.33$

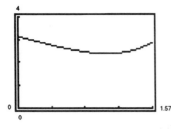

b. $D'(\alpha) = 2 - 4\dfrac{d}{d\alpha}\left[\sin^{-1}\left(\dfrac{1}{k}\sin\alpha\right)\right]$

Let $y = \sin^{-1}[(1/k)\sin\alpha]$, then $\sin y = (1/k)\sin\alpha$ and

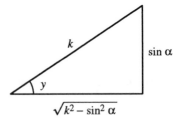

$$\cos y \frac{dy}{d\alpha} = \frac{1}{k}\cos\alpha$$

$$\frac{dy}{d\alpha} = \frac{\frac{1}{k}\cos\alpha}{\cos y} = \frac{\frac{1}{k}\cos\alpha}{\dfrac{\sqrt{k^2 - \sin^2\alpha}}{k}} = \frac{\cos\alpha}{\sqrt{k^2 - \sin^2\alpha}}.$$

Thus, $D'(\alpha) = 2 - 4\left(\dfrac{\cos\alpha}{\sqrt{k^2 - \sin^2\alpha}}\right)$. The critical number occurs when $D'(\alpha) = 0$.

$$2 - 4\left(\frac{\cos\alpha}{\sqrt{k^2 - \sin^2\alpha}}\right) = 0$$

$$\frac{1}{2} = \frac{\cos\alpha}{\sqrt{k^2 - (1 - \cos^2\alpha)}}$$

$$\frac{1}{4} = \frac{\cos^2\alpha}{k^2 - 1 + \cos^2\alpha}$$

$$k^2 - 1 + \cos^2\alpha = 4\cos^2\alpha$$

$$\frac{k^2 - 1}{3} = \cos^2\alpha$$

Thus, $\cos\alpha = \sqrt{(k^2 - 1)/3}$. **Note:** Since $0 \le \alpha \le \pi/2$, $\cos\alpha \ge 0$. $D_{\min}$ occurs when

$$\alpha = \cos^{-1}\sqrt{\frac{k^2 - 1}{3}} = \cos^{-1}\sqrt{\frac{1.33^2 - 1}{3}} \approx 1.04 \text{ radians}$$

$$D_{\min} = D\left(\cos^{-1}\sqrt{\frac{1.33^2 - 1}{3}}\right) \approx 2.4 \text{ radians}$$

$$\pi - D_{\min} \approx 0.74 \text{ radian.}$$

Section 3.4   Concavity and the Second Derivative Test

1. $y = x^2 - x - 2$

Concave upward: $(-\infty, \infty)$

2. $f(x) = \dfrac{24}{x^2 + 12}$

Concave upward: $(-\infty, -2)$, $(2, \infty)$

Concave downward: $(-2, 2)$

3. $y = -x^3 + 3x^2 - 2$

Concave upward: $(-\infty, 1)$

Concave downward: $(1, \infty)$

4. $f(x) = \dfrac{x^2 - 1}{2x + 1}$

Concave upward: $\left(-\infty, -\frac{1}{2}\right)$

Concave downward: $\left(-\frac{1}{2}, \infty\right)$

5. $f(x) = \dfrac{x^2 + 1}{x^2 - 1}$

Concave upward: $(-\infty, -1)$, $(1, \infty)$

Concave downward: $(-1, 1)$

6. $y = \dfrac{1}{270}(-3x^5 + 40x^3 + 135x)$

Concave upward: $(-\infty, -2)$, $(0, 2)$

Concave downward: $(-2, 0)$, $(2, \infty)$

7. $f(x) = 6x - x^2$

$f'(x) = 6 - 2x$

$f''(x) = -2$

Critical number: $x = 3$

$f''(3) < 0$

Therefore, $(3, 9)$ is a relative maximum.

8. $f(x) = x^2 + 3x - 8$

$f'(x) = 2x + 3$

$f''(x) = 2$

Critical number: $x = -\frac{3}{2}$

$f''\left(-\frac{3}{2}\right) > 0$

Therefore, $\left(-\frac{3}{2}, -\frac{41}{4}\right)$ is a relative minimum.

9. $f(x) = (x - 5)^2$

$f'(x) = 2(x - 5)$

$f''(x) = 2$

Critical number: $x = 5$

$f''(5) > 0$

Therefore, $(5, 0)$ is a relative minimum.

10. $f(x) = -(x - 5)^2$

$f'(x) = -2(x - 5)$

$f''(x) = -2$

Critical number: $x = 5$

$f''(5) < 0$

Therefore, $(5, 0)$ is a relative maximum.

11. $f(x) = x^3 - 3x^2 + 3$

$f'(x) = 3x^2 - 6x = 3x(x - 2)$

$f''(x) = 6x - 6 = 6(x - 1)$

Critical numbers: $x = 0$, $x = 2$

$f''(0) = -6 < 0$

Therefore, $(0, 3)$ is a relative maximum.

$f''(2) = 6 > 0$

Therefore, $(2, -1)$ is a relative minimum.

12. $f(x) = 5 + 3x^2 - x^3$

$f'(x) = 6x - 3x^2 = 3x(2 - x)$

$f''(x) = 6 - 6x = 6(1 - x)$

Critical numbers: $x = 0$, $x = 2$

$f''(0) = 6 > 0$

Therefore, $(0, 5)$ is a relative minimum.

$f''(2) = -6 < 0$

Therefore, $(2, 9)$ is a relative maximum.

13. $f(x) = x^4 - 4x^3 + 2$

$f'(x) = 4x^3 - 12x^2 = 4x^2(x - 3)$

$f''(x) = 12x^2 - 24x = 12x(x - 2)$

Critical numbers: $x = 0, \ x = 3$

However, $f''(0) = 0$, so we must use the First Derivative Test. $f'(x) < 0$ on the intervals $(-\infty, \ 0)$ and $(0, 3)$; hence, $(0, 2)$ is not an extremum. $f''(3) > 0$ so $(3, \ -25)$ is a relative minimum.

14. $f(x) = x^3 - 9x^2 + 27x$

$f'(x) = 3x^2 - 18x + 27 = 3(x - 3)^2$

$f''(x) = 6(x - 3)$

Critical number: $x = 3$

However, $f''(3) = 0$, so we must use the First Derivative Test. $f'(x) \geq 0$ for all x and, therefore, there are no relative extrema.

15. $f(x) = x^{2/3} - 3$

$f'(x) = \dfrac{2}{3x^{1/3}}$

$f''(x) = \dfrac{-2}{9x^{4/3}}$

Critical number: $x = 0$

However, $f''(0)$ is undefined, so we must use the First Derivative Test. Since $f'(x) < 0$ on $(-\infty, \ 0)$ and $f'(x) > 0$ on $(0, \ \infty)$, $(0, \ -3)$ is a relative minimum.

16. $f(x) = \sqrt{x^2 + 1}$

$f'(x) = \dfrac{x}{\sqrt{x^2 + 1}}$

Critical number: $x = 0$

$f''(x) = \dfrac{1}{(x^2 + 1)^{3/2}}$

$f''(0) = 1 > 0$

Therefore, $(0, 1)$ is a relative minimum.

17. $f(x) = x + \dfrac{4}{x}$

$f'(x) = 1 - \dfrac{4}{x^2} = \dfrac{x^2 - 4}{x^2}$

$f''(x) = \dfrac{8}{x^3}$

Critical numbers: $x = \pm 2$

$f''(-2) < 0$

Therefore, $(-2, \ -4)$ is a relative maximum.

$f''(2) > 0$

Therefore, $(2, 4)$ is a relative minimum.

18. $f(x) = \dfrac{x}{x - 1}$

$f'(x) = \dfrac{-1}{(x - 1)^2}$

There are no critical numbers and $x = 1$ is not in the domain. There are no relative extrema.

19. $f(x) = \cos x - x, \quad 0 \leq x \leq 4\pi$

$f'(x) = -\sin x - 1 \leq 0$

Therefore, f is non-increasing and there are no relative extrema.

20. $f(x) = 2 \sin x + \cos 2x, \quad 0 \le x \le 2\pi$

$f'(x) = 2 \cos x - 2 \sin 2x = 2 \cos x (1 - 2 \sin x) = 0$ when $x = \dfrac{\pi}{6}, \dfrac{\pi}{2}, \dfrac{5\pi}{6}, \dfrac{3\pi}{2}$.

$f''(x) = -2 \sin x - 4 \cos 2x$

$f''\left(\dfrac{\pi}{6}\right) < 0$

$f''\left(\dfrac{\pi}{2}\right) > 0$

$f''\left(\dfrac{5\pi}{6}\right) < 0$

$f''\left(\dfrac{3\pi}{2}\right) > 0$

Relative maxima: $\left(\dfrac{\pi}{6}, \dfrac{3}{2}\right), \left(\dfrac{5\pi}{6}, \dfrac{3}{2}\right)$

Relative minima: $\left(\dfrac{\pi}{2}, 1\right), \left(\dfrac{3\pi}{2}, -3\right)$

21. $f(x) = x^3 - 12x$

$f'(x) = 3x^2 - 12 = 3(x + 2)(x - 2) = 0$ when $x = \pm 2$.

$f''(x) = 6x$

$f''(-2) = -12 < 0 \implies (-2, \ 16)$ is a relative maximum.

$f''(2) = 12 > 0 \implies (2, \ -16)$ is a relative minimum.

$f''(x) = 6x = 0$ when $x = 0$.

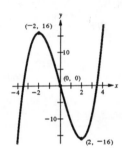

Test interval	$-\infty < x < 0$	$0 < x < \infty$
Sign of $f''(x)$	$f''(x) < 0$	$f''(x) > 0$
Conclusion	Concave downward	Concave upward

Point of inflection: $(0, 0)$

22. $f(x) = x^3 + 1$

$f'(x) = 3x^2 = 0$ when $x = 0$.

$f''(x) = 6x = 0$ when $x = 0$.

Since $f'(x) \ge 0$ for all x and the concavity changes at $x = 0$. $(0, 1)$ is a point of inflection.

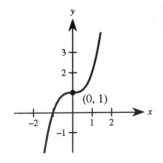

23. $f(x) = x^3 - 6x^2 + 12x - 8$

$f'(x) = 3x^2 - 12x + 12$

$\qquad = 3(x - 2)^2 = 0$ when $x = 2$.

$f''(x) = 6(x - 2) = 0$ when $x = 2$.

Since $f'(x) > 0$ when $x \ne 2$ and the concavity changes at $x = 2$. $(2, 0)$ is a point of inflection.

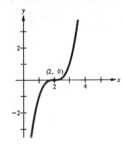

24. $f(x) = 2x^3 - 3x^2 - 12x + 8$

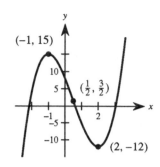

$f'(x) = 6x^2 - 6x - 12 = 6(x - 2)(x + 1) = 0$ when $x = -1, \ 2.$

$f''(x) = 12x - 6$

$f''(-1) = -18 < 0 \ \Rightarrow \ (-1, \ 15)$ is a relative maximum.

$f''(2) = 18 > 0 \ \Rightarrow \ (2, \ -12)$ is a relative minimum.

$f''(x) = 12x - 6 = 0$ when $x = \frac{1}{2}$

Test interval	$-\infty < x < \frac{1}{2}$	$\frac{1}{2} < x < \infty$
Sign of $f''(x)$	$f''(x) < 0$	$f''(x) > 0$
Conclusion	Concave downward	Concave upward

Point of inflection: $\left(\frac{1}{2}, \ \frac{3}{2}\right)$

25. $f(x) = \frac{1}{4}x^4 - 2x^2$

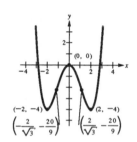

$f'(x) = x^3 - 4x = x(x + 2)(x - 2) = 0$ when $x = 0, \ \pm 2.$

$f''(x) = 3x^2 - 4$

$f''(-2) = 8 > 0 \ \Rightarrow \ (-2, \ -4)$ is a relative minimum.

$f''(0) = -4 < 0 \ \Rightarrow \ (0, \ 0)$ is a relative maximum.

$f''(2) = 8 > 0 \ \Rightarrow \ (2, \ -4)$ is a relative minimum.

$f''(x) = 3x^2 - 4 = 0$ when $x = \pm\dfrac{2}{\sqrt{3}}$

Test interval	$-\infty < x < -\dfrac{2}{\sqrt{3}}$	$-\dfrac{2}{\sqrt{3}} < x < \dfrac{2}{\sqrt{3}}$	$\dfrac{2}{\sqrt{3}} < x < \infty$
Sign of $f''(x)$	$f''(x) > 0$	$f''(x) < 0$	$f''(x) > 0$
Conclusion	Concave upward	Concave downward	Concave upward

Point of inflection: $\left(\pm\dfrac{2}{\sqrt{3}}, \ -\dfrac{20}{9}\right)$

26. $f(x) = 2x^4 - 8x + 3$

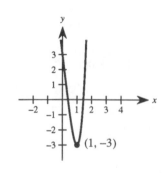

$f'(x) = 8x^3 - 8 = 8(x - 1)(x^2 + x + 1) = 0$ when $x = 1.$

$f''(x) = 24x^2 = 0$ when $x = 0.$

Since $f''(1) > 0, \ (1, \ -3)$ is a relative minimum. However, $(0, 3)$ is not a point of inflection since $f''(x) \geq 0$ for all $x.$

27. $f(x) = x(x-4)^3$

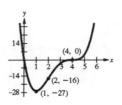

$f'(x) = x[3(x-4)^2] + (x-4)^3$

$\qquad = (x-4)^2(4x-4) = 4(x-1)(x-4)^2 = 0$ when $x = 1, \ 4.$

$f''(x) = 4(x-1)[2(x-4)] + 4(x-4)^2$

$\qquad = 4(x-4)[2(x-1) + (x-4)]$

$\qquad = 4(x-4)(3x-6) = 12(x-4)(x-2)$

$f''(1) = 36 > 0 \ \Rightarrow \ (1, \ -27)$ is a relative minimum.

$f''(4) = 0 \ \Rightarrow \ $ test fails

By the First Derivative Test we see that $x = 4$ does not yield a relative extrema.

$\qquad f''(x) = 12(x-4)(x-2) = 0$ when $x = 2, \ 4.$

Test interval	$-\infty < x < 2$	$2 < x < 4$	$4 < x < \infty$
Sign of $f''(x)$	$f''(x) > 0$	$f''(x) < 0$	$f''(x) > 0$
Conclusion	Concave upward	Concave downward	Concave upward

Points of inflection: $(2, \ -16), \ (4, \ 0)$

28. $f(x) = x^3(x-4)$

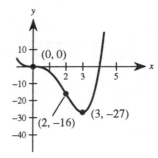

$f'(x) = x^3 + 3x^2(x-4)$

$\qquad = x^2[x + 3(x-4)] = 4x^2(x-3) = 0$ when $x = 0, \ 3.$

$f''(x) = 4x^2 + 8x(x-3) = 4x[x + 2(x-3)] = 12x(x-2) = 0$

$f''(0) = 0 \ \Rightarrow \ $ test fails

$f''(3) = 36 > 0 \ \Rightarrow \ (3, \ -27)$ is a relative minimum.

By the First Derivative Test we see that $x = 0$ does not yield a relative extrema.

$\qquad f''(x) = 12x(x-2) = 0$ when $x = 0, \ 2.$

Test interval	$-\infty < x < 0$	$0 < x < 2$	$2 < x < \infty$
Sign of $f''(x)$	$f''(x) > 0$	$f''(x) < 0$	$f''(x) > 0$
Conclusion	Concave upward	Concave downward	Concave upward

Points of inflection: $(0, \ 0), \ (2, \ -16)$

29. $f(x) = x\sqrt{x+3}, \quad$ Domain: $[-3, \ \infty)$

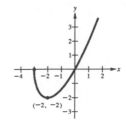

$f'(x) = x\left(\dfrac{1}{2}\right)(x+3)^{-1/2} + \sqrt{x+3} = \dfrac{3(x+2)}{2\sqrt{x+3}} = 0$ when $x = -2.$

$f''(x) = \dfrac{6\sqrt{x+3} - 3(x+2)(x+3)^{-1/2}}{4(x+3)} = \dfrac{3(x+4)}{4(x+3)^{3/2}}$

Since $f''(-2) > 0$, then $(-2, \ -2)$ is a relative minimum. $f''(x) > 0$ on the entire domain of f (except for $x = -3$, for which $f''(x)$ is undefined). There are no points of inflection.

30. $f(x) = x\sqrt{x+1}$,   Domain: $[-1, \infty)$

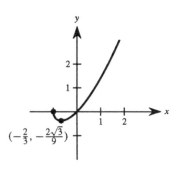

$$f'(x) = (x)\frac{1}{2}(x+1)^{-1/2} + \sqrt{x+1} = \frac{3x+2}{2\sqrt{x+1}} = 0 \text{ when } x = -\frac{2}{3}.$$

$$f''(x) = \frac{6\sqrt{x+1} - (3x+2)(x+1)^{-1/2}}{4(x+1)} = \frac{3x+4}{4(x+1)^{3/2}}$$

Since $f'\left(\frac{2}{3}\right) > 0$, then $\left(-\frac{2}{3}, -\frac{2\sqrt{3}}{9}\right)$ is a relative minimum. $f''(x) > 0$ on the entire domain of f (except for $x = -1$, for which $f''(x)$ is undefined). There are no points of inflection.

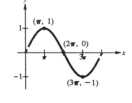

31. $f(x) = \sin\left(\frac{x}{2}\right)$, $0 \le x \le 4\pi$

$$f'(x) = \frac{1}{2}\cos\left(\frac{x}{2}\right) = 0 \text{ when } x = \pi, \ 3\pi.$$

$$f''(x) = -\frac{1}{4}\sin\left(\frac{\pi}{2}\right)$$

$f''(\pi) < 0 \Rightarrow (\pi, \ 1)$ is a relative maximum.

$f''(3\pi) > 0 \Rightarrow (3\pi, \ -1)$ is a relative minimum.

$f''(x) = 0$ when $x = 0, \ 2\pi, \ 4\pi$.

Test interval	$0 < x < 2\pi$	$2\pi < x < 4\pi$
Sign of $f''(x)$	$f'' < 0$	$f'' > 0$
Conclusion	Concave downward	Concave upward

Point of inflection: $(2\pi, \ 0)$

32. $f(x) = 2\csc\frac{3x}{2}$, $0 < x < 2\pi$

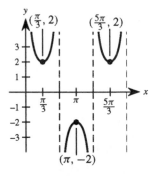

$$f'(x) = -3\csc\frac{3x}{2}\cot\frac{3x}{2} = 0 \text{ when } x = \frac{\pi}{3}, \ \pi, \ \frac{5\pi}{3}.$$

$$f''(x) = \frac{9}{2}\left(\csc^3\frac{3x}{2} + \csc\frac{3x}{2}\cot^2\frac{3x}{2}\right) \ne 0 \text{ for any } x \text{ in the domain of } f.$$

$f''\left(\frac{\pi}{3}\right) > 0 \Rightarrow \left(\frac{\pi}{3}, \ 2\right)$ is a relative minimum.

$f''(\pi) < 0 \Rightarrow (\pi, \ -2)$ is a relative maximum.

$f''\left(\frac{5\pi}{3}\right) > 0 \Rightarrow \left(\frac{5\pi}{3}, \ 2\right)$ is a relative minimum.

33. $f(x) = \sec\left(x - \dfrac{\pi}{2}\right), \quad 0 < x < 4\pi$

$f'(x) = \sec\left(x - \dfrac{\pi}{2}\right)\tan\left(x - \dfrac{\pi}{2}\right) = 0$ when $x = \dfrac{\pi}{2}, \dfrac{3\pi}{2}, \dfrac{5\pi}{2}, \dfrac{7\pi}{2}.$

$f''(x) = \sec^3\left(x - \dfrac{\pi}{2}\right) + \sec\left(x - \dfrac{\pi}{2}\right)\tan^2\left(x - \dfrac{\pi}{2}\right) \neq 0$ for any x in the domain of f.

$f''\left(\dfrac{\pi}{2}\right) > 0 \Rightarrow \left(\dfrac{\pi}{2}, \; 1\right)$ is a relative minimum.

$f''\left(\dfrac{3\pi}{2}\right) < 0 \Rightarrow \left(\dfrac{3\pi}{2}, \; -1\right)$ is a relative maximum.

$f''\left(\dfrac{5\pi}{2}\right) > 0 \Rightarrow \left(\dfrac{5\pi}{2}, \; 1\right)$ is a relative minimum.

$f''\left(\dfrac{7\pi}{2}\right) < 0 \Rightarrow \left(\dfrac{7\pi}{2}, \; -1\right)$ is a relative maximum.

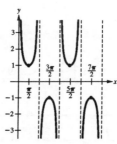

34. $f(x) = \sin x + \cos x, \quad 0 \le x \le 2\pi$

$f'(x) = \cos x - \sin x = 0$ when $x = \dfrac{\pi}{4}, \dfrac{5\pi}{4}.$

$f''(x) = -\sin x - \cos x$

$f''\left(\dfrac{\pi}{4}\right) < 0 \Rightarrow \left(\dfrac{\pi}{4}, \; \sqrt{2}\right)$ is a relative maximum.

$f''\left(\dfrac{5\pi}{4}\right) > 0 \Rightarrow \left(\dfrac{5\pi}{4}, \; -\sqrt{2}\right)$ is a relative minimum.

$f''(x) = 0$ when $x = \dfrac{3\pi}{4}, \dfrac{7\pi}{4}.$

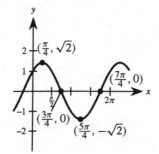

Test interval	$0 < x < \dfrac{3\pi}{4}$	$\dfrac{3\pi}{4} < x < \dfrac{7\pi}{4}$	$\dfrac{7\pi}{4} < x < 2\pi$
Sign of $f''(x)$	$f'' < 0$	$f'' > 0$	$f'' < 0$
Conclusion	Concave downward	Concave upward	Concave downward

Points of inflection: $\left(\dfrac{3\pi}{4}, \; 0\right), \left(\dfrac{7\pi}{4}, \; 0\right)$

35. $f(x) = 2\sin x + \sin 2x, \quad 0 \le x \le 2\pi$

$f'(x) = 2\cos x + 2\cos 2x = 0$

$2(2\cos^2 x + \cos x - 1) = 0$

$2(\cos x + 1)(2\cos x - 1) = 0$ when $\cos x = \dfrac{1}{2}, \; -1$ or $x = \dfrac{\pi}{3}, \; \pi, \; \dfrac{5\pi}{3}.$

$f''(x) = -2\sin x - 4\sin 2x = -2\sin x(1 + 4\cos x)$

$f''\left(\dfrac{\pi}{3}\right) < 0 \Rightarrow \left(\dfrac{\pi}{3}, \; 2.598\right)$ is a relative maximum.

$f''\left(\dfrac{5\pi}{3}\right) > 0 \Rightarrow \left(\dfrac{5\pi}{3}, \; -2.598\right)$ is a relative minimum.

$f''(x) = 0$ when $x = 0, \; 1.823, \; \pi, \; 4.460.$

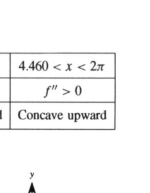

Test interval	$0 < x < 1.823$	$1.823 < x < \pi$	$\pi < x < 4.460$	$4.460 < x < 2\pi$
Sign of $f''(x)$	$f'' < 0$	$f'' > 0$	$f'' < 0$	$f'' > 0$
Conclusion	Concave downward	Concave upward	Concave downward	Concave upward

Points of inflection: $(1.823, \; 1.452), \; (\pi, \; 0), \; (4.46, \; -1.452)$

36. $f(x) = x - \sin x, \quad 0 \le x \le 4\pi$

$f'(x) = 1 - \cos x \ge 0$

Therefore, f is nondecreasing, and there are no relative extrema.

$f''(x) = \sin x = 0$ when $x = 0, \; \pi, \; 2\pi, \; 3\pi, \; 4\pi.$

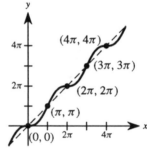

Test interval	$0 < x < \pi$	$\pi < x < 2\pi$	$2\pi < x < 3\pi$	$3\pi < x < 4\pi$
Sign of $f''(x)$	$f'' > 0$	$f'' < 0$	$f'' > 0$	$f'' < 0$
Conclusion	Concave upward	Concave downward	Concave upward	Concave downward

Points of inflection: $(\pi, \; \pi), \; (2\pi, \; 2\pi), \; (3\pi, \; 3\pi)$

Note: $(0, \; 0)$ and $(4\pi, \; 4\pi)$ are not points of inflection since they are endpoints.

37. $f(x) = 0.2x^2(x - 3)^3, \quad [-1, \; 4]$

   a. $f'(x) = 0.2x(5x - 6)(x - 3)^2$

       $f''(x) = (x - 3)(4x^2 - 9.6x + 3.6)$

           $= 0.4(x - 3)(10x^2 - 24x + 9)$

   b. $f''(0) < 0 \Rightarrow (0, \; 0)$ is a relative maximum.

       $f''\left(\dfrac{6}{5}\right) > 0 \Rightarrow (1.2, \; -1.6796)$ is a relative minimum.

       Points of inflection:

       $(3, \; 0), \; (0.4652, \; -0.7049), \; (1.9348, \; -0.9049)$

c.

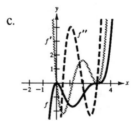

f is increasing when $f' > 0$ and decreasing when $f' < 0$. f is concave upward when $f'' > 0$ and concave downward when $f'' < 0$.

38. $f(x) = x^2\sqrt{6-x^2}, \quad [-\sqrt{6}, \sqrt{6}]$

a. $f'(x) = \dfrac{3x(4-x^2)}{\sqrt{6-x^2}}$

$f'(x) = 0$ when $x = 0, \; x = \pm 2$.

$f''(x) = \dfrac{6(x^4 - 9x^2 + 12)}{(6-x^2)^{3/2}}$

$f''(x) = 0$ when $x = \pm\sqrt{\dfrac{9-\sqrt{33}}{2}}$.

b. $f''(0) > 0 \Rightarrow (0, \; 0)$ is a relative minimum.
$f''(\pm 2) < 0 \Rightarrow (\pm 2, \; 4\sqrt{2})$ are relative maxima.

Points of inflection: $(\pm 1.2758, \; 3.4035)$

c.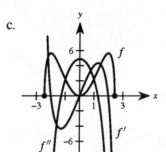

The graph of f is increasing when $f' > 0$ and decreasing when $f' < 0$. f is concave upward when $f'' > 0$ and concave downward when $f'' < 0$.

39. $f(x) = \sin x - \dfrac{1}{3}\sin 3x + \dfrac{1}{5}\sin 5x, \quad [0, \; \pi]$

a. $f'(x) = \cos x - \cos 3x + \cos 5x$

$f'(x) = 0$ when $x = \dfrac{\pi}{6}, \; x = \dfrac{\pi}{2}, \; x = \dfrac{5\pi}{6}$.

$f''(x) = -\sin x + 3\sin 3x - 5\sin 5x$

$f''(x) = 0$ when $x = \dfrac{\pi}{6}, \; x = \dfrac{5\pi}{6}, \; x \approx 1.1731, \; x \approx 1.9685$.

b. $f''\left(\dfrac{\pi}{2}\right) < 0 \Rightarrow \left(\dfrac{\pi}{2}, \; 1.53333\right)$ is a relative maximum.

Points of inflection: $\left(\dfrac{\pi}{6}, \; 0.2667\right)$, $(1.1731, \; 0.9638)$,

$(1.9685, \; 0.9637)$, $\left(\dfrac{5\pi}{6}, \; 0.2667\right)$

Note: $(0, \; 0)$ and $(\pi, \; 0)$ are not points of inflection since they are endpoints.

c.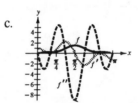

The graph of f is increasing when $f' > 0$ and decreasing when $f' < 0$. f is concave upward when $f'' > 0$ and concave downward when $f'' < 0$.

40. $f(x) = \sqrt{2x}\sin x, \quad [0, \; 2\pi]$

a. $f'(x) = \sqrt{2x}\cos x + \dfrac{\sin x}{\sqrt{2x}}$

Critical numbers: $x \approx 1.84, \; 4.82$

$f''(x) = -\sqrt{2x}\sin x + \dfrac{\cos x}{\sqrt{2x}} + \dfrac{\cos x}{\sqrt{2x}} - \dfrac{\sin x}{2x\sqrt{2x}}$

$= \dfrac{2\cos x}{\sqrt{2x}} - \dfrac{(4x^2 + 1)\sin x}{2x\sqrt{2x}}$

$= \dfrac{4x\cos x - (4x^2 + 1)\sin x}{2x\sqrt{2x}}$

b. Relative maximum: $(1.84, \; 1.85)$
Relative minimum: $(4.82, \; -3.09)$
Points of inflection: $(0.75, \; 0.83), \; (3.42, \; -0.72)$

c.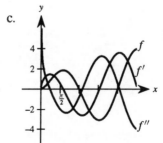

f is increasing when $f' > 0$ and decreasing when $f' < 0$. f is concave upward when $f'' > 0$ and concave downward when $f'' < 0$.

41.

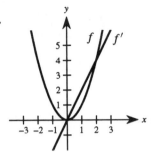

42.

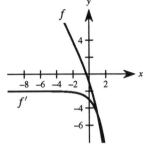

43.

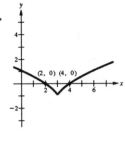

44.

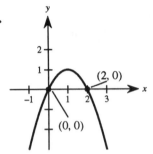

45.

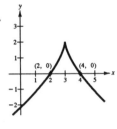

46.

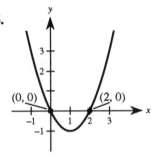

47.

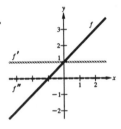

48.

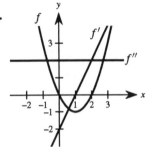

49.

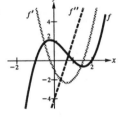

50.

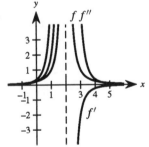

51.

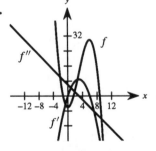

52. a.

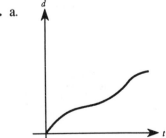

b. Since the depth d is always increasing, there are no relative extrema. $f'(x) > 0$

c. The rate of change of d is decreasing until you reach the widest point of the jug, then the rate increases until you reach the narrowest part of the jug's neck, then the rate decreases until you reach the top of the jug.

53.

$n = 1$:

$f(x) = x - c$

$f'(x) = 1$

$f''(x) = 0$

No inflection points

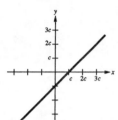

$n = 2$:

$f(x) = (x - c)^2$

$f'(x) = 2(x - c)$

$f''(x) = 2$

No inflection points
Relative minimum:
    $(c, 0)$

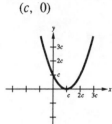

$n = 3$:

$f(x) = (x - c)^3$

$f'(x) = 3(x - c)^2$

$f''(x) = 6(x - c)$

Inflection point $(c, 0)$

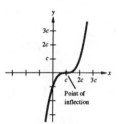

$n = 4$:

$f(x) = (x - c)^4$

$f'(x) = 4(x - c)^3$

$f''(x) = 12(x - c)^2$

No inflection points
Relative minimum:
    $(c, 0)$

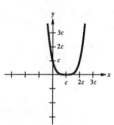

Conclusion: If $n \geq 3$ and n is odd, then $(c, 0)$ is an inflection point. If $n \geq 2$ and n is even, then $(c, 0)$ is a relative minimum.

54. a. $f(x) = \sqrt[3]{x}$

$f'(x) = \frac{1}{3}x^{-2/3}$

$f''(x) = -\frac{2}{9}x^{-5/3}$

Inflection point: $(0, 0)$

b. $f''(x)$ does not exist at $x = 0$.

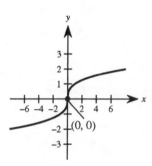

55. $f(x) = ax^3 + bx^2 + cx + d$

Relative maximum: $(3, 3)$

Relative minimum: $(5, 1)$

Point of inflection: $(4, 2)$

$f'(x) = 3ax^2 + 2bx + c, \quad f''(x) = 6ax + 2b$

$\left.\begin{array}{l} f(3) = 27a + 9b + 3c + d = 3 \\ f(5) = 125a + 25b + 5c + d = 1 \end{array}\right\} 98a + 16b + 2c = -2 \Rightarrow 49a + 8b + c = -1$

$f'(3) = 27a + 6b + c = 0, \quad f''(4) = 24a + 2b = 0$

$\begin{array}{ll} 49a + 8b + c = -1 & 24a + 2b = 0 \\ \underline{27a + 6b + c = 0} & \underline{22a + 2b = -1} \\ 22a + 2b = -1 & 2a = 1 \end{array}$

$a = \frac{1}{2}, \quad b = -6, \quad c = \frac{45}{2}, \quad d = -24$

$f(x) = \frac{1}{2}x^3 - 6x^2 + \frac{45}{2}x - 24$

56. $f(x) = ax^3 + bx^2 + cx + d$

Relative maximum: (2, 4)

Relative minimum: (4, 2)

Point of inflection: (3, 3)

$f'(x) = 3ax^2 + 2bx + c, \quad f''(x) = 6ax + 2b$

$\left. \begin{array}{l} f(2) = 8a + 4b + 2c + d = 4 \\ f(4) = 64a + 16b + 4c + d = 2 \end{array} \right\} \; 56a + 12b + 2c = -2 \Rightarrow 28a + 6b + c = -1$

$f'(2) = 12a + 4b + c = 0, \quad f'(4) = 48a + 8b + c = 0, \quad f''(3) = 18a + 2b = 0$

$$\begin{array}{cc} \begin{array}{l} 28a + 6b + c = -1 \\ 12a + 4b + c = 0 \\ \hline 16a + 2b = -1 \end{array} & \begin{array}{l} 18a + 2b = 0 \\ 16a + 2b = -1 \\ \hline 2a = 1 \end{array} \end{array}$$

$a = \frac{1}{2}, \quad b = -\frac{9}{2}, \quad c = 12, \quad d = -6$

$f(x) = \frac{1}{2}x^3 - \frac{9}{2}x^2 + 12x - 6$

57. $f(x) = ax^3 + bx^2 + cx + d$

Maximum: (−4, 1)

Minimum: (0, 0)

Point of inflection: $\left(-2, \frac{1}{2}\right)$

a. $f'(x) = 3ax^2 + 2bx + c, \quad f''(x) = 6ax + 2b$

$f(0) = 0 \Rightarrow d = 0$

$f(-4) = 1 \Rightarrow -64a + 16b - 4c = 1$

$f'(-4) = 0 \Rightarrow 48a - 8b + c = 0$

$f'(0) = 0 \Rightarrow \phantom{48a - 8b + {}} c = 0$

$f''(-2) = 0 \Rightarrow -12a + 2b = 0$

Solving this system yields $a = \frac{1}{32}$ and $b = 6a = \frac{3}{16}$.

$f(x) = \frac{1}{32}x^3 + \frac{3}{16}x^2$

b. The plane would be descending at the greatest rate at the point of inflection, or two miles from touchdown.

58. $f(x) = x(x - 6)^2 = x^3 - 12x^2 + 36x$

$f'(x) = 3x^2 - 24x + 36 = 3(x - 2)(x - 6) = 0$

$f''(x) = 6x - 24 = 6(x - 4) = 0$

Relative extrema: (2, 32) and (6, 0)

Point of inflection (4, 16) is midway between the relative extrema of f.

59. Assume the zeros of f are all real. Then express the function as $f(x) = a(x - r_1)(x - r_2)(x - r_3)$ where r_1, r_2, and r_3 are the distinct zeros of f. From the Product Rule for a function involving three factors, we have

$$f'(x) = a[(x - r_1)(x - r_2) + (x - r_1)(x - r_3) + (x - r_2)(x - r_3)]$$

$$f''(x) = a[(x - r_1) + (x - r_2) + (x - r_1) + (x - r_3) + (x - r_2) + (x - r_3)]$$

$$= a[6x - 2(r_1 + r_2 + r_3)].$$

Consequently, $f''(x) = 0$ if

$$x = \frac{2(r_1 + r_2 + r_3)}{6} = \frac{r_1 + r_2 + r_3}{3} = \text{(Average of } r_1, \ r_2, \ \text{and } r_3).$$

60. $S = \dfrac{5.755T^3}{10^8} - \dfrac{8.521T^2}{10^6} + \dfrac{0.654T}{10^4} + 0.99987, \quad 0 < T < 25$

a. The maximum occurs when $T \approx 4°$ and $S \approx 0.999999$.

b. $S(20°) \approx 0.9982$

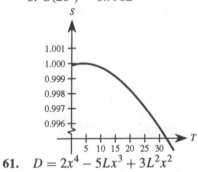

61. $D = 2x^4 - 5Lx^3 + 3L^2x^2$

$$D' = 8x^3 - 15Lx^2 + 6L^2x = x(8x^2 - 15Lx + 6L^2) = 0$$

$$x = 0 \text{ or } x = \frac{15L \pm \sqrt{33}\,L}{16} = \left(\frac{15 \pm \sqrt{33}}{16}\right)L$$

By the Second Derivative Test, the deflection is maximum when

$$x = \left(\frac{15 - \sqrt{33}}{16}\right)L \approx 0.578L.$$

62. $E = kqx(x^2 + a^2)^{-3/2}$

$$E' = kqx\left[-\frac{3}{2}(x^2 + a^2)^{-5/2}(2x)\right] + kq(x^2 + a^2)^{-3/2}$$

$$= kq(x^2 + a^2)^{-5/2}[-3x^2 + (x^2 + a^2)] = \frac{kq(a^2 - 2x^2)}{(x^2 + a^2)^{5/2}}$$

By the First Derivative Test, E is maximized when $x = (a\sqrt{2})/2$.

63. $C = 0.5x^2 + 15x + 5000$

$$\overline{C} = \frac{C}{x} = 0.5x + 15 + \frac{5000}{x}$$

$\overline{C} = $ average cost per unit

$$\frac{d\overline{C}}{dx} = 0.5 - \frac{5000}{x^2} = 0 \text{ when } x = 100$$

By the First Derivative Test, $\overline{C}$ is minimized when $x = 100$ units.

64. $C = 2x + \dfrac{300,000}{x}$

$$C' = 2 - \frac{300,000}{x^2} = 0 \text{ when } x = 100\sqrt{15} \approx 387$$

By the First Derivative Test, C is minimized when $x \approx 387$ units.

65. $v = -1200\pi \sin\theta$

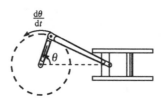

$$\frac{d\theta}{dt} = (200)(2\pi) = 400\pi \text{ rad/min}$$

$$\frac{dv}{dt} = -1200\pi \cos\theta \frac{d\theta}{dt}$$

$$= -1200\pi(400\pi)\cos\theta$$

$$= 0 \text{ when } \theta = \frac{(2n+1)\pi}{2}, \quad n \text{ is any integer.}$$

By the First Derivative Test, the velocity v is maximum when $\theta = (3\pi/2) + 2n\pi$. The speed $|v|$ is maximum when

$$\theta = \frac{\pi}{2} + 2n\pi \text{ and } \frac{3\pi}{2} + 2n\pi.$$

66. $f(x) = x\sin\left(\dfrac{1}{x}\right)$

$$f'(x) = x\left[-\frac{1}{x^2}\cos\left(\frac{1}{x}\right)\right] + \sin\left(\frac{1}{x}\right) = -\frac{1}{x}\cos\left(\frac{1}{x}\right) + \sin\left(\frac{1}{x}\right)$$

$$f''(x) = -\frac{1}{x}\left[\frac{1}{x^2}\sin\left(\frac{1}{x}\right)\right] + \frac{1}{x^2}\cos\left(\frac{1}{x}\right) - \frac{1}{x^2}\cos\left(\frac{1}{x}\right) = -\frac{1}{x^3}\sin\left(\frac{1}{x}\right) = 0$$

$$x = \frac{1}{\pi}$$

Point of inflection: $\left(\dfrac{1}{\pi}, 0\right)$

When $x > 1/\pi$, $f'' < 0$, so the graph is concave downward.

67. $f(x) = 2(\sin x + \cos x)$, $f(0) = 2$

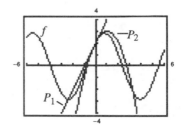

$f'(x) = 2(\cos x - \sin x)$, $f'(0) = 2$

$f''(x) = 2(-\sin x - \cos x)$, $f''(0) = -2$

$P_1(x) = 2 + 2(x - 0) = 2(1 + x)$

$P_1'(x) = 2$

$P_2(x) = 2 + 2(x - 0) + \frac{1}{2}(-2)(x - 0)^2 = 2 + 2x - x^2$

$P_2'(x) = 2 - 2x$

$P_2''(x) = -2$

The values of f, P_1, P_2, and their first derivatives are equal at $x = 0$. The values of the second derivatives of f and P_2 are equal at $x = 0$.

68. $f(x) = \sqrt{1-x}$, $f(0) = 1$

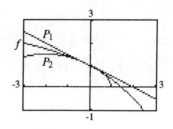

$$f'(x) = -\frac{1}{2\sqrt{1-x}}, \qquad f'(0) = -\frac{1}{2}$$

$$f''(x) = -\frac{1}{4(1-x)^{3/2}}, \qquad f''(0) = -\frac{1}{4}$$

$$P_1(x) = 1 + \left(-\frac{1}{2}\right)(x-0) = 1 - \frac{x}{2}$$

$$P_1'(x) = -\frac{1}{2}$$

$$P_2(x) = 1 + \left(-\frac{1}{2}\right)(x-0) + \frac{1}{2}\left(-\frac{1}{4}\right)(x-0)^2 = 1 - \frac{x}{2} - \frac{x^2}{8}$$

$$P_2'(x) = -\frac{1}{2} - \frac{x}{4}$$

$$P_2''(x) = -\frac{1}{4}$$

The values of f, P_1, P_2, and their first derivatives are equal at $x = 0$. The values of the second derivatives of f and P_2 are equal at $x = 0$.

69. $f(x) = 2(\sin x + \cos x)$, $f\left(\frac{\pi}{4}\right) = 2\sqrt{2}$

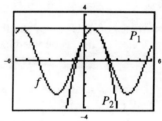

$$f'(x) = 2(\cos x - \sin x), \qquad f'\left(\frac{\pi}{4}\right) = 0$$

$$f''(x) = 2(-\sin x - \cos x), \qquad f''\left(\frac{\pi}{4}\right) = -2\sqrt{2}$$

$$P_1(x) = 2\sqrt{2} + 0\left(x - \frac{\pi}{4}\right) = 2\sqrt{2}$$

$$P_1'(x) = 0$$

$$P_2(x) = 2\sqrt{2} + 0\left(x - \frac{\pi}{4}\right) + \frac{1}{2}(-2\sqrt{2})\left(x - \frac{\pi}{4}\right)^2 = 2\sqrt{2} - \sqrt{2}\left(x - \frac{\pi}{4}\right)^2$$

$$P_2'(x) = -2\sqrt{2}\left(x - \frac{\pi}{4}\right)$$

$$P_2''(x) = -2\sqrt{2}$$

The values of f, P_1, P_2, and their first derivatives are equal at $x = \pi/4$. The values of the second derivatives of f and P_2 are equal at $x = \pi/4$.

70. $f(x) = \dfrac{\sqrt{x}}{x-1}$, $f(2) = \sqrt{2}$

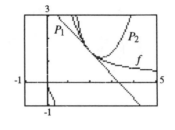

$f'(x) = \dfrac{-(x+1)}{2\sqrt{x}(x-1)^2}$, $f'(2) = -\dfrac{3}{2\sqrt{2}} = -\dfrac{3\sqrt{2}}{4}$

$f''(x) = \dfrac{3x^2 + 6x - 1}{4x^{3/2}(x-1)^3}$, $f''(2) = \dfrac{23}{8\sqrt{2}} = \dfrac{23\sqrt{2}}{16}$

$P_1(x) = \sqrt{2} + \left(-\dfrac{3\sqrt{2}}{4}\right)(x-2) = -\dfrac{3\sqrt{2}}{4}x + \dfrac{5\sqrt{2}}{2}$

$P_1'(x) = -\dfrac{3\sqrt{2}}{4}$

$P_2(x) = \sqrt{2} + \left(-\dfrac{3\sqrt{2}}{4}\right)(x-2) + \dfrac{1}{2}\left(\dfrac{23\sqrt{2}}{16}\right)(x-2)^2$

$\quad = \sqrt{2} - \dfrac{3\sqrt{2}}{4}(x-2) + \dfrac{23\sqrt{2}}{32}(x-2)^2$

$P_2'(x) = -\dfrac{3\sqrt{2}}{4} + \dfrac{23\sqrt{2}}{16}(x-2)$

$P_2''(x) = \dfrac{23\sqrt{2}}{16}$

The values of f, P_1, P_2, and their first derivatives are equal at $x = 2$.
The values of the second derivatives of f and P_2 are equal at $x = 2$.

71. a. The rate of change of sales is increasing.

   $S'' > 0$

   b. The rate of change of sales is decreasing.

   $S'' < 0$

   c. The rate of change of sales is constant.

   $S' = C, \quad S'' = 0$

   d. Sales are steady.

   $S = C, \quad S' = 0, \quad S'' = 0$

   e. Sales are declining, but at a lower rate.

   $S' < 0, \quad S'' > 0$

   f. Sales have bottomed out and have started to rise.

   $S' > 0$

72. Let $f(x) = x^4$.

   $f''(x) = 12x^2$

   $f''(0) = 0$, but $(0, 0)$ is not a point of inflection.

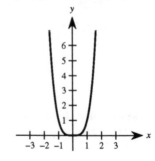

73. True. Let $y = ax^3 + bx^2 + cx + d$, $a \neq 0$. Then $y'' = 6ax + 2b = 0$ when $x = -(b/3a)$, and the concavity changes at this point.

74. False. $f(x) = 1/x$ has a discontinuity at $x = 0$.

75. False

$$f(x) = 3 \sin x + 2 \cos x$$

$$f'(x) = 3 \cos x - 2 \sin x$$

$$3 \cos x - 2 \sin x = 0$$

$$3 \cos x = 2 \sin x$$

$$\tfrac{3}{2} = \tan x$$

Critical number: $x = \tan^{-1}\left(\tfrac{3}{2}\right)$

$f\left(\tan^{-1} \tfrac{3}{2}\right) \approx 3.60555$ is the maximum value of y.

76. True

$$y = \sin(bx)$$

Slope: $y' = b \cos(bx)$

$$-b \le y' \le b$$

77. The First Derivative Test has the following advantages:

1. It does not fail as the Second Derivative Test sometimes does.

Example: $f(x) = x^3$

2. It can be used for critical numbers that make the derivative undefined. The Second Derivative Test can only be used if $f'(c) = 0$.

Example: $f(x) = (x-1)^{2/3}$

3. You do not need to take a second derivative to determine relative extrema.

Example: $f(x) = \dfrac{\sqrt{x+7}}{x^2 - 9}$

The Second Derivative Test usually works best on polynomials or on functions whose derivatives are easily determined. It has the added advantage of providing additional information such as concavity and points of inflection.

78. Suppose $y_1 < d < y_2$ and let $g(x) = f(x) - d(x - a)$. g is continuous on $[a, b]$, and therefore, by the Extreme Value Theorem, attains a minimum value at some point c in $[a, b]$. This point c cannot be the endpoints a or b because

$$g'(a) = f'(a) - d = y_1 - d < 0$$

$$g'(b) = f'(b) - d = y_2 - d > 0.$$

Hence, g attains its minimum value at c, $a < c < b$. This implies that $g'(c) = 0$ and $f'(c) = d$. The proof for $y_2 < d < y_1$ is similar.

79. $p(x) = ax^3 + bx^2 + cx + d$

$p'(x) = 3ax^2 + 2bx + c$

$p''(x) = 6ax + 2b$

$6ax + 2b = 0$

$x = -\dfrac{b}{3a}$

The sign of $p''(x)$ changes at $x = -b/3a$. Therefore, $(-b/3a, \ p(-b/3a))$ is a point of inflection.

$$p\left(-\frac{b}{3a}\right) = a\left(-\frac{b^3}{27a^3}\right) + b\left(\frac{b^2}{9a^2}\right) + c\left(-\frac{b}{3a}\right) + d = \frac{2b^3}{27a^2} - \frac{bc}{3a} + d$$

When $p(x) = x^3 - 3x^2 + 2$, $a = 1$, $b = -3$, $c = 0$, and $d = 2$.

$x_0 = \dfrac{-(-3)}{3(1)} = 1$

$y_0 = \dfrac{2(-3)^3}{27(1)^2} - \dfrac{(-3)(0)}{3(1)} + 2 = -2 - 0 + 2 = 0$

The point of inflection of $p(x) = x^3 - 3x^2 + 2$ is $(x_0, \ y_0) = (1, \ 0)$.

80. $\dfrac{dy}{dt} = Ky(L - y) = KLy - Ky^2$

$\dfrac{d^2y}{dt^2} = KL\dfrac{dy}{dt} - 2Ky\dfrac{dy}{dt} = K\dfrac{dy}{dt}(L - 2y)$

$\dfrac{d^2y}{dt^2} = 0$ when $L = 2y$.

The point of inflection occurs when $y = L/2$.

Section 3.5   Limits at Infinity

1. $f(x) = \dfrac{3x^2}{x^2 + 2}$

No vertical asymptotes

Horizontal asymptote: $y = 3$

Matches (h)

2. $f(x) = \dfrac{2x}{\sqrt{x^2 + 2}}$

No vertical asymptotes

Horizontal asymptotes: $y = \pm 2$

Matches (c)

3. $f(x) = \dfrac{x}{x^2 + 2}$

No vertical asymptotes

Horizontal asymptote: $y = 0$

Matches (e)

4. $f(x) = 2 + \dfrac{x^2}{x^4 + 1}$

No vertical asymptotes

Horizontal asymptote: $y = 2$

Matches (a)

5. $f(x) = \dfrac{-6x}{\sqrt{4x^2 + 5}}$

No vertical asymptotes

Horizontal asymptotes: $y = \pm 3$

Matches (d)

6. $f(x) = 5 - \dfrac{1}{x^2 + 1}$

No vertical asymptotes

Horizontal asymptote: $y = 5$

Matches (g)

7. $f(x) = \dfrac{4\sin x}{x^2+1}$

No vertical asymptotes

Horizontal asymptote: $y=0$

Matches (b)

8. $f(x) = \dfrac{2x^2-3x+5}{x^2+1}$

No vertical asymptotes

Horizontal asymptote: $y=2$

Matches (f)

9. $\displaystyle\lim_{x\to\infty}\frac{2x-1}{3x+2} = \lim_{x\to\infty}\frac{2-(1/x)}{3+(2/x)}$

$\qquad = \dfrac{2-0}{3+0} = \dfrac{2}{3}$

10. $\displaystyle\lim_{x\to\infty}\frac{5x^3+1}{10x^3-3x^2+7} = \lim_{x\to\infty}\frac{5+(1/x^3)}{10-(3/x)+(7/x^3)}$

$\qquad = \dfrac{1}{2}$

11. $\displaystyle\lim_{x\to\infty}\frac{x}{x^2-1} = \lim_{x\to\infty}\frac{1/x}{1-(1/x^2)} = \frac{0}{1} = 0$

12. $\displaystyle\lim_{x\to\infty}\frac{2x^{10}-1}{10x^{11}-3} = \lim_{x\to\infty}\frac{(2/x)-(1/x^{11})}{10-(3/x^{11})} = \frac{0}{10} = 0$

13. $\displaystyle\lim_{x\to-\infty}\frac{5x^2}{x+3} = \lim_{x\to-\infty}\frac{5x}{1+(3/x)} = -\infty$

Limit does not exist.

14. $\displaystyle\lim_{x\to\infty}\frac{x^3-2x^2+3x+1}{x^2-3x+2} = \lim_{x\to\infty}\frac{x-2+(3/x)+(1/x^2)}{1-(3/x)+(2/x^2)} = \infty$

Limit does not exist.

15. $\displaystyle\lim_{x\to\infty}\left(2x-\frac{1}{x^2}\right) = \infty - 0 = \infty$

Limit does not exist.

16. $\displaystyle\lim_{x\to\infty}\frac{1}{(x+3)^2} = 0$

17. $\displaystyle\lim_{x\to-\infty}\left(\frac{2x}{x-1}+\frac{3x}{x+1}\right) = \lim_{x\to-\infty}\left(\frac{2}{1-(1/x)}+\frac{3}{1+(1/x)}\right) = 2+3 = 5$

18. $\displaystyle\lim_{x\to\infty}\left(\frac{2x^2}{x-1}+\frac{3x}{x+1}\right) = \lim_{x\to\infty}\left(\frac{2x}{1-(1/x)}+\frac{3}{1+(1/x)}\right) = \infty+3 = \infty$

Limit does not exist.

19. $\displaystyle\lim_{x\to-\infty}\frac{x}{\sqrt{x^2-x}} = \lim_{x\to-\infty}\frac{1}{\dfrac{\sqrt{x^2-x}}{-\sqrt{x^2}}}$, $\quad\left(\text{for }x<0\text{ we have }x=-\sqrt{x^2}\right)$

$\qquad\qquad = \displaystyle\lim_{x\to-\infty}\frac{-1}{\sqrt{1-(1/x)}} = -1$

20. $\displaystyle\lim_{x\to\infty}\frac{x}{\sqrt{x^2+1}} = \lim_{x\to\infty}\frac{1}{\sqrt{1+(1/x^2)}} = 1$

21. $\displaystyle\lim_{x\to\infty}\frac{2x+1}{\sqrt{x^2-x}} = \lim_{x\to\infty}\frac{2+(1/x)}{\sqrt{1-(1/x)}} = 2$

22. $\displaystyle\lim_{x\to-\infty}\frac{-3x+1}{\sqrt{x^2+x}} = \lim_{x\to-\infty}\frac{-3+(1/x)}{\dfrac{\sqrt{x^2+x}}{-\sqrt{x^2}}}$, $\quad\left(\text{for }x<0\text{ we have }-\sqrt{x^2}=x\right)$

$\qquad\qquad = \displaystyle\lim_{x\to-\infty}\frac{3-(1/x)}{\sqrt{1+(1/x)}} = 3$

23. $\displaystyle\lim_{x\to\infty} \frac{x^2-x}{\sqrt{x^4+x}} = \lim_{x\to\infty} \frac{1-(1/x)}{\sqrt{1+(1/x^3)}} = 1$

24. $\displaystyle\lim_{x\to\infty} \frac{2x}{\sqrt{4x^2+1}} = \lim_{x\to\infty} \frac{2}{\sqrt{4+(1/x^2)}} = 1$

25. Since $-\dfrac{1}{x} \le \dfrac{\sin(2x)}{x} \le \dfrac{1}{x}$ for all $x \ne 0$, we have by the Squeeze Theorem,

$$\lim_{x\to\infty} -\frac{1}{x} \le \lim_{x\to\infty} \frac{\sin(2x)}{x} \le \lim_{x\to\infty} \frac{1}{x}$$

$$0 \le \lim_{x\to\infty} \frac{\sin(2x)}{x} \le 0.$$

Therefore, $\displaystyle\lim_{x\to\infty} \frac{\sin(2x)}{x} = 0.$

26. $\displaystyle\lim_{x\to\infty} \frac{x-\cos x}{x} = \lim_{x\to\infty} \left(1 - \frac{\cos x}{x}\right)$

$$= 1 - 0 = 1$$

Note:

$\displaystyle\lim_{x\to\infty} \frac{\cos x}{x} = 0$ by the Squeeze Theorem since

$$-\frac{1}{x} \le \frac{\cos x}{x} \le \frac{1}{x}.$$

27. $\displaystyle\lim_{x\to\infty} \frac{1}{2x+\sin x} = 0$

28. $\displaystyle\lim_{x\to\infty} \sin \frac{1}{x} = \sin 0 = 0$

29. $\displaystyle\lim_{x\to\infty} x \sin \frac{1}{x} = \lim_{t\to 0^+} \frac{\sin t}{t} = 1$

(Let $x = 1/t$.)

30. $\displaystyle\lim_{x\to\infty} x \tan \frac{1}{x} = \lim_{t\to 0^+} \frac{\tan t}{t} = \lim_{t\to 0^+} \left[\frac{\sin t}{t} \cdot \frac{1}{\cos t}\right]$

$$= (1)(1) = 1$$

(Let $x = 1/t$.)

31. $\displaystyle\lim_{x\to-\infty} \left(x+\sqrt{x^2+3}\right) = \lim_{x\to-\infty} \left[\left(x+\sqrt{x^2+3}\right) \cdot \frac{x-\sqrt{x^2+3}}{x-\sqrt{x^2+3}}\right] = \lim_{x\to-\infty} \frac{-3}{x-\sqrt{x^2+3}} = 0$

32. $\displaystyle\lim_{x\to\infty} \left(2x-\sqrt{4x^2+1}\right) = \lim_{x\to\infty} \left[\left(2x-\sqrt{4x^2+1}\right) \cdot \frac{2x+\sqrt{4x^2+1}}{2x+\sqrt{4x^2+1}}\right] = \lim_{x\to\infty} \frac{-1}{2x+\sqrt{4x^2+1}} = 0$

33. $\displaystyle\lim_{x\to\infty} \left(x-\sqrt{x^2+x}\right) = \lim_{x\to\infty} \left[\left(x-\sqrt{x^2+x}\right) \cdot \frac{x+\sqrt{x^2+x}}{x+\sqrt{x^2+x}}\right]$

$$= \lim_{x\to\infty} \frac{-x}{x+\sqrt{x^2+x}} = \lim_{x\to\infty} \frac{-1}{1+\sqrt{1+(1/x)}} = -\frac{1}{2}$$

34. $\displaystyle\lim_{x\to-\infty} \left(3x+\sqrt{9x^2-x}\right) = \lim_{x\to-\infty} \left[\left(3x+\sqrt{9x^2-x}\right) \cdot \frac{3x-\sqrt{9x^2-x}}{3x-\sqrt{9x^2-x}}\right]$

$$= \lim_{x\to-\infty} \frac{x}{3x-\sqrt{9x^2-x}}$$

$$= \lim_{x\to-\infty} \frac{1}{3 - \dfrac{\sqrt{9x^2-x}}{-\sqrt{x^2}}} \quad \left(\text{for } x < 0 \text{ we have } x = -\sqrt{x^2}\right)$$

$$= \lim_{x\to-\infty} \frac{1}{3+\sqrt{9-(1/x)}} = \frac{1}{6}$$

35. $y = \dfrac{2+x}{1-x}$

Intercepts: $(-2, \ 0), \ (0, \ 2)$

Symmetry: none

Horizontal asymptote: $y = -1$ since

$$\lim_{x \to -\infty} \frac{2+x}{1-x} = -1 = \lim_{x \to \infty} \frac{2+x}{1-x}$$

Discontinuity: $x = 1$ (Vertical asymptote)

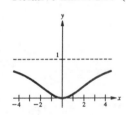

36. $y = \dfrac{x-3}{x-2}$

Intercepts: $(3, \ 0), \ \left(0, \ \dfrac{3}{2}\right)$

Symmetry: none

Horizontal asymptote: $y = 1$ since

$$\lim_{x \to -\infty} \frac{x-3}{x-2} = 1 = \lim_{x \to \infty} \frac{x-3}{x-2}$$

Discontinuity: $x = 2$ (Vertical asymptote)

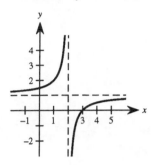

37. $y = \dfrac{x^2}{x^2 + 9}$

Intercept: $(0, \ 0)$

Symmetry: y-axis

Horizontal asymptote: $y = 1$ since

$$\lim_{x \to -\infty} \frac{x^2}{x^2 + 9} = 1 = \lim_{x \to \infty} \frac{x^2}{x^2 + 9}$$

Relative minimum: $(0, 0)$

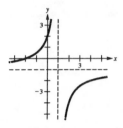

38. $y = \dfrac{x^2}{x^2 - 9}$

Intercept: $(0, \ 0)$

Symmetry: y-axis

Horizontal asymptote: $y = 1$ since

$$\lim_{x \to -\infty} \frac{x^2}{x^2 - 9} = 1 = \lim_{x \to \infty} \frac{x^2}{x^2 - 9}$$

Discontinuities: $x = \pm 3$ (Vertical asymptotes)

Relative maximum: $(0, 0)$

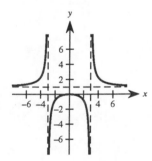

39. $xy^2 = 4$

Domain: $x > 0$

Intercepts: none

Symmetry: x-axis

Horizontal asymptote: $y = 0$ since

$$\lim_{x \to \infty} \frac{2}{\sqrt{x}} = 0 = \lim_{x \to \infty} -\frac{2}{\sqrt{x}}$$

Discontinuity: $x = 0$ (Vertical asymptote)

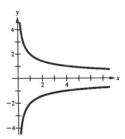

40. $x^2 y = 4$

Intercepts: none

Symmetry: y-axis

Horizontal asymptote: $y = 0$ since

$$\lim_{x \to -\infty} \frac{4}{x^2} = 0 = \lim_{x \to \infty} \frac{4}{x^2}$$

Discontinuity: $x = 0$ (Vertical asymptote)

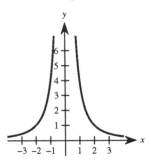

41. $y = \dfrac{2x}{1 - x}$

Intercept: $(0,\ 0)$

Symmetry: none

Horizontal asymptote: $y = -2$ since

$$\lim_{x \to -\infty} \frac{2x}{1 - x} = -2 = \lim_{x \to \infty} \frac{2x}{1 - x}$$

Discontinuity: $x = 1$ (Vertical asymptote)

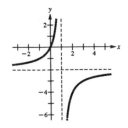

42. $y = \dfrac{2x}{1 - x^2}$

Intercept: $(0,\ 0)$

Symmetry: origin

Horizontal asymptote: $y = 0$ since

$$\lim_{x \to -\infty} \frac{2x}{1 - x^2} = 0 = \lim_{x \to \infty} \frac{2x}{1 - x^2}$$

Discontinuities: $x = \pm 1$ (Vertical asymptotes)

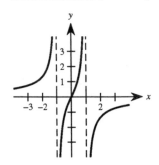

43. $y = 2 - \dfrac{3}{x^2}$

Intercepts: $(\pm\sqrt{3/2},\ 0)$

Symmetry: y-axis

Horizontal asymptote: $y = 2$ since

$$\lim_{x \to -\infty} \left(2 - \frac{3}{x^2}\right) = 2 = \lim_{x \to \infty} \left(2 - \frac{3}{x^2}\right)$$

Discontinuity: $x = 0$ (Vertical asymptote)

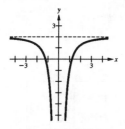

44. $y = 1 + \dfrac{1}{x}$

Intercept: $(-1,\ 0)$

Symmetry: none

Horizontal asymptote: $y = 1$ since

$$\lim_{x \to -\infty} \left(1 + \frac{1}{x}\right) = 1 = \lim_{x \to \infty} \left(1 + \frac{1}{x}\right)$$

Discontinuity: $x = 0$ (Vertical asymptote)

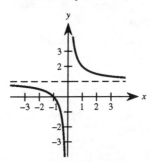

45. $y = \dfrac{x^3}{\sqrt{x^2 - 4}}$

Domain: $(-\infty,\ -2),\ (2,\ \infty)$

Intercepts: none

Symmetry: origin

Horizontal asymptote: none

Vertical asymptotes:

$\quad x = \pm 2$ (discontinuities)

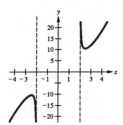

46. $y = \dfrac{x}{\sqrt{x^2 - 4}}$

Domain: $(-\infty,\ -2),\ (2,\ \infty)$

Intercepts: none

Symmetry: origin

Horizontal asymptotes: $y = \pm 1$ since

$$\lim_{x \to \infty} \frac{x}{\sqrt{x^2 - 4}} = 1, \quad \lim_{x \to -\infty} \frac{x}{\sqrt{x^2 - 4}} = -1$$

Vertical asymptotes:

$\quad x = \pm 2$ (discontinuities)

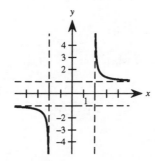

47. $f(x) = 5 - \dfrac{1}{x^2} = \dfrac{5x^2 - 1}{x^2}$

Domain: $(-\infty,\ 0),\ (0,\ \infty)$

$f'(x) = \dfrac{2}{x^3} \Rightarrow$ No relative extrema

$f''(x) = -\dfrac{6}{x^4} \Rightarrow$ No points of inflection

Vertical asymptote: $x = 0$

Horizontal asymptote: $y = 5$

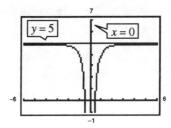

48. $f(x) = \dfrac{x^2}{x^2 - 1}$

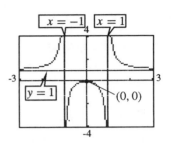

$f'(x) = \dfrac{(x^2 - 1)(2x) - x^2(2x)}{(x^2 - 1)^2} = \dfrac{-2x}{(x^2 - 1)^2} = 0$ when $x = 0$

$f''(x) = \dfrac{(x^2 - 1)^2(-2) + 2x(2)(x^2 - 1)(2x)}{(x^2 - 1)^4} = \dfrac{2(3x^2 + 1)}{(x^2 - 1)^3}$

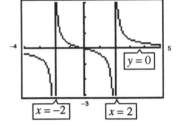

Since $f''(0) < 0$, then $(0, 0)$ is a relative maximum. Since $f''(x) \neq 0$, nor is it undefined in the domain of f, there are no points of inflection.

Vertical asymptotes: $x = \pm 1$
Horizontal asymptote: $y = 1$

49. $f(x) = \dfrac{x}{x^2 - 4}$

$f'(x) = \dfrac{(x^2 - 4) - x(2x)}{(x^2 - 4)^2}$

$ = \dfrac{-(x^2 + 4)}{(x^2 - 4)^2} \neq 0$ for any x in the domain of f

$f''(x) = \dfrac{(x^2 - 4)^2(-2x) + (x^2 + 4)(2)(x^2 - 4)(2x)}{(x^2 - 4)^4}$

$ = \dfrac{2x(x^2 + 12)}{(x^2 - 4)^3} = 0$ when $x = 0$.

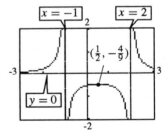

Since $f''(x) > 0$ on $(-2, \ 0)$ and $f''(x) < 0$ on $(0, 2)$, then $(0, 0)$ is a point of inflection.

Vertical asymptotes: $x = \pm 2$
Horizontal asymptote: $y = 0$

50. $f(x) = \dfrac{1}{x^2 - x - 2} = \dfrac{1}{(x + 1)(x - 2)}$

$f'(x) = \dfrac{-(2x - 1)}{(x^2 - x - 2)^2} = 0$ when $x = \dfrac{1}{2}$

$f''(x) = \dfrac{(x^2 - x - 2)^2(-2) + (2x - 1)(2)(x^2 - x - 2)(2x - 1)}{(x^2 - x - 2)^4}$

$ = \dfrac{6(x^2 - x + 1)}{(x^2 - x - 2)^3}$

Since $f''\left(\tfrac{1}{2}\right) < 0$, then $\left(\tfrac{1}{2}, \ -\tfrac{4}{9}\right)$ is a relative maximum. Since $f''(x) \neq 0$, nor is it undefined in the domain of f, there are no points of inflection.

Vertical asymptotes: $x = -1, \ x = 2$
Horizontal asymptote: $y = 0$

51. $f(x) = \dfrac{x - 2}{x^2 - 4x + 3} = \dfrac{x - 2}{(x - 1)(x - 3)}$

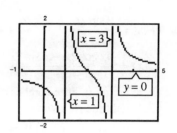

$f'(x) = \dfrac{(x^2 - 4x + 3) - (x - 2)(2x - 4)}{(x^2 - 4x + 3)^2} = \dfrac{-x^2 + 4x - 5}{(x^2 - 4x + 3)^2} \neq 0$

$f''(x) = \dfrac{(x^2 - 4x + 3)^2(-2x + 4) - (-x^2 + 4x - 5)(2)(x^2 - 4x + 3)(2x - 4)}{(x^2 - 4x + 3)^4}$

$= \dfrac{2(x^3 - 6x^2 + 15x - 14)}{(x^2 - 4x + 3)^3} = 0$ when $x = 2$.

Since $f''(x) > 0$ on $(1, 2)$ and $f''(x) < 0$ on $(2, 3)$, then $(2, 0)$ is a point of inflection.

Vertical asymptote: $x = 1$, $x = 3$

Horizontal asymptote: $y = 0$

52. $f(x) = \dfrac{x + 1}{x^2 + x + 1}$

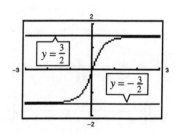

$(-0.6527, 0.4491)$ $(0.5321, 0.8440)$

$f'(x) = \dfrac{-x(x + 2)}{(x^2 + x + 1)^2} = 0$ when $x = 0$, -2.

$f''(x) = \dfrac{2(x^3 + 3x^2 - 1)}{(x^2 + x + 1)^3} = 0$ when $x \approx 0.5321$, -0.6527, -2.8794

$(-2, -\frac{1}{3})$ $(0, 1)$

$f''(0) < 0$

Therefore, $(0, 1)$ is a relative maximum.

$(-2.8794, -0.2931)$

$f''(-2) > 0$

Therefore, $\left(-2, -\frac{1}{3}\right)$ is a relative minimum.

Points of inflection: $(0.5321, 0.8440)$, $(-0.6527, 0.4491)$ and $(-2.8794, -0.2931)$

Horizontal asymptote: $y = 0$

53. $f(x) = \dfrac{3x}{\sqrt{4x^2 + 1}}$

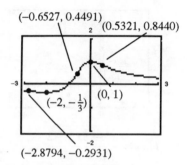

$f'(x) = \dfrac{3}{(4x^2 + 1)^{3/2}} \Rightarrow$ No relative extrema

$f''(x) = \dfrac{-36x}{(4x^2 + 1)^{5/2}} = 0$ when $x = 0$

Point of inflection: $(0, 0)$

Horizontal asymptotes: $y = \pm\dfrac{3}{2}$

No vertical asymptotes

54. $f(x) = \dfrac{2 \sin 2x}{x}$

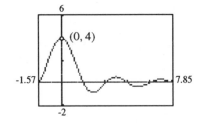

$f'(x) = \dfrac{4x \cos 2x - 2 \sin 2x}{x^2}$

There are an infinite number of relative extrema. In the interval $(-2\pi,\ 2\pi)$, you obtain the following.

Relative minima: $(\pm 2.25,\ -0.869)$, $(\pm 5.45,\ -0.365)$

Relative maxima: $(\pm 3.87,\ 0.513)$

Horizontal asymptote: $y = 0$

No vertical asymptotes

55. $f(x) = \dfrac{x^3 - 3x^2 + 2}{x(x - 3)}$, $\quad g(x) = x + \dfrac{2}{x(x - 3)}$

a.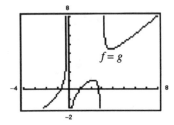

c.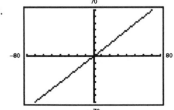

The graph appears as the slant asymptote $y = x$.

b. $f(x) = \dfrac{x^3 - 3x^2 + 2}{x(x - 3)}$

$= \dfrac{x^2(x - 3)}{x(x - 3)} + \dfrac{2}{x(x - 3)}$

$= x + \dfrac{2}{x(x - 3)} = g(x)$

56. $f(x) = -\dfrac{x^3 - 2x^2 + 2}{2x^2}$, $\quad g(x) = -\dfrac{1}{2}x + 1 - \dfrac{1}{x^2}$

a.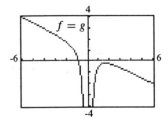

c.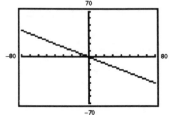

The graph appears as the slant asymptote $y = -\frac{1}{2}x + 1$.

b. $f(x) = -\dfrac{x^3 - 2x^2 + 2}{2x^2}$

$= -\left[\dfrac{x^3}{2x^2} - \dfrac{2x^2}{2x^2} + \dfrac{2}{2x^2} \right]$

$= -\dfrac{1}{2}x + 1 - \dfrac{1}{x^2} = g(x)$

57.

x	10^0	10^1	10^2	10^3	10^4	10^5	10^6
$f(x)$	1	0.513	0.501	0.500	0.500	0.500	0.500

$$\lim_{x \to \infty} (x - \sqrt{x(x-1)}) = \lim_{x \to \infty} \frac{x - \sqrt{x^2 - x}}{1} \cdot \frac{x + \sqrt{x^2 - x}}{x + \sqrt{x^2 - x}}$$

$$= \lim_{x \to \infty} \frac{x}{x + \sqrt{x^2 - x}}$$

$$= \lim_{x \to \infty} \frac{1}{1 + \sqrt{1 - (1/x)}}$$

$$= \frac{1}{2}$$

58.

x	10^0	10^1	10^2	10^3	10^4	10^5	10^6
$f(x)$	1.0	5.1	50.1	500.1	5000.1	50000.1	500000.1

$$\lim_{x \to \infty} \frac{x^2 - x\sqrt{x^2 - x}}{1} \cdot \frac{x^2 + x\sqrt{x^2 - x}}{x^2 + x\sqrt{x^2 - x}} = \lim_{x \to \infty} \frac{x^3}{x^2 + x\sqrt{x^2 - x}} = \infty$$

Limit does not exist

59.

x	10^0	10^1	10^2	10^3	10^4	10^5	10^6
$f(x)$	0.479	0.500	0.500	0.500	0.500	0.500	0.500

Let $x = 1/t$.

$$\lim_{x \to \infty} x \sin\left(\frac{1}{2x}\right) = \lim_{t \to 0^+} \frac{\sin(t/2)}{t} = \lim_{t \to 0^+} \frac{1}{2} \frac{\sin(t/2)}{t/2} = \frac{1}{2}$$

60.

x	10^0	10^1	10^2	10^3	10^4	10^5	10^6
$f(x)$	2.000	0.348	0.101	0.032	0.010	0.003	0.001

$$\lim_{x \to \infty} \frac{x+1}{x\sqrt{x}} = 0$$

61. $C = 0.5x + 500$

$$\overline{C} = \frac{C}{x}$$

$$\overline{C} = 0.5 + \frac{500}{x}$$

$$\lim_{x \to \infty} \left(0.5 + \frac{500}{x}\right) = 0.5$$

62. $\lim\limits_{v \to c} \dfrac{m_0}{\sqrt{1 - (v^2/c^2)}} \to \infty$

Limit does not exist

63. $\lim\limits_{v_1/v_2 \to \infty} 100\left[1 - \dfrac{1}{(v_1/v_2)^c}\right] = 100[1 - 0] = 100\%$

64. $N(t) = \dfrac{109.25 + 4.15t}{1000 - 23.88t + 0.164t^2}$, $\quad 0 \le t \le 89$

a. $N'(t) = -\dfrac{0.6806t^2 + 35.834t - 6758.89}{(1000 - 23.88t + 0.164t^2)^2}$

$N'(t) = 0$ when $t \approx 76.75$ which approximately corresponds to the year 1977.

b. $N''(t) = \dfrac{0.2232368t^3 + 17.630328t^2 - 6650.74776t + 286970.5864}{(1000 - 23.88t + 0.164t^2)^3}$

$N''(t) = 0$ when $t \approx 59.82$ which approximately corresponds to the year 1960.

c. As $t \to \infty$, $\quad N(t) \to 0$.

65. False. Let $f(x) = \dfrac{2x}{\sqrt{x^2 + 2}}$. (See Exercise 2.)

66. False. Let $y_1 = \sqrt{x + 1}$, then $y_1(0) = 1$. Thus, $y_1' = 1/\left(2\sqrt{x + 1}\right)$ and $y_1'(0) = 1/2$. Finally,

$$y_1'' = -\dfrac{1}{4(x + 1)^{3/2}} \text{ and } y_1''(0) = -\dfrac{1}{4}.$$

Let $p = ax^2 + bx + 1$, then $p(0) = 1$. Thus, $p' = 2ax + b$ and $p'(0) = \tfrac{1}{2} \Rightarrow b = \tfrac{1}{2}$. Finally, $p'' = 2a$ and $p''(0) = -\tfrac{1}{4} \Rightarrow a = -\tfrac{1}{8}$. Therefore,

$$f(x) = \begin{cases} -\dfrac{1}{8}x^2 + \dfrac{1}{2}x + 1, & x < 0 \\ \sqrt{x + 1}, & x \ge 0 \end{cases} \quad \text{and} \quad f(0) = 1,$$

$$f'(x) = \begin{cases} \dfrac{1}{2} - \dfrac{1}{4}x, & x < 0 \\ \dfrac{1}{2\sqrt{x + 1}}, & x > 0 \end{cases} \quad \text{and} \quad f'(0) = \dfrac{1}{2}, \quad \text{and}$$

$$f''(x) = \begin{cases} -\dfrac{1}{4}, & x < 0 \\ -\dfrac{1}{4(x + 1)^{3/2}}, & x > 0 \end{cases} \quad \text{and} \quad f''(0) = -\dfrac{1}{4}.$$

$f''(x) < 0$ for all real x, but $f(x)$ *increases* without bound.

67. a. $f(x) = \dfrac{|x|}{x+1}$

$\displaystyle\lim_{x\to\infty} \dfrac{|x|}{x+1} = 1$

$\displaystyle\lim_{x\to-\infty} \dfrac{|x|}{x+1} = -1$

Therefore, $y = 1$ and $y = -1$ are both horizontal asymptotes.

b. $f(x) = \dfrac{2x}{\sqrt{x^2+1}}$

$\displaystyle\lim_{x\to\infty} \dfrac{2x}{\sqrt{x^2+1}} = 2$

$\displaystyle\lim_{x\to-\infty} \dfrac{2x}{\sqrt{x^2+1}} = -2$

Therefore, $y = 2$ and $y = -2$ are both horizontal asymptotes.

68. a. $h(x) = \dfrac{f(x)}{x^2}$

$ = \dfrac{5x^3 - 3x^2 + 10}{x^2} = 5x - 3 + \dfrac{10}{x^2}$

$\displaystyle\lim_{x\to\infty} h(x) = \infty - 3 = \infty$

Limit does not exist.

b. $h(x) = \dfrac{f(x)}{x^3}$

$ = \dfrac{5x^3 - 3x^2 + 10}{x^3} = 5 - \dfrac{3}{x} + \dfrac{10}{x^3}$

$\displaystyle\lim_{x\to\infty} h(x) = 5$

c. $h(x) = \dfrac{f(x)}{x^4}$

$ = \dfrac{5x^3 - 3x^2 + 10}{x^4} = \dfrac{5}{x} - \dfrac{3}{x^2} + \dfrac{10}{x^4}$

$\displaystyle\lim_{x\to\infty} h(x) = 0$

69. $x = 2$ is a critical number.

$f'(x) < 0$ for $x < 2$.

$f'(x) > 0$ for $x > 2$.

$\displaystyle\lim_{x\to-\infty} f(x) = \lim_{x\to\infty} f(x) = 6$

For example, let $f(x) = \dfrac{-6}{0.1(x-2)^2 + 1} + 6$.

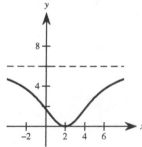

70. For example, let $f(x) = \dfrac{6|x-2|}{\sqrt{(x-2)^2 + 1}}$.

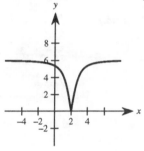

71. $\displaystyle\lim_{x\to\infty} \dfrac{p(x)}{q(x)} = \lim_{x\to\infty} \dfrac{a_n x^n + \cdots + a_1 x + a_0}{b_m x^m + \cdots + b_1 x + b_0}$

Divide $p(x)$ and $q(x)$ by x^m.

Case 1: If $n < m$: $\displaystyle\lim_{x\to\infty} \dfrac{p(x)}{q(x)} = \lim_{x\to\infty} \dfrac{\dfrac{a_n}{x^{m-n}} + \cdots + \dfrac{a_1}{x^{m-1}} + \dfrac{a_0}{x^m}}{b_m + \cdots + \dfrac{b_1}{x^{m-1}} + \dfrac{b_0}{x^m}} = \dfrac{0 + \cdots + 0 + 0}{b_m + \cdots + 0 + 0} = \dfrac{0}{b_m} = 0$

Case 2: If $m = n$: $\displaystyle\lim_{x\to\infty} \dfrac{p(x)}{q(x)} = \lim_{x\to\infty} \dfrac{a_n + \cdots + \dfrac{a_1}{x^{m-1}} + \dfrac{a_0}{x^m}}{b_m + \cdots + \dfrac{b_1}{x^{m-1}} + \dfrac{b_0}{x^m}} = \dfrac{a_n + \cdots + 0 + 0}{b_m + \cdots + 0 + 0} = \dfrac{a_n}{b_m}$

Case 3: If $n > m$: $\displaystyle\lim_{x\to\infty} \dfrac{p(x)}{q(x)} = \lim_{x\to\infty} \dfrac{a_n x^{n-m} + \cdots + \dfrac{a_1}{x^{m-1}} + \dfrac{a_0}{x^m}}{b_m + \cdots + \dfrac{b_1}{x^{m-1}} + \dfrac{b_0}{x^m}} = \dfrac{\pm\infty + \cdots + 0}{b_m + \cdots + 0} = \pm\infty$

Section 3.6   A Summary of Curve Sketching

1. $y = x^3 - 3x^2 + 3$

$y' = 3x^2 - 6x = 3x(x - 2) = 0$ when $x = 0$, $x = 2$

$y'' = 6x - 6 = 6(x - 1) = 0$ when $x = 1$

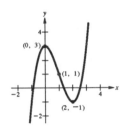

	y	y'	y''	Conclusion
$-\infty < x < 0$		$+$	$-$	Increasing, concave down
$x = 0$	3	0	$-$	Relative maximum
$0 < x < 1$		$-$	$-$	Decreasing, concave down
$x = 1$	1	$-$	0	Point of inflection
$1 < x < 2$		$-$	$+$	Decreasing, concave up
$x = 2$	-1	0	$+$	Relative minimum
$2 < x < \infty$		$+$	$+$	Increasing, concave up

2. $y = -\frac{1}{3}(x^3 - 3x + 2)$

$y' = -x^2 + 1 = 0$ when $x = \pm 1$

$y'' = -2x = 0$ when $x = 0$

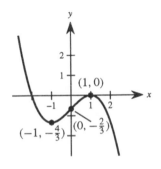

	y	y'	y''	Conclusion
$-\infty < x < -1$		$-$	$+$	Decreasing, concave up
$x = -1$	$-\frac{4}{3}$	0	$+$	Relative minimum
$-1 < x < 0$		$+$	$+$	Increasing, concave up
$x = 0$	$-\frac{2}{3}$	$+$	0	Point of inflection
$0 < x < 1$		$+$	$-$	Increasing, concave down
$x = 1$	0	0	$-$	Relative maximum
$1 < x < \infty$		$-$	$-$	Decreasing, concave down

3. $y = 2 - x - x^3$

$y' = -1 - 3x^2$ No critical numbers

$y'' = -6x = 0$ when $x = 0$

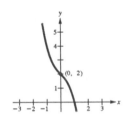

	y	y'	y''	Conclusion
$-\infty < x < 0$		$-$	$+$	Decreasing, concave up
$x = 0$	2	$-$	0	Point of inflection
$0 < x < \infty$		$-$	$-$	Decreasing, concave down

4. $f(x) = \frac{1}{3}(x-1)^3 + 2$

$f'(x) = (x-1)^2 = 0$ when $x = 1$

$f''(x) = 2(x-1) = 0$ when $x = 1$

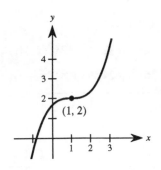

	$f(x)$	$f'(x)$	$f''(x)$	Conclusion
$-\infty < x < 1$		$+$	$-$	Increasing, concave down
$x = 1$	2	0	0	Point of inflection
$1 < x < \infty$		$+$	$+$	Increasing, concave up

5. $f(x) = 3x^3 - 9x + 1$

$f'(x) = 9x^2 - 9 = 9(x^2 - 1) = 0$ when $x = \pm 1$

$f''(x) = 18x = 0$ when $x = 0$

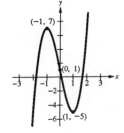

	$f(x)$	$f'(x)$	$f''(x)$	Conclusion
$-\infty < x < -1$		$+$	$-$	Increasing, concave down
$x = -1$	7	0	$-$	Relative maximum
$-1 < x < 0$		$-$	$-$	Decreasing, concave down
$x = 0$	1	$-$	0	Point of inflection
$0 < x < 1$		$-$	$+$	Decreasing, concave up
$x = 1$	-5	0	$+$	Relative minimum
$1 < x < \infty$		$+$	$+$	Increasing, concave up

6. $f(x) = (x+1)(x-2)(x-5)$

$f'(x) = (x+1)(x-2) + (x+1)(x-5) + (x-2)(x-5)$

$\quad = 3(x^2 - 4x + 1) = 0$ when $x = 2 \pm \sqrt{3}$

$f''(x) = 6(x-2) = 0$ when $x = 2$

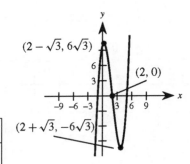

	$f(x)$	$f'(x)$	$f''(x)$	Conclusion
$-\infty < x < 2 - \sqrt{3}$		$+$	$-$	Increasing, concave down
$x = 2 - \sqrt{3}$	$6\sqrt{3}$	0	$-$	Relative maximum
$2 - \sqrt{3} < x < 2$		$-$	$-$	Decreasing, concave down
$x = 2$	0	$-$	0	Point of inflection
$2 < x < 2 + \sqrt{3}$		$-$	$+$	Decreasing, concave up
$x = 2 + \sqrt{3}$	$-6\sqrt{3}$	0	$+$	Relative minimum
$2 + \sqrt{3} < x < \infty$		$+$	$+$	Increasing, concave up

7. $y = 3x^4 + 4x^3$

$y' = 12x^3 + 12x^2 = 12x^2(x + 1) = 0$ when $x = 0$, $x = -1$

$y'' = 36x^2 + 24x = 12x(3x + 2) = 0$ when $x = 0$, $x = -\frac{2}{3}$

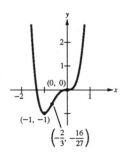

	y	y'	y''	Conclusion
$-\infty < x < -1$		$-$	$+$	Decreasing, concave up
$x = -1$	-1	0	$+$	Relative minimum
$-1 < x < -\frac{2}{3}$		$+$	$+$	Increasing, concave up
$x = -\frac{2}{3}$	$-\frac{16}{27}$	$+$	0	Point of inflection
$-\frac{2}{3} < x < 0$		$+$	$-$	Increasing, concave down
$x = 0$	0	0	0	Point of inflection
$0 < x < \infty$		$+$	$+$	Increasing, concave up

8. $y = 3x^4 - 6x^2$

$y' = 12x^3 - 12x = 12x(x^2 - 1) = 0$ when $x = 0$, $x = \pm 1$

$y'' = 36x^2 - 12 = 12(3x^2 - 1) = 0$ when $x = \pm\dfrac{\sqrt{3}}{3}$

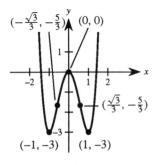

	y	y'	y''	Conclusion
$-\infty < x < -1$		$-$	$+$	Decreasing, concave up
$x = -1$	-3	0	$+$	Relative minimum
$-1 < x < -\frac{\sqrt{3}}{3}$		$+$	$+$	Increasing, concave up
$x = -\frac{\sqrt{3}}{3}$	$-\frac{5}{3}$	$+$	0	Point of inflection
$-\frac{\sqrt{3}}{3} < x < 0$		$+$	$-$	Increasing, concave down
$x = 0$	0	0	$-$	Relative maximum
$0 < x < \frac{\sqrt{3}}{3}$		$-$	$-$	Decreasing, concave down
$x = \frac{\sqrt{3}}{3}$	$-\frac{5}{3}$	$-$	0	Point of inflection
$\frac{\sqrt{3}}{3} < x < 1$		$-$	$+$	Decreasing, concave up
$x = 1$	-3	0	$+$	Relative minimum
$1 < x < \infty$		$+$	$+$	Increasing, concave up

9. $f(x) = x^4 - 4x^3 + 16x$

$f'(x) = 4x^3 - 12x^2 + 16 = 4(x+1)(x-2)^2 = 0$ when $x = -1$, $x = 2$

$f''(x) = 12x^2 - 24x = 12x(x-2) = 0$ when $x = 0$, $x = 2$

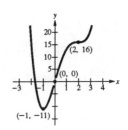

	$f(x)$	$f'(x)$	$f''(x)$	Conclusion
$-\infty < x < -1$		$-$	$+$	Decreasing, concave up
$x = -1$	-11	0	$+$	Relative minimum
$-1 < x < 0$		$+$	$+$	Increasing, concave up
$x = 0$	0	$+$	0	Point of inflection
$0 < x < 2$		$+$	$-$	Increasing, concave down
$x = 2$	16	0	0	Point of inflection
$2 < x < \infty$		$+$	$+$	Increasing, concave up

10. $f(x) = x^4 - 8x^3 + 18x^2 - 16x + 5$

$f'(x) = 4x^3 - 24x^2 + 36x - 16 = 4(x-4)(x-1)^2 = 0$ when $x = 1$, $x = 4$

$f''(x) = 12x^2 - 48x + 36 = 12(x-3)(x-1) = 0$ when $x = 3$, $x = 1$

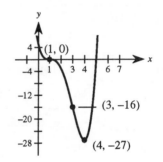

	$f(x)$	$f'(x)$	$f''(x)$	Conclusion
$-\infty < x < 1$		$-$	$+$	Decreasing, concave up
$x = 1$	0	0	0	Point of inflection
$1 < x < 3$		$-$	$-$	Decreasing, concave down
$x = 3$	-16	$-$	0	Point of inflection
$3 < x < 4$		$-$	$+$	Decreasing, concave up
$x = 4$	-27	0	$+$	Relative minimum
$4 < x < \infty$		$+$	$+$	Increasing, concave up

11. $y = x^5 - 5x$

$y' = 5x^4 - 5 = 5(x^4 - 1) = 0$ when $x = \pm 1$

$y'' = 20x^3 = 0$ when $x = 0$

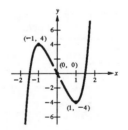

	y	y'	y''	Conclusion
$-\infty < x < -1$		$+$	$-$	Increasing, concave down
$x = -1$	4	0	$-$	Relative maximum
$-1 < x < 0$		$-$	$-$	Decreasing, concave down
$x = 0$	0	$-$	0	Point of inflection
$0 < x < 1$		$-$	$+$	Decreasing, concave up
$x = 1$	-4	0	$+$	Relative minimum
$1 < x < \infty$		$+$	$+$	Increasing, concave up

12. $y = (x - 1)^5$

$y' = 5(x - 1)^4 = 0$ when $x = 1$

$y'' = 20(x - 1)^3 = 0$ when $x = 1$

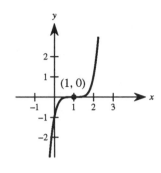

	y	y'	y''	Conclusion
$-\infty < x < 1$		+	−	Increasing, concave down
$x = 1$	0	0	0	Point of inflection
$1 < x < \infty$		+	+	Increasing, concave up

13. $y = |2x - 3|$

$y' = \dfrac{2(2x - 3)}{|2x - 3|}$ undefined at $x = \dfrac{3}{2}$

$y'' = 0$

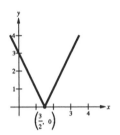

	y	y'	Conclusion
$-\infty < x < \frac{3}{2}$		−	Decreasing
$x = \frac{3}{2}$	0	Undefined	Relative minimum
$\frac{3}{2} < x < \infty$		+	Increasing

14. $y = |x^2 - 6x + 5|$

$y' = \dfrac{2(x - 3)(x^2 - 6x + 5)}{|x^2 - 6x + 5|} = \dfrac{2(x - 3)(x - 5)(x - 1)}{|(x - 5)(x - 1)|}$

$= 0$ when $x = 3$ and undefined when $x = 1$, $x = 5$

$y'' = \dfrac{2(x^2 - 6x + 5)}{|x^2 - 6x + 5|} = \dfrac{2(x - 5)(x - 1)}{|(x - 5)(x - 1)|}$ undefined when $x = 1$, $x = 5$

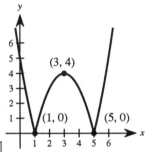

	y	y'	y''	Conclusion
$-\infty < x < 1$		−	+	Decreasing, concave up
$x = 1$	0	Undefined	Undefined	Relative minimum, point of inflection
$1 < x < 3$		+	−	Increasing, concave down
$x = 3$	4	0	−	Relative maximum
$3 < x < 5$		−	−	Decreasing, concave down
$x = 5$	0	Undefined	Undefined	Relative minimum, point of inflection
$5 < x < \infty$		+	+	Increasing, concave up

15. $y = \dfrac{x^2}{x^2 + 3}$

$y' = \dfrac{6x}{(x^2 + 3)^2} = 0$ when $x = 0$

$y'' = \dfrac{18(1 - x^2)}{(x^2 + 3)^3} = 0$ when $x = \pm 1$

Horizontal asymptote: $y = 1$

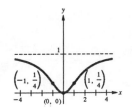

	y	y'	y''	Conclusion
$-\infty < x < -1$		$-$	$-$	Decreasing, concave down
$x = -1$	$\frac{1}{4}$	$-$	0	Point of inflection
$-1 < x < 0$		$-$	$+$	Decreasing, concave up
$x = 0$	0	0	$+$	Relative minimum
$0 < x < 1$		$+$	$+$	Increasing, concave up
$x = 1$	$\frac{1}{4}$	$+$	0	Point of inflection
$1 < x < \infty$		$+$	$-$	Increasing, concave down

16. $y = \dfrac{x}{x^2 + 1}$

$y' = \dfrac{1 - x^2}{(x^2 + 1)^2} = \dfrac{(1 - x)(1 + x)}{(x^2 + 1)^2} = 0$ when $x = \pm 1$

$y'' = -\dfrac{2x(3 - x^2)}{(x^2 + 1)^3} = 0$ when $x = 0,\ \pm\sqrt{3}$

Horizontal asymptote: $y = 0$

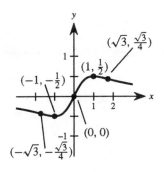

	y	y'	y''	Conclusion
$-\infty < x < -\sqrt{3}$		$-$	$-$	Decreasing, concave down
$x = -\sqrt{3}$	$-\frac{\sqrt{3}}{4}$	$-$	0	Point of inflection
$-\sqrt{3} < x < -1$		$-$	$+$	Decreasing, concave up
$x = -1$	$-\frac{1}{2}$	0	$+$	Relative minimum
$-1 < x < 0$		$+$	$+$	Increasing, concave up
$x = 0$	0	$+$	0	Point of inflection
$0 < x < 1$		$+$	$-$	Increasing, concave down
$x = 1$	$\frac{1}{2}$	0	$-$	Relative maximum
$1 < x < \sqrt{3}$		$-$	$-$	Decreasing, concave down
$x = \sqrt{3}$	$\frac{\sqrt{3}}{4}$	$-$	0	Point of inflection
$\sqrt{3} < x < \infty$		$-$	$+$	Decreasing, concave up

17. $y = x\sqrt{4 - x}$,   Domain: $(-\infty, \ 4]$

$y' = \dfrac{8 - 3x}{2\sqrt{4 - x}} = 0$ when $x = \dfrac{8}{3}$ and undefined when $x = 4$

$y'' = \dfrac{3x - 16}{4(4 - x)^{3/2}} = 0$ when $x = \dfrac{16}{3}$ and undefined when $x = 4$

Note: $x = \dfrac{16}{3}$ is not in the domain.

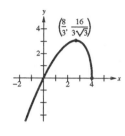

	y	y'	y''	Conclusion
$-\infty < x < \frac{8}{3}$		$+$	$-$	Increasing, concave down
$x = \frac{8}{3}$	$\frac{16}{3\sqrt{3}}$	0	$-$	Relative maximum
$\frac{8}{3} < x < 4$		$-$	$-$	Decreasing, concave down
$x = 4$	0	Undefined	Undefined	Endpoint

18. $y = x\sqrt{4 - x^2}$,   Domain: $[-2, \ 2]$

$y' = \dfrac{4 - 2x^2}{\sqrt{4 - x^2}} = 0$ when $x = \pm\sqrt{2}$ and undefined when $x = \pm 2$

$y'' = \dfrac{2x(x^2 - 6)}{(4 - x^2)^{3/2}} = 0$ when $x = 0, \ \pm 6$ and undefined when $x = \pm 2$

Note: $x = \pm\sqrt{6}$ are not in the domain.

	y	y'	y''	Conclusion
$x = -2$	0	Undefined	Undefined	Endpoint
$-2 < x < -\sqrt{2}$		$-$	$+$	Decreasing, concave up
$x = -\sqrt{2}$	-2	0	$+$	Relative minimum
$-\sqrt{2} < x < 0$		$+$	$+$	Increasing, concave up
$x = 0$	0	$+$	0	Point of inflection
$0 < x < \sqrt{2}$		$+$	$-$	Increasing, concave down
$x = \sqrt{2}$	2	0	$-$	Relative maximum
$\sqrt{2} < x < 2$		$-$	$-$	Decreasing, concave down
$x = 2$	0	Undefined	Undefined	Endpoint

19. $y = 3x^{2/3} - 2x$

$$y' = 2x^{-1/3} - 2 = \frac{2(1 - x^{1/3})}{x^{1/3}}$$

$= 0$ when $x = 1$ and undefined when $x = 0$

$$y'' = \frac{-2}{3x^{4/3}} < 0 \text{ when } x \neq 0$$

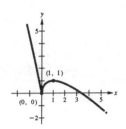

	y	y'	y''	Conclusion
$-\infty < x < 0$		$-$	$-$	Decreasing, concave down
$x = 0$	0	Undefined	Undefined	Relative minimum
$0 < x < 1$		$+$	$-$	Increasing, concave down
$x = 1$	1	0	$-$	Relative maximum
$1 < x < \infty$		$-$	$-$	Decreasing, concave down

20. $y = 3x^{2/3} - x^2$

$$y' = \frac{2}{x^{1/3}} - 2x = \frac{2(1 - x^{4/3})}{x^{1/3}} = 0 \text{ when } x = \pm 1, \text{ undefined when } x = 0$$

$$y'' = -2\left(\frac{1}{3x^{4/3}} + 1\right) < 0 \text{ when } x \neq 0$$

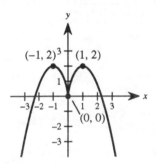

	y	y'	y''	Conclusion
$-\infty < x < -1$		$+$	$-$	Increasing, concave down
$x = -1$	2	0	$-$	Relative maximum
$-1 < x < 0$		$-$	$-$	Decreasing, concave down
$x = 0$	0	Undefined	Undefined	Relative minimum
$0 < x < 1$		$+$	$-$	Increasing, concave down
$x = 1$	2	0	$-$	Relative maximum
$1 < x < \infty$		$-$	$-$	Decreasing, concave down

21. $y = \dfrac{x}{\sqrt{x^2 + 7}}$

$$y' = \frac{7}{(x^2 + 7)^{3/2}} > 0 \text{ for all } x$$

$$y'' = \frac{-21x}{(x^2 + 7)^{5/2}} = 0 \text{ when } x = 0$$

Horizontal asymptotes: $y = \pm 1$

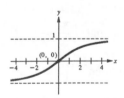

	$f(x)$	$f'(x)$	$f''(x)$	Conclusion
$-\infty < x < 0$		$+$	$+$	Increasing, concave up
$x = 0$	0	$+$	0	Point of inflection
$0 < x < \infty$		$+$	$-$	Increasing, concave down

22. $f(x) = \dfrac{4x}{\sqrt{x^2 + 15}}$

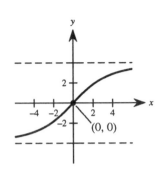

$f'(x) = \dfrac{60}{(x^2 + 15)^{3/2}} > 0$ for all x

$f''(x) = \dfrac{-180x}{(x^2 + 15)^{5/2}} = 0$ when $x = 0$

Horizontal asymptotes: $y = \pm 4$

	$f(x)$	$f'(x)$	$f''(x)$	Conclusion
$-\infty < x < 0$		$+$	$+$	Increasing, concave up
$x = 0$	0	$+$	0	Point of inflection
$0 < x < \infty$		$+$	$-$	Increasing, concave down

23. $y = \sin x - \dfrac{1}{18}\sin 3x, \quad 0 \le x \le 2\pi$

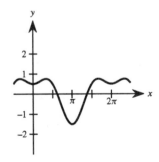

$y' = \cos x - \dfrac{1}{6}\cos 3x = 0$ when $x = \dfrac{\pi}{2}, \dfrac{3\pi}{2}$

$y'' = -\sin x + \dfrac{1}{2}\sin 3x = 0$ when $x = 0, \dfrac{\pi}{6}, \dfrac{5\pi}{6}, \pi, \dfrac{7\pi}{6}, \dfrac{11\pi}{6}$

Relative maximum: $\left(\dfrac{\pi}{2}, \dfrac{19}{18}\right)$

Relative minimum: $\left(\dfrac{3\pi}{2}, -\dfrac{19}{18}\right)$

Inflection points: $\left(\dfrac{\pi}{6}, \dfrac{4}{9}\right), \left(\dfrac{5\pi}{6}, \dfrac{4}{9}\right), (\pi, 0), \left(\dfrac{7\pi}{6}, -\dfrac{4}{9}\right), \left(\dfrac{11\pi}{6}, -\dfrac{4}{9}\right)$

24. $y = \cos x - \dfrac{1}{2}\cos 2x, \quad 0 \le x \le 2\pi$

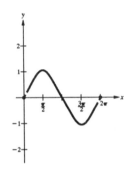

$y' = -\sin x + \sin 2x = -\sin x(1 - 2\cos x) = 0$ when $x = 0, \pi, \dfrac{\pi}{3}, \dfrac{5\pi}{3}$

$y'' = -\cos x + 2\cos 2x = -\cos x + 2(2\cos^2 x - 1)$

$\qquad = 4\cos^2 x - \cos x - 2 = 0$ when $\cos x = \dfrac{1 \pm \sqrt{33}}{8} \approx 0.8431, \ -0.5931$

Therefore, $x \approx 0.5678$ or 5.7154, $x \approx 2.2057$ or 4.0775.

Relative maxima: $\left(\dfrac{\pi}{3}, \dfrac{3}{4}\right), \left(\dfrac{5\pi}{3}, \dfrac{3}{4}\right)$

Relative minimum: $\left(\pi, -\dfrac{3}{2}\right)$

Inflection points: $(0.5678, 0.6323), (2.2057, -0.4449), (5.7154, 0.6323), (4.0775, -0.4449)$

25. $y = 2x - \tan x$, $-\dfrac{\pi}{2} < x < \dfrac{\pi}{2}$

$y' = 2 - \sec^2 x = 0$ when $x = \pm\dfrac{\pi}{4}$

$y'' = -2\sec^2 x \tan x = 0$ when $x = 0$

Relative maximum: $\left(\dfrac{\pi}{4},\ \dfrac{\pi}{2} - 1\right)$

Relative minimum: $\left(-\dfrac{\pi}{4},\ 1 - \dfrac{\pi}{2}\right)$

Inflection point: $(0, 0)$

Vertical asymptotes: $x = \pm\dfrac{\pi}{2}$

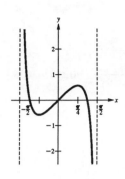

26. $y = 2x + \cot x$, $0 < x < \pi$

$y' = 2 - \csc^2 x = 0$ when $x = \dfrac{\pi}{4},\ \dfrac{3\pi}{4}$

$y'' = 2\csc^2 x \cot x = 0$ when $x = \dfrac{\pi}{2}$

Relative maximum: $\left(\dfrac{3\pi}{4},\ \dfrac{3\pi}{2} - 1\right)$

Relative minimum: $\left(\dfrac{\pi}{4},\ \dfrac{\pi}{2} + 1\right)$

Inflection point: $\left(\dfrac{\pi}{2},\ \pi\right)$

Vertical asymptotes: $x = 0,\ \pi$

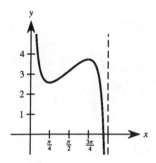

27. $y = \dfrac{1}{x - 2} - 3$

$y' = -\dfrac{1}{(x - 2)^2} < 0$ when $x \neq 2$

$y'' = \dfrac{2}{(x - 2)^3}$

No relative extrema, no points of inflection

Intercepts: $\left(\dfrac{7}{3},\ 0\right),\ \left(0,\ -\dfrac{7}{2}\right)$

Vertical asymptote: $x = 2$

Horizontal asymptote: $y = -3$

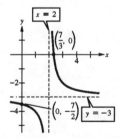

28. $y = \dfrac{x^2 + 1}{x^2 - 2}$

$y' = \dfrac{-6x}{(x^2 - 2)^2} = 0$ when $x = 0$

$y'' = \dfrac{6(3x^2 + 2)}{(x^2 - 2)^3} < 0$ when $x = 0$

Therefore, $\left(0, \ -\frac{1}{2}\right)$ is a relative maximum.

Intercept: $\left(0, \ -\frac{1}{2}\right)$

Vertical asymptotes: $x = \pm\sqrt{2}$

Horizontal asymptote: $y = 1$

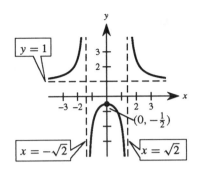

29. $y = \dfrac{2x}{x^2 - 1}$

$y' = \dfrac{-2(x^2 + 1)}{(x^2 - 1)^2} < 0$ if $x \neq \pm1$

$y'' = \dfrac{4x(x^2 + 3)}{(x^2 - 1)^3} = 0$ if $x = 0$

Inflection point: $(0, 0)$

Intercept: $(0, 0)$

Vertical asymptotes: $x = \pm1$

Horizontal asymptote: $y = 0$

Symmetry with respect to the origin

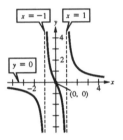

30. $f(x) = \dfrac{x + 2}{x} = 1 + \dfrac{2}{x}$

$f'(x) = \dfrac{-2}{x^2} < 0$ when $x \neq 0$

$f''(x) = \dfrac{4}{x^3} \neq 0$

Intercept: $(-2, \ 0)$

Vertical asymptote: $x = 0$

Horizontal asymptote: $y = 1$

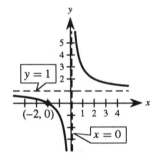

31. $g(x) = x + \dfrac{4}{x^2 + 1}$

$g'(x) = 1 - \dfrac{8x}{(x^2 + 1)^2} = \dfrac{x^4 + 2x^2 - 8x + 1}{(x^2 + 1)^2} = 0$ when $x \approx 0.1292, \ 1.6085$

$g''(x) = \dfrac{8(3x^2 - 1)}{(x^2 + 1)^3} = 0$ when $x = \pm\dfrac{\sqrt{3}}{3}$

$g''(0.1292) < 0$, therefore, $(0.1292, \ 4.064)$ is a relative maximum.

$g''(1.6085) > 0$, therefore, $(1.6085, \ 2.724)$ is a relative minimum.

Points of inflection: $\left(-\dfrac{\sqrt{3}}{3}, \ 2.423\right), \ \left(\dfrac{\sqrt{3}}{3}, \ 3.577\right)$

Intercepts: $(0, \ 4), \ (-1.3788, \ 0)$

Slant asymptote: $y = x$

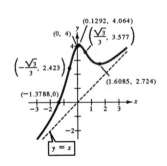

32. $f(x) = x + \dfrac{32}{x^2}$

$f'(x) = 1 - \dfrac{64}{x^3} = \dfrac{(x-4)(x^2+4x+16)}{x^3} = 0$ when $x = 4$

$f''(x) = \dfrac{192}{x^4} > 0$ if $x \neq 0$

Therefore, $(4, 6)$ is a relative minimum.

Intercept: $(-2\sqrt[3]{4},\ 0)$

Vertical asymptote: $x = 0$

Slant asymptote: $y = x$

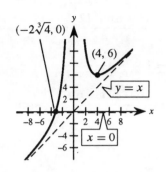

33. $f(x) = \dfrac{x^2+1}{x} = x + \dfrac{1}{x}$

$f'(x) = 1 - \dfrac{1}{x^2} = 0$ when $x = \pm 1$

$f''(x) = \dfrac{2}{x^3} \neq 0$

Relative maximum: $(-1,\ -2)$

Relative minimum: $(1, 2)$

Vertical asymptote: $x = 0$

Slant asymptote: $y = x$

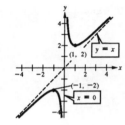

34. $f(x) = \dfrac{x^3}{x^2-1} = x + \dfrac{x}{x^2-1}$

$f'(x) = \dfrac{x^2(x^2-3)}{(x^2-1)^2} = 0$ when $x = 0,\ \pm\sqrt{3}$

$f''(x) = \dfrac{2x(x^2+3)}{(x^2-1)^3} = 0$ when $x = 0$

Relative maximum: $\left(-\sqrt{3},\ -\dfrac{3\sqrt{3}}{2}\right)$

Relative minimum: $\left(\sqrt{3},\ \dfrac{3\sqrt{3}}{2}\right)$

Inflection point: $(0, 0)$

Vertical asymptotes: $x = \pm 1$

Slant asymptote: $y = x$

Origin symmetry

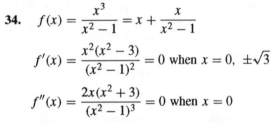
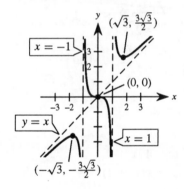

35. $y = \dfrac{x^2 - 6x + 12}{x - 4} = x - 2 + \dfrac{4}{x - 4}$

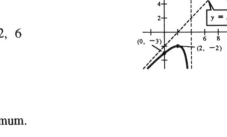

$y' = 1 - \dfrac{4}{(x - 4)^2}$

$ = \dfrac{(x - 2)(x - 6)}{(x - 4)^2} = 0$ when $x = 2, \ 6$

$y'' = \dfrac{8}{(x - 4)^3}$

$y'' < 0$ when $x = 2$.

Therefore, $(2, \ -2)$ is a relative maximum.

$y'' > 0$ when $x = 6$.

Therefore, $(6, 6)$ is a relative minimum.

Vertical asymptote: $x = 4$

Slant asymptote: $y = x - 2$

---

36. $y = \dfrac{2x^2 - 5x + 5}{x - 2} = 2x - 1 + \dfrac{3}{x - 2}$

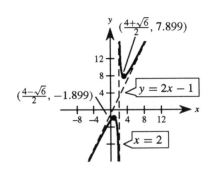

$y' = 2 - \dfrac{3}{(x - 2)^2} = \dfrac{2x^2 - 8x + 5}{(x - 2)^2} = 0$ when $x = \dfrac{4 \pm \sqrt{6}}{2}$

$y'' = \dfrac{6}{(x - 2)^3} \neq 0$

Relative maximum: $\left(\dfrac{4 - \sqrt{6}}{2}, \ -1.8990 \right)$

Relative minimum: $\left(\dfrac{4 + \sqrt{6}}{2}, \ 7.8990 \right)$

Vertical asymptote: $x = 2$

Slant asymptote: $y = 2x - 1$

---

37. $f(x) = \dfrac{20x}{x^2 + 1} - \dfrac{1}{x} = \dfrac{19x^2 - 1}{x(x^2 + 1)}$

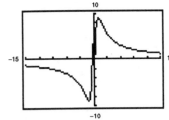

38. $f(x) = 5 \left(\dfrac{1}{x - 4} - \dfrac{1}{x + 2} \right)$

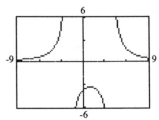

39. $f(x) = \dfrac{4(x-1)^2}{x^2 - 4x + 5}$

Vertical asymptote: none

Horizontal asymptote: $y = 4$

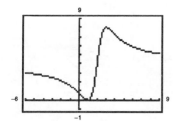

The graph crosses the horizontal asymptote $y = 4$. If a function has a vertical asymptote at $x = c$, the graph would not cross it since $f(c)$ is undefined.

40. $g(x) = \dfrac{3x^4 - 5x + 3}{x^4 + 1}$

Vertical asymptote: none

Horizontal asymptote: $y = 3$

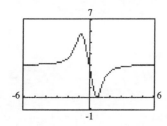

The graph crosses the horizontal asymptote $y = 3$. If a function has a vertical asymptote at $x = c$, the graph would not cross it since $f(c)$ is undefined.

41. $h(x) = \dfrac{6 - 2x}{3 - x}$

$= \dfrac{2(3 - x)}{3 - x} = \begin{cases} 2, & \text{if } x \neq 3 \\ \text{Undefined}, & \text{if } x = 3 \end{cases}$

The rational function is not reduced to lowest terms.

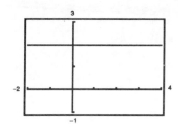
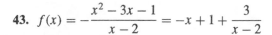

42. $g(x) = \dfrac{x^2 + x - 2}{x - 1}$

$= \dfrac{(x + 2)(x - 1)}{x - 1} = \begin{cases} x + 2, & \text{if } x \neq 1 \\ \text{Undefined}, & \text{if } x = 1 \end{cases}$

The rational function is not reduced to lowest terms.

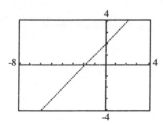

43. $f(x) = -\dfrac{x^2 - 3x - 1}{x - 2} = -x + 1 + \dfrac{3}{x - 2}$

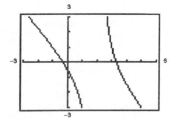

The graph appears to approach the slant asymptote $y = -x + 1$.

44. $g(x) = \dfrac{2x^2 - 8x - 15}{x - 5} = 2x + 2 - \dfrac{5}{x - 5}$

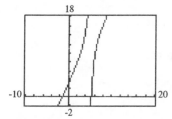

The graph appears to approach the slant asymptote $y = 2x + 2$.

45. Vertical asymptote: $x = 5$

Horizontal asymptote: $y = 0$

$y = \dfrac{1}{x - 5}$

46. Vertical asymptote: $x = -3$

Horizontal asymptote: none

$y = \dfrac{x^2}{x + 3}$

47. Vertical asymptote: $x = 5$

Slant asymptote: $y = 3x + 2$

$$y = 3x + 2 + \frac{1}{x - 5} = \frac{3x^2 - 13x - 9}{x - 5}$$

48. Vertical asymptote: $x = 0$

Slant asymptote: $y = -x$

$$y = -x + \frac{1}{x} = \frac{1 - x^2}{x}$$

49. $f(x) = \dfrac{ax}{x - b}$

   a. The horizontal asymptote is $y = a$.

   b. The vertical asymptote is $x = b$.

50. $f(x) = \dfrac{1}{2}(ax)^2 - (ax) = \dfrac{1}{2}(ax)(ax - 2), \quad a \neq 0$

$f'(x) = a^2 x - a = a(ax - 1) = 0$ when $x = \dfrac{1}{a}$

$f''(x) = a^2 > 0$ for all x

   a. Intercepts: $(0, 0)$, $\left(\dfrac{2}{a}, 0\right)$

      Relative minimum: $\left(\dfrac{1}{a}, -\dfrac{1}{2}\right)$

      Points of inflection: none

   b.

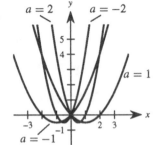

In Exercises 51–56:

$$f(x) = ax^3 + bx^2 + cx + d$$

$$f'(x) = 3ax^2 + 2bx + c$$

$$f''(x) = 6ax + 2b$$

$$f'(x) = 0 \text{ when } x = \frac{-2b \pm \sqrt{4b^2 - 12ac}}{6a} = \frac{-b \pm \sqrt{b^2 - 3ac}}{3a}$$

Furthermore,

$$\lim_{x \to \infty} f(x) = \infty \text{ if and only if } a > 0$$

$$\lim_{x \to \infty} f(x) = -\infty \text{ if and only if } a < 0.$$

51. Since $\lim\limits_{x \to \infty} f(x) = -\infty$, $a < 0$.

Also, $f'(x) < 0$ for all x. Therefore, the discriminant is negative.

$b^2 - 3ac < 0 \Rightarrow b^2 < 3ac$

52. Since $\lim\limits_{x \to \infty} f(x) = \infty$, $a > 0$.

Also, $f'(x) > 0$ for all x. Therefore, the discriminant is negative.

$b^2 - 3ac < 0 \Rightarrow b^2 < 3ac$

53. Since $\lim\limits_{x \to \infty} f(x) = -\infty$, $a < 0$.

Also, since there is only one critical point, the discriminant is zero.

$b^2 - 3ac = 0 \Rightarrow b^2 = 3ac$

54. Since $\lim\limits_{x \to \infty} f(x) = \infty$, $a > 0$.

Also, since there is only one critical point, the discriminant is zero.

$b^2 - 3ac = 0 \Rightarrow b^2 = 3ac$

55. Since $\lim_{x \to \infty} f(x) = -\infty$, $a < 0$.

Also, since there are two critical points, the discriminant is positive.

$b^2 - 3ac > 0 \Rightarrow b^2 > 3ac$

56. Since $\lim_{x \to \infty} f(x) = \infty$, $a > 0$.

Also, since there are two critical points, the discriminant is positive.

$b^2 - 3ac > 0 \Rightarrow b^2 > 3ac$

57. $f'(x) = -(x - 1)(x - 3) = -x^2 + 4x - 3$

Relative minimum when $x = 1$

Relative maximum when $x = 3$

$f(x) = -\dfrac{x^3}{3} + 2x^2 - 3x + C$, C is any constant

Let $C = 0$; $f(x) = -\dfrac{x^3}{3} + 2x^2 - 3x$

$f(1) = -\dfrac{4}{3}$

$f(3) = 0$

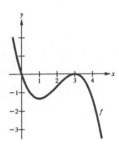

58. $f' > 0$ for $x > 0 \Rightarrow f(x)$ is increasing.

$f' < 0$ for $x < 0 \Rightarrow f(x)$ is decreasing.

f' is undefined when $x = 0$.

One such function is

$f(x) = -\dfrac{1}{x^2}$.

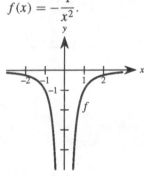

59. $f'(x) = 2$

$f(x) = 2x + C$, C is any constant

If $C = 0$, $f(x) = 2x$.

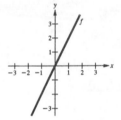

60. f' is linear $\Rightarrow f$ is quadratic.

$f' < 0$ when $x < 0 \Rightarrow f$ is decreasing. $f' > 0$ when $x > 0 \Rightarrow f$ is increasing.

Therefore, $(0, \ f(0))$ is a relative minimum.

$f(x) = ax^2 + b$, $a > 0$

If $a = \dfrac{5}{2}$ and $b = 0$, then

$f(x) = \dfrac{5}{2}x^2$.

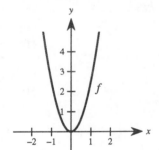

61. $f''(x) = 2 > 0 \Rightarrow f$ is concave up.

$f'(x) = 2x + C_1$

$f(x) = x^2 + C_1 x + C_2$

Let $C_1 = C_2 = 0$, then $f(x) = x^2$.

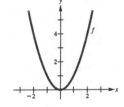

62. f'' is linear $\Rightarrow f$ is cubic.

$f(x) = ax^3 + bx^2 + cx + d$

$f'' < 0$ when $x < 0 \Rightarrow$ concave down.

$f'' > 0$ when $x > 0 \Rightarrow$ concave up.

Therefore, $(0, \ f(0))$ is a point of inflection. Since f is concave up to the right and concave down to the left $\Rightarrow a > 0$. Let $f(x) = x^3$.

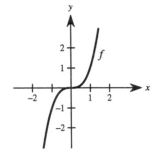

63. f is cubic.

f' is quadratic.

f'' is linear.

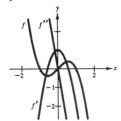

64. a. f is decreasing on $(-\pi, \ -\pi/2), \ (\pi/2, \ \pi) \Rightarrow f'$ is negative.

f is increasing on $(-\pi/2, \ \pi/2) \Rightarrow f'$ is positive.

Matches B.

b. f is increasing on $(-\pi, \ 0) \Rightarrow f'$ is positive.

f is decreasing on $(0, \ \pi) \Rightarrow f'$ is negative.

Matches C.

c. f is increasing on $(-2, \ 2) \Rightarrow f'$ is positive.

Matches A.

d. f is decreasing on $(-2, \ 0) \Rightarrow f'$ is negative.

f is increasing on $(0, \ 2) \Rightarrow f'$ is positive.

Matches D.

Section 3.7   Optimization Problems

1. a.

First number, x	Second number	Product, P
10	$110 - 10$	$10(110 - 10) = 1000$
20	$110 - 20$	$20(110 - 20) = 1800$
30	$110 - 30$	$30(110 - 30) = 2400$
40	$110 - 40$	$40(110 - 40) = 2800$
50	$110 - 50$	$50(110 - 50) = 3000$
60	$110 - 60$	$60(110 - 60) = 3000$

b. $P = x(110 - x) = 110x - x^2$

c. $\dfrac{dP}{dx} = 110 - 2x = 0$ when $x = 55$.

$\dfrac{d^2P}{dx^2} = -2 < 0$

P is a maximum when $x = 110 - x = 55$.

d.

First number, x	Second number	Product, P
10	$110 - 10$	$10(110 - 10) = 1000$
20	$110 - 20$	$20(110 - 20) = 1800$
30	$110 - 30$	$30(110 - 30) = 2400$
40	$110 - 40$	$40(110 - 40) = 2800$
50	$110 - 50$	$50(110 - 50) = 3000$
60	$110 - 60$	$60(110 - 60) = 3000$
70	$110 - 70$	$70(110 - 70) = 2800$
80	$110 - 80$	$80(110 - 80) = 2400$
90	$110 - 90$	$90(110 - 90) = 1800$
100	$110 - 100$	$100(110 - 100) = 1000$

e.

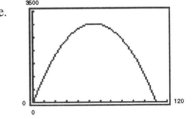

2. Let x and y be two positive numbers such that $x + y = S$.

$$P = xy = x(S - x) = Sx - x^2$$

$$\frac{dP}{dx} = S - 2x = 0 \text{ when } x = \frac{S}{2}.$$

$$\frac{d^2P}{dx^2} = -2 < 0 \text{ when } x = \frac{S}{2}.$$

P is a maximum when $x = y = S/2$.

3. Let x and y be two positive numbers such that $xy = 192$.

$$S = x + y = x + \frac{192}{x}$$

$$\frac{dS}{dx} = 1 - \frac{192}{x^2} = 0 \text{ when } x = \sqrt{192}.$$

$$\frac{d^2S}{dx^2} = \frac{384}{x^3} > 0 \text{ when } x = \sqrt{192}.$$

S is a minimum when $x = y = \sqrt{192}$.

4. Let x and y be two positive numbers such that $xy = 192$.

$$S = x + 3y = \frac{192}{y} + 3y$$

$$\frac{dS}{dy} = 3 - \frac{192}{y^2} = 0 \text{ when } y = 8.$$

$$\frac{d^2S}{dy^2} = \frac{384}{y^3} > 0 \text{ when } y = 8.$$

S is a minimum when $y = 8$ and $x = 24$.

5. Let x be the positive number.

$$S = x + \frac{1}{x}$$

$$\frac{dS}{dx} = 1 - \frac{1}{x^2} = 0 \text{ when } x = 1.$$

$$\frac{d^2S}{dx^2} = \frac{2}{x^3} > 0 \text{ when } x = 1.$$

The sum is a minimum when $x = 1$ and $1/x = 1$.

6. Let x and y be two positive numbers such that $x + 2y = 100$.

$$P = xy = y(100 - 2y) = 100y - 2y^2$$

$$\frac{dP}{dy} = 100 - 4y = 0 \text{ when } y = 25.$$

$$\frac{d^2P}{dy^2} = -4 < 0 \text{ when } y = 25.$$

P is a maximum when $x = 50$ and $y = 25$.

7. Let x be the length and y the width of the rectangle.

$$2x + 2y = 100$$

$$y = 50 - x$$

$$A = xy = x(50 - x)$$

$$\frac{dA}{dx} = 50 - 2x = 0 \text{ when } x = 25.$$

$$\frac{d^2A}{dx^2} = -2 < 0 \text{ when } x = 25.$$

A is maximum when $x = y = 25$ feet.

8. Let x be the length and y the width of the rectangle.

$$2x + 2y = P$$

$$y = \frac{P - 2x}{2} = \frac{P}{2} - x$$

$$A = xy = x\left(\frac{P}{2} - x\right) = \frac{P}{2}x - x^2$$

$$\frac{dA}{dx} = \frac{P}{2} - 2x = 0 \text{ when } x = \frac{P}{4}.$$

$$\frac{d^2A}{dx^2} = -2 < 0 \text{ when } x = \frac{P}{4}.$$

A is maximum when $x = y = P/4$ units.

9. Let x be the length and y the width of the rectangle.

$$xy = 64$$

$$y = \frac{64}{x}$$

$$P = 2x + 2y = 2x + 2\left(\frac{64}{x}\right) = 2x + \frac{128}{x}$$

$$\frac{dP}{dx} = 2 - \frac{128}{x^2} = 0 \text{ when } x = 8.$$

$$\frac{d^2P}{dx^2} = \frac{256}{x^3} > 0 \text{ when } x = 8.$$

P is minimum when $x = y = 8$ feet.

10. Let x be the length and y the width of the rectangle.

$$xy = A$$

$$y = \frac{A}{x}$$

$$P = 2x + 2y = 2x + 2\left(\frac{A}{x}\right) = 2x + \frac{2A}{x}$$

$$\frac{dP}{dx} = 2 - \frac{2A}{x^2} = 0 \text{ when } x = \sqrt{A}.$$

$$\frac{d^2P}{dx^2} = \frac{4A}{x^3} > 0 \text{ when } x = \sqrt{A}.$$

P is minimum when $x = y = \sqrt{A}$ feet.

11. $d = \sqrt{(x-4)^2 + (\sqrt{x} - 0)^2}$

$$= \sqrt{x^2 - 7x + 16}$$

Since d is smallest when the expression inside the radical is smallest, you need only find the critical numbers of $f(x) = x^2 - 7x + 16$.

$$f'(x) = 2x - 7 = 0$$

$$x = \tfrac{7}{2}$$

By the First Derivative Test, the point nearest to $(4, 0)$ is $\left(7/2, \sqrt{7/2}\right)$.

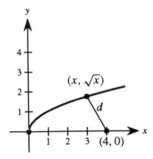

12. $d = \sqrt{(x-2)^2 + [x^2 - (1/2)]^2}$

$$= \sqrt{x^4 - 4x + (17/4)}$$

Since d is smallest when the expression inside the radical is smallest, you need only find the critical numbers of $f(x) = x^4 - 4x + \frac{17}{4}$.

$$f'(x) = 4x^3 - 4 = 0$$

$$x = 1$$

By the First Derivative Test, the point nearest to $(2, 1/2)$ is $(1, 1)$.

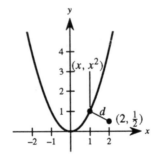

13. $$\frac{dQ}{dx} = kx(Q_0 - x) = kQ_0 x - kx^2$$

$$\frac{d^2Q}{dx^2} = kQ_0 - 2kx$$

$$= k(Q_0 - 2x) = 0 \text{ when } x = \frac{Q_0}{2}.$$

$$\frac{d^3Q}{dx^3} = -2k < 0 \text{ when } x = \frac{Q_0}{2}.$$

dQ/dx is maximum when $x = Q_0/2$.

14. $$F = \frac{v}{22 + 0.02v^2}$$

$$\frac{dF}{dv} = \frac{22 - 0.02v^2}{(22 + 0.02v^2)^2}$$

$$= 0 \text{ when } v = \sqrt{1100} \approx 33.166.$$

By the First Derivative Test, the flow rate on the road is maximized when $v \approx 33$ mph.

15. $xy = 180,000$ (see figure)

$$S = x + 2y = \left(x + \frac{360,000}{x}\right) \text{ where}$$

S is the length of fence needed.

$$\frac{dS}{dx} = 1 - \frac{360,000}{x^2} = 0 \text{ when } x = 600.$$

$$\frac{d^2S}{dx^2} = \frac{720,000}{x^3} > 0 \text{ when } x = 600.$$

S is a minimum when $x = 600$ meters and $y = 300$ meters.

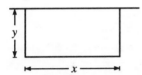

16. $4x + 3y = 200$ is the perimeter. (see figure)

$$A = 2xy = 2x\left(\frac{200 - 4x}{3}\right) = \frac{8}{3}(50x - x^2)$$

$$\frac{dA}{dx} = \frac{8}{3}(50 - 2x) = 0 \text{ when } x = 25.$$

$$\frac{d^2A}{dx^2} = -\frac{16}{3} < 0 \text{ when } x = 25.$$

A is a maximum when $x = 25$ feet and $y = \frac{100}{3}$ feet.

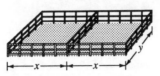

17. $A = 4(\text{Area of side}) + 2(\text{Area of top})$

$V = (\text{Length})(\text{Width})(\text{Height})$

a. $A = 4[(3)(11)] + 2[(3)(3)] = 150$ square inches

$V = (3)(3)(11) = 99$ cubic inches

b. $A = 4[(5)(5)] + 2[(5)(5)] = 150$ square inches

$V = (5)(5)(5) = 125$ cubic inches

c. $A = 4[(3.25)(6)] + 2[(6)(6)]$

$= 150$ square inches

$V = (6)(6)(3.25) = 117$ cubic inches

18. $A = 2\pi rh + 2\pi r^2$

$V = \pi r^2 h$

a. $A = 2\pi(1)(11) + 2\pi(1)^2 = 24\pi$ square inches

$V = \pi(1)^2(11) = 11\pi$ cubic inches

b. $A = 2\pi(2)(4) + 2\pi(2)^2 = 24\pi$ square inches

$V = \pi(2)^2(4) = 16\pi$ cubic inches

c. $A = 2\pi(3)(1) + 2\pi(3)^2 = 24\pi$ square inches

$V = \pi(3)^2(1) = 9\pi$ cubic inches

19. a.

Height, x	Length & Width	Volume
1	$24 - 2(1)$	$1[24 - 2(1)]^2 = 484$
2	$24 - 2(2)$	$2[24 - 2(2)]^2 = 800$
3	$24 - 2(3)$	$3[24 - 2(3)]^2 = 972$
4	$24 - 2(4)$	$4[24 - 2(4)]^2 = 1024$
5	$24 - 2(5)$	$5[24 - 2(5)]^2 = 980$
6	$24 - 2(6)$	$6[24 - 2(6)]^2 = 864$

b. $V = x(24 - 2x)^2, \quad 0 < x < 12$

c. $\dfrac{dV}{dx} = 2x(24 - 2x)(-2) + (24 - 2x)^2 = (24 - 2x)(24 - 6x)$

$= 12(12 - x)(4 - x) = 0$ when $x = 12, 4$ (12 is not in the domain).

$\dfrac{d^2V}{dx^2} = 12(2x - 16)$

$\dfrac{d^2V}{dx^2} < 0$ when $x = 4$.

When $x = 4, \quad V = 1024$ is maximum.

d.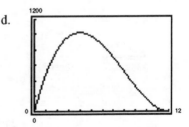

20. a. $V = x(s - 2x)^2, \ 0 < x < \dfrac{s}{2}$

$\dfrac{dV}{dx} = 2x(s - 2x)(-2) + (s - 2x)^2$

$\qquad = (s - 2x)(s - 6x) = 0$ when $x = \dfrac{s}{2}, \ \dfrac{s}{6}$ ($s/2$ is not in the domain).

$\dfrac{d^2V}{dx^2} = 24x - 8s$

$\dfrac{d^2V}{dx^2} < 0$ when $x = \dfrac{s}{6}$.

$V = \dfrac{2s^3}{27}$ is maximum when $x = \dfrac{s}{6}$.

b. If the length is doubled, $V = \frac{2}{27}(2s)^3 = 8\left(\frac{2}{27}s^3\right)$. Volume is increased by a factor of 8.

21. $V = x(3 - 2x)(2 - 2x)$ (see figure), $0 < x < 1$

$\dfrac{dV}{dx} = (3 - 2x)(2 - 2x) + x(2 - 2x)(-2) + x(3 - 2x)(-2)$

$\qquad = 12x^2 - 20x + 6 = 0$ when $x = \dfrac{5 \pm \sqrt{7}}{6}$.

$\dfrac{d^2V}{dx^2} = 24x - 20 < 0$ when $x = \dfrac{5 - \sqrt{7}}{6}$.

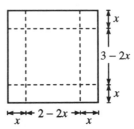

The dimensions are $\left(\dfrac{5 - \sqrt{7}}{6} \ \text{ft}\right)$ by $\left(\dfrac{1 + \sqrt{7}}{3} \ \text{ft}\right)$ by $\left(\dfrac{4 + \sqrt{7}}{3} \ \text{ft}\right)$

or (0.392 ft) by (1.215 ft) by (2.215 ft) yield a maximum volume.

22. $S = x^2 + 3xy$ (see figure)

$x^2 y = \dfrac{250}{3}$

$y = \dfrac{250}{3x^2}$

$S = x^2 + \dfrac{250}{x}$

$\dfrac{dS}{dx} = 2x - \dfrac{250}{x^2} = \dfrac{2x^3 - 250}{x^2} = 0$ when $x = 5$.

$\dfrac{d^2S}{dx^2} = 2 + \dfrac{500}{x^3} > 0$ when $x = 5$.

S is a minimum when $x = 5$ and $y = 10/3$.

23. $16 = 2y + x + \pi\left(\dfrac{x}{2}\right)$

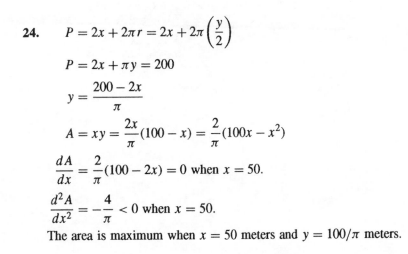

$32 = 4y + 2x + \pi x$

$y = \dfrac{32 - 2x - \pi x}{4}$

$A = xy + \dfrac{\pi}{2}\left(\dfrac{x}{2}\right)^2 = \left(\dfrac{32 - 2x - \pi x}{4}\right)x + \dfrac{\pi x^2}{8}$

$= 8x - \dfrac{1}{2}x^2 - \dfrac{\pi}{4}x^2 + \dfrac{\pi}{8}x^2$

$\dfrac{dA}{dx} = 8 - x - \dfrac{\pi}{2}x + \dfrac{\pi}{4}x = 8 - x\left(1 + \dfrac{\pi}{4}\right)$

$= 0$ when $x = \dfrac{8}{1 + (\pi/4)} = \dfrac{32}{4 + \pi}$.

$\dfrac{d^2A}{dx^2} = -\left(1 + \dfrac{\pi}{4}\right) < 0$ when $x = \dfrac{32}{4 + \pi}$

$y = \dfrac{32 - 2[32/(4 + \pi)] - \pi[32/(4 + \pi)]}{4} = \dfrac{16}{4 + \pi}$

The area is maximum when $y = \dfrac{16}{4 + \pi}$ feet and $x = \dfrac{32}{4 + \pi}$ feet.

24. $P = 2x + 2\pi r = 2x + 2\pi\left(\dfrac{y}{2}\right)$

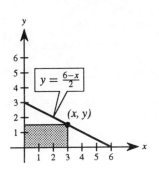

$P = 2x + \pi y = 200$

$y = \dfrac{200 - 2x}{\pi}$

$A = xy = \dfrac{2x}{\pi}(100 - x) = \dfrac{2}{\pi}(100x - x^2)$

$\dfrac{dA}{dx} = \dfrac{2}{\pi}(100 - 2x) = 0$ when $x = 50$.

$\dfrac{d^2A}{dx^2} = -\dfrac{4}{\pi} < 0$ when $x = 50$.

The area is maximum when $x = 50$ meters and $y = 100/\pi$ meters.

25. You can see from the figure that $A = xy$ and since $y = \dfrac{6 - x}{2}$,

$A = x\left(\dfrac{6 - x}{2}\right) = \dfrac{1}{2}(6x - x^2)$

$\dfrac{dA}{dx} = \dfrac{1}{2}(6 - 2x) = 0$ when $x = 3$.

$\dfrac{d^2A}{dx^2} = -1 < 0$ when $x = 3$.

A is a maximum when $x = 3$ and $y = 3/2$.

26. a.
$$\frac{y-2}{0-1} = \frac{0-2}{x-1}$$

$$y = 2 + \frac{2}{x-1}$$

$$L = \sqrt{x^2 + y^2} = \sqrt{x^2 + \left(2 + \frac{2}{x-1}\right)^2}$$

$$= \sqrt{x^2 + 4 + \frac{8}{x-1} + \frac{4}{(x-1)^2}}, \quad x > 1$$

b.

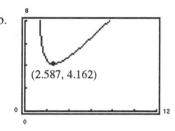

$(2.587, 4.162)$

L is minimum when $x \approx 2.587$.

27. $A = \dfrac{xy}{2} = \dfrac{1}{2}x\left(3 + \dfrac{6}{x-2}\right) = \dfrac{3x^2}{2(x-2)}, \quad x > 2$

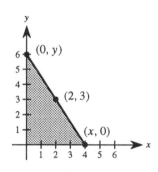

$\dfrac{dA}{dx} = \dfrac{3x(x-4)}{2(x-2)^2} = 0$ when $x = 0, \ 4$.

By the First Derivative Test, A is a minimum when $x = 4$ and $y = 6$. The vertices of the triangle are $(0, 0)$, $(4, 0)$, and $(0, 6)$.

28. $A = \dfrac{1}{2}\text{Base} \times \text{height}$

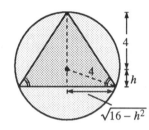

$$= \frac{1}{2}\left(2\sqrt{16 - h^2}\right)(4 + h) = \sqrt{16 - h^2}(4 + h)$$

$$\frac{dA}{dh} = \frac{1}{2}(16 - h^2)^{-1/2}(-2h)(4 + h) + (16 - h^2)^{1/2}$$

$$= (16 - h^2)^{-1/2}[-h(4 + h) + (16 - h^2)]$$

$$= \frac{-2[h^2 + 2h - 8]}{\sqrt{16 - h^2}} = \frac{-2(h + 4)(h - 2)}{\sqrt{16 - h^2}}$$

$dA/dh = 0$ when $h = 2$, which is a maximum by the First Derivative Test. Hence, the sides of the equilateral triangle are $2\sqrt{16 - h^2} = 2\sqrt{16 - 4} = 4\sqrt{3}$.

29. $A = 2xy = 2x\sqrt{25 - x^2}$ (see figure)

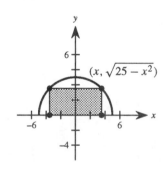

$$\frac{dA}{dx} = 2x\left(\frac{1}{2}\right)\left(\frac{-2x}{\sqrt{25 - x^2}}\right) + 2\sqrt{25 - x^2}$$

$$= 2\left(\frac{25 - 2x^2}{\sqrt{25 - x^2}}\right) = 0 \text{ when } x = y = \frac{5\sqrt{2}}{2} \approx 3.54.$$

By the First Derivative Test, the inscribed rectangle of maximum area has vertices

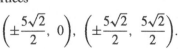

Width: $\dfrac{5\sqrt{2}}{2}$; Length: $5\sqrt{2}$

30. $A = 2xy = 2x\sqrt{r^2 - x^2}$ (see figure)

$\dfrac{dA}{dx} = \dfrac{2(r^2 - 2x^2)}{\sqrt{r^2 - x^2}} = 0$ when $x = \dfrac{\sqrt{2}\,r}{2}$.

By the First Derivative Test, A is maximum when the rectangle has dimensions $\sqrt{2}\,r$ by $(\sqrt{2}\,r)/2$.

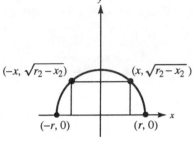

31. $A = \dfrac{1}{2}(2r + 2x)\sqrt{r^2 - x^2} = (r + x)\sqrt{r^2 - x^2}$ (see figure)

$\dfrac{dA}{dx} = (r + x)\left(\dfrac{1}{2}\right)(r^2 - x^2)^{-1/2}(-2x) + \sqrt{r^2 - x^2}$

$\qquad = \dfrac{-x(r + x)}{\sqrt{r^2 - x^2}} + \sqrt{r^2 - x^2} = \dfrac{r^2 - rx - 2x^2}{\sqrt{r^2 - x^2}}$

$\qquad = \dfrac{(r - 2x)(r + x)}{\sqrt{r^2 - x^2}} = 0$ when $x = \dfrac{r}{2}$.

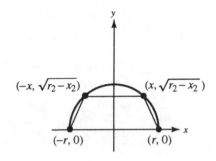

By the First Derivative Test, A will be a maximum when the trapezoid bases are r and $2r$, and the altitude is $(\sqrt{3}\,r)/2$.

32. $xy = 30 \Rightarrow y = \dfrac{30}{x}$

$A = (x + 2)\left(\dfrac{30}{x} + 2\right)$ (see figure)

$\dfrac{dA}{dx} = (x + 2)\left(\dfrac{-30}{x^2}\right) + \left(\dfrac{30}{x} + 2\right) = \dfrac{2(x^2 - 30)}{x^2} = 0$ when $x = \sqrt{30}$.

$y = \dfrac{30}{\sqrt{30}} = \sqrt{30}$

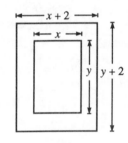

By the First Derivative Test, the dimensions $(x + 2)$ by $(y + 2)$ are $(2 + \sqrt{30})$ by $(2 + \sqrt{30})$ (approximately 7.477 by 7.477). These dimensions yield a minimum area.

33. $V = \pi r^2 h = 22$ cubic inches or $h = \dfrac{22}{\pi r^2}$

e.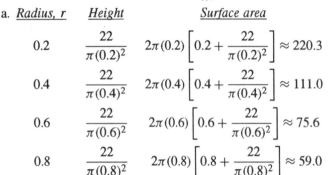

a.

Radius, r	Height	Surface area
0.2	$\dfrac{22}{\pi (0.2)^2}$	$2\pi(0.2)\left[0.2 + \dfrac{22}{\pi (0.2)^2}\right] \approx 220.3$
0.4	$\dfrac{22}{\pi (0.4)^2}$	$2\pi(0.4)\left[0.4 + \dfrac{22}{\pi (0.4)^2}\right] \approx 111.0$
0.6	$\dfrac{22}{\pi (0.6)^2}$	$2\pi(0.6)\left[0.6 + \dfrac{22}{\pi (0.6)^2}\right] \approx 75.6$
0.8	$\dfrac{22}{\pi (0.8)^2}$	$2\pi(0.8)\left[0.8 + \dfrac{22}{\pi (0.8)^2}\right] \approx 59.0$

b. $S = 2\pi r^2 + 2\pi rh = 2\pi r(r + h) = 2\pi r\left[r + \dfrac{22}{\pi r^2}\right] = 2\pi r^2 + \dfrac{44}{r}$

c. $\dfrac{dS}{dr} = 4\pi r - \dfrac{44}{r^2} = 0$ when $r = \sqrt[3]{11/\pi} \approx 1.52$ in.

$h = \dfrac{22}{\pi r^2} \approx 3.04$ in.

Note: Notice that $h = \dfrac{22}{\pi r^2} = \dfrac{22}{\pi (11/\pi)^{2/3}} = 2 \cdot \dfrac{11^{1/3}}{\pi^{1/3}} = 2r.$

d.

Radius, r	Height	Surface Area
0.2	$\dfrac{22}{\pi (0.2)^2}$	$2\pi(0.2)\left[0.2 + \dfrac{22}{\pi (0.2)^2}\right] \approx 220.3$
0.4	$\dfrac{22}{\pi (0.4)^2}$	$2\pi(0.4)\left[0.4 + \dfrac{22}{\pi (0.4)^2}\right] \approx 111.0$
0.6	$\dfrac{22}{\pi (0.6)^2}$	$2\pi(0.6)\left[0.6 + \dfrac{22}{\pi (0.6)^2}\right] \approx 75.6$
0.8	$\dfrac{22}{\pi (0.8)^2}$	$2\pi(0.8)\left[0.8 + \dfrac{22}{\pi (0.8)^2}\right] \approx 59.0$
1.0	$\dfrac{22}{\pi (1.0)^2}$	$2\pi(1.0)\left[1.0 + \dfrac{22}{\pi (1.0)^2}\right] \approx 50.3$
1.2	$\dfrac{22}{\pi (1.2)^2}$	$2\pi(1.2)\left[1.2 + \dfrac{22}{\pi (1.2)^2}\right] \approx 45.7$
1.4	$\dfrac{22}{\pi (1.4)^2}$	$2\pi(1.4)\left[1.4 + \dfrac{22}{\pi (1.4)^2}\right] \approx 43.7$
1.6	$\dfrac{22}{\pi (1.6)^2}$	$2\pi(1.6)\left[1.6 + \dfrac{22}{\pi (1.6)^2}\right] \approx 43.6$
1.8	$\dfrac{22}{\pi (1.8)^2}$	$2\pi(1.8)\left[1.8 + \dfrac{22}{\pi (1.8)^2}\right] \approx 44.8$
2.0	$\dfrac{22}{\pi (2.0)^2}$	$2\pi(2.0)\left[2.0 + \dfrac{22}{\pi (2.0)^2}\right] \approx 47.1$

34. $V = \pi r^2 h = V_0$ cubic units or $h = \dfrac{V_0}{\pi r^2}$

$$S = 2\pi r^2 + 2\pi rh = 2\left(\pi r^2 + \dfrac{V_0}{r}\right)$$

$$\dfrac{dS}{dr} = 2\left(2\pi r - \dfrac{V_0}{r^2}\right) = 0 \text{ when } r = \sqrt[3]{\dfrac{V_0}{2\pi}} \text{ units.}$$

$$h = \dfrac{V_0}{\pi\left(\sqrt[3]{V_0/2\pi}\right)^2} = \dfrac{V_0(2\pi)^{2/3}}{\pi V_0^{2/3}} = \dfrac{2V_0^{1/3}}{(2\pi)^{1/3}} = 2r$$

By the First Derivative Test, this will yield the minimum surface area.

35. Let x be the sides of the square ends and y the length of the package.

$$P = 4x + y = 108$$

$$V = x^2 y = x^2(108 - 4x) = 108x^2 - 4x^3$$

$$\dfrac{dV}{dx} = 216x - 12x^2$$

$$= 12x(18 - x) = 0 \text{ when } x = 18.$$

$$\dfrac{d^2V}{dx^2} = 216 - 24x = -216 \text{ when } x = 18.$$

The volume is maximum when $x = 18$ inches and $y = 108 - 4(18) = 36$ inches.

36. $V = \pi r^2 x$

$$x + 2\pi r = 108 \Rightarrow x = 108 - 2\pi r \quad \text{(see figure)}$$

$$V = \pi r^2(108 - 2\pi r) = \pi(108r^2 - 2\pi r^3)$$

$$\dfrac{dV}{dr} = \pi(216r - 6\pi r^2) = 6\pi r(36 - \pi r)$$

$$= 0 \text{ when } r = \dfrac{36}{\pi} \text{ and } x = 36.$$

$$\dfrac{d^2V}{dr^2} = \pi(216 - 12\pi r) < 0 \text{ when } r = \dfrac{36}{\pi}.$$

Volume is maximum when $x = 36$ inches and $r = 36/\pi \approx 11.459$ inches.

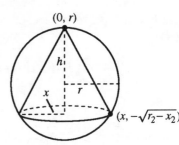

37. $V = \dfrac{1}{3}\pi x^2 h = \dfrac{1}{3}\pi x^2(r + \sqrt{r^2 - x^2})$ (see figure)

$$\dfrac{dV}{dx} = \dfrac{1}{3}\pi\left[\dfrac{-x^3}{\sqrt{r^2 - x^2}} + 2x(r + \sqrt{r^2 - x^2})\right] = \dfrac{\pi x}{3\sqrt{r^2 - x^2}}(2r^2 + 2r\sqrt{r^2 - x^2} - 3x^2) = 0$$

$$2r^2 + 2r\sqrt{r^2 - x^2} - 3x^2 = 0$$

$$2r\sqrt{r^2 - x^2} = 3x^2 - 2r^2$$

$$4r^2(r^2 - x^2) = 9x^4 - 12x^2r^2 + 4r^4$$

$$0 = 9x^4 - 8x^2r^2 = x^2(9x^2 - 8r^2)$$

$$x = 0, \ \dfrac{2\sqrt{2}\,r}{3}$$

By the First Derivative Test, the volume is a maximum when $x = \dfrac{2\sqrt{2}\,r}{3}$ and $h = r + \sqrt{r^2 - x^2} = \dfrac{4r}{3}$. Thus, the maximum volume is $V = \dfrac{1}{3}\pi\left(\dfrac{8r^2}{9}\right)\left(\dfrac{4r}{3}\right) = \dfrac{32\pi r^3}{81}$ cubic units.

38. $V = \pi r^2 h = \pi x^2 (2\sqrt{r^2 - x^2}) = 2\pi x^2 \sqrt{r^2 - x^2}$ (see figure)

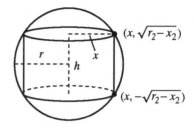

$$\frac{dV}{dx} = 2\pi \left[x^2 \left(\frac{1}{2}\right)(r^2 - x^2)^{-1/2}(-2x) + 2x\sqrt{r^2 - x^2} \right]$$

$$= \frac{2\pi x}{\sqrt{r^2 - x^2}}(2r^2 - 3x^2)$$

$= 0$ when $x = 0$ and $x^2 = \dfrac{2r^2}{3} \Rightarrow x = \dfrac{\sqrt{6}\,r}{3}$.

By the First Derivative Test, the volume is a maximum when

$$x = \frac{\sqrt{6}\,r}{3} \text{ and } h = \frac{2r}{\sqrt{3}}.$$

Thus, the maximum volume is

$$V = \pi \left(\frac{2}{3}r^2\right)\left(\frac{2r}{\sqrt{3}}\right) = \frac{4\pi r^3}{3\sqrt{3}}.$$

39. $12 = \dfrac{4}{3}\pi r^3 + \pi r^2 h$

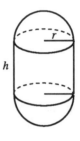

$$h = \frac{12 - (4/3)\pi r^3}{\pi r^2} = \frac{12}{\pi r^2} - \frac{4}{3}r$$

$$S = 4\pi r^2 + 2\pi r h = 4\pi r^2 + 2\pi r\left(\frac{12}{\pi r^2} - \frac{4}{3}r\right)$$

$$= 4\pi r^2 + \frac{24}{r} - \frac{8}{3}\pi r^2 = \frac{4}{3}\pi r^2 + \frac{24}{r}$$

$$\frac{dS}{dr} = \frac{8}{3}\pi r - \frac{24}{r^2} = 0 \text{ when } r = \sqrt[3]{9/\pi} \approx 1.42 \text{ inches.}$$

$$\frac{d^2S}{dr^2} = \frac{8}{3}\pi + \frac{48}{r^3} > 0 \text{ when } r = \sqrt[3]{9/\pi} \text{ inches.}$$

The surface area is minimum when $r = \sqrt[3]{9/\pi}$ inches and $h = 0$. The resulting solid is a sphere of radius $r \approx 1.42$ inches.

40. $3000 = \dfrac{4}{3}\pi r^3 + \pi r^2 h$

$$h = \frac{3000}{\pi r^2} - \frac{4}{3}r$$

Let k = cost per square foot of the surface area of the sides, then $2k$ = cost per square foot of the hemispherical ends.

$$C = 2k(4\pi r^2) + k(2\pi r h) = k\left[8\pi r^2 + 2\pi r\left(\frac{3000}{\pi r^2} - \frac{4}{3}r\right)\right] = k\left[\frac{16}{3}\pi r^2 + \frac{6000}{r}\right]$$

$$\frac{dC}{dr} = k\left[\frac{32}{3}\pi r - \frac{6000}{r^2}\right] = 0 \text{ when } r = \sqrt[3]{\frac{1125}{2\pi}} \approx 5.636 \text{ feet and } h \approx 22.545 \text{ feet.}$$

By the Second Derivative Test, we have

$$\frac{d^2C}{dr^2} = k\left[\frac{32}{3}\pi + \frac{12000}{r^3}\right] > 0 \text{ when } r = \sqrt[3]{\frac{1125}{2\pi}}.$$

Therefore, these dimensions will produce a minimum cost.

41. Let x be the length of a side of the square and y the length of a side of the triangle.

$$4x + 3y = 10$$

$$A = x^2 + \frac{1}{2}y\left(\frac{\sqrt{3}}{2}y\right)$$

$$= \frac{(10 - 3y)^2}{16} + \frac{\sqrt{3}}{4}y^2$$

$$\frac{dA}{dy} = \frac{1}{8}(10 - 3y)(-3) + \frac{\sqrt{3}}{2}y = 0$$

$$-30 + 9y + 4\sqrt{3}\,y = 0$$

$$y = \frac{30}{9 + 4\sqrt{3}}$$

$$\frac{d^2A}{dy^2} = \frac{9 + 4\sqrt{3}}{8} > 0$$

A is minimum when

$$y = \frac{30}{9 + 4\sqrt{3}} \text{ and } x = \frac{10\sqrt{3}}{9 + 4\sqrt{3}}.$$

42. Let x be the length of a side of the square and r the radius of the circle.

$$4x + 2\pi r = 16$$

$$x = 4 - \frac{1}{2}\pi r$$

$$A = x^2 + \pi r^2 = \left(4 - \frac{1}{2}\pi r\right)^2 + \pi r^2$$

$$\frac{dA}{dr} = 2\left(4 - \frac{1}{2}\pi r\right)\left(-\frac{1}{2}\pi\right) + 2\pi r$$

$$-8\pi + (\pi^2 + 4\pi)r = 0 \text{ when } r = \frac{8}{4 + \pi}.$$

$$\frac{d^2A}{dr^2} = \frac{\pi^2}{2} + 2\pi > 0$$

A is minimum when $r = \dfrac{8}{4 + \pi}$ and $x = \dfrac{16}{4 + \pi}$.

43. Let x be the length of the sides of the isosceles triangle (the hypotenuse will be $\sqrt{2}x$) and r the radius of the circle.

$$(2x + \sqrt{2}x) + 2\pi r = 10$$

$$x = \frac{10 - 2\pi r}{2 + \sqrt{2}}$$

$$A = \frac{1}{2}x^2 + \pi r^2 = \frac{1}{2}\left(\frac{10 - 2\pi r}{2 + \sqrt{2}}\right)^2 + \pi r^2$$

$$= \frac{1}{3 + 2\sqrt{2}}(25 - 10\pi r + \pi^2 r^2) + \pi r^2, \quad 0 \le r \le \frac{10}{2\pi}$$

$$\frac{dA}{dr} = \frac{1}{3 + 2\sqrt{2}}(-10\pi + 2\pi^2 r) + 2\pi r = 0 \Rightarrow r = \frac{5}{\pi + 3 + 2\sqrt{2}}$$

a. Minimum: $2\pi r = 2\pi\left(\dfrac{5}{\pi + 3 + 2\sqrt{2}}\right) = \dfrac{10\pi}{\pi + 3 + 2\sqrt{2}} \approx 3.5$ feet

b. Maximum: $2\pi r = 10$ feet (endpoint extrema; that is, use all of the wire for the circle.)

44. a. Let x be the side of the triangle and y the side of the square.

$$A = \frac{3}{4}\left(\cot\frac{\pi}{3}\right)x^2 + \frac{4}{4}\left(\cot\frac{\pi}{4}\right)y^2 \text{ where } 3x + 4y = 20$$

$$= \frac{\sqrt{3}}{4}x^2 + \left(5 - \frac{3}{4}x\right)^2, \quad 0 \le x \le \frac{20}{3}$$

$$A' = \frac{\sqrt{3}}{2}x + 2\left(5 - \frac{3}{4}x\right)\left(-\frac{3}{4}\right) = 0$$

$$x = \frac{60}{4\sqrt{3} + 9}$$

When $x = 0$, $A = 25$, when $x = 60/(4\sqrt{3} + 9)$, $A \approx 10.847$, and when $x = 20/3$, $A \approx 19.245$. Area is maximum when all 20 feet are used on the square.

b. Let x be the side of the square and y the side of the pentagon.

$$A = \frac{4}{4}\left(\cot\frac{\pi}{4}\right)x^2 + \frac{5}{4}\left(\cot\frac{\pi}{5}\right)y^2 \text{ where } 4x + 5y = 20$$

$$= x^2 + 1.7204774\left(4 - \frac{4}{5}x\right)^2, \quad 0 \le x \le 5$$

$$A' = 2x - 2.75276384\left(4 - \frac{4}{5}x\right) = 0$$

$$x \approx 2.62$$

When $x = 0$, $A \approx 27.528$, when $x \approx 2.62$, $A \approx 13.102$, and when $x = 5$, $A \approx 25$. Area is maximum when all 20 feet are used on the pentagon.

c. Let x be the side of the pentagon and y the side of the hexagon.

$$A = \frac{5}{4}\left(\cot\frac{\pi}{5}\right)x^2 + \frac{6}{4}\left(\cot\frac{\pi}{6}\right)y^2 \text{ where } 5x + 6y = 20$$

$$= \frac{5}{4}\left(\cot\frac{\pi}{5}\right)x^2 + \frac{3}{2}(\sqrt{3})\left(\frac{20 - 5x}{6}\right)^2, \quad 0 \le x \le 4$$

$$A' = \frac{5}{2}\left(\cot\frac{\pi}{5}\right)x + 3\sqrt{3}\left(-\frac{5}{6}\right)\left(\frac{20 - 5x}{6}\right) = 0$$

$$x \approx 2.0475$$

When $x = 0$, $A \approx 28.868$, when $x \approx 2.0475$, $A \approx 14.091$, and when $x = 4$, $A \approx 27.528$. Area is maximum when all 20 feet are used on the hexagon.

d. Let x be the side of the hexagon and r the radius of the circle.

$$A = \frac{6}{4}\left(\cot\frac{\pi}{6}\right)x^2 + \pi r^2 \text{ where } 6x + 2\pi r = 20$$

$$= \frac{3\sqrt{3}}{2}x^2 + \pi\left(\frac{10}{\pi} - \frac{3x}{\pi}\right)^2, \quad 0 \le x \le \frac{10}{3}$$

$$A' = 3\sqrt{3}x - 6\left(\frac{10}{\pi} - \frac{3x}{\pi}\right) = 0$$

$$x \approx 1.748$$

When $x = 0$, $A \approx 31.831$, when $x \approx 1.748$, $A \approx 15.138$, and when $x = 10/3$, $A \approx 28.868$. Area is maximum when all 20 feet are used on the circle. In general, using all of the wire for the figure with more sides will enclose the most area.

45. Let S be the strength and k the constant of proportionality. Given $h^2 + w^2 = 24^2$, $h^2 = 24^2 - w^2$,

$$S = kwh^2$$

$$S = kw(576 - w^2) = k(576w - w^3)$$

$$\frac{dS}{dw} = k(576 - 3w^2) = 0 \text{ when } w = 8\sqrt{3}, \quad h = 8\sqrt{6}.$$

$$\frac{d^2S}{dw^2} = -6kw < 0 \text{ when } w = 8\sqrt{3}.$$

These values yield a maximum.

46. Let A be the amount of the power line.

$$A = h - y + 2\sqrt{x^2 + y^2}$$

$$\frac{dA}{dy} = -1 + \frac{2y}{\sqrt{x^2 + y^2}} = 0 \text{ when } y = \frac{x}{\sqrt{3}}.$$

$$\frac{d^2A}{dy^2} = \frac{2x^2}{(x^2 + y^2)^{3/2}} > 0 \text{ for } y = \frac{x}{\sqrt{3}}.$$

The amount of power line is minimum when $y = x/\sqrt{3}$.

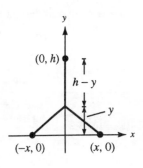

47. $\sin \alpha = \dfrac{h}{s} \Rightarrow s = \dfrac{h}{\sin \alpha}, \quad 0 < \alpha < \dfrac{\pi}{2}$

$$\tan \alpha = \frac{h}{2} \Rightarrow h = 2 \tan \alpha \Rightarrow s = \frac{2 \tan \alpha}{\sin \alpha} = 2 \sec \alpha$$

$$I = \frac{k \sin \alpha}{s^2} = \frac{k \sin \alpha}{4 \sec^2 \alpha} = \frac{k}{4} \sin \alpha \cos^2 \alpha$$

$$\frac{dI}{d\alpha} = \frac{k}{4} [\sin \alpha (-2 \sin \alpha \cos \alpha) + \cos^2 \alpha (\cos \alpha)]$$

$$= \frac{k}{4} \cos \alpha [\cos^2 \alpha - 2 \sin^2 \alpha]$$

$$= \frac{k}{4} \cos \alpha [1 - 3 \sin^2 \alpha]$$

$$= 0 \text{ when } \alpha = \frac{\pi}{2}, \frac{3\pi}{2}, \text{ or when } \sin \alpha = \pm \frac{1}{\sqrt{3}}.$$

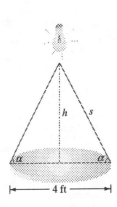

Since α is acute, we have

$$\sin \alpha = \frac{1}{\sqrt{3}} \Rightarrow h = 2 \tan \alpha = 2 \left(\frac{1}{\sqrt{2}} \right) = \sqrt{2} \text{ feet.}$$

Since $(d^2 I)/(d\alpha^2) = (k/4) \sin \alpha (9 \sin^2 \alpha - 7) < 0$ when $\sin \alpha = 1/\sqrt{3}$, this yields a maximum.

48. Let F be the illumination at point P which is x units from source 1.

$$F = \frac{kI_1}{x^2} + \frac{kI_2}{(d-x)^2}$$

$$\frac{dF}{dx} = \frac{-2kI_1}{x^3} + \frac{2kI_2}{(d-x)^3} = 0 \text{ when } \frac{2kI_1}{x^3} = \frac{2kI_2}{(d-x)^3}.$$

$$\frac{\sqrt[3]{I_1}}{\sqrt[3]{I_2}} = \frac{x}{d-x}$$

$$(d-x)\sqrt[3]{I_1} = x\sqrt[3]{I_2}$$

$$d\sqrt[3]{I_1} = x\left(\sqrt[3]{I_1} + \sqrt[3]{I_2}\right)$$

$$x = \frac{d\sqrt[3]{I_1}}{\sqrt[3]{I_1} + \sqrt[3]{I_2}}$$

$$\frac{d^2 F}{dx^2} = \frac{6kI_1}{x^4} + \frac{6kI_2}{(d-x)^4} > 0 \text{ when } x = \frac{d\sqrt[3]{I_1}}{\sqrt[3]{I_1} + \sqrt[3]{I_2}}.$$

This is the minimum point.

49.

$$S = \sqrt{x^2 + 4}, \quad L = \sqrt{1 + (3-x)^2}$$

$$T = \frac{\sqrt{x^2 + 4}}{2} + \frac{\sqrt{x^2 - 6x + 10}}{4}$$

$$\frac{dT}{dx} = \frac{x}{2\sqrt{x^2 + 4}} + \frac{x-3}{4\sqrt{x^2 - 6x + 10}} = 0$$

$$\frac{x^2}{x^2 + 4} = \frac{9 - 6x + x^2}{4(x^2 - 6x + 10)}$$

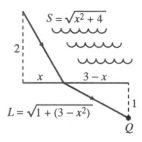

$$x^4 - 6x^3 + 9x^2 + 8x - 12 = 0$$

You need to find the roots of this equation in the interval $[0, 3]$. By using a computer or graphics calculator, you can determine that this equation has only one root in this interval ($x = 1$). Testing at this value and at the endpoints, you see that $x = 1$ yields the minimum time. Thus, the man should row to a point 1 mile from the nearest point on the coast.

50.

$$T = \frac{\sqrt{x^2 + 4}}{4} + \frac{\sqrt{x^2 - 6x + 10}}{4}$$

$$\frac{dT}{dx} = \frac{1}{4}\left[\frac{x}{\sqrt{x^2 + 4}} + \frac{x-3}{\sqrt{x^2 - 6x + 10}}\right] = 0 \text{ when } \frac{x^2}{x^2 + 4} = \frac{(x-3)^2}{x^2 - 6x + 10}.$$

Thus, $x^2 - 8x + 12 = 0 \Rightarrow x = 2, 6$. In the interval $[0, 3]$, we have $x = 2$ miles. Testing at $x = 2$ and at the endpoints, we see that $x = 2$ yields a minimum time.

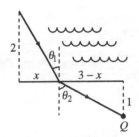

51. $T = \dfrac{\sqrt{x^2 + 4}}{v_1} + \dfrac{\sqrt{x^2 - 6x + 10}}{v_2}$

$\dfrac{dT}{dx} = \dfrac{x}{v_1\sqrt{x^2 + 4}} + \dfrac{x - 3}{v_2\sqrt{x^2 - 6x + 10}} = 0$

Since

$$\dfrac{x}{\sqrt{x^2 + 4}} = \sin\theta_1 \quad \text{and} \quad \dfrac{x - 3}{\sqrt{x^2 - 6x + 10}} = -\sin\theta_2$$

we have

$\dfrac{\sin\theta_1}{v_1} - \dfrac{\sin\theta_2}{v_2} = 0 \Rightarrow \dfrac{\sin\theta_1}{v_1} = \dfrac{\sin\theta_2}{v_2}.$

Since

$$\dfrac{d^2T}{dx^2} = \dfrac{4}{v_1(x^2 + 4)^{3/2}} + \dfrac{1}{v_2(x^2 - 6x + 10)^{3/2}} > 0$$

this condition yields a minimum time.

52. $T = \dfrac{\sqrt{x^2 + d_1{}^2}}{v_1} + \dfrac{\sqrt{d_2{}^2 + (a - x)^2}}{v_2}$

$\dfrac{dT}{dx} = \dfrac{x}{v_1\sqrt{x^2 + d_1{}^2}} + \dfrac{x - a}{v_2\sqrt{d_2{}^2 + (a - x)^2}} = 0$

Since

$$\dfrac{x}{\sqrt{x^2 + d_1{}^2}} = \sin\theta_1 \quad \text{and} \quad \dfrac{x - a}{\sqrt{d_2{}^2 + (a - x)^2}} = -\sin\theta_2$$

we have

$\dfrac{\sin\theta_1}{v_1} - \dfrac{\sin\theta_2}{v_2} = 0 \Rightarrow \dfrac{\sin\theta_1}{v_1} = \dfrac{\sin\theta_2}{v_2}.$

Since

$$\dfrac{d^2T}{dx^2} = \dfrac{d_1{}^2}{v_1(x^2 + d_1{}^2)^{3/2}} + \dfrac{d_2{}^2}{v_2[d_2{}^2 + (a - x)^2]^{3/2}} > 0$$

this condition yields a minimum time.

53. $F\cos\theta = k(W - F\sin\theta)$

$F = \dfrac{kW}{\cos\theta + k\sin\theta}$

$\dfrac{dF}{d\theta} = \dfrac{-kW(k\cos\theta - \sin\theta)}{(\cos\theta + k\sin\theta)^2} = 0$

$k\cos\theta = \sin\theta \Rightarrow k = \tan\theta \Rightarrow \theta = \arctan k$

Since

$$\cos\theta + k\sin\theta = \dfrac{1}{\sqrt{k^2 + 1}} + \dfrac{k^2}{\sqrt{k^2 + 1}} = \sqrt{k^2 + 1},$$

the minimum force is

$$F = \dfrac{kW}{\cos\theta + k\sin\theta} = \dfrac{kW}{\sqrt{k^2 + 1}}.$$

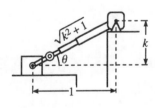

54. $R = \dfrac{v_0{}^2}{g} \sin 2\theta$

$\dfrac{dR}{d\theta} = \dfrac{2v_0{}^2}{g} \cos 2\theta = 0$ when $\theta = \dfrac{\pi}{4},\ \dfrac{3\pi}{4}$.

$\dfrac{d^2R}{d\theta^2} = -\dfrac{4v_0{}^2}{g} \sin 2\theta < 0$ when $\theta = \dfrac{\pi}{4}$.

By the Second Derivative Test, R is maximum when $\theta = \pi/4$.

55. $V = \dfrac{1}{3}\pi r^2 h = \dfrac{1}{3}\pi r^2 \sqrt{144 - r^2}$

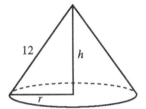

$\dfrac{dV}{dr} = \dfrac{1}{3}\pi \left[r^2 \left(\dfrac{1}{2}\right)(144 - r^2)^{-1/2}(-2r) + 2r\sqrt{144 - r^2} \right]$

$= \dfrac{1}{3}\pi \left[\dfrac{288r - 3r^3}{\sqrt{144 - r^2}} \right] = \pi \left[\dfrac{r(96 - r^2)}{\sqrt{144 - r^2}} \right] = 0$ when $r = 0,\ 4\sqrt{6}$.

By the First Derivative Test, V is maximum when $r = 4\sqrt{6}$ and $h = 4\sqrt{3}$.

Area of circle: $A = \pi(12)^2 = 144\pi$

Lateral surface area of cone: $S = \pi(4\sqrt{6})\sqrt{(4\sqrt{6})^2 + (4\sqrt{3})^2} = 48\sqrt{6}\,\pi$

Area of sector: $144\pi - 48\sqrt{6}\,\pi = \dfrac{1}{2}\theta r^2 = 72\theta$

$$\theta = \dfrac{144\pi - 48\sqrt{6}\,\pi}{72} = \dfrac{2\pi}{3}(3 - \sqrt{6}) \approx 1.153 \text{ radians or } 66°$$

56. a.

Base 1	Base 2	Altitude	Area
8	$8 + 16\cos 10°$	$8\sin 10°$	≈ 22.1
8	$8 + 16\cos 20°$	$8\sin 20°$	≈ 42.5
8	$8 + 16\cos 30°$	$8\sin 30°$	≈ 59.7
8	$8 + 16\cos 40°$	$8\sin 40°$	≈ 72.7
8	$8 + 16\cos 50°$	$8\sin 50°$	≈ 80.5
8	$8 + 16\cos 60°$	$8\sin 60°$	≈ 83.1

b. $A = (a + b)\dfrac{h}{2}$

$= [8 + (8 + 16\cos\theta)]\dfrac{8\sin\theta}{2}$

$= 64(1 + \cos\theta)\sin\theta, \quad 0° < \theta < 90°$

c. $\dfrac{dA}{d\theta} = 64(1 + \cos\theta)\cos\theta + (-64\sin\theta)\sin\theta$

$= 64(\cos\theta + \cos^2\theta - \sin^2\theta)$

$= 64(2\cos^2\theta + \cos\theta - 1)$

$= 64(2\cos\theta - 1)(\cos\theta + 1)$

$= 0$ when $\theta = 60°,\ 180°,\ 300°$.

The maximum occurs when $\theta = 60°$.

d.

Base 1	Base 2	Altitude	Area
8	$8 + 16\cos 10°$	$8\sin 10°$	≈ 22.1
8	$8 + 16\cos 20°$	$8\sin 20°$	≈ 42.5
8	$8 + 16\cos 30°$	$8\sin 30°$	≈ 59.7
8	$8 + 16\cos 40°$	$8\sin 40°$	≈ 72.7
8	$8 + 16\cos 50°$	$8\sin 50°$	≈ 80.5
8	$8 + 16\cos 60°$	$8\sin 60°$	≈ 83.1
8	$8 + 16\cos 70°$	$8\sin 70°$	≈ 80.7
8	$8 + 16\cos 80°$	$8\sin 80°$	≈ 74.0
8	$8 + 16\cos 90°$	$8\sin 90°$	≈ 64.0

The maximum cross-sectional area is approximately 83.1 square feet.

e.

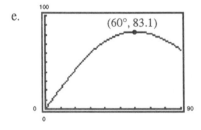

57. $A = 2x\sqrt{100 - x^2}$

$\dfrac{dA}{dx} = \dfrac{2(100 - 2x^2)}{\sqrt{100 - x^2}} = 0$ (from Exercise 30)

$x = 5\sqrt{2}$

Dimensions: $10\sqrt{2} \times 5\sqrt{2} \approx 14.1421 \times 7.0711$

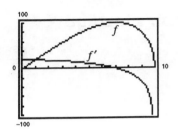

58. $\dfrac{y - 8}{0 - 5} = \dfrac{8 - 0}{5 - x} \Rightarrow y = \dfrac{8x}{x - 5}$

$L^2 = x^2 + y^2 = x^2 + \left(\dfrac{8x}{x - 5}\right)^2$

$2L\dfrac{dL}{dx} = 2x + 2\left(\dfrac{8x}{x - 5}\right)\left[\dfrac{-40}{(x - 5)^2}\right]$

$L\dfrac{dL}{dx} = x - \dfrac{320x}{(x - 5)^3}$

$\dfrac{dL}{dx} = \dfrac{1}{L}\left[x - \dfrac{320x}{(x - 5)^3}\right] = 0$

When $x[(x - 5)^3 - 320] = 0$:

$x = \sqrt[3]{320} + 5 \approx 11.84$

$L = \sqrt{(\sqrt[3]{320} + 5)^2 + \left[\dfrac{8(\sqrt[3]{320} + 5)}{\sqrt[3]{320}}\right]^2} \approx 18.22$ feet

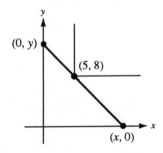
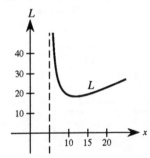
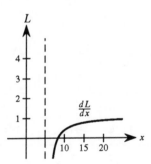

59. $E = \dfrac{\tan\phi(1 - 0.1\tan\phi)}{0.1 + \tan\phi}$

$\dfrac{dE}{d\phi} = \dfrac{(0.1 + \tan\phi)[\sec^2\phi - 0.2\tan\phi\sec^2\phi] - (\tan\phi - 0.1\tan^2\phi)(\sec^2\phi)}{(0.1 + \tan\phi)^2}$

$= \dfrac{0.1\sec^2\phi(1 - 0.2\tan\phi - \tan^2\phi)}{(0.1 + \tan\phi)^2} = 0$

When $1 - 0.2\tan\phi - \tan^2\phi = 0$:

$\tan\phi = \dfrac{-1 \pm \sqrt{101}}{10}$

$\phi \approx 42.1°, \ 137.9°.$

By the First Derivative Test, the maximum occurs when $\phi \approx 42.1°$.

Section 3.8   Newton's Method

1. $f(x) = x^2 - 3$

$f'(x) = 2x$

$x_1 = 1.7$

n	x_n	$f(x_n)$	$f'(x_n)$	$\dfrac{f(x_n)}{f'(x_n)}$	$x_n - \dfrac{f(x_n)}{f'(x_n)}$
1	1.7000	−0.1100	3.4000	−0.0324	1.7324
2	1.7324	0.0012	3.4648	0.0003	1.7321

2. $f(x) = 3x^2 - 2$

$f'(x) = 6x$

$x_1 = 1$

n	x_n	$f(x_n)$	$f'(x_n)$	$\dfrac{f(x_n)}{f'(x_n)}$	$x_n - \dfrac{f(x_n)}{f'(x_n)}$
1	1.0000	1.0000	6.0000	0.1667	0.8333
2	0.8333	0.0832	4.9998	0.0166	0.8166

3. $f(x) = \sin x$

$f'(x) = \cos x$

$x_1 = 3$

n	x_n	$f(x_n)$	$f'(x_n)$	$\dfrac{f(x_n)}{f'(x_n)}$	$x_n - \dfrac{f(x_n)}{f'(x_n)}$
1	3.0000	0.1411	−0.9900	−0.1425	3.1425
2	3.1425	−0.0009	−1.0000	0.0009	3.1416

4. $f(x) = \tan x$

$f'(x) = \sec^2 x$

$x_1 = 0.1$

n	x_n	$f(x_n)$	$f'(x_n)$	$\dfrac{f(x_n)}{f'(x_n)}$	$x_n - \dfrac{f(x_n)}{f'(x_n)}$
1	0.1000	0.1003	1.0101	0.0993	0.0007
2	0.0007	0.0007	1.0000	0.0007	0.0000

5. $f(x) = x^3 + x - 1$

$f'(x) = 3x^2 + 1$

n	x_n	$f(x_n)$	$f'(x_n)$	$\dfrac{f(x_n)}{f'(x_n)}$	$x_n - \dfrac{f(x_n)}{f'(x_n)}$
1	0.5000	-0.3750	1.7500	-0.2143	0.7143
2	0.7143	0.0788	2.5307	0.0311	0.6832
3	0.6832	0.0021	2.4003	0.0009	0.6823

Approximation of the zero of f is 0.682.

6. $f(x) = x^5 + x - 1$

$f'(x) = 5x^4 + 1$

n	x_n	$f(x_n)$	$f'(x_n)$	$\dfrac{f(x_n)}{f'(x_n)}$	$x_n - \dfrac{f(x_n)}{f'(x_n)}$
1	0.5000	-0.4688	1.3125	-0.3571	0.8571
2	0.8571	0.3196	3.6983	0.0864	0.7707
3	0.7707	0.0426	2.7641	0.0154	0.7553
4	0.7553	0.0011	2.6272	0.0004	0.7549

Approximation of the zero of f is 0.755.

7. $f(x) = 3\sqrt{x-1} - x$

$f'(x) = \dfrac{3}{2\sqrt{x-1}} - 1$

n	x_n	$f(x_n)$	$f'(x_n)$	$\dfrac{f(x_n)}{f'(x_n)}$	$x_n - \dfrac{f(x_n)}{f'(x_n)}$
1	1.2000	0.1416	2.3541	0.0602	1.1398
2	1.1398	-0.0181	3.0118	-0.0060	1.1458
3	1.1458	-0.0003	2.9284	-0.0001	1.1459

Approximation of the zero of f is 1.146.

8. $f(x) = x^3 - 3.9x^2 + 4.79x - 1.881$

$f'(x) = 3x^2 - 7.8x + 4.79$

n	x_n	$f(x_n)$	$f'(x_n)$	$\dfrac{f(x_n)}{f'(x_n)}$	$x_n - \dfrac{f(x_n)}{f'(x_n)}$
1	0.5000	−0.3360	1.6400	−0.2049	0.7049
2	0.7049	−0.0921	0.7824	−0.1177	0.8226
3	0.8226	−0.0231	0.4037	−0.0573	0.8799
4	0.8799	−0.0045	0.2495	−0.0181	0.8980
5	0.8980	−0.0004	0.2048	−0.0020	0.9000
6	0.9000	0.0000	0.2000	0.0000	0.9000

Approximation of the zero of f is 0.900.

n	x_n	$f(x_n)$	$f'(x_n)$	$\dfrac{f(x_n)}{f'(x_n)}$	$x_n - \dfrac{f(x_n)}{f'(x_n)}$
1	1.1	0.0000	−0.1600	−0.0000	1.1000

Approximation of the zero of f is 1.100.

n	x_n	$f(x_n)$	$f'(x_n)$	$\dfrac{f(x_n)}{f'(x_n)}$	$x_n - \dfrac{f(x_n)}{f'(x_n)}$
1	1.9	0.0000	0.8000	0.0000	1.9000

Approximation of the zero of f is 1.900.

9. $f(x) = x^4 - 10x^2 - 11$

$f'(x) = 4x^3 - 20x$

n	x_n	$f(x_n)$	$f'(x_n)$	$\dfrac{f(x_n)}{f'(x_n)}$	$x_n - \dfrac{f(x_n)}{f'(x_n)}$
1	3.5000	16.5625	101.5000	0.1632	3.3368
2	3.3368	1.6288	81.8749	0.0199	3.3169
3	3.3169	0.0219	79.6298	0.0003	3.3166

Approximation of the zero of f is 3.317.

10. $f(x) = x^3 + 3$

$f'(x) = 3x^2$

n	x_n	$f(x_n)$	$f'(x_n)$	$\dfrac{f(x_n)}{f'(x_n)}$	$x_n - \dfrac{f(x_n)}{f'(x_n)}$
1	−1.5000	−0.3750	6.7500	−0.0556	−1.4444
2	−1.4444	−0.0134	6.2589	−0.0021	−1.4423
3	−1.4423	−0.0003	6.2407	−0.0001	−1.4422

Approximation of the zero of f is −1.442.

11. $f(x) = x + \sin(x + 1)$

$f'(x) = 1 + \cos(x + 1)$

n	x_n	$f(x_n)$	$f'(x_n)$	$\dfrac{f(x_n)}{f'(x_n)}$	$x_n - \dfrac{f(x_n)}{f'(x_n)}$
1	-0.5000	-0.0206	1.8776	-0.0110	-0.4890
2	-0.4890	0.0000	1.8723	0.0000	-0.4890

Approximation of the zero of f is -0.489.

12. $f(x) = x^3 - \cos x$

$f'(x) = 3x^2 + \sin x$

n	x_n	$f(x_n)$	$f'(x_n)$	$\dfrac{f(x_n)}{f'(x_n)}$	$x_n - \dfrac{f(x_n)}{f'(x_n)}$
1	0.9000	0.1074	3.2133	0.0334	0.8666
2	0.8666	0.0034	3.0151	0.0011	0.8655
3	0.8655	0.0001	3.0087	0.0000	0.8655

Approximation of the zero of f is 0.866.

13. $h(x) = f(x) - g(x) = 2x + 1 - \sqrt{x + 4}$

$h'(x) = 2 - \dfrac{1}{2\sqrt{x + 4}}$

n	x_n	$h(x_n)$	$h'(x_n)$	$\dfrac{h(x_n)}{h'(x_n)}$	$x_n - \dfrac{h(x_n)}{h'(x_n)}$
1	0.6000	0.0552	1.7669	0.0313	0.5687
2	0.5687	-0.0001	1.7661	0.0000	0.5687

Point of intersection of the graphs of f and g occurs when $x \approx 0.569$.

14. $h(x) = f(x) - g(x) = 3 - x - \dfrac{1}{x^2 + 1}$

$h'(x) = -1 + \dfrac{2x}{(x^2 + 1)^2}$

n	x_n	$h(x_n)$	$h'(x_n)$	$\dfrac{h(x_n)}{h'(x_n)}$	$x_n - \dfrac{h(x_n)}{h'(x_n)}$
1	2.9000	-0.0063	-0.9345	0.0067	2.8933
2	2.8933	0.0000	-0.9341	0.0000	2.8933

Point of intersection of the graphs of f and g occurs when $x \approx 2.893$.

15. $h(x) = f(x) - g(x) = x - \tan x$

$h'(x) = 1 - \sec^2 x$

n	x_n	$h(x_n)$	$h'(x_n)$	$\dfrac{h(x_n)}{h'(x_n)}$	$x_n - \dfrac{h(x_n)}{h'(x_n)}$
1	4.5000	-0.1373	-21.5048	0.0064	4.4936
2	4.4936	-0.0039	-20.2271	0.0002	4.4934

Point of intersection of the graphs of f and g occurs when $x \approx 4.493$.

16. $h(x) = f(x) - g(x) = x^2 - \cos x$

$h'(x) = 2x + \sin x$

n	x_n	$h(x_n)$	$h'(x_n)$	$\dfrac{h(x_n)}{h'(x_n)}$	$x_n - \dfrac{h(x_n)}{h'(x_n)}$
1	0.8000	-0.0567	2.3174	-0.0245	0.8245
2	0.8245	0.0009	2.3832	0.0004	0.8241

One point of intersection of the graphs of f and g occurs when $x \approx 0.824$. Since $f(x) = x^2$ and $g(x) = \cos x$ are both symmetric with respect to the y-axis, the other point of intersection occurs when $x \approx -0.824$.

17. Let $g(x) = f(x) - x = \cos x - x$

$g'(x) = -\sin x - 1.$

n	x_n	$g(x_n)$	$g'(x_n)$	$\dfrac{g(x_n)}{g'(x_n)}$	$x_n - \dfrac{g(x_n)}{g'(x_n)}$
1	1.0000	-0.4597	-1.8415	0.2496	0.7504
2	0.7504	-0.0190	-1.6819	0.0113	0.7391
3	0.7391	0.0000	-1.6736	0.0000	0.7391

The fixed point is approximately 0.74.

18. Let $g(x) = f(x) - x = \cot x - x$

$g'(x) = -\csc^2 x - 1.$

n	x_n	$g(x_n)$	$g'(x_n)$	$\dfrac{g(x_n)}{g'(x_n)}$	$x_n - \dfrac{g(x_n)}{g'(x_n)}$
1	1.0000	-0.3579	-2.4123	0.1484	0.8516
2	0.8516	0.0240	-2.7668	-0.0087	0.8603
3	0.8603	0.0001	-2.7403	0.0000	0.8603

The fixed point is approximately 0.86.

19. $f(x) = x^3 - 3x^2 + 3, \quad f'(x) = 3x^2 - 6x$

a.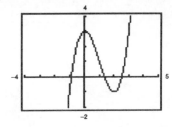

b. $x_1 = 1$

$$x_2 = x_1 - \frac{f(x_1)}{f'(x_1)} \approx 1.333$$

c. $x_1 = \frac{1}{4}$

$$x_2 = x_1 - \frac{f(x_1)}{f'(x_1)} \approx 2.405$$

d.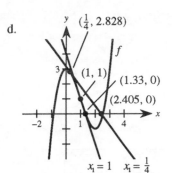

The x-intercepts correspond to the values resulting from the first iteration of Newton's Method.

e. If the initial guess x_1 is not "close to" the desired zero of the function, the x-intercept of the tangent line may approximate another zero of the function.

20. $f(x) = \sin x, \quad f'(x) = \cos x$

a.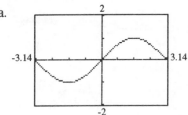

b. $x_1 = 1.8$

$$x_2 = x_1 - \frac{f(x_1)}{f'(x_1)} \approx 6.086$$

c. $x_1 = 3$

$$x_2 = x_1 - \frac{f(x_1)}{f'(x_1)} \approx 3.143$$

d.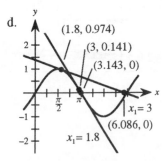

The x-intercepts correspond to the values resulting from the first iteration of Newton's Method.

e. If the initial guess x_1 is not "close to" the desired zero of the function, the x-intercept of the tangent line may approximate another zero of the function.

21. $y = 2x^3 - 6x^2 + 6x - 1 = f(x)$

$y' = 6x^2 - 12x + 6 = f'(x)$

$x_1 = 1$

n	x_n	$f(x_n)$	$f'(x_n)$
1	1	1	0

$f'(x_1) = 0$; therefore, the method fails.

22.　$y = 4x^3 - 12x^2 + 12x - 3 = f(x)$

　　$y' = 12x^2 - 24x + 12 = f'(x)$

　　$x_1 = \dfrac{3}{2}$

n	x_n	$f(x_n)$	$f'(x_n)$	$\dfrac{f(x_n)}{f'(x_n)}$	$x_n - \dfrac{f(x_n)}{f'(x_n)}$
1	$\dfrac{3}{2}$	$\dfrac{3}{2}$	3	$\dfrac{1}{2}$	1
2	1	1	0	—	—

　　$f'(x_2) = 0$; therefore, the method fails.

23.　$y = -x^3 + 3x^2 - x + 1 = f(x)$

　　$y' = -3x^2 + 6x - 1 = f'(x)$

　　$x_1 = 1$

n	x_n	$f(x_n)$	$f'(x_n)$	$\dfrac{f(x_n)}{f'(x_n)}$	$x_n - \dfrac{f(x_n)}{f'(x_n)}$
1	1	2	2	1	0
2	0	1	-1	-1	1
3	1	2	2	1	0

　　Fails to converge

24.　$f(x) = 2\sin x + \cos 2x$

　　$f'(x) = 2\cos x - 2\sin 2x$

　　$x_1 = \dfrac{3\pi}{2}$

n	x_n	$f(x_n)$	$f'(x_n)$
1	$\dfrac{3\pi}{2}$	-3	0

　　Fails because $f'(x_1) = 0$

25.　$f(x) = x^2 - a = 0$

　　$f'(x) = 2x$

　　$\begin{aligned} x_{i+1} &= x_i - \dfrac{x_i^2 - a}{2x_i} \\[2mm] &= \dfrac{2x_i^2 - x_i^2 + a}{2x_i} = \dfrac{x_i^2 + a}{2x_i} \end{aligned}$

26.　$f(x) = x^n - a = 0$

　　$f'(x) = nx^{n-1}$

　　$\begin{aligned} x_{i+1} &= x_i - \dfrac{x_i^n - a}{nx_i^{n-1}} \\[2mm] &= \dfrac{nx_i^n - x_i^n + a}{nx_i^{n-1}} = \dfrac{(n-1)x_i^n + a}{nx_i^{n-1}} \end{aligned}$

27. $x_{i+1} = \dfrac{x_i{}^2 + 7}{2x_i}$

i	1	2	3	4	5
x_i	2.0000	2.7500	2.6477	2.6458	2.6458

$\sqrt{7} \approx 2.646$

28. $x_{i+1} = \dfrac{x_i{}^2 + 5}{2x_i}$

i	1	2	3	4
x_i	2.0000	2.2500	2.2361	2.2361

$\sqrt{5} \approx 2.236$

29. $x_{i+1} = \dfrac{3x_i{}^4 + 6}{4x_i{}^3}$

i	1	2	3	4
x_i	1.5000	1.5694	1.5651	1.5651

$\sqrt[4]{6} \approx 1.565$

30. $x_{i+1} = \dfrac{2x_i{}^3 + 15}{3x_i{}^2}$

i	1	2	3	4
x_i	2.5000	2.4667	2.4662	2.4662

$\sqrt[3]{15} \approx 2.466$

31. $f(x) = \dfrac{1}{x} - a = 0$

$f'(x) = -\dfrac{1}{x^2}$

$x_{n+1} = x_n - \dfrac{(1/x_n) - a}{-1/x_n{}^2} = x_n + x_n{}^2\left(\dfrac{1}{x_n} - a\right) = x_n + x_n - x_n{}^2 a = 2x_n - x_n{}^2 a = x_n(2 - ax_n)$

32. a. $x_{n+1} = x_n(2 - 3x_n)$

i	1	2	3	4
x_i	0.3000	0.3300	0.3333	0.3333

$\frac{1}{3} \approx 0.333$

b. $x_{n+1} = x_n(2 - 11x_n)$

i	1	2	3	4
x_i	0.1000	0.0900	0.0909	0.0909

$\frac{1}{11} \approx 0.091$

33. $f(x) = 1 + \cos x$

   $f'(x) = -\sin x$

n	x_n	$f(x_n)$	$f'(x_n)$	$\dfrac{f(x_n)}{f'(x_n)}$	$x_n - \dfrac{f(x_n)}{f'(x_n)}$
1	3.0000	0.0100	−0.1411	−0.0709	3.0709
2	3.0709	0.0025	−0.0706	−0.0354	3.1063
3	3.1063	0.0006	−0.0353	−0.0176	3.1239
4	3.1239	0.0002	−0.0177	−0.0088	3.1327
5	3.1327	0.0000	−0.0089	−0.0044	3.1371
6	3.1371	0.0000	−0.0045	−0.0022	3.1393
7	3.1393	0.0000	−0.0023	−0.0011	3.1404
8	3.1404	0.0000	−0.0012	−0.0006	3.1410

Approximation of the zero: 3.141

34. $f(x) = \tan x$

   $f'(x) = \sec^2 x$

n	x_n	$f(x_n)$	$f'(x_n)$	$\dfrac{f(x_n)}{f'(x_n)}$	$x_n - \dfrac{f(x_n)}{f'(x_n)}$
1	3.0000	−0.1425	1.0203	−0.1397	3.1397
2	3.1397	−0.0019	1.0000	−0.0019	3.1416
3	3.1416	0.0000	1.0000	0.0000	3.1416

Approximation of the zero: 3.142

35. $f(x) = x \cos x$

   $f'(x) = -x \sin x + \cos x = 0$

Letting $F(x) = f'(x)$, we can use Newton's Method as follows.

   $[F'(x) = -2 \sin x - x \cos x]$

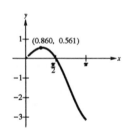
(0.860, 0.561)

n	x_n	$F(x_n)$	$F'(x_n)$	$\dfrac{F(x_n)}{F'(x_n)}$	$x_n - \dfrac{F(x_n)}{F'(x_n)}$
1	0.9000	−0.0834	−2.1261	0.0392	0.8608
2	0.8608	−0.0010	−2.0778	0.0005	0.8603

Approximation to the critical number: 0.860

36. $f(x) = x \sin x$

$f'(x) = x \cos x + \sin x = 0$

The simple root is $x = 0$. Letting $F(x) = f'(x)$, we can use Newton's Method as follows.

$$[F'(x) = 2 \cos x - x \sin x]$$

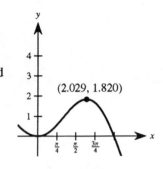

(2.029, 1.820)

n	x_n	$F(x_n)$	$F'(x_n)$	$\dfrac{F(x_n)}{F'(x_n)}$	$x_n - \dfrac{F(x_n)}{F'(x_n)}$
1	2.0000	0.0770	−2.6509	−0.0290	2.0290
2	2.0290	−0.0007	−2.7044	0.0002	2.0288

Approximation to the critical number: 2.029

37. $y = f(x) = 4 - x^2$, $\quad (1, \ 0)$

$d = \sqrt{(x - 1)^2 + (y - 0)^2} = \sqrt{(x - 1)^2 + (4 - x^2)^2} = \sqrt{x^4 - 7x^2 - 2x + 17}$

d is minimized when $d_i = x^4 - 7x^2 - 2x + 17$ is a minimum.

$g(x) = d_i' = 4x^3 - 14x - 2$

$g'(x) = 12x^2 - 14$

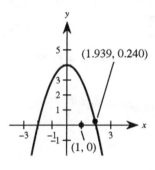

(1.939, 0.240)

(1, 0)

n	x_n	$g(x_n)$	$g'(x_n)$	$\dfrac{g(x_n)}{g'(x_n)}$	$x_n - \dfrac{g(x_n)}{g'(x_n)}$
1	2.0000	2.0000	34.0000	0.0588	1.9412
2	1.9412	0.0830	31.2191	0.0027	1.9385
3	1.9385	−0.0012	31.0934	0.0000	1.9385

$x \approx 1.939$

Point closest to (1, 0) is $\approx$ (1.939, 0.240).

38. $y = f(x) = x^2$, $\quad (4, \ -3)$

$d = \sqrt{(x - 4)^2 + (y + 3)^2} = \sqrt{(x - 4)^2 + (x^2 + 3)^2} = \sqrt{x^4 + 7x^2 - 8x + 25}$

d is minimum when $d_i = x^4 + 7x^2 - 8x + 25$ is minimum.

$g(x) = d_i' = 4x^3 + 14x - 8$

$g'(x) = 12x^2 + 14$

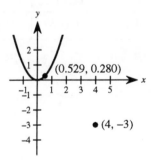

(0.529, 0.280)

• (4, −3)

n	x_n	$g(x_n)$	$g'(x_n)$	$\dfrac{g(x_n)}{g'(x_n)}$	$x_n - \dfrac{g(x_n)}{g'(x_n)}$
1	0.5000	−0.5000	17.0000	−0.0294	0.5294
2	0.5294	0.0051	17.3632	0.0003	0.5291
3	0.5291	−0.0001	17.3594	0.0000	0.5291

$x \approx 0.529$

Point closest to (4, −3) is approximately (0.529, 0.280).

39.

$$\text{Minimize: } T = \frac{\text{Distance rowed}}{\text{Rate rowed}} + \frac{\text{Distance walked}}{\text{Rate walked}}$$

$$T = \frac{\sqrt{x^2 + 4}}{3} + \frac{\sqrt{x^2 - 6x + 10}}{4}$$

$$T' = \frac{x}{3\sqrt{x^2 + 4}} + \frac{x - 3}{4\sqrt{x^2 - 6x + 10}} = 0$$

$$4x\sqrt{x^2 - 6x + 10} = -3(x - 3)\sqrt{x^2 + 4}$$

$$16x^2(x^2 - 6x + 10) = 9(x - 3)^2(x^2 + 4)$$

$$7x^4 - 42x^3 + 43x^2 + 216x - 324 = 0$$

Let $f(x) = 7x^4 - 42x^3 + 43x^2 + 216x - 324$ and $f'(x) = 28x^3 - 126x^2 + 86x + 216$. Since $f(1) = -100$ and $f(2) = 56$, the solution is in the interval $(1, 2)$.

n	x_n	$f(x_n)$	$f'(x_n)$	$\dfrac{f(x_n)}{f'(x_n)}$	$x_n - \dfrac{f(x_n)}{f'(x_n)}$
1	1.7000	19.5887	135.6240	0.1444	1.5556
2	1.5556	−1.0480	150.2780	−0.0070	1.5626
3	1.5626	0.0014	49.5591	0.0000	1.5626

Approximation: $x \approx 1.563$ miles

40. Maximize: $\quad C = \dfrac{3t^2 + t}{50 + t^3}$

$$C' = \frac{-3t^4 - 2t^3 + 300t + 50}{(50 + t^3)^2} = 0$$

Let $\quad f(x) = 3t^4 + 2t^3 - 300t - 50$

$$f'(x) = 12t^3 + 6t^2 - 300.$$

Since $f(4) = -354$ and $f(5) = 575$, the solution is in the interval $(4, 5)$.

n	x_n	$f(x_n)$	$f'(x_n)$	$\dfrac{f(x_n)}{f'(x_n)}$	$x_n - \dfrac{f(x_n)}{f'(x_n)}$
1	4.5000	12.4375	915.0000	0.0136	4.4864
2	4.4864	0.0658	904.3822	0.0001	4.4863

Approximation: $t \approx 4.486$ hours

41. $2,500,000 = -76x^3 + 4830x^2 - 320,000$

$76x^3 - 4830x^2 + 2,820,000 = 0$

Let $f(x) = 76x^3 - 4830x^2 + 2,820,000$

$f'(x) = 228x^2 - 9660x.$

From the graph, choose $x_1 = 40$.

n	x_n	$f(x_n)$	$f'(x_n)$	$\dfrac{f(x_n)}{f'(x_n)}$	$x_n - \dfrac{f(x_n)}{f'(x_n)}$
1	40.0000	−44000.0000	−21600.0000	2.0370	37.9630
2	37.9630	17157.6209	−38131.4039	−0.4500	38.4130
3	38.4130	780.0914	−34642.2263	−0.0225	38.4355
4	38.4355	2.6308	−34465.3435	−0.0001	38.4356

The zero occurs when $x \approx 38.4356$ which corresponds to \$384,356.

42. $170 = 0.808x^3 - 17.974x^2 + 71.248x + 110.843, \quad 1 \le x \le 5$

Let $f(x) = 0.808x^3 - 17.974x^2 + 71.248x - 59.157$

$f'(x) = 2.424x^2 - 35.948x + 71.248.$

From the graph, choose $x_1 = 1$ and $x_1 = 3.5$. Apply Newton's Method.

n	x_n	$f(x_n)$	$f'(x_n)$	$\dfrac{f(x_n)}{f'(x_n)}$	$x_n - \dfrac{f(x_n)}{f'(x_n)}$
1	1.0000	−5.0750	37.7240	−0.1345	1.1345
2	1.1345	−0.2805	33.5849	−0.0084	1.1429
3	1.1429	0.0006	33.3293	0.0000	1.1429

n	x_n	$f(x_n)$	$f'(x_n)$	$\dfrac{f(x_n)}{f'(x_n)}$	$x_n - \dfrac{f(x_n)}{f'(x_n)}$
1	3.5000	4.6725	−24.8760	−0.1878	3.6878
2	3.6878	−0.3286	−28.3550	0.0116	3.6762
3	3.6762	−0.0009	−28.1450	0.0000	3.6762

The zeros occur when $x \approx 1.1429$ and $x \approx 3.6762$. These approximately correspond to engine speeds of 1143 rev/min and 3676 rev/min.

43. $f(x) = \frac{1}{4}x^3 - 3x^2 + \frac{3}{4}x - 2$

$f'(x) = \frac{3}{4}x^2 - 6x + \frac{3}{4}$

Let $x_1 = 12$.

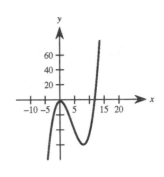

n	x_n	$f(x_n)$	$f'(x_n)$	$\dfrac{f(x_n)}{f'(x_n)}$	$x_n - \dfrac{f(x_n)}{f'(x_n)}$
1	12.0000	7.0000	36.7500	0.1905	11.8095
2	11.8095	0.2151	34.4912	0.0062	11.8033
3	11.8033	0.0015	34.4186	0.0000	11.8033

Approximation: $x \approx 11.803$

44. $f(x) = \sqrt{4 - x^2}\sin(x - 2)$ Domain: $[-2, 2]$

$x = -2$ and $x = 2$ are both zeros.

$$f'(x) = \sqrt{4 - x^2}\cos(x - 2) - \frac{x}{\sqrt{4 - x^2}}\sin(x - 2)$$

Let $x_1 = -1$.

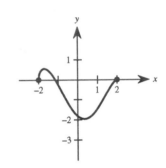

n	x_n	$f(x_n)$	$f'(x_n)$	$\dfrac{f(x_n)}{f'(x_n)}$	$x_n - \dfrac{f(x_n)}{f'(x_n)}$
1	-1.0000	-0.2444	-1.7962	0.1361	-1.1361
2	-1.1361	-0.0090	-1.6498	0.0055	-1.1416
3	-1.1416	0.0000	-1.6422	0.0000	-1.1416

Zeros: $x = \pm 2$, $x \approx -1.142$

45. False. Let $f(x) = \dfrac{x^2 - 1}{x - 1}$. $x = 1$ is a discontinuity. It is not a zero of $f(x)$. This statement would be true if $f(x) = p(x)/q(x)$ is given in **reduced** form.

46. True

47. True

48. True

49. $x = \tan x$

$x = 0$ is a solution.

Let $f(x) = \tan x - x$

$f'(x) = \sec^2 x - 1$.

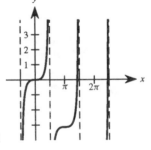

From the graph, it would appear that the second positive solution is in the interval $(7, 8)$. Also, from the graph, you can see that $f(x) > 0$ in the interval $(0, \pi/2)$ and has no solution there. In fact, $f'(x) = \sec^2 x - 1 > 0$ for $x > 0$ implies f is increasing and hence, $\tan x$ is greater than x. Use Newton's Method to find the desired zero. Let $x_1 = 7.8$.

i	1	2	3	4	5	6
x_i	7.8	7.7687	7.7400	7.7270	7.7253	7.7253

Zero: $x \approx 7.7253$

Section 3.9 Differentials

1. $f(x) = x^2$

$f'(x) = 2x$

Tangent line at $(2, 4)$:

$y - f(2) = f'(2)(x - 2)$

$y - 4 = 4(x - 2)$

$y = 4x - 4$

x	1.9	1.99	2	2.01	2.1
$f(x) = x^2$	3.6100	3.9601	4	4.0401	4.4100
$T(x) = 4x - 4$	3.6000	3.9600	4	4.0400	4.4000

2. $f(x) = \dfrac{1}{x^2}$

$f'(x) = -\dfrac{2}{x^3}$

Tangent line at $\left(2, \frac{1}{4}\right)$:

$y - f(2) = f'(2)(x - 2)$

$y - \dfrac{1}{4} = -\dfrac{1}{4}(x - 2)$

$y = -\dfrac{1}{4}x + \dfrac{3}{4}$

x	1.9	1.99	2	2.01	2.1
$f(x) = \dfrac{1}{x^2}$	0.2770	0.2525	0.2500	0.2475	0.2268
$T(x) = -\dfrac{1}{4}x + \dfrac{3}{4}$	0.2750	0.2525	0.2500	0.2475	0.2250

3. $f(x) = x^5$

$f'(x) = 5x^4$

Tangent line at $(2, 32)$:

$y - f(2) = f'(2)(x - 2)$

$y - 32 = 80(x - 2)$

$y = 80x - 128$

x	1.9	1.99	2	2.01	2.1
$f(x) = x^5$	24.7610	31.2080	32	32.8080	40.8410
$T(x) = 80x - 128$	24.0000	31.2000	32	32.8000	40.0000

4. $f(x) = \sqrt{x}$

$f'(x) = \dfrac{1}{2\sqrt{x}}$

Tangent line at $(2, \sqrt{2})$:

$y - f(2) = f'(2)(x - 2)$

$y - \sqrt{2} = \dfrac{1}{2\sqrt{2}}(x - 2)$

$y = \dfrac{x}{2\sqrt{2}} + \dfrac{1}{\sqrt{2}}$

x	1.9	1.99	2	2.01	2.1
$f(x) = \sqrt{x}$	1.3784	1.4107	1.4142	1.4177	1.4491
$T(x) = \dfrac{x}{2\sqrt{2}} + \dfrac{1}{\sqrt{2}}$	1.3789	1.4107	1.4142	1.4177	1.4496

5. $f(x) = \sin x$

$f'(x) = \cos x$

Tangent line at $(2, \sin 2)$:

$y - f(2) = f'(2)(x - 2)$

$y - \sin 2 = (\cos 2)(x - 2)$

$y = (\cos 2)(x - 2) + \sin 2$

x	1.9	1.99	2	2.01	2.1
$f(x) = \sin x$	0.9463	0.9134	0.9093	0.9051	0.8632
$T(x) = (\cos 2)(x - 2) + \sin 2$	0.9509	0.9135	0.9093	0.9051	0.8677

6. $f(x) = \csc x$

$f'(x) = -\csc x \cot x$

Tangent line at $(2, \csc 2)$:

$y - f(2) = f'(2)(x - 2)$

$y - \csc 2 = (-\csc 2 \cot 2)(x - 2)$

$y = (-\csc 2 \cot 2)(x - 2) + \csc 2$

x	1.9	1.99	2	2.01	2.1
$f(x) = \csc x$	1.0567	1.0948	1.0998	1.1049	1.1585
$T(x) = (-\csc 2 \cot 2)(x - 2) + \csc 2$	1.0494	1.0947	1.0998	1.1048	1.1501

7. $y = f(x) = x^3$, $f'(x) = 3x^2$, $x = 1$, $\Delta x = dx = 0.1$

$\Delta y = f(x + \Delta x) - f(x)$ $dy = f'(x)\,dx$

$= f(1.1) - f(1)$ $= f'(1)(0.1)$

$= (1.1)^3 - (1)^3$ $= 3(0.1)$

$= 0.331$ $= 0.3$

8. $y = f(x) = 1 - 2x^2, \; f'(x) = -4x, \; x = 0, \; \Delta x = dx = -0.1$

$\Delta y = f(x + \Delta x) - f(x)$ $dy = f'(x)\,dx$

     $= f(-0.1) - f(0)$ $= f'(0)(-0.1)$

     $= [1 - 2(-0.1)^2] - [1 - 2(0)^2] = -0.02$ $= (0)(-0.1) = 0$

9. $y = f(x) = x^4 + 1, \; f'(x) = 4x^3, \; x = -1, \; \Delta x = dx = 0.01$

$\Delta y = f(x + \Delta x) - f(x)$ $dy = f'(x)\,dx$

     $= f(-0.99) - f(-1)$ $= f'(-1)(0.01)$

     $= [(-0.99)^4 + 1] - [(-1)^4 + 1] \approx -0.0394$ $= (-4)(0.01) = -0.04$

10. $y = f(x) = 2x + 1, \; f'(x) = 2, \; x = 2, \; \Delta x = dx = 0.01$

$\Delta y = f(x + \Delta x) - f(x)$ $dy = f'(x)\,dx$

     $= f(2.01) - f(2)$ $= f'(2)(0.01)$

     $= [2(2.01) + 1] - [2(2) + 1] = 0.02$ $= 2(0.01) = 0.02$

11. $y = 3x^2 - 4$ **12.** $y = 2x^{3/2}$ **13.** $y = \dfrac{x + 1}{2x - 1}$ **14.** $y = \sqrt{x^2 - 4}$

    $dy = 6x\,dx$ $dy = 3\sqrt{x}\,dx$ $dy = \dfrac{-3}{(2x - 1)^2}\,dx$ $dy = \dfrac{x}{\sqrt{x^2 - 4}}\,dx$

15. $y = x\sqrt{1 - x^2}$ **16.** $y = \sqrt{x} + \dfrac{1}{\sqrt{x}}$

    $dy = \left(x\,\dfrac{-x}{\sqrt{1 - x^2}} + \sqrt{1 - x^2}\right) dx = \dfrac{1 - 2x^2}{\sqrt{1 - x^2}}\,dx$ $dy = \left(\dfrac{1}{2\sqrt{x}} - \dfrac{1}{2x\sqrt{x}}\right) dx = \dfrac{x - 1}{2x\sqrt{x}}\,dx$

17. $y = \dfrac{\sec^2 x}{x^2 + 1}$ **18.** $y = x \sin x$

    $dy = \left[\dfrac{(x^2 + 1)2\sec^2 x \tan x - \sec^2 x(2x)}{(x^2 + 1)^2}\right] dx$ $dy = (x \cos x + \sin x)\,dx$

       $= \left[\dfrac{2\sec^2 x(x^2 \tan x + \tan x - x)}{(x^2 + 1)^2}\right] dx$

19. $y = \dfrac{1}{3}\cos\left(\dfrac{6\pi x - 1}{2}\right)$ **20.** $y = x - \tan^2 x$

    $dy = -\pi \sin\left(\dfrac{6\pi x - 1}{2}\right) dx$ $dy = (1 - 2\tan x \sec^2 x)\,dx$

21. $A = x^2$ **22.** $A = \frac{1}{2}bh, \; b = 36, \; h = 50$

      $x = 12$ $db = dh = \pm 0.25$

      $\Delta x = dx = \pm\frac{1}{64}$ $dA = \frac{1}{2}b\,dh + \frac{1}{2}h\,db$

      $dA = 2x\,dx$ $\Delta A \approx dA = \frac{1}{2}(36)(\pm 0.25) + \frac{1}{2}(50)(\pm 0.25)$

      $\Delta A \approx dA = 2(12)\left(\pm\frac{1}{64}\right)$ $= \pm 10.75$ square centimeters

           $= \pm\frac{3}{8}$ square inches

23. $A = \pi r^2$

$r = 14$

$\Delta r = dr = \pm\frac{1}{4}$

$\Delta A \approx dA = 2\pi r \, dr = \pi(28)\left(\pm\frac{1}{4}\right)$

$\quad = \pm 7\pi$ square inches

24. $x = 12$ inches

$\Delta x = dx = \pm 0.03$ inch

a. $V = x^3$

$dV = 3x^2 \, dx = 3(12)^2(\pm 0.03)$

$\quad = \pm 12.96$ cubic inches

b. $S = 6x^2$

$dS = 12x \, dx = 12(12)(\pm 0.03)$

$\quad = \pm 4.32$ square inches

25. a. $x = 15$ centimeter

$\Delta x = dx = \pm 0.05$ centimeters

$A = x^2$

$dA = 2x \, dx = 2(15)(\pm 0.05)$

$\quad = \pm 1.5$ square centimeters

Percentage error:

$\frac{dA}{A} = \frac{\pm 1.5}{(15)^2} = 0.00666\ldots = \frac{2}{3}\%$

b. $\frac{dA}{A} = \frac{2x \, dx}{x^2} = \frac{2 \, dx}{x} \le 0.025$

$\frac{dx}{x} \le \frac{0.025}{2} = 0.0125 = 1.25\%$

26. a. $C = 56$ centimeters

$\Delta C = dC = \pm 1.2$ centimeters

$C = 2\pi r \Rightarrow r = \frac{C}{2\pi}$

$A = \pi r^2 = \pi\left(\frac{C}{2\pi}\right)^2 = \frac{1}{4\pi}C^2$

$dA = \frac{1}{2\pi}C \, dC = \frac{1}{2\pi}(56)(\pm 1.2) = \frac{33.6}{\pi}$

$\frac{dA}{A} = \frac{33.6/\pi}{[1/(4\pi)](56)^2} \approx 0.042857 = 4.2857\%$

b. $\frac{dA}{A} = \frac{(1/2\pi)C \, dC}{(1/4\pi)C^2} = \frac{2 \, dC}{C} \le 0.03$

$\frac{dC}{C} \le \frac{0.03}{2} = 0.015 = 1.5\%$

27. $r = 6$ inches

$\Delta r = dr = \pm 0.02$ inches

a. $V = \frac{4}{3}\pi r^3$

$dV = 4\pi r^2 \, dr = 4\pi(6)^2(\pm 0.02) = \pm 2.88\pi$ cubic inches

b. $S = 4\pi r^2$

$dS = 8\pi r \, dr = 8\pi(6)(\pm 0.02) = \pm 0.96\pi$ square inches

c. Relative error: $\frac{dV}{V} = \frac{4\pi r^2 \, dr}{(4/3)\pi r^3} = \frac{3 \, dr}{r} = \frac{3}{6}(0.02) = 0.01 = 1\%$

Relative error: $\frac{dS}{S} = \frac{8\pi r \, dr}{4\pi r^2} = \frac{2 \, dr}{r} = \frac{2(0.02)}{6} = 0.00666\ldots = \frac{2}{3}\%$

28. $P = (500x - x^2) - \left(\frac{1}{2}x^2 - 77x + 3000\right)$, x changes from 115 to 120

$dP = (500 - 2x - x + 77) \, dx = (577 - 3x) \, dx = [577 - 3(115)](120 - 115) = 1160$

Approximate percentage change: $\frac{dP}{P}(100) = \frac{1160}{43517.50}(100) \approx 2.7\%$

29. $V = \pi r^2 h = 40\pi r^2$, $r = 5$ cm, $h = 40$ cm, $dr = 0.2$ cm

$\Delta V \approx dV = 80\pi r \, dr = 80\pi(5)(0.2) = 80\pi$ cm^3

30. $V = \frac{4}{3}\pi r^3$, $r = 100$ cm, $dr = 0.2$ cm

$\Delta V \approx dV = 4\pi r^2 \, dr = 4\pi (100)^2 (0.2) = 8000\pi$ cm^3

31. a. $T = 2\pi \sqrt{L/g}$

$dT = \dfrac{\pi}{g\sqrt{L/g}} dL$

Relative error:

$\dfrac{dT}{T} = \dfrac{(\pi \, dL)/(g\sqrt{L/g})}{2\pi \sqrt{L/g}}$

$= \dfrac{dL}{2L}$

$= \dfrac{1}{2}$(relative error in L)

$= \dfrac{1}{2}(0.005) = 0.0025$

Percentage error:

$\dfrac{dT}{T}(100) = 0.25\% = \dfrac{1}{4}\%$

b. $(0.0025)(3600)(24) = 216$ seconds

$= 3.6$ minutes

32. $T = 2\pi \sqrt{L/g}$

$T^2 = 4\pi^2 \left(\dfrac{L}{g} \right)$

$g = \dfrac{4\pi^2 L}{T^2}$

$dg = 4\pi^2 L \dfrac{(-2)}{T^3} dT$

Relative error in g:

$\dfrac{dg}{g} = \dfrac{(-8\pi^2 L \, dT)/T^3}{(4\pi^2 L)/T^2}$

$= -2\dfrac{dT}{T} = -2$(relative error in period)

$= -2(0.001) = -0.002$

Percentage error in g:

$\left| \dfrac{dg}{g} \right| (100) = \dfrac{1}{5}\%$

33. $E = IR$

$R = \dfrac{E}{I}$

$dR = -\dfrac{E}{I^2} dI$

$\dfrac{dR}{R} = \dfrac{-(E/I^2)\, dI}{E/I} = -\dfrac{dI}{I}$

$\left| \dfrac{dR}{R} \right| = \left| -\dfrac{dI}{I} \right| = \left| \dfrac{dI}{I} \right|$

34. $C = \dfrac{80,000p}{100 - p}$, $0 \le p < 100$

a. $p = 40$

$\Delta p = dp = 2$

$dC = \dfrac{8,000,000}{(100 - p)^2} dp$

$= \dfrac{8,000,000}{(100 - 40)^2}(2) \approx \$4444.44$

b. $p = 75$

$\Delta p = dp = 2$

$dC = \dfrac{8,000,000}{(100 - 75)^2}(2) = \$25,600.00$

35. $\theta = 26°45' = 26.75°$

$d\theta = \pm 15' = \pm 0.25°$

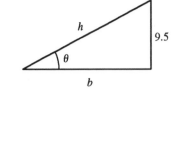

a. $h = 9.5 \csc \theta$

$dh = -9.5 \csc \theta \cot \theta \, d\theta$

$\dfrac{dh}{h} = -\cot \theta \, d\theta$

$\left| \dfrac{dh}{h} \right| = (\cot 26.75°)(0.25°)$

Converting to radians, $(\cot 0.4669)(0.0044) \approx 0.0087 = 0.87\%$ (in radians).

b. $\left| \dfrac{dh}{h} \right| = \cot \theta \, d\theta \le 0.02$

$\dfrac{d\theta}{\theta} \le \dfrac{0.02}{\theta(\cot \theta)} = \dfrac{0.02 \tan \theta}{\theta}$

$\dfrac{d\theta}{\theta} \le \dfrac{0.02 \tan 26.75°}{26.75°} \approx \dfrac{0.02 \tan 0.4669}{0.4669} \approx 0.0216 = 2.16\%$ (in radians)

36. See Exercise 35.

$A = \dfrac{1}{2}(\text{base})(\text{height}) = \dfrac{1}{2}(9.5 \cot \theta)(9.5) = 45.125 \cot \theta$

$dA = -45.125 \csc^2 \theta \, d\theta$

$\left| \dfrac{dA}{A} \right| = \dfrac{\csc^2 \theta \, d\theta}{\cot \theta} = \dfrac{d\theta}{\sin \theta \cos \theta}$

$= \dfrac{0.25°}{(\sin 26.75°)(\cos 26.75°)} \approx \dfrac{0.0044}{(\sin 0.4669)(\cos 0.4669)} \approx 0.0109 = 1.09\%$ (in radians)

37. $r = \dfrac{v_0{}^2}{32}(\sin 2\theta)$

$v_0 = 2200$ ft/sec

θ changes from 10° to 11°

$dr = \dfrac{(2200)^2}{16}(\cos 2\theta) \, d\theta$

$\theta = 10\left(\dfrac{\pi}{180}\right)$

$d\theta = (11 - 10)\dfrac{\pi}{180}$

$\Delta r \approx dr$

$= \dfrac{(2200)^2}{16} \cos\left(\dfrac{20\pi}{180}\right)\left(\dfrac{\pi}{180}\right)$

≈ 4961 feet

38. $h = 50 \tan \theta$

$\theta = 71.5°$ (see figure)

$dh = 50 \sec^2 \theta \, d\theta$

$\dfrac{dh}{h} = \dfrac{\sec^2 71.5°}{\tan 71.5°} \, d\theta \le \pm 0.06$

$d\theta \le \pm 0.018°$

39. Let $f(x) = \sqrt{x}$, $x = 100$, $dx = -0.6$.

$$f(x + \Delta x) \approx f(x) + f'(x)\,dx$$

$$= \sqrt{x} + \frac{1}{2\sqrt{x}}\,dx$$

$$f(x + \Delta x) = \sqrt{99.4}$$

$$\approx \sqrt{100} + \frac{1}{2\sqrt{100}}(-0.6) = 9.97$$

Using a calculator: $\sqrt{99.4} \approx 9.96995$

40. Let $f(x) = \sqrt[3]{x}$, $x = 27$, $dx = 1$.

$$f(x + \Delta x) \approx f(x) + f'(x)\,dx$$

$$= \sqrt[3]{x} + \frac{1}{3\sqrt[3]{x^2}}\,dx$$

$$f(x + \Delta x) = \sqrt[3]{28}$$

$$\approx \sqrt[3]{27} + \frac{1}{3(\sqrt[3]{27})^2}(1)$$

$$= 3 + \frac{1}{27} \approx 3.0370$$

Using a calculator: $\sqrt[3]{28} \approx 3.0366$

41. Let $f(x) = \sqrt[4]{x}$, $x = 625$, $dx = -1$.

$$f(x + \Delta x) \approx f(x) + f'(x)\,dx$$

$$= \sqrt[4]{x} + \frac{1}{4\sqrt[4]{x^3}}\,dx$$

$$f(x + \Delta x) = \sqrt[4]{624}$$

$$\approx \sqrt[4]{625} + \frac{1}{4(\sqrt[4]{625})^3}(-1)$$

$$= 5 - \frac{1}{500} = 4.998$$

Using a calculator: $\sqrt[4]{624} \approx 4.9980$

42. Let $f(x) = x^3$, $x = 3$, $dx = -0.01$.

$$f(x + \Delta x) \approx f(x) + f'(x)\,dx$$

$$= x^3 + 3x^2\,dx$$

$$f(x + \Delta x) = (2.99)^3 \approx 3^3 + 3(3)^2(-0.01)$$

$$= 27 - 0.27 = 26.73$$

Using a calculator: $(2.99)^3 \approx 26.7309$

43. Let $f(x) = x^4$, $x = 1$, $dx = -0.01$, $f'(x) = 4x^3$. Then

$$f(0.99) \approx f(1) + f'(1)\,dx$$

$$(0.99)^4 \approx (1)^4 + 4(1)^3(-0.01) = 1 - 4(0.01).$$

44. Let $f(x) = \sqrt{x}$, $x = 4$, $dx = 0.02$, $f'(x) = 1/(2\sqrt{x})$. Then

$$f(4.02) \approx f(4) + f'(4)\,dx$$

$$\sqrt{4.02} \approx \sqrt{4} + \frac{1}{2\sqrt{4}}(0.02) = 2 + \frac{1}{4}(0.02).$$

45. Let $f(x) = \sec x$, $x = 0$, $dx = 0.03$, $f'(x) = \sec x \tan x$. Then

$$f(0.03) \approx f(0) + f'(0)\,dx$$

$$\sec 0.03 \approx \sec 0 + (\sec 0 \tan 0)(0.03)$$

$$= 1 + 0(0.03).$$

46. Let $f(x) = \tan x$, $x = 0$, $dx = 0.05$ $f'(x) = \sec^2 x$. Then

$$f(0.05) \approx f(0) + f'(0)\,dx$$

$$\tan 0.05 \approx \tan 0 + \sec^2 0(0.05) = 0 + 1(0.05).$$

47. True

48. True, $\dfrac{\Delta y}{\Delta x} = \dfrac{dy}{dx} = a$

49. True

50. False

Let $f(x) = \sqrt{x}$, $x = 1$, and $\Delta x = dx = 3$. Then

$$\Delta y = f(x + \Delta x) - f(x) = f(4) - f(1) = 1$$

and

$$dy = f'(x)\,dx = \frac{1}{2\sqrt{1}}(3) = \frac{3}{2}.$$

Thus, $dy > \Delta y$ in this example.

51. $A(x) = x^2$

a. $dA = 2x\,dx = 2x\Delta x$

$$\Delta A = (x + \Delta x)^2 - x^2$$

$$= 2x\Delta x + (\Delta x)^2$$

b.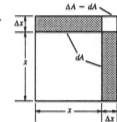

c. $\Delta A - dA = (\Delta x)^2$

52. Let $\epsilon = (\Delta y/\Delta x) - f'(x)$. Then $\Delta x \epsilon = \Delta y - f'(x)\Delta x$. Since $\Delta x = dx$, we have $\Delta y - f'(x)\,dx = \Delta x \epsilon$ or $\Delta y - dy = \epsilon \Delta x$ where $\epsilon \to 0$ as $\Delta x \to 0$.

53. $f(x) = \dfrac{1}{1 - x}$

$$f'(x) = \frac{1}{(1-x)^2}$$

Tangent line at $(0, 1)$:

$$y - f(0) = f'(0)(x - 0)$$

$$y - 1 = 1(x - 0)$$

$$y = 1 + x$$

$g(x) = 1 + x$ is the tangent line approximation of $f(x)$ for x near 0.

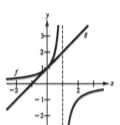

Section 3.10   Business and Economics Applications

1. a. $C(0)$ represents the fixed costs.

c. The marginal cost function has a relative minimum. It occurs when production costs are increasing at their slowest rate.

b.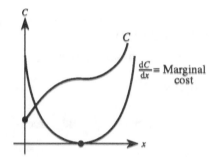

2. a.

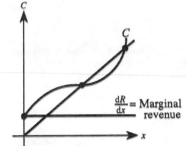

$\dfrac{dR}{dx}$ = Marginal revenue

b.

Maximum profit

3. $R = 900x - 0.1x^2$

$\dfrac{dR}{dx} = 900 - 0.2x = 0$ when $x = 4500$.

By the First Derivative Test, $x = 4500$ is a maximum.

4. $R = 600x^2 - 0.02x^3$

$\dfrac{dR}{dx} = 1200x - 0.06x^2 = 6x(200 - 0.01x) = 0$ when $x = 0, \quad 20{,}000$.

By the First Derivative Test, $x = 20{,}000$ is a maximum.

5.
$$R = \frac{1{,}000{,}000x}{0.02x^2 + 1800}$$

$$\frac{dR}{dx} = 1{,}000{,}000 \left[\frac{0.02x^2 + 1800 - x(0.04x)}{(0.02x^2 + 1800)^2} \right] = 0$$

$1800 - 0.02x^2 = 0$ when $x = 300$.

By the First Derivative Test, $x = 300$ is a maximum.

6. $R = 30x^{2/3} - 2x$

$\dfrac{dR}{dx} = 20x^{-1/3} - 2 = \dfrac{20}{x^{1/3}} - 2 = 0$ when $x = 1000$.

By the First Derivative Test, $x = 1000$ is a maximum.

7. $\overline{C} = 0.125x + 20 + \dfrac{5000}{x}$

$\dfrac{d\overline{C}}{dx} = 0.125 - \dfrac{5000}{x^2} = 0$ when $x = 200$.

By the First Derivative Test, $x = 200$ yields the minimum average cost.

8. $\overline{C} = 0.001x^2 - 5 + \dfrac{250}{x}$

$\dfrac{d\overline{C}}{dx} = 0.002x - \dfrac{250}{x^2} = 0$ when $x = 50$.

By the First Derivative Test, $x = 50$ yields the minimum average cost.

9. $\overline{C} = 3000 - x(300 - x)^{1/2}$

$$\frac{d\overline{C}}{dx} = -x\left(\frac{1}{2}\right)(300 - x)^{-1/2}(-1) - (300 - x)^{1/2} = -\frac{3}{2}(300 - x)^{-1/2}(200 - x) = 0 \text{ when } x = 200.$$

By the First Derivative Test, $x = 200$ yields the minimum average cost.

10. $\overline{C} = \dfrac{2x^2 - x + 5000}{x^2 + 2500}$

$$\frac{d\overline{C}}{dx} = \frac{(x^2 + 2500)(4x - 1) - (2x^2 - x + 5000)(2x)}{(x^2 + 2500)^2} = \frac{x^2 - 2500}{(x^2 + 2500)^2} = 0 \text{ when } x = 50.$$

By the First Derivative Test, $x = 50$ yields the minimum average cost.

11. $C = 100 + 30x$

    $p = 90 - x$

    $P = xp - C$

      $= 90x - x^2 - 30x - 100$

      $= -x^2 + 60x - 100$

  $\dfrac{dP}{dx} = -2x + 60$

      $= 0$ when $x = 30$, so $p = 60$.

By the First Derivative Test, $x = 30$ is a maximum.

12. $C = 2400x + 5200$

    $p = 6000 - 0.4x^2$

    $P = xp - C$

      $= (6000x - 0.4x^3) - (2400x + 5200)$

      $= -0.4x^3 + 3600x - 5200$

  $\dfrac{dP}{dx} = -1.2x^2 + 3600$

      $= 0$ when $x \approx 55$, so $p \approx 4800$.

By the First Derivative Test, $x \approx 55$ is a maximum.

13. $C = 4000 - 40x + 0.02x^2$

    $p = 50 - 0.01x$

    $P = xp - C = 50x - 0.01x^2 - 4000 + 40x - 0.02x^2 = -0.03x^2 + 90x - 4000$

  $\dfrac{dP}{dx} = -0.06x + 90 = 0$ when $x = 1500$, so $p = 35$.

By the First Derivative Test, $x = 1500$ is a maximum.

14. $C = 35x + 2\sqrt{x - 1}$

    $p = 40 - \sqrt{x - 1}$

    $P = xp - C = 40x - x\sqrt{x - 1} - 35x - 2\sqrt{x - 1} = 5x - (x + 2)\sqrt{x - 1}$

  $\dfrac{dP}{dx} = 5 - \left[(x + 2)\left(\dfrac{1}{2}\right)(x - 1)^{-1/2} + (x - 1)^{1/2}\right]$

      $= 5 - \left[\dfrac{x + 2 + 2(x - 1)}{2\sqrt{x - 1}}\right] = 5 - \dfrac{3x}{2\sqrt{x - 1}} = \dfrac{10\sqrt{x - 1} - 3x}{2\sqrt{x - 1}} = 0$ when $x = 10$, so $p = 37$.

By the First Derivative Test, $x = 10$ is a maximum.

15.
$$C = 2x^2 + 5x + 18$$

Average cost $= \dfrac{C}{x} = \overline{C} = 2x + 5 + \dfrac{18}{x}$

$\dfrac{d\overline{C}}{dx} = 2 - \dfrac{18}{x^2} = 0$ when $x = 3$

$\overline{C}(3) = 6 + 5 + 6 = 17$

By the First Derivative Test, $x = 3$ is a minimum.

Marginal cost: $\dfrac{dC}{dx} = 4x + 5$

At $x = 3$: $\dfrac{dC}{dx} = 17 = \overline{C}(3)$

16.
$$C = x^3 - 6x^2 + 13x$$

Average cost $= \dfrac{C}{x} = \overline{C} = x^2 - 6x + 13$

$\dfrac{d\overline{C}}{dx} = 2x - 6 = 0$ when $x = 3$

$\dfrac{dC}{dx} = 3x^2 - 12x + 13$ when $x = 3$

Marginal cost $= \dfrac{dC}{dx} = 27 - 36 + 13 = 4$

Average cost $= \overline{C} = 9 - 18 + 13 = 4$

17. Average cost: $\overline{C}(x) = \dfrac{C(x)}{x}$

$\dfrac{d\overline{C}}{dx} = \dfrac{xC'(x) - C(x)}{x^2} = 0 \Rightarrow xC'(x) - C(x) = 0$ when $C'(x) = \dfrac{C(x)}{x} = \overline{C}(x)$.

Marginal cost = average cost

This condition will yield a minimum (if it exists).

18. $P = 230 + 20s - \dfrac{1}{2}s^2$

$\dfrac{dP}{ds} = 20 - s = 0$ when $s = 20$.

P is maximum when advertising is \$2000 since $d^2P/ds^2 = -1 < 0$.

19. a.

Order size, x	Price	Profit
102	$90 - 2(0.15)$	$102[90 - 2(0.15)] - 102(60) = 3029.40$
104	$90 - 4(0.15)$	$104[90 - 4(0.15)] - 104(60) = 3057.60$
106	$90 - 6(0.15)$	$106[90 - 6(0.15)] - 106(60) = 3084.60$
108	$90 - 8(0.15)$	$108[90 - 8(0.15)] - 108(60) = 3110.40$

e.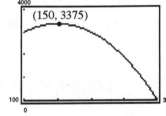
(150, 3375)

b. $P = x[90 - (x - 100)(0.15)] - x(60)$

$= x(90 - 0.15x + 15 - 60) = x(45 - 0.15x), \ x \geq 100$

c. $\dfrac{dP}{dx} = x(-0.15) + (45 - 0.15x)(1)$

$= 45 - 0.30x = 0$ when $x = 150$.

Since $d^2P/dx^2 = -0.30 < 0$, an order size of $x = 150$ units yields a maximum profit.

d.

Order size, x	Price	Profit
146	$90 - 46(0.15)$	$146[90 - 46(0.15)] - 146(60) = 3372.60$
148	$90 - 48(0.15)$	$148[90 - 48(0.15)] - 148(60) = 3374.40$
150	$90 - 50(0.15)$	$150[90 - 50(0.15)] - 150(60) = 3375.00$
152	$90 - 52(0.15)$	$152[90 - 52(0.15)] - 152(60) = 3374.40$
154	$90 - 54(0.15)$	$154[90 - 54(0.15)] - 154(60) = 3372.60$

20. x = Number of \$30 increases in rent

Profit = Revenue − Cost

P = (Number of apartments rented)(Rent/unit) − (Number of apartments rented)(\$36)

$= (50 - x)(540 + 30x) - (50 - x)(36) = (50 - x)(504 + 30x)$

$\dfrac{dP}{dx} = (50 - x)(30) + (504 + 30x)(-1)$

$\phantom{\dfrac{dP}{dx}} = 996 - 60x = 0$ when $x = 16.6 \approx 17$ $\left(\text{maximum since } \dfrac{d^2 P}{dx^2} = -60 < 0\right)$

Maximum profit when there are 17 increases of \$30 in rent and the rent is $540 + 30(17) = \$1050$.

21. Total cost = (Cost per hour)(Number of hours)

$$T = \left(\frac{v^2}{600} + 5\right)\left(\frac{110}{v}\right) = \frac{11v}{60} + \frac{550}{v}$$

$$\frac{dT}{dv} = \frac{11}{60} - \frac{550}{v^2} = \frac{11v^2 - 33,000}{60v^2}$$

$= 0$ when $v = \sqrt{3000} = 10\sqrt{30} \approx 54.8$ mph.

$$\frac{d^2 T}{dv^2} = \frac{1100}{v^3} > 0 \text{ when } v = 10\sqrt{30} \text{ so this value yields a minimum.}$$

22. Total cost = (Cost per hour)(Number of hours)

$$T = \left(\frac{v^2}{500} + 7.50\right)\left(\frac{110}{v}\right) = \frac{11v}{50} + \frac{825}{v}$$

$$\frac{dT}{dv} = \frac{11}{50} - \frac{825}{v^2} = \frac{11v^2 - 41,250}{50v^2}$$

$= 0$ when $v = \sqrt{3750} = 25\sqrt{6} \approx 61.2$ mph.

$$\frac{d^2 T}{dv^2} = \frac{1650}{v^3} > 0 \text{ when } v = 25\sqrt{6} \text{ so this value yields a minimum.}$$

23. Let T be the total cost.

$$T = 8(5280)\sqrt{x^2 + (1/4)} + 6(5280)(6 - x)$$

$$= 2(5280)[4\sqrt{x^2 + (1/4)} + 18 - 3x]$$

$$\frac{dT}{dx} = 2(5280)\left[\frac{4(2x)}{2\sqrt{x^2 + (1/4)}} - 3\right]$$

$$= 2(5280)\left[\frac{4x - 3\sqrt{x^2 + (1/4)}}{\sqrt{x^2 + (1/4)}}\right] = 0$$

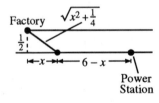

When $4x - 3\sqrt{x^2 + (1/4)} = 0$, $x = 3/(2\sqrt{7}) \approx 0.57$ mile.

By the First Derivative Test, $x \approx 0.57$ is a minimum.

24. Let K be the cost per mile for laying pipe on land and T be the total cost.

$$T = 2K\sqrt{x^2 + 1} + K(2 - x) = K(2\sqrt{x^2 + 1} + 2 - x)$$

$$\frac{dT}{dx} = K\left[\frac{2x}{\sqrt{x^2 + 1}} - 1\right]$$

$$= 0 \text{ when } 2x - \sqrt{x^2 + 1} = 0, \quad x = \frac{1}{\sqrt{3}} \approx 0.58 \text{ mile}$$

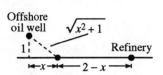

By the First Derivative Test, $x = 1/\sqrt{3}$ is a minimum.

25. $S_1 = (4m - 1)^2 + (5m - 6)^2 + (10m - 3)^2$

$$\frac{dS_1}{dm} = 2(4m - 1)(4) + 2(5m - 6)(5) + 2(10m - 3)(10) = 282m - 128 = 0 \text{ when } m = \frac{64}{141}.$$

Line: $y = \frac{64}{141}x$

$$S = \left|4\left(\frac{64}{141}\right) - 1\right| + \left|5\left(\frac{64}{141}\right) - 6\right| + \left|10\left(\frac{64}{141}\right) - 3\right|$$

$$= \left|\frac{256}{141} - 1\right| + \left|\frac{320}{141} - 6\right| + \left|\frac{640}{141} - 3\right| = \frac{858}{141} \approx 6.1 \text{ mi}$$

26. $S_2 = |4m - 1| + |5m - 6| + |10m - 3|$

Using a graphing utility, you can see that the minimum occurs when $m = 0.3$.

Line $y = 0.3x$

$$S_2 = |4(0.3) - 1| + |5(0.3) - 6| + |10(0.3) - 3| = 4.7 \text{ mi}.$$

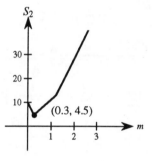

27. $S_3 = \dfrac{|4m - 1|}{\sqrt{m^2 + 1}} + \dfrac{|5m - 6|}{\sqrt{m^2 + 1}} + \dfrac{|10m - 3|}{\sqrt{m^2 + 1}}$

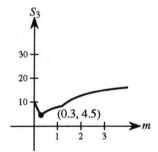

Using a graphing utility, you can see that the minimum occurs when $x \approx 0.3$.

Line: $y \approx 0.3x$

$$S_3 = \frac{|4(0.3) - 1| + |5(0.3) - 6| + |10(0.3) - 3|}{\sqrt{(0.3)^2 + 1}}$$

$$\approx 4.5 \text{ mi}.$$

28. By the methods of calculus, finding the minimum of the sums of the squares of the lengths of the vertical feeder lines is easiest (see Exercise 25). The minimum of the sum of the perpendicular distances best meets the objective of minimizing the amount of feeder line required (see Exercise 27). This minimum is easily found by using a graphing utility.

29. Let d be the amount deposited in the bank, i be the interest rate paid by the bank, and P be the profit.

$$P = (0.12)d - id$$

$$d = ki^2 \text{ (since } d \text{ is proportional to } i^2)$$

$$P = (0.12)(ki^2) - i(ki^2) = k(0.12i^2 - i^3)$$

$$\frac{dP}{di} = k(0.24i - 3i^2) = 0 \text{ when } i = \frac{0.24}{3} = 0.08$$

$$\frac{d^2 P}{di^2} = k(0.24 - 6i) < 0 \text{ when } i = 0.08 \quad (\textbf{Note: } k > 0)$$

The profit is a maximum when $i = 8\%$.

30. For the linear demand function, the rate of change (slope) is $m = -\frac{25}{5}$. Therefore, the demand x is

$$x = 800 - \frac{25}{5}(p - 25) = 925 - 5p$$

$$R = xp = -5p^2 + 925p$$

$$\frac{dR}{dp} = -10p + 925 = 0 \text{ when } p = \$92.50.$$

31. $C = 100\left(\dfrac{200}{x^2} + \dfrac{x}{x + 30}\right), \quad 1 \le x$

$C' = 100\left(-\dfrac{400}{x^3} + \dfrac{30}{(x + 30)^2}\right)$

$\overline{C}' = f(x) = 100\left(-\dfrac{400}{x^3} + \dfrac{30}{(x + 30)^2}\right)$

$f'(x) = 100\left(\dfrac{1200}{x^4} - \dfrac{60}{(x + 30)^3}\right)$

Approximation: $x \approx 40$ units

n	x_n	$f(x_n)$	$f'(x_n)$	$\dfrac{f(x_n)}{f'(x_n)}$	$x_n - \dfrac{f(x_n)}{f'(x_n)}$
1	40.0000	−0.0128	0.0294	−0.4341	40.4341
2	40.4341	−0.0004	0.0277	−0.0131	40.4472
3	40.4472	0.0000	0.0277	0.0000	40.4472

32. $\overline{C} = \dfrac{C}{x} = \dfrac{800}{x} + 0.4 + 0.02x + 0.0001x^2$

$\overline{C}' = -\dfrac{800}{x^2} + 0.02 + 0.0002x$

$\overline{C}' = f(x) = -\dfrac{800}{x^2} + 0.02 + 0.0002x$

$f'(x) = \dfrac{1600}{x^3} + 0.0002$

Approximation: $x \approx 131$ units

n	x_n	$f(x_n)$	$f'(x_n)$	$\dfrac{f(x_n)}{f'(x_n)}$	$x_n - \dfrac{f(x_n)}{f'(x_n)}$
1	130.0000	−0.0013	0.0009	−1.4406	131.4406
2	131.4406	0.0000	0.0009	0.0000	131.4406

33. $R = 900x - 0.1x^2$

$x = 3000$

$dx = 100$

$dR = (900 - 0.2x)\,dx$

$\quad = [900 - 0.2(3000)](100)$

$\quad = \$30,000$

34. $P = (500x - x^2) - \left(\dfrac{1}{2}x^2 - 77x + 3000\right)$

$x = 175$

$dx = 5$

$dP = [(500 - 2x) - (x - 77)]\,dx$

$\quad = (-3x + 577)\,dx$

$\quad = [-3(175) + 577](5)$

$\quad = \$260$

$\dfrac{dP}{P} = \dfrac{260}{52037.5} \approx 0.004996 \approx 0.5\%$ or $\dfrac{1}{2}\%$

35. $F = 100,000\left(1 + \sin\left[\dfrac{2\pi(t - 60)}{365}\right]\right)$

a. $\dfrac{dF}{dt} = 100,000\left(\dfrac{2\pi}{365}\cos\left[\dfrac{2\pi(t - 60)}{365}\right]\right)$

$\quad = 0$ when $\dfrac{2\pi(t - 60)}{365} = \dfrac{\pi}{2}$ or $\dfrac{3\pi}{2}$.

$t = 151.25$ or 333.75

The maximum occurs when $t = 151.25$ which corresponds to May 31.

b.
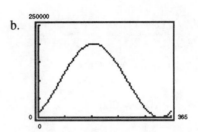

Sales are minimum when $t = 333.75$ which corresponds to November 30.

36. $G = 1600 + 418\sin\left(\dfrac{\pi t}{6} + 7\right)$

a. August ($t = 8$ and $t = 20$)

b. The temperature causes the seasonal variation in the sales of natural gas.

c. $G = 1600 + 10t + 418\sin\left(\dfrac{\pi t}{6} + 7\right)$, $\quad t = 1$ corresponds to January 1991

The inclusion of the term $10t$ means that natural gas consumption is predicted to rise. In the year 2000, we have $109 \le t \le 120$.

$\dfrac{dG}{dt} = 10 + 418\left(\dfrac{\pi}{6}\right)\cos\left(\dfrac{\pi t}{6} + 7\right) = 0$ when $t \approx 109.72$ and 115.54.

The maximum monthly consumption would occur when $t \approx 109.72$.

$G(109.72) \approx 3114.75$ billion cubic feet

37. $\eta = \dfrac{p/x}{dp/dx} = \dfrac{(400 - 3x)/x}{-3} = 1 - \dfrac{400}{3x}$

When $x = 20$, we have

$\eta = 1 - \dfrac{400}{3(20)} = -\dfrac{17}{3}$.

Since $|\eta| = \dfrac{17}{3} > 1$, the demand is elastic.

38. $\eta = \dfrac{p/x}{dp/dx} = \dfrac{(5 - 0.03x)/x}{-0.03} = 1 - \dfrac{5}{0.03x}$

When $x = 100$, we have

$$\eta = 1 - \frac{5}{0.03(100)} = -\frac{2}{3}.$$

Since $|\eta| = \frac{2}{3} < 1$, the demand is inelastic.

39. $\eta = \dfrac{p/x}{dp/dx} = \dfrac{(400 - 0.5x^2)/x}{-x} = \dfrac{1}{2} - \dfrac{400}{x^2}$

When $x = 20$, we have

$$\eta = \frac{1}{2} - \frac{400}{(20)^2} = -\frac{1}{2}.$$

Since $|\eta| = \frac{1}{2} < 1$, the demand is inelastic.

40. $\eta = \dfrac{p/x}{dp/dx} = \dfrac{500/[x(x + 2)]}{-500/[(x + 2)^2]} = -\dfrac{x + 2}{x}$

When $x = 23$, we have $\eta = -\frac{25}{23}$. Since $|\eta| = \frac{25}{23} > 1$, the demand is elastic.

41. $P = -2s^3 + 35s^2 - 100s + 200$

$\dfrac{dP}{ds} = -6s^2 + 70s - 100 = -2(3s^2 - 35s + 50) = -2(3s - 5)(s - 10) = 0$ when $s = \dfrac{5}{3}$, $s = 10$.

$\dfrac{d^2P}{ds^2} = -12s + 70$

$\dfrac{d^2P}{ds^2}\left(\dfrac{5}{3}\right) > 0 \Rightarrow s = \dfrac{5}{3}$ yields a minimum.

$\dfrac{d^2P}{ds^2}(10) < 0 \Rightarrow s = 10$ yields a maximum.

$\dfrac{d^2P}{ds^2} = -12s + 70 = 0$ when $s = \dfrac{35}{6}$.

The maximum profit occurs when $s = 10$, which corresponds to $10,000 ($P = \$700,000$). The point of diminishing returns occurs when $s = \frac{35}{6}$, which corresponds to \$5833.33 being spent on advertising.

42. $P = -\dfrac{1}{10}s^3 + 6s^2 + 400$

$\dfrac{dP}{ds} = -\dfrac{3}{10}s^2 + 12s = -\dfrac{3}{10}s(s - 40) = 0$ when $s = 0$, $s = 40$.

$\dfrac{d^2P}{ds^2} = -\dfrac{3}{5}s + 12$

$\dfrac{d^2P}{ds^2}(0) > 0 \Rightarrow s = 0$ yields a minimum.

$\dfrac{d^2P}{ds^2}(40) < 0 \Rightarrow s = 40$ yields a maximum.

$\dfrac{d^2P}{ds^2} = -\dfrac{3}{5}s + 12 = 0$ when $s = 20$.

The maximum profit occurs when $s = 40$, which corresponds to $40,000 ($P = \$3,600,000$). The point of diminishing returns occurs when $s = 20$, which corresponds to \$20,000 being spent on advertising.

Chapter 3   Review Exercises

1. $f(x) = (x-1)^2(x-3)$

$f'(x) = (x-1)^2(1) + (x-3)(2)(x-1)$

$= (x-1)(3x-7)$

Critical numbers: $x = 1$ and $x = \frac{7}{3}$

Interval	$-\infty < x < 1$	$1 < x < \frac{7}{3}$	$\frac{7}{3} < x < \infty$
Sign of $f'(x)$	$f'(x) > 0$	$f'(x) < 0$	$f'(x) > 0$
Conclusion	Increasing	Decreasing	Increasing

2. $g(x) = (x+1)^3$

$g'(x) = 3(x+1)^2$

Critical number: $x = -1$

Interval	$-\infty < x < -1$	$-1 < x < \infty$
Sign of $g'(x)$	$g'(x) > 0$	$g'(x) > 0$
Conclusion	Increasing	Increasing

3. $h(x) = \sqrt{x}(x-3) = x^{3/2} - 3x^{1/2}$

Domain: $[0, \infty)$

$h'(x) = \frac{3}{2}x^{1/2} - \frac{3}{2}x^{-1/2}$

$= \frac{3}{2}x^{-1/2}(x-1) = \frac{3(x-1)}{2\sqrt{x}}$

Critical numbers: $x = 0$, $x = 1$

Interval	$0 < x < 1$	$1 < x < \infty$
Sign of $h'(x)$	$h'(x) < 0$	$h'(x) > 0$
Conclusion	Decreasing	Increasing

4. $f(x) = \sin x + \cos x$, $0 \le x \le 2\pi$

$f'(x) = \cos x - \sin x$

Critical numbers: $x = \frac{\pi}{4}$, $x = \frac{5\pi}{4}$

Interval	$0 < x < \frac{\pi}{4}$	$\frac{\pi}{4} < x < \frac{5\pi}{4}$	$\frac{5\pi}{4} < x < 2\pi$
Sign of $f'(x)$	$f'(x) > 0$	$f'(x) < 0$	$f'(x) > 0$
Conclusion	Increasing	Decreasing	Increasing

5. $g(x) = 2x + 5\cos x$, $[0, 2\pi]$

$g'(x) = 2 - 5\sin x$

$= 0$ when $\sin x = \frac{2}{5}$.

Critical numbers: $x \approx 0.41$, $x \approx 2.73$

Left endpoint: $(0, 5)$

Critical number: $(0.41, 5.41)$

Critical number: $(2.73, 0.88)$ Minimum

Right endpoint: $(2\pi, 17.57)$ Maximum

6. $f(x) = \dfrac{x}{\sqrt{x^2+1}}$, $[0, 2]$

$f'(x) = x\left[-\frac{1}{2}(x^2+1)^{-3/2}(2x)\right] + (x^2+1)^{-1/2}$

$= \dfrac{1}{(x^2+1)^{3/2}}$

No critical numbers

Left endpoint: $(0, 0)$ Minimum

Right endpoint: $(2, 2/\sqrt{5})$ Maximum

7. $h(t) = \frac{1}{4}t^4 - 8t$

$h'(t) = t^3 - 8 = 0$ when $t = 2$.

Relative minimum: $(2, -12)$

Test interval	$-\infty < t < 2$	$2 < t < \infty$
Sign of $h'(t)$	$h'(t) < 0$	$h'(t) > 0$
Conclusion	Decreasing	Increasing

8. $g(\theta) = \dfrac{3}{2} \sin\left(\dfrac{\pi\theta}{2} - 1\right),$ [0, 4]

$g'(\theta) = \dfrac{3}{2}\left(\dfrac{\pi}{2}\right) \cos\left(\dfrac{\pi\theta}{2} - 1\right)$

$= 0$ when $\theta = 1 + \dfrac{2}{\pi},\ 3 + \dfrac{2}{\pi}$

Relative maximum: $\left(1 + \dfrac{2}{\pi},\ \dfrac{3}{2}\right)$

Relative minimum: $\left(3 + \dfrac{2}{\pi},\ -\dfrac{3}{2}\right)$

Test interval	$0 < \theta < 1 + \dfrac{2}{\pi}$	$1 + \dfrac{2}{\pi} < \theta < 3 + \dfrac{2}{\pi}$	$3 + \dfrac{2}{\pi} < \theta < 4$
Sign of $g'(\theta)$	$g'(\theta) > 0$	$g'(\theta) < 0$	$g'(\theta) > 0$
Conclusion	Increasing	Decreasing	Increasing

9. $f(x) = x + \cos x,\ \ 0 \le x \le 2\pi$

$f'(x) = 1 - \sin x$

$f''(x) = -\cos x = 0$ when $x = \dfrac{\pi}{2},\ \dfrac{3\pi}{2}.$

Test interval	$0 < x < \dfrac{\pi}{2}$	$\dfrac{\pi}{2} < x < \dfrac{3\pi}{2}$	$\dfrac{3\pi}{2} < x < 2\pi$
Sign of $f''(x)$	$f''(x) < 0$	$f''(x) > 0$	$f''(x) < 0$
Conclusion	Concave downward	Concave upward	Concave downward

Points of inflection: $\left(\dfrac{\pi}{2},\ \dfrac{\pi}{2}\right),\ \left(\dfrac{3\pi}{2},\ \dfrac{3\pi}{2}\right)$

10. $f(x) = (x + 2)^2(x - 4) = x^3 - 12x - 16$

$f'(x) = 3x^2 - 12$

$f''(x) = 6x = 0$ when $x = 0.$

Test interval	$-\infty < x < 0$	$0 < x < \infty$
Sign of $f''(x)$	$f''(x) < 0$	$f''(x) > 0$
Conclusion	Concave downward	Concave upward

Point of inflection: $(0,\ -16)$

11. $h(x) = \dfrac{2x + 3}{x - 4}$

Discontinuity: $x = 4$

$\displaystyle \lim_{x \to \infty} \dfrac{2x + 3}{x - 4} = \lim_{x \to \infty} \dfrac{2 + (3/x)}{1 - (4/x)} = 2$

Vertical asymptote: $x = 4$

Horizontal asymptote: $y = 2$

12. $g(x) = \dfrac{5x^2}{x^2 + 2}$

$\displaystyle \lim_{x \to \infty} \dfrac{5x^2}{x^2 + 2} = \lim_{x \to \infty} \dfrac{5}{1 + (2/x^2)} = 5$

Horizontal asymptote: $y = 5$

13. $f(x) = \dfrac{3}{x} - 2$

Discontinuity: $x = 0$

$\displaystyle\lim_{x \to \infty} \left(\dfrac{3}{x} - 2 \right) = -2$

Vertical asymptote: $x = 0$

Horizontal asymptote: $y = -2$

14. $f(x) = \dfrac{3x}{\sqrt{x^2 + 2}}$

$\displaystyle\lim_{x \to \infty} \dfrac{3x}{\sqrt{x^2 + 2}} = \lim_{x \to \infty} \dfrac{3x/x}{\sqrt{x^2 + 2}/\sqrt{x^2}}$

$\qquad = \displaystyle\lim_{x \to \infty} \dfrac{3}{\sqrt{1 + (2/x^2)}} = 3$

$\displaystyle\lim_{x \to -\infty} \dfrac{3x}{\sqrt{x^2 + 2}} = \lim_{x \to -\infty} \dfrac{3x/x}{\sqrt{x^2 + 2}/(-\sqrt{x^2})}$

$\qquad = \displaystyle\lim_{x \to -\infty} \dfrac{3}{-\sqrt{1 + (2/x^2)}} = -3$

Horizontal asymptotes: $y = \pm 3$

15. $f(x) = 4x - x^2 = x(4 - x)$, Domain: $(-\infty, \infty)$, Range: $(-\infty, 4]$

$f'(x) = 4 - 2x = 0$ when $x = 2$.

$f''(x) = -2$

Therefore, $(2, 4)$ is a relative maximum.

Intercepts: $(0, 0)$, $(4, 0)$

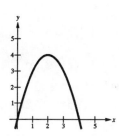

16. $f(x) = 4x^3 - x^4 = x^3(4 - x)$, Domain: $(-\infty, \infty)$, Range: $(-\infty, 27]$

$f'(x) = 12x^2 - 4x^3 = 4x^2(3 - x) = 0$ when $x = 0, 3$.

$f''(x) = 24x - 12x^2 = 12x(2 - x) = 0$ when $x = 0, 2$.

$f''(3) < 0$

Therefore, $(3, 27)$ is a relative maximum.

Points of inflection: $(0, 0)$, $(2, 16)$

Intercepts: $(0, 0)$, $(4, 0)$

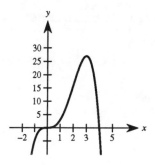

17. $f(x) = x\sqrt{16 - x^2}$, Domain: $[-4, 4]$, Range: $[-8, 8]$

$f'(x) = \dfrac{16 - 2x^2}{\sqrt{16 - x^2}} = 0$ when $x = \pm 2\sqrt{2}$ and undefined when $x = \pm 4$.

$f''(x) = \dfrac{2x(x^2 - 24)}{(16 - x^2)^{3/2}}$

$f''(-2\sqrt{2}) > 0$

Therefore, $(-2\sqrt{2}, -8)$ is a relative minimum.

$f''(2\sqrt{2}) < 0$

Therefore, $(2\sqrt{2}, 8)$ is a relative maximum.

Point of inflection: $(0, 0)$

Intercepts: $(-4, 0)$, $(0, 0)$, $(4, 0)$

Symmetry with respect to origin

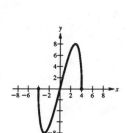

18. $f(x) = (x^2 - 4)^2$, Domain: $(-\infty, \infty)$, Range: $[0, \infty)$

$f'(x) = 4x(x^2 - 4) = 0$ when $x = 0, \pm 2$.

$f''(x) = 4(3x^2 - 4) = 0$ when $x = \pm\dfrac{2\sqrt{3}}{3}$.

$f''(0) < 0$

Therefore, $(0, 16)$ is a relative maximum.

$f''(\pm 2) > 0$

Therefore, $(\pm 2, 0)$ are relative minima.

Points of inflection: $(\pm 2\sqrt{3}/3, \ 64/9)$
Intercepts: $(-2, 0), \ (0, 16), \ (2, 0)$
Symmetry with respect to y-axis

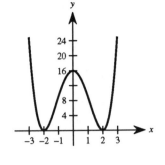

19. $f(x) = (x - 1)^3(x - 3)^2$, Domain: $(-\infty, \infty)$, Range: $(-\infty, \infty)$

$f'(x) = (x - 1)^2(x - 3)(5x - 11) = 0$ when $x = 1, \ \dfrac{11}{5}, \ 3$.

$f''(x) = 4(x - 1)(5x^2 - 22x + 23) = 0$ when $x = 1, \ \dfrac{11 \pm \sqrt{6}}{5}$.

$f''(3) > 0$

Therefore, $(3, 0)$ is a relative minimum.

$f''\left(\dfrac{11}{5}\right) < 0$

Therefore, $\left(\frac{11}{5}, \ \frac{3456}{3125}\right)$ is a relative maximum.

Points of inflection: $(1, \ 0), \ \left(\dfrac{11 - \sqrt{6}}{5}, \ 0.60\right), \ \left(\dfrac{11 + \sqrt{6}}{5}, \ 0.46\right)$

Intercepts: $(0, -9), \ (1, 0), \ (3, 0)$

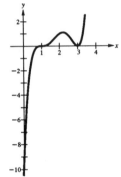

20. $f(x) = (x - 3)(x + 2)^3$, Domain: $(-\infty, \infty)$, Range: $\left[-\dfrac{16,875}{256}, \ \infty\right)$

$f'(x) = (x - 3)(3)(x + 2)^2 + (x + 2)^3$

$\qquad = (4x - 7)(x + 2)^2 = 0$ when $x = -2, \ \frac{7}{4}$.

$f''(x) = (4x - 7)(2)(x + 2) + (x + 2)^2(4)$

$\qquad = 6(2x - 1)(x + 2) = 0$ when $x = -2, \ \frac{1}{2}$.

$f''\left(\frac{7}{4}\right) > 0$

Therefore, $\left(\frac{7}{4}, \ -\dfrac{16,875}{256}\right)$ is a relative minimum.

Points of inflection: $(-2, \ 0), \ \left(\frac{1}{2}, \ -\dfrac{625}{16}\right)$
Intercepts: $(-2, 0), \ (0, -24), \ (3, 0)$

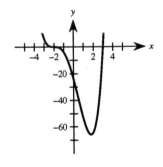

21. $f(x) = x^{1/3}(x+3)^{2/3}$, Domain: $(-\infty, \infty)$, Range: $(-\infty, \infty)$

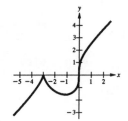

$f'(x) = \dfrac{x+1}{(x+3)^{1/3}x^{2/3}} = 0$ when $x = -1$ and undefined when $x = -3, 0$.

$f''(x) = \dfrac{-2}{x^{5/3}(x+3)^{4/3}}$ is undefined when $x = 0, -3$.

By the First Derivative Test $(-3, 0)$ is a relative maximum and $\left(-1, -\sqrt[3]{4}\right)$ is a relative minimum. $(0, 0)$ is a point of inflection.
Intercepts: $(-3, 0)$, $(0, 0)$

22. $f(x) = (x-2)^{1/3}(x+1)^{2/3}$

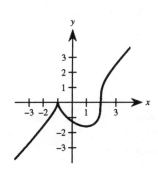

Graph of Exercise 21 translated 2 units to the right (x replaces by $x - 2$).

$(-1, 0)$ is a relative maximum.

$\left(1, -\sqrt[3]{4}\right)$ is a relative minimum.

$(2, 0)$ is a point of inflection.

Intercepts: $(-1, 0)$, $(2, 0)$

23. $f(x) = \dfrac{x+1}{x-1}$, Domain: $(-\infty, 1)$, $(1, \infty)$, Range: $(-\infty, 1)$, $(1, \infty)$

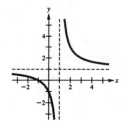

$f'(x) = \dfrac{-2}{(x-1)^2} < 0$ if $x \neq 1$

$f''(x) = \dfrac{4}{(x-1)^3}$

Horizontal asymptote: $y = 1$
Vertical asymptote: $x = 1$
Intercepts: $(-1, 0)$, $(0, -1)$

24. $f(x) = \dfrac{2x}{1+x^2}$, Domain: $(-\infty, \infty)$, Range: $[-1, 1]$

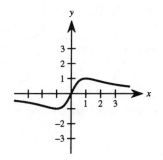

$f'(x) = \dfrac{2(1-x)(1+x)}{(1+x^2)^2} = 0$ when $x = \pm 1$.

$f''(x) = \dfrac{-2x(3-x^2)}{(1+x^2)^3} = 0$ when $x = 0, \pm\sqrt{3}$.

$f''(1) < 0$
Therefore, $(1, 1)$ is a relative maximum.

$f''(-1) > 0$
Therefore, $(-1, -1)$ is a relative minimum.

Points of inflection: $(-\sqrt{3}, -\sqrt{3}/2)$, $(0, 0)$, $(\sqrt{3}, \sqrt{3}/2)$
Intercept: $(0, 0)$
Symmetric with respect to the origin
Horizontal asymptote: $y = 0$

25. $f(x) = \dfrac{4}{1 + x^2}$,   Domain: $(-\infty, \infty)$,   Range: $(0, 4]$

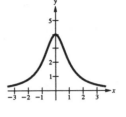

$f'(x) = \dfrac{-8x}{(1 + x^2)^2} = 0$ when $x = 0$.

$f''(x) = \dfrac{-8(1 - 3x^2)}{(1 + x^2)^3} = 0$ when $x = \pm\dfrac{\sqrt{3}}{3}$.

$f''(0) < 0$

Therefore, $(0, 4)$ is a relative maximum.

Points of inflection: $(\pm\sqrt{3}/3, \ 3)$

Intercept: $(0, 4)$

Symmetric to the y-axis

Horizontal asymptote: $y = 0$

26. $f(x) = \dfrac{x^2}{1 + x^4}$,   Domain: $(-\infty, \infty)$,   Range: $\left[0, \dfrac{1}{2}\right]$

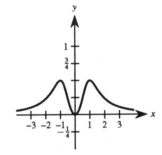

$f'(x) = \dfrac{(1 + x^4)(2x) - x^2(4x^3)}{(1 + x^4)^2}$

$\qquad = \dfrac{2x(1 - x)(1 + x)(1 + x^2)}{(1 + x^4)^2} = 0$ when $x = 0, \ \pm 1$.

$f''(x) = \dfrac{(1 + x^4)^2(2 - 10x^4) - (2x - 2x^5)(2)(1 + x^4)(4x^3)}{(1 + x^4)^4}$

$\qquad = \dfrac{2(1 - 12x^4 + 3x^8)}{(1 + x^4)^3} = 0$ when $x = \pm\sqrt[4]{\dfrac{6 \pm \sqrt{33}}{3}}$.

$f''(\pm 1) < 0$

Therefore, $\left(\pm 1, \ \tfrac{1}{2}\right)$ are relative maxima.

$f''(0) > 0$

Therefore, $(0, 0)$ is a relative minimum.

Points of inflection: $\left(\pm\sqrt[4]{\dfrac{6 - \sqrt{33}}{3}}, \ 0.29\right), \ \left(\pm\sqrt[4]{\dfrac{6 + \sqrt{33}}{3}}, \ 0.40\right)$

Intercept: $(0, 0)$

Symmetric to the y-axis

Horizontal asymptote: $y = 0$

27. $f(x) = x^3 + x + \dfrac{4}{x}$

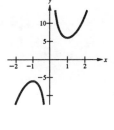

Domain: $(-\infty, 0), (0, \infty)$, Range: $(-\infty, -6], [6, \infty)$

$$f'(x) = 3x^2 + 1 - \frac{4}{x^2} = \frac{3x^4 + x^2 - 4}{x^2} = 0 \text{ when } x = \pm 1.$$

$$f''(x) = 6x + \frac{8}{x^3} = \frac{6x^4 + 8}{x^3} \neq 0$$

$f''(-1) < 0$
Therefore, $(-1, -6)$ is a relative maximum.

$f''(1) > 0$
Therefore, $(1, 6)$ is a relative minimum.

Vertical asymptote: $x = 0$
Symmetric with respect to origin

28. $f(x) = x^2 + \dfrac{1}{x} = \dfrac{x^3 + 1}{x}$

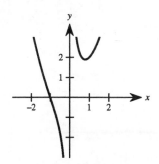

Domain: $(-\infty, 0), (0, \infty)$, Range: $(-\infty, \infty)$

$$f'(x) = 2x - \frac{1}{x^2} = \frac{2x^3 - 1}{x^2} = 0 \text{ when } x = \frac{1}{\sqrt[3]{2}}.$$

$$f''(x) = 2 + \frac{2}{x^3} = \frac{2(x^3 + 1)}{x^3} = 0 \text{ when } x = -1.$$

$f''(1/\sqrt[3]{2}) > 0$
Therefore, $(1/\sqrt[3]{2},\ 3/\sqrt[3]{4})$ is a relative minimum.

Point of inflection: $(-1,\ 0)$
Intercept: $(-1,\ 0)$
Vertical asymptote: $x = 0$

29. $f(x) = |x^2 - 9|$, Domain: $(-\infty, \infty)$, Range: $[0, \infty)$

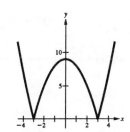

$$f'(x) = \frac{2x(x^2 - 9)}{|x^2 - 9|} = 0 \text{ when } x = 0 \text{ and is undefined when } x = \pm 3.$$

$$f''(x) = \frac{2(x^2 - 9)}{|x^2 - 9|} \text{ is undefined at } x = \pm 3.$$

$f''(0) < 0$
Therefore, $(0, 9)$ is a relative maximum.

Relative minima: $(\pm 3,\ 0)$
Points of inflection: $(\pm 3,\ 0)$
Intercepts: $(\pm 3,\ 0),\ (0,\ 9)$
Symmetric to the y-axis

30. $f(x) = |x - 1| + |x - 3| = \begin{cases} -2x + 4, & x \le 1 \\ 2, & 1 < x \le 3 \\ 2x - 4, & x > 3 \end{cases}$

Domain: $(-\infty, \infty)$

Range: $[2, \infty)$

Intercept: $(0, 4)$

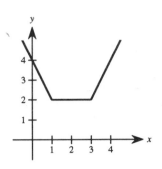

31. $f(x) = x + \cos x$,   Domain: $[0, 2\pi]$,   Range: $[1, 1 + 2\pi]$

$f'(x) = 1 - \sin x \ge 0$, f is increasing

$f''(x) = -\cos x = 0$ when $x = \dfrac{\pi}{2}, \dfrac{3\pi}{2}$

Points of inflection: $\left(\dfrac{\pi}{2}, \dfrac{\pi}{2}\right), \left(\dfrac{3\pi}{2}, \dfrac{3\pi}{2}\right)$

Intercept: $(0, 1)$

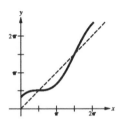

32. $f(x) = \dfrac{1}{\pi}(2 \sin \pi x - \sin 2\pi x)$,   Domain: $[-1, 1]$,   Range: $\left[\dfrac{-3\sqrt{3}}{2\pi}, \dfrac{3\sqrt{3}}{2\pi}\right]$

$f'(x) = 2(\cos \pi x - \cos 2\pi x) = -2(2\cos \pi x + 1)(\cos \pi x - 1) = 0$,   Critical numbers: $x = \pm\dfrac{2}{3}$, 0

$f''(x) = 2\pi(-\sin \pi x + 2\sin 2\pi x) = 2\pi \sin \pi x(-1 + 4\cos \pi x) = 0$ when $x = 0, \pm 1, \pm 0.420$

By the First Derivative Test:

$\left(-\dfrac{2}{3}, \dfrac{-3\sqrt{3}}{2\pi}\right)$ is a relative minimum .

$\left(\dfrac{2}{3}, \dfrac{3\sqrt{3}}{2\pi}\right)$ is a relative maximum.

Points of inflection: $(-0.420, -0.462)$, $(0.420, 0.462)$, $(\pm 1, 0)$, $(0, 0)$

Intercepts: $(-1, 0)$, $(0, 0)$, $(1, 0)$

Symmetric with respect to the origin

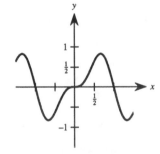

33. $f(x) = x^3 + \dfrac{243}{x}$

Relative minimum: $(3, 108)$

Relative maximum: $(-3, -108)$

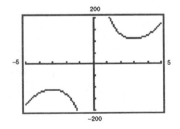

34. $f(x) = |x^3 - 3x^2 + 2x| = |x(x - 1)(x - 2)|$

Relative minima: $(0, 0)$, $(1, 0)$, $(2, 0)$

Relative maxima: $(1.577, 0.38)$, $(0.423, 0.38)$

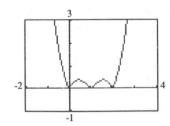

35. $f(x) = \dfrac{x-1}{1+3x^2}$

Relative minimum: $(-0.155, \ -1.077)$

Relative maximum: $(2.155, \ 0.077)$

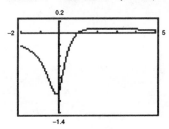

36. $g(x) = \dfrac{\pi^2}{3} - 4\cos x + \cos 2x$

Relative minima: $(2\pi k, \ 0.29)$ where k is any integer.

Relative maxima: $((2k-1)\pi, \ 8.29)$ where k is any integer.

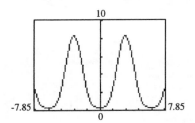

37.
$$f(x) = x^{2/3}, \quad 1 \le x \le 8$$
$$f'(x) = \frac{2}{3}x^{-1/3}$$
$$\frac{f(b) - f(a)}{b - a} = \frac{4 - 1}{8 - 1} = \frac{3}{7}$$
$$f'(c) = \frac{2}{3}c^{-1/3} = \frac{3}{7}$$
$$c = \left(\frac{14}{9}\right)^3 = \frac{2744}{729} \approx 3.764$$

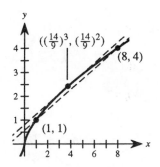

38.
$$f(x) = \frac{1}{x}, \quad 1 \le x \le 4$$
$$f'(x) = -\frac{1}{x^2}$$
$$\frac{f(b) - f(a)}{b - a} = \frac{(1/4) - 1}{4 - 1} = \frac{-3/4}{3} = -\frac{1}{4}$$
$$f'(c) = \frac{-1}{c^2} = -\frac{1}{4}$$
$$c = 2$$

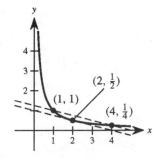

39.
$$f(x) = x - \cos x, \quad -\frac{\pi}{2} \le x \le \frac{\pi}{2}$$
$$f'(x) = 1 + \sin x$$
$$\frac{f(b) - f(a)}{b - a} = \frac{(\pi/2) - (-\pi/2)}{(\pi/2) - (-\pi/2)} = 1$$
$$f'(c) = 1 + \sin c = 1$$
$$c = 0$$

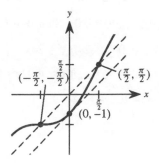

40. $f(x) = \sqrt{x} - 2x, \quad 0 \le x \le 4$

$$f'(x) = \frac{1}{2\sqrt{x}} - 2$$

$$\frac{f(b) - f(a)}{b - a} = \frac{-6 - 0}{4 - 0} = -\frac{3}{2}$$

$$f'(c) = \frac{1}{2\sqrt{c}} - 2 = -\frac{3}{2}$$

$$c = 1$$

41. No; the function is discontinuous at $x = 0$ which is in the interval $[-2, \ 1]$.

42. $f(x) = 3 - |x - 4|$

$$f(1) = 3 - |1 - 4| = 0$$

$$f(7) = 3 - |7 - 4| = 0$$

$f'(4)$ does not exist.

Rolle's Theorem requires that the function be differentiable in $(1, \ 7)$, but this function is not differentiable at $x = 4$.

43. $f(x) = Ax^2 + Bx + C$

$$f'(x) = 2Ax + B$$

$$\frac{f(x_2) - f(x_1)}{x_2 - x_1} = \frac{A(x_2{}^2 - x_1{}^2) + B(x_2 - x_1)}{x_2 - x_1}$$

$$= A(x_1 + x_2) + B$$

$$f'(c) = 2Ac + B = A(x_1 + x_2) + B$$

$$2Ac = A(x_1 + x_2)$$

$$c = \frac{x_1 + x_2}{2} = \text{ Midpoint of } [x_1, \ x_2]$$

44. $f(x) = 2x^2 - 3x + 1$

$$f'(x) = 4x - 3$$

$$\frac{f(b) - f(a)}{b - a} = \frac{21 - 1}{4 - 0} = 5$$

$$f'(c) = 4c - 3 = 5$$

$$c = 2 = \text{ Midpoint of } [0, \ 4]$$

45. Let $t = 0$ at noon.

$$L = d^2 = (100 - 12t)^2 + (-10t)^2 = 10,000 - 2400t + 244t^2$$

$$\frac{dL}{dt} = -2400 + 488t = 0 \text{ when } t = \frac{300}{61} \approx 4.92 \text{ hr}$$

Ship A at $(40.98, 0)$; Ship B at $(0, \ -49.18)$

$$d^2 = 10,000 - 2400t + 244t^2$$

$$\approx 4098.36 \text{ when } t \approx 4.92 \approx 4{:}55 \text{ P.M.}$$

$$d \approx 64 \text{ miles}$$

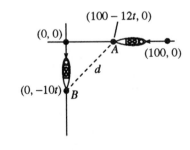

46. Ellipse: $\dfrac{x^2}{144} + \dfrac{y^2}{16} = 1$, $y = \dfrac{1}{3}\sqrt{144 - x^2}$

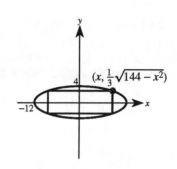

$$A = (2x)\left(\dfrac{2}{3}\sqrt{144 - x^2}\right) = \dfrac{4}{3}x\sqrt{144 - x^2}$$

$$\dfrac{dA}{dx} = \dfrac{4}{3}\left[\dfrac{-x^2}{\sqrt{144 - x^2}} + \sqrt{144 - x^2}\right]$$

$$= \dfrac{4}{3}\left[\dfrac{144 - 2x^2}{\sqrt{144 - x^2}}\right] = 0 \text{ when } x = \sqrt{72} = 6\sqrt{2}$$

The dimensions of the rectangle are $2x = 12\sqrt{2}$ by $y = \dfrac{2}{3}\sqrt{144 - 72} = 4\sqrt{2}$.

47. We have points $(0, \ y)$, $(x, \ 0)$, and $(1, 8)$. Thus:

$$m = \dfrac{y - 8}{0 - 1} = \dfrac{0 - 8}{x - 1} \text{ or } y = \dfrac{8x}{x - 1}.$$

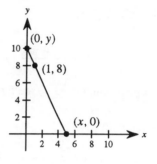

Let $f(x) = L^2 = x^2 + \left(\dfrac{8x}{x - 1}\right)^2$.

$$f'(x) = 2x + 128\left(\dfrac{x}{x - 1}\right)\left[\dfrac{(x - 1) - x}{(x - 1)^2}\right] = 0$$

$$x - \dfrac{64x}{(x - 1)^3} = 0$$

$x[(x - 1)^3 - 64] = 0$ when $x = 0, \ 5$ (minimum).

Vertices of triangle: $(0, 0)$, $(5, 0)$, $(0, 10)$

48. We have points $(0, \ y)$, $(x, \ 0)$, and $(4, 5)$. Thus:

$$m = \dfrac{y - 5}{0 - 4} = \dfrac{5 - 0}{4 - x} \text{ or } y = \dfrac{5x}{x - 4}.$$

Let $f(x) = L^2 = x^2 + \left(\dfrac{5x}{x - 4}\right)^2$.

$$f'(x) = 2x + 50\left(\dfrac{x}{x - 4}\right)\left[\dfrac{x - 4 - x}{(x - 4)^2}\right] = 0$$

$$x - \dfrac{100x}{(x - 4)^3} = 0$$

$x[(x - 4)^3 - 100] = 0$ when $x = 0$ or $x = 4 + \sqrt[3]{100}$.

$$L = \sqrt{x^2 + \dfrac{25x^2}{(x - 4)^2}} = \dfrac{x}{x - 4}\sqrt{(x - 4)^2 + 25} = \dfrac{\sqrt[3]{100} + 4}{\sqrt[3]{100}}\sqrt{100^{2/3} + 25} \approx 12.7 \text{ feet}$$

49. $A = $ (Average of bases)(Height)

$$= \left(\dfrac{x + s}{2}\right)\dfrac{\sqrt{3s^2 + 2sx - x^2}}{2} \quad \text{(see figure)}$$

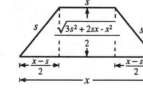

$$\dfrac{dA}{dx} = \dfrac{1}{4}\left[\dfrac{(s - x)(s + x)}{\sqrt{3s^2 + 2sx - x^2}} + \sqrt{3s^2 + 2sx - x^2}\right]$$

$$= \dfrac{2(2s - x)(s + x)}{4\sqrt{3s^2 + 2sx - x^2}} = 0 \text{ when } x = 2s.$$

A is a maximum when $x = 2s$.

50. Label triangle with vertices $(0, 0)$, $(a, 0)$, and (b, c). The equations of the sides of the triangle are $y = (c/b)x$ and $y = [c/(b - a)](x - a)$. Let $(x, 0)$ be a vertex of the inscribed rectangle. The coordinates of the upper left vertex are $(x, (c/b)x)$. The y-coordinate of the upper right vertex of the rectangle is $(c/b)x$. Solving for the x-coordinate $\bar{x}$ of the rectangle's upper right vertex, you get

$$\frac{c}{b}x = \frac{c}{b - a}(\bar{x} - a)$$

$$(b - a)x = b(\bar{x} - a)$$

$$\bar{x} = \frac{b - a}{b}x + a = a - \frac{a - b}{b}x.$$

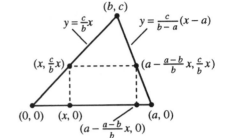

Finally, the lower right vertex is

$$\left(a - \frac{a - b}{b}x, \ 0\right).$$

Width of rectangle: $a - \dfrac{a - b}{b}x - x$

Height of rectangle: $\dfrac{c}{b}x$ (see figure)

$$A = \text{(Width)(Height)} = \left(a - \frac{a - b}{b}x - x\right)\left(\frac{c}{b}x\right) = \left(a - \frac{a}{b}x\right)\frac{c}{b}x$$

$$\frac{dA}{dx} = \left(a - \frac{a}{b}x\right)\frac{c}{b} + \left(\frac{c}{b}x\right)\left(-\frac{a}{b}\right) = \frac{ac}{b} - \frac{2ac}{b^2}x = 0 \text{ when } x = \frac{b}{2}.$$

$$A\left(\frac{b}{2}\right) = \left(a - \frac{a}{b}\frac{b}{2}\right)\left(\frac{c}{b}\frac{b}{2}\right) = \left(\frac{a}{2}\right)\left(\frac{c}{2}\right) = \frac{1}{4}ac = \frac{1}{2}\left(\frac{1}{2}ac\right) = \frac{1}{2}\text{(Area of triangle)}$$

51. $m = \dfrac{y - 6}{0 - 4} = \dfrac{6 - 0}{4 - x}$ or $y = \dfrac{6x}{x - 4}$

Let $f(x) = L^2 = x^2 + \left(\dfrac{6x}{x - 4}\right)^2$.

$$f'(x) = 2x + 72\left(\frac{x}{x - 4}\right)\left[\frac{-4}{(x - 4)^2}\right] = 0$$

$x[(x - 4)^3 - 144] = 0$ when $x = 0$ or $x = 4 + \sqrt[3]{144}$.

$$L \approx 14.05 \text{ feet}$$

(See solution to Exercise 48.)

52. You can form a right triangle with vertices $(0, y)$, $(0, 0)$, and $(x, 0)$. Choosing a point (a, b) on the hypotenuse (assuming the triangle is in the first quadrant), the slope is

$$m = \frac{y - b}{0 - a} = \frac{b - 0}{a - x} \Rightarrow y = \frac{-bx}{a - x}.$$

Let $f(x) = L^2 = x^2 + y^2 = x^2 + \left(\dfrac{-bx}{a - x}\right)^2$.

$$f'(x) = 2x + 2\left(\frac{-bx}{a - x}\right)\left[\frac{-ab}{(a - x)^2}\right]$$

$$\frac{2x[(a - x)^3 + ab^2]}{(a - x)^3} = 0 \text{ when } x = 0, \ a + \sqrt[3]{ab^2}.$$

Choosing the nonzero value, we have $y = b + \sqrt[3]{a^2 b}$.

$$L = \sqrt{(a + \sqrt[3]{ab^2})^2 + (b + \sqrt[3]{a^2 b})^2} = (a^2 + 3a^{4/3}b^{2/3} + 3a^{2/3}b^{4/3} + b^2)^{1/2} = (a^{2/3} + b^{2/3})^{3/2} \text{ feet}$$

53. $\csc\theta = \dfrac{L_1}{6}$ or $L_1 = 6\csc\theta$ (see figure)

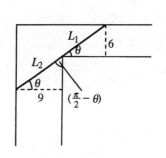

$$\csc\left(\frac{\pi}{2}-\theta\right) = \frac{L_2}{9} \text{ or } L_2 = 9\csc\left(\frac{\pi}{2}-\theta\right)$$

$$L = L_1 + L_2 = 6\csc\theta + 9\csc\left(\frac{\pi}{2}-\theta\right) = 6\csc\theta + 9\sec\theta$$

$$\frac{dL}{d\theta} = -6\csc\theta\cot\theta + 9\sec\theta\tan\theta = 0$$

$$\tan^3\theta = \frac{2}{3} \Rightarrow \tan\theta = \frac{\sqrt[3]{2}}{\sqrt[3]{3}}$$

$$\sec\theta = \sqrt{1+\tan^2\theta} = \sqrt{1+\left(\frac{2}{3}\right)^{2/3}} = \frac{\sqrt{3^{2/3}+2^{2/3}}}{3^{1/3}}$$

$$\csc\theta = \frac{\sec\theta}{\tan\theta} = \frac{\sqrt{3^{2/3}+2^{2/3}}}{2^{1/3}}$$

$$L = 6\frac{(3^{2/3}+2^{2/3})^{1/2}}{2^{1/3}} + 9\frac{(3^{2/3}+2^{2/3})^{1/2}}{3^{1/3}} = 3(3^{2/3}+2^{2/3})^{3/2} \text{ ft} \approx 21.07 \text{ ft}$$

54. Using Exercise 53 as a guide we have $L_1 = a\csc\theta$ and $L_2 = b\sec\theta$.
Then $dL/d\theta = -a\csc\theta\cot\theta + b\sec\theta\tan\theta = 0$ when

$$\tan\theta = \sqrt[3]{a/b}, \quad \sec\theta = \frac{\sqrt{a^{2/3}+b^{2/3}}}{b^{1/3}}, \quad \csc\theta = \frac{\sqrt{a^{2/3}+b^{2/3}}}{a^{1/3}} \text{ and}$$

$$L = L_1 + L_2 = a\csc\theta + b\sec\theta = a\frac{(a^{2/3}+b^{2/3})^{1/2}}{a^{1/3}} + b\frac{(a^{2/3}+b^{2/3})^{1/2}}{b^{1/3}} = (a^{2/3}+b^{2/3})^{3/2}.$$

This matches the result of Exercise 52.

55. $y = \dfrac{1}{3}\cos(12t) - \dfrac{1}{4}\sin(12t)$

$v = y' = -4\sin(12t) - 3\cos(12t)$

a. When $t = \dfrac{\pi}{8}$, $y = \dfrac{1}{4}$ inch and $v = y' = 4$ inches/second.

b. $y' = -4\sin(12t) - 3\cos(12t) = 0$ when $\dfrac{\sin(12t)}{\cos(12t)} = -\dfrac{3}{4} \Rightarrow \tan(12t) = -\dfrac{3}{4}$.

Therefore, $\sin(12t) = -\dfrac{3}{5}$ and $\cos(12t) = \dfrac{4}{5}$. The maximum displacement is

$$y = \left(\frac{1}{3}\right)\left(\frac{4}{5}\right) - \frac{1}{4}\left(-\frac{3}{5}\right) = \frac{5}{12} \text{ inch.}$$

c. Period: $\dfrac{2\pi}{12} = \dfrac{\pi}{6}$

Frequency: $\dfrac{1}{\pi/6} = \dfrac{6}{\pi}$

56. a. $y = A\sin(\sqrt{k/m}\ t) + B\cos(\sqrt{k/m}\ t)$

$y' = A\sqrt{k/m}\ \cos(\sqrt{k/m}\ t) - B\sqrt{k/m}\ \sin(\sqrt{k/m}\ t)$

$= 0$ when $\dfrac{\sin\sqrt{k/m}\ t}{\cos\sqrt{k/m}\ t} = \dfrac{A}{B} \Rightarrow \tan(\sqrt{k/m}\ t) = \dfrac{A}{B}$

Therefore,

$\sin(\sqrt{k/m}\ t) = \dfrac{A}{\sqrt{A^2 + B^2}}$

$\cos(\sqrt{k/m}\ t) = \dfrac{B}{\sqrt{A^2 + B^2}}.$

When $v = y' = 0$,

$y = A\left(\dfrac{A}{\sqrt{A^2 + B^2}}\right) + B\left(\dfrac{B}{\sqrt{A^2 + B^2}}\right) = \sqrt{A^2 + B^2}.$

b. Period: $\dfrac{2\pi}{\sqrt{k/m}}$

Frequency: $\dfrac{1}{2\pi/\sqrt{k/m}} = \dfrac{1}{2\pi}\sqrt{k/m}$

c. The frequency is changed as the square root of k.

d. The frequency is inversely proportional to the square root of the mass.

57. $f(x) = x^3 - 3x - 1$

From the graph you can see that $f(x)$ has three real zeros.

$f'(x) = 3x^2 - 3$

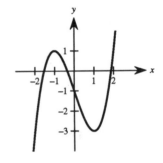

n	x_n	$f(x_n)$	$f'(x_n)$	$\dfrac{f(x_n)}{f'(x_n)}$	$x_n - \dfrac{f(x_n)}{f'(x_n)}$
1	−1.5000	0.1250	3.7500	0.0333	−1.5333
2	−1.5333	−0.0049	4.0530	−0.0012	−1.5321

n	x_n	$f(x_n)$	$f'(x_n)$	$\dfrac{f(x_n)}{f'(x_n)}$	$x_n - \dfrac{f(x_n)}{f'(x_n)}$
1	−0.5000	0.3750	−2.2500	−0.1667	−0.3333
2	−0.3333	−0.0371	−2.6667	0.0139	−0.3472
3	−0.3472	−0.0003	−2.6384	0.0001	−0.3473

n	x_n	$f(x_n)$	$f'(x_n)$	$\dfrac{f(x_n)}{f'(x_n)}$	$x_n - \dfrac{f(x_n)}{f'(x_n)}$
1	−1.9000	0.1590	7.8300	0.0203	1.8797
2	1.8797	0.0024	7.5998	0.0003	1.8794

The three real zeros of $f(x)$ are $x \approx -1.532$, $x \approx -0.347$, and $x \approx 1.879$.

58. $f(x) = x^3 + 2x + 1$

From the graph, you can see that $f(x)$ has one real zero.

$f'(x) = 3x^2 + 2$

f changes sign in $[-1, \ 0]$.

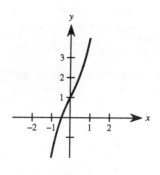

n	x_n	$f(x_n)$	$f'(x_n)$	$\dfrac{f(x_n)}{f'(x_n)}$	$x_n - \dfrac{f(x_n)}{f'(x_n)}$
1	-0.5000	-0.1250	2.7500	-0.0455	-0.4545
2	-0.4545	-0.0029	2.6197	-0.0011	-0.4534

On the interval $[-1, \ 0]$: $x \approx -0.453$.

59. Find the zeros of $f(x) = x^4 - x - 3$.

$f'(x) = 4x^3 - 1$

From the graph you can see that $f(x)$ has two real zeros.

f changes sign in $[-2, \ -1]$.

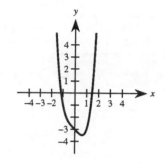

n	x_n	$f(x_n)$	$f'(x_n)$	$\dfrac{f(x_n)}{f'(x_n)}$	$x_n - \dfrac{f(x_n)}{f'(x_n)}$
1	-1.2000	0.2736	-7.9120	-0.0346	-1.1654
2	-1.1654	0.0100	-7.3312	-0.0014	-1.1640

On the interval $[-2, \ -1]$: $x \approx -1.164$

f changes sign in $[1, 2]$.

n	x_n	$f(x_n)$	$f'(x_n)$	$\dfrac{f(x_n)}{f'(x_n)}$	$x_n - \dfrac{f(x_n)}{f'(x_n)}$
1	1.5000	0.5625	12.5000	0.0450	1.4550
2	1.4550	0.0268	11.3211	0.0024	1.4526
3	1.4526	-0.0003	11.2602	0.0000	1.4526

On the interval $[1, \ 2]$: $x \approx 1.453$.

60. Find the zeros of $f(x) = \sin \pi x + x - 1$.

$$f'(x) = \pi \cos \pi x + 1$$

From the graph you can see that $f(x)$ has three real zeros.

n	x_n	$f(x_n)$	$f'(x_n)$	$\dfrac{f(x_n)}{f'(x_n)}$	$x_n - \dfrac{f(x_n)}{f'(x_n)}$
1	0.2000	−0.2122	3.5416	−0.0599	0.2599
2	0.2599	−0.0113	3.1513	−0.0036	0.2635
3	0.2635	0.0000	3.1253	0.0000	0.2635

n	x_n	$f(x_n)$	$f'(x_n)$	$\dfrac{f(x_n)}{f'(x_n)}$	$x_n - \dfrac{f(x_n)}{f'(x_n)}$
1	1.0000	0.0000	−2.1416	0.0000	1.0000

n	x_n	$f(x_n)$	$f'(x_n)$	$\dfrac{f(x_n)}{f'(x_n)}$	$x_n - \dfrac{f(x_n)}{f'(x_n)}$
1	1.8000	0.2122	3.5416	0.0599	1.7401
2	1.7401	0.0113	3.1513	0.0036	1.7365
3	1.7365	0.0000	3.1253	0.0000	1.7365

The three real zeros of $f(x)$ are $x \approx 0.264$, $x = 1$, and $x \approx 1.737$.

61.
$$S = 4\pi r^2 \cdot dr = \Delta r = \pm 0.025$$
$$dS = 8\pi r\, dr = 8\pi(9)(\pm 0.025)$$
$$= \pm 1.8\pi \text{ square inches}$$
$$\frac{dS}{S}(100) = \frac{8\pi r\, dr}{4\pi r^2}(100) = \frac{2\, dr}{r}(100)$$
$$= \frac{2(\pm 0.025)}{9}(100) \approx \pm 0.56\%$$
$$V = \frac{4}{3}\pi r^3$$
$$dV = 4\pi r^2\, dr = 4\pi(9)^2(\pm 0.025)$$
$$= \pm 8.1\pi \text{ cubic inches}$$
$$\frac{dV}{V}(100) = \frac{4\pi r^2\, dr}{(4/3)\pi r^3}(100) = \frac{3\, dr}{r}(100)$$
$$= \frac{3(\pm 0.025)}{9}(100) \approx \pm 0.83\%$$

62.
$$p = 75 - \frac{1}{4}x$$
$$\Delta p = p(8) - p(7)$$
$$= \left(75 - \frac{8}{4}\right) - \left(75 - \frac{7}{4}\right) = -\frac{1}{4}$$
$$dp = -\frac{1}{4}\, dx = -\frac{1}{4}(1) = -\frac{1}{4}$$

REVIEW EXERCISES for Chapter 3   315

63. $p = 36 - 4x$

$C = 2x^2 + 6$

$R = xp$

$P = R - C = x(36 - 4x) - (2x^2 + 6) = -6(x^2 - 6x + 1)$

$\dfrac{dP}{dx} = -6(2x - 6) = 0$ when $x = 3$.

$\dfrac{d^2P}{dx^2} = -12 < 0, \quad x = 3$ is a maximum.

For $x = 3$, the profit is $P = \$48$.

64. $C = \dfrac{1}{4}x^2 + 62x + 125$

$p = 75 - \dfrac{1}{3}x$

$R = xp$

a. $P = R - C = x\left(75 - \dfrac{1}{3}x\right) - \left(\dfrac{1}{4}x^2 + 62x + 125\right)$

$\dfrac{dP}{dx} = -\dfrac{1}{3}x + 75 - \dfrac{1}{3}x - \dfrac{1}{2}x - 62 = -\dfrac{7}{6}x + 13 = 0$ when $x = \dfrac{78}{7}$ (11 units).

b. $\overline{C}(x) = \dfrac{C}{x} = \dfrac{1}{4}x + 62 + \dfrac{125}{x}$ (average cost)

$\dfrac{d\overline{C}}{dx} = \dfrac{1}{4} - \dfrac{125}{x^2} = 0$

$x^2 - 500 = 0$ when $x = 10\sqrt{5}$ (22 units).

c. $\eta = \dfrac{p/x}{dp/dx} = \dfrac{(75/x) - (1/3)}{-1/3} = 1 - \dfrac{225}{x} = \dfrac{x - 255}{x}$

65. $R = $ (Number of people)(Rate per person)

$= n[8.00 - 0.05(n - 80)], \quad n \geq 80$

$= 12n - 0.05n^2$

$\dfrac{dR}{dn} = 12 - 0.10n = 0$ when $n = 120$.

$\dfrac{d^2R}{dn^2} = -0.10 < 0 \Rightarrow$ Revenue will be maximum when 120 people go on the bus.

66. The cost of fuel is proportional to $s^{3/2}$, $C_F = 50$ when $s = 25$, and fixed costs are \$100.

$C_F = ks^{3/2}$ where $50 = k(25)^{3/2} \Rightarrow k = \dfrac{2}{5}$.

The total cost $C = \dfrac{2}{5}s^{3/2} + 100$.

$\overline{C} = \dfrac{C}{s} = \dfrac{(2/5)s^{3/2} + 100}{s} = \dfrac{2}{5}s^{1/2} + \dfrac{100}{s}$

$\dfrac{d\overline{C}}{ds} = \dfrac{1}{5s^{1/2}} - \dfrac{100}{s^2} = 0$ when $s^{3/2} = 500$.

$s = 500^{2/3} \approx 63$ mi/hr

67. $C = \left(\dfrac{Q}{x}\right)s + \left(\dfrac{x}{2}\right)r$

$\dfrac{dC}{dx} = -\dfrac{Qs}{x^2} + \dfrac{r}{2} = 0$

$\dfrac{Qs}{x^2} = \dfrac{r}{2}$

$x^2 = \dfrac{2Qs}{r}$

$x = \sqrt{\dfrac{2Qs}{r}}$

68. $p = 600 - 3x$

$C = 0.3x^2 + 6x + 600$

$P = xp - C - xt$

$\quad = x(600 - 3x) - (0.3x^2 + 600) - xt$

$\quad = -3.3x^2 + (594 - t)x - 600$

$\dfrac{dP}{dx} = -6.6x + 594 - t = 0$

$x = \dfrac{594 - t}{6.6}$

a. $t = 5$, $x = 89$, $P = 25{,}681.70$

b. $t = 10$, $x = 88$, $P = 25{,}236.80$

c. $t = 20$, $x = 87$, $P = 24{,}360.30$

69. $x^2 + 4y^2 - 2x - 16y + 13 = 0$

a. $(x^2 - 2x + 1) + 4(y^2 - 4y + 4) = -13 + 1 + 16$

$\qquad (x - 1)^2 + 4(y - 2)^2 = 4$

$\qquad \dfrac{(x - 1)^2}{4} + \dfrac{(y - 2)^2}{1} = 1$

The graph is an ellipse.

Maximum: $(1, 3)$

Minimum: $(1, 1)$

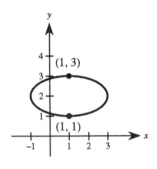

b. $x^2 + 4y^2 - 2x - 16y + 13 = 0$

$\qquad 2x + 8y\dfrac{dy}{dx} - 2 - 16\dfrac{dy}{dx} = 0$

$\qquad\qquad \dfrac{dy}{dx}(8y - 16) = 2 - 2x$

$\qquad\qquad\qquad \dfrac{dy}{dx} = \dfrac{2 - 2x}{8y - 16} = \dfrac{1 - x}{4y - 8}$

The critical numbers are $x = 1$ and $y = 2$. These correspond to the points $(1, 1)$, $(1, 3)$, $(2, -1)$, and $(2, 3)$. Hence, the maximum is $(1, 3)$ and the minimum is $(1, 1)$.

70. $f(x) = x^n$, n is a positive integer.

a. $f'(x) = nx^{n-1}$

The function has a relative minimum at $(0, 0)$ when n is even.

b. $f''(x) = n(n - 1)x^{n-2}$

The function has a point of inflection at $(0, 0)$ when n is odd and $n \geq 3$.

71. False

Let $f(x) = x^3$, $c = 0$.

72. False

Let $f(x) = \dfrac{2x}{\sqrt{x^2 + 2}}$.

f has horizontal asymptotes at $y = 2$ and $y = -2$.

73. The first and second derivatives are both positive. The graph is increasing and is concave upward.

CHAPTER 4
Integration

Section 4.1   Antiderivatives and Indefinite Integration

1. $\dfrac{dy}{dt} = 3t^2$

$y = t^3 + C$

Check: $\dfrac{d}{dt}[t^3 + C] = 3t^2$

2. $\dfrac{dr}{d\theta} = \pi$

$r = \pi\theta + C$

Check: $\dfrac{d}{d\theta}[\pi\theta + C] = \pi$

3. $\dfrac{dy}{dx} = x^{3/2}$

$y = \dfrac{2}{5}x^{5/2} + C$

Check: $\dfrac{d}{dx}\left[\dfrac{2}{5}x^{5/2} + C\right] = x^{3/2}$

4. $\dfrac{dy}{dx} = 3x^{-4}$

$y = -x^{-3} + C = -\dfrac{1}{x^3} + C$

Check: $\dfrac{d}{dx}[-x^{-3} + C] = 3x^{-4}$

	Given	*Rewrite*	*Integrate*	*Simplify*
5.	$\displaystyle\int \sqrt[3]{x}\, dx$	$\displaystyle\int x^{1/3}\, dx$	$\dfrac{x^{4/3}}{4/3} + C$	$\dfrac{3}{4}x^{4/3} + C$
6.	$\displaystyle\int \dfrac{1}{x^2}\, dx$	$\displaystyle\int x^{-2}\, dx$	$\dfrac{x^{-1}}{-1} + C$	$-\dfrac{1}{x} + C$
7.	$\displaystyle\int \dfrac{1}{x\sqrt{x}}\, dx$	$\displaystyle\int x^{-3/2}\, dx$	$\dfrac{x^{-1/2}}{-1/2} + C$	$-\dfrac{2}{\sqrt{x}} + C$
8.	$\displaystyle\int x(x^2 + 3)\, dx$	$\displaystyle\int (x^3 + 3x)\, dx$	$\dfrac{x^4}{4} + 3\left(\dfrac{x^2}{2}\right) + C$	$\dfrac{1}{4}x^4 + \dfrac{3}{2}x^2 + C$
9.	$\displaystyle\int \dfrac{1}{2x^3}\, dx$	$\dfrac{1}{2}\displaystyle\int x^{-3}\, dx$	$\dfrac{1}{2}\left(\dfrac{x^{-2}}{-2}\right) + C$	$-\dfrac{1}{4x^2} + C$
10.	$\displaystyle\int \dfrac{1}{(2x)^3}\, dx$	$\dfrac{1}{8}\displaystyle\int x^{-3}\, dx$	$\dfrac{1}{8}\left(\dfrac{x^{-2}}{-2}\right) + C$	$-\dfrac{1}{16x^2} + C$

11. $\displaystyle\int (x^3 + 2)\, dx = \dfrac{1}{4}x^4 + 2x + C$

Check: $\dfrac{d}{dx}\left(\dfrac{1}{4}x^4 + 2x + C\right) = x^3 + 2$

12. $\displaystyle\int (x^2 - 2x + 3)\, dx = \dfrac{1}{3}x^3 - x^2 + 3x + C$

Check: $\dfrac{d}{dx}\left(\dfrac{1}{3}x^3 - x^2 + 3x + C\right) = x^2 - 2x + 3$

13. $\displaystyle\int (x^{3/2} + 2x + 1)\, dx = \dfrac{2}{5}x^{5/2} + x^2 + x + C$

Check: $\dfrac{d}{dx}\left(\dfrac{2}{5}x^{5/2} + x^2 + x + C\right) = x^{3/2} + 2x + 1$

14. $\int \left(\sqrt{x} + \dfrac{1}{2\sqrt{x}} \right) dx = \int \left(x^{1/2} + \dfrac{1}{2} x^{-1/2} \right) dx = \dfrac{x^{3/2}}{3/2} + \dfrac{1}{2}\left(\dfrac{x^{1/2}}{1/2} \right) + C = \dfrac{2}{3} x^{3/2} + x^{1/2} + C$

Check: $\dfrac{d}{dx}\left(\dfrac{2}{3} x^{3/2} + x^{1/2} + C \right) = x^{1/2} + \dfrac{1}{2} x^{-1/2} = \sqrt{x} + \dfrac{1}{2\sqrt{x}}$

15. $\int \sqrt[3]{x^2}\, dx = \int x^{2/3}\, dx = \dfrac{x^{5/3}}{5/3} + C = \dfrac{3}{5} x^{5/3} + C$

Check: $\dfrac{d}{dx}\left(\dfrac{3}{5} x^{5/3} + C \right) = x^{2/3} = \sqrt[3]{x^2}$

16. $\int (\sqrt[4]{x^3} + 1)\, dx = \int (x^{3/4} + 1)\, dx = \dfrac{4}{7} x^{7/4} + x + C$

Check: $\dfrac{d}{dx}\left(\dfrac{4}{7} x^{7/4} + x + C \right) = x^{3/4} + 1 = \sqrt[4]{x^3} + 1$

17. $\int \dfrac{1}{x^3}\, dx = \int x^{-3}\, dx = \dfrac{x^{-2}}{-2} + C = -\dfrac{1}{2x^2} + C$

Check: $\dfrac{d}{dx}\left(-\dfrac{1}{2x^2} + C \right) = \dfrac{1}{x^3}$

18. $\int \dfrac{1}{x^4}\, dx = \int x^{-4}\, dx = \dfrac{x^{-3}}{-3} + C = -\dfrac{1}{3x^3} + C$

Check: $\dfrac{d}{dx}\left(-\dfrac{1}{3x^3} + C \right) = \dfrac{1}{x^4}$

19. $\int \dfrac{x^2 + x + 1}{\sqrt{x}}\, dx = \int (x^{3/2} + x^{1/2} + x^{-1/2})\, dx = \dfrac{2}{5} x^{5/2} + \dfrac{2}{3} x^{3/2} + 2x^{1/2} + C = \dfrac{2}{15} x^{1/2}(3x^2 + 5x + 15) + C$

Check: $\dfrac{d}{dx}\left(\dfrac{2}{5} x^{5/2} + \dfrac{2}{3} x^{3/2} + 2x^{1/2} + C \right) = x^{3/2} + x^{1/2} + x^{-1/2} = \dfrac{x^2 + x + 1}{\sqrt{x}}$

20. $\int \dfrac{x^2 + 1}{x^2}\, dx = \int (1 + x^{-2})\, dx = x + \dfrac{x^{-1}}{-1} + C = x - \dfrac{1}{x} + C$

Check: $\dfrac{d}{dx}\left(x - \dfrac{1}{x} + C \right) = 1 + \dfrac{1}{x^2} = \dfrac{x^2 + 1}{x^2}$

21. $\int (x + 1)(3x - 2)\, dx = \int (3x^2 + x - 2)\, dx = x^3 + \dfrac{1}{2} x^2 - 2x + C$

Check: $\dfrac{d}{dx}\left(x^3 + \dfrac{1}{2} x^2 - 2x + C \right) = 3x^2 + x - 2 = (x + 1)(3x - 2)$

22. $\int (2t^2 - 1)^2\, dt = \int (4t^4 - 4t^2 + 1)\, dt = \dfrac{4}{5} t^5 - \dfrac{4}{3} t^3 + t + C$

Check: $\dfrac{d}{dt}\left(\dfrac{4}{5} t^5 - \dfrac{4}{3} t^3 + t + C \right) = 4t^4 - 4t^2 + 1 = (2t^2 - 1)^2$

23. $\int y^2 \sqrt{y}\, dy = \int y^{5/2}\, dy = \dfrac{2}{7} y^{7/2} + C$

Check: $\dfrac{d}{dy}\left(\dfrac{2}{7} y^{7/2} + C \right) = y^{5/2} = y^2 \sqrt{y}$

24. $\int (1 + 3t)t^2\, dt = \int (t^2 + 3t^3)\, dt = \dfrac{1}{3} t^3 + \dfrac{3}{4} t^4 + C$

Check: $\dfrac{d}{dt}\left(\dfrac{1}{3} t^3 + \dfrac{3}{4} t^4 + C \right) = t^2 + 3t^3 = (1 + 3t)t^2$

25. $\int dx = \int 1\, dx = x + C$

Check: $\dfrac{d}{dx}(x + C) = 1$

26. $\int 3\, dt = 3t + C$

Check: $\dfrac{d}{dt}(3t + C) = 3$

27. $\int (2\sin x + 3\cos x)\, dx = -2\cos x + 3\sin x + C$

Check: $\dfrac{d}{dx}(-2\cos x + 3\sin x + C) = 2\sin x + 3\cos x$

28. $\int (t^2 - \sin t)\, dt = \dfrac{1}{3} t^3 + \cos t + C$

Check: $\dfrac{d}{dt}\left(\dfrac{1}{3} t^3 + \cos t + C \right) = t^2 - \sin t$

29. $\int (1 - \csc t \cot t)\, dt = t + \csc t + C$

Check: $\dfrac{d}{dt}(t + \csc t + C) = 1 - \csc t \cot t$

30. $\int (\theta^2 + \sec^2 \theta)\, d\theta = \dfrac{1}{3}\theta^3 + \tan \theta + C$

Check: $\dfrac{d}{d\theta}\left(\dfrac{1}{3}\theta^3 + \tan \theta + C\right) = \theta^2 + \sec^2 \theta$

31. $\int (\sec^2 \theta - \sin \theta)\, d\theta = \tan \theta + \cos \theta + C$

Check: $\dfrac{d}{d\theta}(\tan \theta + \cos \theta + C) = \sec^2 \theta - \sin \theta$

32. $\int \sec y(\tan y - \sec y)\, dy = \int (\sec y \tan y - \sec^2 y)\, dy = \sec y - \tan y + C$

Check: $\dfrac{d}{dy}(\sec y - \tan y + C) = \sec y \tan y - \sec^2 y = \sec y(\tan y - \sec y)$

33. $\int (\tan^2 y + 1)\, dy = \int \sec^2 y\, dy = \tan y + C$

Check: $\dfrac{d}{dy}(\tan y + C) = \sec^2 y = \tan^2 y + 1$

34. $\int \dfrac{\sin x}{1 - \sin^2 x}\, dx = \int \dfrac{\sin x}{\cos^2 x}\, dx$

$\qquad = \int \left(\dfrac{1}{\cos}\right)\left(\dfrac{\sin x}{\cos x}\right)\, dx$

$\qquad = \int \sec x \tan x\, dx = \sec x + C$

Check: $\dfrac{d}{dx}(\sec x + C) = \sec x \tan x = \dfrac{\sin x}{\cos^2 x}$

$\qquad = \dfrac{\sin x}{1 - \sin^2 x}$

35. $f(x) = \cos x$

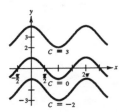

36. $f(x) = \sqrt{x}$

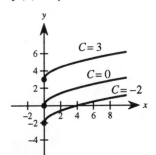

37. $f'(x) = 2$

$f(x) = 2x + C$

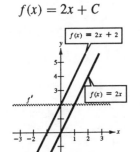

38. $f'(x) = x$

$f(x) = \dfrac{x^2}{2} + C$

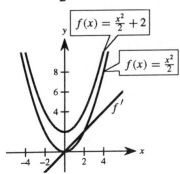

39. $f'(x) = 1 - x^2$

$f(x) = x - \dfrac{x^3}{3} + C$

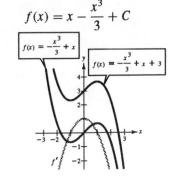

40. $f'(x) = \dfrac{1}{x^2}$

$f(x) = -\dfrac{1}{x} + C$

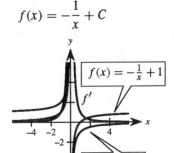

41. $\dfrac{dy}{dx} = 2x - 1$,   (1, 1)

$y = \displaystyle\int (2x-1)\,dx = x^2 - x + C$

$1 = (1)^2 - (1) + C \Rightarrow C = 1$

$y = x^2 - x + 1$

42. $\dfrac{dy}{dx} = 2(x-1) = 2x - 2$,   (3, 2)

$y = \displaystyle\int 2(x-1)\,dx = x^2 - 2x + C$

$2 = (3)^2 - 2(3) + C \Rightarrow C = -1$

$y = x^2 - 2x - 1$

43. $\dfrac{dy}{dx} = \cos x$,   (0, 4)

$y = \displaystyle\int \cos x\,dx = \sin x + C$

$4 = \sin 0 + C \Rightarrow C = 4$

$y = \sin x + 4$

44. $\dfrac{dy}{dx} = -\dfrac{1}{x^2} = -x^{-2}$,   (1, 3)

$y = \displaystyle\int -x^{-2}\,dx = \dfrac{1}{x} + C$

$3 = \dfrac{1}{1} + C \Rightarrow C = 2$

$y = \dfrac{1}{x} + 2$

45. $f''(x) = 2$

$f'(2) = 5$

$f(2) = 10$

$f'(x) = \displaystyle\int 2\,dx = 2x + C_1$

$f'(2) = 4 + C_1 = 5 \Rightarrow C_1 = 1$

$f'(x) = 2x + 1$

$f(x) = \displaystyle\int (2x+1)\,dx = x^2 + x + C_2$

$f(2) = 6 + C_2 = 10 \Rightarrow C_2 = 4$

$f(x) = x^2 + x + 4$

46. $f''(x) = x^2$

$f'(0) = 6$

$f(0) = 3$

$f'(x) = \displaystyle\int x^2\,dx = \tfrac{1}{3}x^3 + C_1$

$f'(0) = 0 + C_1 = 6 \Rightarrow C_1 = 6$

$f'(x) = \tfrac{1}{3}x^3 + 6$

$f(x) = \displaystyle\int \left(\tfrac{1}{3}x^3 + 6\right)dx = \tfrac{1}{12}x^4 + 6x + C_2$

$f(0) = 0 + 0 + C_2 = 3 \Rightarrow C_2 = 3$

$f(x) = \tfrac{1}{12}x^4 + 6x + 3$

47. $f''(x) = x^{-3/2}$

$f'(4) = 2$

$f(0) = 0$

$f'(x) = \displaystyle\int x^{-3/2}\,dx$

$\qquad = -2x^{-1/2} + C_1 = -\dfrac{2}{\sqrt{x}} + C_1$

$f'(4) = -\dfrac{2}{2} + C_1 = 2 \Rightarrow C_1 = 3$

$f'(x) = -\dfrac{2}{\sqrt{x}} + 3$

$f(x) = \displaystyle\int (-2x^{-1/2} + 3)\,dx$

$\qquad = -4x^{1/2} + 3x + C_2$

$f(0) = 0 + 0 + C_2 = 0 \Rightarrow C_2 = 0$

$f(x) = -4x^{1/2} + 3x = -4\sqrt{x} + 3x$

48. $f''(x) = \sin x$

$f'(0) = 1$

$f(0) = 6$

$f'(x) = \displaystyle\int \sin x\,dx = -\cos x + C_1$

$f'(0) = -1 + C_1 = 1 \Rightarrow C_1 = 2$

$f'(x) = -\cos x + 2$

$f(x) = \displaystyle\int (-\cos x + 2)\,dx$

$\qquad = -\sin x + 2x + C_2$

$f(0) = 0 + 0 + C_2 = 6 \Rightarrow C_2 = 6$

$f(x) = -\sin x + 2x + 6$

49. **a.** $h(t) = \int (0.5t + 2) \, dt = 0.25t^2 + 2t + C$

$h(0) = 0 + 0 + C = 5 \Rightarrow C = 5$

$h(t) = 0.25t^2 + 2t + 5 = \dfrac{t^2}{4} + 2t + 5$

b. $h(6) = 0.25(6)^2 + 2(6) + 5 = 26$ inches

50. $\dfrac{dP}{dt} = k\sqrt{t}, \quad 0 \le t \le 10$

$P(t) = \int kt^{1/2} \, dt = \dfrac{2}{3} kt^{3/2} + C$

$P(0) = 0 + C = 500 \Rightarrow C = 500$

$P(1) = \dfrac{2}{3} k + 500 = 600 \Rightarrow k = 150$

$P(t) = \dfrac{2}{3}(150)t^{3/2} + 500 = 100t^{3/2} + 500$

$P(7) = 100(7)^{3/2} + 500 \approx 2352$ bacteria

51. $f''(t) = a(t) = -32$ ft/sec^2

$f'(0) = v_0$

$f(0) = s_0$

$f'(t) = v(t) = \int -32 \, dt = -32t + C_1$

$f'(0) = 0 + C_1 = v_0 \Rightarrow C_1 = v_0$

$f'(t) = -32t + v_0$

$f(t) = s(t) = \int (-32t + v_0) \, dt$

$\quad = -16t^2 + v_0 t + C_2$

$f(0) = 0 + 0 + C_2 = s_0 \Rightarrow C_2 = s_0$

$f(t) = -16t^2 + v_0 t + s_0$

52. From Exercise 51, we have

$s(t) = -16t^2 + 60t$

$s'(t) = v(t) = -32t + 60 = 0$ when $t = 1.875$.

$s''(t) = -32 < 0.$

Maximum height at:

$s(1.875) = -56.25 + 112.50 = 56.25$ feet

53. From Exercise 51, we have:

$s(t) = -16t^2 + v_0 t$

$s'(t) = -32t + v_0 = 0$ when $t = \dfrac{v_0}{32} = $ time to reach maximum height.

$s\left(\dfrac{v_0}{32}\right) = -16\left(\dfrac{v_0}{32}\right)^2 + v_0\left(\dfrac{v_0}{32}\right) = 550$

$-\dfrac{v_0^2}{64} + \dfrac{v_0^2}{32} = 550$

$v_0^2 = 35,200$

$v_0 \approx 187.617$ ft/sec

54. $v_0 = 16$ ft/sec

$s_0 = 64$ ft

a. $\qquad s(t) = -16t^2 + 16t + 64 = 0$

$-16(t^2 - t - 4) = 0$

$$t = \frac{1 \pm \sqrt{17}}{2}$$

Choosing the positive value,

$t = \dfrac{1 + \sqrt{17}}{2} \approx 2.562$ seconds.

b. $\qquad v(t) = s'(t) = -32t + 16$

$$v\left(\frac{1 + \sqrt{17}}{2}\right) = -32\left(\frac{1 + \sqrt{17}}{2}\right) + 16$$

$$= -16\sqrt{17}$$

$$\approx -65.970 \text{ ft/sec}$$

55. $v(0) = 15 \text{ mph} = 15 \dfrac{5280 \text{ ft}}{3600 \text{ sec}} = 22$ ft/sec

$v(13) = 50 \text{ mph} = 50 \dfrac{5280 \text{ ft}}{3600 \text{ sec}} = \dfrac{220}{3}$ ft/sec

$s(0) = 0$

a. $a(t) = a$(constant acceleration)

$v(t) = at + C$

$v(0) = C = 22$ ft/sec

$v(t) = at + 22$

$v(13) = 13a + 22 = \dfrac{220}{3}$ when

$a = \dfrac{154}{39} \approx 3.95$ ft/sec^2.

b. $v(t) = \dfrac{154}{39}t + 22$

$s(t) = \left(\dfrac{77}{39}\right)t^2 + 22t$ since $s(0) = 0$.

$s(13) = \dfrac{77}{39}(13)^2 + 22(13)$

$= \dfrac{1859}{3}$ ft ≈ 619.67 ft

56. $v(0) = 45 \text{ mph} = 66$ ft/sec

$30 \text{ mph} = 44$ ft/sec

$15 \text{ mph} = 22$ ft/sec

$a(t) = -a$

$v(t) = -at + 66$

$s(t) = -\dfrac{a}{2}t^2 + 66t$ (Let $s(0) = 0$.)

$v(t) = 0$ after car moves 132 ft.

$-at + 66 = 0$ when $t = \dfrac{66}{a}$.

$s\left(\dfrac{66}{a}\right) = -\dfrac{a}{2}\left(\dfrac{66}{a}\right)^2 + 66\left(\dfrac{66}{a}\right)$

$= 132$ when $a = \dfrac{33}{2} = 16.5$.

$a(t) = -16.5$

$v(t) = -16.5t + 66$

$s(t) = -8.25t^2 + 66t$

a. $-16.5t + 66 = 44$

$$t = \frac{22}{16.5}$$

$s\left(\dfrac{22}{16.5}\right) \approx 73.33$ ft

b. $-16.5t + 66 = 22$

$$t = \frac{44}{16.5}$$

$s\left(\dfrac{44}{16.5}\right) \approx 117.33$ ft

c.

It takes 1.333 seconds to reduce the speed from 45 mph to 30 mph, 1.333 seconds to reduce the speed from 30 mph to 15 mph, and 1.333 seconds to reduce the speed from 15 mph to 0 mph.. Each time, less distance is needed to reach the next speed reduction.

57. Truck:

$v(t) = 30$

$s(t) = 30t$ (Let $s(0) = 0$.)

Automobile:

$a(t) = 6$

$v(t) = 6t$ (Let $v(0) = 0$.)

$s(t) = 3t^2$ (Let $s(0) = 0$.)

At the point where the automobile overtakes the truck:

$30t = 3t^2$

$0 = 3t^2 - 30t$

$0 = 3t(t - 10)$ when $t = 10$ sec

a. $s(10) = 3(10)^2 = 300$ ft

b. $v(10) = 6(10) = 60$ ft/sec ≈ 41 mph

58. $a(t) = k$

$v(t) = kt$

$s(t) = \dfrac{k}{2}t^2$ since $v(0) = s(0) = 0$.

At the time of lift-off, $kt = 160$ and $(k/2)t^2 = 0.7$. Since $(k/2)t^2 = 0.7$,

$$t = \sqrt{\frac{1.4}{k}}$$

$$v\left(\sqrt{\frac{1.4}{k}}\right) = k\sqrt{\frac{1.4}{k}} = 160$$

$$1.4k = 160^2 \Rightarrow k = \frac{160^2}{1.4}$$

$$\approx 18285.714 \text{ mi/hr}^2$$

$$\approx 7.45 \text{ ft/sec}^2.$$

59. Let d be the distance traversed and a be the uniform acceleration. Furthermore, note that $v(0) = 0$ and let $s(0) = 0$.

$a(t) = a$

$v(t) = at$

$s(t) = \dfrac{at^2}{2} = d$ when $t = \sqrt{\dfrac{2d}{a}}$.

The highest speed is

$$v\left(\sqrt{\frac{2d}{a}}\right) = a\sqrt{\frac{2d}{a}} = \sqrt{2ad}$$

and the mean speed is

$$\frac{\sqrt{2ad}}{2} = \sqrt{\frac{ad}{2}}.$$

The time necessary to traverse the distance at the mean speed must satisfy the equation

$$\sqrt{\frac{ad}{2}}\,t = d \text{ or } t = \sqrt{\frac{2d}{a}}.$$

This is the same time traversed under uniform acceleration.

60. $\displaystyle\int v\,dv = -GM\int \frac{1}{y^2}\,dy$

$$\frac{1}{2}v^2 = \frac{GM}{y} + C$$

When $y = R$, $v = v_0$.

$$\frac{1}{2}v_0^2 = \frac{GM}{R} + C$$

$$C = \frac{1}{2}v_0^2 - \frac{GM}{R}$$

$$\frac{1}{2}v^2 = \frac{GM}{y} + \frac{1}{2}v_0^2 - \frac{GM}{R}$$

$$v^2 = \frac{2GM}{y} + v_0^2 - \frac{2GM}{R}$$

$$v^2 = v_0^2 + 2GM\left(\frac{1}{y} - \frac{1}{R}\right)$$

61. $\dfrac{dC}{dx} = 2x - 12$

$C(0) = 50$

$C(x) = x^2 - 12x + C$

$C(0) = 0 + C = 50 \Rightarrow C = 50$

Cost: $C(x) = x^2 - 12x + 50$

Average cost:

$$\overline{C}(x) = \dfrac{C(x)}{x} = x - 12 + \dfrac{50}{x}$$

62. $\dfrac{dC}{dx} = \dfrac{\sqrt[4]{x}}{10} + 10$

$C(0) = 2300$

$C(x) = \displaystyle\int \left(\dfrac{1}{10} x^{1/4} + 10 \right) dx$

$\qquad = \dfrac{2}{25} x^{5/4} + 10x + C$

$C(0) = 0 + 0 + C = 2300 \Rightarrow C = 2300$

Cost: $C(x) = \dfrac{2}{25} x^{5/4} + 10x + 2300$

Average cost:

$$\overline{C}(x) = \dfrac{C(x)}{x} = \dfrac{2}{25} x^{1/4} + 10 + \dfrac{2300}{x}$$

63. $\dfrac{dR}{dx} = 100 - 5x$

$R = 100x - \dfrac{5}{2} x^2 + C$

$(C = 0$; if no units are sold, no revenue is generated.$)$

Revenue: $R = 100x - \dfrac{5}{2} x^2$

$R = xp = x \left(100 - \dfrac{5}{2} x \right)$

Demand: $p = 100 - \dfrac{5}{2} x$

64. $\dfrac{dR}{dx} = 100 - 6x - 2x^2$

$R = 100x - 3x^2 - \dfrac{2}{3} x^3 + C$

$(C = 0$; if no units are sold, no revenue is generated.$)$

Revenue: $R = 100x - 3x^2 - \dfrac{2}{3} x^3$

$R = xp = x \left(100 - 3x - \dfrac{2}{3} x^2 \right)$

Demand: $p = 100 - 3x - \dfrac{2}{3} x^2$

65. True

66. True

67. True

68. True

69. $\qquad a = -1.6$

$\quad v(t) = \displaystyle\int -1.6 \, dt = -1.6t + v_0, \quad$ since the stone was dropped, $v_0 = 0$.

$\quad s(t) = \displaystyle\int (-1.6t) \, dt = -0.8t^2 + s_0$

$\quad s(20) = 0 \Rightarrow -0.8(20)^2 + s_0 = 0$

$$s_0 = 320$$

Since the stone was dropped, $v_0 = 0$. Thus, the height of the cliff is 320 meters.

$\quad v(t) = -1.6t$

$\quad v(20) = -32$ meters/second

70. $\dfrac{d}{dx} \left[[s(x)]^2 + [c(x)]^2 \right] = 2s(x)s'(x) + 2c(x)c'(x)$

$\qquad\qquad\qquad\qquad\quad = 2s(x)c(x) - 2c(x)s(x)$

$\qquad\qquad\qquad\qquad\quad = 0$

Thus, $[s(x)]^2 + [c(x)]^2 = k$ for some constant k. Since, $s(0) = 0$ and $c(0) = 1$, $k = 1$. Therefore, $[s(x)]^2 + [c(x)]^2 = 1$.

71.

Since f'' is negative on $(-\infty, 0)$, f' is decreasing on $(-\infty, 0)$. Since f'' is positive on $(0, \infty)$, f' is increasing on $(0, \infty)$. f' has a relative minimum at $(0, 0)$. Since f' is positive on $(-\infty, \infty)$, f is increasing on $(-\infty, \infty)$.

Section 4.2   Area

1. $\displaystyle\sum_{i=1}^{5} (2i + 1) = 2\sum_{i=1}^{5} i + \sum_{i=1}^{5} 1 = 2(1 + 2 + 3 + 4 + 5) + 5 = 35$

2. $\displaystyle\sum_{k=2}^{5} (k + 1)(k - 3) = (3)(-1) + (4)(0) + (5)(1) + (6)(2) = 14$

3. $\displaystyle\sum_{k=0}^{4} \frac{1}{k^2 + 1} = 1 + \frac{1}{2} + \frac{1}{5} + \frac{1}{10} + \frac{1}{17} = \frac{158}{85}$

4. $\displaystyle\sum_{j=3}^{5} \frac{1}{j} = \frac{1}{3} + \frac{1}{4} + \frac{1}{5} = \frac{47}{60}$

5. $\displaystyle\sum_{k=1}^{4} c = c + c + c + c = 4c$

6. $\displaystyle\sum_{i=1}^{4} [(i - 1)^2 + (i + 1)^3] = (0 + 8) + (1 + 27) + (4 + 64) + (9 + 125) = 238$

7. $\displaystyle\sum_{i=1}^{9} \frac{1}{3i}$

8. $\displaystyle\sum_{i=1}^{15} \frac{5}{1 + i}$

9. $\displaystyle\sum_{j=1}^{8} \left[2\left(\frac{j}{8}\right) + 3 \right]$

10. $\displaystyle\sum_{j=1}^{4} \left[1 - \left(\frac{j}{4}\right)^2 \right]$

11. $\displaystyle\frac{2}{n} \sum_{i=1}^{n} \left[\left(\frac{2i}{n}\right)^3 - \left(\frac{2i}{n}\right) \right]$

12. $\displaystyle\frac{2}{n} \sum_{i=1}^{n} \left[1 - \left(\frac{2i}{n} - 1\right)^2 \right]$

13. $\displaystyle\frac{3}{n} \sum_{i=1}^{n} \left[2\left(1 + \frac{3i}{n}\right)^2 \right]$

14. $\displaystyle\frac{1}{n} \sum_{i=0}^{n-1} \sqrt{1 - \left(\frac{i}{n}\right)^2}$

15. $\displaystyle\sum_{i=1}^{20} 2i = 2\sum_{i=1}^{20} i$

$= 2\left[\frac{20(21)}{2}\right] = 420$

16. $\displaystyle\sum_{i=1}^{15} (2i - 3) = 2\sum_{i=1}^{15} i - 3(15)$

$= 2\left[\frac{15(16)}{2}\right] - 45 = 195$

17. $\displaystyle\sum_{i=1}^{20} (i - 1)^2 = \sum_{i=1}^{19} i^2$

$= \left[\frac{19(20)(39)}{6}\right] = 2470$

18. $\displaystyle\sum_{i=1}^{10} (i^2 - 1) = \sum_{i=1}^{10} i^2 - \sum_{i=1}^{10} 1$

$= \left[\frac{10(11)(21)}{6}\right] - 10 = 375$

19. $\displaystyle\sum_{i=1}^{15} \frac{1}{n^3}(i-1)^2 = \frac{1}{n^3}\sum_{i=1}^{14} i^2$

$\displaystyle\qquad\qquad = \frac{1}{n^3}\left[\frac{14(15)(29)}{6}\right]$

$\displaystyle\qquad\qquad = \frac{1015}{n^3}$

20. $\displaystyle\sum_{i=1}^{10} i(i^2+1) = \sum_{i=1}^{10} i^3 + \sum_{i=1}^{10} i$

$\displaystyle\qquad\qquad = \frac{10^2(11)^2}{4} + \left[\frac{10(11)}{2}\right]$

$\displaystyle\qquad\qquad = 3080$

21. $\displaystyle\lim_{n\to\infty}\left[\left(\frac{4}{3n^3}\right)(2n^3+3n^2+n)\right] = \lim_{n\to\infty}\left[\frac{8}{3}+\frac{4}{n}+\frac{4}{3n^2}\right] = \frac{8}{3}$

22. $\displaystyle\lim_{n\to\infty}\left(\frac{8}{3}+\frac{4}{n}+\frac{4}{3n^2}\right) = \frac{8}{3}$

23. $\displaystyle\lim_{n\to\infty}\left[\left(\frac{81}{n^4}\right)\frac{n^2(n+1)^2}{4}\right] = \frac{81}{4}\lim_{n\to\infty}\left[\frac{n^4+2n^3+n^2}{n^4}\right] = \frac{81}{4}(1) = \frac{81}{4}$

24. $\displaystyle\lim_{n\to\infty}\left[\left(\frac{64}{n^3}\right)\frac{n(n+1)(2n+1)}{6}\right] = \frac{64}{6}\lim_{n\to\infty}\left[\frac{2n^3+3n^2+n}{n^3}\right] = \frac{64}{6}(2) = \frac{64}{3}$

25. $\displaystyle\lim_{n\to\infty}\left[\left(\frac{18}{n^2}\right)\frac{n(n+1)}{2}\right] = \frac{18}{2}\lim_{n\to\infty}\left[\frac{n^2+n}{n^2}\right] = \frac{18}{2}(1) = 9$

26. $\displaystyle\lim_{n\to\infty}\left[\left(\frac{1}{n^2}\right)\frac{n(n+1)}{2}\right] = \frac{1}{2}\lim_{n\to\infty}\left[\frac{n^2+n}{n^2}\right] = \frac{1}{2}(1) = \frac{1}{2}$

27. $\displaystyle\lim_{n\to\infty}\sum_{i=1}^{n}\left(\frac{16i}{n^2}\right) = \lim_{n\to\infty}\frac{16}{n^2}\sum_{i=1}^{n} i = \lim_{n\to\infty}\frac{16}{n^2}\left(\frac{n(n+1)}{2}\right) = \lim_{n\to\infty}\left[8\left(\frac{n^2+n}{n^2}\right)\right] = 8\lim_{n\to\infty}\left(1+\frac{1}{n}\right) = 8$

28. $\displaystyle\lim_{n\to\infty}\sum_{i=1}^{n}\left(\frac{2i}{n}\right)\left(\frac{2}{n}\right) = \lim_{n\to\infty}\frac{4}{n^2}\sum_{i=1}^{n} i = \lim_{n\to\infty}\frac{4}{n^2}\left(\frac{n(n+1)}{2}\right) = \lim_{n\to\infty}\frac{4}{2}\left(1+\frac{1}{n}\right) = 2$

29. $\displaystyle\lim_{n\to\infty}\sum_{i=1}^{n}\frac{1}{n^3}(i-1)^2 = \lim_{n\to\infty}\frac{1}{n^3}\sum_{i=1}^{n-1} i^2 = \lim_{n\to\infty}\frac{1}{n^3}\left[\frac{(n-1)(n)(2n-1)}{6}\right]$

$\displaystyle\qquad\qquad = \lim_{n\to\infty}\frac{1}{6}\left[\frac{2n^3-3n^2+n}{n^3}\right] = \lim_{n\to\infty}\left[\frac{1}{6}\left(\frac{2-(3/n)+(1/n^2)}{1}\right)\right] = \frac{1}{3}$

30. $\displaystyle\lim_{n\to\infty}\sum_{i=1}^{n}\left(1+\frac{2i}{n}\right)^2\left(\frac{2}{n}\right) = \lim_{n\to\infty}\frac{2}{n^3}\sum_{i=1}^{n}(n+2i)^2$

$\displaystyle\qquad\qquad = \lim_{n\to\infty}\frac{2}{n^3}\left[\sum_{i=1}^{n} n^2 + 4n\sum_{i=1}^{n} i + 4\sum_{i=1}^{n} i^2\right]$

$\displaystyle\qquad\qquad = \lim_{n\to\infty}\frac{2}{n^3}\left[n^3 + (4n)\left(\frac{n(n+1)}{2}\right) + \frac{4(n)(n+1)(2n+1)}{6}\right]$

$\displaystyle\qquad\qquad = 2\lim_{n\to\infty}\left[1+2+\frac{2}{n}+\frac{4}{3}+\frac{2}{n}+\frac{2}{3n^2}\right]$

$\displaystyle\qquad\qquad = 2\left(1+2+\frac{4}{3}\right) = \frac{26}{3}$

31. $\displaystyle\lim_{n\to\infty}\sum_{i=1}^{n}\left(1+\frac{2i}{n}\right)^3\left(\frac{2}{n}\right)=2\lim_{n\to\infty}\frac{1}{n^4}\sum_{i=1}^{n}(n+2i)^3$

$$=2\lim_{n\to\infty}\frac{1}{n^4}\sum_{i=1}^{n}(n^3+6n^2i+12ni^2+8i^3)$$

$$=2\lim_{n\to\infty}\frac{1}{n^4}\left[n^4+6n^2\left(\frac{n(n+1)}{2}\right)+12n\left(\frac{n(n+1)(2n+1)}{6}\right)+8\left(\frac{n^2(n+1)^2}{4}\right)\right]$$

$$=2\lim_{n\to\infty}\left(1+3+\frac{3}{n}+4+\frac{6}{n}+\frac{2}{n^2}+2+\frac{4}{n}+\frac{2}{n^2}\right)$$

$$=2\lim_{n\to\infty}\left(10+\frac{13}{n}+\frac{4}{n^2}\right)=20$$

32. $\displaystyle\lim_{n\to\infty}\sum_{i=1}^{n}\left(1+\frac{i}{n}\right)\left(\frac{2}{n}\right)=2\lim_{n\to\infty}\frac{1}{n}\left[\sum_{i=1}^{n}1+\frac{1}{n}\sum_{i=1}^{n}i\right]=2\lim_{n\to\infty}\frac{1}{n}\left[n+\frac{1}{n}\left(\frac{n(n+1)}{2}\right)\right]$

$$=2\lim_{n\to\infty}\left[1+\frac{n^2+n}{2n^2}\right]=2\left(1+\frac{1}{2}\right)=3$$

33. $S(4)=\sqrt{\dfrac{1}{4}}\left(\dfrac{1}{4}\right)+\sqrt{\dfrac{1}{2}}\left(\dfrac{1}{4}\right)+\sqrt{\dfrac{3}{4}}\left(\dfrac{1}{4}\right)+\sqrt{1}\left(\dfrac{1}{4}\right)=\dfrac{3+\sqrt{2}+\sqrt{3}}{8}\approx0.768$

$s(4)=0\left(\dfrac{1}{4}\right)+\sqrt{\dfrac{1}{4}}\left(\dfrac{1}{4}\right)+\sqrt{\dfrac{1}{2}}\left(\dfrac{1}{4}\right)+\sqrt{\dfrac{3}{4}}\left(\dfrac{1}{4}\right)=\dfrac{1+\sqrt{2}+\sqrt{3}}{8}\approx0.518$

34. $S(8)=\left(\sqrt{\dfrac{1}{4}}+1\right)\dfrac{1}{4}+\left(\sqrt{\dfrac{1}{2}}+1\right)\dfrac{1}{4}+\left(\sqrt{\dfrac{3}{4}}+1\right)\dfrac{1}{4}+(\sqrt{1}+1)\dfrac{1}{4}$

$$+\left(\sqrt{\dfrac{5}{4}}+1\right)\dfrac{1}{4}+\left(\sqrt{\dfrac{3}{2}}+1\right)\dfrac{1}{4}+\left(\sqrt{\dfrac{7}{4}}+1\right)\dfrac{1}{4}+(\sqrt{2}+1)\dfrac{1}{4}$$

$$=\dfrac{1}{4}\left(8+\dfrac{1}{2}+\dfrac{\sqrt{2}}{2}+\dfrac{\sqrt{3}}{2}+1+\dfrac{\sqrt{5}}{2}+\dfrac{\sqrt{6}}{2}+\dfrac{\sqrt{7}}{2}+\sqrt{2}\right)\approx4.038$$

$s(8)=(0+1)\dfrac{1}{4}+\left(\sqrt{\dfrac{1}{4}}+1\right)\dfrac{1}{4}+\left(\sqrt{\dfrac{1}{2}}+1\right)\dfrac{1}{4}+\cdots+\left(\sqrt{\dfrac{7}{4}}+1\right)\dfrac{1}{4}\approx3.685$

35. $S(5)=1\left(\dfrac{1}{5}\right)+\dfrac{1}{6/5}\left(\dfrac{1}{5}\right)+\dfrac{1}{7/5}\left(\dfrac{1}{5}\right)+\dfrac{1}{8/5}\left(\dfrac{1}{5}\right)+\dfrac{1}{9/5}\left(\dfrac{1}{5}\right)=\dfrac{1}{5}+\dfrac{1}{6}+\dfrac{1}{7}+\dfrac{1}{8}+\dfrac{1}{9}\approx0.746$

$s(5)=\dfrac{1}{6/5}\left(\dfrac{1}{5}\right)+\dfrac{1}{7/5}\left(\dfrac{1}{5}\right)+\dfrac{1}{8/5}\left(\dfrac{1}{5}\right)+\dfrac{1}{9/5}\left(\dfrac{1}{5}\right)+\dfrac{1}{2}\left(\dfrac{1}{5}\right)=\dfrac{1}{6}+\dfrac{1}{7}+\dfrac{1}{8}+\dfrac{1}{9}+\dfrac{1}{10}\approx0.646$

36. $S(4)=\dfrac{1}{4-2}\left(\dfrac{1}{2}\right)+\dfrac{1}{4.5-2}\left(\dfrac{1}{2}\right)+\dfrac{1}{5-2}\left(\dfrac{1}{2}\right)+\dfrac{1}{5.5-2}\left(\dfrac{1}{2}\right)=\dfrac{1}{2}\left(\dfrac{1}{2}+\dfrac{2}{5}+\dfrac{1}{3}+\dfrac{2}{7}\right)\approx0.760$

$s(4)=\dfrac{1}{4.5-2}\left(\dfrac{1}{2}\right)+\dfrac{1}{5-2}\left(\dfrac{1}{2}\right)+\dfrac{1}{5.5-2}\left(\dfrac{1}{2}\right)+\dfrac{1}{6-2}\left(\dfrac{1}{2}\right)=\dfrac{1}{2}\left(\dfrac{2}{5}+\dfrac{1}{3}+\dfrac{2}{7}+\dfrac{1}{4}\right)\approx0.635$

37. $S(5)=1\left(\dfrac{1}{5}\right)+\sqrt{1-\left(\dfrac{1}{5}\right)^2}\left(\dfrac{1}{5}\right)+\sqrt{1-\left(\dfrac{2}{5}\right)^2}\left(\dfrac{1}{5}\right)+\sqrt{1-\left(\dfrac{3}{5}\right)^2}\left(\dfrac{1}{5}\right)+\sqrt{1-\left(\dfrac{4}{5}\right)^2}\left(\dfrac{1}{5}\right)$

$$=\dfrac{1}{5}\left[1+\dfrac{\sqrt{24}}{5}+\dfrac{\sqrt{21}}{5}+\dfrac{\sqrt{16}}{5}+\dfrac{\sqrt{9}}{5}\right]\approx0.859$$

$s(5)=\sqrt{1-\left(\dfrac{1}{5}\right)^2}\left(\dfrac{1}{5}\right)+\sqrt{1-\left(\dfrac{2}{5}\right)^2}\left(\dfrac{1}{5}\right)+\sqrt{1-\left(\dfrac{3}{5}\right)^2}\left(\dfrac{1}{5}\right)+\sqrt{1-\left(\dfrac{4}{5}\right)^2}\left(\dfrac{1}{5}\right)+0\approx0.659$

38. $y = x^2 - 1 \Rightarrow x = \sqrt{y+1}$

$$S(4) = \sqrt{\frac{1}{4}+1}\left(\frac{1}{4}\right) + \sqrt{\frac{1}{2}+1}\left(\frac{1}{4}\right) + \sqrt{\frac{3}{4}+1}\left(\frac{1}{4}\right) + \sqrt{2}\left(\frac{1}{4}\right) = \frac{\sqrt{5}+\sqrt{6}+\sqrt{7}+\sqrt{8}}{8} \approx 1.270$$

$$s(4) = 1\left(\frac{1}{4}\right) + \sqrt{\frac{1}{4}+1}\left(\frac{1}{4}\right) + \sqrt{\frac{1}{2}+1}\left(\frac{1}{4}\right) + \sqrt{\frac{3}{4}+1}\left(\frac{1}{4}\right) = \frac{2+\sqrt{5}+\sqrt{6}+\sqrt{7}}{8} \approx 1.166$$

39. a.

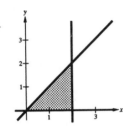

b. $\Delta x = \dfrac{2-0}{n} = \dfrac{2}{n}$

Endpoints:

$$0 < 1\left(\frac{2}{n}\right) < 2\left(\frac{2}{n}\right) < \cdots < (n-1)\left(\frac{2}{n}\right) < n\left(\frac{2}{n}\right) = 2$$

c. Since $y = x$ is increasing, $f(m_i) = f(x_{i-1})$ on $[x_{i-1}, \ x_i]$.

$$s(n) = \sum_{i=1}^{n} f(x_i)\Delta x$$

$$= \sum_{i=1}^{n} f\left(\frac{2i-2}{n}\right)\left(\frac{2}{n}\right) = \sum_{i=1}^{n} \left[(i-1)\left(\frac{2}{n}\right)\right]\left(\frac{2}{n}\right)$$

d. $f(M_i) = f(x_i)$ on $[x_{i-1}, \ x_i]$

$$S(n) = \sum_{i=1}^{n} f(x_i)\Delta x = \sum_{i=1}^{n} f\left(\frac{2i}{n}\right)\frac{2}{n} = \sum_{i=1}^{n} \left[i\left(\frac{2}{n}\right)\right]\left(\frac{2}{n}\right)$$

e.

n	5	10	50	100
$s(n)$	1.6	1.8	1.96	1.98
$S(n)$	2.4	2.2	2.04	2.02

f. $\displaystyle \lim_{n\to\infty} \sum_{i=1}^{n} \left[(i-1)\left(\frac{2}{n}\right)\right]\left(\frac{2}{n}\right) = \lim_{n\to\infty} \frac{4}{n^2} \sum_{i=1}^{n} (i-1)$

$$= \lim_{n\to\infty} \frac{4}{n^2}\left[\frac{n(n+1)}{2} - n\right] = \lim_{n\to\infty} \left[\frac{2(n+1)}{n} - \frac{4}{n}\right] = 2$$

$$\lim_{n\to\infty} \sum_{i=1}^{n} \left[i\left(\frac{2}{n}\right)\right]\left(\frac{2}{n}\right) = \lim_{n\to\infty} \frac{4}{n^2}\sum_{i=1}^{n} i = \lim_{n\to\infty} \left(\frac{4}{n^2}\right)\frac{n(n+1)}{2} = \lim_{n\to\infty} \frac{2(n+1)}{n} = 2$$

40. a.

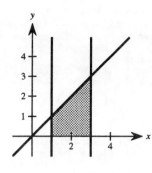

b. $\Delta x = \dfrac{3-1}{n} = \dfrac{2}{n}$

Endpoints:

$$1 < 1 + \frac{2}{n} < 1 + \frac{4}{n} < \cdots < 1 + \frac{2n}{n} = 3$$

$$1 < 1 + 1\left(\frac{2}{n}\right) < 1 + 2\left(\frac{2}{n}\right) < \cdots < 1 + (n-1)\left(\frac{2}{n}\right) < 1 + n\left(\frac{2}{n}\right)$$

c. Since $y = x$ is increasing, $f(m_i) = f(x_{i-1})$ on $[x_{i-1},\ x_i]$.

$$s(n) = \sum_{i=1}^{n} f(x_{i-1})\Delta x$$

$$= \sum_{i=1}^{n} f\left[1 + (i-1)\left(\frac{2}{n}\right)\right]\left(\frac{2}{n}\right) = \sum_{i=1}^{n}\left[1 + (i-1)\left(\frac{2}{n}\right)\right]\left(\frac{2}{n}\right)$$

d. $f(M_i) = f(x_i)$ on $[x_{i-1},\ x_i]$

$$S(n) = \sum_{i=1}^{n} f(x_i)\Delta x = \sum_{i=1}^{n} f\left[1 + i\left(\frac{2}{n}\right)\right]\left(\frac{2}{n}\right) = \sum_{i=1}^{n}\left[1 + i\left(\frac{2}{n}\right)\right]\left(\frac{2}{n}\right)$$

e.

n	5	10	50	100
$s(n)$	3.6	3.8	3.96	3.98
$S(n)$	4.4	4.2	4.04	4.02

f. $\displaystyle\lim_{n\to\infty}\sum_{i=1}^{n}\left[1 + (i-1)\left(\frac{2}{n}\right)\right]\left(\frac{2}{n}\right) = \lim_{n\to\infty}\left(\frac{2}{n}\right)\left[n + \frac{2}{n}\left(\frac{n(n-1)}{2} - n\right)\right]$

$$= \lim_{n\to\infty}\left[2 + \frac{2n+2}{n} - \frac{4}{n}\right] = \lim_{n\to\infty}\left[4 - \frac{2}{n}\right] = 4$$

$$\lim_{n\to\infty}\sum_{i=1}^{n}\left[1 + i\left(\frac{2}{n}\right)\right]\left(\frac{2}{n}\right) = \lim_{n\to\infty}\frac{2}{n}\left[n + \left(\frac{2}{n}\right)\frac{n(n+1)}{2}\right]$$

$$= \lim_{n\to\infty}\left[2 + \frac{2(n+1)}{n}\right] = \lim_{n\to\infty}\left[4 + \frac{2}{n}\right] = 4$$

41. $y = -2x + 3$ on $[0,\ 1]$. $\left(\text{Note: } \Delta x = \dfrac{1-0}{n} = \dfrac{1}{n}\right)$

$$s(n) = \sum_{i=1}^{n} f\left(\frac{i}{n}\right)\left(\frac{1}{n}\right) = \sum_{i=1}^{n}\left[-2\left(\frac{i}{n}\right) + 3\right]\left(\frac{1}{n}\right)$$

$$= 3 - \frac{2}{n^2}\sum_{i=1}^{n} i = 3 - \frac{2(n+1)n}{2n^2} = 2 - \frac{1}{n}$$

Area $= \displaystyle\lim_{n\to\infty} s(n) = 2$

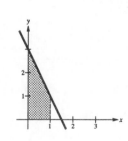

42. $y = 3x - 4$ on [2, 5]. $\left(\textbf{Note: } \Delta x = \dfrac{5-2}{n} = \dfrac{3}{n}\right)$

$$S(n) = \sum_{i=1}^{n} f\left(2 + \frac{3i}{n}\right)\left(\frac{3}{n}\right)$$

$$= \sum_{i=1}^{n} \left[3\left(2 + \frac{3i}{n}\right) - 4\right]\left(\frac{3}{n}\right) = 18 + 3\left(\frac{3}{n}\right)^2 \sum_{i=1}^{n} i - 12$$

$$= 6 + \frac{27}{n^2}\left(\frac{(n+1)n}{2}\right) = 6 + \frac{27}{2}\left(1 + \frac{1}{n}\right)$$

Area $= \displaystyle\lim_{n\to\infty} S(n) = 6 + \frac{27}{2} = \frac{39}{2}$

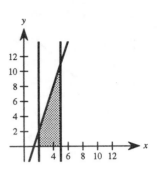

43. $y = x^2 + 2$ on [0, 1]. $\left(\textbf{Note: } \Delta x = \dfrac{1}{n}\right)$

$$S(n) = \sum_{i=1}^{n} f\left(\frac{i}{n}\right)\left(\frac{1}{n}\right) = \sum_{i=1}^{n} \left[\left(\frac{i}{n}\right)^2 + 2\right]\left(\frac{1}{n}\right)$$

$$= \left[\frac{1}{n^3}\sum_{i=1}^{n} i^2\right] + 2 = \frac{n(n+1)(2n+1)}{6n^3} + 2 = \frac{1}{6}\left(2 + \frac{3}{n} + \frac{1}{n^2}\right) + 2$$

Area $= \displaystyle\lim_{n\to\infty} S(n) = \frac{7}{3}$

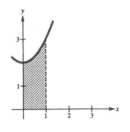

44. $y = 1 - x^2$ on [−1, 1]. Find area of region over the interval [0, 1].

$\left(\textbf{Note: } \Delta x = \dfrac{1}{n}\right)$

$$s(n) = \sum_{i=1}^{n} f\left(\frac{i}{n}\right)\left(\frac{1}{n}\right) = \sum_{i=1}^{n}\left[1 - \left(\frac{i}{n}\right)^2\right]\frac{1}{n}$$

$$= 1 - \frac{1}{n^3}\sum_{i=1}^{n} i^2 = 1 - \frac{n(n+1)(2n+1)}{6n^3} = 1 - \frac{1}{6}\left(2 + \frac{3}{n} + \frac{1}{n^2}\right)$$

$\dfrac{1}{2}$Area $= \displaystyle\lim_{n\to\infty} s(n) = 1 - \frac{1}{3} = \frac{2}{3}$

Area $= \dfrac{4}{3}$

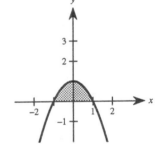

45. $y = 27 - x^3$ on [1, 3]. $\left(\textbf{Note:}\quad \Delta x = \dfrac{3-1}{n} = \dfrac{2}{n}\right)$

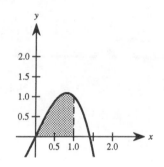

$$s(n) = \sum_{i=1}^{n} f\left(1 + \frac{2i}{n}\right)\left(\frac{2}{n}\right)$$

$$= \sum_{i=1}^{n}\left[27 - \left(1 + \frac{2i}{n}\right)^3\right]\left(\frac{2}{n}\right)$$

$$= \frac{2}{n}\sum_{i=1}^{n}\left(26 - \frac{6i}{n} - \frac{12i^2}{n^2} - \frac{8i^3}{n^3}\right)$$

$$= \frac{2}{n}\left[26n - \frac{6}{n}\left(\frac{n(n+1)}{2}\right) - \frac{12}{n^2}\left(\frac{n(n+1)(2n+1)}{6}\right) - \frac{8}{n^3}\left(\frac{n^2(n+1)^2}{4}\right)\right]$$

$$= 34 - \frac{26}{n} - \frac{8}{n^2}$$

Area $= \lim_{n\to\infty} s(n) = 34$

46. $y = 2x - x^3$ on [0, 1]. $\left(\textbf{Note:}\ \Delta x = \dfrac{1-0}{n} = \dfrac{1}{n}\right)$

Since y both increases and decreases on [0, 1], $T(n)$ is neither an upper nor a lower sum.

$$T(n) = \sum_{i=1}^{n} f\left(\frac{i}{n}\right)\left(\frac{1}{n}\right)$$

$$= \sum_{i=1}^{n}\left[2\left(\frac{i}{n}\right) - \left(\frac{i}{n}\right)^3\right]\left(\frac{1}{n}\right)$$

$$= \frac{2}{n^2}\sum_{i=1}^{n} i - \frac{1}{n^4}\sum_{i=1}^{n} i^3$$

$$= \frac{n(n+1)}{n^2} - \frac{1}{n^4}\left[\frac{n^2(n+1)^2}{4}\right]$$

$$= 1 + \frac{1}{n} - \frac{1}{4} - \frac{2}{4n} - \frac{1}{4n^2}$$

Area $= \lim_{n\to\infty} T(n) = 1 - \frac{1}{4} = \frac{3}{4}$

47. $y = x^2 - x^3$ on $[-1, 1]$. $\left(\text{Note: } \Delta x = \dfrac{1 - (-1)}{n} = \dfrac{2}{n}\right)$

Again, $T(n)$ is neither an upper nor a lower sum.

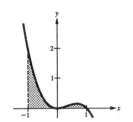

$$T(n) = \sum_{i=1}^{n} f\left(-1 + \frac{2i}{n}\right)\left(\frac{2}{n}\right)$$

$$= \sum_{i=1}^{n} \left[\left(-1 + \frac{2i}{n}\right)^2 - \left(-1 + \frac{2i}{n}\right)^3\right]\left(\frac{2}{n}\right)$$

$$= \sum_{i=1}^{n} \left[\left(1 - \frac{4i}{n} + \frac{4i^2}{n^2}\right) - \left(-1 + \frac{6i}{n} - \frac{12i^2}{n^2} + \frac{8i^3}{n^3}\right)\right]\left(\frac{2}{n}\right)$$

$$= \sum_{i=1}^{n} \left[2 - \frac{10i}{n} + \frac{16i^2}{n^2} - \frac{8i^3}{n^3}\right]\left(\frac{2}{n}\right)$$

$$= \frac{4}{n} \sum_{i=1}^{n} 1 - \frac{20}{n^2} \sum_{i=1}^{n} i + \frac{32}{n^3} \sum_{i=1}^{n} i^2 - \frac{16}{n^4} \sum_{i=1}^{n} i^3$$

$$= \frac{4}{n}(n) - \frac{20}{n^2} \cdot \frac{n(n+1)}{2} + \frac{32}{n^3} \cdot \frac{n(n+1)(2n+1)}{6} - \frac{16}{n^4} \cdot \frac{n^2(n+1)^2}{4}$$

$$= 4 - 10\left(1 + \frac{1}{n}\right) + \frac{16}{3}\left(2 + \frac{3}{n} + \frac{1}{n^2}\right) - 4\left(1 + \frac{2}{n} + \frac{1}{n^2}\right)$$

$$\text{Area} = \lim_{n \to \infty} T(n) = 4 - 10 + \frac{32}{3} - 4 = \frac{2}{3}$$

48. $y = x^2 - x^3$ on $[-1, 0]$. $\left(\text{Note: } \Delta x = \dfrac{0 - (-1)}{n} = \dfrac{1}{n}\right)$

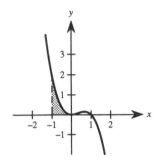

$$s(n) = \sum_{i=1}^{n} f\left(-1 + \frac{i}{n}\right)\left(\frac{1}{n}\right)$$

$$= \sum_{i=1}^{n} \left[\left(-1 + \frac{i}{n}\right)^2 - \left(-1 + \frac{i}{n}\right)^3\right]\left(\frac{1}{n}\right)$$

$$= \sum_{i=1}^{n} \left[2 - \frac{5i}{n} + \frac{4i^2}{n^2} - \frac{i^3}{n^3}\right]\left(\frac{1}{n}\right)$$

$$= 2 - \frac{5}{n^2} \sum_{i=1}^{n} i + \frac{4}{n^3} \sum_{i=1}^{n} i^2 - \frac{1}{n^4} \sum_{i=1}^{n} i^3$$

$$= 2 - \frac{5}{2} - \frac{5}{2n} + \frac{4}{3} + \frac{2}{n} + \frac{2}{3n^3} - \frac{1}{4} - \frac{1}{2n} - \frac{1}{4n^2}$$

$$\text{Area} = \lim_{n \to \infty} s(n) = 2 - \frac{5}{2} + \frac{4}{3} - \frac{1}{4} = \frac{7}{12}$$

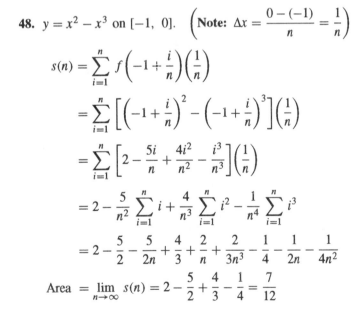

49. $f(y) = 3y$ on $[0, 2]$. $\left(\textbf{Note: } \Delta y = \dfrac{2-0}{n} = \dfrac{2}{n}\right)$

$$S(n) = \sum_{i=1}^{n} f(m_i)\Delta y = \sum_{i=1}^{n} f\left(\frac{2i}{n}\right)\left(\frac{2}{n}\right) = \sum_{i=1}^{n} 3\left(\frac{2i}{n}\right)\left(\frac{2}{n}\right)$$

$$= \frac{12}{n^2}\sum_{i=1}^{n} i = \left(\frac{12}{n^2}\right)\cdot\frac{n(n+1)}{2} = \frac{6(n+1)}{n} = 6 + \frac{6}{n}$$

Area $= \lim_{n\to\infty} S(n) = \lim_{n\to\infty}\left(6 + \dfrac{6}{n}\right) = 6$

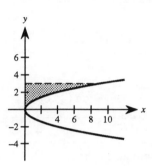

50. $f(y) = y^2$ on $[0, 3]$. $\left(\textbf{Note: } \Delta y = \dfrac{3-0}{n} = \dfrac{3}{n}\right)$

$$S(n) = \sum_{i=1}^{n} f\left(\frac{3i}{n}\right)\left(\frac{3}{n}\right) = \sum_{i=1}^{n}\left(\frac{3i}{n}\right)^2\left(\frac{3}{n}\right) = \frac{27}{n^3}\sum_{i=1}^{n} i^2$$

$$= \frac{27}{n^3}\cdot\frac{n(n+1)(2n+1)}{6} = \frac{9}{n^2}\left(\frac{2n^2+3n+1}{2}\right) = 9 + \frac{27}{2n} + \frac{9}{2n^2}$$

Area $= \lim_{n\to\infty} S(n) = \lim_{n\to\infty}\left(9 + \dfrac{27}{2n} + \dfrac{9}{2n^2}\right) = 9$

51. $f(x) = x^2 + 3,\ \ 0 \le x \le 2,\ \ n = 4$

Let $c_i = \dfrac{x_i + x_{i-1}}{2}$.

$\Delta x = \dfrac{1}{2},\ \ c_1 = \dfrac{1}{4},\ \ c_2 = \dfrac{3}{4},\ \ c_3 = \dfrac{5}{4},\ \ c_4 = \dfrac{7}{4}$

Area $\approx \displaystyle\sum_{i=1}^{n} f(c_i)\Delta x = \sum_{i=1}^{4} [c_i^2 + 3]\left(\frac{1}{2}\right) = \frac{1}{2}\left[\left(\frac{1}{16}+3\right) + \left(\frac{9}{16}+3\right) + \left(\frac{25}{16}+3\right) + \left(\frac{49}{16}+3\right)\right] = \frac{69}{8}$

52. $f(x) = x^2 + 4x,\ \ 0 \le x \le 4,\ \ n = 4$

Let $c_i = \dfrac{x_i + x_{i-1}}{2}$.

$\Delta x = 1,\ \ c_1 = \dfrac{1}{2},\ \ c_2 = \dfrac{3}{2},\ \ c_3 = \dfrac{5}{2},\ \ c_4 = \dfrac{7}{2}$

Area $\approx \displaystyle\sum_{i=1}^{n} f(c_i)\Delta x = \sum_{i=1}^{4} [c_i^2 + 4c_i](1) = \left[\left(\frac{1}{4}+2\right) + \left(\frac{9}{4}+6\right) + \left(\frac{25}{4}+10\right) + \left(\frac{49}{4}+14\right)\right] = 53$

53. $f(x) = \tan x,\ \ 0 \le x \le \dfrac{\pi}{4},\ \ n = 4$

Let $c_i = \dfrac{x_i + x_{i-1}}{2}$.

$\Delta x = \dfrac{\pi}{16},\ \ c_1 = \dfrac{\pi}{32},\ \ c_2 = \dfrac{3\pi}{32},\ \ c_3 = \dfrac{5\pi}{32},\ \ c_4 = \dfrac{7\pi}{32}$

Area $\approx \displaystyle\sum_{i=1}^{n} f(c_i)\Delta x = \sum_{i=1}^{4} (\tan c_i)\left(\frac{\pi}{16}\right) = \frac{\pi}{16}\left(\tan\frac{\pi}{32} + \tan\frac{3\pi}{32} + \tan\frac{5\pi}{32} + \tan\frac{7\pi}{32}\right) \approx 0.345$

54. $f(x) = \sin x, \quad 0 \le x \le \dfrac{\pi}{2}, \quad n = 4$

Let $c_i = \dfrac{x_i + x_{i-1}}{2}$.

$\Delta x = \dfrac{\pi}{8}, \quad c_1 = \dfrac{\pi}{16}, \quad c_2 = \dfrac{3\pi}{16}, \quad c_3 = \dfrac{5\pi}{16}, \quad c_4 = \dfrac{7\pi}{16}$

$\text{Area} \approx \sum_{i=1}^{n} f(c_i)\Delta x = \sum_{i=1}^{4} (\sin c_i)\left(\dfrac{\pi}{8}\right) = \dfrac{\pi}{8}\left(\sin\dfrac{\pi}{16} + \sin\dfrac{3\pi}{16} + \sin\dfrac{5\pi}{16} + \sin\dfrac{7\pi}{16}\right) \approx 1.006$

55. $f(x) = \sqrt{x}$ on $[0, \ 4]$.

n	4	8	12	16	20
Approximate area	5.3838	5.3523	5.3439	5.3403	5.3384

56. $f(x) = \dfrac{8}{x^2 + 1}$ on $[2, \ 6]$.

n	4	8	12	16	20
Approximate area	2.3397	2.3755	2.3824	2.3848	2.3860

57. $f(x) = \tan\left(\dfrac{\pi x}{8}\right)$ on $[1, \ 3]$.

n	4	8	12	16	20
Approximate area	2.2223	2.2387	2.2418	2.2430	2.2435

58. $f(x) = \cos\sqrt{x}$ on $[0, \ 2]$.

n	4	8	12	16	20
Approximate area	1.1041	1.1053	1.1055	1.1056	1.1056

59.

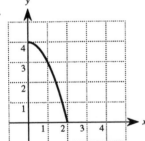

b. $A \approx 6$ square units

60.

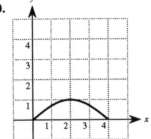

a. $A \approx 3$ square units

61. True

62. True

63. $f(x) = \sin x,$ $\left[0, \dfrac{\pi}{2}\right]$

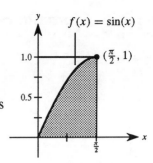

Let A_1 = area bounded by $f(x) = \sin x$, the x-axis, $x = 0$ and $x = \pi/2$. Let A_2 = area of the rectangle bounded by $y = 1$, $y = 0$, $x = 0$, and $x = \pi/2$. Thus, $A_2 = (\pi/2)(1) \approx 1.570796$. In this program, the computer is generating N_2 pairs of random points in the rectangle whose area is represented by A_2. It is keeping track of how many of these points, N_1, lie in the region whose area is represented by A_1. Since the points are randomly generated, we assume that

$$\frac{A_1}{A_2} \approx \frac{N_1}{N_2} \Rightarrow A_1 \approx \frac{N_1}{N_2} A_2.$$

The larger N_2 is, the better the approximation to A_1.

64. a. $\theta = \dfrac{2\pi}{n}$

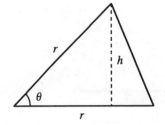

b. $\sin \theta = \dfrac{h}{r}$

$h = r \sin \theta$

$A = \dfrac{1}{2}bh = \dfrac{1}{2}r(r \sin \theta) = \dfrac{1}{2}r^2 \sin \theta$

c. $A_n = n \left(\dfrac{1}{2}r^2 \sin \dfrac{2\pi}{n}\right) = \dfrac{r^2 n}{2} \sin \dfrac{2\pi}{n} = \pi r^2 \left(\dfrac{\sin(2\pi/n)}{2\pi/n}\right)$

Let $x = 2\pi/n$. As $n \to \infty$, $x \to 0$.

$\displaystyle \lim_{n \to \infty} A_n = \lim_{x \to 0} \pi r^2 \left(\frac{\sin x}{x}\right) = \pi r^2(1) = \pi r^2$

65. a. $(i + 1)^3 - i^3 = (i^3 + 3i^2 + 3i + 1) - i^3 = 3i^2 + 3i + 1$

b. $\displaystyle \sum_{i=1}^{n} (3i^2 + 3i + 1) = \sum_{i=1}^{n} \left[(i + 1)^3 - i^3\right]$

$$= (2^3 - 1^3) + (3^3 - 2^3) + (4^3 - 3^3) + \cdots + (n^3 - (n - 1)^3) + ((n + 1)^3 - n^3)$$

$$= -1 + (n + 1)^3$$

c. $\displaystyle \sum_{i=1}^{n} (3i^2 + 3i + 1) = 3\sum_{i=1}^{n} i^2 + 3\sum_{i=1}^{n} i + \sum_{i=1}^{n} 1$

$$-1 + (n + 1)^3 = 3\sum_{i=1}^{n} i^2 + \frac{3n(n + 1)}{2} + n$$

$$n^3 + 3n^2 + 3n = 3\sum_{i=1}^{n} i^2 + \frac{3}{2}n^2 + \frac{5}{2}n$$

$$\frac{n^3 + (3/2)n^2 + (1/2)n}{3} = \sum_{i=1}^{n} i^2$$

$$\frac{2n^3 + 3n^2 + n}{6} = \sum_{i=1}^{n} i^2$$

$$\frac{n(2n + 1)(n + 1)}{6} = \sum_{i=1}^{n} i^2$$

Section 4.3   Riemann Sums and Definite Integrals

1. $\displaystyle\int_0^5 3\,dx$

2. $\displaystyle\int_0^2 (4-2x)\,dx$

3. $\displaystyle\int_{-4}^4 (4-|x|)\,dx$

4. $\displaystyle\int_0^2 x^2\,dx$

5. $\displaystyle\int_{-2}^2 (4-x^2)\,dx$

6. $\displaystyle\int_{-1}^1 \frac{1}{x^2+1}\,dx$

7. $\displaystyle\int_0^\pi \sin x\,dx$

8. $\displaystyle\int_0^{\pi/4} \tan x\,dx$

9. $\displaystyle\int_0^2 y^3\,dy$

10. $\displaystyle\int_0^2 (y-2)^2\,dy$

11. Rectangle

$A = bh = 3(4)$

$A = \displaystyle\int_0^3 4\,dx = 12$

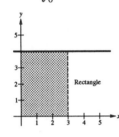

12. Rectangle

$A = bh = 2(4)(a)$

$A = \displaystyle\int_{-a}^a 4\,dx = 8a$

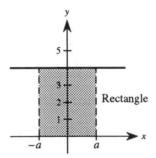

13. Triangle

$A = \dfrac{1}{2}bh = \dfrac{1}{2}(4)(4)$

$A = \displaystyle\int_0^4 x\,dx = 8$

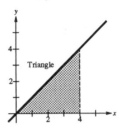

14. Triangle

$A = \dfrac{1}{2}bh = \dfrac{1}{2}(4)(2)$

$A = \displaystyle\int_0^4 \frac{x}{2}\,dx = 4$

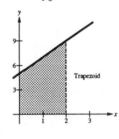

15. Trapezoid

$A = \dfrac{b_1+b_2}{2}h = 2\left(\dfrac{5+9}{2}\right)$

$A = \displaystyle\int_0^2 (2x+5)\,dx = 14$

16. Triangle

$A = \dfrac{1}{2}bh = \dfrac{1}{2}(5)(5)$

$A = \displaystyle\int_0^5 (5-x)\,dx = \dfrac{25}{2}$

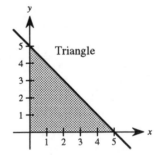

17. Triangle

$$a = \frac{1}{2}bh = \frac{1}{2}(2)(1)$$

$$A = \int_{-1}^{1} (1 - |x|)\,dx = 1$$

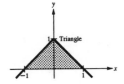

18. Triangle

$$A = \frac{1}{2}bh = \frac{1}{2}(2a)a$$

$$A = \int_{-a}^{a} (a - |x|)\,dx = a^2$$

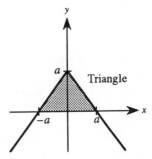

Triangle

19. Semicircle

$$A = \frac{1}{2}\pi r^2 = \frac{1}{2}\pi(3)^2$$

$$A = \int_{-3}^{3} \sqrt{9 - x^2}\,dx = \frac{9\pi}{2}$$

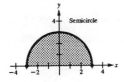

20. Semicircle

$$A = \frac{1}{2}\pi r^2$$

$$A = \int_{-r}^{r} \sqrt{r^2 - x^2}\,dx = \frac{1}{2}\pi r^2$$

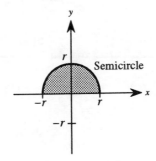

Semicircle

21. a. $\displaystyle\int_{0}^{7} f(x)\,dx = \int_{0}^{5} f(x)\,dx + \int_{5}^{7} f(x)\,dx = 10 + 3 = 13$

b. $\displaystyle\int_{5}^{0} f(x)\,dx = -\int_{0}^{5} f(x)\,dx = -10$

c. $\displaystyle\int_{5}^{5} f(x)\,dx = 0$

d. $\displaystyle\int_{0}^{5} 3f(x)\,dx = 3\int_{0}^{5} f(x)\,dx = 3(10) = 30$

22. a. $\displaystyle\int_{0}^{6} f(x)\,dx = \int_{0}^{3} f(x)\,dx + \int_{3}^{6} f(x)\,dx = 4 + (-1) = 3$

b. $\displaystyle\int_{6}^{3} f(x)\,dx = -\int_{3}^{6} f(x)\,dx = -(-1) = 1$

c. $\displaystyle\int_{4}^{4} f(x)\,dx = 0$

d. $\displaystyle\int_{3}^{6} -5f(x)\,dx = -5\int_{3}^{6} f(x)\,dx = -5(-1) = 5$

23. a. $\displaystyle\int_2^6 [f(x)+g(x)]\,dx = \int_2^6 f(x)\,dx + \int_2^6 g(x)\,dx = 10+(-2)=8$

b. $\displaystyle\int_2^6 [g(x)-f(x)]\,dx = \int_2^6 g(x)\,dx - \int_2^6 f(x)\,dx = -2-10=-12$

c. $\displaystyle\int_2^6 2g(x)\,dx = 2\int_2^6 g(x)\,dx = 2(-2)=-4$

d. $\displaystyle\int_2^6 3f(x)\,dx = 3\int_2^6 f(x)\,dx = 3(10)=30$

24. a. $\displaystyle\int_{-1}^0 f(x)\,dx = \int_{-1}^1 f(x)\,dx - \int_0^1 f(x)\,dx = 0-5=-5$

b. $\displaystyle\int_0^1 f(x)\,dx - \int_{-1}^0 f(x)\,dx = 5-(-5)=10$

c. $\displaystyle\int_{-1}^1 3f(x)\,dx = 3\int_{-1}^1 f(x)\,dx = 3(0)=0$

d. $\displaystyle\int_0^1 3f(x)\,dx = 3\int_0^1 f(x)\,dx = 3(5)=15$

25. $y=6$ on $[4,\ 10]$. $\left(\textbf{Note: } \Delta x = \dfrac{10-4}{n} = \dfrac{6}{n},\ \|\Delta\|\to 0 \text{ as } n\to\infty\right)$

$$\sum_{i=1}^n f(c_i)\Delta x_i = \sum_{i=1}^n f\left(4+\frac{6i}{n}\right)\left(\frac{6}{n}\right) = \sum_{i=1}^n 6\left(\frac{6}{n}\right) = \sum_{i=1}^n \frac{36}{n} = 36$$

$$\int_4^{10} 6\,dx = \lim_{n\to\infty} 36 = 36$$

26. $y=x$ on $[-2,\ 3]$. $\left(\textbf{Note: } \Delta x = \dfrac{3-(-2)}{n} = \dfrac{5}{n},\ \|\Delta\|\to 0 \text{ as } n\to\infty\right)$

$$\sum_{i=1}^n f(c_i)\Delta x_i = \sum_{i=1}^n f\left(-2+\frac{5i}{n}\right)\left(\frac{5}{n}\right) = \sum_{i=1}^n \left(-2+\frac{5i}{n}\right)\left(\frac{5}{n}\right) = -10+\frac{25}{n^2}\sum_{i=1}^n i$$

$$= -10+\left(\frac{25}{n^2}\right)\frac{n(n+1)}{2} = -10+\frac{25}{2}\left(1+\frac{1}{n}\right) = \frac{5}{2}+\frac{25}{2n}$$

$$\int_{-2}^3 x\,dx = \lim_{n\to\infty}\left(\frac{5}{2}+\frac{25}{2n}\right) = \frac{5}{2}$$

27. $y=x^3$ on $[-1,\ 1]$. $\left(\textbf{Note: } \Delta x = \dfrac{1-(-1)}{n} = \dfrac{2}{n},\ \|\Delta\|\to 0 \text{ as } n\to\infty\right)$

$$\sum_{i=1}^n f(c_i)\Delta x_i = \sum_{i=1}^n f\left(-1+\frac{2i}{n}\right)\left(\frac{2}{n}\right) = \sum_{i=1}^n\left(-1+\frac{2i}{n}\right)^3\left(\frac{2}{n}\right) = \sum_{i=1}^n\left[-1+\frac{6i}{n}-\frac{12i^2}{n^2}+\frac{8i^3}{n^3}\right]\left(\frac{2}{n}\right)$$

$$= -2+\frac{12}{n^2}\sum_{i=1}^n i - \frac{24}{n^3}\sum_{i=1}^n i^2 + \frac{16}{n^4}\sum_{i=1}^n i^3$$

$$= -2+6\left(1+\frac{1}{n}\right) - 4\left(2+\frac{3}{n}+\frac{1}{n^2}\right) + 4\left(1+\frac{2}{n}+\frac{1}{n^2}\right) = \frac{2}{n}$$

$$\int_{-1}^1 x^3\,dx = \lim_{n\to\infty}\frac{2}{n} = 0$$

28. $y = x^3$ on $[0,\ 1]$. $\left(\textbf{Note: } \Delta x = \dfrac{1-0}{n} = \dfrac{1}{n},\ \|\Delta\| \to 0 \text{ as } n \to \infty\right)$

$$\sum_{i=1}^{n} f(c_i)\Delta x_i = \sum_{i=1}^{n} f\left(\frac{i}{n}\right)\left(\frac{1}{n}\right) = \sum_{i=1}^{n}\left(\frac{i}{n}\right)^3\left(\frac{1}{n}\right) = \frac{1}{n^4}\sum_{i=1}^{n}i^3 = \frac{1}{n^4}\left[\frac{n^2(n+1)^2}{4}\right] = \frac{1}{4}\left(1 + \frac{2}{n} + \frac{1}{n^2}\right)$$

$$\int_0^1 x^3\,dx = \frac{1}{4}\lim_{n\to\infty}\left(1 + \frac{2}{n} + \frac{1}{n^2}\right) = \frac{1}{4}$$

29. $y = x^2 + 1$ on $[1,\ 2]$. $\left(\textbf{Note: } \Delta x = \dfrac{2-1}{n} = \dfrac{1}{n},\ \|\Delta\| \to 0 \text{ as } n \to \infty\right)$

$$\sum_{i=1}^{n} f(c_i)\Delta x_i = \sum_{i=1}^{n} f\left(1 + \frac{i}{n}\right)\left(\frac{1}{n}\right) = \sum_{i=1}^{n}\left[\left(1 + \frac{i}{n}\right)^2 + 1\right]\left(\frac{1}{n}\right) = \sum_{i=1}^{n}\left[1 + \frac{2i}{n} + \frac{i^2}{n^2} + 1\right]\left(\frac{1}{n}\right)$$

$$= 2 + \frac{2}{n^2}\sum_{i=1}^{n}i + \frac{1}{n^3}\sum_{i=1}^{n}i^2 = 2 + \left(1 + \frac{1}{n}\right) + \frac{1}{6}\left(2 + \frac{3}{n} + \frac{1}{n^2}\right) = \frac{10}{3} + \frac{3}{2n} + \frac{1}{6n^2}$$

$$\int_1^2 (x^2 + 1)\,dx = \lim_{n\to\infty}\left(\frac{10}{3} + \frac{3}{2n} + \frac{1}{6n^2}\right) = \frac{10}{3}$$

30. $y = 4x^2$ on $[1,\ 2]$. $\left(\textbf{Note: } \Delta x = \dfrac{2-1}{n} = \dfrac{1}{n},\ \|\Delta\| \to 0 \text{ as } n \to \infty\right)$

$$\sum_{i=1}^{n} f(c_i)\Delta x_i = \sum_{i=1}^{n} f\left(1 + \frac{i}{n}\right)\left(\frac{1}{n}\right) = \sum_{i=1}^{n}\left[4\left(1 + \frac{i}{n}\right)^2\right]\left(\frac{1}{n}\right) = 4\sum_{i=1}^{n}\left[1 + \frac{2i}{n} + \frac{i^2}{n^2}\right]\left(\frac{1}{n}\right)$$

$$= 4\left[1 + \frac{1}{n^2}\sum_{i=1}^{n}i + \frac{1}{n^3}\sum_{i=1}^{n}i^2\right] = 4\left[1 + \left(\frac{2}{n^2}\right)\frac{n(n+1)}{2} + \left(\frac{1}{n^3}\right)\frac{n(n+1)(2n+1)}{6}\right]$$

$$= 4\left[1 + \left(1 + \frac{1}{n}\right) + \frac{1}{6}\left(2 + \frac{3}{n} + \frac{1}{n^2}\right)\right] = 4\left(\frac{7}{3} + \frac{3}{2n} + \frac{1}{6n^2}\right)$$

$$\int_1^2 4x^2\,dx = 4\lim_{n\to\infty}\left(\frac{7}{3} + \frac{3}{2n} + \frac{1}{6n^2}\right) = \frac{28}{3}$$

31. $\displaystyle\lim_{\|\Delta\|\to 0}\sum_{i=1}^{n}(3c_i + 10)\Delta x_i = \int_{-1}^{5}(3x + 10)\,dx$ on the interval $[-1,\ 5]$.

32. $\displaystyle\lim_{\|\Delta\|\to 0}\sum_{i=1}^{n}6c_i(4 - c_i)^2\Delta x_i = \int_{0}^{4}6x(4 - x)^2\,dx$ on the interval $[0,\ 4]$.

33. $\displaystyle\lim_{\|\Delta\|\to 0}\sum_{i=1}^{n}\sqrt{c_i^2 + 4}\,\Delta x_i = \int_{0}^{3}\sqrt{x^2 + 4}\,dx$ on the interval $[0,\ 3]$.

34. $\displaystyle\lim_{\|\Delta\|\to 0}\sum_{i=1}^{n}\left(\frac{3}{c^2}\right)\Delta x_i = \int_{1}^{3}\frac{3}{x^2}\,dx$ on the interval $[1,\ 3]$.

35. $\displaystyle\int_0^3 x\sqrt{3-x}\,dx$

n	4	8	12	16	20
L(n)	3.6830	3.9956	4.0707	4.1016	4.1177
M(n)	4.3082	4.2076	4.1838	4.1740	4.1690
R(n)	3.6830	3.9956	4.0707	4.1016	4.1177

36. $\displaystyle\int_0^3 \frac{5}{x^2+1}\,dx$

n	4	8	12	16	20
L(n)	7.9224	7.0855	6.8062	6.6662	6.5822
M(n)	6.2485	6.2470	6.2460	6.2457	6.2455
R(n)	4.5474	5.3980	5.6812	5.8225	5.9072

37. $\displaystyle\int_0^{\pi/2} \sin^2 x\,dx$

n	4	8	12	16	20
L(n)	0.5890	0.6872	0.7199	0.7363	0.7461
M(n)	0.7854	0.7854	0.7854	0.7854	0.7854
R(n)	0.9817	0.8836	0.8508	0.8345	0.8247

38. $\displaystyle\int_0^3 x\sin x\,dx$

n	4	8	12	16	20
L(n)	2.8186	2.9985	3.0434	3.0631	3.0740
M(n)	3.1784	3.1277	3.1185	3.1152	3.1138
R(n)	3.1361	3.1573	3.1493	3.1425	3.1375

39.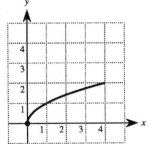

a. $A \approx 5$ square units

40.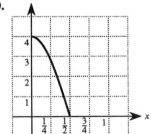

b. $A \approx \frac{4}{3}$ square units

41. $f(x) = x^2 + 3x$, $[0, 8]$

$x_0 = 0$, $x_1 = 1$, $x_2 = 3$, $x_3 = 7$, $x_4 = 8$

$\Delta x_1 = 1$, $\Delta x_2 = 2$, $\Delta x_3 = 4$, $\Delta x_4 = 1$

$C_1 = 1$, $C_2 = 2$, $C_3 = 5$, $C_4 = 8$

$$\sum_{i=1}^{4} f(c_i)\Delta x_i = f(1)\Delta x_1 + f(2)\Delta x_2 + f(5)\Delta x_3 + f(8)\Delta x_4 = (4)(1) + (10)(2) + (40)(4) + (88)(1) = 272$$

42. $f(x) = \sin x$, $[0, 2\pi]$

$x_0 = 0$, $x_1 = \dfrac{\pi}{4}$, $x_2 = \dfrac{\pi}{3}$, $x_3 = \pi$, $x_4 = 2\pi$

$\Delta x_1 = \dfrac{\pi}{4}$, $\Delta x_2 = \dfrac{\pi}{12}$, $\Delta x_3 = \dfrac{2\pi}{3}$, $\Delta x_4 = \pi$

$C_1 = \dfrac{\pi}{6}$, $C_2 = \dfrac{\pi}{3}$, $C_3 = \dfrac{2\pi}{3}$, $C_4 = \dfrac{3\pi}{2}$

$$\sum_{i=1}^{4} f(c_i)\Delta x_i = f\left(\frac{\pi}{6}\right)\Delta x_1 + f\left(\frac{\pi}{3}\right)\Delta x_2 + f\left(\frac{2\pi}{3}\right)\Delta x_3 + f\left(\frac{3\pi}{2}\right)\Delta x_4$$

$$= \left(\frac{1}{2}\right)\left(\frac{\pi}{4}\right) + \left(\frac{\sqrt{3}}{2}\right)\left(\frac{\pi}{12}\right) + \left(\frac{\sqrt{3}}{2}\right)\left(\frac{2\pi}{3}\right) + (-1)(\pi) \approx -0.708$$

43. True

44. False

$$\int_0^1 x\sqrt{x}\,dx$$

$$\neq \left(\int_0^1 x\,dx\right)\left(\int_0^1 \sqrt{x}\,dx\right)$$

45. True

46. True

47. False

$$\int_0^2 (-x)\,dx = -2$$

48. False

$$\int_{-2}^4 x\,dx = 6$$

49. $f(x) = \sqrt{x}$, $y = 0$, $x = 0$, $x = 2$, $c_i = \dfrac{2i^2}{n^2}$

$$\Delta x_i = \frac{2i^2}{n^2} - \frac{2(i-1)^2}{n^2} = \frac{2(2i-1)}{n^2}$$

$$\lim_{n\to\infty} \sum_{i=1}^{n} f(c_i)\Delta x_i = \lim_{n\to\infty} \sum_{i=1}^{n} \sqrt{\frac{2i^2}{n^2}}\left[\frac{2(2i-1)}{n^2}\right]$$

$$= \lim_{n\to\infty} \frac{2\sqrt{2}}{n^3} \sum_{i=1}^{n} (2i^2 - i)$$

$$= \lim_{n\to\infty} \frac{2\sqrt{2}}{n^3}\left[2\left(\frac{n(n+1)(2n+1)}{6}\right) - \frac{n(n+1)}{2}\right]$$

$$= \lim_{n\to\infty} \frac{2\sqrt{2}}{n^3}\left[\frac{4n^3 + 3n^2 - n}{6}\right] = \lim_{n\to\infty} \sqrt{2}\left[\frac{4}{3} + \frac{1}{n} - \frac{2}{n^2}\right] = \frac{4\sqrt{2}}{3}$$

50. $f(x) = \sqrt[3]{x}$, $y = 0$, $x = 0$, $x = 1$, $c_i = \dfrac{i^3}{n^3}$

$$\Delta x_i = \frac{i^3}{n^3} - \frac{(i-1)^3}{n^3} = \frac{3i^2 - 3i + 1}{n^3}$$

$$\lim_{n \to \infty} \sum_{i=1}^{n} f(c_i)\Delta x_i = \lim_{n \to \infty} \sum_{i=1}^{n} \sqrt[3]{\frac{i^3}{n^3}} \left[\frac{3i^2 - 3i + 1}{n^3}\right]$$

$$= \lim_{n \to \infty} \frac{1}{n^4} \sum_{i=1}^{n} (3i^3 - 3i^2 + i)$$

$$= \lim_{n \to \infty} \frac{1}{n^4} \left[3\left(\frac{n^2(n+1)^2}{4}\right) - 3\left(\frac{n(n+1)(2n+1)}{6}\right) + \frac{n(n+1)}{2}\right]$$

$$= \lim_{n \to \infty} \frac{1}{n^4} \left[\frac{3n^4 + 6n^3 + 3n^2}{4} - \frac{2n^3 + 3n^2 + n}{2} + \frac{n^2 + n}{2}\right]$$

$$= \lim_{n \to \infty} \frac{1}{n^4} \left[\frac{3n^4}{4} + \frac{n^3}{2} - \frac{n^2}{4}\right] = \lim_{n \to \infty} \left[\frac{3}{4} + \frac{1}{2n} - \frac{1}{4n^2}\right] = \frac{3}{4}$$

51. Since $-|f(x)| \le f(x) \le |f(x)|$,

$$-\int_a^b |f(x)| \, dx \le \int_a^b f(x) \, dx \le \int_a^b |f(x)| \, dx \Rightarrow \left|\int_a^b f(x) \, dx\right| \le \int_a^b |f(x)| \, dx.$$

52. $f(x) = \dfrac{1}{x - 4}$ is not integrable on the interval $[3, 5]$ since f has a discontinuity at $x = 4$.

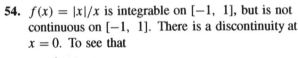

$$\int_3^5 \frac{1}{x - 4} \, dx = \int_3^4 \frac{1}{x - 4} \, dx + \int_4^5 \frac{1}{x - 4} \, dx \text{ which is infinite.}$$

53. $f(x) = \begin{cases} 1, & x \text{ is rational} \\ 0, & x \text{ is irrational} \end{cases}$

is not integrable on the interval $[0, 1]$. As $\|\Delta\| \to 0$, $f(c_i) = 1$ or $f(c_i) = 0$ in each subinterval since there are an infinite number of both rational and irrational numbers in any interval, no matter how small.

54. $f(x) = |x|/x$ is integrable on $[-1, 1]$, but is not continuous on $[-1, 1]$. There is a discontinuity at $x = 0$. To see that

$$\int_{-1}^{1} \frac{|x|}{x} \, dx$$

is integrable, sketch a graph of the region bounded by $f(x) = |x|/x$ and the x-axis for $-1 \le x \le 1$.

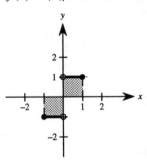

55. To find $\int_0^2 [\![x]\!]\, dx$, use a geometric approach.

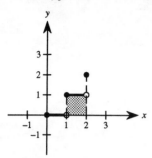

Thus,

$$\int_0^2 [\![x]\!]\, dx = 1(2-1) = 1.$$

56.

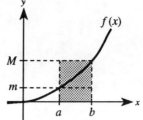

From the graph, the area of the rectangle with width m and length $b-a$ is $m(b-a)$, which is less than the area given by $\int_a^b f(x)\,dx$. The area of the rectangle with width M and length $b-a$ is $M(b-a)$, which is greater than the area given by $\int_a^b f(x)\,dx$. Thus,

$$m(b-a) \le \int_a^b f(x)\,dx \le M(b-a)$$

$$1 \le \sqrt{1+x^4} \le \sqrt{2} \text{ on } [0,\ 1]$$

$$\int_0^1 dx \le \int_0^1 \sqrt{1+x^4}\,dx \le \int_0^1 \sqrt{2}\,dx$$

$$1 \le \int_0^1 \sqrt{1+x^4}\,dx \le \sqrt{2}$$

57. $\displaystyle \lim_{n\to\infty} \frac{1}{n^3}[1^2 + 2^2 + 3^2 + \cdots + n^2] = \lim_{n\to\infty} \frac{1}{n^3} \cdot \frac{n(2n+1)(n+1)}{6}$

$$= \lim_{n\to\infty} \frac{2n^2 + 3n + 1}{6n^2} = \lim_{n\to\infty} \left(\frac{1}{3} + \frac{1}{2n} + \frac{1}{6n^2}\right) = \frac{1}{3}$$

Section 4.4   The Fundamental Theorem of Calculus

1. $\displaystyle \int_0^1 2x\, dx = \left[x^2\right]_0^1 = 1 - 0 = 1$

2. $\displaystyle \int_2^7 3\, dv = \left[3v\right]_2^7 = 3(7) - 3(2) = 15$

3. $\displaystyle \int_{-1}^0 (x-2)\, dx = \left[\frac{x^2}{2} - 2x\right]_{-1}^0 = 0 - \left(\frac{1}{2} + 2\right) = -\frac{5}{2}$

4. $\displaystyle \int_2^5 (-3v+4)\, dv = \left[-\frac{3}{2}v^2 + 4v\right]_2^5 = \left(-\frac{75}{2} + 20\right) - (-6+8) = -\frac{39}{2}$

5. $\displaystyle \int_{-1}^1 (t^2 - 2)\, dt = \left[\frac{t^3}{3} - 2t\right]_{-1}^1 = \left(\frac{1}{3} - 2\right) - \left(-\frac{1}{3} + 2\right) = -\frac{10}{3}$

6. $\displaystyle \int_0^3 (3x^2 + x - 2)\, dx = \left[x^3 + \frac{x^2}{2} - 2x\right]_0^3 = \left(27 + \frac{9}{2} - 6\right) - 0 = \frac{51}{2}$

7. $\displaystyle \int_0^1 (2t-1)^2\, dt = \int_0^1 (4t^2 - 4t + 1)\, dt = \left[\frac{4}{3}t^3 - 2t^2 + t\right]_0^1 = \frac{4}{3} - 2 + 1 = \frac{1}{3}$

8. $\displaystyle \int_{-1}^1 (t^3 - 9t)\, dt = \left[\frac{1}{4}t^4 - \frac{9}{2}t^2\right]_{-1}^1 = \left(\frac{1}{4} - \frac{9}{2}\right) - \left(\frac{1}{4} - \frac{9}{2}\right) = 0$

9. $\int_1^2 \left(\frac{3}{x^2}-1\right)dx = \left[-\frac{3}{x}-x\right]_1^2 = \left(-\frac{3}{2}-2\right)-(-3-1) = \frac{1}{2}$

10. $\int_{-2}^{-1}\left(u-\frac{1}{u^2}\right)du = \left[\frac{u^2}{2}+\frac{1}{u}\right]_{-2}^{-1} = \left(\frac{1}{2}-1\right)-\left(2-\frac{1}{2}\right) = -2$

11. $\int_1^4 \frac{u-2}{\sqrt{u}}\,du = \int_1^4 (u^{1/2}-2u^{-1/2})\,du = \left[\frac{2}{3}u^{3/2}-4u^{1/2}\right]_1^4 = \left[\frac{2}{3}(\sqrt{4})^3-4\sqrt{4}\right]-\left[\frac{2}{3}-4\right] = \frac{2}{3}$

12. $\int_{-3}^3 v^{1/3}\,dv = \left[\frac{3}{4}v^{4/3}\right]_{-3}^3 = \frac{3}{4}[(\sqrt[3]{3})^4-(\sqrt[3]{-3})^4] = 0$

13. $\int_{-1}^1 (\sqrt[3]{t}-2)\,dt = \left[\frac{3}{4}t^{4/3}-2t\right]_{-1}^1 = \left(\frac{3}{4}-2\right)-\left(\frac{3}{4}+2\right) = -4$

14. $\int_1^8 \sqrt{\frac{2}{x}}\,dx = \sqrt{2}\int_1^8 x^{-1/2}\,dx = \left[\sqrt{2}(2)x^{1/2}\right]_1^8 = \left[2\sqrt{2x}\right]_1^8 = 8-2\sqrt{2}$

15. $\int_0^1 \frac{x-\sqrt{x}}{3}\,dx = \frac{1}{3}\int_0^1 (x-x^{1/2})\,dx = \frac{1}{3}\left[\frac{x^2}{2}-\frac{2}{3}x^{3/2}\right]_0^1 = \frac{1}{3}\left(\frac{1}{2}-\frac{2}{3}\right) = -\frac{1}{18}$

16. $\int_0^2 (2-t)\sqrt{t}\,dt = \int_0^2 (2t^{1/2}-t^{3/2})\,dt = \left[\frac{4}{3}t^{3/2}-\frac{2}{5}t^{5/2}\right]_0^2 = \left[\frac{t\sqrt{t}}{15}(20-6t)\right]_0^2 = \frac{2\sqrt{2}}{15}(20-12) = \frac{16\sqrt{2}}{15}$

17. $\int_{-1}^0 (t^{1/3}-t^{2/3})\,dt = \left[\frac{3}{4}t^{4/3}-\frac{3}{5}t^{5/3}\right]_{-1}^0 = 0-\left(\frac{3}{4}+\frac{3}{5}\right) = -\frac{27}{20}$

18. $\int_{-8}^{-1} \frac{x-x^2}{2\sqrt[3]{x}}\,dx = \frac{1}{2}\int_{-8}^{-1}(x^{2/3}-x^{5/3})\,dx$

$= \frac{1}{2}\left[\frac{3}{5}x^{5/3}-\frac{3}{8}x^{8/3}\right]_{-8}^{-1} = \left[\frac{x^{5/3}}{80}(24-15x)\right]_{-8}^{-1} = -\frac{1}{80}(39)+\frac{32}{80}(144) = \frac{4569}{80}$

19. $\int_0^3 |2x-3|\,dx = \int_0^{3/2}(3-2x)\,dx + \int_{3/2}^3(2x-3)\,dx = \left[3x-x^2\right]_0^{3/2}+\left[x^2-3x\right]_{3/2}^3$

$= \left(\frac{9}{2}-\frac{9}{4}\right)-0+(9-9)-\left(\frac{9}{4}-\frac{9}{2}\right)$

$= 2\left(\frac{9}{2}-\frac{9}{4}\right) = \frac{9}{2}$

20. $\int_0^4 |x^2-4x+3|\,dx = \int_0^1 (x^2-4x+3)\,dx - \int_1^3 (x^2-4x+3)\,dx + \int_3^4 (x^2-4x+3)\,dx$

$= \left[\frac{x^3}{3}-2x^2+3x\right]_0^1 - \left[\frac{x^3}{3}-2x^2+3x\right]_1^3 + \left[\frac{x^3}{3}-2x^2+3x\right]_3^4$

$= \left(\frac{1}{3}-2+3\right)+\left(\frac{1}{3}-2+3\right)+\left(\frac{64}{3}-32+12\right)-(9-18+9)$

$= \frac{8}{3}+\frac{64}{3}-20 = 4$

21. $\int_0^\pi (1+\sin x)\,dx = \left[x-\cos x\right]_0^\pi = (\pi+1)-(0-1) = 2+\pi$

22. $\displaystyle\int_0^{\pi/4} \frac{1 - \sin^2\theta}{\cos^2\theta}\, d\theta = \int_0^{\pi/4} d\theta = \Big[\theta\Big]_0^{\pi/4} = \frac{\pi}{4}$

23. $\displaystyle\int_{-\pi/6}^{\pi/6} \sec^2 x\, dx = \Big[\tan x\Big]_{-\pi/6}^{\pi/6} = \frac{\sqrt{3}}{3} - \left(-\frac{\sqrt{3}}{3}\right) = \frac{2\sqrt{3}}{3}$

24. $\displaystyle\int_{\pi/4}^{\pi/2} (2 - \csc^2 x)\, dx = \Big[2x + \cot x\Big]_{\pi/4}^{\pi/2} = (\pi + 0) - \left(\frac{\pi}{2} + 1\right) = \frac{\pi}{2} - 1 = \frac{\pi - 2}{2}$

25. $\displaystyle\int_{-\pi/3}^{\pi/3} 4\sec\theta\tan\theta\, d\theta = \Big[4\sec\theta\Big]_{-\pi/3}^{\pi/3} = 4(2) - 4(2) = 0$

26. $\displaystyle\int_{-\pi/2}^{\pi/2} (2t + \cos t)\, dt = \Big[t^2 + \sin t\Big]_{-\pi/2}^{\pi/2} = \left(\frac{\pi^2}{4} + 1\right) - \left(\frac{\pi^2}{4} - 1\right) = 2$

27. $\displaystyle\int_0^3 10{,}000(t - 6)\, dt = 10{,}000\Big[\frac{t^2}{2} - 6t\Big]_0^3 = -\$135{,}000$

28. $\displaystyle P = \frac{2}{\pi}\int_0^{\pi/2} \sin\theta\, d\theta = \Big[-\frac{2}{\pi}\cos\theta\Big]_0^{\pi/2} = -\frac{2}{\pi}(0 - 1) = \frac{2}{\pi} \approx 63.7\%$

29. $\displaystyle A = \int_0^1 (x - x^2)\, dx = \Big[\frac{x^2}{2} - \frac{x^3}{3}\Big]_0^1 = \frac{1}{6}$

30. $\displaystyle A = \int_{-1}^1 (1 - x^4)\, dx = \Big[x - \frac{1}{5}x^5\Big]_{-1}^1 = \frac{8}{5}$

31. $\displaystyle A = \int_0^3 (3 - x)\sqrt{x}\, dx = \int_0^3 (3x^{1/2} - x^{3/2})\, dx = \Big[2x^{3/2} - \frac{2}{5}x^{5/2}\Big]_0^3 = \Big[\frac{x\sqrt{x}}{5}(10 - 2x)\Big]_0^3 = \frac{12\sqrt{3}}{5}$

32. $\displaystyle A = \int_1^2 \frac{1}{x^2}\, dx = \Big[-\frac{1}{x}\Big]_1^2 = -\frac{1}{2} + 1 = \frac{1}{2}$

33. $\displaystyle A = \int_0^{\pi/2} \cos x\, dx = \Big[\sin x\Big]_0^{\pi/2} = 1$

34. $\displaystyle A = \int_0^{\pi} (x + \sin x)\, dx = \Big[\frac{x^2}{2} - \cos x\Big]_0^{\pi} = \frac{\pi^2}{2} + 2 = \frac{\pi^2 + 4}{2}$

35. Since $y \geq 0$ on $[0, 2]$, $\displaystyle A = \int_0^2 (3x^2 + 1)\, dx = \Big[x^3 + x\Big]_0^2 = 8 + 2 = 10.$

36. Since $y \geq 0$ on $[0, 4]$, $\displaystyle A = \int_0^4 (1 + \sqrt{x})\, dx = \Big[x + \frac{2}{3}x^{3/2}\Big]_0^4 = 4 + \frac{2}{3}(8) = \frac{28}{3}.$

37. Since $y \geq 0$ on $[0, 2]$, $\displaystyle A = \int_0^2 (x^3 + x)\, dx = \Big[\frac{x^4}{4} + \frac{x^2}{2}\Big]_0^2 = 4 + 2 = 6.$

38. Since $y \geq 0$ on $[0, 3]$, $\displaystyle A = \int_0^3 (3x - x^2)\, dx = \Big[\frac{3}{2}x^2 - \frac{x^3}{3}\Big]_0^3 = \frac{9}{2}.$

39. $\displaystyle\int_0^2 (x - 2\sqrt{x})\,dx = \left[\frac{x^2}{2} - \frac{4x^{3/2}}{3}\right]_0^2 = 2 - \frac{8\sqrt{2}}{3}$

$$f(c)(2 - 0) = \frac{6 - 8\sqrt{2}}{3}$$

$$c - 2\sqrt{c} = \frac{3 - 4\sqrt{2}}{3}$$

$$c - 2\sqrt{c} + 1 = \frac{3 - 4\sqrt{2}}{3} + 1$$

$$(\sqrt{c} - 1)^2 = \frac{6 - 4\sqrt{2}}{3}$$

$$\sqrt{c} - 1 = \pm\sqrt{\frac{6 - 4\sqrt{2}}{3}}$$

$$c = \left[1 \pm \sqrt{\frac{6 - 4\sqrt{2}}{3}}\right]^2$$

$$c \approx 0.4380 \text{ or } c \approx 1.7908$$

40. $\displaystyle\int_1^3 \frac{9}{x^3}\,dx = \left[-\frac{9}{2x^2}\right]_1^3 = -\frac{1}{2} + \frac{9}{2} = 4$

$$f(c)(3 - 1) = 4$$

$$\frac{9}{c^3} = 2$$

$$c^3 = \frac{9}{2}$$

$$c = \sqrt[3]{\frac{9}{2}} \approx 1.6510$$

41. $\displaystyle\int_{-\pi/4}^{\pi/4} 2\sec^2 x\,dx = \Big[2\tan x\Big]_{-\pi/4}^{\pi/4}$

$$= 2(1) - 2(-1)$$

$$= 4$$

$$f(c)\left[\frac{\pi}{4} - \left(-\frac{\pi}{4}\right)\right] = 4$$

$$2\sec^2 c = \frac{8}{\pi}$$

$$\sec^2 c = \frac{4}{\pi}$$

$$\sec c = \pm\frac{2}{\sqrt{\pi}}$$

$$c = \pm\operatorname{arcsec}\left(\frac{2}{\sqrt{\pi}}\right)$$

$$= \pm\arccos\frac{\sqrt{\pi}}{2} \approx \pm 0.4817$$

42. $\displaystyle\int_{-\pi/3}^{\pi/3} \cos x\,dx = \Big[\sin x\Big]_{-\pi/3}^{\pi/3} = \sqrt{3}$

$$f(c)\left[\frac{\pi}{3} - \left(-\frac{\pi}{3}\right)\right] = \sqrt{3}$$

$$\cos c = \frac{3\sqrt{3}}{2\pi}$$

$$c \approx \pm 0.5971$$

43. $\displaystyle\frac{1}{2 - (-2)}\int_{-2}^2 (4 - x^2)\,dx = \frac{1}{4}\left[4x - \frac{1}{3}x^3\right]_{-2}^2$

$$= \frac{1}{4}\left[\left(8 - \frac{8}{3}\right) - \left(-8 + \frac{8}{3}\right)\right] = \frac{8}{3}$$

Average value $= \dfrac{8}{3}$

$4 - x^2 = \dfrac{8}{3}$ when $x^2 = 4 - \dfrac{8}{3}$ or $x = \pm\dfrac{2\sqrt{3}}{3} \approx \pm 1.155.$

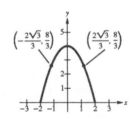

44. $\dfrac{1}{2-(1/2)} \displaystyle\int_{1/2}^{2} \dfrac{x^2+1}{x^2}\,dx = \dfrac{2}{3} \displaystyle\int_{1/2}^{2} (1+x^{-2})\,dx$

$$= \dfrac{2}{3}\left[x - \dfrac{1}{x}\right]_{1/2}^{2} = \dfrac{2}{3}\left(\dfrac{3}{2}+\dfrac{3}{2}\right) = 2$$

Average value $= 2$

In the interval $\left[\dfrac{1}{2},\ 2\right]$, $\dfrac{x^2+1}{x^2} = 2$ when $x^2 + 1 = 2x^2$ or $x = 1$.

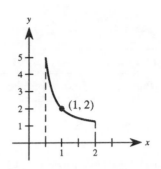

45. $\dfrac{1}{\pi - 0} \displaystyle\int_{0}^{\pi} \sin x\,dx = \left[-\dfrac{1}{\pi}\cos x\right]_{0}^{\pi} = \dfrac{2}{\pi}$

Average value $= \dfrac{2}{\pi}$

$\sin x = \dfrac{2}{\pi}$

$x \approx 0.690,\ 2.451$

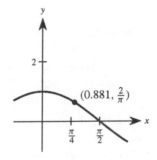

46. $\dfrac{1}{(\pi/2) - 0} \displaystyle\int_{0}^{\pi/2} \cos x\,dx = \left[\dfrac{2}{\pi}\sin x\right]_{0}^{\pi/2} = \dfrac{2}{\pi}$

Average value $= \dfrac{2}{\pi}$

$\cos x = \dfrac{2}{\pi}$

$x \approx 0.881$

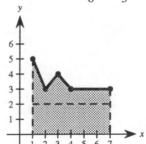

47. a. $\displaystyle\int_{1}^{7} f(x)\,dx$

    $=$ Sum of the areas

    $= A_1 + A_2 + A_3 + A_4$

    $= \dfrac{1}{2}(3+1) + \dfrac{1}{2}(1+2) + \dfrac{1}{2}(2+1) + (3)(1)$

    $= 8$

b. Average value $= \dfrac{\displaystyle\int_{1}^{7} f(x)\,dx}{7-1} = \dfrac{8}{6} = \dfrac{4}{3}$

c. $A = 8 + (6)(2) = 20$

    Average value $= \dfrac{20}{6} = \dfrac{10}{3}$

48. $P = 5(\sqrt{t} + 30)$

a.

t	1	2	3	4	5	6
P	155	157.071	158.660	160	161.180	162.247

Average profit $\approx \dfrac{1}{6}(155 + 157.071 + 158.660 + 160 + 161.180 + 162.247) = \dfrac{954.158}{6} \approx 159.026$

b. $\dfrac{1}{6}\displaystyle\int_{0.5}^{6.5} 5(\sqrt{t} + 30)\, dt = \dfrac{1}{6}\left[5\left(\dfrac{2}{3}t^{3/2} + 30t\right)\right]_{0.5}^{6.5} \approx \dfrac{954.061}{6} \approx 159.010$

c. The definite integral yields a better approximation.

49. $R = -90.767 - 6.434x + 0.037x^2 + 46.004\sqrt{x},\quad 20 \le x \le 60$

a.

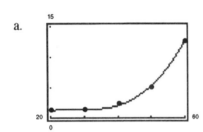

b. $R'(x) = -6.434 + 0.074x + \dfrac{23.002}{\sqrt{x}}$

$R'(40) \approx 0.16$

$R'(50) \approx 0.52$

c. $\dfrac{1}{10}\displaystyle\int_{30}^{40} (-90.767 - 6.434x + 0.037x^2 + 46.004\sqrt{x})\, dx$

$= \dfrac{1}{10}\left[-90.767x - 3.217x^2 + \dfrac{0.037}{3}x^3 + \dfrac{92.008x\sqrt{x}}{3}\right]_{30}^{40} \approx 1.61$

$\dfrac{1}{10}\displaystyle\int_{50}^{60} (-90.767 - 6.434x + 0.037x^2 + 46.004\sqrt{x})\, dx$

$= \dfrac{1}{10}\left[-90.767x - 3.217x^2 + \dfrac{0.037}{3}x^3 + \dfrac{92.008x\sqrt{x}}{3}\right]_{50}^{60} \approx 8.65$

50. $\dfrac{1}{R-0}\displaystyle\int_0^R k(R^2 - r^2)\, dr = \dfrac{k}{R}\left[R^2 r - \dfrac{r^3}{3}\right]_0^R = \dfrac{2kR^2}{3}$

51. a. $F(x) = k\sec^2 x$

$F(0) = k = 100$

$F(x) = 100\sec^2 x$

b. $\dfrac{1}{(\pi/3) - 0}\displaystyle\int_0^{\pi/3} 100\sec^2 x\, dx = \left[\dfrac{300}{\pi}\tan x\right]_0^{\pi/3} = \dfrac{300}{\pi}(\sqrt{3} - 0) \approx 165.4 \text{ lb}$

52. $\dfrac{1}{5-0}\displaystyle\int_0^5 (0.1729t + 0.1522t^2 - 0.0374t^3)\, dt \approx \dfrac{1}{5}\left[0.08645t^2 + 0.05073t^3 - 0.00935t^4\right]_0^5 \approx 0.5318 \text{ liter}$

53. a. $\displaystyle\int_0^x (t + 2)\, dt = \left[\dfrac{t^2}{2} + 2t\right]_0^x = \dfrac{1}{2}x^2 + 2x$

b. $\dfrac{d}{dx}\left[\dfrac{1}{2}x^2 + 2x\right] = x + 2$

54. a. $\int_0^x t(t^2+1)\,dt = \int_0^x (t^3+t)\,dt = \left[\frac{1}{4}t^4 + \frac{1}{2}t^2\right]_0^x = \frac{1}{4}x^4 + \frac{1}{2}x^2 = \frac{x^2}{4}(x^2+2)$

b. $\dfrac{d}{dx}\left[\dfrac{1}{4}x^4 + \dfrac{1}{2}x^2\right] = x^3 + x = x(x^2+1)$

55. a. $\int_8^x \sqrt[3]{t}\,dt = \left[\frac{3}{4}t^{4/3}\right]_8^x = \frac{3}{4}(x^{4/3} - 16) = \frac{3}{4}x^{4/3} - 12$

b. $\dfrac{d}{dx}\left[\dfrac{3}{4}x^{4/3} - 12\right] = x^{1/3} = \sqrt[3]{x}$

56. a. $\int_4^x \sqrt{t}\,dt = \left[\frac{2}{3}t^{3/2}\right]_4^x = \frac{2}{3}x^{3/2} - \frac{16}{3} = \frac{2}{3}(x^{3/2} - 8)$

b. $\dfrac{d}{dx}\left[\dfrac{2}{3}x^{3/2} - \dfrac{16}{3}\right] = x^{1/2} = \sqrt{x}$

57. a. $\int_{\pi/4}^x \sec^2 t\,dt = \left[\tan t\right]_{\pi/4}^x = \tan x - 1$

b. $\dfrac{d}{dx}[\tan x - 1] = \sec^2 x$

58. a. $\int_{\pi/3}^x \sec t \tan t\,dt = \left[\sec t\right]_{\pi/3}^x = \sec x - 2$

b. $\dfrac{d}{dx}[\sec x - 2] = \sec x \tan x$

59. $F(x) = \int_{-2}^x (t^2 - 2t + 5)\,dt$

$F'(x) = x^2 - 2x + 5$

60. $F(x) = \int_1^x \sqrt[4]{t}\,dt$

$F'(x) = \sqrt[4]{x}$

61. $F(x) = \int_{-1}^x \sqrt{t^4 + 1}\,dt$

$F'(x) = \sqrt{x^4 + 1}$

62. $F(x) = \int_0^x \tan^4 t\,dt$

$F'(x) = \tan^4 x$

63. $F(x) = \int_0^x t \cos t\,dt$

$F'(x) = x \cos x$

64. $F(x) = \int_1^x \dfrac{t^2}{t^2+1}\,dt$

$F'(x) = \dfrac{x^2}{x^2+1}$

65. $F(x) = \int_x^{x+2} (4t+1)\,dt$

$\qquad = \left[2t^2 + t\right]_x^{x+2}$

$\qquad = [2(x+2)^2 + (x+2)] - [2x^2 + x]$

$\qquad = 8x + 10$

$F'(x) = 8$

Alternate solution:

$F(x) = \int_x^{x+2} (4t+1)\,dt$

$\qquad = \int_x^0 (4t+1)\,dt + \int_0^{x+2} (4t+1)\,dt$

$\qquad = -\int_0^x (4t+1)\,dt + \int_0^{x+2} (4t+1)\,dt$

$F'(x) = -(4x+1) + 4(x+2) + 1 = 8$

66. $F(x) = \int_{-x}^x t^3\,dt = \dfrac{t^4}{4}\Big]_{-x}^x = 0$

$F'(x) = 0$

Alternate solution:

$F(x) = \int_{-x}^x t^3\,dt$

$\qquad = \int_{-x}^0 t^3\,dt + \int_0^x t^3\,dt$

$\qquad = -\int_0^{-x} t^3\,dt + \int_0^x t^3\,dt$

$F'(x) = -(-x)^3(-1) + (x^3) = 0$

67. $F(x) = \displaystyle\int_0^{\sin x} \sqrt{t}\, dt$

$ = \dfrac{2}{3} t^{3/2} \Big]_0^{\sin x} = \dfrac{2}{3}(\sin x)^{3/2}$

$F'(x) = (\sin x)^{1/2} \cos x = \cos x \sqrt{\sin x}$

Alternate solution:

$F(x) = \displaystyle\int_0^{\sin x} \sqrt{t}\, dt$

$F'(x) = \sqrt{\sin x}\,(\cos x)$

68. $F(x) = \displaystyle\int_2^{x^2} \dfrac{1}{t^2}\, dt = \left[-\dfrac{1}{t}\right]_2^{x^2} = -\dfrac{1}{x^2} + \dfrac{1}{2}$

Alternate solution:

$F(x) = \displaystyle\int_2^{x^2} \dfrac{1}{t^2}\, dt$

$F'(x) = \dfrac{1}{(x^2)^2}(2x) = \dfrac{2}{x^3}$

69. $F(x) = \displaystyle\int_0^{x^3} \sin t^2 \, dt$

$F'(x) = \sin(x^3)^2 \cdot 3x^2 = 3x^2 \sin x^6$

70. $F(x) = \displaystyle\int_0^{3x} \sqrt{1 + t^3}\, dt$

$F'(x) = \sqrt{1 + (3x)^3} \cdot 3 = 3\sqrt{1 + 27x^3}$

71.

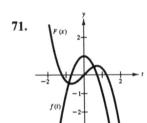

The extrema of F correspond to the zeros of f and the inflection point of F corresponds to the extrema of f.

72.

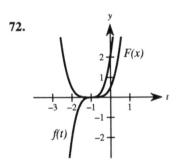

The extrema of F corresponds to the zero of f. Since f has no extrema, F has no points of inflection.

73. a. $C(x) = 5000\left(25 + 3\displaystyle\int_0^x t^{1/4}\, dt\right) = 5000\left(25 + 3\left[\dfrac{4}{5}t^{5/4}\right]_0^x\right) = 5000\left(25 + \dfrac{12}{5}x^{5/4}\right) = 1000(125 + 12x^{5/4})$

b. $C(1) = 1000(125 + 12(1)) = \$137{,}000$

$C(5) = 1000(125 + 12(5)^{5/4}) \approx \$214{,}721$

$C(10) = 1000(125 + 12(10)^{5/4}) \approx \$338{,}394$

74. a. $g(t) = 4 - \dfrac{4}{t^2}$

$\displaystyle\lim_{t\to\infty} g(t) = 4$

Horizontal asymptote: $y = 4$

b. $A(x) = \displaystyle\int_1^x \left(4 - \dfrac{4}{t^2}\right) dt = \left[4t + \dfrac{4}{t}\right]_1^x = 4x + \dfrac{4}{x} - 8 = \dfrac{4x^2 - 8x + 4}{x} = \dfrac{4(x-1)^2}{x}$

$\displaystyle\lim_{x\to\infty} A(x) = \lim_{x\to\infty}\left(4x + \dfrac{4}{x} - 8\right) = \infty + 0 - 8 = \infty$

The graph of $A(x)$ does not have a horizontal asymptote.

75. True

76. True

77. False; $\displaystyle\int_{-1}^{1} x^{-2}\,dx = \int_{-1}^{0} x^{-2}\,dx + \int_{0}^{1} x^{-2}\,dx$

Each of these integrals is infinite. $f(x) = x^{-2}$ has a nonremovable discontinuity at $x = 0$.

78. Let $F(t)$ be an antiderivative of $f(t)$. Then,

$$\int_{u(x)}^{v(x)} f(t)\,dt = \left[F(t)\right]_{u(x)}^{v(x)} = F(v(x)) - F(u(x))$$

$$\frac{d}{dx}\left[\int_{u(x)}^{v(x)} f(t)\,dt\right] = \frac{d}{dx}[F(v(x)) - F(u(x))]$$

$$= F'(v(x))v'(x) - F'(u(x))u'(x) = f(v(x))v'(x) - f(u(x))u'(x)$$

79. $\displaystyle f(x) = \int_{0}^{1/x} \frac{1}{t^2+1}\,dt + \int_{0}^{x} \frac{1}{t^2+1}\,dt$

By the Second Fundamental Theorem of Calculus, we have

$$f'(x) = \frac{1}{(1/x)^2 + 1}\left(-\frac{1}{x^2}\right) + \frac{1}{x^2+1}$$

$$= -\frac{1}{1+x^2} + \frac{1}{x^2+1} = 0.$$

Since $f'(x) = 0$, $f(x)$ must be constant.

80. $\displaystyle G(x) = \int_{0}^{x}\left[s\int_{0}^{s} f(t)\,dt\right]ds$

a. $\displaystyle G(0) = \int_{0}^{0}\left[s\int_{0}^{s} f(t)\,dt\right]ds = 0$

b. Let $\displaystyle F(s) = s\int_{0}^{s} f(t)\,dt.$

$$G(x) = \int_{0}^{x} F(s)\,ds$$

$$G'(x) = F(x) = x\int_{0}^{x} f(t)\,dt$$

c. $\displaystyle G'(0) = 0\int_{0}^{0} f(t)\,dt = 0$

d. $\displaystyle G''(x) = x \cdot f(x) + \int_{0}^{x} f(t)\,dt$

e. $\displaystyle G''(0) = 0 \cdot f(0) + \int_{0}^{0} f(t)\,dt = 0$

Section 4.5 Integration by Substitution

$\displaystyle\int f(g(x))g'(x)\,dx$	$u = g(x)$	$du = g'(x)\,dx$
1. $\displaystyle\int (5x^2 + 1)^2(10x)\,dx$	$5x^2 + 1$	$10x\,dx$
2. $\displaystyle\int x^2\sqrt{x^3 + 1}\,dx$	$x^3 + 1$	$3x^2\,dx$
3. $\displaystyle\int \frac{x}{\sqrt{x^2+1}}\,dx$	$x^2 + 1$	$2x\,dx$
4. $\displaystyle\int \sec 2x \tan 2x\,dx$	$2x$	$2\,dx$
5. $\displaystyle\int \tan^2 x \sec^2 x\,dx$	$\tan x$	$\sec^2 x\,dx$
6. $\displaystyle\int \frac{\cos x}{\sin^2 x}\,dx$	$\sin x$	$\cos x\,dx$

7. $\int (1+2x)^4 2\,dx = \dfrac{(1+2x)^5}{5} + C$

Check: $\dfrac{d}{dx}\left[\dfrac{(1+2x)^5}{5} + C\right] = 2(1+2x)^4$

8. $\int (x^2-1)^3 2x\,dx = \dfrac{(x^2-1)^4}{4} + C$

Check: $\dfrac{d}{dx}\left[\dfrac{(x^2-1)^4}{4} + C\right] = 2x(x^2-1)^3$

9. $\int (9-x^2)^{1/2}(-2x)\,dx = \dfrac{(9-x^2)^{3/2}}{3/2} + C = \dfrac{2}{3}(9-x^2)^{3/2} + C$

Check: $\dfrac{d}{dx}\left[\dfrac{2}{3}(9-x^2)^{3/2} + C\right] = \dfrac{2}{3}\cdot\dfrac{3}{2}(9-x^2)^{1/2}(-2x) = \sqrt{9-x^2}(-2x)$

10. $\int (1-2x^2)^3(-4x)\,dx = \dfrac{(1-2x^2)^4}{4} + C$

Check: $\dfrac{d}{dx}\left[\dfrac{(1-2x^2)^4}{4} + C\right] = \dfrac{4(1-2x^2)^3(-4x)}{4} = (1-2x^2)^3(-4x)$

11. $\int x^2(x^3-1)^4\,dx = \dfrac{1}{3}\int (x^3-1)^4(3x^2)\,dx = \dfrac{1}{3}\left[\dfrac{(x^3-1)^5}{5}\right] + C = \dfrac{(x^3-1)^5}{15} + C$

Check: $\dfrac{d}{dx}\left[\dfrac{(x^3-1)^5}{15} + C\right] = \dfrac{5(x^3-1)^4(3x^2)}{15} = x^2(x^3-1)^4$

12. $\int x(4x^2+3)^3\,dx = \dfrac{1}{8}\int (4x^2+3)^3(8x)\,dx = \dfrac{1}{8}\left[\dfrac{(4x^2+3)^4}{4}\right] + C = \dfrac{(4x^2+3)^4}{32} + C$

Check: $\dfrac{d}{dx}\left[\dfrac{(4x^2+3)^4}{32} + C\right] = \dfrac{4(4x^2+3)^3(8x)}{32} = x(4x^2+3)^3$

13. $\int 5x(1-x^2)^{1/3}\,dx = -\dfrac{5}{2}\int (1-x^2)^{1/3}(-2x)\,dx = -\dfrac{5}{2}\cdot\dfrac{(1-x^2)^{4/3}}{4/3} + C = -\dfrac{15}{8}(1-x^2)^{4/3} + C$

Check: $\dfrac{d}{dx}\left[-\dfrac{15}{8}(1-x^2)^{4/3} + C\right] = -\dfrac{15}{8}\cdot\dfrac{4}{3}(1-x^2)^{1/3}(-2x) = 5x(1-x^2)^{1/3} = 5x\sqrt[3]{1-x^2}$

14. $\int u^3\sqrt{u^4+2}\,du = \dfrac{1}{4}\int (u^4+2)^{1/2}(4u^3)\,du = \dfrac{1}{4}\left[\dfrac{(u^4+2)^{3/2}}{3/2}\right] + C = \dfrac{1}{6}(u^4+2)^{3/2} + C$

Check: $\dfrac{d}{dx}\left[\dfrac{1}{6}(u^4+2)^{3/2} + C\right] = \dfrac{1}{6}\cdot\dfrac{3}{2}(u^4+2)^{1/2}(4u^3) = u^3\sqrt{u^4+2}$

15. $\int \dfrac{x^2}{(1+x^3)^2}\,dx = \dfrac{1}{3}\int (1+x^3)^{-2}(3x^2)\,dx = \dfrac{1}{3}\left[\dfrac{(1+x^3)^{-1}}{-1}\right] + C = -\dfrac{1}{3(1+x^3)} + C$

Check: $\dfrac{d}{dx}\left[-\dfrac{1}{3(1+x^3)} + C\right] = -\dfrac{1}{3}(-1)(1+x^3)^{-2}(3x^2) = \dfrac{x^2}{(1+x^3)^2}$

16. $\int \dfrac{x^2}{(16-x^3)^2}\,dx = -\dfrac{1}{3}\int (16-x^3)^{-2}(-3x^2)\,dx = -\dfrac{1}{3}\left[\dfrac{(16-x^3)^{-1}}{-1}\right] + C = \dfrac{1}{3(16-x^3)} + C$

Check: $\dfrac{d}{dx}\left[\dfrac{1}{3(16-x^3)} + C\right] = \dfrac{1}{3}(-1)(16-x^3)^{-2}(-3x^2) = \dfrac{x^2}{(16-x^3)^2}$

17. $\int \left(1+\dfrac{1}{t}\right)^3\left(\dfrac{1}{t^2}\right)\,dt = -\int \left(1+\dfrac{1}{t}\right)^3\left(-\dfrac{1}{t^2}\right)\,dt = -\dfrac{[1+(1/t)]^4}{4} + C$

Check: $\dfrac{d}{dt}\left[-\dfrac{[1+(1/t)]^4}{4} + C\right] = -\dfrac{1}{4}(4)\left(1+\dfrac{1}{t}\right)^3\left(-\dfrac{1}{t^2}\right) = \dfrac{1}{t^2}\left(1+\dfrac{1}{t}\right)^3$

18. $\int \left[x^2 + \dfrac{1}{(3x)^2}\right] dx = \int \left(x^2 + \dfrac{1}{9}x^{-2}\right) dx = \dfrac{x^3}{3} + \dfrac{1}{9}\left(\dfrac{x^{-1}}{-1}\right) + C = \dfrac{x^3}{3} - \dfrac{1}{9x} + C = \dfrac{3x^4 - 1}{9x} + C$

Check: $\dfrac{d}{dx}\left[\dfrac{1}{3}x^3 - \dfrac{1}{9}x^{-1} + C\right] = x^2 + \dfrac{1}{9}x^{-2} = x^2 + \dfrac{1}{(3x)^2}$

19. $\int \dfrac{1}{\sqrt{2x}} dx = \dfrac{1}{2} \int (2x)^{-1/2} 2\, dx = \dfrac{1}{2}\left[\dfrac{(2x)^{1/2}}{1/2}\right] + C = \sqrt{2x} + C$

Check: $\dfrac{d}{dx}[\sqrt{2x} + C] = \dfrac{1}{2}(2x)^{-1/2}(2) = \dfrac{1}{\sqrt{2x}}$

20. $\int \dfrac{1}{2\sqrt{x}} dx = \dfrac{1}{2} \int x^{-1/2}\, dx = \dfrac{1}{2}\left(\dfrac{x^{1/2}}{1/2}\right) + C = \sqrt{x} + C$

Check: $\dfrac{d}{dx}[\sqrt{x} + C] = \dfrac{1}{2\sqrt{x}}$

21. $\int \dfrac{x^2 + 3x + 7}{\sqrt{x}} dx = \int (x^{3/2} + 3x^{1/2} + 7x^{-1/2})\, dx = \dfrac{2}{5}x^{5/2} + 2x^{3/2} + 14x^{1/2} + C = \dfrac{2}{5}\sqrt{x}(x^2 + 5x + 35) + C$

Check: $\dfrac{d}{dx}\left[\dfrac{2}{5}x^{5/2} + 2x^{3/2} + 14x^{1/2} + C\right] = \dfrac{x^2 + 3x + 7}{\sqrt{x}}$

22. $\int \dfrac{t + 2t^2}{\sqrt{t}} dt = \int (t^{1/2} + 2t^{3/2})\, dt = \dfrac{2}{3}t^{3/2} + \dfrac{4}{5}t^{5/2} + C = \dfrac{2}{15}t^{3/2}(5 + 6t) + C$

Check: $\dfrac{d}{dt}\left[\dfrac{2}{3}t^{3/2} + \dfrac{4}{5}t^{5/2} + C\right] = t^{1/2} + 2t^{3/2} = \dfrac{t + 2t^2}{\sqrt{t}}$

23. $\int t^2\left(t - \dfrac{2}{t}\right) dt = \int (t^3 - 2t)\, dt = \dfrac{1}{4}t^4 - t^2 + C$

Check: $\dfrac{d}{dt}\left[\dfrac{1}{4}t^4 - t^2 + C\right] = t^3 - 2t = t^2\left(t - \dfrac{2}{t}\right)$

24. $\int \left(\dfrac{t^3}{3} + \dfrac{1}{4t^2}\right) dt = \int \left(\dfrac{1}{3}t^3 + \dfrac{1}{4}t^{-2}\right) dt = \dfrac{1}{3}\left(\dfrac{t^4}{4}\right) + \dfrac{1}{4}\left(\dfrac{t^{-1}}{-1}\right) + C = \dfrac{1}{12}t^4 - \dfrac{1}{4t} + C$

Check: $\dfrac{d}{dt}\left[\dfrac{1}{12}t^4 - \dfrac{1}{4t} + C\right] = \dfrac{1}{3}t^3 + \dfrac{1}{4t^2}$

25. $\int (9 - y)\sqrt{y}\, dy = \int (9y^{1/2} - y^{3/2})\, dy = 9\left(\dfrac{2}{3}y^{3/2}\right) - \dfrac{2}{5}y^{5/2} + C = \dfrac{2}{5}y^{3/2}(15 - y) + C$

Check: $\dfrac{d}{dy}\left[\dfrac{2}{5}y^{3/2}(15 - y) + C\right] = \dfrac{d}{dy}\left[6y^{3/2} - \dfrac{2}{5}y^{5/2} + C\right] = 9y^{1/2} - y^{3/2} = (9 - y)\sqrt{y}$

26. $\int 2\pi y(8 - y^{3/2})\, dy = 2\pi \int (8y - y^{5/2})\, dy = 2\pi\left(4y^2 - \dfrac{2}{7}y^{7/2}\right) + C = \dfrac{4\pi y^2}{7}(14 - y^{3/2}) + C$

Check: $\dfrac{d}{dy}\left[\dfrac{4\pi y^2}{7}(14 - y^{3/2}) + C\right] = \dfrac{d}{dy}\left[2\pi\left(4y^2 - \dfrac{2}{7}y^{7/2}\right) + C\right] = 16\pi y - 2\pi y^{5/2} = (2\pi y)(8 - y^{3/2})$

27. $y = \int \left[4x + \dfrac{4x}{\sqrt{16 - x^2}}\right] dx$

$= 4 \int x\, dx - 2 \int (16 - x^2)^{-1/2}(-2x)\, dx = 4\left(\dfrac{x^2}{2}\right) - 2\left[\dfrac{(16 - x^2)^{1/2}}{1/2}\right] + C = 2x^2 - 4\sqrt{16 - x^2} + C$

28. $y = \displaystyle\int \frac{10x^2}{\sqrt{1+x^3}}\,dx = \frac{10}{3}\int (1+x^3)^{-1/2}(3x^2)\,dx = \frac{10}{3}\left[\frac{(1+x^3)^{1/2}}{1/2}\right] + C = \frac{20}{3}\sqrt{1+x^3} + C$

29. $y = \displaystyle\int \frac{x+1}{(x^2+2x-3)^2}\,dx$

$\qquad = \dfrac{1}{2}\displaystyle\int (x^2+2x-3)^{-2}(2x+2)\,dx = \dfrac{1}{2}\left[\dfrac{(x^2+2x-3)^{-1}}{-1}\right] + C = -\dfrac{1}{2(x^2+2x-3)} + C$

30. $y = \displaystyle\int \frac{x-4}{\sqrt{x^2-8x+1}}\,dx = \frac{1}{2}\int (x^2-8x+1)^{-1/2}(2x-8)\,dx = \frac{1}{2}\left[\frac{(x^2-8x+1)^{1/2}}{1/2}\right] + C = \sqrt{x^2-8x+1} + C$

31. $\displaystyle\int \sin 2x\,dx = \frac{1}{2}\int (\sin 2x)(2)\,dx = -\frac{1}{2}\cos 2x + C$

32. $\displaystyle\int x\sin x^2\,dx = \frac{1}{2}\int (\sin x^2)(2x)\,dx = -\frac{1}{2}\cos x^2 + C$

33. $\displaystyle\int \frac{1}{\theta^2}\cos\frac{1}{\theta}\,d\theta = -\int \cos\frac{1}{\theta}\left(-\frac{1}{\theta^2}\right)d\theta = -\sin\frac{1}{\theta} + C$

34. $\displaystyle\int \cos 6x\,dx = \frac{1}{6}\int (\cos 6x)(6)\,dx = \frac{1}{6}\sin 6x + C$

35. $\displaystyle\int \sin 2x\cos 2x\,dx = \frac{1}{2}\int (\sin 2x)(2\cos 2x)\,dx = \frac{1}{2}\frac{(\sin 2x)^2}{2} + C = \frac{1}{4}\sin^2 2x + C$

$\qquad$ OR

$\qquad \displaystyle\int \sin 2x\cos 2x\,dx = -\frac{1}{2}\int (\cos 2x)(-2\sin 2x)\,dx = -\frac{1}{2}\frac{(\cos 2x)^2}{2} + C_1 = -\frac{1}{4}\cos^2 2x + C_1$

36. $\displaystyle\int \sec(1-x)\tan(1-x)\,dx = -\int \left[\sec(1-x)\tan(1-x)\right](-1)\,dx = -\sec(1-x) + C$

37. $\displaystyle\int \tan^4 x\sec^2 x\,dx = \frac{\tan^5 x}{5} + C = \frac{1}{5}\tan^5 x + C$

38. $\displaystyle\int \sqrt{\cot x}\,\csc^2 x\,dx = -\int (\cot x)^{1/2}(-\csc^2 x)\,dx = -\frac{2}{3}(\cot x)^{3/2} + C$

39. $\displaystyle\int \frac{\csc^2 x}{\cot^3 x}\,dx = -\int (\cot x)^{-3}(-\csc^2 x)\,dx$

$\qquad\qquad = -\dfrac{(\cot x)^{-2}}{-2} + C = \dfrac{1}{2\cot^2 x} + C = \dfrac{1}{2}\tan^2 x + C = \dfrac{1}{2}(\sec^2 x - 1) + C = \dfrac{1}{2}\sec^2 x + C_1$

40. $\displaystyle\int \frac{\sin x}{\cos^2 x}\,dx = \int \left(\frac{1}{\cos x}\right)\left(\frac{\sin x}{\cos x}\right)dx = \int \sec x\tan x\,dx = \sec x + C$

41. $\displaystyle\int \cot^2 x\,dx = \int (\csc^2 x - 1)\,dx = -\cot x - x + C$

42. $\displaystyle\int \csc^2\left(\frac{x}{2}\right)dx = 2\int \csc^2\left(\frac{x}{2}\right)\left(\frac{1}{2}\right)dx = -2\cot\left(\frac{x}{2}\right) + C$

43. $u = x + 2, \ x = u - 2, \ dx = du$

$$\int x\sqrt{x+2}\,dx = \int (u-2)\sqrt{u}\,du$$

$$= \int (u^{3/2} - 2u^{1/2})\,du$$

$$= \frac{2}{5}u^{5/2} - \frac{4}{3}u^{3/2} + C$$

$$= \frac{2u^{3/2}}{15}(3u - 10) + C$$

$$= \frac{2}{15}(x+2)^{3/2}[3(x+2) - 10] + C$$

$$= \frac{2}{15}(x+2)^{3/2}(3x - 4) + C$$

44. $u = 2x + 1, \ x = \frac{1}{2}(u - 1), \ dx = \frac{1}{2}du$

$$\int x\sqrt{2x+1}\,dx = \int \frac{1}{2}(u-1)\sqrt{u}\frac{1}{2}\,du$$

$$= \frac{1}{4}\int (u^{3/2} - u^{1/2})\,du$$

$$= \frac{1}{4}\left(\frac{2}{5}u^{5/2} - \frac{2}{3}u^{3/2}\right) + C$$

$$= \frac{u^{3/2}}{30}(3u - 5) + C$$

$$= \frac{1}{30}(2x+1)^{3/2}[3(2x+1) - 5] + C$$

$$= \frac{1}{30}(2x+1)^{3/2}(6x - 2) + C$$

$$= \frac{1}{15}(2x+1)^{3/2}(3x - 1) + C$$

45. $u = 1 - x, \ x = 1 - u, \ dx = -du$

$$\int x^2\sqrt{1-x}\,dx = -\int (1-u)^2\sqrt{u}\,du$$

$$= -\int (u^{1/2} - 2u^{3/2} + u^{5/2})\,du$$

$$= -\left(\frac{2}{3}u^{3/2} - \frac{4}{5}u^{5/2} + \frac{2}{7}u^{7/2}\right) + C$$

$$= -\frac{2u^{3/2}}{105}(35 - 42u + 15u^2) + C$$

$$= -\frac{2}{105}(1-x)^{3/2}[35 - 42(1-x) + 15(1-x)^2] + C$$

$$= -\frac{2}{105}(1-x)^{3/2}(15x^2 + 12x + 8) + C$$

46. $u = 2 - x, \ x = 2 - u, \ dx = -du$

$$\int (x+1)\sqrt{2-x}\,dx = -\int (3-u)\sqrt{u}\,du$$

$$= -\int (3u^{1/2} - u^{3/2})\,du$$

$$= -\left(2u^{3/2} - \frac{2}{5}u^{5/2}\right) + C$$

$$= -\frac{2u^{3/2}}{5}(5 - u) + C$$

$$= -\frac{2}{5}(2-x)^{3/2}[5 - (2-x)] + C$$

$$= -\frac{2}{5}(2-x)^{3/2}(x+3) + C$$

47. $u = 2x - 1$, $x = \dfrac{1}{2}(u + 1)$, $dx = \dfrac{1}{2}du$

$$\int \frac{x^2 - 1}{\sqrt{2x - 1}}\,dx = \int \frac{[(1/2)(u + 1)]^2 - 1}{\sqrt{u}}\,\frac{1}{2}\,du$$

$$= \frac{1}{8}\int u^{-1/2}[(u^2 + 2u + 1) - 4]\,du$$

$$= \frac{1}{8}\int (u^{3/2} + 2u^{1/2} - 3u^{-1/2})\,du$$

$$= \frac{1}{8}\left(\frac{2}{5}u^{5/2} + \frac{4}{3}u^{3/2} - 6u^{1/2}\right) + C$$

$$= \frac{u^{1/2}}{60}(3u^2 + 10u - 45) + C$$

$$= \frac{\sqrt{2x - 1}}{60}[3(2x - 1)^2 + 10(2x - 1) - 45] + C$$

$$= \frac{1}{60}\sqrt{2x - 1}(12x^2 + 8x - 52) + C$$

$$= \frac{1}{15}\sqrt{2x - 1}(3x^2 + 2x - 13) + C$$

48. $u = x + 3$, $x = u - 3$, $dx = du$

$$\int \frac{2x - 1}{\sqrt{x + 3}}\,dx = \int \frac{2(u - 3) - 1}{\sqrt{u}}\,du$$

$$= \int (2u^{1/2} - 7u^{-1/2})\,du$$

$$= \frac{4}{3}u^{3/2} - 14u^{1/2} + C$$

$$= \frac{2u^{1/2}}{3}(2u - 21) + C$$

$$= \frac{2}{3}\sqrt{x + 3}[2(x + 3) - 21] + C$$

$$= \frac{2}{3}\sqrt{x + 3}(2x - 15) + C$$

49. $u = x + 1$, $x = u - 1$, $dx = du$

$$\int \frac{-x}{(x + 1) - \sqrt{x + 1}}\,dx = \int \frac{-(u - 1)}{u - \sqrt{u}}\,du$$

$$= -\int \frac{(\sqrt{u} + 1)(\sqrt{u} - 1)}{\sqrt{u}(\sqrt{u} - 1)}\,du$$

$$= -\int (1 + u^{-1/2})\,du$$

$$= -(u + 2u^{1/2}) + C$$

$$= -u - 2\sqrt{u} + C$$

$$= -(x + 1) - 2\sqrt{x + 1} + C$$

$$= -x - 2\sqrt{x + 1} - 1 + C$$

$$= (x + 2\sqrt{x + 1}) + C_1 \text{ where}$$

$$C_1 = -1 + C$$

50. $u = t - 4$, $t = u + 4$, $dt = du$

$$\int t\sqrt[3]{t - 4}\,dt = \int (u + 4)u^{1/3}\,du$$

$$= \int (u^{4/3} + 4u^{1/3})\,du$$

$$= \frac{3}{7}u^{7/3} + 3u^{4/3} + C$$

$$= \frac{3u^{4/3}}{7}(u + 7) + C$$

$$= \frac{3}{7}(t - 4)^{4/3}[(t - 4) + 7] + C$$

$$= \frac{3}{7}(t - 4)^{4/3}(t + 3) + C$$

51. Let $u = x^2 + 1$, $du = 2x\,dx$

$$\int_{-1}^{1} x(x^2+1)^3\,dx = \frac{1}{2}\int_{-1}^{1}(x^2+1)^3(2x)\,dx = \left[\frac{1}{8}(x^2+1)^4\right]_{-1}^{1} = 0$$

52. Let $u = 1 - x^2$, $du = -2x\,dx$

$$\int_{0}^{1} x\sqrt{1-x^2}\,dx = -\frac{1}{2}\int_{0}^{1}(1-x^2)^{1/2}(-2x)\,dx = \left[-\frac{1}{3}(1-x^2)^{3/2}\right]_{0}^{1} = 0 + \frac{1}{3} = \frac{1}{3}$$

53. Let $u = 2x + 1$, $du = 2\,dx$

$$\int_{0}^{4} \frac{1}{\sqrt{2x+1}}\,dx = \frac{1}{2}\int_{0}^{4}(2x+1)^{-1/2}(2)\,dx = \left[\sqrt{2x+1}\right]_{0}^{4} = \sqrt{9} - \sqrt{1} = 2$$

54. Let $u = 1 + 2x^2$, $du = 4x\,dx$

$$\int_{0}^{2} \frac{x}{\sqrt{1+2x^2}}\,dx = \frac{1}{4}\int_{0}^{2}(1+2x^2)^{-1/2}(4x)\,dx = \left[\frac{1}{2}\sqrt{1+2x^2}\right]_{0}^{2} = \frac{3}{2} - \frac{1}{2} = 1$$

55. Let $u = 1 + \sqrt{x}$, $du = \frac{1}{2\sqrt{x}}\,dx$

$$\int_{1}^{9} \frac{1}{\sqrt{x}(1+\sqrt{x})^2}\,dx = 2\int_{1}^{9}(1+\sqrt{x})^{-2}\left(\frac{1}{2\sqrt{x}}\right)\,dx = \left[-\frac{2}{1+\sqrt{x}}\right]_{1}^{9} = -\frac{1}{2} + 1 = \frac{1}{2}$$

56. Let $u = 4 + x^2$, $du = 2x\,dx$

$$\int_{0}^{2} x\sqrt[3]{4+x^2}\,dx = \frac{1}{2}\int_{0}^{2}(4+x^2)^{1/3}(2x)\,dx = \left[\frac{3}{8}(4+x^2)^{4/3}\right]_{0}^{2} = \frac{3}{8}(8^{4/3} - 4^{4/3}) = 6 - \frac{3}{2}\sqrt[3]{4} \approx 3.619$$

57. $u = 2 - x$, $x = 2 - u$, $dx = -du$
When $x = 1$, $u = 1$. When $x = 2$, $u = 0$.

$$\int_{1}^{2} (x-1)\sqrt{2-x}\,dx = \int_{1}^{0} -[(2-u)-1]\sqrt{u}\,du$$

$$= \int_{1}^{0}(u^{3/2} - u^{1/2})\,du = \left[\frac{2}{5}u^{5/2} - \frac{2}{3}u^{3/2}\right]_{1}^{0} = \left[\frac{2}{5} - \frac{2}{3}\right] = \frac{4}{15}$$

58. $u = 2x + 1$, $x = \frac{1}{2}(u-1)$, $dx = \frac{1}{2}\,du$
When $x = 0$, $u = 1$. When $x = 4$, $u = 9$.

$$\int_{0}^{4} \frac{x}{\sqrt{2x+1}}\,dx = \int_{1}^{9} \frac{(1/2)(u-1)}{\sqrt{u}}\,\frac{1}{2}\,du$$

$$= \frac{1}{4}\int_{1}^{9}(u^{1/2} - u^{-1/2})\,du = \frac{1}{4}\left[\frac{2}{3}u^{3/2} - 2u^{1/2}\right]_{1}^{9} = \frac{1}{4}(18-6) - \frac{1}{4}\left(\frac{2}{3}-2\right) = \frac{10}{3}$$

59. $u = x + 1$, $x = u - 1$, $dx = du$
When $x = 0$, $u = 1$. When $x = 7$, $u = 8$.

$$\int_{0}^{7} x\sqrt[3]{x+1}\,dx = \int_{1}^{8}(u-1)\sqrt[3]{u}\,du$$

$$= \int_{1}^{8}(u^{4/3} - u^{1/3})\,du = \left[\frac{3}{7}u^{7/3} - \frac{3}{4}u^{4/3}\right]_{1}^{8} = \left(\frac{384}{7} - 12\right) - \left(\frac{3}{7} - \frac{3}{4}\right) = \frac{1209}{28}$$

60. $u = x + 2$, $x = u - 2$, $dx = du$

When $x = -2$, $u = 0$. When $x = 6$, $u = 8$.

$$\int_{-2}^{6} x^2 \sqrt[3]{x+2}\, dx = \int_{0}^{8} (u-2)^2 \sqrt[3]{u}\, du$$

$$= \int_{0}^{8} (u^{7/3} - 4u^{4/3} + 4u^{1/3})\, du = \left[\frac{3}{10} u^{10/3} - \frac{12}{7} u^{7/3} + 3u^{4/3}\right]_0^8 = \frac{4752}{35}$$

61. $\displaystyle\int_0^{\pi/2} \cos\left(\frac{2}{3}x\right) dx = \left[\frac{3}{2}\sin\left(\frac{2}{3}x\right)\right]_0^{\pi/2} = \frac{3}{2}\left(\frac{\sqrt{3}}{2}\right) = \frac{3\sqrt{3}}{4}$

62. $\displaystyle\int_{\pi/3}^{\pi/2} (x + \cos x)\, dx = \left[\frac{x^2}{2} + \sin x\right]_{\pi/3}^{\pi/2} = \left(\frac{\pi^2}{8} + 1\right) - \left(\frac{\pi^2}{18} + \frac{\sqrt{3}}{2}\right) = \frac{5\pi^2}{72} + \frac{2-\sqrt{3}}{2}$

63. $\displaystyle\int_{\pi/2}^{2\pi/3} \sec^2\left(\frac{x}{2}\right) dx = 2 \int_{\pi/2}^{2\pi/3} \sec^2\left(\frac{x}{2}\right)\left(\frac{1}{2}\right) dx = \left[2\tan\left(\frac{x}{2}\right)\right]_{\pi/2}^{2\pi/3} = 2(\sqrt{3}-1)$

64. Let $u = 2x$, $du = 2\,dx$

$$\int_{\pi/12}^{\pi/4} \csc 2x \cot 2x\, dx = \frac{1}{2}\int_{\pi/12}^{\pi/4} \csc 2x \cot 2x (2)\, dx = \left[-\frac{1}{2}\csc 2x\right]_{\pi/12}^{\pi/4} = \frac{1}{2}$$

65. $\displaystyle\int \frac{x}{\sqrt{2x+1}}\, dx = \frac{1}{3}\sqrt{2x+1}(x-1) + C$

66. $\displaystyle\int x^3\sqrt{x+2}\, dx = \frac{2}{315}(x+2)^{3/2}(35x^3 - 60x^2 + 96x - 128) + C$

67. $\displaystyle\int_3^7 x\sqrt{x-3}\, dx = \frac{144}{5}$

68. $\displaystyle\int_0^1 \frac{1}{\sqrt{x}+\sqrt{x+1}}\, dx = \frac{4}{3}(\sqrt{2}-1)$

69. $\displaystyle\int_1^5 x^2\sqrt{x-1}\, dx = \frac{7088}{105}$

70. $\displaystyle\int_0^{\pi/2} \sin 2x\, dx = 1$

71. $\displaystyle\int (2x-1)^2\, dx = \frac{1}{2}\int (2x-1)^2 2\, dx = \frac{1}{6}(2x-1)^3 + C_1 = \frac{4}{3}x^3 - 2x^2 + x - \frac{1}{6} + C_1$

$\displaystyle\int (2x-1)^2\, dx = \int (4x^2 - 4x + 1)\, dx = \frac{4}{3}x^3 - 2x^2 + x + C_2$

They differ by a constant: $C_2 = C_1 - \dfrac{1}{6}$.

72. $\displaystyle\int \sin x \cos x\, dx = \int (\sin x)^1(\cos x\, dx) = \frac{\sin^2 x}{2} + C_1$

$\displaystyle\int \sin x \cos x\, dx = -\int (\cos x)^1(-\sin x\, dx) = -\frac{\cos^2 x}{2} + C_2$

$-\dfrac{\cos^2 x}{2} + C_2 = -\dfrac{(1-\sin^2 x)}{2} + C_2 = \dfrac{\sin^2 x}{2} - \dfrac{1}{2} + C_2$

They differ by a constant: $C_2 = C_1 + \dfrac{1}{2}$.

73. $A = \displaystyle\int_0^{\pi} (2\sin x + \sin 2x)\, dx = -\left[2\cos x + \frac{1}{2}\cos 2x\right]_0^{\pi} = 4$

74. $A = \int_0^{\pi} (\sin x + \cos 2x)\, dx = \left[-\cos x + \frac{1}{2} \sin 2x \right]_0^{\pi} = 2$

75. $\int_0^2 x^2\, dx = \frac{8}{3}$; the function x^2 is an even function.

   a. $\int_{-2}^0 x^2\, dx = \int_0^2 x^2\, dx = \frac{8}{3}$ b. $\int_{-2}^2 x^2\, dx = 2\int_0^2 x^2\, dx = \frac{16}{3}$

   c. $\int_0^2 (-x^2)\, dx = -\int_0^2 x^2\, dx = -\frac{8}{3}$ d. $\int_{-2}^0 3x^2\, dx = 3\int_0^2 x^2\, dx = 8$

76. a. $\int_{-\pi/4}^{\pi/4} \sin x\, dx = 0$ since $\sin x$ is symmetric to the origin.

   b. $\int_{-\pi/4}^{\pi/4} \cos x\, dx = 2\int_0^{\pi/4} \cos x\, dx = \left[2\sin x \right]_0^{\pi/4} = \sqrt{2}$ since $\cos x$ is symmetric to the y-axis.

   c. $\int_{-\pi/2}^{\pi/2} \cos x\, dx = 2\int_0^{\pi/2} \cos x\, dx = \left[2\sin x \right]_0^{\pi/2} = 2$

   d. $\int_{-\pi/2}^{\pi/2} \sin x \cos x\, dx = 0$ since $\sin(-x)\cos(-x) = -\sin x \cos x$ and hence, is symmetric to the origin.

77. $\int_{-4}^4 (x^3 + 6x^2 - 2x - 3)\, dx = \int_{-4}^4 (x^3 - 2x)\, dx + \int_{-4}^4 (6x^2 - 3)\, dx$

$$= 0 + 2\int_0^4 (6x^2 - 3)\, dx = 2\left[2x^3 - 3x \right]_0^4 = 232$$

78. $\int_{-\pi}^{\pi} (\sin 3x + \cos 3x)\, dx = \int_{-\pi}^{\pi} \sin 3x\, dx + \int_{-\pi}^{\pi} \cos 3x\, dx$

$$= 0 + 2\int_0^{\pi} \cos 3x\, dx = \left[\frac{2}{3} \sin 3x \right]_0^{\pi} = 0$$

79. $f'(x) = x\sqrt{1 - x^2}, \quad \left(0, \frac{7}{3} \right)$

$f(x) = \int x\sqrt{1 - x^2}\, dx$

$\quad = -\frac{1}{2} \int (1 - x^2)^{1/2}(-2x)\, dx$

$\quad = -\frac{1}{3}(1 - x^2)^{3/2} + C$

$f(0) = -\frac{1}{3} + C = \frac{7}{3} \Rightarrow C = \frac{8}{3}$

$f(x) = -\frac{1}{3}(1 - x^2)^{3/2} + \frac{8}{3} = -\frac{1}{3}[(1 - x^2)^{3/2} - 8]$

80. $f'(x) = \pi \sec \pi x \tan \pi x, \quad \left(\frac{1}{3}, 1 \right)$

$f(x) = \int \pi \sec \pi x \tan \pi x\, dx = \sec \pi x + C$

$f\left(\frac{1}{3} \right) = 2 + C = 1 \Rightarrow C = -1$

$f(x) = \sec \pi x - 1$

81. $\dfrac{dV}{dt} = \dfrac{k}{(t+1)^2}$

$V(t) = \displaystyle\int \dfrac{k}{(t+1)^2}\, dt = -\dfrac{k}{t+1} + C$

$V(0) = -k + C = 500{,}000$

$V(1) = -\dfrac{1}{2}k + C = 400{,}000$

Solving this system yields $k = -200{,}000$ and $C = 300{,}000$. Thus,

$V(t) = \dfrac{200{,}000}{t+1} + 300{,}000.$

When $t = 4, \quad V(4) = \$340{,}000.$

82. $\dfrac{dQ}{dt} = k(100 - t)^2$

$Q(t) = \displaystyle\int k(100 - t)^2\, dt = -\dfrac{k}{3}(100 - t)^3 + C$

$Q(100) = C = 0$

$Q(t) = -\dfrac{k}{3}(100 - t)^3$

$Q(0) = -\dfrac{k}{3}(100)^3 = 2{,}000{,}000 \Rightarrow k = -6$

Thus, $Q(t) = 2(100 - t)^3$. When $t = 50, \quad Q(50) = \$250{,}000.$

83. $\dfrac{dC}{dx} = \dfrac{12}{\sqrt[3]{12x + 1}}$

a. $C(x) = \displaystyle\int \dfrac{12}{\sqrt[3]{12x + 1}}\, dx$

$= \displaystyle\int (12x + 1)^{-1/3} 12\, dx = \dfrac{3}{2}(12x + 1)^{2/3} + C_1$

$C(13) \approx 43.65 + C_1 = 100$

$C_1 = 56.35$

$C(x) = \dfrac{3}{2}(12x + 1)^{2/3} + 56.35$

b.

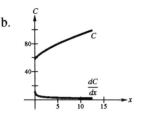

84. $\dfrac{1}{b-a}\displaystyle\int_a^b \left[217 + 13\cos\dfrac{\pi(t-3)}{6}\right] dt = \dfrac{1}{b-a}\left[217t + \dfrac{78}{\pi}\sin\left(\dfrac{\pi(t-3)}{6}\right)\right]_a^b$

a. $\dfrac{1}{3}\left[217t + \dfrac{78}{\pi}\sin\left(\dfrac{\pi(t-3)}{6}\right)\right]_0^3 = \dfrac{1}{3}\left(651 + \dfrac{78}{\pi}\right) \approx 225.28$ million barrels

b. $\dfrac{1}{3}\left[217t + \dfrac{78}{\pi}\sin\left(\dfrac{\pi(t-3)}{6}\right)\right]_3^6 = \dfrac{1}{3}\left(1302 + \dfrac{78}{\pi} - 651\right) \approx 225.28$ million barrels

c. $\dfrac{1}{12}\left[217t + \dfrac{78}{\pi}\sin\left(\dfrac{\pi(t-3)}{6}\right)\right]_0^{12} = \dfrac{1}{12}\left(2604 - \dfrac{78}{\pi} + \dfrac{78}{\pi}\right) = 217$ million barrels

85. $\dfrac{1}{b-a}\displaystyle\int_a^b \left[74.50 + 43.75\sin\dfrac{\pi t}{6}\right] dt = \dfrac{1}{b-a}\left[74.50t - \dfrac{262.5}{\pi}\cos\dfrac{\pi t}{6}\right]_a^b$

a. $\dfrac{1}{3}\left[74.50t - \dfrac{262.5}{\pi}\cos\dfrac{\pi t}{6}\right]_0^3 = \dfrac{1}{3}\left(223.5 + \dfrac{262.5}{\pi}\right) \approx 102.352$ thousand units

b. $\dfrac{1}{3}\left[74.50t - \dfrac{262.5}{\pi}\cos\dfrac{\pi t}{6}\right]_3^6 = \dfrac{1}{3}\left(447 + \dfrac{262.5}{\pi} - 223.5\right) \approx 102.352$ thousand units

c. $\dfrac{1}{12}\left[74.50t - \dfrac{262.5}{\pi}\cos\dfrac{\pi t}{6}\right]_0^{12} = \dfrac{1}{12}\left(894 - \dfrac{262.5}{\pi} + \dfrac{262.5}{\pi}\right) = 74.5$ thousand units

86. $\dfrac{1}{b-a} \displaystyle\int_a^b \left[2\sin(60\pi t) + \cos(120\pi t) \right] dt = \dfrac{1}{b-a} \left[-\dfrac{1}{30\pi}\cos(60\pi t) + \dfrac{1}{120\pi}\sin(120\pi t) \right]_a^b$

a. $\dfrac{1}{(1/60) - 0} \left[-\dfrac{1}{30\pi}\cos(60\pi t) + \dfrac{1}{120\pi}\sin(120\pi t) \right]_0^{1/60} = 60\left[\left(\dfrac{1}{30\pi} + 0 \right) - \left(-\dfrac{1}{30\pi} \right) \right] = \dfrac{4}{\pi} \approx 1.273$ amps

b. $\dfrac{1}{(1/240) - 0} \left[-\dfrac{1}{30\pi}\cos(60\pi t) + \dfrac{1}{120\pi}\sin(120\pi t) \right]_0^{1/240} = 240\left[\left(-\dfrac{1}{30\sqrt{2}\,\pi} + \dfrac{1}{120\pi} \right) - \left(-\dfrac{1}{30\pi} \right) \right]$

$\qquad\qquad\qquad\qquad\qquad\qquad\qquad\qquad\qquad\qquad\qquad = \dfrac{2}{\pi}(5 - 2\sqrt{2}) \approx 1.382$ amps

c. $\dfrac{1}{(1/30) - 0} \left[-\dfrac{1}{30\pi}\cos(60\pi t) + \dfrac{1}{120\pi}\sin(120\pi t) \right]_0^{1/30} = 30\left[\left(-\dfrac{1}{30\pi} \right) - \left(-\dfrac{1}{30\pi} \right) \right] = 0$ amps

87. False

$$\int (2x+1)^2 \, dx = \dfrac{1}{2}\int (2x+1)^2 \, 2\, dx = \dfrac{1}{6}(2x+1)^3 + C$$

88. False

$$\int x(x^2+1) \, dx = \dfrac{1}{2}\int (x^2+1)(2x) \, dx = \dfrac{1}{4}(x^2+1)^2 + C$$

89. True

$$\int_{-10}^{10} (ax^3 + bx^2 + cx + d) \, dx = \underbrace{\int_{-10}^{10} (ax^3 + cx) \, dx}_{\text{Odd}} + \underbrace{\int_{-10}^{10} (bx^2 + d) \, dx}_{\text{Even}}$$

$$= 0 + 2\int_0^{10} (bx^2 + d) \, dx$$

90. True

$$\int_a^b \sin x \, dx = \left[-\cos x \right]_a^b = -\cos b + \cos a = -\cos(b + 2\pi) + \cos a = \int_a^{b+2\pi} \sin x \, dx$$

91. True

$$4\int \sin x \cos x \, dx = 2\sin^2 x + C_1$$

$$= 2\left(\dfrac{1 - \cos 2x}{2} \right) + C_1 = 1 - \cos 2x + C_1 = -\cos 2x + C \text{ where } C = 1 + C_1.$$

92. False

$$\int \sin^2 2x \cos 2x \, dx = \dfrac{1}{2}\int (\sin 2x)^2 (2\cos 2x) \, dx = \dfrac{1}{2}\dfrac{(\sin 2x)^3}{3} + C = \dfrac{1}{6}\sin^3 2x + C$$

93. Let $u = x + h$, then $du = dx$. When $x = a$, $u = a + h$. When $x = b$, $u = b + h$. Thus,

$$\int_a^b f(x+h) \, dx = \int_{a+h}^{b+h} f(u) \, du = \int_{a+h}^{b+h} f(x) \, dx.$$

94. $\displaystyle\int_0^b \frac{f(x)}{f(x) + f(b-x)}\,dx$

Let $u = b - x$, then $du = -dx$. When $x = 0$, $u = b$. When $x = b$, $u = 0$.

$$\int_0^b \frac{f(x)}{f(x) + f(b-x)}\,dx = -\int_b^0 \frac{f(b-u)}{f(b-u) + f(u)}\,du = \int_0^b \frac{f(b-x)}{f(b-x) + f(x)}\,dx$$

Also, by division,

$$\int_0^b \frac{f(x)}{f(x) + f(b-x)}\,dx = \int_0^b \left[1 - \frac{f(b-x)}{f(x) + f(b-x)}\right] dx.$$

Thus,

$$\int_0^b \frac{f(x)}{f(x) + f(b-x)}\,dx = \int_0^b 1\,dx - \int_0^b \frac{f(b-x)}{f(x) + f(b-x)}\,dx$$

$$\int_0^b \frac{f(x)}{f(x) + f(b-x)}\,dx + \int_0^b \frac{f(b-x)}{f(x) + f(b-x)}\,dx = \int_0^b 1\,dx$$

$$2\int_0^b \frac{f(x)}{f(x) + f(b-x)}\,dx = \Big[x\Big]_0^b = b.$$

Therefore,

$$\int_0^b \frac{f(x)}{f(x) + f(b-x)}\,dx = \frac{b}{2}$$

$$\int_0^1 \frac{\sin x}{\sin x + \sin(1-x)}\,dx = \frac{1}{2}\quad \text{(Let } f(x) = \sin x \text{ and } b = 1.)$$

Section 4.6   Numerical Integration

1. Exact:
$$\int_0^2 x^2\,dx = \left[\frac{1}{3}x^3\right]_0^2 = \frac{8}{3} \approx 2.6667$$

Trapezoidal:
$$\int_0^2 x^2\,dx \approx \frac{1}{4}\left[0 + 2\left(\frac{1}{2}\right)^2 + 2(1)^2 + 2\left(\frac{3}{2}\right)^2 + (2)^2\right] = \frac{11}{4} = 2.7500$$

Simpson's:
$$\int_0^2 x^2\,dx \approx \frac{1}{6}\left[0 + 4\left(\frac{1}{2}\right)^2 + 2(1)^2 + 4\left(\frac{3}{2}\right)^2 + (2)^2\right] = \frac{8}{3} \approx 2.6667$$

2. Exact:
$$\int_0^1 \left(\frac{x^2}{2} + 1\right) dx = \left[\frac{x^3}{6} + x\right]_0^1 = \frac{7}{6} \approx 1.1667$$

Trapezoidal:
$$\int_0^1 \left(\frac{x^2}{2} + 1\right) dx \approx \frac{1}{8}\left[1 + 2\left(\frac{(1/4)^2}{2} + 1\right) + 2\left(\frac{(1/2)^2}{2} + 1\right) + 2\left(\frac{(3/4)^2}{2} + 1\right) + \left(\frac{1^2}{2} + 1\right)\right]$$

$$= \frac{75}{64} \approx 1.1719$$

Simpson's:
$$\int_0^1 \left(\frac{x^2}{2} + 1\right) dx \approx \frac{1}{12}\left[1 + 4\left(\frac{(1/4)^2}{2} + 1\right) + 2\left(\frac{(1/2)^2}{2} + 1\right) + 4\left(\frac{(3/4)^2}{2} + 1\right) + \left(\frac{1^2}{2} + 1\right)\right]$$

$$= \frac{7}{6} \approx 1.1667$$

3. Exact: $\displaystyle\int_0^2 x^3\,dx = \left[\frac{x^4}{4}\right]_0^2 = 4.0000$

Trapezoidal: $\displaystyle\int_0^2 x^3\,dx \approx \frac{1}{4}\left[0 + 2\left(\frac{1}{2}\right)^3 + 2(1)^3 + 2\left(\frac{3}{2}\right)^3 + (2)^3\right] = \frac{17}{4} = 4.2500$

Simpson's: $\displaystyle\int_0^2 x^3\,dx \approx \frac{1}{6}\left[0 + 4\left(\frac{1}{2}\right)^3 + 2(1)^3 + 4\left(\frac{3}{2}\right)^3 + (2)^3\right] = \frac{24}{6} = 4.0000$

4. Exact: $\displaystyle\int_1^2 \frac{1}{x^2}\,dx = \left[\frac{-1}{x}\right]_1^2 = 0.5000$

Trapezoidal: $\displaystyle\int_1^2 \frac{1}{x^2}\,dx \approx \frac{1}{8}\left[1 + 2\left(\frac{4}{5}\right)^2 + 2\left(\frac{4}{6}\right)^2 + 2\left(\frac{4}{7}\right)^2 + \frac{1}{4}\right] \approx 0.5090$

Simpson's: $\displaystyle\int_1^2 \frac{1}{x^2}\,dx \approx \frac{1}{12}\left[1 + 4\left(\frac{4}{5}\right)^2 + 2\left(\frac{4}{6}\right)^2 + 4\left(\frac{4}{7}\right)^2 + \frac{1}{4}\right] \approx 0.5004$

5. Exact: $\displaystyle\int_0^2 x^3\,dx = \left[\frac{1}{4}x^4\right]_0^2 = 4.0000$

Trapezoidal: $\displaystyle\int_0^2 x^3\,dx \approx \frac{1}{8}\left[0 + 2\left(\frac{1}{4}\right)^3 + 2\left(\frac{2}{4}\right)^3 + 2\left(\frac{3}{4}\right)^3 + 2(1)^3 + 2\left(\frac{5}{4}\right)^3 + 2\left(\frac{6}{4}\right)^3 + 2\left(\frac{7}{4}\right)^3 + 8\right]$
$= 4.0625$

Simpson's: $\displaystyle\int_0^2 x^3\,dx \approx \frac{1}{12}\left[0 + 4\left(\frac{1}{4}\right)^3 + 2\left(\frac{2}{4}\right)^3 + 4\left(\frac{3}{4}\right)^3 + 2(1)^3 + 4\left(\frac{5}{4}\right)^3 + 2\left(\frac{6}{4}\right)^3 + 4\left(\frac{7}{4}\right)^3 + 8\right]$
$= 4.0000$

6. Exact: $\displaystyle\int_0^8 \sqrt[3]{x}\,dx = \left[\frac{3}{4}x^{4/3}\right]_0^8 = 12.0000$

Trapezoidal: $\displaystyle\int_0^8 \sqrt[3]{x}\,dx \approx \frac{1}{2}\left[0 + 2 + 2\sqrt[3]{2} + 2\sqrt[3]{3} + 2\sqrt[3]{4} + 2\sqrt[3]{5} + 2\sqrt[3]{6} + 2\sqrt[3]{7} + 2\right] \approx 11.7296$

Simpson's: $\displaystyle\int_0^8 \sqrt[3]{x}\,dx \approx \frac{1}{3}\left[0 + 4 + 2\sqrt[3]{2} + 4\sqrt[3]{3} + 2\sqrt[3]{4} + 4\sqrt[3]{5} + 2\sqrt[3]{6} + 4\sqrt[3]{7} + 2\right] \approx 11.8632$

7. Exact: $\displaystyle\int_4^9 \sqrt{x}\,dx = \left[\frac{2}{3}x^{3/2}\right]_4^9 = 18 - \frac{16}{3} = \frac{38}{3} \approx 12.6667$

Trapezoidal: $\displaystyle\int_4^9 \sqrt{x}\,dx \approx \frac{5}{16}\left[2 + 2\sqrt{\frac{37}{8}} + 2\sqrt{\frac{21}{4}} + 2\sqrt{\frac{47}{8}} + 2\sqrt{\frac{26}{4}} + 2\sqrt{\frac{57}{8}} + 2\sqrt{\frac{31}{4}} + 2\sqrt{\frac{67}{8}} + 3\right] \approx 12.6640$

Simpson's: $\displaystyle\int_4^9 \sqrt{x}\,dx \approx \frac{5}{24}\left[2 + 4\sqrt{\frac{37}{8}} + \sqrt{21} + 4\sqrt{\frac{47}{8}} + \sqrt{26} + 4\sqrt{\frac{57}{8}} + \sqrt{31} + 4\sqrt{\frac{67}{8}} + 3\right] \approx 12.6667$

8. Exact: $\displaystyle\int_1^3 (4 - x^2)\,dx = \left[4x - \frac{x^3}{3}\right]_1^3 = 3 - \frac{11}{3} = -\frac{2}{3} \approx -0.6667$

Trapezoidal: $\displaystyle\int_1^3 (4 - x^2)\,dx \approx \frac{1}{4}\left\{3 + 2\left[4 - \left(\frac{3}{2}\right)^2\right] + 2(0) + 2\left[4 - \left(\frac{5}{2}\right)^2\right] - 5\right\} = -0.7500$

Simpson's: $\displaystyle\int_1^3 (4 - x^2)\,dx \approx \frac{1}{6}\left[3 + 4\left(4 - \frac{9}{4}\right) + 0 + 4\left(4 - \frac{25}{4}\right) - 5\right] \approx -0.6667$

9. Exact: $\int_1^2 \frac{1}{(x+1)^2}\,dx = \left[-\frac{1}{x+1}\right]_1^2 = -\frac{1}{3} + \frac{1}{2} = \frac{1}{6} \approx 0.1667$

Trapezoidal: $\int_1^2 \frac{1}{(x+1)^2}\,dx \approx \frac{1}{8}\left[\frac{1}{4} + 2\left(\frac{1}{(\frac{5}{4}+1)^2}\right) + 2\left(\frac{1}{(\frac{3}{2}+1)^2}\right) + 2\left(\frac{1}{(\frac{7}{4}+1)^2}\right) + \frac{1}{9}\right]$

$= \frac{1}{8}\left(\frac{1}{4} + \frac{32}{81} + \frac{8}{25} + \frac{32}{121} + \frac{1}{9}\right) \approx 0.1676$

Simpson's: $\int_1^2 \frac{1}{(x+1)^2}\,dx \approx \frac{1}{12}\left[\frac{1}{4} + 4\left(\frac{1}{(\frac{5}{4}+1)^2}\right) + 2\left(\frac{1}{(\frac{3}{2}+1)^2}\right) + 4\left(\frac{1}{(\frac{7}{4}+1)^2}\right) + \frac{1}{9}\right]$

$= \frac{1}{12}\left(\frac{1}{4} + \frac{64}{81} + \frac{8}{25} + \frac{64}{121} + \frac{1}{9}\right) \approx 0.1667$

10. Exact: $\int_0^2 x\sqrt{x^2+1}\,dx = \frac{1}{3}(x^2+1)^{3/2}\Big]_0^2 = \frac{1}{3}(5^{3/2}-1) \approx 3.393$

Trapezoidal: $\int_0^2 x\sqrt{x^2+1}\,dx \approx \frac{1}{4}\left[0 + 2\left(\frac{1}{2}\right)\sqrt{(1/2)^2+1} + 2(1)\sqrt{1^2+1} + 2\left(\frac{3}{2}\right)\sqrt{(3/2)^2+1} + 2\sqrt{2^2+1}\right]$

≈ 3.457

Simpson's: $\int_0^2 x\sqrt{x^2+1}\,dx \approx \frac{1}{6}\left[0 + 4\left(\frac{1}{2}\right)\sqrt{(1/2)^2+1} + 2(1)\sqrt{1^2+1} + 4\left(\frac{3}{2}\right)\sqrt{(3/2)^2+1} + 2\sqrt{2^2+1}\right]$

≈ 3.392

11. Trapezoidal: $\int_0^2 \sqrt{1+x^3}\,dx \approx \frac{1}{4}[1 + 2\sqrt{1+(1/8)} + 2\sqrt{2} + 2\sqrt{1+(27/8)} + 3] \approx 3.283$

Simpson's: $\int_0^2 \sqrt{1+x^3}\,dx \approx \frac{1}{6}[1 + 4\sqrt{1+(1/8)} + 2\sqrt{2} + 4\sqrt{1+(27/8)} + 3] \approx 3.240$

12. Trapezoidal: $\int_0^2 \frac{1}{\sqrt{1+x^3}}\,dx \approx \frac{1}{4}\left[1 + 2\left(\frac{1}{\sqrt{1+(1/2)^3}}\right) + 2\left(\frac{1}{\sqrt{1+1^3}}\right) + 2\left(\frac{1}{\sqrt{1+(3/2)^3}}\right) + \frac{1}{3}\right] \approx 1.397$

Simpson's: $\int_0^2 \frac{1}{\sqrt{1+x^3}}\,dx \approx \frac{1}{6}\left[1 + 4\left(\frac{1}{\sqrt{1+(1/2)^3}}\right) + 2\left(\frac{1}{\sqrt{1+1^3}}\right) + 4\left(\frac{1}{\sqrt{1+(3/2)^3}}\right) + \frac{1}{3}\right] \approx 1.405$

13. $\int_0^1 \sqrt{x}\sqrt{1-x}\,dx = \int_0^1 \sqrt{x(1-x)}\,dx$

Trapezoidal: $\int_0^1 \sqrt{x(1-x)}\,dx \approx \frac{1}{8}\left[0 + 2\sqrt{\frac{1}{4}\left(1-\frac{1}{4}\right)} + 2\sqrt{\frac{1}{2}\left(1-\frac{1}{2}\right)} + 2\sqrt{\frac{3}{4}\left(1-\frac{3}{4}\right)}\right] \approx 0.342$

Simpson's: $\int_0^1 \sqrt{x(1-x)}\,dx \approx \frac{1}{12}\left[0 + 4\sqrt{\frac{1}{4}\left(1-\frac{1}{4}\right)} + 2\sqrt{\frac{1}{2}\left(1-\frac{1}{2}\right)} + 4\sqrt{\frac{3}{4}\left(1-\frac{3}{4}\right)}\right] \approx 0.372$

14. Trapezoidal:

$$\int_{\pi/2}^{\pi} \sqrt{x} \sin x \, dx \approx \frac{\pi}{16}\left[\sqrt{\frac{\pi}{2}}(1) + 2\sqrt{\frac{5\pi}{8}}\sin\left(\frac{5\pi}{8}\right) + 2\sqrt{\frac{3\pi}{4}}\sin\left(\frac{3\pi}{4}\right) + 2\sqrt{\frac{7\pi}{8}}\sin\left(\frac{7\pi}{8}\right) + 0\right] \approx 1.430$$

Simpson's:

$$\int_{\pi/2}^{\pi} \sqrt{x} \sin x \, dx \approx \frac{\pi}{24}\left[\sqrt{\frac{\pi}{2}} + 4\sqrt{\frac{5\pi}{8}}\sin\left(\frac{5\pi}{8}\right) + 2\sqrt{\frac{3\pi}{4}}\sin\left(\frac{3\pi}{4}\right) + 4\sqrt{\frac{7\pi}{8}}\sin\left(\frac{7\pi}{8}\right) + 0\right] \approx 1.458$$

15. Trapezoidal:

$$\int_{0}^{\sqrt{\pi/2}} \cos(x^2) \, dx \approx \frac{\sqrt{\pi/2}}{8}\left[\cos 0 + 2\cos\left(\frac{\sqrt{\pi/2}}{4}\right)^2 + 2\cos\left(\frac{\sqrt{\pi/2}}{2}\right)^2 + 2\cos\left(\frac{3\sqrt{\pi/2}}{4}\right)^2 + \cos\left(\sqrt{\frac{\pi}{2}}\right)^2\right]$$

$$\approx 0.957$$

Simpson's:

$$\int_{0}^{\sqrt{\pi/2}} \cos(x^2) \, dx \approx \frac{\sqrt{\pi/2}}{12}\left[\cos 0 + 4\cos\left(\frac{\sqrt{\pi/2}}{4}\right)^2 + 2\cos\left(\frac{\sqrt{\pi/2}}{2}\right)^2 + 4\cos\left(\frac{3\sqrt{\pi/2}}{4}\right)^2 + \cos\left(\sqrt{\frac{\pi}{2}}\right)^2\right]$$

$$\approx 0.978$$

16. Trapezoidal:

$$\int_{0}^{\sqrt{\pi/4}} \tan(x^2) \, dx \approx \frac{\sqrt{\pi/4}}{8}\left[\tan 0 + 2\tan\left(\frac{\sqrt{\pi/4}}{4}\right)^2 + 2\tan\left(\frac{\sqrt{\pi/4}}{2}\right)^2 + 2\tan\left(\frac{3\sqrt{\pi/4}}{4}\right)^2 + \tan\left(\sqrt{\frac{\pi}{4}}\right)^2\right]$$

$$\approx 0.271$$

Simpson's:

$$\int_{0}^{\sqrt{\pi/4}} \tan(x^2) \, dx \approx \frac{\sqrt{\pi/4}}{12}\left[\tan 0 + 4\tan\left(\frac{\sqrt{\pi/4}}{4}\right)^2 + 2\tan\left(\frac{\sqrt{\pi/4}}{2}\right)^2 + 4\tan\left(\frac{3\sqrt{\pi/4}}{4}\right)^2 + \tan\left(\sqrt{\frac{\pi}{4}}\right)^2\right]$$

$$\approx 0.257$$

17. Trapezoidal: $\displaystyle\int_{1}^{1.1} \sin x^2 \, dx \approx \frac{1}{80}\left[\sin(1) + 2\sin(1.025)^2 + 2\sin(1.05)^2 + 2\sin(1.075)^2 + \sin(1.1)^2\right]$

$$\approx 0.089$$

Simpson's: $\displaystyle\int_{1}^{1.1} \sin x^2 \, dx \approx \frac{1}{120}\left[\sin(1) + 4\sin(1.025)^2 + 2\sin(1.05)^2 + 4\sin(1.075)^2 + \sin(1.1)^2\right]$

$$\approx 0.089$$

18. Trapezoidal: $\displaystyle\int_{0}^{\pi/2} \sqrt{1 + \cos^2 x} \, dx \approx \frac{\pi}{16}\left[\sqrt{2} + 2\sqrt{1 + \cos^2(\pi/8)} + 2\sqrt{1 + \cos^2(\pi/4)} + 2\sqrt{1 + \cos^2(3\pi/8)} + 1\right]$

$$\approx 1.910$$

Simpson's: $\displaystyle\int_{0}^{\pi/2} \sqrt{1 + \cos^2 x} \, dx \approx \frac{\pi}{24}\left[\sqrt{2} + 4\sqrt{1 + \cos^2(\pi/8)} + 2\sqrt{1 + \cos^2(\pi/4)} + 4\sqrt{1 + \cos^2(3\pi/8)} + 1\right]$

$$\approx 1.910$$

19. Trapezoidal: $\displaystyle\int_0^{\pi/4} x \tan x \, dx \approx \frac{\pi}{32}\left[0 + 2\left(\frac{\pi}{16}\right)\tan\left(\frac{\pi}{16}\right) + 2\left(\frac{2\pi}{16}\right)\tan\left(\frac{2\pi}{16}\right) + 2\left(\frac{3\pi}{16}\right)\tan\left(\frac{3\pi}{16}\right) + \frac{\pi}{4}\right]$

≈ 0.194

Simpson's: $\displaystyle\int_0^{\pi/4} x \tan x \, dx \approx \frac{\pi}{48}\left[0 + 4\left(\frac{\pi}{16}\right)\tan\left(\frac{\pi}{16}\right) + 2\left(\frac{2\pi}{16}\right)\tan\left(\frac{2\pi}{16}\right) + 4\left(\frac{3\pi}{16}\right)\tan\left(\frac{3\pi}{16}\right) + \frac{\pi}{4}\right]$

≈ 0.186

20. Trapezoidal: $\displaystyle\int_0^{\pi} \frac{\sin x}{x} \, dx \approx \frac{\pi}{8}\left[1 + \frac{2\sin(\pi/4)}{\pi/4} + \frac{2\sin(\pi/2)}{\pi/2} + \frac{2\sin(3\pi/4)}{3\pi/4} + 0\right] \approx 1.836$

Simpson's: $\displaystyle\int_0^{\pi} \frac{\sin x}{x} \, dx \approx \frac{\pi}{12}\left[1 + \frac{4\sin(\pi/4)}{\pi/4} + \frac{2\sin(\pi/2)}{\pi/2} + \frac{4\sin(3\pi/4)}{3\pi/4} + 0\right] \approx 1.852$

21. $f(x) = x^3$

$f'(x) = 3x^2$

$f''(x) = 6x$

$f'''(x) = 6$

$f^{(4)}(x) = 0$

Trapezoidal:

Error $\leq \dfrac{(2-0)^3}{12(4^2)}(12) = 0.5$ since

$f''(x)$ is maximum in $[0, 2]$ when $x = 2$.

Simpson's:

Error $\leq \dfrac{(2-0)^5}{180(4^4)}(0) = 0$ since

$f^{(4)}(x) = 0$.

22. $f(x) = \dfrac{1}{x + 1}$

$f'(x) = \dfrac{-1}{(x+1)^2}$

$f''(x) = \dfrac{2}{(x+1)^3}$

$f'''(x) = \dfrac{-6}{(x+1)^4}$

$f^{(4)}(x) = \dfrac{24}{(x+1)^5}$

Trapezoidal:

Error $\leq \dfrac{(1-0)^2}{12(4^2)}(2) = \dfrac{1}{96} \approx 0.01$ since

$f''(x)$ is maximum in $[0, 1]$ when $x = 0$.

Simpson's:

Error $\leq \dfrac{(1-0)^5}{180(4^4)}(24) = \dfrac{1}{1920} \approx 0.0005$ since

$f^{(4)}(x)$ is maximum in $[0, 1]$ when $x = 0$.

23. $f''(x) = \dfrac{2}{x^3}$ in $[1, 3]$

$|f''(x)|$ is maximum when $x = 1$ and $|f''(1)| = 2$.

Trapezoidal: Error $\leq \dfrac{2^3}{12n^2}(2) < 0.00001, \quad n^2 > 133333.33, \quad n > 365.15$; let $n = 366$.

$f^{(4)}(x) = \dfrac{24}{x^5}$ in $[1, 3]$

$|f^{(4)}(x)|$ is maximum when $x = 1$ and when $|f^{(4)}(1)| = 24$.

Simpson's: Error $\leq \dfrac{2^5}{180n^4}(24) < 0.00001, \quad n^4 > 426666.67, \quad n > 25.56$; let $n = 26$.

24. $f''(x) = \dfrac{2}{(1 + x)^3}$ in $[0, 1]$

$|f''(x)|$ is maximum when $x = 0$ and $|f''(0)| = 2$.

Trapezoidal: Error $\leq \dfrac{1}{12n^2}(2) < 0.00001$, $\quad n^2 > 16666.67$, $\quad n > 129.10$; let $n = 130$.

$f^{(4)}(x) = \dfrac{24}{(1 + x)^5}$ in $[0, 1]$

$|f^{(4)}(x)|$ is maximum when $x = 0$ and $|f^{(4)}(0)| = 24$.

Simpson's: Error $\leq \dfrac{1}{180n^4}(24) < 0.00001$, $\quad n^4 > 13333.33$, $\quad n > 10.75$; let $n = 12$.

(In Simpson's Rule n must be even.)

25. $f(x) = \sqrt{1 + x}$

$f''(x) = -\dfrac{1}{4(1 + x)^{3/2}}$ in $[0, 2]$

$|f''(x)|$ is maximum when $x = 0$ and $|f''(0)| = \dfrac{1}{4}$.

Trapezoidal: Error $\leq \dfrac{8}{12n^2}\left(\dfrac{1}{4}\right) < 0.00001$, $\quad n^2 > 16666.67$, $\quad n > 129.10$; let $n = 130$.

$f^{(4)}(x) = \dfrac{-15}{16(1 + x)^{7/2}}$ in $[0, 2]$

$|f^{(4)}(x)|$ is maximum when $x = 0$ and $|f^{(4)}(0)| = \dfrac{15}{16}$.

Simpson's: Error $\leq \dfrac{32}{180n^4}\left(\dfrac{15}{16}\right) < 0.00001$, $\quad n^4 > 16666.67$, $\quad n > 11.36$; let $n = 12$.

26. $f(x) = (x + 1)^{2/3}$

$f''(x) = -\dfrac{2}{9(x + 1)^{4/3}}$ in $[0, 2]$

$|f''(x)|$ is maximum when $x = 0$ and $|f''(0)| = \dfrac{2}{9}$.

Trapezoidal: Error $\leq \dfrac{8}{12n^2}\left(\dfrac{2}{9}\right) < 0.00001$, $\quad n^2 > 14814.81$, $\quad n > 121.72$; let $n = 122$.

$f^{(4)}(x) = -\dfrac{56}{81(x + 1)^{10/3}}$ in $[0, 2]$

$|f^{(4)}(x)|$ is maximum when $x = 0$ and $|f^{(4)}(0)| = \dfrac{56}{81}$.

Simpson's: Error $\leq \dfrac{32}{180n^4}\left(\dfrac{56}{81}\right) < 0.00001$, $\quad n^4 > 12290.81$, $\quad n > 10.53$; let $n = 12$.

(In Simpson's Rule n must be even.)

27. $f(x) = \tan(x^2)$

$f''(x) = 2\sec^2(x^2)[1 + 4x^2\tan(x^2)]$ in $[0, 1]$.
$|f''(x)|$ is maximum when $x = 1$ and $|f''(1)| \approx 49.5305$.
$f^{(4)}(x) = 8\sec^2(x^2)[12x^2 + (3 + 32x^4)\tan(x^2) + 36x^2\tan^2(x^2) + 48x^4\tan^3(x^2)]$ in $[0, 1]$.
$|f^{(4)}(x)|$ is maximum when $x = 1$ and $|f^{(4)}(1)| \approx 9184.4734$.
Trapezoidal:

$$\text{Error} \le \frac{(1-0)^3}{12n^2}(49.5305) < 0.00001, \quad n^2 > 412754.17, \quad n > 642.46; \text{ let } n = 643.$$

Simpson's:

$$\text{Error} \le \frac{(1-0)^5}{180n^4}(9184.4734) < 0.00001, \quad n^4 > 5102485.22, \quad n > 47.53; \text{ let } n = 48.$$

28. $f(x) = \sin(x^2)$

$f''(x) = 2[-2x^2\sin(x^2) + \cos(x^2)]$ in $[0, 1]$.
$|f''(x)|$ is maximum when $x = 1$ and $|f''(1)| \approx 2.2853$.
$f^{(4)}(x) = (16x^4 - 12)\sin(x^2) - 48x^2\cos(x^2)$ in $[0, 1]$.
$|f^{(4)}(x)|$ is maximum when $x \approx 0.852$ and $|f^{(4)}(0.852)| \approx 28.4285$.
Trapezoidal:

$$\text{Error} \le \frac{(1-0)^3}{12n^2}(2.2853) < 0.00001, \quad n^2 > 19044.17, \quad n > 138.00; \text{ let } n = 139.$$

Simpson's:

$$\text{Error} \le \frac{(1-0)^5}{180n^4}(28.4285) < 0.00001, \quad n^4 > 15793.61, \quad n > 11.21; \text{ let } n = 12.$$

29. Let $f(x) = Ax^3 + Bx^2 + Cx + D$. Then $f^{(4)}(x) = 0$.

Simpson's: $\text{Error} \le \dfrac{(b-a)^5}{180n^4}(0) = 0$

Therefore, Simpson's Rule is exact when approximating the integral of a cubic polynomial.

Example: $\displaystyle\int_0^1 x^3\,dx = \frac{1}{6}\left[0 + 4\left(\frac{1}{2}\right)^3 + 1\right] = \frac{1}{4}$

This is the exact value of the integral.

30. The program will vary depending upon the computer or programmable calculator that you use.

31. $f(x) = \sqrt{2 + 3x^2}$ on $[0, 4]$

n	$L(n)$	$M(n)$	$R(n)$	$T(n)$	$S(n)$
4	12.7771	15.3965	18.4340	15.6055	15.4845
8	14.0868	15.4480	16.9152	15.5010	15.4662
10	14.3569	15.4544	16.6197	15.4883	15.4658
12	14.5386	15.4578	16.4242	15.4814	15.4657
16	14.7674	15.4613	16.1816	15.4745	15.4657
20	14.9056	15.4628	16.0370	15.4713	15.4657

32. $f(x) = \sqrt{1 - x^2}$ on $[0, 1]$

n	$L(n)$	$M(n)$	$R(n)$	$T(n)$	$S(n)$
4	0.8739	0.7960	0.6239	0.7489	0.7709
8	0.8350	0.7892	0.7100	0.7725	0.7803
10	0.8261	0.7881	0.7261	0.7761	0.7818
12	0.8200	0.7875	0.7367	0.7783	0.7826
16	0.8121	0.7867	0.7496	0.7808	0.7836
20	0.8071	0.7864	0.7571	0.7821	0.7841

33. $f(x) = \sin \sqrt{x}$ on $[0, 4]$

n	$L(n)$	$M(n)$	$R(n)$	$T(n)$	$S(n)$
4	2.8163	3.5456	3.7256	3.2709	3.3996
8	3.1809	3.5053	3.6356	3.4083	3.4541
10	3.2478	3.4990	3.6115	3.4296	3.4624
12	3.2909	3.4952	3.5940	3.4425	3.4674
16	3.3431	3.4910	3.5704	3.4568	3.4730
20	3.3734	3.4888	3.5552	3.4643	3.4759

34. $f(x) = \dfrac{\sin x}{x}$ on $[1, 2]$

n	$L(n)$	$M(n)$	$R(n)$	$T(n)$	$S(n)$
4	0.7070	0.6597	0.6103	0.6586	0.6593
8	0.6833	0.6594	0.6350	0.6592	0.6593
10	0.6786	0.6594	0.6399	0.6592	0.6593
12	0.6754	0.6594	0.6431	0.6593	0.6593
16	0.6714	0.6594	0.6472	0.6593	0.6593
20	0.6690	0.6593	0.6496	0.6593	0.6593

35. $A = \displaystyle\int_0^{\pi/2} \sqrt{x} \cos x \, dx$

Simpson's Rule: $n = 14$

$$\int_0^{\pi/2} \sqrt{x} \cos x \, dx$$

$$\approx \frac{\pi}{84} \left[\sqrt{0} \cos 0 + 4\sqrt{\frac{\pi}{28}} \cos \frac{\pi}{28} + 2\sqrt{\frac{\pi}{14}} \cos \frac{\pi}{14} \right.$$

$$\left. + 4\sqrt{\frac{3\pi}{28}} \cos \frac{3\pi}{28} + \cdots + \sqrt{\frac{\pi}{2}} \cos \frac{\pi}{2} \right] \approx 0.701$$

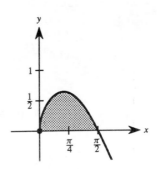

36. Simpson's Rule: $n = 8$

$$8\sqrt{3} \int_0^{\pi/2} \sqrt{1 - \frac{2}{3} \sin^2 \theta}\, d\theta$$

$$\approx \frac{\sqrt{3}\,\pi}{6} \left[\sqrt{1 - \frac{2}{3} \sin^2 0} + 4\sqrt{1 - \frac{2}{3} \sin^2 \frac{\pi}{16}} + 2\sqrt{1 - \frac{2}{3} \sin^2 \frac{\pi}{8}} + \cdots + \sqrt{1 - \frac{2}{3} \sin^2 \frac{\pi}{2}} \right]$$

$$\approx 17.476$$

37. Simpson's Rule: $n = 6$

$$\pi = 4 \int_0^1 \frac{1}{1+x^2}\, dx \approx \frac{4}{3(6)} \left[1 + \frac{4}{1 + (1/6)^2} + \frac{2}{1 + (2/6)^2} + \frac{4}{1 + (3/6)^2} + \frac{2}{1 + (4/6)^2} + \frac{4}{1 + (5/6)^2} + \frac{1}{2} \right]$$

$$\approx 3.14159$$

38. $W = \int_0^5 100x\sqrt{125 - x^3}\, dx$

Simpson's Rule: $n = 12$

$$\int_0^5 100x\sqrt{125 - x^3}\, dx$$

$$\approx \frac{5}{3(12)} \left[0 + 400 \left(\frac{5}{12} \right) \sqrt{125 - \left(\frac{5}{12} \right)^3} + 200 \left(\frac{10}{12} \right) \sqrt{125 - \left(\frac{10}{12} \right)^3} \right.$$

$$\left. + 400 \left(\frac{15}{12} \right) \sqrt{125 - \left(\frac{15}{12} \right)^3} + \cdots + 0 \right] \approx 10233.58 \text{ ft} \cdot \text{lb}$$

39. Area $\approx \dfrac{1000}{2(10)} [125 + 2(125) + 2(120) + 2(112) + 2(90) + 2(90) + 2(95) + 2(88) + 2(75) + 2(35)] = 89250$ sq ft

40. Area $\approx \dfrac{120}{2(12)} [75 + 2(81) + 2(84) + 2(76) + 2(67) + 2(68) + 2(69) + 2(72) + 2(68) + 2(56)$

$$+ 2(42) + 2(23) + 0] = 7435 \text{ sq ft}$$

41. a. Trapezoidal:

Area $\approx \dfrac{160}{2(8)} [0 + 2(50) + 2(54) + 2(82) + 2(82) + 2(73) + 2(75) + 2(80) + 0] = 9920$ sq ft

b. Simpson's:

Area $\approx \dfrac{160}{3(8)} [0 + 4(50) + 2(54) + 4(82) + 2(82) + 4(73) + 2(75) + 4(80) + 0] = 10413\frac{1}{3}$ sq ft

42. Trapezoidal:

$$\int_0^2 f(x)\, dx \approx \frac{2}{2(8)} [4.32 + 2(4.36) + 2(4.58) + 2(5.79) + 2(6.14) + 2(7.25) + 2(7.64) + 2(8.08) + 8.14] \approx 12.518$$

Simpson's:

$$\int_0^2 f(x)\, dx \approx \frac{2}{3(8)} [4.32 + 4(4.36) + 2(4.58) + 4(5.79) + 2(6.14) + 4(7.25) + 2(7.64) + 4(8.08) + 8.14] \approx 12.592$$

43. $\int_0^t \sin \sqrt{x}\, dx = 2, \quad n = 10$

By trial and error, we obtain $t \approx 2.478$.

44. a.

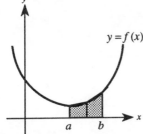

The Trapezoidal Rule overestimates the area if the graph of the integrand is concave up.

b.

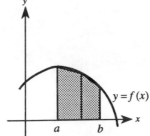

The Trapezoidal Rule underestimates the area if the graph of the integrand is concave down.

45. $L(x) = \int_1^x \frac{1}{t}\, dt$

a. $L(1) = \int_1^1 \frac{1}{t}\, dt = 0$

b. $L'(x) = \frac{1}{x}$

$L'(1) = 1$

c. By trial and error, $x \approx 2.718$.

d. $L'(x) = \frac{1}{x} > 0$ on $(0, \infty)$. Thus, L is increasing on $(0, \infty)$.

e. $L(x_1 x_2) = \int_1^{x_1 x_2} \frac{1}{t}\, dt = \int_1^{x_1} \frac{1}{t}\, dt + \int_{x_1}^{x_1 x_2} \frac{1}{t}\, dt$

But, this second integral is equal to $\int_1^{x_2} (1/t)\, dt$, by substitution $u = t/x_1$, $du = dt/x_1$.

Chapter 4 Review Exercises

1.

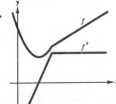

2.

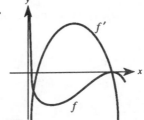

3. $\displaystyle\int (2x^2 + x - 1)\, dx = \frac{2}{3}x^3 + \frac{1}{2}x^2 - x + C$

4. $u = 3x$

$du = 3\, dx$

$\displaystyle\int \frac{2}{\sqrt[3]{3x}}\, dx = \frac{2}{3} \int (3x)^{-1/3}(3)\, dx = (3x)^{2/3} + C$

5. $\displaystyle\int (x^2 + 1)^3\, dx = \int (x^6 + 3x^4 + 3x^2 + 1)\, dx = \frac{1}{7}x^7 + \frac{3}{5}x^5 + x^3 + x + C$

6. $\displaystyle\int \left(x + \frac{1}{x}\right)^2 dx = \int (x^2 + 2 + x^{-2})\, dx = \frac{1}{3}x^3 + 2x - \frac{1}{x} + C$

7. $\displaystyle\int \frac{x^3 + 1}{x^2}\, dx = \int \left(x + \frac{1}{x^2}\right) dx = \frac{1}{2}x^2 - \frac{1}{x} + C$

8. $\displaystyle\int \frac{x^3 - 2x^2 + 1}{x^2}\, dx = \int (x - 2 + x^{-2})\, dx = \frac{1}{2}x^2 - 2x - \frac{1}{x} + C$

9. $u = x^3 + 3$

$du = 3x^2\,dx$

$$\int \frac{x^2}{\sqrt{x^3+3}}\,dx = \frac{1}{3}\int (x^3+3)^{-1/2}(3x^2)\,dx = \frac{2}{3}\sqrt{x^3+3} + C$$

10. $u = x^3 + 3$

$du = 3x^2\,dx$

$$\int x^2\sqrt{x^3+3}\,dx = \frac{1}{3}\int (x^3+3)^{1/2}(3x^2)\,dx = \frac{2}{9}(x^3+3)^{3/2} + C$$

11. $u = 1 - 3x^2$

$du = -6x\,dx$

$$\int x(1-3x^2)^4\,dx = -\frac{1}{6}\int (1-3x^2)^4(-6x)\,dx = -\frac{1}{30}(1-3x^2)^5 + C = \frac{(3x^2-1)^5}{30} + C$$

12. $u = x^2 + 6x - 5$

$du = (2x+6)\,dx = 2(x+3)\,dx$

$$\int \frac{x+3}{(x^2+6x-5)^2}\,dx = \frac{1}{2}\int (x^2+6x-5)^{-2}2(x+3)\,dx = -\frac{1}{2(x^2+6x-5)} + C$$

13. $\displaystyle\int \sin^3 x \cos x\,dx = \frac{1}{4}\sin^4 x + C$ **14.** $\displaystyle\int x\sin 3x^2\,dx = \frac{1}{6}\int (\sin 3x^2)(6x)\,dx$

$$= -\frac{1}{6}\cos 3x^2 + C$$

15. $\displaystyle\int \frac{\sin\theta}{\sqrt{1-\cos\theta}}\,d\theta = \int (1-\cos\theta)^{-1/2}\sin\theta\,d\theta = 2(1-\cos\theta)^{1/2} + C = 2\sqrt{1-\cos\theta} + C$

16. $\displaystyle\int \frac{\cos x}{\sqrt{\sin x}}\,dx = \int (\sin x)^{-1/2}\cos x\,dx = 2(\sin x)^{1/2} + C = 2\sqrt{\sin x} + C$

17. $\displaystyle\int \tan^n x \sec^2 x\,dx = \frac{\tan^{n+1} x}{n+1} + C, \quad n \neq -1$

18. $\displaystyle\int \sec 2x \tan 2x\,dx = \frac{1}{2}\int (\sec 2x \tan 2x)(2)\,dx = \frac{1}{2}\sec 2x + C$

19. $\displaystyle\int (1+\sec \pi x)^2 \sec \pi x \tan \pi x\,dx = \frac{1}{\pi}\int (1+\sec \pi x)^2(\pi \sec \pi x \tan \pi x)\,dx = \frac{1}{3\pi}(1+\sec \pi x)^3 + C$

20. $\displaystyle\int \cot^4 \alpha \csc^2 \alpha\,d\alpha = -\int (\cot\alpha)^4(-\csc^2\alpha)\,d\alpha = -\frac{1}{5}\cot^5\alpha + C$

21. $f'(x) = -2x, \quad (-1, \ 1)$

$$f(x) = \int -2x\,dx = -x^2 + C$$

When $x = -1$:

$y = -1 + C = 1$

$C = 2$

$y = 2 - x^2$

22. $f''(x) = 6(x - 1)$

Solution of $f''(x)$ tangent to $3x - y - 5 = 0$ at $(2, \ 1)$:

$$f'(x) = \int 6(x - 1)\,dx = 3(x - 1)^2 + C_1$$

Since the slope of the tangent line at $(2, \ 1)$ is 3, it follows that $f'(2) = 3 + C_1 = 3$ when $C_1 = 0$.

$f'(x) = 3(x - 1)^2$

$$f(x) = \int 3(x - 1)^2\,dx = (x - 1)^3 + C_2$$

$f(2) = 1 + C_2 = 1$ when $C_2 = 0$.

$f(x) = (x - 1)^3$

23. $a(t) = a$

$$v(t) = \int a\,dt = at + C_1$$

$v(0) = 0 + C_1 = 0$ when $C_1 = 0$.

$v(t) = at$

$$s(t) = \int at\,dt = \frac{a}{2}t^2 + C_2$$

$s(0) = 0 + C_2 = 0$ when $C_2 = 0$.

$s(t) = \frac{a}{2}t^2$

$s(30) = \frac{a}{2}(30)^2 = 3600$ or

$a = \dfrac{2(3600)}{(30)^2} = 8$ ft/sec^2

$v(30) = 8(30) = 240$ ft/sec

24. 45 mph $= 66$ ft/sec

30 mph $= 44$ ft/sec

$a(t) = -a$

$v(t) = -at + 66$ since $v(0) = 66$ ft/sec.

$s(t) = -\dfrac{a}{2}t^2 + 66t$ since $s(0) = 0$.

Solving the system

$v(t) = -at + 66 = 44$

$s(t) = -\dfrac{a}{2}t^2 + 66t = 264$

we obtain $t = \frac{24}{5}$ and $a = \frac{55}{12}$. We now solve $-(\frac{55}{12})t + 66 = 0$ and get $t = \frac{72}{5}$. Thus,

$$s\left(\frac{72}{5}\right) = -\frac{55/12}{2}\left(\frac{72}{5}\right)^2 + 66\left(\frac{72}{5}\right) \approx 475.2 \text{ ft.}$$

Stopping distance from 30 mph to rest is $475.2 - 264 = 211.2$ ft.

25. $a(t) = -32$

$v(t) = -32t + 96$

$s(t) = -16t^2 + 96t$

a. $v(t) = -32t + 96 = 0$ when $t = 3$ sec.

b. $s(3) = -144 + 288 = 144$ ft

c. $v(t) = -32t + 96 = \frac{96}{2}$ when $t = \frac{3}{2}$ sec.

d. $s(\frac{3}{2}) = -16(\frac{9}{4}) + 96(\frac{3}{2}) = 108$ ft

26. $a(t) = -32$

$v(t) = -32t + 128$

$s(t) = -16t^2 + 128t$

a. $v(t) = -32t + 128 = 0$ when $t = 4$ sec.

b. $s(4) = -256 + 512 = 256$ ft

c. $v(t) = -32t + 128 = \frac{128}{2}$ when $t = 2$ sec.

d. $s(2) = -16(2)^2 + 128(2) = 192$ ft

27. a. $\displaystyle\sum_{i=1}^{10}(2i - 1)$ b. $\displaystyle\sum_{i=1}^{n}i^3$ c. $\displaystyle\sum_{i=1}^{10}(4i + 2)$

28. $x_1 = 2$, $x_2 = -1$, $x_3 = 5$, $x_4 = 3$, $x_5 = 7$

a. $\dfrac{1}{5}\displaystyle\sum_{i=1}^{5} x_i = \dfrac{1}{5}(2 - 1 + 5 + 3 + 7) = \dfrac{16}{5}$

b. $\displaystyle\sum_{i=1}^{5} \dfrac{1}{x_i} = \dfrac{1}{2} - 1 + \dfrac{1}{5} + \dfrac{1}{3} + \dfrac{1}{7} = \dfrac{37}{210}$

c. $\displaystyle\sum_{i=1}^{5} (2x_i - x_i{}^2) = [2(2) - (2)^2] + [2(-1) - (-1)^2] + [2(5) - (5)^2] + [2(3) - (3)^2] + [2(7) - (7)^2] = -56$

d. $\displaystyle\sum_{i=2}^{5} (x_i - x_{i-1}) = (-1 - 2) + [5 - (-1)] + (3 - 5) + (7 - 3) = 5$

29. a. $S = m\left(\dfrac{b}{4}\right)\left(\dfrac{b}{4}\right) + m\left(\dfrac{2b}{4}\right)\left(\dfrac{b}{4}\right) + m\left(\dfrac{3b}{4}\right)\left(\dfrac{b}{4}\right) + m\left(\dfrac{4b}{4}\right)\left(\dfrac{b}{4}\right) = \dfrac{mb^2}{16}(1 + 2 + 3 + 4) = \dfrac{5mb^2}{8}$

$s = m(0)\left(\dfrac{b}{4}\right) + m\left(\dfrac{b}{4}\right)\left(\dfrac{b}{4}\right) + m\left(\dfrac{2b}{4}\right)\left(\dfrac{b}{4}\right) + m\left(\dfrac{3b}{4}\right)\left(\dfrac{b}{4}\right) = \dfrac{mb^2}{16}(1 + 2 + 3) = \dfrac{3mb^2}{8}$

b. $S(n) = \displaystyle\sum_{i=1}^{n} f\left(\dfrac{bi}{n}\right)\left(\dfrac{b}{n}\right) = \sum_{i=1}^{n}\left(\dfrac{mbi}{n}\right)\left(\dfrac{b}{n}\right) = m\left(\dfrac{b}{n}\right)^2 \sum_{i=1}^{n} i$

$= \dfrac{mb^2}{n^2}\left(\dfrac{n(n+1)}{2}\right) = \dfrac{mb^2(n+1)}{2n}$

$s(n) = \displaystyle\sum_{i=0}^{n-1} f\left(\dfrac{bi}{n}\right)\left(\dfrac{b}{n}\right) = \sum_{i=0}^{n-1} m\left(\dfrac{bi}{n}\right)\left(\dfrac{b}{n}\right) = m\left(\dfrac{b}{n}\right)^2 \sum_{i=0}^{n-1} i$

$= \dfrac{mb^2}{n^2}\left(\dfrac{(n-1)n}{2}\right) = \dfrac{mb^2(n-1)}{2n}$

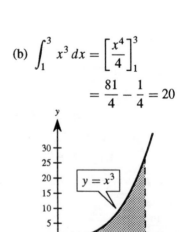

c. Area $= \displaystyle\lim_{n\to\infty} \dfrac{mb^2(n+1)}{2n} = \lim_{n\to\infty}\dfrac{mb^2(n-1)}{2n} = \dfrac{1}{2}mb^2$

d. $\displaystyle\int_0^b mx\,dx = \left[\dfrac{1}{2}mx^2\right]_0^b = \dfrac{1}{2}mb^2$

30. a. $S(n) = \displaystyle\sum_{i=1}^{n} f\left(1 + \dfrac{2i}{n}\right)\left(\dfrac{2}{n}\right)$

$= \displaystyle\sum_{i=1}^{n}\left(1 + \dfrac{2i}{n}\right)^3\left(\dfrac{2}{n}\right)$

$= \displaystyle\sum_{i=1}^{n}\left(1 + \dfrac{6i}{n} + \dfrac{12i^2}{n^2} + \dfrac{8i^3}{n^3}\right)\left(\dfrac{2}{n}\right)$

$= 2 + \dfrac{12}{n^2}\displaystyle\sum_{i=1}^{n} i + \dfrac{24}{n^3}\sum_{i=1}^{n} i^2 + \dfrac{16}{n^4}\sum_{i=1}^{n} i^3$

$= 2 + \dfrac{12}{n^2}[n(n+1)] + \dfrac{24}{n^3}\left[\dfrac{n(n+1)(2n+1)}{6}\right] + \dfrac{16}{n^4}\left[\dfrac{n^2(n+1)^2}{4}\right]$

$= 2 + 6\left(1 + \dfrac{1}{n}\right) + 4\left(2 + \dfrac{3}{n} + \dfrac{1}{n^2}\right) + 4\left(1 + \dfrac{2}{n} + \dfrac{1}{n^2}\right)$

Area $= \displaystyle\lim_{n\to\infty} S(n) = 2 + 6 + 8 + 4 = 20$

(b) $\displaystyle\int_1^3 x^3\,dx = \left[\dfrac{x^4}{4}\right]_1^3$

$= \dfrac{81}{4} - \dfrac{1}{4} = 20$

31. a. $\displaystyle\int_2^6 [f(x) + g(x)]\,dx = \int_2^6 f(x)\,dx + \int_2^6 g(x)\,dx = 10 + 3 = 13$

b. $\displaystyle\int_2^6 [f(x) - g(x)]\,dx = \int_2^6 f(x)\,dx - \int_2^6 g(x)\,dx = 10 - 3 = 7$

c. $\displaystyle\int_2^6 [2f(x) - 3g(x)]\,dx = 2\int_2^6 f(x)\,dx - 3\int_2^6 g(x)\,dx = 2(10) - 3(3) = 11$

d. $\displaystyle\int_2^6 5f(x)\,dx = 5\int_2^6 f(x)\,dx = 5(10) = 50$

32. a. $\displaystyle\int_0^6 f(x)\,dx = \int_0^3 f(x)\,dx + \int_3^6 f(x)\,dx = 4 + (-1) = 3$

b. $\displaystyle\int_6^3 f(x)\,dx = -\int_3^6 f(x)\,dx = -(-1) = 1$

c. $\displaystyle\int_4^4 f(x)\,dx = 0$

d. $\displaystyle\int_3^6 -10f(x)\,dx = -10\int_3^6 f(x)\,dx = -10(-1) = 10$

33. $\displaystyle\int_0^4 (2 + x)\,dx = \left[2x + \frac{x^2}{2}\right]_0^4 = 8 + \frac{16}{2} = 16$

34. $\displaystyle\int_{-1}^1 (t^2 + 2)\,dt = \left[\frac{t^3}{3} + 2t\right]_{-1}^1 = \frac{14}{3}$

35. $\displaystyle\int_{-1}^1 (4t^3 - 2t)\,dt = \left[t^4 - t^2\right]_{-1}^1 = 0$

36. $\displaystyle\int_3^6 \frac{x}{3\sqrt{x^2 - 8}}\,dx = \frac{1}{6}\int_3^6 (x^2 - 8)^{-1/2}(2x)\,dx$

$$= \left[\frac{1}{3}(x^2 - 8)^{1/2}\right]_3^6$$

$$= \frac{1}{3}(2\sqrt{7} - 1)$$

37. $\displaystyle\int_0^3 \frac{1}{\sqrt{1+x}}\,dx = \int_0^3 (1 + x)^{-1/2}\,dx = \left[2(1 + x)^{1/2}\right]_0^3 = 4 - 2 = 2$

38. $\displaystyle\int_0^1 x^2(x^3 + 1)^3\,dx = \frac{1}{3}\int_0^1 (x^3 + 1)^3(3x^2)\,dx = \frac{1}{12}\left[(x^3 + 1)^4\right]_0^1 = \frac{1}{12}(16 - 1) = \frac{5}{4}$

39. $\displaystyle\int_4^9 x\sqrt{x}\,dx = \int_4^9 x^{3/2}\,dx = \left[\frac{2}{5}x^{5/2}\right]_4^9 = \frac{2}{5}[(\sqrt{9})^5 - (\sqrt{4})^5] = \frac{2}{5}(243 - 32) = \frac{422}{5}$

40. $\displaystyle\int_1^2 \left(\frac{1}{x^2} - \frac{1}{x^3}\right)dx = \int_1^2 (x^{-2} - x^{-3})\,dx = \left[-\frac{1}{x} + \frac{1}{2x^2}\right]_1^2 = \left(-\frac{1}{2} + \frac{1}{8}\right) - \left(-1 + \frac{1}{2}\right) = \frac{1}{8}$

41. $u = 1 - y$, $y = 1 - u$, $dy = -du$

When $y = 0$, $u = 1$. When $y = 1$, $u = 0$.

$$2\pi \int_0^1 (y + 1)\sqrt{1 - y}\,dy = 2\pi \int_1^0 -[(1 - u) + 1]\sqrt{u}\,du$$

$$= 2\pi \int_1^0 (u^{3/2} - 2u^{1/2})\,du = 2\pi\left[\frac{2}{5}u^{5/2} - \frac{4}{3}u^{3/2}\right]_1^0 = \frac{28\pi}{15}$$

42. $u = x + 1, \ x = u - 1, \ dx = du$

When $x = -1, \ u = 0$. When $x = 0, \ u = 1$.

$$2\pi \int_{-1}^{0} x^2 \sqrt{x+1} \, dx = 2\pi \int_{0}^{1} (u-1)^2 \sqrt{u} \, du$$

$$= 2\pi \int_{0}^{1} (u^{5/2} - 2u^{3/2} + u^{1/2}) \, du = 2\pi \left[\frac{2}{7} u^{7/2} - \frac{4}{5} u^{5/2} + \frac{2}{3} u^{3/2} \right]_{0}^{1} = \frac{32\pi}{105}$$

43. $\displaystyle\int_{0}^{\pi} \cos\left(\frac{x}{2}\right) dx = 2 \int_{0}^{\pi} \cos\left(\frac{x}{2}\right) 2 \, dx = \left[2 \sin\left(\frac{x}{2}\right) \right]_{0}^{\pi} = 2$

44. $\displaystyle\int_{-\pi/4}^{\pi/4} \sin 2x \, dx = 0$ since $\sin 2x$ is an odd function.

45. $\displaystyle\int_{1}^{3} (2x - 1) \, dx = \left[x^2 - x \right]_{1}^{3} = 6$

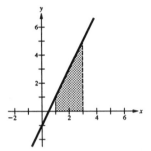

46. $\displaystyle\int_{0}^{2} (x + 4) \, dx = \left[\frac{x^2}{2} + 4x \right]_{0}^{2} = 10$

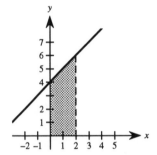

47. $\displaystyle\int_{3}^{4} (x^2 - 9) \, dx = \left[\frac{x^3}{3} - 9x \right]_{3}^{4}$

$$= \left(\frac{64}{3} - 36 \right) - (9 - 27)$$

$$= \frac{64}{3} - \frac{54}{3} = \frac{10}{3}$$

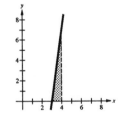

48. $\displaystyle\int_{-1}^{2} (-x^2 + x + 2) \, dx$

$$= \left[-\frac{x^3}{3} + \frac{x^2}{2} + 2x \right]_{-1}^{2}$$

$$= \left(-\frac{8}{3} + 2 + 4 \right) - \left(\frac{1}{3} + \frac{1}{2} - 2 \right)$$

$$= \frac{10}{3} + \frac{7}{6} = \frac{9}{2}$$

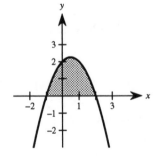

49. $\displaystyle\int_0^1 (x - x^3)\, dx = \left[\frac{x^2}{2} - \frac{x^4}{4}\right]_0^1$

$$= \frac{1}{2} - \frac{1}{4} = \frac{1}{4}$$

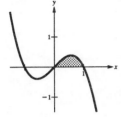

50. $\displaystyle\int_0^1 \sqrt{x}(1 - x)\, dx = \int_0^1 (x^{1/2} - x^{3/2})\, dx$

$$= \left[\frac{2}{3}x^{3/2} - \frac{2}{5}x^{5/2}\right]_0^1$$

$$= \frac{2}{3} - \frac{2}{5} = \frac{4}{15}$$

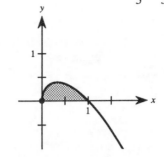

51. $\displaystyle\int_0^8 \frac{4}{\sqrt{x+1}}\, dx = 4\int_0^8 (x+1)^{-1/2}\, dx$

$$= \left[8\sqrt{x+1}\right]_0^8 = 16$$

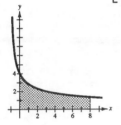

52. $\displaystyle\int_{-1}^0 (x^5 - x)\, dx + \int_0^1 (x - x^5)\, dx$

$$= 2\int_0^1 (x - x^5)\, dx$$

$$= 2\left[\frac{x^2}{2} - \frac{x^6}{6}\right]_0^1 = \frac{2}{3}$$

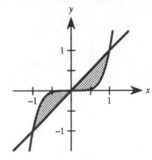

53. $\displaystyle\int_0^{\pi/3} \sec^2 x\, dx = \left[\tan x\right]_0^{\pi/3} = \sqrt{3}$

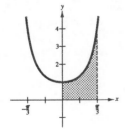

54. $\displaystyle\int_{-\pi/4}^{\pi/4} \cos x\, dx = 2\int_0^{\pi/4} \cos x\, dx$

$$= \left[2\sin x\right]_0^{\pi/4} = \sqrt{2}$$

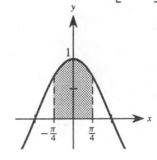

55. $\dfrac{1}{10-5} \displaystyle\int_5^{10} \dfrac{1}{\sqrt{x-1}}\, dx = \left[\dfrac{2}{5}\sqrt{x-1}\right]_5^{10} = \dfrac{2}{5}$

$\dfrac{1}{\sqrt{x-1}} = \dfrac{2}{5}$

$\sqrt{x-1} = \dfrac{5}{2}$

$x - 1 = \dfrac{25}{4}$

$x = \dfrac{29}{4}$

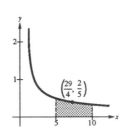

56. $\dfrac{1}{2-0} \displaystyle\int_0^2 x^3\, dx = \left[\dfrac{x^4}{8}\right]_0^2 = 2$

$x^3 = 2$

$x = \sqrt[3]{2}$

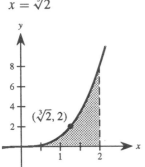

57. $\dfrac{1}{4-0} \displaystyle\int_0^4 x\, dx = \left[\dfrac{x^2}{8}\right]_0^4 = 2$

$x = 2$

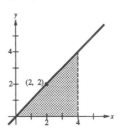

58. $\dfrac{1}{2-1} \displaystyle\int_1^2 \left(x^2 - \dfrac{1}{x^2}\right) dx = \left[\dfrac{x^3}{3} + \dfrac{1}{x}\right]_1^2 = \left(\dfrac{8}{3} + \dfrac{1}{2}\right) - \left(\dfrac{1}{3} + 1\right) = \dfrac{11}{6}$

$x^2 - \dfrac{1}{x^2} = \dfrac{11}{6}$

$6x^4 - 6 = 11x^2$

$6x^4 - 11x^2 - 6 = 0$ when $x^2 = \dfrac{11 \pm \sqrt{265}}{12}$.

In the interval $[1, 2]$: $x = \sqrt{\dfrac{11 + \sqrt{265}}{12}} \approx 1.508$.

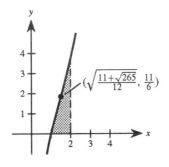

59. Simpson's Rule ($n = 4$):

$\displaystyle\int_1^2 \dfrac{1}{1+x^3}\, dx \approx \dfrac{1}{12}\left[\dfrac{1}{1+1^3} + \dfrac{4}{1+(1.25)^3} + \dfrac{2}{1+(1.5)^3} + \dfrac{4}{1+(1.75)^3} + \dfrac{1}{1+2^3}\right] \approx 0.254$

60. Simpson's Rule ($n = 4$):

$\displaystyle\int_0^1 \dfrac{x^{3/2}}{3-x^2}\, dx \approx \dfrac{1}{12}\left[0 + \dfrac{4(1/4)^{3/2}}{3-(1/4)^2} + \dfrac{2(1/2)^{3/2}}{3-(1/2)^2} + \dfrac{4(3/4)^{3/2}}{3-(3/4)^2} + \dfrac{1}{2}\right] \approx 0.166$

61. $V = \displaystyle\int_0^3 0.85 \sin\dfrac{\pi t}{3}\, dt = -\dfrac{3}{\pi}\left[0.85 \cos\dfrac{\pi t}{3}\right]_0^3 = -\dfrac{2.55}{\pi}(-1 - 1) = \dfrac{5.1}{\pi} \approx 1.6234$ liters

62. $\displaystyle\int_0^2 1.75\sin\frac{\pi t}{2}\,dt = -\frac{2}{\pi}\left[1.75\cos\frac{\pi t}{2}\right]_0^2 = -\frac{2}{\pi}(1.75)(-1-1) = \frac{7}{\pi} \approx 2.2282$ liters

Increase is $\dfrac{7}{\pi} - \dfrac{5.1}{\pi} = \dfrac{1.9}{\pi} \approx 0.6048$ liters.

63. (a) $\displaystyle C = 0.1\int_8^{20}\left[12\sin\frac{\pi(t-8)}{12}\right]dt = \left[-\frac{14.4}{\pi}\cos\frac{\pi(t-8)}{12}\right]_8^{20} = \frac{-14.4}{\pi}(-1-1) \approx \$9.17$

(b) $\displaystyle C = 0.1\int_{10}^{18}\left[12\sin\frac{\pi(t-8)}{12} - 6\right]dt = \left[\frac{-14.4}{\pi}\cos\frac{\pi(t-8)}{12} - 0.6t\right]_{10}^{18}$

$\displaystyle = \left[\frac{-14.4}{\pi}\left(\frac{-\sqrt{3}}{2}\right) - 10.8\right] - \left[\frac{-14.4}{\pi}\left(\frac{\sqrt{3}}{2}\right) - 6\right] \approx \$3.14$

Savings $\approx 9.17 - 3.14 = \$6.03$.

64. $\displaystyle\frac{1}{365}\int_0^{365}100{,}000\left[1+\sin\frac{2\pi(t-60)}{365}\right]dt = \frac{100{,}000}{365}\left[t - \frac{365}{2\pi}\cos\frac{2\pi(t-60)}{365}\right]_0^{365} = 100{,}000$ lbs.

65. $p = 1.10 + 0.04t$

$\displaystyle C = \frac{15{,}000}{M}\int_t^{t+1}(1.10 + 0.04s)\,ds = \frac{15{,}000}{M}\left[1.10s + 0.02s^2\right]_t^{t+1}$

a. 1995 is represented by $t = 5$.

$\displaystyle C = \frac{15{,}000}{M}\left[1.10s + 0.02s^2\right]_5^6 = \frac{19{,}800}{M}$

b. 2000 is represented by $t = 10$.

$\displaystyle C = \frac{15{,}000}{M}\left[1.10s + 0.02s^2\right]_{10}^{11} = \frac{22{,}800}{M}$

66. $u = 1 - x,\ \ x = 1 - u,\ \ dx = -du$

When $x = a,\ u = 1 - a$. When $x = b,\ u = 1 - b$.

$\displaystyle P_{a,b} = \int_a^b \frac{15}{4}x\sqrt{1-x}\,dx$

$\displaystyle = \frac{15}{4}\int_{1-a}^{1-b} -(1-u)\sqrt{u}\,du = \frac{15}{4}\int_{1-a}^{1-b}(u^{3/2} - u^{1/2})\,du$

$\displaystyle = \frac{15}{4}\left[\frac{2}{5}u^{5/2} - \frac{2}{3}u^{3/2}\right]_{1-a}^{1-b} = \frac{15}{4}\left[\frac{2u^{3/2}}{15}(3u-5)\right]_{1-a}^{1-b} = \left[-\frac{(1-x)^{3/2}}{2}(3x+2)\right]_a^b$

a. $\displaystyle P_{0.50,0.75} = \left[-\frac{(1-x)^{3/2}}{2}(3x+2)\right]_{0.50}^{0.75} = 35.3\%$

b. $\displaystyle P_{0,b} = \left[-\frac{(1-x)^{3/2}}{2}(3x+2)\right]_0^b = -\frac{(1-b)^{3/2}}{2}(3b+2) + 1 = 0.5$

$(1-b)^{3/2}(3b+2) = 1$

$b \approx 0.586 \approx 58.6\%$

67. $u = 1 - x, \; x = 1 - u, \; dx = -du$

When $x = a, \; u = 1 - a$. When $x = b, \; u = 1 - b$.

$$P_{a,b} = \int_a^b \frac{1155}{32} x^3 (1-x)^{3/2} \, dx = \frac{1155}{32} \int_{1-a}^{1-b} -(1-u)^3 u^{3/2} \, du$$

$$= \frac{1155}{32} \int_{1-a}^{1-b} (u^{9/2} - 3u^{7/2} + 3u^{5/2} - u^{3/2}) \, du = \frac{1155}{32} \left[\frac{2}{11} u^{11/2} - \frac{2}{3} u^{9/2} + \frac{6}{7} u^{7/2} - \frac{2}{5} u^{5/2} \right]_{1-a}^{1-b}$$

$$= \frac{1155}{32} \left[\frac{2u^{5/2}}{1155} (105u^3 - 385u^2 + 495u - 231) \right]_{1-a}^{1-b} = \left[\frac{u^{5/2}}{16} (105u^3 - 385u^2 + 495u - 231) \right]_{1-a}^{1-b}$$

a. $P_{0,0.25} = \left[\frac{u^{5/2}}{16} (105u^3 - 385u^2 + 495u - 231) \right]_1^{0.75} \approx 0.025 = 2.5\%$

b. $P_{0.5,1} = \left[\frac{u^{5/2}}{16} (105u^3 - 385u^2 + 495u - 231) \right]_{0.5}^{0} \approx 0.736 = 73.6\%$

68. False. This property only holds for **constants.**

69. False

$$\int -\frac{1}{x^2} \, dx = \frac{1}{x} + C \text{ or } \frac{d}{dx} \left[-\frac{1}{x^2} \right] \neq \frac{1}{x}.$$

70. True. If $f(x) = -f(-x)$, then $f(-x) = -f(x)$ and $f(x)$ is odd.

71. True

$$\frac{1}{2\pi - 0} \int_0^{2\pi} \sin x \, dx = \left[-\frac{1}{2\pi} \cos x \right]_0^{2\pi} = 0$$

72. False

$$\int \sec^2 x \, dx = \tan x + C \text{ or } \frac{d}{dx} [\sec^2 x] \neq \tan x$$

CHAPTER 5
Logarithmic, Exponential, and Other Transcendental Functions

Section 5.1   The Natural Logarithmic Function and Differentiation

1. Simpson's Rule: $n = 10$

x	0.5	1.5	2	2.5	3	3.5	4
$\int_1^x \frac{1}{t}\, dt$	−0.6932	0.4055	0.6932	0.9163	1.0987	1.2529	1.3865

Note: $\displaystyle\int_1^{0.5} \frac{1}{t}\, dt = -\int_{0.5}^1 \frac{1}{t}\, dt$

2. The curve generated in Exercise 1 is very close to the graph of $y = \ln x$.

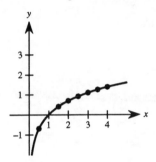

3. $f(x) = \ln x + 2$
   Vertical shift 2 units upward
   Matches (b)

4. $f(x) = -\ln x$
   Reflection in the x-axis
   Matches (d)

5. $f(x) = \ln(x - 1)$
   Horizontal shift 1 unit to the right
   Matches (a)

6. $f(x) = -\ln(-x)$
   Reflection in the y-axis and the x-axis
   Matches (c)

7. $f(x) = 3 \ln x$
   Domain: $x > 0$

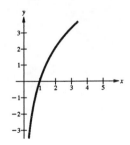

8. $f(x) = -2 \ln x$
   Domain: $x > 0$

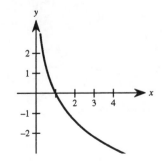

9. $f(x) = \ln 2x$
   Domain: $x > 0$

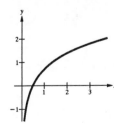

10. $f(x) = \ln |x|$
Domain: $x \neq 0$

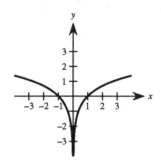

11. $f(x) = \ln(x - 1)$
Domain: $x > 1$

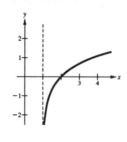

12. $g(x) = 2 + \ln x$
Domain: $x > 0$

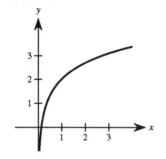

13. a. $\ln 6 = \ln 2 + \ln 3 \approx 1.7917$
b. $\ln \frac{2}{3} = \ln 2 - \ln 3 \approx -0.4055$
c. $\ln 81 = 4 \ln 3 \approx 4.3944$
d. $\ln \sqrt{3} = \frac{1}{2} \ln 3 \approx 0.5493$

14. a. $\ln 0.25 = \ln \frac{1}{4} = \ln 1 - 2 \ln 2 \approx -1.3862$
b. $\ln 24 = 3 \ln 2 + \ln 3 \approx 3.1779$
c. $\ln \sqrt[3]{12} = \frac{1}{3}(2 \ln 2 + \ln 3) \approx 0.8283$
d. $\ln \frac{1}{72} = \ln 1 - (3 \ln 2 + 2 \ln 3) \approx -4.2765$

15. $\ln \frac{2}{3} = \ln 2 - \ln 3$

16. $\ln \frac{1}{5} = \ln 1 - \ln 5 = -\ln 5$

17. $\ln \frac{xy}{z} = \ln x + \ln y - \ln z$

18. $\ln xyz = \ln x + \ln y + \ln z$

19. $\ln \sqrt{2^3} = \ln 2^{3/2} = \frac{3}{2} \ln 2$

20. $\ln \sqrt{a - 1} = \ln(a - 1)^{1/2} = \left(\frac{1}{2}\right) \ln(a - 1)$

21. $\ln \left(\frac{x^2 - 1}{x^3}\right)^3 = 3[\ln(x^2 - 1) - \ln x^3]$
$= 3[\ln(x + 1) + \ln(x - 1) - 3 \ln x]$

22. $\ln 3e^2 = \ln 3 + 2 \ln e = 2 + \ln 3$

23. $\ln z(z - 1)^2 = \ln z + \ln(z - 1)^2$
$= \ln z + 2 \ln(z - 1)$

24. $\ln \frac{1}{e} = \ln 1 - \ln e = -1$

25. $\ln(x - 2) - \ln(x + 2) = \ln \frac{x - 2}{x + 2}$

26. $3 \ln x + 2 \ln y - 4 \ln z = \ln x^3 + \ln y^2 - \ln z^4$
$= \ln \frac{x^3 y^2}{z^4}$

27. $\frac{1}{3}[2 \ln(x + 3) + \ln x - \ln(x^2 - 1)] = \frac{1}{3} \ln \frac{x(x + 3)^2}{x^2 - 1} = \ln \sqrt[3]{\frac{x(x + 3)^2}{x^2 - 1}}$

28. $2[\ln x - \ln(x + 1) - \ln(x - 1)] = 2 \ln \frac{x}{(x + 1)(x - 1)} = \ln \left(\frac{x}{x^2 - 1}\right)^2$

29. $2 \ln 3 - \frac{1}{2} \ln(x^2 + 1) = \ln \frac{9}{\sqrt{x^2 + 1}}$

30. $\frac{3}{2}[\ln(x^2 + 1) - \ln(x + 1) - \ln(x - 1)] = \frac{3}{2} \ln \frac{x^2 + 1}{(x + 1)(x - 1)} = \ln \sqrt{\left(\frac{x^2 + 1}{x^2 - 1}\right)^3}$

31. $y = \ln x^3 = 3\ln x$

$y' = \dfrac{3}{x}$

At $(1, 0)$, $y' = 3$.

32. $y = \ln x^{3/2} = \dfrac{3}{2}\ln x$

$y' = \dfrac{3}{2x}$

At $(1, 0)$, $y' = \frac{3}{2}$.

33. $y = \ln x^2 = 2\ln x$

$y' = \dfrac{2}{x}$

At $(1, 0)$, $y' = 2$.

34. $y = \ln x^{1/2} = \dfrac{1}{2}\ln x$

$y' = \dfrac{1}{2x}$

At $(1, 0)$, $y' = \frac{1}{2}$.

35. $g(x) = \ln x^2 = 2\ln x$

$g'(x) = \dfrac{2}{x}$

36. $h(x) = \ln(x^2 + 3)$

$h'(x) = \dfrac{2x}{x^2 + 3}$

37. $y = (\ln x)^4$

$\dfrac{dy}{dx} = 4(\ln x)^3\left(\dfrac{1}{x}\right) = \dfrac{4(\ln x)^3}{x}$

38. $y = x\ln x$

$\dfrac{dy}{dx} = x\left(\dfrac{1}{x}\right) + \ln x = 1 + \ln x$

39. $y = \ln x\sqrt{x^2 - 1} = \ln x + \dfrac{1}{2}\ln(x^2 - 1)$

$\dfrac{dy}{dx} = \dfrac{1}{x} + \dfrac{1}{2}\left(\dfrac{2x}{x^2 - 1}\right) = \dfrac{2x^2 - 1}{x(x^2 - 1)}$

40. $y = \ln\sqrt{x^2 - 4} = \dfrac{1}{2}\ln(x^2 - 4)$

$\dfrac{dy}{dx} = \dfrac{1}{2}\left(\dfrac{2x}{x^2 - 4}\right) = \dfrac{x}{x^2 - 4}$

41. $f(x) = \ln\dfrac{x}{x^2 + 1} = \ln x - \ln(x^2 + 1)$

$f'(x) = \dfrac{1}{x} - \dfrac{2x}{x^2 + 1} = \dfrac{1 - x^2}{x(x^2 + 1)}$

42. $f(x) = \ln\dfrac{x}{x + 1} = \ln x - \ln(x + 1)$

$f'(x) = \dfrac{1}{x} - \dfrac{1}{x + 1} = \dfrac{1}{x^2 + x}$

43. $g(t) = \dfrac{\ln t}{t^2}$

$g'(t) = \dfrac{t^2(1/t) - 2t\ln t}{t^4} = \dfrac{1 - 2\ln t}{t^3}$

44. $h(t) = \dfrac{\ln t}{t}$

$h'(t) = \dfrac{t(1/t) - \ln t}{t^2} = \dfrac{1 - \ln t}{t^2}$

45. $y = \ln(\ln x^2)$

$\dfrac{dy}{dx} = \dfrac{(2x/x^2)}{\ln x^2} = \dfrac{2}{x\ln x^2} = \dfrac{1}{x\ln x}$

46. $y = \ln(\ln x)$

$\dfrac{dy}{dx} = \dfrac{1/x}{\ln x} = \dfrac{1}{x\ln x}$

47. $y = \ln\sqrt{\dfrac{x + 1}{x - 1}} = \dfrac{1}{2}[\ln(x + 1) - \ln(x - 1)]$

$\dfrac{dy}{dx} = \dfrac{1}{2}\left[\dfrac{1}{x + 1} - \dfrac{1}{x - 1}\right] = \dfrac{1}{1 - x^2}$

48. $y = \ln\sqrt{\dfrac{x - 1}{x + 1}} = \dfrac{1}{2}[\ln(x - 1) - \ln(x + 1)]$

$\dfrac{dy}{dx} = \dfrac{1}{2}\left[\dfrac{1}{x - 1} - \dfrac{1}{x + 1}\right] = \dfrac{1}{x^2 - 1}$

49. $f(x) = \ln\dfrac{\sqrt{4 + x^2}}{x} = \dfrac{1}{2}\ln(4 + x^2) - \ln x$

$f'(x) = \dfrac{x}{4 + x^2} - \dfrac{1}{x} = \dfrac{-4}{x(x^2 + 4)}$

50. $f(x) = \ln\left(x + \sqrt{4 + x^2}\right)$

$f'(x) = \dfrac{1}{x + \sqrt{4 + x^2}}\left(1 + \dfrac{x}{\sqrt{4 + x^2}}\right)$

$= \dfrac{1}{\sqrt{4 + x^2}}$

51. $y = \dfrac{-\sqrt{x^2+1}}{x} + \ln\left(x + \sqrt{x^2+1}\right)$

$\dfrac{dy}{dx} = \dfrac{-x\left(x/\sqrt{x^2+1}\right) + \sqrt{x^2+1}}{x^2} + \left(\dfrac{1}{x+\sqrt{x^2+1}}\right)\left(1 + \dfrac{x}{\sqrt{x^2+1}}\right)$

$= \dfrac{1}{x^2\sqrt{x^2+1}} + \left(\dfrac{1}{x+\sqrt{x^2+1}}\right)\left(\dfrac{\sqrt{x^2+1}+x}{\sqrt{x^2+1}}\right)$

$= \dfrac{1}{x^2\sqrt{x^2+1}} + \dfrac{1}{\sqrt{x^2+1}} = \dfrac{1+x^2}{x^2\sqrt{x^2+1}} = \dfrac{\sqrt{x^2+1}}{x^2}$

52. $y = \dfrac{-\sqrt{x^2+4}}{2x^2} - \dfrac{1}{4}\ln\left(\dfrac{2+\sqrt{x^2+4}}{x}\right) = \dfrac{-\sqrt{x^2+4}}{2x^2} - \dfrac{1}{4}\ln\left(2+\sqrt{x^2+4}\right) + \dfrac{1}{4}\ln x$

$\dfrac{dy}{dx} = \dfrac{-2x^2\left(x/\sqrt{x^2+4}\right) + 4x\sqrt{x^2+4}}{4x^4} - \dfrac{1}{4}\left(\dfrac{1}{2+\sqrt{x^2+4}}\right)\left(\dfrac{x}{\sqrt{x^2+4}}\right) + \dfrac{1}{4x}$

Note that:

$\dfrac{1}{2+\sqrt{x^2+4}} = \dfrac{1}{2+\sqrt{x^2+4}} \cdot \dfrac{2-\sqrt{x^2+4}}{2-\sqrt{x^2+4}} = \dfrac{2-\sqrt{x^2+4}}{-x^2}$

$\dfrac{dy}{dx} = \dfrac{-1}{2x\sqrt{x^2+4}} + \dfrac{\sqrt{x^2+4}}{x^3} - \dfrac{1}{4}\dfrac{\left(2-\sqrt{x^2+4}\right)}{-x^2}\left(\dfrac{x}{\sqrt{x^2+4}}\right) + \dfrac{1}{4x}$

$= \dfrac{-1+(1/2)\left(2-\sqrt{x^2+4}\right)}{2x\sqrt{x^2+4}} + \dfrac{\sqrt{x^2+4}}{x^3} + \dfrac{1}{4x}$

$= \dfrac{-\sqrt{x^2+4}}{4x\sqrt{x^2+4}} + \dfrac{\sqrt{x^2+4}}{x^3} + \dfrac{1}{4x} = \dfrac{\sqrt{x^2+4}}{x^3}$

53. $y = \ln|\sin x|$

$\dfrac{dy}{dx} = \dfrac{\cos x}{\sin x} = \cot x$

54. $y = \ln|\sec x|$

$\dfrac{dy}{dx} = \dfrac{\sec x \tan x}{\sec x} = \tan x$

55. $y = \ln\left|\dfrac{\cos x}{\cos x - 1}\right|$

$= \ln|\cos x| - \ln|\cos x - 1|$

$\dfrac{dy}{dx} = \dfrac{-\sin x}{\cos x} - \dfrac{-\sin x}{\cos x - 1} = -\tan x + \dfrac{\sin x}{\cos x - 1}$

56. $y = \ln|\sec x + \tan x|$

$\dfrac{dy}{dx} = \dfrac{\sec x \tan x + \sec^2 x}{\sec x + \tan x}$

$= \dfrac{\sec x(\sec x + \tan x)}{\sec x + \tan x} = \sec x$

57. $y = \ln\left|\dfrac{-1+\sin x}{2+\sin x}\right|$

$= \ln|-1+\sin x| - \ln|2+\sin x|$

$\dfrac{dy}{dx} = \dfrac{\cos x}{-1+\sin x} - \dfrac{\cos x}{2+\sin x}$

$= \dfrac{3\cos x}{(\sin x - 1)(\sin x + 2)}$

58. $y = \ln\sqrt{1+\sin^2 x} = \dfrac{1}{2}\ln(1+\sin^2 x)$

$\dfrac{dy}{dx} = \left(\dfrac{1}{2}\right)\dfrac{2\sin x \cos x}{1+\sin^2 x} = \dfrac{\sin x \cos x}{1+\sin^2 x}$

59. $f(x) = \sin 2x \ln x^2 = 2 \sin 2x \ln x$

$f'(x) = (2 \sin 2x)\left(\dfrac{1}{x}\right) + 4 \cos 2x \ln x$

$\qquad = \dfrac{2}{x}(\sin 2x + 2x \cos 2x \ln x)$

$\qquad = \dfrac{2}{x}(\sin 2x + x \cos 2x \ln x^2)$

60. $g(x) = \displaystyle\int_{1}^{\ln x} (t^2 + 3)\, dt$

$g'(x) = [(\ln x)^2 + 3]\dfrac{d}{dx}(\ln x) = \dfrac{(\ln x)^2 + 3}{x}$

61. $x^2 - 3 \ln y + y^2 = 10$

$2x - \dfrac{3}{y}\dfrac{dy}{dx} + 2y\dfrac{dy}{dx} = 0$

$\qquad 2x = \dfrac{dy}{dx}\left(\dfrac{3}{y} - 2y\right)$

$\qquad \dfrac{dy}{dx} = \dfrac{2x}{(3/y) - 2y} = \dfrac{2xy}{3 - 2y^2}$

62. $\ln(xy) + 5x = 30$

$\ln x + \ln y + 5x = 30$

$\dfrac{1}{x} + \dfrac{1}{y}\dfrac{dy}{dx} + 5 = 0$

$\qquad \dfrac{1}{y}\dfrac{dy}{dx} = -\dfrac{1}{x} - 5$

$\qquad \dfrac{dy}{dx} = -\dfrac{y}{x} - 5y = -\left(\dfrac{y + 5xy}{x}\right)$

63. $y = 3x^2 - \ln x, \quad (1, \ 3)$

$\dfrac{dy}{dx} = 6x - \dfrac{1}{x}$

When $x = 1, \quad \dfrac{dy}{dx} = 5.$

Tangent line: $y - 3 = 5(x - 1)$

$\qquad\qquad\qquad y = 5x - 2$

$\qquad\qquad\quad 0 = 5x - y - 2$

64. $x^2 + \ln(x + 1) + y^2 = 4, \quad (0, \ 2)$

$2x + \dfrac{1}{x + 1} + 2y\dfrac{dy}{dx} = 0$

$\qquad\qquad \dfrac{dy}{dx} = -\dfrac{2x^2 + 2x + 1}{2y(x + 1)}$

At $(0, 2), \ \dfrac{dy}{dx} = -\dfrac{1}{4}.$

Tangent line: $y = -\dfrac{1}{4}x + 2$

$\qquad\qquad\quad x + 4y - 8 = 0$

65. $\qquad y = 2(\ln x) + 3$

$\qquad y' = \dfrac{2}{x}$

$\qquad y'' = -\dfrac{2}{x^2}$

$xy'' + y' = x\left(-\dfrac{2}{x^2}\right) + \dfrac{2}{x} = 0$

66. $\qquad y = x(\ln x) - 4x$

$\qquad y' = x\left(\dfrac{1}{x}\right) + \ln x - 4 = -3 + \ln x$

$(x + y) - xy' = x + x \ln x - 4x - x(-3 + \ln x)$

$\qquad\qquad\qquad\quad = 0$

67. $y = \dfrac{x^2}{2} - \ln x$

Domain: $x > 0$

$y' = x - \dfrac{1}{x} = \dfrac{(x + 1)(x - 1)}{x} = 0$ when $x = 1.$

$y'' = 1 + \dfrac{1}{x^2} > 0$

Relative minimum: $\left(1, \ \tfrac{1}{2}\right)$

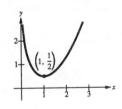

68. $y = x - \ln x$

Domain: $x > 0$

$y' = 1 - \dfrac{1}{x} = 0$ when $x = 1$.

$y'' = \dfrac{1}{x^2} > 0$

Relative minimum: $(1, 1)$

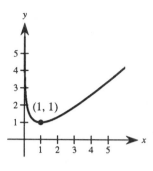

69. $y = x \ln x$

Domain: $x > 0$

$y' = x\left(\dfrac{1}{x}\right) + \ln x = 1 + \ln x = 0$ when $x = e^{-1}$.

$y'' = \dfrac{1}{x} > 0$

Relative minimum: $(e^{-1}, -e^{-1})$

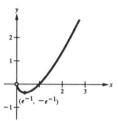

70. $y = \dfrac{\ln x}{x}$

Domain: $x > 0$

$y' = \dfrac{x(1/x) - \ln x}{x^2} = \dfrac{1 - \ln x}{x^2} = 0$ when $x = e$.

$y'' = \dfrac{x^2(-1/x) - (1 - \ln x)(2x)}{x^4} = \dfrac{2(\ln x) - 3}{x^3} = 0$ when $x = e^{3/2}$

Relative maximum: $(e, \ e^{-1})$

Point of inflection: $\left(e^{3/2}, \ \frac{3}{2}e^{-3/2}\right)$

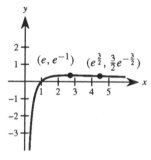

71. $y = \dfrac{x}{\ln x}$

Domain: $0 < x < 1, \ x > 1$

$y' = \dfrac{(\ln x)(1) - (x)(1/x)}{(\ln x)^2} = \dfrac{\ln x - 1}{(\ln x)^2} = 0$ when $x = e$.

$y'' = \dfrac{2 - \ln x}{x(\ln x)^3} = 0$ when $x = e^2$

Relative minimum: $(e, \ e)$

Point of inflection: $(e^2, \ e^2/2)$

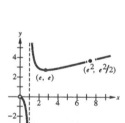

72. $y = x^2 \ln x$

Domain: $x > 0$

$y' = x^2\left(\dfrac{1}{x}\right) + 2x(\ln x) = x(1 + 2\ln x) = 0$ when $x = e^{-1/2}$.

$y'' = x\left(\dfrac{2}{x}\right) + (1 + 2\ln x) = 3 + 2(\ln x) = 0$ when $x = e^{-3/2}$

Relative minimum: $\left(e^{-1/2}, \ -\frac{1}{2}e^{-1}\right)$

Point of inflection: $\left(e^{-3/2}, \ -\frac{3}{2}e^{-3}\right)$

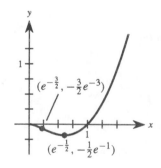

73. $f(x) = \ln x, \quad f(1) = 0$

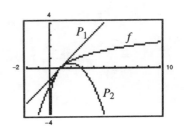

$f'(x) = \dfrac{1}{x}, \quad f'(1) = 1$

$f''(x) = -\dfrac{1}{x^2}, \quad f''(1) = -1$

$P_1(x) = f(1) + f'(1)(x - 1) = x - 1, \quad P_1(1) = 0$

$P_2(x) = f(1) + f'(1)(x - 1) + \dfrac{1}{2}f''(1)(x - 1)^2$

$\quad\quad = (x - 1) - \dfrac{1}{2}(x - 1)^2, \quad P_2(1) = 0$

$P_1'(x) = 1, \quad P_1'(1) = 1$

$P_2'(x) = 1 - (x - 1) = 2 - x, \quad P_2'(1) = 1$

$P_2''(x) = -1, \quad P_2''(1) = -1$

The values of f, P_1, P_2, and their first derivatives agree at $x = 1$. The values of the second derivatives of f and P_2 agree at $x = 1$.

74. $f(x) = x \ln x, \quad f(1) = 0$

$f'(x) = 1 + \ln x, \quad f'(1) = 1$

$f''(x) = \dfrac{1}{x}, \quad f''(1) = 1$

$P_1(x) = f(1) + f'(1)(x - 1) = x - 1, \quad P_1(1) = 0$

$P_2(x) = f(1) + f'(1)(x - 1) + \dfrac{1}{2}f''(1)(x - 1)^2$

$\quad\quad = (x - 1) + \dfrac{1}{2}(x - 1)^2, \quad P_2(1) = 0$

$P_1'(x) = 1, \quad P_1'(1) = 1$

$P_2'(x) = 1 + (x - 1) = x, \quad P_2'(1) = 1$

$P_2''(x) = x, \quad P_2''(1) = 1$

The values of f, P_1, P_2, and their first derivatives agree at $x = 1$. The values of the second derivatives of f and P_2 agree at $x = 1$.

75. Find x such that $\ln x = -x$.

$f(x) = (\ln x) + x = 0$

$f'(x) = \dfrac{1}{x} + 1$

$x_{n+1} = x_n - \dfrac{f(x_n)}{f'(x_n)} = x_n \left[\dfrac{1 - \ln x_n}{1 + x_n} \right]$

n	1	2	3
x_n	0.5	0.5644	0.5671
$f(x_n)$	-0.1931	-0.0076	-0.0001

Approximate root: $x = 0.567$

76. Find x such that $\ln x = 3 - x$.

$f(x) = x + (\ln x) - 3 = 0$

$f'(x) = 1 + \dfrac{1}{x}$

$x_{n+1} = x_n - \dfrac{f(x_n)}{f'(x_n)} = x_n \left[\dfrac{4 - \ln x_n}{1 + x_n} \right]$

n	1	2	3
x_n	2	2.2046	2.2079
$f(x_n)$	-0.3069	-0.0049	0.0000

Approximate root: $x = 2.208$

77.
$$y = x\sqrt{x^2 - 1}$$

$$\ln y = \ln x + \frac{1}{2}\ln(x^2 - 1)$$

$$\frac{1}{y}\left(\frac{dy}{dx}\right) = \frac{1}{x} + \frac{x}{x^2 - 1}$$

$$\frac{dy}{dx} = y\left[\frac{2x^2 - 1}{x(x^2 - 1)}\right] = \frac{2x^2 - 1}{\sqrt{x^2 - 1}}$$

78.
$$y = \sqrt{(x - 1)(x - 2)(x - 3)}$$

$$\ln y = \frac{1}{2}[\ln(x - 1) + \ln(x - 2) + \ln(x - 3)]$$

$$\frac{1}{y}\left(\frac{dy}{dx}\right) = \frac{1}{2}\left[\frac{1}{x - 1} + \frac{1}{x - 2} + \frac{1}{x - 3}\right]$$

$$= \frac{1}{2}\left[\frac{3x^2 - 12x + 11}{(x - 1)(x - 2)(x - 3)}\right]$$

$$\frac{dy}{dx} = \frac{3x^2 - 12x + 11}{2y}$$

$$= \frac{3x^2 - 12x + 11}{2\sqrt{(x - 1)(x - 2)(x - 3)}}$$

79.
$$y = \frac{x^2\sqrt{3x - 2}}{(x - 1)^2}$$

$$\ln y = 2\ln x + \frac{1}{2}\ln(3x - 2) - 2\ln(x - 1)$$

$$\frac{1}{y}\left(\frac{dy}{dx}\right) = \frac{2}{x} + \frac{3}{2(3x - 2)} - \frac{2}{x - 1}$$

$$\frac{dy}{dx} = y\left[\frac{3x^2 - 15x + 8}{2x(3x - 2)(x - 1)}\right]$$

$$= \frac{3x^3 - 15x^2 + 8x}{2(x - 1)^3\sqrt{3x - 2}}$$

80.
$$y = \sqrt[3]{\frac{x^2 + 1}{x^2 - 1}}$$

$$\ln y = \frac{1}{3}[\ln(x^2 + 1) - \ln(x + 1) - \ln(x - 1)]$$

$$\frac{1}{y}\left(\frac{dy}{dx}\right) = \frac{1}{3}\left[\frac{2x}{x^2 + 1} - \frac{1}{x + 1} - \frac{1}{x - 1}\right]$$

$$\frac{dy}{dx} = \frac{-4xy}{3(x^4 - 1)} = -\frac{4x}{3(x^4 - 1)}\sqrt[3]{\frac{x^2 + 1}{x^2 - 1}}$$

$$= \frac{-4x}{3(x^2 + 1)^{2/3}(x^2 - 1)^{4/3}}$$

81.
$$y = \frac{x(x - 1)^{3/2}}{\sqrt{x + 1}}$$

$$\ln y = \ln x + \frac{3}{2}\ln(x - 1) - \frac{1}{2}\ln(x + 1)$$

$$\frac{1}{y}\left(\frac{dy}{dx}\right) = \frac{1}{x} + \frac{3}{2}\left(\frac{1}{x - 1}\right) - \frac{1}{2}\left(\frac{1}{x + 1}\right)$$

$$\frac{dy}{dx} = \frac{y}{2}\left[\frac{2}{x} + \frac{3}{x - 1} - \frac{1}{x + 1}\right]$$

$$= \frac{y}{2}\left[\frac{4x^2 + 4x - 2}{x(x^2 - 1)}\right] = \frac{(2x^2 + 2x - 1)\sqrt{x - 1}}{(x + 1)^{3/2}}$$

82.
$$y = \frac{(x + 1)(x + 2)}{(x - 1)(x - 2)}$$

$$\ln y = \ln(x + 1) + \ln(x + 2) - \ln(x - 1) - \ln(x - 2)$$

$$\frac{1}{y}\left(\frac{dy}{dx}\right) = \frac{1}{x + 1} + \frac{1}{x + 2} - \frac{1}{x - 1} - \frac{1}{x - 2}$$

$$\frac{dy}{dx} = y\left[\frac{-2}{x^2 - 1} + \frac{-4}{x^2 - 4}\right] = y\left[\frac{-6x^2 + 12}{(x^2 - 1)(x^2 - 4)}\right]$$

$$= \frac{(x + 1)(x + 2)}{(x - 1)(x - 2)} \cdot \frac{-6(x^2 - 2)}{(x + 1)(x - 1)(x + 2)(x - 2)} = -\frac{6(x^2 - 2)}{(x - 1)^2(x - 2)^2}$$

83. $$\beta = \frac{10}{\ln 10} \ln \left(\frac{I}{10^{-16}} \right) = \frac{10}{\ln 10} [\ln I + 16 \ln 10]$$

$$\beta(10^{-10}) = \frac{10}{\ln 10} [\ln 10^{-10} + 16 \ln 10] = \frac{10}{\ln 10} [-10 \ln 10 + 16 \ln 10] = \frac{10}{\ln 10} [6 \ln 10] = 60 \text{ decibels}$$

84. $t = \dfrac{5.315}{-6.7968 + \ln x}, \quad 1000 < x$

a.

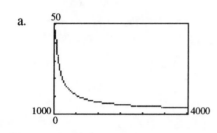

b. $t(1167.41) \approx 20$ years

$$T = (1167.41)(20)(12)$$

$$= \$280,178.40$$

c. $t(1068.45) \approx 30$ years

$$T = (1068.45)(30)(12)$$

$$= \$384,642.00$$

d. $\dfrac{dt}{dx} = -5.315(-6.7968 + \ln x)^{-2} \left(\dfrac{1}{x} \right) = -\dfrac{5.315}{x(-6.7968 + \ln x)^2}$

When $x = 1167.41$, $dt/dx \approx -0.0645$. When $x = 1068.45$, $dt/dx \approx -0.1585$.

e. There are two obvious benefits to paying a higher monthly payment:

1. The term is lower.
2. The total amount paid is lower.

85. a.

T

Temperature (°C)

350 ┤
300 ┤
250 ┤
200 ┤
150 ┤

20 40 60 80 100 → *p*

Pressure (in psi)

b. $$T'(t) = \frac{34.96}{p} + \frac{3.955}{\sqrt{p}}$$

$$T'(10) \approx 4.75 \text{ deg/lb/in}^2$$

$$T'(70) \approx 0.97 \text{ deg/lb/in}^2$$

86. $\displaystyle\int_{1}^{2.7} \frac{1}{t}\, dt \approx \frac{2.7 - 1}{3(4)} \left[\frac{1}{1} + \frac{4}{1.425} + \frac{2}{1.85} + \frac{4}{2.275} + \frac{1}{2.7} \right] \approx 0.9940$

$\displaystyle\int_{1}^{2.8} \frac{1}{t}\, dt \approx \frac{2.8 - 1}{3(4)} \left[\frac{1}{1} + \frac{4}{1.45} + \frac{2}{1.9} + \frac{4}{2.35} + \frac{1}{2.8} \right] \approx 1.0306$

Therefore, $\displaystyle\int_{1}^{2.7} \frac{1}{t}\, dt < 1 < \int_{1}^{2.8} \frac{1}{t}\, dt$. Since $\displaystyle\int_{1}^{2.7} \frac{1}{t}\, dt = \ln 2.7$, $\displaystyle\int_{1}^{2.8} \frac{1}{t}\, dt = \ln 2.8$, and $\ln e = 1$

we have $\ln 2.7 < \ln e < \ln 2.8$. Also, since $y = \ln x$ is an increasing function, we have $2.7 < e < 2.8$.

87. $g(x) = \ln f(x), \quad f(x) > 0$

$$g'(x) = \frac{f'(x)}{f(x)}$$

a. Yes. If the graph of g is increasing, then $g'(x) > 0$. Since $f(x) > 0$, you know that $f'(x) = g'(x) f(x)$ and thus, $f'(x) > 0$. Therefore, the graph of f is increasing.

b. Yes. If the graph of f is increasing, then $f'(x) > 0$. Since $f(x) > 0$ and $g'(x) = f'(x)/f(x)$, $g'(x) > 0$. Therefore, the graph of g is increasing.

88. $y = 10\ln\left(\dfrac{10 + \sqrt{100 - x^2}}{x}\right) - \sqrt{100 - x^2} = 10[\ln(10 + \sqrt{100 - x^2}) - \ln x] - \sqrt{100 - x^2}$

a.
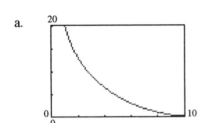

b. $\dfrac{dy}{dx} = 10\left[\dfrac{-x}{\sqrt{100 - x^2}(10 + \sqrt{100 - x^2})} - \dfrac{1}{x}\right] + \dfrac{x}{\sqrt{100 - x^2}}$

$= \dfrac{x}{\sqrt{100 - x^2}}\left[\dfrac{-10}{10 + \sqrt{100 - x^2}}\right] - \dfrac{10}{x} + \dfrac{x}{\sqrt{100 - x^2}}$

$= \dfrac{x}{\sqrt{100 - x^2}}\left[\dfrac{-10}{10 + \sqrt{100 - x^2}} + 1\right] - \dfrac{10}{x}$

$= \dfrac{x}{\sqrt{1000 - x^2}}\left[\dfrac{\sqrt{100 - x^2}}{10 + \sqrt{100 - x^2}}\right] - \dfrac{10}{x}$

$= \dfrac{x}{10 + \sqrt{100 - x^2}} - \dfrac{10}{x}$

$= \dfrac{x\left(10 - \sqrt{100 - x^2}\right)}{x^2} - \dfrac{10}{x} = -\dfrac{\sqrt{100 - x^2}}{x}$

When $x = 5$, $dy/dx = -\sqrt{3}$. When $x = 9$, $dy/dx = -\sqrt{19}/9$.

c. $\lim\limits_{x \to 10^-} \dfrac{dy}{dx} = 0$

---

89. a. $f(x) = \ln x$, $g(x) = \sqrt{x}$

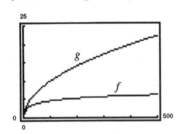

$f'(x) = \dfrac{1}{x}$, $g'(x) = \dfrac{1}{2\sqrt{x}}$

For $x > 4$, $g'(x) > f'(x)$. g is increasing at a higher rate than f for "large" values of x.

b. $f(x) = \ln x$, $g(x) = \sqrt[4]{x}$

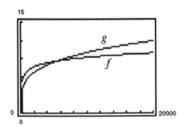

$f'(x) = \dfrac{1}{x}$, $g'(x) = \dfrac{1}{4\sqrt[4]{x^3}}$

For $x > 256$, $g'(x) > f'(x)$. g is increasing at a higher rate than f for "large" values of x. $f(x) = \ln x$ increases very slowly for "large" values of x.

---

90. When $x = 1000$,

$\dfrac{x}{\ln x} \approx 144.7648$.

When $x = 10^6$,

$\dfrac{x}{\ln x} \approx 72{,}382.4137$.

When $x = 10^9$,

$\dfrac{x}{\ln x} \approx 48{,}254{,}942.43$.

$\dfrac{p(1000)}{144.7648} \approx 1.1605$

$\dfrac{p(10^6)}{72{,}382.4137} \approx 1.0845$

$\dfrac{p(10^9)}{48{,}254{,}942.43} \approx 1.0537$

91.
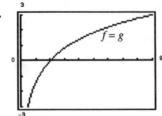

92.
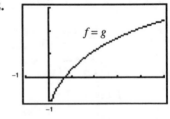

93. False

$\ln x + \ln 25 = \ln(25x) \neq \ln(x + 25)$

94. True

$f(x) = \ln x, \quad x > 0$

$f'(x) = \dfrac{1}{x} > 0 \text{ for } x > 0$

95. True

$f(x) = \ln(ax), \quad f'(x) = \dfrac{a}{ax} = \dfrac{1}{x}$

$g(x) = \ln(bx), \quad g'(x) = \dfrac{b}{bx} = \dfrac{1}{x}$

96. False

π is a constant.

$\dfrac{d}{dx}[\ln \pi] = 0$

97. $y = \ln x$

$y' = \dfrac{1}{x} > 0 \text{ for } x > 0$

Since $\ln x$ is increasing on its entire domain $(0, \infty)$, it is a strictly monotonic function and therefore, is one-to-one.

98. $\ln\left(x + \sqrt{x^2 - 1}\right) + \ln\left(x - \sqrt{x^2 - 1}\right) = \ln\left[\left(x + \sqrt{x^2 - 1}\right)\left(x - \sqrt{x^2 - 1}\right)\right]$

$= \ln[x^2 - (x^2 - 1)]$

$= \ln 1$

$= 0$

Section 5.2   The Natural Logarithmic Function and Integration

1. $u = x + 1, \quad du = dx$

$\displaystyle \int \frac{1}{x+1}\, dx = \ln|x+1| + C$

2. $u = x - 5, \quad du = dx$

$\displaystyle \int \frac{1}{x-5}\, dx = \ln|x-5| + C$

3. $u = 3 - 2x, \quad du = -2\, dx$

$\displaystyle \int \frac{1}{3-2x}\, dx = -\frac{1}{2}\int \frac{1}{3-2x}(-2)\, dx$

$= -\dfrac{1}{2}\ln|3 - 2x| + C$

4. $u = 6x + 1, \quad du = 6\, dx$

$\displaystyle \int \frac{1}{6x+1}\, dx = \frac{1}{6}\int \frac{1}{6x+1}(6)\, dx$

$= \dfrac{1}{6}\ln|6x + 1| + C$

5. $u = x^2 + 1, \quad du = 2x\, dx$

$\displaystyle \int \frac{x}{x^2+1}\, dx = \frac{1}{2}\int \frac{1}{x^2+1}(2x)\, dx$

$= \dfrac{1}{2}\ln(x^2 + 1) + C$

$= \ln\sqrt{x^2 + 1} + C$

6. $u = 3 - x^3, \quad du = -3x^2\, dx$

$\displaystyle \int \frac{x^2}{3-x^3}\, dx = -\frac{1}{3}\int \frac{1}{3-x^3}(-3x^2)\, dx$

$= -\dfrac{1}{3}\ln|3 - x^3| + C$

7. $\displaystyle \int \frac{x^2 - 4}{x}\, dx = \int \left(x - \frac{4}{x}\right) dx$

$= \dfrac{x^2}{2} - 4\ln|x| + C$

8. $u = 9 - x^2, \quad du = -2x\, dx$

$\displaystyle \int \frac{x}{\sqrt{9-x^2}}\, dx = -\frac{1}{2}\int (9 - x^2)^{-1/2}(-2x)\, dx$

$= -\sqrt{9 - x^2} + C$

9. $u = x^3 + 3x^2 + 9x, \quad du = 3(x^2 + 2x + 3)\, dx$

$$\int \frac{x^2 + 2x + 3}{x^3 + 3x^2 + 9x}\, dx = \frac{1}{3} \int \frac{3(x^2 + 2x + 3)}{x^3 + 3x^2 + 9x}\, dx$$

$$= \frac{1}{3} \ln |x^3 + 3x^2 + 9x| + C$$

10. $u = x^2 + 6x + 7, \quad du = 2(x + 3)\, dx$

$$\int \frac{x + 3}{x^2 + 6x + 7}\, dx = \frac{1}{2} \int \frac{2x + 6}{x^2 + 6x + 7}\, dx$$

$$= \frac{1}{2} \ln |x^2 + 6x + 7| + C$$

11. $u = \ln x, \quad du = \frac{1}{x}\, dx$

$$\int \frac{(\ln x)^2}{x}\, dx = \frac{1}{3} (\ln x)^3 + C$$

12. $u = \ln x, \quad du = \frac{1}{x}\, dx$

$$\int \frac{1}{x \ln x^2}\, dx = \frac{1}{2} \int \frac{1}{x \ln x}\, dx$$

$$= \frac{1}{2} \ln |\ln |x|| + C$$

13. $u = x + 1, \quad du = dx$

$$\int \frac{1}{\sqrt{x + 1}}\, dx = \int (x + 1)^{-1/2}\, dx$$

$$= 2(x + 1)^{1/2} + C$$

$$= 2\sqrt{x + 1} + C$$

14. $u = 1 + x^{1/3}, \quad du = \frac{1}{3x^{2/3}}\, dx$

$$\int \frac{1}{x^{2/3}(1 + x^{1/3})}\, dx = 3 \int \frac{1}{1 + x^{1/3}} \left(\frac{1}{3x^{2/3}} \right)\, dx$$

$$= 3 \ln |1 + x^{1/3}| + C$$

15. $u = \sqrt{x} - 3, \quad du = \frac{1}{2\sqrt{x}}\, dx \Rightarrow 2(u + 3)\, du = dx$

$$\int \frac{\sqrt{x}}{\sqrt{x} - 3}\, dx = 2 \int \frac{(u + 3)^2}{u}\, du = 2 \int \frac{u^2 + 6u + 9}{u}\, du = 2 \int \left(u + 6 + \frac{9}{u} \right) du$$

$$= 2 \left[\frac{u^2}{2} + 6u + 9 \ln |u| \right] + C_1 = u^2 + 12u + 18 \ln |u| + C_1$$

$$= \left(\sqrt{x} - 3 \right)^2 + 12 \left(\sqrt{x} - 3 \right) + 18 \ln |\sqrt{x} - 3| + C_1$$

$$= x + 6\sqrt{x} + 18 \ln |\sqrt{x} - 3| + C \text{ where } C = C_1 - 27.$$

16. $u = 1 + \sqrt{2x}, \quad du = \frac{1}{\sqrt{2x}}\, dx \Rightarrow (u - 1)\, du = dx$

$$\int \frac{1}{1 + \sqrt{2x}}\, dx = \int \frac{(u - 1)}{u}\, du = \int \left(1 - \frac{1}{u} \right) du = u - \ln |u| + C_1 = (1 + \sqrt{2x}) - \ln |1 + \sqrt{2x}| + C_1$$

$$= \sqrt{2x} - \ln (1 + \sqrt{2x}) + C \text{ where } C = C_1 + 1.$$

17. $\displaystyle \int \frac{2x}{(x - 1)^2}\, dx = \int \frac{2x - 2 + 2}{(x - 1)^2}\, dx = \int \frac{2(x - 1)}{(x - 1)^2}\, dx + 2 \int \frac{1}{(x - 1)^2}\, dx$

$$= 2 \int \frac{1}{x - 1}\, dx + 2 \int \frac{1}{(x - 1)^2}\, dx = 2 \ln |x - 1| - \frac{2}{(x - 1)} + C$$

18. $\displaystyle \int \frac{x(x - 2)}{(x - 1)^3}\, dx = \int \frac{x^2 - 2x + 1 - 1}{(x - 1)^3}\, dx = \int \frac{(x - 1)^2}{(x - 1)^3}\, dx - \int \frac{1}{(x - 1)^3}\, dx$

$$= \int \frac{1}{x - 1}\, dx - \int \frac{1}{(x - 1)^3}\, dx = \ln |x - 1| + \frac{1}{2(x - 1)^2} + C$$

19. $\displaystyle \int \frac{\cos \theta}{\sin \theta}\, d\theta = \ln |\sin \theta| + C$

20. $\displaystyle \int \tan 5\theta\, d\theta = \frac{1}{5} \int \frac{5 \sin 5\theta}{\cos 5\theta}\, d\theta$

$$= -\frac{1}{5} \ln |\cos 5\theta| + C$$

21. $\displaystyle\int \csc 2x \, dx = \frac{1}{2}\int (\csc 2x)(2)\, dx$

$\displaystyle = -\frac{1}{2}\ln|\csc 2x + \cot 2x| + C$

22. $\displaystyle\int \sec\frac{x}{2}\, dx = 2\int \sec\frac{x}{2}\left(\frac{1}{2}\right)dx$

$\displaystyle = 2\ln\left|\sec\frac{x}{2} + \tan\frac{x}{2}\right| + C$

23. $\displaystyle\int \frac{\cos t}{1 + \sin t}\, dt = \ln|1 + \sin t| + C$

24. $\displaystyle\int \frac{\sin x}{1 + \cos x}\, dx = -\int \frac{-\sin x}{1 + \cos x}\, dx$

$\displaystyle = -\ln|1 + \cos x| + C$

25. $\displaystyle\int \frac{\sec x \tan x}{\sec x - 1}\, dx = \ln|\sec x - 1| + C$

26. $\displaystyle\int (\sec t + \tan t)\, dt = \ln|\sec t + \tan t| - \ln|\cos t| + C$

$\displaystyle = \ln\left|\frac{\sec t + \tan t}{\cos t}\right| + C = \ln|\sec t(\sec t + \tan t)| + C$

27. $\displaystyle y = \int \frac{3}{2 - x}\, dx$

$\displaystyle = -3\int \frac{-1}{2-x}\, dx = -3\ln|2 - x| + C$

28. $\displaystyle y = \int \frac{2x}{x^2 - 9}\, dx = \ln|x^2 - 9| + C$

29. $\displaystyle s = \int \tan 2\theta \, d\theta$

$\displaystyle = \frac{1}{2}\int \tan 2\theta (2)\, d\theta$

$\displaystyle = -\frac{1}{2}\ln|\cos 2\theta| + C$

$\displaystyle = -\ln\sqrt{|\cos 2\theta|} + C$

30. $\displaystyle r = \int \frac{\sec^2 t}{\tan t + 1}\, dt = \ln|\tan t + 1| + C$

31. $\displaystyle\int_0^4 \frac{5}{3x + 1}\, dx = \left[\frac{5}{3}\ln|3x + 1|\right]_0^4$

$\displaystyle = \frac{5}{3}\ln 13 \approx 4.275$

32. $\displaystyle\int_{-1}^1 \frac{1}{5 - 2x}\, dx = \left[-\frac{1}{2}\ln|5 - 2x|\right]_{-1}^1$

$\displaystyle = -\frac{1}{2}\ln 3 + \frac{1}{2}\ln 7$

$\displaystyle = \frac{1}{2}\ln\frac{7}{3} \approx 0.424$

33. $u = 1 + \ln x, \quad du = \dfrac{1}{x}\, dx$

$\displaystyle\int_1^e \frac{(1 + \ln x)^2}{x}\, dx = \left[\frac{1}{3}(1 + \ln|x|)^3\right]_1^e$

$\displaystyle = \frac{7}{3}$

34. $u = \ln x, \quad du = \dfrac{1}{x}\, dx$

$\displaystyle\int_e^{e^2} \frac{1}{x \ln x}\, dx = \int_e^{e^2}\left(\frac{1}{\ln x}\right)\frac{1}{x}\, dx$

$\displaystyle = \Big[\ln|\ln|x||\Big]_e^{e^2} = \ln 2$

35. $\displaystyle\int_0^2 \frac{x^2 - 2}{x + 1}\, dx = \int_0^2\left(x - 1 - \frac{1}{x + 1}\right)dx$

$\displaystyle = \left[\frac{1}{2}x^2 - x - \ln|x + 1|\right]_0^2 = -\ln 3$

36. $\displaystyle\int_0^1 \frac{x - 1}{x + 1}\, dx = \int_0^1 1\, dx + \int_0^1 \frac{-2}{x + 1}\, dx$

$\displaystyle = \Big[x - 2\ln|x + 1|\Big]_0^1 = 1 - 2\ln 2$

37. $\int_1^2 \frac{1 - \cos\theta}{\theta - \sin\theta}\, d\theta = \left[\ln|\theta - \sin\theta|\right]_1^2 = \ln\left|\frac{2 - \sin 2}{1 - \sin 1}\right| \approx 1.929$

38. $\int_{0.1}^{0.2} (\csc 2\theta - \cot 2\theta)^2\, d\theta = \int_{0.1}^{0.2} (\csc^2 2\theta - 2\csc 2\theta \cot 2\theta + \cot^2 2\theta)\, d\theta$

$$= \int_{0.1}^{0.2} (2\csc^2 2\theta - 2\csc 2\theta \cot 2\theta - 1)\, d\theta$$

$$= \left[-\cot 2\theta + \csc 2\theta - \theta\right]_{0.1}^{0.2} \approx 0.002$$

39. $\int \frac{1}{1 + \sqrt{x}}\, dx = 2(1 + \sqrt{x}) - 2\ln(1 + \sqrt{x}) + C_1 = 2[\sqrt{x} - \ln(1 + \sqrt{x})] + C$ where $C = C_1 + 2$.

40. $\int \frac{1 - \sqrt{x}}{1 + \sqrt{x}}\, dx = -(1 + \sqrt{x})^2 + 6(1 + \sqrt{x}) - 4\ln(1 + \sqrt{x}) + C_1 = 4\sqrt{x} - x - 4\ln(1 + \sqrt{x}) + C$ where $C = C_1 + 5$.

41. $\int \cos(1 - x)\, dx = -\sin(1 - x) + C$

42. $\int \frac{\tan^2 2x}{\sec 2x}\, dx = \frac{1}{2}[\ln|\sec 2x + \tan 2x| - \sin 2x] + C$

43. $\int_{\pi/4}^{\pi/2} (\csc x - \sin x)\, dx = \left[-\ln|\csc x + \cot x| + \cos x\right]_{\pi/4}^{\pi/2} = \ln(\sqrt{2} + 1) - \frac{\sqrt{2}}{2} \approx 0.174$

44. $\int_{-\pi/4}^{\pi/4} \frac{\sin^2 x - \cos^2 x}{\cos x}\, dx = \left[\ln|\sec x + \tan x| - 2\sin x\right]_{-\pi/4}^{\pi/4} = \ln\left(\frac{\sqrt{2} + 1}{\sqrt{2} - 1}\right) - 2\sqrt{2} \approx -1.066$

45. $-\ln|\cos x| + C = \ln\left|\frac{1}{\cos x}\right| + C = \ln|\sec x| + C$ **46.** $\ln|\sin x| + C = \ln\left|\frac{1}{\csc x}\right| + C = -\ln|\csc x| + C$

47. $\ln|\sec x + \tan x| + C = \ln\left|\frac{(\sec x + \tan x)(\sec x - \tan x)}{(\sec x - \tan x)}\right| + C$

$$= \ln\left|\frac{\sec^2 x - \tan^2 x}{\sec x - \tan x}\right| + C = \ln\left|\frac{1}{\sec x - \tan x}\right| + C = -\ln|\sec x - \tan x| + C$$

48. $-\ln|\csc x + \cot x| + C = -\ln\left|\frac{(\csc x + \cot x)(\csc x - \cot x)}{(\csc x - \cot x)}\right| + C$

$$= -\ln\left|\frac{\csc^2 x - \cot^2 x}{\csc x - \cot x}\right| + C = -\ln\left|\frac{1}{\csc x - \cot x}\right| + C = \ln|\csc x - \cot x| + C$$

49. $F(x) = \int_1^x \frac{1}{t}\, dt$ **50.** $F(x) = \int_0^x \tan t\, dt$ **51.** $F(x) = \int_x^{3x} \frac{1}{t}\, dt$ **52.** $F(x) = \int_1^{x^2} \frac{1}{t}\, dt$

$F'(x) = \frac{1}{x}$ $F'(x) = \tan x$ $F'(x) = \frac{3}{3x} - \frac{1}{x}$ $F'(x) = \frac{2x}{x^2} = \frac{2}{x}$

$= 0$

53.

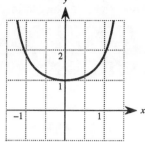

$A \approx 1.25$

Matches (d)

54.

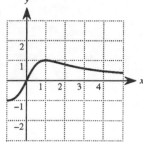

$A \approx 3$

Matches (a)

55. $A = \displaystyle\int_1^4 \frac{x^2+4}{x}\,dx = \int_1^4 \left(x + \frac{4}{x}\right) dx = \left[\frac{x^2}{2} + 4\ln x\right]_1^4 = (8 + 4\ln 4) - \frac{1}{2} = \frac{15}{2} + 8\ln 2 \approx 13.045$ square units

56. $A = \displaystyle\int_1^5 \frac{x+5}{x}\,dx = \int_1^5 \left(1 + \frac{5}{x}\right) dx = \left[x + 5\ln x\right]_1^5 = 4 + 5\ln 5 \approx 12.047$ square units

57. $P(t) = \displaystyle\int \frac{3000}{1+0.25t}\,dt = (3000)(4)\int \frac{0.25}{1+0.25t}\,dt = 12{,}000\ln|1 + 0.25t| + C$

$\quad P(0) = 12{,}000\ln|1 + 0.25(0)| + C = 1000$

$\qquad C = 1000$

$\quad P(t) = 12{,}000\ln|1 + 0.25t| + 1000 = 1000[12\ln|1+0.25t| + 1]$

$\quad P(3) = 1000[12(\ln 1.75) + 1] \approx 7715$

58. $t = \dfrac{10}{\ln 2} \displaystyle\int_{250}^{300} \frac{1}{T-100}\,dT$

$\quad = \dfrac{10}{\ln 2}\left[\ln(T-100)\right]_{250}^{300} = \dfrac{10}{\ln 2}[\ln 200 - \ln 150] = \dfrac{10}{\ln 2}\left[\ln\left(\frac{4}{3}\right)\right] \approx 4.1504$ units of time

59. $\dfrac{1}{50-40}\displaystyle\int_{40}^{50} \frac{90{,}000}{400+3x}\,dx = \left[3000\ln|400+3x|\right]_{40}^{50} \approx \$168.27$

60. $\dfrac{dS}{dt} = \dfrac{k}{t}$

$\quad S(t) = \displaystyle\int \frac{k}{t}\,dt = k\ln|t| + C = k\ln t + C$ since $t > 1$.

$\quad S(2) = k\ln 2 + C = 200$

$\quad S(4) = k\ln 4 + C = 300$

Solving this system yields $k = 100/\ln 2$ and $C = 100$. Thus,

$$S(t) = \frac{100\ln t}{\ln 2} + 100 = 100\left[\frac{\ln t}{\ln 2} + 1\right].$$

61. False

$\quad \dfrac{1}{2}(\ln x) = \ln(x^{1/2}) \neq (\ln x)^{1/2}$

62. False

$\quad \dfrac{d}{dx}[\ln x] = \dfrac{1}{x}$

63. True

$$\int \frac{1}{x}\,dx = \ln|x| + C_1$$

$$= \ln|x| + \ln|C| = \ln|Cx|, \quad C \neq 0$$

64. False; the integrand has a nonremovable discontinuity at $x = 0$.

65.

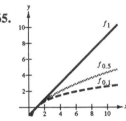

$k = 1: \ f_1(x) = x - 1$

$k = 0.5: \ f_{0.5}(x) = \dfrac{\sqrt{x} - 1}{0.5} = 2\left(\sqrt{x} - 1\right)$

$k = 0.1: \ f_{0.1}(x) = \dfrac{\sqrt[10]{x} - 1}{0.1} = 10\left(\sqrt[10]{x} - 1\right)$

$\displaystyle \lim_{k \to 0} \ f_k(x) = \ln x$

Section 5.3   Inverse Functions

1. $f(x) = 5x + 1$

$g(x) = \dfrac{x - 1}{5}$

$f(g(x)) = f\left(\dfrac{x-1}{5}\right) = 5\left(\dfrac{x-1}{5}\right) + 1 = x$

$g(f(x)) = g(5x + 1) = \dfrac{(5x+1) - 1}{5} = x$

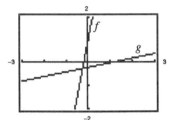

2. $f(x) = 3 - 4x$

$g(x) = \dfrac{3 - x}{4}$

$f(g(x)) = f\left(\dfrac{3-x}{4}\right) = 3 - 4\left(\dfrac{3-x}{4}\right) = x$

$g(f(x)) = g(3 - 4x) = \dfrac{3 - (3 - 4x)}{4} = x$

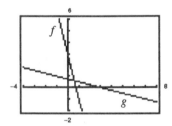

3. $f(x) = x^3$

$g(x) = \sqrt[3]{x}$

$f(g(x)) = f\left(\sqrt[3]{x}\right) = \left(\sqrt[3]{x}\right)^3 = x$

$g(f(x)) = g(x^3) = \sqrt[3]{x^3} = x$

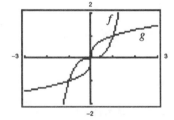

4. $f(x) = 1 - x^3$

$g(x) = \sqrt[3]{1 - x}$

$f(g(x)) = f\left(\sqrt[3]{1-x}\right) = 1 - \left(\sqrt[3]{1-x}\right)^3$

$\qquad\qquad = 1 - (1 - x) = x$

$g(f(x)) = g(1 - x^3)$

$\qquad = \sqrt[3]{1 - (1 - x^3)} = \sqrt[3]{x^3} = x$

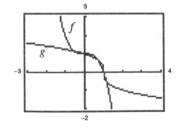

5. $f(x) = \sqrt{x-4}$

$g(x) = x^2 + 4, \quad x \geq 0$

$f(g(x)) = f(x^2 + 4)$

$\qquad = \sqrt{(x^2 + 4) - 4} = \sqrt{x^2} = x$

$g(f(x)) = g\left(\sqrt{x-4}\right)$

$\qquad = \left(\sqrt{x-4}\right)^2 + 4 = x - 4 + 4 = x$

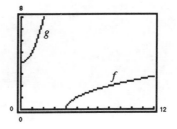

6. $f(x) = 9 - x^2, \quad x \geq 0$

$g(x) = \sqrt{9-x}, \quad x \leq 9$

$f(g(x)) = f\left(\sqrt{9-x}\right)$

$\qquad = 9 - \left(\sqrt{9-x}\right)^2 = 9 - (9 - x) = x$

$g(f(x)) = g(9 - x^2)$

$\qquad = \sqrt{9 - (9 - x^2)} = \sqrt{x^2} = x$

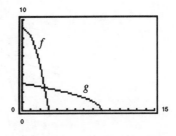

7. $f(x) = \dfrac{1}{x}$

$g(x) = \dfrac{1}{x}$

$f(g(x)) = g(f(x)) = \dfrac{1}{1/x} = x$

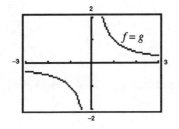

8. $f(x) = \dfrac{1}{1+x}, \quad x \geq 0$

$g(x) = \dfrac{1-x}{x}, \quad 0 < x \leq 1$

$f(g(x)) = f\left(\dfrac{1-x}{x}\right) = \dfrac{1}{1 + \dfrac{1-x}{x}} = \dfrac{1}{\dfrac{1}{x}} = x$

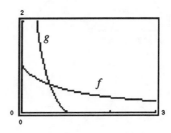

$g(f(x)) = g\left(\dfrac{1}{1+x}\right) = \dfrac{1 - \dfrac{1}{1+x}}{\dfrac{1}{1+x}} = \dfrac{x}{1+x} \cdot \dfrac{1+x}{1} = x$

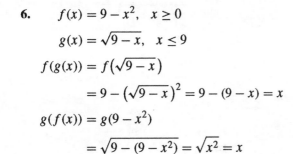
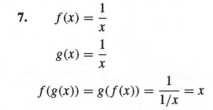
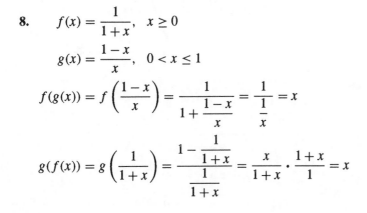

9. $f(x) = 2x - 3 = y$

$x = \dfrac{y+3}{2}$

$y = \dfrac{x+3}{2}$

$f^{-1}(x) = \dfrac{x+3}{2}$

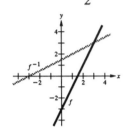

10. $f(x) = 3x = y$

$x = \dfrac{y}{3}$

$y = \dfrac{x}{3}$

$f^{-1}(x) = \dfrac{x}{3}$

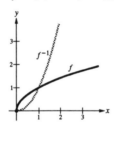

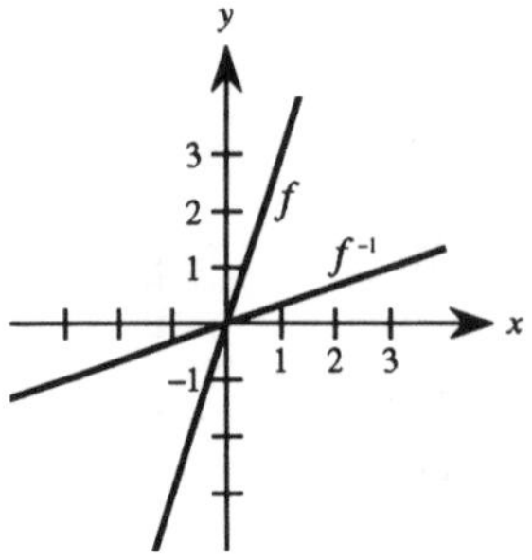

11. $f(x) = x^5 = y$

$x = \sqrt[5]{y}$

$y = \sqrt[5]{x}$

$f^{-1}(x) = \sqrt[5]{x} = x^{1/5}$

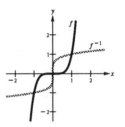

12. $f(x) = x^3 + 1 = y$

$x = \sqrt[3]{y - 1}$

$y = \sqrt[3]{x - 1}$

$f^{-1}(x) = \sqrt[3]{x - 1}$

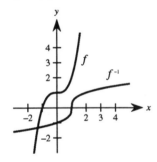

13. $f(x) = \sqrt{x} = y$

$x = y^2$

$y = x^2$

$f^{-1}(x) = x^2, \quad x \geq 0$

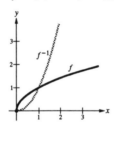

14. $f(x) = x^2 = y, \quad 0 \leq x$

$x = \sqrt{y}$

$y = \sqrt{x}$

$f^{-1}(x) = \sqrt{x}$

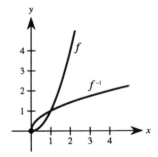

15. $f(x) = \sqrt{4 - x^2} = y, \quad 0 \leq x \leq 2$

$x = \sqrt{4 - y^2}$

$y = \sqrt{4 - x^2}$

$f^{-1}(x) = \sqrt{4 - x^2}, \quad 0 \leq x \leq 2$

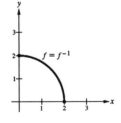

16. $f(x) = \sqrt{x^2 - 4} = y, \quad x \geq 2$

$x = \sqrt{y^2 + 4}$

$y = \sqrt{x^2 + 4}$

$f^{-1}(x) = \sqrt{x^2 + 4}, \quad x \geq 0$

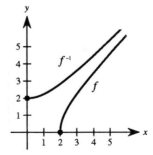

17. $f(x) = \sqrt[3]{x-1} = y$

$x = y^3 + 1$

$y = x^3 + 1$

$f^{-1}(x) = x^3 + 1$

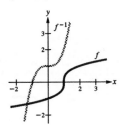

18. $f(x) = 3\sqrt[5]{2x-1} = y$

$x = \dfrac{y^5 + 243}{486}$

$y = \dfrac{x^5 + 243}{486}$

$f^{-1}(x) = \dfrac{x^5 + 243}{486}$

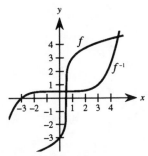

19. $f(x) = x^{2/3} = y, \quad x \geq 0$

$x = y^{3/2}$

$y = x^{3/2}$

$f^{-1}(x) = x^{3/2}, \quad x \geq 0$

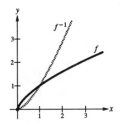

20. $f(x) = x^{3/5} = y$

$x = y^{5/3}$

$y = x^{5/3}$

$f^{-1}(x) = x^{5/3}$

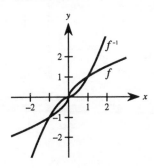

21. $f(x) = \dfrac{x}{\sqrt{x^2 + 7}} = y$

$x = \dfrac{\sqrt{7}\, y}{\sqrt{1 - y^2}}$

$y = \dfrac{\sqrt{7}\, x}{\sqrt{1 - x^2}}$

$f^{-1}(x) = \dfrac{\sqrt{7}\, x}{\sqrt{1 - x^2}}, \quad -1 < x < 1$

22. $f(x) = \dfrac{x+2}{x} = y$

$x = \dfrac{2}{y-1}$

$y = \dfrac{2}{x-1}$

$f^{-1}(x) = \dfrac{2}{x-1}$

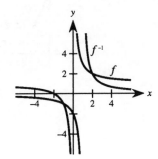

23. $f(x) = \dfrac{x}{x^2 - 4} = y$ on $(-2,\ 2)$

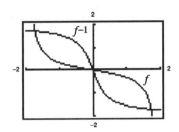

$x^2 y - 4y = x$

$x^2 y - x - 4y = 0$

$a = y,\quad b = -1,\quad c = -4y$

$x = \dfrac{1 \pm \sqrt{1 - 4(y)(-4y)}}{2y} = \dfrac{1 \pm \sqrt{1 + 16y^2}}{2y}$

On $(-2,\ 2)$: $y = \begin{cases} \dfrac{1 - \sqrt{1 + 16x^2}}{2x}, & \text{if } x \neq 0 \\ 0, & \text{if } x = 0 \end{cases}$

$f^{-1}(x) = \begin{cases} \dfrac{1 - \sqrt{1 + 16x^2}}{2x}, & \text{if } x \neq 0 \\ 0, & \text{if } x = 0 \end{cases}$

24. $f(x) = 2 - \dfrac{3}{x^2} = y$ on $(0,\ 10)$

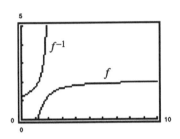

$2x^2 - 3 = x^2 y$

$x^2(2 - y) = 3$

$x = \pm\sqrt{\dfrac{3}{2 - y}}$

$y = \pm\sqrt{\dfrac{3}{2 - x}}$

On $(0,\ 10)$, $f^{-1}(x) = \sqrt{\dfrac{3}{2 - x}},\quad x < 2.$

25.

x	1	2	3	4
$f^{-1}(x)$	0	1	2	4

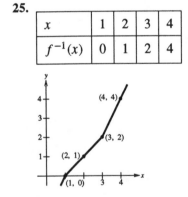

26.

x	0	2	4
$f^{-1}(x)$	6	2	0

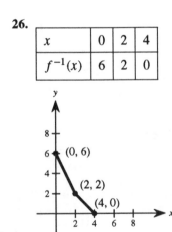

27. $f(x) = \frac{3}{4}x + 6$
One-to-one; has an inverse

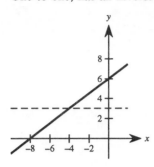

28. $f(x) = 5x - 3$
One-to-one; has an inverse

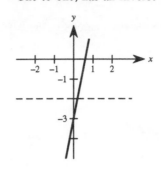

29. $f(\theta) = \sin\theta$
Not one-to-one; does not have an inverse

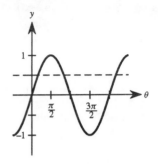

30. $F(x) = \dfrac{x^2}{x^2 + 4}$
Not one-to-one; does not have an inverse

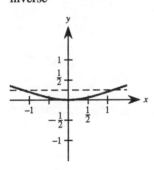

31. $h(s) = \dfrac{1}{s - 2} - 3$
One-to-one; has an inverse

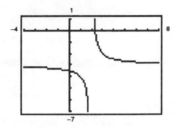

32. $g(t) = \dfrac{1}{\sqrt{t^2 + 1}}$
Not one-to-one; does not have an inverse

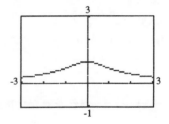

33. $f(x) = \ln x$
One-to-one; has an inverse

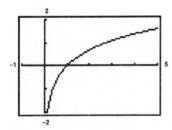

34. $f(x) = 3x\sqrt{x + 1}$
Not one-to-one; does not have an inverse

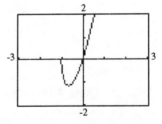

35. $h(x) = |x + 4| - |x - 4|$
Not one-to-one; does not have an inverse

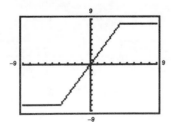

36. $g(x) = (x + 5)^3$
One-to-one; has an inverse

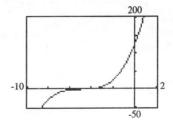

37. $f(x) = (x + a)^3 + b$

$f'(x) = 3(x + a)^2 \geq 0$ for all x

f is increasing on $(-\infty, \infty)$. Therefore, f is strictly monotonic and has an inverse.

38. $f(x) = \cos \dfrac{3x}{2}$

$f'(x) = -\dfrac{3}{2} \sin \dfrac{3x}{2} = 0$ when $x = 0, \dfrac{2\pi}{3}, \dfrac{4\pi}{3}, \ldots$

f is not strictly monotonic on $(-\infty, \infty)$. Therefore, f does not have an inverse.

39. $f(x) = \dfrac{x^4}{4} - 2x^2$

$f'(x) = x^3 - 4x = 0$ when $x = 0, 2, -2$

f is not strictly monotonic on $(-\infty, \infty)$. Therefore, f does not have an inverse.

40. $f(x) = x^3 - 6x^2 + 12x$

$f'(x) = 3x^2 - 12x + 12 = 3(x - 2)^2 \geq 0$ for all x

f is increasing on $(-\infty, \infty)$. Therefore, f is strictly monotonic and has an inverse.

41. $f(x) = 2 - x - x^3$

$f'(x) = -1 - 3x^2 < 0$ for all x

f is decreasing on $(-\infty, \infty)$. Therefore, f is strictly monotonic and has an inverse.

42. $f(x) = \ln(x - 3), \quad x > 3$

$f'(x) = \dfrac{1}{x - 3} > 0$ for $x > 3$

f is increasing on $(3, \infty)$. Therefore, f is strictly monotonic and has an inverse.

43. $f(x) = (x - 4)^2$ on $[4, \infty)$

$f'(x) = 2(x - 4) > 0$ on $(4, \infty)$

f is increasing on $[4, \infty)$. Therefore, f is strictly monotonic and has an inverse.

44. $f(x) = |x + 2|$ on $[-2, \infty)$

$f'(x) = \dfrac{|x + 2|}{x + 2}(1) = 1 > 0$ on $(-2, \infty)$

f is increasing on $[-2, \infty)$. Therefore, f is strictly monotonic and has an inverse.

45. $f(x) = \dfrac{4}{x^2}$ on $(0, \infty)$

$f'(x) = -\dfrac{8}{x^3} < 0$ on $(0, \infty)$

f is decreasing on $(0, \infty)$. Therefore, f is strictly monotonic and has an inverse.

46. $f(x) = \tan x$ on $\left(-\dfrac{\pi}{2}, \dfrac{\pi}{2}\right)$

$f'(x) = \sec^2 x > 0$ on $\left(-\dfrac{\pi}{2}, \dfrac{\pi}{2}\right)$

f is increasing on $(-\pi/2, \pi/2)$. Therefore, f is strictly monotonic and has an inverse.

47. $f(x) = \cos x$ on $[0, \pi]$

$f'(x) = -\sin x < 0$ on $(0, \pi)$

f is decreasing on $[0, \pi]$. Therefore, f is strictly monotonic and has an inverse.

48. $f(x) = \sec x$ on $\left[0, \dfrac{\pi}{2}\right)$

$f'(x) = \sec x \tan x > 0$ on $\left(0, \dfrac{\pi}{2}\right)$

f is increasing on $[0, \pi/2)$. Therefore, f is strictly monotonic and has an inverse.

49. f is not one-to-one because many different x-values yield the same y-value.

Example: $f(0) = f(\pi) = 0$

Not continuous at $\dfrac{(2n-1)\pi}{2}$

50. f is not one-to-one because many different x-values yield the same y-value.

Example: $f(3) = f\left(-\dfrac{4}{3}\right) = \dfrac{3}{5}$

Not continuous at ± 2

51. $f(x) = \sqrt{x-2}$, Domain: $x \geq 2$

$f'(x) = \dfrac{1}{2\sqrt{x-2}} > 0$ for $x > 2$

f is one-to-one; has an inverse

$\sqrt{x-2} = y$

$x - 2 = y^2$

$x = y^2 + 2$

$y = x^2 + 2$

$f^{-1}(x) = x^2 + 2, \quad x \geq 0$

52. $f(x) = -3$

Not one-to-one; does not have an inverse

53. $f(x) = |x-2|, \quad x \leq 2$

$= -(x-2)$

$= 2 - x$

f is one-to-one; has an inverse

$2 - x = y$

$2 - y = x$

$f^{-1}(x) = 2 - x, \quad x \geq 0$

54. $f(x) = ax + b$

f is one-to-one; has an inverse

$ax + b = y$

$x = \dfrac{y-b}{a}$

$y = \dfrac{x-b}{a}$

$f^{-1}(x) = \dfrac{x-b}{a}, \quad a \neq 0$

55. $f(x) = (x-3)^2$ is one-to-one for $x \geq 3$.

$(x-3)^2 = y$

$x - 3 = \sqrt{y}$

$x = \sqrt{y} + 3$

$y = \sqrt{x} + 3$

$f^{-1}(x) = \sqrt{x} + 3, \quad x \geq 0$

56. $f(x) = 16 - x^4$ is one-to-one for $x \geq 0$.

$16 - x^4 = y$

$16 - y = x^4$

$\sqrt[4]{16-y} = x$

$\sqrt[4]{16-x} = y$

$f^{-1}(x) = \sqrt[4]{16-x}, \quad x \leq 16$

57. $f(x) = |x+3|$ is one-to-one for $x \geq -3$.

$x + 3 = y$

$x = y - 3$

$y = x - 3$

$f^{-1}(x) = x - 3, \quad x \geq 0$

58. $f(x) = |x-3|$ is one-to-one for $x \geq 3$.

$x - 3 = y$

$x = y + 3$

$y = x + 3$

$f^{-1}(x) = x + 3, \quad x \geq 0$

59. $\quad f(x) = x^3 + 2x - 1, \quad f(1) = 2 = a$

$f'(x) = 3x^2 + 2$

$(f^{-1})'(2) = \dfrac{1}{f'(f^{-1}(2))} = \dfrac{1}{f'(1)} = \dfrac{1}{3(1)^2 + 2} = \dfrac{1}{5}$

60. $\quad f(x) = 2x^5 + x^3 + 1, \quad f(-1) = -2 = a$

$f'(x) = 10x^4 + 3x^2$

$(f^{-1})'(-2) = \dfrac{1}{f'(f^{-1}(-2))} = \dfrac{1}{f'(-1)} = \dfrac{1}{10(-1)^4 + 3(-1)^2} = \dfrac{1}{13}$

61. $\quad f(x) = \sin x, \quad f\left(\dfrac{\pi}{6}\right) = \dfrac{1}{2} = a$

$f'(x) = \cos x$

$(f^{-1})'\left(\dfrac{1}{2}\right) = \dfrac{1}{f'(f^{-1}(1/2))} = \dfrac{1}{f'(\pi/6)} = \dfrac{1}{\cos(\pi/6)} = \dfrac{1}{\sqrt{3}/2} = \dfrac{2\sqrt{3}}{3}$

62. $\quad f(x) = \cos 2x, \quad f(0) = 1 = a$

$f'(x) = -2\sin 2x$

$(f^{-1})'(1) = \dfrac{1}{f'(f^{-1}(1))} = \dfrac{1}{f'(0)} = \dfrac{1}{-2\sin 0} = \dfrac{1}{0}$ which is undefined.

63. $\quad f(x) = x^3 - \dfrac{4}{x}, \quad f(2) = 6 = a$

$f'(x) = 3x^2 + \dfrac{4}{x^2}$

$(f^{-1})'(6) = \dfrac{1}{f'(f^{-1}(6))} = \dfrac{1}{f'(2)} = \dfrac{1}{3(2)^2 + (4/2^2)} = \dfrac{1}{13}$

64. $\quad f(x) = \sqrt{x-4}, \quad f(8) = 2 = a$

$f'(x) = \dfrac{1}{2\sqrt{x-4}}$

$(f^{-1})'(2) = \dfrac{1}{f'(f^{-1}(2))} = \dfrac{1}{f'(8)} = \dfrac{1}{1/(2\sqrt{8-4})} = \dfrac{1}{1/4} = 4$

65. $f(x) = x^3,$ $\left(\dfrac{1}{2}, \dfrac{1}{8}\right)$

$f'(x) = 3x^2$

$f'\left(\dfrac{1}{2}\right) = \dfrac{3}{4}$

$f^{-1}(x) = \sqrt[3]{x},$ $\left(\dfrac{1}{8}, \dfrac{1}{2}\right)$

$(f^{-1})'(x) = \dfrac{1}{3\sqrt[3]{x^2}}$

$(f^{-1})'\left(\dfrac{1}{8}\right) = \dfrac{4}{3}$

66. $f(x) = 3 - 4x,$ $(1, -1)$

$f'(x) = -4$

$f'(1) = -4$

$f^{-1}(x) = \dfrac{3 - x}{4},$ $(-1, 1)$

$(f^{-1})'(x) = -\dfrac{1}{4}$

$(f^{-1})'(-1) = -\dfrac{1}{4}$

67. $f(x) = \sqrt{x - 4},$ $(5, 1)$

$f'(x) = \dfrac{1}{2\sqrt{x - 4}}$

$f'(5) = \dfrac{1}{2}$

$f^{-1}(x) = x^2 + 4,$ $(1, 5)$

$(f^{-1})'(x) = 2x$

$(f^{-1})'(1) = 2$

68. $f(x) = \dfrac{1}{1 + x^2},$ $\left(1, \dfrac{1}{2}\right)$

$f'(x) = \dfrac{-2x}{(1 + x^2)^2}$

$f'(1) = -\dfrac{1}{2}$

$f^{-1}(x) = \sqrt{\dfrac{1 - x}{x}},$ $\left(\dfrac{1}{2}, 1\right)$

$(f^{-1})'(x) = -\dfrac{1}{2x^2}\sqrt{\dfrac{x}{1 - x}}$

$(f^{-1})'\left(\dfrac{1}{2}\right) = -2$

In Exercises 69–72, use the following.

$f(x) = \tfrac{1}{8}x - 3$ and $g(x) = x^3$

$f^{-1}(x) = 8(x + 3)$ and $g^{-1}(x) = \sqrt[3]{x}$

69. $(f^{-1} \circ g^{-1})(1) = f^{-1}(g^{-1}(1)) = f^{-1}(1) = 32$

70. $(g^{-1} \circ f^{-1})(-3) = g^{-1}(f^{-1}(-3)) = g^{-1}(0) = 0$

71. $(f^{-1} \circ f^{-1})(6) = f^{-1}(f^{-1}(6)) = f^{-1}(72) = 600$

72. $(g^{-1} \circ g^{-1})(-4) = g^{-1}(g^{-1}(-4)) = g^{-1}(\sqrt[3]{-4})$

$= \sqrt[3]{\sqrt[3]{-4}} = -\sqrt[9]{4}$

In Exercises 73–76, use the following.

$f(x) = x + 4$ and $g(x) = 2x - 5$

$f^{-1}(x) = x - 4$ and $g^{-1}(x) = \dfrac{x + 5}{2}$

73. $(g^{-1} \circ f^{-1})(x) = g^{-1}(f^{-1}(x))$

$= g^{-1}(x - 4)$

$= \dfrac{(x - 4) + 5}{2}$

$= \dfrac{x + 1}{2}$

74. $(f^{-1} \circ g^{-1})(x) = f^{-1}(g^{-1}(x))$

$= f^{-1}\left(\dfrac{x + 5}{2}\right)$

$= \dfrac{x + 5}{2} - 4$

$= \dfrac{x - 3}{2}$

75. $(f \circ g)^{-1} = [f(g(x))]^{-1}$

$\qquad = [f(2x - 5)]^{-1}$

$\qquad = [2x - 1]^{-1}$

$\qquad = \dfrac{x + 1}{2}$

76. $(g \circ f)^{-1} = [g(f(x))]^{-1}$

$\qquad = [g(x + 4)]^{-1}$

$\qquad = [2(x + 4) - 5]^{-1}$

$\qquad = [2x + 3]^{-1}$

$\qquad = \dfrac{x - 3}{2}$

77. Let $(f \circ g)(x) = y$ then $x = (f \circ g)^{-1}(y)$. Also,

$\qquad (f \circ g)(x) = y$

$\qquad f(g(x)) = y$

$\qquad g(x) = f^{-1}(y)$

$\qquad x = g^{-1}(f^{-1}(y))$

$\qquad = (g^{-1} \circ f^{-1})(y)$

Since f and g are one-to-one functions,
$(f \circ g)^{-1} = g^{-1} \circ f^{-1}$.

78. If f has an inverse, then f and f^{-1} are both one-to-one. Let $(f^{-1})^{-1}(x) = y$ then $x = f^{-1}(y)$ and $f(x) = y$. Thus, $(f^{-1})^{-1} = f$.

79. Suppose $g(x)$ and $h(x)$ are both inverses of $f(x)$. Then the graph of $f(x)$ contains the point $(a, \ b)$ if and only if the graphs of $g(x)$ and $h(x)$ contain the point $(b, \ a)$. Since the graphs of $g(x)$ and $h(x)$ are the same, $g(x) = h(x)$. Therefore, the inverse of $f(x)$ is unique.

80. If f has an inverse and $f(x_1) = f(x_2)$, then $f^{-1}(f(x_1)) = f^{-1}(f(x_2)) \Rightarrow x_1 = x_2$. Therefore, f is one-to-one. If $f(x)$ is one-to-one, then for every value b in the range, there corresponds exactly one value a in the domain. Define $g(x)$ such that the domain of g equals the range of f and $g(b) = a$. By the reflexive property of inverses, $g = f^{-1}$.

81. False

Let $f(x) = x^2$.

82. True; if f has a y-intercept.

83. True

84. False

Let $f(x) = x$ or $g(x) = 1/x$.

85. Not true

Let $f(x) = \begin{cases} x, & 0 \le x \le 1 \\ 1 - x, & 1 < x \le 2. \end{cases}$

86. From Theorem 5.9, we have:

$g'(x) = \dfrac{1}{f'(g(x))}$

$g''(x) = \dfrac{f'(g(x))(0) - f''(g(x))g'(x)}{[f'(g(x))]^2}$

$\qquad = -\dfrac{f''(g(x)) \cdot [1/(f'(g(x)))]}{[f'(g(x))]^2}$

$\qquad = -\dfrac{f''(g(x))}{[f'(g)]^3}$

If f is increasing and concave down, then $f' > 0$ and $f'' < 0$ which implies that g is increasing and concave up.

87. $f(x) = \displaystyle\int_2^x \dfrac{dt}{\sqrt{1 + t^4}}, \quad f(2) = 0$

$f'(x) = \dfrac{1}{\sqrt{1 + x^4}}$

$(f^{-1})'(0) = \dfrac{1}{f'(2)} = \dfrac{1}{1/\sqrt{17}} = \sqrt{17}$

Section 5.4   Exponential Functions: Differentiation and Integration

1. a. $e^0 = 1$

      $\ln 1 = 0$

  b. $e^2 = 7.389\ldots$

      $\ln 7.389\ldots = 2$

2. a. $e^{-2} = 0.1353\ldots$

      $\ln 0.1353\ldots = -2$

  b. $e^{-1} = 0.3679\ldots$

      $\ln 0.3679\ldots = -1$

3. a. $\ln 2 = 0.6931\ldots$

      $e^{0.6931\ldots} = 2$

  b. $\ln 8.4 = 2.128\ldots$

      $e^{2.128\ldots} = 8.4$

4. a. $\ln 0.5 = -0.6931\ldots$

      $e^{-0.6931\ldots} = \frac{1}{2}$

  b. $\ln 1 = 0$

      $e^0 = 1$

5. a. $e^{\ln x} = 4$

      $x = 4$

  b. $\ln e^{2x} = 3$

      $2x = 3$

      $x = \frac{3}{2}$

6. a. $e^{\ln 2x} = 12$

      $2x = 12$

      $x = 6$

  b. $\ln e^{-x} = 0$

      $-x = 0$

      $x = 0$

7. a. $\ln x = 2$

      $x = e^2 \approx 7.3891$

  b. $e^x = 4$

      $x = \ln 4 \approx 1.3863$

8. a. $\ln x^2 = 10$

      $x^2 = e^{10}$

      $x = \pm e^5 \approx \pm 148.4132$

  b. $e^{-4x} = 5$

      $-4x = \ln 5$

      $x = -\frac{1}{4} \ln 5 \approx -0.4024$

9. $y = e^{-x}$

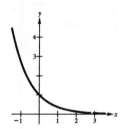

10. $y = \frac{1}{2} e^x$

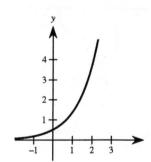

11. $y = e^{-x^2}$

Symmetric with respect to the y-axis

Horizontal asymptote

$y = 0$

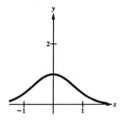

12. $y = e^{-x/2}$

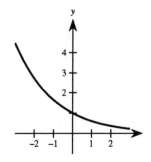

13. a.

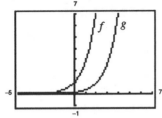

Horizontal shift 2 units to the right

b.

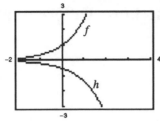

A reflection in the *x*-axis and a vertical shrink

c.

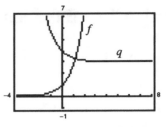

Vertical shift 3 units upward and a reflection in the *y*-axis

14. a.

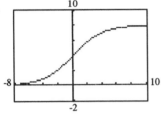

Horizontal asymptotes: $y = 0$ and $y = 8$

b.

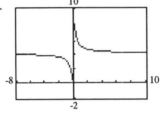

Horizontal asymptote: $y = 4$

15. $f(x) = e^{2x}$

$g(x) = \ln \sqrt{x} = \frac{1}{2} \ln x$

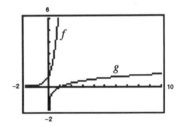

16. $f(x) = e^{x/3}$

$g(x) = \ln x^3 = 3 \ln x$

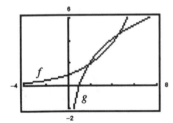

17. $f(x) = e^x - 1$

$g(x) = \ln(x + 1)$

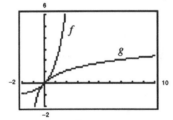

18. $f(x) = e^{x-1}$

$g(x) = 1 + \ln x$

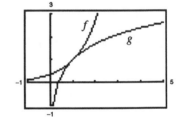

19. $y = Ce^{ax}$
Horizontal asymptote: $y = 0$
Matches (c)

20. $y = Ce^{-ax}$
Horizontal asymptote: $y = 0$
Reflection in the *y*-axis
Matches (d)

21. $y = C(1 - e^{-ax})$
Vertical shift C units
Reflection in both the *x*- and *y*-axes
Matches (a)

22. $y = \dfrac{C}{1 + e^{-ax}}$

$\lim\limits_{x \to \infty} \dfrac{C}{1 + e^{-ax}} = C$

$\lim\limits_{x \to -\infty} \dfrac{C}{1 + e^{-ax}} = 0$

Horizontal asymptotes: $y = C$ and $y = 0$
Matches (b)

23. $\left(1 + \dfrac{1}{1,000,000}\right)^{1,000,000} \approx 2.718280469$

$$e \approx 2.718281828$$

$$e > \left(1 + \dfrac{1}{1,000,000}\right)^{1,000,000}$$

24. $1 + 1 + \dfrac{1}{2} + \dfrac{1}{6} + \dfrac{1}{24} + \dfrac{1}{120} + \dfrac{1}{720} + \dfrac{1}{5040} = 2.71\overline{825396}$

$$e \approx 2.718281828$$

$$e > 1 + 1 + \dfrac{1}{2} + \dfrac{1}{6} + \dfrac{1}{24} + \dfrac{1}{120} + \dfrac{1}{720} + \dfrac{1}{5040}$$

25. a. $y = e^{3x}$

$y' = 3e^{3x}$

At $(0, 1)$, $y' = 3$.

b. $y = e^{-3x}$

$y' = -3e^{-3x}$

At $(0, 1)$, $y' = -3$.

26. a. $y = e^{2x}$

$y' = 2e^{2x}$

At $(0, 1)$, $y' = 2$.

b. $y = e^{-2x}$

$y' = -2e^{-2x}$

At $(0, 1)$, $y' = -2$.

27. $f(x) = e^{2x}$

$f'(x) = 2e^{2x}$

28. $f(x) = e^{1-x}$

$f'(x) = -e^{1-x}$

29. $y = e^{-2x+x^2}$

$\dfrac{dy}{dx} = 2(x - 1)e^{-2x+x^2}$

30. $y = e^{-x^2}$

$\dfrac{dy}{dx} = -2xe^{-x^2}$

31. $y = e^{\sqrt{x}}$

$\dfrac{dy}{dx} = \dfrac{e^{\sqrt{x}}}{2\sqrt{x}}$

32. $y = x^2 e^{-x}$

$\dfrac{dy}{dx} = -x^2 e^{-x} + 2xe^{-x} = xe^{-x}(2 - x)$

33. $g(t) = (e^{-t} + e^t)^3$

$g'(t) = 3(e^{-t} + e^t)^2(e^t - e^{-t})$

34. $g(t) = e^{-1/t^2}$

$g'(t) = \dfrac{2e^{-1/t^2}}{t^3}$

35. $y = \ln e^{x^2} = x^2$

$\dfrac{dy}{dx} = 2x$

36. $y = \ln\left(\dfrac{1 + e^x}{1 - e^x}\right) = \ln(1 + e^x) - \ln(1 - e^x)$

$\dfrac{dy}{dx} = \dfrac{e^x}{1 + e^x} + \dfrac{e^x}{1 - e^x} = \dfrac{2e^x}{1 - e^{2x}}$

37. $y = \ln(1 + e^{2x})$

$\dfrac{dy}{dx} = \dfrac{2e^{2x}}{1 + e^{2x}}$

38. $y = \ln\left(\dfrac{e^x + e^{-x}}{2}\right) = \ln(e^x + e^{-x}) - \ln 2$

$\dfrac{dy}{dx} = \dfrac{e^x - e^{-x}}{e^x + e^{-x}} = \dfrac{e^{2x} - 1}{e^{2x} + 1}$

39. $y = \dfrac{2}{e^x + e^{-x}} = 2(e^x + e^{-x})^{-1}$

$\dfrac{dy}{dx} = -2(e^x + e^{-x})^{-2}(e^x - e^{-x})$

$\quad = \dfrac{-2(e^x - e^{-x})}{(e^x + e^{-x})^2}$

40. $y = \dfrac{e^x - e^{-x}}{2}$

$\dfrac{dy}{dx} = \dfrac{e^x + e^{-x}}{2}$

41. $y = x^2 e^x - 2xe^x + 2e^x = e^x(x^2 - 2x + 2)$

$\dfrac{dy}{dx} = e^x(2x - 2) + e^x(x^2 - 2x + 2) = x^2 e^x$

42. $y = xe^x - e^x = e^x(x - 1)$

$\dfrac{dy}{dx} = e^x + e^x(x - 1) = xe^x$

43. $f(x) = e^{-x} \ln x$

$f'(x) = e^{-x}\left(\dfrac{1}{x}\right) - e^{-x} \ln x = e^{-x}\left(\dfrac{1}{x} - \ln x\right)$

44. $f(x) = e^3 \ln x$

$f'(x) = \dfrac{e^3}{x}$

45. $y = e^x(\sin x + \cos x)$

$\dfrac{dy}{dx} = e^x(\cos x - \sin x) + (\sin x + \cos x)(e^x)$

$\quad = e^x(2 \cos x) = 2e^x \cos x$

46. $y = \ln e^x = x$

$\dfrac{dy}{dx} = 1$

47. $\qquad xe^y - 10x + 3y = 0$

$xe^y \dfrac{dy}{dx} + e^y - 10 + 3\dfrac{dy}{dx} = 0$

$\qquad \dfrac{dy}{dx}(xe^y + 3) = 10 - e^y$

$\qquad\qquad \dfrac{dy}{dx} = \dfrac{10 - e^y}{xe^y + 3}$

48. $\qquad e^{xy} + x^2 - y^2 = 10$

$\left(x\dfrac{dy}{dx} + y\right)e^{xy} + 2x - 2y\dfrac{dy}{dx} = 0$

$\qquad \dfrac{dy}{dx}(xe^{xy} - 2y) = -ye^{xy} - 2x$

$\qquad\qquad \dfrac{dy}{dx} = -\dfrac{ye^{xy} + 2x}{xe^{xy} - 2y}$

49. $f(x) = (3 + 2x)e^{-3x}$

$f'(x) = (3 + 2x)(-3e^{-3x}) + 2e^{-3x}$

$\quad = (-7 - 6x)e^{-3x}$

$f''(x) = (-7 - 6x)(-3e^{-3x}) - 6e^{-3x}$

$\quad = 3(6x + 5)e^{-3x}$

50. $g(x) = \sqrt{x} + e^x \ln x$

$g'(x) = \dfrac{1}{2\sqrt{x}} + \dfrac{e^x}{x} + e^x \ln x$

$g''(x) = -\dfrac{1}{4x^{3/2}} + \dfrac{xe^x - e^x}{x^2} + \dfrac{e^x}{x} + e^x \ln x$

$\quad = -\dfrac{1}{4x\sqrt{x}} + \dfrac{e^x(2x - 1)}{x^2} + e^x \ln x$

51. $\qquad y = e^x\left(\cos \sqrt{2}\,x + \sin \sqrt{2}\,x\right)$

$y' = e^x\left(-\sqrt{2} \sin \sqrt{2}\,x + \sqrt{2} \cos \sqrt{2}\,x\right) + e^x\left(\cos \sqrt{2}\,x + \sin \sqrt{2}\,x\right)$

$\quad = e^x\left[(1 + \sqrt{2}) \cos \sqrt{2}\,x + (1 - \sqrt{2}) \sin \sqrt{2}\,x\right]$

$y'' = e^x\left[-(\sqrt{2} + 2) \sin \sqrt{2}\,x + (\sqrt{2} - 2) \cos \sqrt{2}\,x\right] + e^x\left[(1 + \sqrt{2}) \cos \sqrt{2}\,x + (1 - \sqrt{2}) \sin \sqrt{2}\,x\right]$

$\quad = e^x\left[(-1 - 2\sqrt{2}) \sin \sqrt{2}\,x + (-1 + 2\sqrt{2}) \cos \sqrt{2}\,x\right]$

$-2y' + 3y = -2e^x\left[(1 + \sqrt{2}) \cos \sqrt{2}\,x + (1 - \sqrt{2}) \sin \sqrt{2}\,x\right] + 3e^x\left[\cos \sqrt{2}\,x + \sin \sqrt{2}\,x\right]$

$\quad = e^x\left[(1 - 2\sqrt{2}) \cos \sqrt{2}\,x + (1 + 2\sqrt{2}) \sin \sqrt{2}\,x\right] = -y''$

Therefore, $-2y' + 3y = -y'' \Rightarrow y'' - 2y' + 3y = 0.$

52. $y = e^x(3\cos 2x - 4\sin 2x)$

$\quad\quad y' = e^x(-6\sin 2x - 8\cos 2x) + e^x(3\cos 2x - 4\sin 2x)$

$\quad\quad\quad = e^x(-10\sin 2x - 5\cos 2x) = -5e^x(2\sin 2x + \cos 2x)$

$\quad\quad y'' = -5e^x(4\cos 2x - 2\sin 2x) - 5e^x(2\sin 2x + \cos 2x) = -5e^x(5\cos 2x) = -25e^x\cos 2x$

$\quad y'' - 2y' = -5e^x(5\cos 2x) - 2(-5e^x)(2\sin 2x + \cos 2x) = -5e^x(3\cos 2x - 4\sin 2x) = -5y$

$\quad$ Therefore, $y'' - 2y' = -5y \Rightarrow y'' - 2y' + 5y = 0.$

53. $f(x) = \dfrac{1}{\sqrt{2\pi}} e^{-x^2/2}$

$\quad\quad f'(x) = \dfrac{1}{\sqrt{2\pi}}(-xe^{-x^2/2}) = 0$ when $x = 0$

$\quad\quad f''(x) = \dfrac{1}{\sqrt{2\pi}}(x^2 e^{-x^2/2} - e^{-x^2/2})$

$\quad\quad\quad = \dfrac{e^{-x^2/2}}{\sqrt{2\pi}}(x^2 - 1) = 0$ when $x = \pm 1$

$\quad$ Relative maximum: $\left(0, \ \dfrac{1}{\sqrt{2\pi}}\right)$

$\quad$ Points of inflection: $\left(\pm 1, \ \dfrac{1}{\sqrt{2e\pi}}\right)$

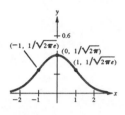

54. $f(x) = \dfrac{e^x - e^{-x}}{2}$

$\quad\quad f'(x) = \dfrac{e^x + e^{-x}}{2} > 0$

$\quad\quad f''(x) = \dfrac{e^x - e^{-x}}{2} = 0$ when $x = 0$

$\quad$ Point of inflection: $(0, 0)$

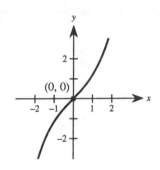

55. $f(x) = \dfrac{e^x + e^{-x}}{2}$

$\quad\quad f'(x) = \dfrac{e^x - e^{-x}}{2} = 0$ when $x = 0$

$\quad\quad f''(x) = \dfrac{e^x + e^{-x}}{2} > 0$

$\quad$ Relative minimum: $(0, 1)$

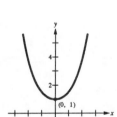

56. $f(x) = xe^{-x}$

$\quad\quad f'(x) = -xe^{-x} + e^{-x} = e^{-x}(1 - x) = 0$ when $x = 1$

$\quad\quad f''(x) = -e^{-x} + (-e^{-x})(1 - x) = e^{-x}(x - 2) = 0$ when $x = 2$

$\quad$ Relative maximum: $(1, \ e^{-1})$

$\quad$ Point of inflection: $(2, \ 2e^{-2})$

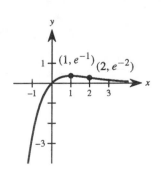

57. $f(x) = x^2 e^{-x}$

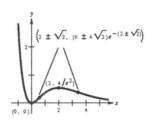

$f'(x) = -x^2 e^{-x} + 2xe^{-x} = xe^{-x}(2 - x) = 0$ when $x = 0,\ 2$

$f''(x) = -e^{-x}(2x - x^2) + e^{-x}(2 - 2x)$

$\qquad = e^{-x}(x^2 - 4x + 2) = 0$ when $x = 2 \pm \sqrt{2}$

Relative minimum: $(0, 0)$

Relative maximum: $(2,\ 4e^{-2})$

$x = 2 \pm \sqrt{2}$

$y = \left(2 \pm \sqrt{2}\right)^2 e^{-(2\pm\sqrt{2})}$

Points of inflection: $(3.414, 0.384)$, $(0.586, 0.191)$

58. $f(x) = -2 + e^{3x}(4 - 2x)$

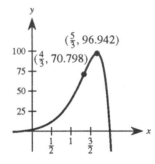

$f'(x) = e^{3x}(-2) + 3e^{3x}(4 - 2x) = e^{3x}(10 - 6x) = 0$ when $x = \frac{5}{3}$

$f''(x) = e^{3x}(-6) + 3e^{3x}(10 - 6x) = e^{3x}(24 - 18x) = 0$ when $x = \frac{4}{3}$

Relative maximum: $\left(\frac{5}{3},\ 96.942\right)$

Point of inflection: $\left(\frac{4}{3},\ 70.798\right)$

59. $f(x) = e^{x/2}$, $f(0) = 1$

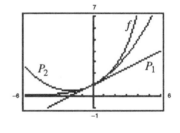

$f'(x) = \frac{1}{2}e^{x/2}$, $f'(0) = \frac{1}{2}$

$f''(x) = \frac{1}{4}e^{x/2}$, $f''(0) = \frac{1}{4}$

$P_1(x) = 1 + \frac{1}{2}(x - 0) = \frac{x}{2} + 1$, $P_1(0) = 1$

$P_1'(x) = \frac{1}{2}$, $P_1'(0) = \frac{1}{2}$

$P_2(x) = 1 + \frac{1}{2}(x - 0) + \frac{1}{8}(x - 0)^2 = \frac{x^2}{8} + \frac{x}{2} + 1$, $P_2(0) = 1$

$P_2'(x) = \frac{1}{4}x + \frac{1}{2}$, $P_2'(0) = \frac{1}{2}$

$P_2''(x) = \frac{1}{4}$, $P_2''(0) = \frac{1}{4}$

The values of f, P_1, P_2 and their first derivatives agree at $x = 0$. The values of the second derivatives of f and P_2 agree at $x = 0$.

60. $f(x) = e^{-x^2/2}, \quad f(0) = 1$

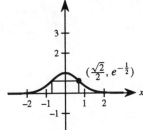

$f'(x) = -xe^{-x^2/2}, \quad f'(0) = 0$

$f''(x) = x^2 e^{-x^2/2} - e^{-x^2/2} = e^{-x^2/2}(x^2 - 1), \quad f''(0) = -1$

$P_1(x) = 1 + 0(x - 0) = 1, \quad P_1(0) = 1$

$P_1'(x) = 0, \quad P_1'(0) = 0$

$P_2(x) = 1 + 0(x - 0) - \dfrac{1}{2}(x - 0)^2 = 1 - \dfrac{x^2}{2}, \quad P_2(0) = 1$

$P_2'(x) = -x, \quad P_2'(0) = 0$

$P_2''(x) = -1, \quad P_2''(0) = -1$

The values of f, P_1, P_2 and their first derivatives agree at $x = 0$. The values of the second derivatives of f and P_2 agree at $x = 0$.

61. $A = 2xe^{-x^2}$

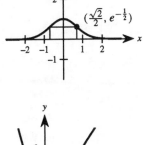

$\dfrac{dA}{dx} = -4x^2 e^{-x^2} + 2e^{-x^2}$

$\quad = 2e^{-x^2}(1 - 2x^2) = 0 \text{ when } x = \dfrac{\sqrt{2}}{2}$

$A = \sqrt{2}\, e^{-1/2}$

62. $y = e^{-x}$

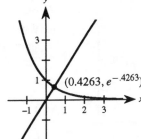

$y' = -e^{-x} \quad$ (Slope of tangent line)

$-\dfrac{1}{y'} = e^x \quad$ (Slope of normal line)

$y - e^{-x_0} = e^{x_0}(x - x_0)$

We want $(0, 0)$ to satisfy the equation:

$-e^{-x_0} = -x_0 e^{x_0}$

$1 = x_0 e^{2x_0}$

$x_0 e^{2x_0} - 1 = 0$

Solving by Newton's Method or using a computer, the solution is $x_0 \approx 0.4263$.

$(0.4263, \ e^{-0.4263})$

63. $e^{-x} = x \Rightarrow f(x) = x - e^{-x}$

$f'(x) = 1 + e^{-x}$

$$x_{n+1} = x_n - \frac{f(x_n)}{f'(x_n)} = x_n - \frac{x_n - e^{-x_n}}{1 + e^{-x_n}}$$

$x_1 = 1$

$$x_2 = x_1 - \frac{f(x_1)}{f'(x_1)} \approx 0.5379$$

$$x_3 = x_2 - \frac{f(x_2)}{f'(x_2)} \approx 0.5670$$

$$x_4 = x_3 - \frac{f(x_3)}{f'(x_3)} \approx 0.5671$$

We approximate the root of f to be $x = 0.567$.

64. $V = 15{,}000e^{-0.6286t}, \quad 0 \le t \le 10$

a.
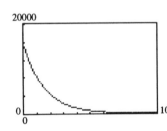

b. $\dfrac{dV}{dt} = -9429e^{-0.6286t}$

When $t = 1$, $\dfrac{dV}{dt} \approx -5028.84$.

When $t = 5$, $\dfrac{dV}{dt} \approx -406.89$.

c.
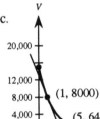

65. a.
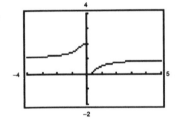

b. When x increases without bound, $1/x$ approaches zero, and $e^{1/x}$ approaches 1. Therefore, $f(x)$ approaches $2/(1 + 1) = 1$. Thus, $f(x)$ has a horizontal asymptote at $y = 1$. As x approaches zero from the right, $1/x$ approaches ∞, $e^{1/x}$ approaches ∞ and $f(x)$ approaches zero. As x approaches zero from the left, $1/x$ approaches $-\infty$, $e^{1/x}$ approaches zero, and $f(x)$ approaches 2. The limit does not exist since the left limit does not equal the right limit. Therefore, $x = 0$ is a nonremovable discontinuity.

66. $1.56e^{-0.22t} \cos 4.9t \le 0.25$ (3 inches equals one-fourth foot.) Using a graphing utility with Newton's Method, we have $t \ge 7.79$ seconds.

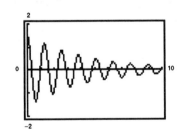

67. Let $u = 5x, \quad du = 5\,dx$.

$$\int e^{5x}5\,dx = e^{5x} + C$$

68. Let $u = -x^4, \quad du = -4x^3\,dx$.

$$\int e^{-x^4}(-4x^3)\,dx = e^{-x^4} + C$$

69. Let $u = -2x$, $du = -2\,dx$.

$$\int_0^1 e^{-2x}\,dx = -\frac{1}{2}\int_0^1 e^{-2x}(-2)\,dx = \left[-\frac{1}{2}e^{-2x}\right]_0^1 = \frac{1}{2}(1 - e^{-2}) = \frac{e^2 - 1}{2e^2}$$

70. Let $u = 1 - x$, $du = -dx$.

$$\int_1^2 e^{1-x}\,dx = -\int_1^2 e^{1-x}(-1)\,dx = \left[-e^{1-x}\right]_1^2 = 1 - e^{-1} = \frac{e - 1}{e}$$

71. Let $u = 1 + e^{-x}$, $du = -e^{-x}\,dx$.

$$\int \frac{e^{-x}}{1 + e^{-x}}\,dx = -\int \frac{-e^{-x}}{1 + e^{-x}}\,dx = -\ln(1 + e^{-x}) + C = \ln\left(\frac{e^x}{e^x + 1}\right) + C = x - \ln(e^x + 1) + C$$

72. Let $u = 1 + e^{2x}$, $du = 2e^{2x}\,dx$.

$$\int \frac{e^{2x}}{1 + e^{2x}}\,dx = \frac{1}{2}\int \frac{2e^{2x}}{1 + e^{2x}}\,dx = \frac{1}{2}\ln(1 + e^{2x}) + C$$

73. Let $u = \dfrac{3}{x}$, $du = -\dfrac{3}{x^2}\,dx$.

$$\int_1^3 \frac{e^{3/x}}{x^2}\,dx = -\frac{1}{3}\int_1^3 e^{3/x}\left(-\frac{3}{x^2}\right)dx = \left[-\frac{1}{3}e^{3/x}\right]_1^3 = \frac{e}{3}(e^2 - 1)$$

74. Let $u = \dfrac{-x^2}{2}$, $du = -x\,dx$.

$$\int_0^{\sqrt{2}} xe^{-x^2/2}\,dx = -\int_0^{\sqrt{2}} e^{-x^2/2}(-x)\,dx = \left[-e^{-x^2/2}\right]_0^{\sqrt{2}} = 1 - e^{-1} = \frac{e - 1}{e}$$

75. Let $u = 1 - e^x$, $du = -e^x\,dx$.

$$\int e^x\sqrt{1 - e^x}\,dx = -\int (1 - e^x)^{1/2}(-e^x)\,dx = -\frac{2}{3}(1 - e^x)^{3/2} + C$$

76. Let $u = e^x + e^{-x}$, $du = (e^x - e^{-x})\,dx$.

$$\int \frac{e^x - e^{-x}}{e^x + e^{-x}}\,dx = \ln(e^x + e^{-x}) + C$$

77. Let $u = e^x - e^{-x}$, $du = (e^x + e^{-x})\,dx$.

$$\int \frac{e^x + e^{-x}}{e^x - e^{-x}}\,dx = \ln|e^x - e^{-x}| + C$$

78. Let $u = e^x + e^{-x}$, $du = (e^x - e^{-x})\,dx$.

$$\int \frac{2e^x - 2e^{-x}}{(e^x + e^{-x})^2}\,dx = 2\int (e^x + e^{-x})^{-2}(e^x - e^{-x})\,dx$$

$$= \frac{-2}{e^x + e^{-x}} + C$$

79. $\displaystyle\int \frac{5 - e^x}{e^{2x}}\,dx = \int 5e^{-2x}\,dx - \int e^{-x}\,dx$

$$= -\frac{5}{2}e^{-2x} + e^{-x} + C$$

80. $\displaystyle\int \frac{e^{2x} + 2e^x + 1}{e^x}\,dx = \int (e^x + 2 + e^{-x})\,dx$

$$= e^x + 2x - e^{-x} + C$$

81. $\displaystyle\int e^{\sin \pi x} \cos \pi x\,dx = \frac{1}{\pi}\int e^{\sin \pi x}(\pi \cos \pi x)\,dx$

$$= \frac{1}{\pi}e^{\sin \pi x} + C$$

82. $\displaystyle\int e^{\tan 2x} \sec^2 2x\,dx = \frac{1}{2}\int e^{\tan 2x}(2\sec^2 2x)\,dx$

$$= \frac{1}{2}e^{\tan 2x} + C$$

83. $\displaystyle\int e^{-x}\tan(e^{-x})\,dx = -\int [\tan(e^{-x})](-e^{-x})\,dx$

$\qquad = \ln|\cos(e^{-x})| + C$

84. $\displaystyle\int \ln(e^{2x-1})\,dx = \int (2x-1)\,dx$

$\qquad = x^2 - x + C$

85. Let $u = ax^2$, $\ du = 2ax\,dx$.

$\quad y = \displaystyle\int xe^{ax^2}\,dx$

$\quad = \dfrac{1}{2a}\displaystyle\int e^{ax^2}(2ax)\,dx = \dfrac{1}{2a}e^{ax^2} + C$

86. $y = \displaystyle\int (e^x - e^{-x})^2\,dx$

$\quad = \displaystyle\int (e^{2x} - 2 + e^{-2x})\,dx$

$\quad = \dfrac{1}{2}e^{2x} - 2x - \dfrac{1}{2}e^{-2x} + C$

87. $f'(x) = \displaystyle\int \dfrac{1}{2}(e^x + e^{-x})\,dx = \dfrac{1}{2}(e^x - e^{-x}) + C_1$

$\quad f'(0) = C_1 = 0$

$\quad f(x) = \displaystyle\int \dfrac{1}{2}(e^x - e^{-x})\,dx = \dfrac{1}{2}(e^x + e^{-x}) + C_2$

$\quad f(0) = 1 + C_2 = 1 \Rightarrow C_2 = 0$

$\quad f(x) = \dfrac{1}{2}(e^x + e^{-x})$

88. $f'(x) = \displaystyle\int (\sin x + e^{2x})\,dx = -\cos x + \dfrac{1}{2}e^{2x} + C_1$

$\quad f'(0) = -1 + \dfrac{1}{2} + C_1 = \dfrac{1}{2} \Rightarrow C_1 = 1$

$\quad f'(x) = -\cos x + \dfrac{1}{2}e^{2x} + 1$

$\quad f(x) = \displaystyle\int \left(-\cos x + \dfrac{1}{2}e^{2x} + 1\right)dx$

$\qquad = -\sin x + \dfrac{1}{4}e^{2x} + x + C_2$

$\quad f(0) = \dfrac{1}{4} + C_2 = \dfrac{1}{4} \Rightarrow C_2 = 0$

$\quad f(x) = x - \sin x + \dfrac{1}{4}e^{2x}$

89. $\displaystyle\int_0^5 e^x\,dx = \left[e^x\right]_0^5 = e^5 - 1 \approx 147.413$

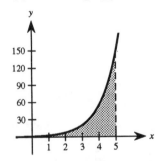

90. $\displaystyle\int_a^b e^{-x}\,dx = \left[-e^{-x}\right]_a^b = e^{-a} - e^{-b}$

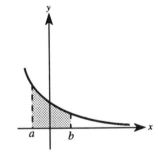

91. $\displaystyle\int_0^{\sqrt{2}} x e^{-(x^2/2)}\, dx = \left[-e^{-(x^2/2)} \right]_0^{\sqrt{2}}$

$$= -e^{-1} + 1 \approx 0.632$$

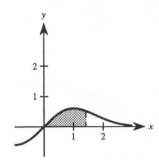

92. $\displaystyle\int_0^2 (e^{-2x} + 2)\, dx = \left[-\frac{1}{2} e^{-2x} + 2x \right]_0^2$

$$= -\frac{1}{2} e^{-4} + 4 + \frac{1}{2} \approx 4.491$$

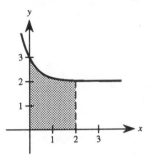

93. a. $\displaystyle\int_0^4 \sqrt{x}\, e^x\, dx, \quad n = 12$

    Midpoint Rule: 92.1898
    Trapezoidal Rule: 93.8371
    Simpson's Rule: 92.7385

b. $\displaystyle\int_0^2 2x e^{-x}\, dx, \quad n = 12$

    Midpoint Rule: 1.1906
    Trapezoidal Rule: 1.1827
    Simpson's Rule: 1.1880

94. $\displaystyle 0.0665 \int_{48}^{60} e^{-0.0139(t-48)^2}\, dt, \quad n = 24$

    Simpson's Rule: $0.4772 = 47.72\%$

95. $\displaystyle\int_0^x e^t\, dt \geq \int_0^x 1\, dt$

$$\left[e^t \right]_0^x \geq \left[t \right]_0^x$$

$$e^x - 1 \geq x \Rightarrow e^x \geq 1 + x \text{ for } x \geq 0$$

96. a. $e^x \geq 1 + x$

$$\int_0^x e^t\, dt \geq \int_0^x (1 + t)\, dt$$

$$\left[e^t \right]_0^x \geq \left[t + \left(\frac{t^2}{2} \right) \right]_0^x$$

$$e^x - 1 \geq x + \frac{x^2}{2} \Rightarrow e^x \geq 1 + x + \frac{x^2}{2}$$

$$x \geq 0$$

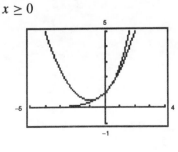

b. $e^x \geq 1 + x + \dfrac{x^2}{2}$

$$\int_0^x e^t\, dt \geq \int_0^x \left(1 + t + \frac{t^2}{2} \right) dt$$

$$\left[e^t \right]_0^x \geq t + \frac{t^2}{2} + \frac{t^3}{6} \Big]_0^x$$

$$e^x - 1 \geq x + \frac{x^2}{2} + \frac{x^3}{6} \Rightarrow$$

$$e^x \geq 1 + x + \frac{x^2}{2} + \frac{x^3}{6}$$

$$x \geq 0$$

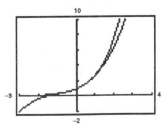

96. —CONTINUED—

c. $e^x \geq 1 + x + \dfrac{x^2}{2} + \dfrac{x^3}{6}$

$$\int_0^x e^t \, dt > \int_0^x \left(1 + t + \frac{t^2}{2} + \frac{t^3}{6}\right) dt$$

$$e^x - 1 \geq x + \frac{x^2}{2} + \frac{x^3}{6} + \frac{x^4}{24} \Rightarrow$$

$$e^x \geq 1 + x + \frac{x^2}{2} + \frac{x^3}{6} + \frac{x^4}{24}$$

$x \geq 0$

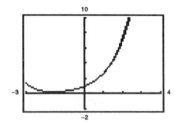

d. $e^x \geq 1 + x + \dfrac{x^2}{2} + \dfrac{x^3}{6} + \dfrac{x^4}{24}$

$$\int_0^x e^t \, dt \geq \int_0^x \left(1 + t + \frac{t^2}{2} + \frac{t^3}{6} + \frac{t^4}{24}\right) dt$$

$$e^x - 1 \geq x + \frac{x^2}{2} + \frac{x^3}{6} + \frac{x^4}{24} + \frac{x^5}{120}$$

$$e^x \geq 1 + x + \frac{x^2}{2} + \frac{x^3}{6} + \frac{x^4}{24} + \frac{x^5}{120}$$

$x \geq 0$

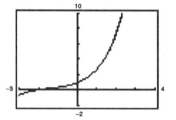

The higher the degree of the polynomial, the better the approximation to $y = e^x$ for values of x near zero.

97. $\ln \dfrac{e^a}{e^b} = \ln e^a - \ln e^b = a - b$

$\ln e^{a-b} = a - b$

Therefore, $\ln \dfrac{e^a}{e^b} = \ln e^{a-b}$ and since $y = \ln x$ is one-to-one, we have $\dfrac{e^a}{e^b} = e^{a-b}$

98. $f(x) = \dfrac{\ln x}{x}$

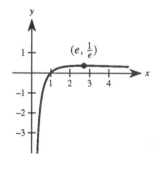

a. $f'(x) = \dfrac{1 - \ln x}{x^2} = 0$ when $x = e$.

On $(0, e)$, $f'(x) > 0 \Rightarrow f$ is increasing.
On (e, ∞), $f'(x) < 0 \Rightarrow f$ is decreasing.

b. For $e \leq A < B$, we have:

$$\frac{\ln A}{A} > \frac{\ln B}{B}$$

$$B \ln A > A \ln B$$

$$\ln A^B > \ln B^A$$

$$A^B > B^A$$

c. Since $e < \pi$, from part b we have $e^\pi > \pi^e$.

Section 5.5   Bases Other than *e* and Applications

1. $\log_2 \frac{1}{8} = \log_2 2^{-3} = -3$

2. $\log_{27} 9 = \log_{27} 27^{2/3} = \frac{2}{3}$

3. $\log_7 1 = 0$

4. $\log_a \dfrac{1}{a} = \log_a 1 - \log_a a = -1$

5. a. $2^3 = 8$

$\log_2 8 = 3$

b. $3^{-1} = \frac{1}{3}$

$\log_3 \frac{1}{3} = -1$

6. a. $27^{2/3} = 9$

$\log_{27} 9 = \frac{2}{3}$

b. $16^{3/4} = 8$

$\log_{16} 8 = \frac{3}{4}$

7. a. $\log_{10} 0.01 = -2$

$10^{-2} = 0.01$

b. $\log_{0.5} 8 = -3$

$0.5^{-3} = 8$

$\left(\dfrac{1}{2}\right)^{-3} = 8$

8. a. $\log_3 \frac{1}{9} = -2$

$3^{-2} = \frac{1}{9}$

b. $49^{1/2} = 7$

$\log_{49} 7 = \frac{1}{2}$

9. a. $\log_{10} 1000 = x$

$10^x = 1000$

$x = 3$

b. $\log_{10} 0.1 = x$

$10^x = 0.1$

$x = -1$

10. a. $\log_4 \frac{1}{64} = x$

$4^x = \frac{1}{64}$

$x = -3$

b. $\log_5 25 = x$

$5^x = 25$

$x = 2$

11. a. $\log_3 x = -1$

$3^{-1} = x$

$x = \frac{1}{3}$

b. $\log_2 x = -4$

$2^{-4} = x$

$x = \frac{1}{16}$

12. a. $\log_b 27 = 3$

$b^3 = 27$

$b = 3$

b. $\log_b 125 = 3$

$b^3 = 125$

$b = 5$

13. a.

$$x^2 - x = \log_5 25$$
$$x^2 - x = 2$$
$$x^2 - x - 2 = 0$$
$$(x + 1)(x - 2) = 0$$
$$x = -1 \text{ OR } x = 2$$

b. $3x + 5 = \log_2 64$

$$3x + 5 = 6$$
$$3x = 1$$
$$x = \frac{1}{3}$$

14. a. $\log_3 x + \log_3(x - 2) = 1$

$$\log_3[x(x - 2)] = 1$$
$$x(x - 2) = 3^1$$
$$x^2 - 2x - 3 = 0$$
$$(x + 1)(x - 3) = 0$$
$$x = -1 \text{ OR } x = 3$$

$x = 3$ is the only solution since the domain of the logarithmic function is the set of all *positive* real numbers.

b. $\log_{10}(x + 3) - \log_{10} x = 1$

$$\log_{10} \frac{x + 3}{x} = 1$$
$$\frac{x + 3}{x} = 10^1$$
$$x + 3 = 10x$$
$$3 = 9x$$
$$x = \frac{1}{3}$$

15. $y = 3^x$

x	-2	-1	0	1	2
y	$\frac{1}{9}$	$\frac{1}{3}$	1	3	9

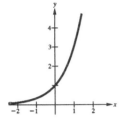

16. $y = 3^{x-1}$

x	-1	0	1	2	3
y	$\frac{1}{9}$	$\frac{1}{3}$	1	3	9

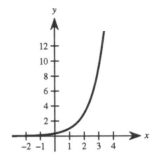

17. $y = \left(\dfrac{1}{3}\right)^x = 3^{-x}$

x	-2	-1	0	1	2
y	9	3	1	$\frac{1}{3}$	$\frac{1}{9}$

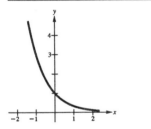

18. $y = 2^{x^2}$

x	-2	-1	0	1	2
y	16	2	1	2	16

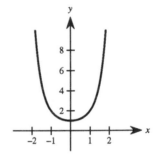

19. $h(x) = 5^{x-2}$

x	-1	0	1	2	3
y	$\frac{1}{125}$	$\frac{1}{25}$	$\frac{1}{5}$	1	5

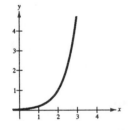

20. $y = 3^{-|x|}$

x	0	±1	±2
y	1	$\frac{1}{3}$	$\frac{1}{9}$

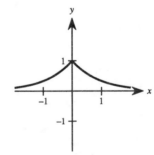

21. $f(x) = 4^x$

$g(x) = \log_4 x$

x	-2	-1	0	$\frac{1}{2}$	1
$f(x)$	$\frac{1}{16}$	$\frac{1}{4}$	1	2	4

x	$\frac{1}{16}$	$\frac{1}{4}$	1	2	4
$g(x)$	-2	-1	0	$\frac{1}{2}$	1

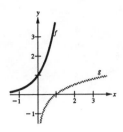

22. $f(x) = 3^x$

$g(x) = \log_3 x$

x	-2	-1	0	1	2
$f(x)$	$\frac{1}{9}$	$\frac{1}{3}$	1	3	9

x	$\frac{1}{9}$	$\frac{1}{3}$	1	3	9
$g(x)$	-2	-1	0	1	2

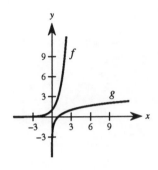

23.

x	1	2	8
y	0	1	3

a. y is an exponential function of x: False
b. y is a logarithmic function of x: True; $y = \log_2 x$
c. x is an exponential function of y: True; $2^y = x$
d. y is a linear function of x: False

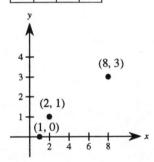

24. $f(x) = \log_{10} x$

a. Domain: $x > 0$

b. $\quad y = \log_{10} x$

$\quad\quad 10^y = x$

$\quad f^{-1}(x) = 10^x$

c. $\quad \log_{10} 1000 = \log_{10} 10^3 = 3$

$\quad \log_{10} 10{,}000 = \log_{10} 10^4 = 4$

If $1000 \le x \le 10{,}000$, then $3 \le f(x) \le 4$.

d. If $f(x) < 0$, then $0 < x < 1$.

e. $f(x) + 1 = \log_{10} x + \log_{10} 10$

$\quad\quad\quad\quad = \log_{10}(10x)$

x must have been increased by a factor of 10.

f. $\log_{10}\left(\dfrac{x_1}{x_2}\right) = \log_{10} x_1 - \log_{10} x_2$

$\quad\quad\quad\quad = 3n - n = 2n$

Thus, $x_1/x_2 = 10^{2n} = 100^n$.

25. $f(x) = 4^x$

$f'(x) = (\ln 4)4^x$

26. $g(x) = 2^{-x}$

$g'(x) = -(\ln 2)2^{-x}$

27. $y = 5^{x-2}$

$\dfrac{dy}{dx} = (\ln 5)5^{x-2}$

28. $y = x(7^{-3x})$

$\dfrac{dy}{dx} = x\left[-3(\ln 7)7^{-3x}\right] + 7^{-3x}$

$= 7^{-3x}[-3x(\ln 7) + 1]$

$= 7^{-3x}(1 - 3x \ln 7)$

29. $g(t) = t^2 2^t$

$g'(t) = t^2(\ln 2)2^t + (2t)2^t$

$= t2^t(t \ln 2 + 2)$

$= 2^t t(2 + t \ln 2)$

30. $f(t) = \dfrac{3^{2t}}{t}$

$f'(t) = \dfrac{t(2\ln 3)3^{2t} - 3^{2t}}{t^2}$

$= \dfrac{3^{2t}(2t \ln 3 - 1)}{t^2}$

31. $y = \log_3 x$

$\dfrac{dy}{dx} = \dfrac{1}{x \ln 3}$

32. $y = \log_{10}(2x) = \log_{10} 2 + \log_{10} x$

$\dfrac{dy}{dx} = 0 + \dfrac{1}{x \ln 10} = \dfrac{1}{x \ln 10}$

33. $f(x) = \log_2 \dfrac{x^2}{x-1}$

$= 2\log_2 x - \log_2(x-1)$

$f'(x) = \dfrac{2}{x \ln 2} - \dfrac{1}{(x-1)\ln 2}$

$= \dfrac{x-2}{(\ln 2)x(x-1)}$

34. $h(x) = \log_3 \dfrac{x\sqrt{x-1}}{2}$

$= \log_3 x + \dfrac{1}{2}\log_3(x-1) - \log_3 2$

$h'(x) = \dfrac{1}{x \ln 3} + \dfrac{1}{2}\cdot\dfrac{1}{(x-1)\ln 3} - 0$

$= \dfrac{1}{\ln 3}\left[\dfrac{1}{x} + \dfrac{1}{2(x-1)}\right]$

$= \dfrac{1}{\ln 3}\left[\dfrac{3x-2}{2x(x-1)}\right]$

35. $y = \log_5 \sqrt{x^2 - 1} = \dfrac{1}{2}\log_5(x^2 - 1)$

$\dfrac{dy}{dx} = \dfrac{1}{2}\cdot\dfrac{2x}{(x^2-1)\ln 5} = \dfrac{x}{(x^2-1)\ln 5}$

36. $y = \log_{10}\dfrac{x^2 - 1}{x}$

$= \log_{10}(x^2 - 1) - \log_{10} x$

$\dfrac{dy}{dx} = \dfrac{2x}{(x^2-1)\ln 10} - \dfrac{1}{x \ln 10}$

$= \dfrac{1}{\ln 10}\left[\dfrac{2x}{x^2-1} - \dfrac{1}{x}\right]$

$= \dfrac{1}{\ln 10}\left[\dfrac{x^2 + 1}{x(x^2 - 1)}\right]$

37. $y = x^{2/x}$

$\ln y = \dfrac{2}{x}\ln x$

$\dfrac{1}{y}\left(\dfrac{dy}{dx}\right) = \dfrac{2}{x}\left(\dfrac{1}{x}\right) + \ln x\left(-\dfrac{2}{x^2}\right)$

$\dfrac{dy}{dx} = \dfrac{2y}{x^2}(1 - \ln x) = 2x^{(2/x)-2}(1 - \ln x)$

38. $y = x^{x-1}$

$\ln y = (x-1)(\ln x)$

$\dfrac{1}{y}\left(\dfrac{dy}{dx}\right) = (x-1)\left(\dfrac{1}{x}\right) + \ln x$

$\dfrac{dy}{dx} = y\left[\dfrac{x-1}{x} + \ln x\right]$

$= x^{x-2}(x - 1 + x \ln x)$

39.
$$y = (x-2)^{x+1}$$
$$\ln y = (x+1)\ln(x-2)$$
$$\frac{1}{y}\left(\frac{dy}{dx}\right) = (x+1)\left(\frac{1}{x-2}\right) + \ln(x-2)$$
$$\frac{dy}{dx} = y\left[\frac{x+1}{x-2} + \ln(x-2)\right]$$
$$= (x-2)^{x+1}\left[\frac{x+1}{x-2} + \ln(x-2)\right]$$

40.
$$y = (1+x)^{1/x}$$
$$\ln y = \frac{1}{x}\ln(1+x)$$
$$\frac{1}{y}\left(\frac{dy}{dx}\right) = \frac{1}{x}\left(\frac{1}{1+x}\right) + \ln(1+x)\left(-\frac{1}{x^2}\right)$$
$$\frac{dy}{dx} = \frac{y}{x}\left[\frac{1}{x+1} - \frac{\ln(x+1)}{x}\right]$$
$$= \frac{(1+x)^{1/x}}{x}\left[\frac{1}{x+1} - \frac{\ln(x+1)}{x}\right]$$

41. $f(x) = \log_2 x \Rightarrow f'(x) = \dfrac{1}{x\ln 2}$

$g(x) = x^x \Rightarrow g'(x) = x^x(1 + \ln x)$

(**Note:** Let $y = g(x)$. Then: $\ln y = \ln x^x$
$$\ln y = x \ln x$$
$$\frac{1}{y}y' = x\cdot\frac{1}{x} + \ln x$$
$$y' = y(1 + \ln x)$$
$$y' = x^x(1 + \ln x) = g'(x).)$$

$h(x) = x^2 \Rightarrow h'(x) = 2x$

$k(x) = 2^x \Rightarrow k'(x) = (\ln 2)2^x$

From greatest to smallest rate of growth: $g(x),\ k(x),\ h(x),\ f(x)$

42. $f(x) = a^x$

a. $f(u+v) = a^{u+v} = a^u a^v = f(u)f(v)$

b. $f(2x) = a^{2x} = (a^x)^2 = [f(x)]^2$

43. $C(t) = P(1.05)^t$

a. $C(10) = 24.95(1.05)^{10}$
$$\approx \$40.64$$

b. $\dfrac{dC}{dt} = P(\ln 1.05)(1.05)^t$

When $t = 1$: $\dfrac{dC}{dt} \approx 0.051P$

When $t = 8$: $\dfrac{dC}{dt} \approx 0.072P$

c. $\dfrac{dC}{dt} = (\ln 1.05)[P(1.05)^t]$
$$= (\ln 1.05)C(t)$$
The constant of proportionality is $\ln 1.05$.

44. $V(t) = 20,000 \left(\dfrac{3}{4}\right)^t$

a.

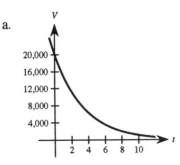

$$V(2) = 20,000 \left(\dfrac{3}{4}\right)^2 = \$11,250$$

b. $\dfrac{dV}{dt} = 20,000 \left(\ln \dfrac{3}{4}\right) \left(\dfrac{3}{4}\right)^t$

When $t = 1$: $\dfrac{dV}{dt} \approx -4315.23$

When $t = 4$: $\dfrac{dV}{dt} \approx -1820.49$

c.

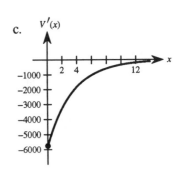

Horizontal asymptote: $y = 0$

As the car ages, it is worth less each year and depreciates less each year, but the value of the car will never reach \$0.

45. $1000(1 + 0.075)^{10} \approx \$2061.03$

$1000 \left(1 + \dfrac{0.075}{2}\right)^{20} \approx \$2088.15$

$1000 \left(1 + \dfrac{0.075}{4}\right)^{40} \approx \$2102.35$

$1000 \left(1 + \dfrac{0.075}{12}\right)^{120} \approx \$2112.06$

$1000 \left(1 + \dfrac{0.075}{365}\right)^{3650} \approx \$2116.84$

$1000e^{0.75} \approx \$2117.00$

n	1	2	4	12	365	Continuous compounding
A	\$2061.03	\$2088.15	\$2102.35	\$2112.06	\$2116.84	\$2117.00

46. $2500(1 + 0.12)^{20} \approx \$24,115.73$

$2500 \left(1 + \dfrac{0.12}{2}\right)^{40} \approx \$25,714.29$

$2500 \left(1 + \dfrac{0.12}{4}\right)^{80} \approx \$26,602.23$

$2500 \left(1 + \dfrac{0.12}{12}\right)^{240} \approx \$27,231.38$

$2500 \left(1 + \dfrac{0.12}{365}\right)^{7300} \approx \$27,547.07$

$2500e^{2.4} \approx \$27,557.94$

n	1	2	4	12	365	Continuous compounding
A	\$24,115.73	\$25,714.29	\$26,602.23	\$27,231.38	\$27,547.07	\$27,557.94

47. $1000(1 + 0.10)^{30} \approx \$17,449.40$

$$1000\left(1 + \frac{0.10}{2}\right)^{60} \approx \$18,679.19$$

$$1000\left(1 + \frac{0.10}{4}\right)^{120} \approx \$19,358.15$$

$$1000\left(1 + \frac{0.10}{12}\right)^{360} \approx \$19,837.40$$

$$1000\left(1 + \frac{0.10}{365}\right)^{10,950} \approx \$20,077.29$$

$1000e^3 \approx \$20,085.54$

n	1	2	4	12	365	Continuous compounding
A	\$17,449.40	\$18,679.19	\$19,358.15	\$19,837.40	\$20,077.29	\$20,085.54

48. $2500(1 + 0.10)^{40} \approx \$113,148.14$

$$2500\left(1 + \frac{0.10}{2}\right)^{80} \approx \$123,903.60$$

$$2500\left(1 + \frac{0.10}{4}\right)^{160} \approx \$129,944.67$$

$$2500\left(1 + \frac{0.10}{12}\right)^{480} \approx \$134,251.66$$

$$2500\left(1 + \frac{0.10}{365}\right)^{14,600} \approx \$136,420.62$$

$2500e^4 \approx \$136,495.38$

n	1	2	4	12	365	Continuous compounding
A	\$113,148.14	\$123,903.60	\$129,944.67	\$134,251.66	\$136,420.62	\$136,495.38

49. $100,000 = Pe^{0.09t}$

$P = 100,000e^{-0.09t}$

t	1	10	20	30	40	50
P	\$91,393.12	\$40,656.97	\$16,529.89	\$6720.55	\$2732.37	\$1110.90

50. $100,000 = Pe^{0.12t}$

$P = 100,000e^{-0.12t}$

t	1	10	20	30	40	50
P	\$88,692.04	\$30,119.42	\$9071.80	\$2732.37	\$822.97	\$247.88

51. $100{,}000 = P\left(1 + \dfrac{0.10}{12}\right)^{12t}$

$P = 100{,}000\left(1 + \dfrac{0.10}{12}\right)^{-12t}$

t	1	10	20	30	40	50
P	\$90,521.24	\$36,940.70	\$13,646.15	\$5040.98	\$1862.17	\$687.90

52. $100{,}000 = P\left(1 + \dfrac{0.07}{365}\right)^{365t}$

$P = 100{,}000\left(1 + \dfrac{0.07}{365}\right)^{-365t}$

t	1	10	20	30	40	50
P	\$93,240.01	\$49,661.86	\$24,663.01	\$12,248.11	\$6082.64	\$3020.75

53. a. $A = 20{,}000\left(1 + \dfrac{0.09}{365}\right)^{(365)(8)} \approx \$41{,}085.02$

b. $A = \$30{,}000$

c. $A = 8000\left(1 + \dfrac{0.09}{365}\right)^{(365)(8)} + 20{,}000\left(1 + \dfrac{0.09}{365}\right)^{(365)(4)} \approx \$45{,}099.32$

d. $A = 9000\left[\left(1 + \dfrac{0.09}{365}\right)^{(365)(8)} + \left(1 + \dfrac{0.09}{365}\right)^{(365)(4)} + 1\right] \approx \$40{,}387.65$

Take option c.

54. $A = 25{,}000e^{(0.0875)(25)} \approx \$222{,}822.57$

55. Let $P = \$100, \quad 0 \le t \le 20.$

a. $\quad A = 100e^{0.06t}$

$A(20) \approx \$332.01$

b. $\quad A = 100e^{0.09t}$

$A(20) \approx \$604.96$

c. $\quad A = 100e^{0.12t}$

$A(20) \approx \$1102.32$

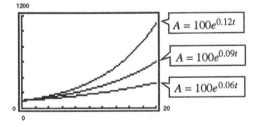

56. a. $\displaystyle\lim_{t \to \infty} 6.7e^{(-48.1)/t} = 6.7e^0 = 6.7$ million ft^3

b. $\quad V' = \dfrac{322.27}{t^2}e^{-(48.1)/t}$

$V'(20) \approx 0.073$ million ft^3/yr

$V'(60) \approx 0.040$ million ft^3/yr

57. $p(t) = \dfrac{10,000}{1 + 19e^{-t/5}}$

$p'(t) = \dfrac{e^{-t/5}}{(1 + 19e^{t/5})^2}\left(\dfrac{19}{5}\right)(10,000)$

$= \dfrac{38,000e^{-t/5}}{(1 + 19e^{-t/5})^2}$

$p'(1) \approx 113.5$ fish/month

$p'(10) \approx 403.2$ fish/month

$p''(t) = -\dfrac{38,000}{5}(e^{-t/5})\left[\dfrac{1 - 19e^{-t/5}}{(1 + 19e^{-t/5})^3}\right] = 0$

$19e^{-t/5} = 1$

$\dfrac{t}{5} = \ln 19$

$t = 5 \ln 19 \approx 14.72$

58. a. $\displaystyle\lim_{n\to\infty} \dfrac{0.83}{1 + e^{-0.2n}} = 0.83 = 83\%$

b. $P' = \dfrac{0.166e^{-0.2n}}{(1 + e^{-0.2n})^2}$

$P'(3) \approx 0.038$

$P'(10) \approx 0.017$

59. $y = \dfrac{300}{3 + 17e^{-1.57x}}, \quad x \ge 0$

a.

b. If $x = 2$, then $y \approx 80.3\%$.

c. $66\frac{2}{3} = \dfrac{300}{3 + 17e^{-1.57x}}$

$3 + 17e^{-1.57x} = 4.5$

$e^{-1.57x} = \dfrac{1.5}{17}$

$x = \dfrac{\ln(1.5/17)}{-1.57}$

≈ 1.546 or 1546 egg masses

d. $y = 300(3 + 17e^{-1.57x})^{-1}$

$y' = -300(3 + 17e^{-1.57x})^{-2}(-26.69e^{-1.57x}) = 8007e^{-1.57x}(3 + 17e^{-1.57x})^{-2}$

$y'' = 8007\left[-2e^{-1.57x}(3 + 17e^{-1.57x})^{-3}(-26.69e^{-1.57x}) + (3 + 17e^{-1.57x})^{-2}(-1.57e^{-1.57x})\right]$

$= 8007e^{-1.57x}(3 + 17e^{-1.57x})^{-3}[53.38e^{-1.57x} - 1.57(3 + 17e^{-1.57x})]$

$= \dfrac{8007e^{-1.57x}(26.69e^{-1.57x} - 4.71)}{(3 + 17e^{-1.57x})^3} = 0$

$26.69e^{-1.57x} - 4.71 = 0$

$e^{-1.57x} = \dfrac{4.71}{26.69}$

$x = \dfrac{\ln(4.71/26.69)}{-1.57} \approx 1.105$ or 1105 egg masses

60. $\displaystyle\int 3^x\, dx = \dfrac{3^x}{\ln 3} + C$

61. $\displaystyle\int 4^{-x}\, dx = -\dfrac{4^{-x}}{\ln 4} + C$

62. $\displaystyle\int_{-1}^{2} 2^x\, dx = \left[\dfrac{2^x}{\ln 2}\right]_{-1}^{2}$

$= \dfrac{1}{\ln 2}\left[4 - \dfrac{1}{2}\right]$

$= \dfrac{7}{2 \ln 2} = \dfrac{7}{\ln 4}$

63.
$$\int_{-2}^{0} (3^3 - 5^2)\, dx = \int_{-2}^{0} (27 - 25)\, dx$$
$$= \int_{-2}^{0} 2\, dx$$
$$= \Big[2x \Big]_{-2}^{0} = 4$$

64.
$$\int x5^{-x^2}\, dx = -\frac{1}{2} \int 5^{-x^2}(-2x)\, dx$$
$$= -\left(\frac{1}{2} \right) \frac{5^{-x^2}}{\ln 5} + C$$
$$= \frac{-1}{2\ln 5}(5^{-x^2}) + C$$

65.
$$\int (3 - x)7^{(3-x)^2}\, dx = -\frac{1}{2} \int -2(3 - x)7^{(3-x)^2}\, dx$$
$$= -\frac{1}{2\ln 7}\Big[7^{(3-x)^2} \Big] + C$$

66.
$$A = \int_{0}^{3} 3^x\, dx = \left[\frac{3^x}{\ln 3} \right]_{0}^{3} = \frac{26}{\ln 3} \approx 23.666$$

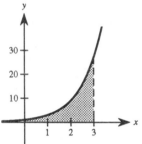

67. Average
$$= \frac{1}{2 - 0} \int_{0}^{2} 2500e^{0.12t}\, dt$$
$$= \left[\frac{1250}{0.12}e^{0.12t} \right]_{0}^{2}$$
$$\approx \$2825.51$$

68.
$$P = \int_{0}^{10} 2000e^{-0.10t}\, dt$$
$$= \left[\frac{2000}{-0.10}e^{-0.10t} \right]_{0}^{10}$$
$$\approx \$12,642.41$$

69.

x	1	10^{-1}	10^{-2}	10^{-4}	10^{-6}
$(1 + x)^{1/x}$	2	2.594	2.705	2.718	2.718

70.

t	0	1	2	3	4
y	1200	720	432	259.20	155.52

$y = C(k^t)$
When $t = 0$, $y = 1200 \Rightarrow C = 1200$.
$y = 1200(k^t)$
$$\frac{720}{1200} = 0.6, \quad \frac{432}{720} = 0.6, \quad \frac{259.20}{432} = 0.6, \quad \frac{155.52}{259.20} = 0.6$$
Let $k = 0.6$.
$y = 1200(0.6)^t$

71.

t	0	1	2	3	4
y	600	630	661.50	694.58	729.30

$y = C(k^t)$

When $t = 0$, $y = 600 \Rightarrow C = 600$.

$y = 600(k^t)$

$\dfrac{630}{600} = 1.05, \quad \dfrac{661.50}{630} = 1.05, \quad \dfrac{694.58}{661.50} \approx 1.05, \quad \dfrac{729.30}{694.58} \approx 1.05$

Let $k = 1.05$.

$y = 600(1.05)^t$

72. False. e is an irrational number.

73. True

$$f(e^{n+1}) - f(e^n) = \ln e^{n+1} - \ln e^n$$
$$= n + 1 - n$$
$$= 1$$

74. True

$$f(g(x)) = 2 + e^{\ln(x-2)}$$
$$= 2 + x - 2 = x$$
$$g(f(x)) = \ln(2 + e^x - 2)$$
$$= \ln e^x = x$$

75. True

$$\frac{d^n y}{dx^n} = Ce^x$$
$$= y \text{ for } n = 1, 2, 3, \ldots$$

76. True

$$\frac{d}{dx}[e^x] = e^x \text{ and}$$
$$\frac{d}{dx}[e^{-x}] = -e^{-x}$$
$$e^x = e^{-x} \text{ when } x = 0.$$
$$(e^0)(-e^{-0}) = -1$$

77. True

$$f(x) = g(x)e^x = 0 \Rightarrow$$
$$g(x) = 0 \text{ since } e^x > 0$$
$$\text{for all } x.$$

78.

$$\frac{dy}{dt} = \frac{8}{25} y \left(\frac{5}{4} - y \right)$$

$$\frac{dy}{y[(5/4) - y]} = \frac{8}{25} dt \Rightarrow \frac{4}{5} \int \left(\frac{1}{y} + \frac{1}{(5/4) - y} \right) dy = \int \frac{8}{25} dt \Rightarrow$$

$$\ln y - \ln \left(\frac{5}{4} - y \right) = \frac{2}{5} t + C$$

$$\ln \left(\frac{y}{(5/4) - y} \right) = \frac{2}{5} t + C$$

$$\frac{y}{(5/4) - y} = e^{(2/5)t + C} = C_1 e^{(2/5)t}$$

$$y(0) = 1 \Rightarrow C_1 = 4 \Rightarrow 4e^{(2/5)t} = \frac{y}{(5/4) - y}$$

$$\Rightarrow 4e^{(2/5)t} \left(\frac{5}{4} - y \right) = y \Rightarrow 5e^{(2/5)t} = 4e^{(2/5)t} y + y = (4e^{(2/5)t} + 1)y$$

$$\Rightarrow y = \frac{5e^{(2/5)t}}{4e^{(2/5)t} + 1} = \frac{5}{4 + e^{-0.4t}} = \frac{1.25}{1 + 0.25e^{-0.4t}}$$

Section 5.6   Differential Equations: Growth and Decay

1. $y' = \dfrac{5x}{y}$

$yy' = 5x$

$\displaystyle \int yy' \, dx = \int 5x \, dx$

$\displaystyle \int y \, dy = \int 5x \, dx$

$\dfrac{1}{2}y^2 = \dfrac{5}{2}x^2 + C_1$

$y^2 - 5x^2 = C$

2. $y' = \dfrac{\sqrt{x}}{2y}$

$2yy' = \sqrt{x}$

$\displaystyle \int 2yy' \, dx = \int \sqrt{x} \, dx$

$\displaystyle \int 2y \, dy = \int \sqrt{x} \, dx$

$y^2 = \dfrac{2}{3}x^{3/2} + C_1$

$3y^2 - 2x^{3/2} = C$

3. $y' = \sqrt{x}\, y$

$\dfrac{y'}{y} = \sqrt{x}$

$\displaystyle \int \dfrac{y'}{y} \, dx = \int \sqrt{x} \, dx$

$\displaystyle \int \dfrac{dy}{y} = \int \sqrt{x} \, dx$

$\ln y = \dfrac{2}{3}x^{3/2} + C_1$

$y = e^{(2/3)x^{3/2}+C_1}$

$\quad = e^{C_1}e^{(2/3)x^{3/2}}$

$\quad = Ce^{(2/3)x^{3/2}}$

4. $y' = x(1+y)$

$\dfrac{y'}{1+y} = x$

$\displaystyle \int \dfrac{y'}{1+y} \, dx = \int x \, dx$

$\displaystyle \int \dfrac{dy}{1+y} = \int x \, dx$

$\ln(1+y) = \dfrac{x^2}{2} + C_1$

$1+y = e^{(x^2/2)+C_1}$

$y = e^{C_1}e^{x^2/2} - 1$

$\quad = Ce^{x^2/2} - 1$

5. $(1+x^2)y' - 2xy = 0$

$y' = \dfrac{2xy}{1+x^2}$

$\dfrac{y'}{y} = \dfrac{2x}{1+x^2}$

$\displaystyle \int \dfrac{y'}{y} \, dx = \int \dfrac{2x}{1+x^2} \, dx$

$\displaystyle \int \dfrac{dy}{y} = \int \dfrac{2x}{1+x^2} \, dx$

$\ln y = \ln(1+x^2) + C_1$

$\ln y = \ln(1+x^2) + \ln C$

$\ln y = \ln C(1+x^2)$

$y = C(1+x^2)$

6. $xy + y' = 100x$

$y' = 100x - xy = x(100 - y)$

$\dfrac{y'}{100 - y} = x$

$\displaystyle \int \dfrac{y'}{100 - y} \, dx = \int x \, dx$

$\displaystyle \int \dfrac{1}{100 - y} \, dy = \int x \, dx$

$-\ln(100 - y) = \dfrac{x^2}{2} + C_1$

$\ln(100 - y) = -\dfrac{x^2}{2} - C_1$

$100 - y = e^{-(x^2/2)-C_1}$

$-y = e^{-C_1}e^{-x^2/2} - 100$

$y = 100 - Ce^{-x^2/2}$

7.
$$\frac{dQ}{dt} = \frac{k}{t^2}$$

$$\int \frac{dQ}{dt}\, dt = \int \frac{k}{t^2}\, dt$$

$$\int dQ = -\frac{k}{t} + C$$

$$Q = -\frac{k}{t} + C$$

8.
$$\frac{dP}{dt} = k(10 - t)$$

$$\int \frac{dP}{dt}\, dt = \int k(10 - t)\, dt$$

$$\int dP = -\frac{k}{2}(10 - t)^2 + C$$

$$P = -\frac{k}{2}(10 - t)^2 + C$$

9.
$$\frac{dN}{ds} = k(250 - s)$$

$$\int \frac{dN}{ds}\, ds = \int k(250 - s)\, ds$$

$$\int dN = -\frac{k}{2}(250 - s)^2 + C$$

$$N = -\frac{k}{2}(250 - s)^2 + C$$

10.
$$\frac{dy}{dx} = kx(L - y)$$

$$\frac{1}{L - y}\frac{dy}{dx} = kx$$

$$\int \frac{1}{L - y}\frac{dy}{dx}\, dx = \int kx\, dx$$

$$\int \frac{1}{L - y}\, dy = \frac{kx^2}{2} + C_1$$

$$-\ln(L - y) = \frac{kx^2}{2} + C_1$$

$$L - y = e^{-(kx^2/2) - C_1}$$

$$-y = -L + e^{-C_1}e^{-kx^2/2}$$

$$y = L - Ce^{-kx^2/2}$$

11. $\dfrac{dy}{dt} = \dfrac{1}{2}t$, (0, 10)

$$\int dy = \int \frac{1}{2}t\, dt$$

$$y = \frac{1}{4}t^2 + C$$

$$10 = \frac{1}{4}(0)^2 + C \Rightarrow C = 10$$

$$y = \frac{1}{4}t^2 + 10$$

12. $\dfrac{dy}{dt} = -\dfrac{3}{4}\sqrt{t}$, (0, 10)

$$\int dy = \int -\frac{3}{4}\sqrt{t}\, dt$$

$$y = -\frac{1}{2}t^{3/2} + C$$

$$10 = -\frac{1}{2}(0)^{3/2} + C \Rightarrow C = 10$$

$$y = -\frac{1}{2}t^{3/2} + 10$$

13. $\dfrac{dy}{dt} = -\dfrac{1}{2}y$, (0, 10)

$$\int \frac{dy}{y} = \int -\frac{1}{2}\, dt$$

$$\ln y = -\frac{1}{2}t + C_1$$

$$y = e^{-(t/2) + C_1} = e^{C_1}e^{-t/2} = Ce^{-t/2}$$

$$10 = Ce^0 \Rightarrow C = 10$$

$$y = 10e^{-t/2}$$

14. $\dfrac{dy}{dt} = \dfrac{3}{4}y$, (0, 10)

$$\int \frac{dy}{y} = \int \frac{3}{4}\, dt$$

$$\ln y = \frac{3}{4}t + C_1$$

$$y = e^{(3/4)t + C_1}$$

$$= e^{C_1}e^{(3/4)t} = Ce^{3t/4}$$

$$10 = Ce^0 \Rightarrow C = 10$$

$$y = 10e^{3t/4}$$

15. $y = Ce^{kt}$, $\left(0, \frac{1}{2}\right)$, $(5, 5)$

$$C = \frac{1}{2}$$

$$y = \frac{1}{2}e^{kt}$$

$$5 = \frac{1}{2}e^{5k}$$

$$k = \frac{\ln 10}{5} \approx 0.4605$$

$$y = \frac{1}{2}e^{0.4605t}$$

16. $y = Ce^{kt}$, $(0, 4)$, $\left(5, \frac{1}{2}\right)$

$$C = 4$$

$$y = 4e^{kt}$$

$$\frac{1}{2} = 4e^{5k}$$

$$k = \frac{\ln(1/8)}{5} \approx -0.4159$$

$$y = 4e^{-0.4159t}$$

17. $y = Ce^{kt}$, $(1, 1)$, $(5, 5)$

$$1 = Ce^{k}$$

$$5 = Ce^{5k}$$

$$5Ce^{k} = Ce^{5k}$$

$$5e^{k} = e^{5k}$$

$$5 = e^{4k}$$

$$k = \frac{\ln 5}{4} \approx 0.4024$$

$$y = Ce^{0.4024t}$$

$$1 = Ce^{0.4024}$$

$$C \approx 0.6687$$

$$y = 0.6687e^{0.4024t}$$

18. $y = Ce^{kt}$, $\left(3, \frac{1}{2}\right)$, $(4, 5)$

$$\tfrac{1}{2} = Ce^{3k}$$

$$5 = Ce^{4k}$$

$$2Ce^{3k} = \tfrac{1}{5}Ce^{4k}$$

$$10e^{3k} = e^{4k}$$

$$10 = e^{k}$$

$$k = \ln 10 \approx 2.3026$$

$$y = Ce^{2.3026t}$$

$$5 = Ce^{2.3026(4)}$$

$$C \approx 0.0005$$

$$y = 0.0005e^{2.3026t}$$

19. Since the initial quantity is 10 grams, $y = 10e^{[\ln(1/2)/1620]t}$. When $t = 1000$, $y = 10e^{[\ln(1/2)/1620](1000)} \approx 6.52$ grams. When $t = 10{,}000$, $y = 10e^{[\ln(1/2)/1620](10{,}000)} \approx 0.14$ gram.

20. Since $y = Ce^{[\ln(1/2)/1620]t}$, we have $1.5 = Ce^{[\ln(1/2)/1620](1000)} \Rightarrow C \approx 2.30$ which implies that the initial quantity is 2.30 grams. When $t = 10{,}000$, we have $y = 2.30e^{[\ln(1/2)/1620](10{,}000)} \approx 0.03$ gram.

21. Since $y = Ce^{[\ln(1/2)/5730]t}$, we have $2.0 = Ce^{[\ln(1/2)/5730](10{,}000)} \Rightarrow C \approx 6.70$ which implies that the initial quantity is 6.70 grams. When $t = 1000$, we have $y = 6.70e^{[\ln(1/2)/5730](1000)} \approx 5.94$ grams.

22. Since the initial quantity is 3.0 grams, we have $y = 3.0e^{[\ln(1/2)/5730]t}$. When $t = 1000$, $y = 3.0e^{[\ln(1/2)/5730](1000)} \approx$ 2.66 grams. When $t = 10{,}000$, $y = 3.0e^{[\ln(1/2)/5730](10{,}000)} \approx 0.89$ gram.

23. Since $y = Ce^{[\ln(1/2)/24{,}360]t}$, we have $2.1 = Ce^{[\ln(1/2)/24{,}360](1000)} \Rightarrow C \approx 2.16$. Thus, the initial quantity is 2.16 grams. When $t = 10{,}000$, $y = 2.16e^{[\ln(1/2)/24{,}360](10{,}000)} \approx 1.63$ grams.

24. Since $y = Ce^{[\ln(1/2)/24{,}360]t}$, we have $0.4 = Ce^{[\ln(1/2)/24{,}360](10{,}000)} \Rightarrow C \approx 0.53$ which implies that the initial quantity is 0.53 gram. When $t = 1000$, we have $y = 0.53e^{[\ln(1/2)/24{,}360](1000)} \approx 0.52$ gram.

25. Since $\dfrac{dy}{dx} = ky$, $y = Ce^{kt}$ or $y = y_0 e^{kt}$.

$$\frac{1}{2} y_0 = y_0 e^{1620k}$$

$$k = \frac{-\ln 2}{1620}$$

$$y = y_0 e^{-(\ln 2)t/1620}$$

When $t = 100$, $y = y_0 e^{-(\ln 2)/16.2} \approx y_0(0.9581)$.
Therefore, 95.81% of the present amount still exists.

27. Since $A = 1000e^{0.12t}$, the time to double is given by $2000 = 1000e^{0.12t}$ and we have

$$t = \frac{\ln 2}{0.12} \approx 5.78 \text{ years.}$$

Amount after 10 years:

$$A = 1000e^{1.2} \approx \$3320.12$$

29. Since $A = 750e^{rt}$ and $A = 1500$ when $t = 7.75$, we have the following.

$$1500 = 750e^{7.75r}$$

$$r = \frac{\ln 2}{7.75} \approx 0.0894 = 8.94\%$$

Amount after 10 years:

$$A = 750e^{0.0894(10)} \approx \$1833.67$$

31. Since $A = 500e^{rt}$ and $A = 1292.85$ when $t = 10$, we have the following.

$$1292.85 = 500e^{10r}$$

$$r = \frac{\ln(1292.85/500)}{10} \approx 0.0950 = 9.50\%$$

The time to double is given by

$$1000 = 500e^{0.0950t}$$

$$t = \frac{\ln 2}{0.095} \approx 7.30 \text{ years}$$

33. $500,000 = P\left(1 + \dfrac{0.075}{12}\right)^{(12)(20)}$

$$P = 500,000\left(1 + \frac{0.075}{12}\right)^{-240}$$

$$\approx \$112,087.09$$

26. Since $\dfrac{dy}{dx} = ky$, $y = Ce^{kt}$ or $y = y_0 e^{kt}$.

$$\frac{1}{2} y_0 = y_0 e^{5730k}$$

$$k = -\frac{\ln 2}{5730}$$

$$0.15 y_0 = y_0 e^{(-\ln 2/5730)t}$$

$$\ln 0.15 = -\frac{(\ln 2)t}{5730}$$

$$t = -\frac{5730 \ln 0.15}{\ln 2} \approx 15,682.813 \text{ years.}$$

28. Since $A = 20,000e^{0.105t}$, the time to double is given by the following.

$$40,000 = 20,000e^{0.105t}$$

$$\ln 2 = 0.105t$$

$$t = \frac{\ln 2}{0.105} \approx 6.6 = 6\tfrac{3}{5} \text{ years}$$

Amount after 10 years:

$$A = 20,000e^{0.105(10)} \approx \$57,153.02$$

30. Since $A = 10,000e^{rt}$ and $A = 20,000$ when $t = 5$, we have the following.

$$20,000 = 10,000e^{5r}$$

$$r = \frac{\ln 2}{5} \approx 0.1386 = 13.86\%$$

Amount after 10 years:

$$A = 10,000e^{[(\ln 2)/5](10)} = \$40,000$$

32. Since $A = 2000e^{rt}$ and $A = 5436.56$ when $t = 10$, we have the following.

$$5436.56 = 2000e^{10r}$$

$$r = \frac{\ln(5436.56/2000)}{10} \approx 0.10 = 10\%$$

The time to double is given by

$$4000 = 2000e^{0.10t}$$

$$t = \frac{\ln 2}{0.10} \approx 6.93 \text{ years}$$

34. $500,000 = P\left(1 + \dfrac{0.12}{12}\right)^{(12)(40)}$

$$P = 500,000(1.01)^{-480}$$

$$\approx \$4214.16$$

35. a. $2000 = 1000(1 + 0.11)^t$

$2 = 1.11^t$

$\ln 2 = t \ln 1.11$

$t = \dfrac{\ln 2}{\ln 1.11} \approx 6.64$ years

b. $2000 = 1000\left(1 + \dfrac{0.11}{12}\right)^{12t}$

$2 = \left(1 + \dfrac{0.11}{12}\right)^{12t}$

$\ln 2 = 12t \ln\left(1 + \dfrac{0.11}{12}\right)$

$t = \dfrac{\ln 2}{12 \ln\left(1 + \dfrac{0.11}{12}\right)} \approx 6.33$ years

c. $200 = 100\left(1 + \dfrac{0.11}{365}\right)^{365t}$

$2 = \left(1 + \dfrac{0.11}{365}\right)^{365t}$

$\ln 2 = 365t \ln\left(1 + \dfrac{0.11}{365}\right)$

$t = \dfrac{\ln 2}{365 \ln\left(1 + \dfrac{0.11}{365}\right)} \approx 6.30$ years

d. $2000 = 1000e^{0.11t}$

$2 = e^{0.11t}$

$\ln 2 = 0.11t$

$t = \dfrac{\ln 2}{0.11} \approx 6.30$ years

36. a. $2000 = 1000(1 + 0.105)^t$

$2 = 1.105^t$

$\ln 2 = t \ln 1.105$

$t = \dfrac{\ln 2}{\ln 1.105} \approx 6.94$ years

b. $2000 = 1000\left(1 + \dfrac{0.105}{12}\right)^{12t}$

$2 = \left(1 + \dfrac{0.105}{12}\right)^{12t}$

$\ln 2 = 12t \ln\left(1 + \dfrac{0.105}{12}\right)$

$t = \dfrac{\ln 2}{12 \ln\left(1 + \dfrac{0.105}{12}\right)} \approx 6.63$ years

c. $2000 = 1000\left(1 + \dfrac{0.105}{365}\right)^{365t}$

$2 = \left(1 + \dfrac{0.105}{365}\right)^{365t}$

$\ln 2 = 365t \ln\left(1 + \dfrac{0.105}{365}\right)$

$t = \dfrac{\ln 2}{365 \ln\left(1 + \dfrac{0.105}{365}\right)} \approx 6.60$ years

d. $2000 = 1000e^{0.105t}$

$2 = e^{0.105t}$

$\ln 2 = 0.105t$

$t = \dfrac{\ln 2}{0.105} \approx 6.60$ years

37. Let $t = 0$ represent 1990.

$$y = Ce^{kt}, \quad (0, \ 4.22), \ (10, \ 6.49)$$

$$C = 4.22$$

$$6.49 = 4.22e^{10k}$$

$$k = \frac{\ln(6.49/4.22)}{10} \approx 0.0430$$

$$y \approx 4.22e^{0.0430t}$$

When $t = 20$ (for 2010), $y \approx 9.97$ million.

38. Let $t = 0$ represent 1990.

$$y = Ce^{kt}, \quad (0, \ 2.30), \ (10, \ 2.65)$$

$$C = 2.30$$

$$2.65 = 2.30e^{10k}$$

$$k = \frac{\ln(2.65/2.30)}{10} \approx 0.0142$$

$$y \approx 2.30e^{0.0142t}$$

When $t = 20$ (for 2010), $y \approx 3.06$ million.

39. k; the larger the value of k, the greater the rate of growth of the population.

40. $y = Ce^{kt}, \quad (0, \ 100), \ (5, \ 300)$

$$C = 100$$

$$300 = 100e^{5k}$$

$$k = \frac{\ln 3}{5} \approx 0.2197$$

$$y \approx 100e^{0.2197t}$$

When $t = 10$, $y \approx 900$.

$$200 = 100e^{0.2197t}$$

$$t = \frac{\ln 2}{0.2197} \approx 3.15 \text{ hours}$$

41. $P = Ce^{kx}, \quad (0, \ 760), \ (1000, \ 672.71)$

$$C = 760$$

$$672.71 = 760e^{1000x}$$

$$x = \frac{\ln(672.71/760)}{1000} \approx -0.000122$$

$$P \approx 760e^{-0.000122x}$$

When $x = 3000$, $P \approx 527.06$ mm Hg.

42. $y = Ce^{kt}, \quad (0, \ 742{,}000), \ (2, \ 632{,}000)$

$$C = 742{,}000$$

$$632{,}000 = 742{,}000e^{2k}$$

$$k = \frac{\ln(632/742)}{2} \approx -0.0802$$

$$y \approx 742{,}000e^{-0.0802t}$$

When $t = 3$, $y \approx \$583{,}327.77$.

43. a. $\quad 19 = 30(1 - e^{20k})$

$$30e^{20k} = 11$$

$$k = \frac{\ln(11/30)}{20} \approx -0.0502$$

$$N \approx 30(1 - e^{-0.0502t})$$

b. $\quad 25 = 30(1 - e^{-0.0502t})$

$$e^{-0.0502t} = \frac{1}{6}$$

$$t = \frac{-\ln 6}{-0.0502} \approx 36 \text{ days}$$

44. a. $\quad 20 = 30(1 - e^{30k})$

$$30e^{30k} = 10$$

$$k = \frac{\ln(1/3)}{30} = \frac{-\ln 3}{30} \approx -0.0366$$

$$N \approx 30(1 - e^{-0.0366t})$$

b. $\quad 25 = 30(1 - e^{-0.0366t})$

$$e^{-0.0366t} = \frac{1}{6}$$

$$t = \frac{-\ln 6}{-0.0366} \approx 49 \text{ days}$$

45. $S = Ce^{k/t}$

    a. $S = 5$ when $t = 1$

                $5 = Ce^k$

      $\lim\limits_{t \to \infty} Ce^{k/t} = C = 30$

              $5 = 30e^k$

              $k = \ln \frac{1}{6} \approx -1.7918$

              $S \approx 30e^{-1.7918/t}$

    b. When $t = 5$, $S \approx 20.9646$ which is 20,965 units.

c.
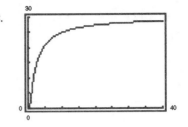

46. $S = 30(1 - e^{kt})$

    a. $5 = 30(1 - e^{k(1)}) \Rightarrow k = \ln \frac{5}{6} \approx -0.1823$

       $S \approx 30(1 - e^{-0.1823t})$

    b. 30,000 units

    c. When $t = 5$, $S \approx 17.9424$ which is 17,942 units.

d.
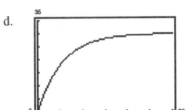

47. $A(t) = V(t)e^{-0.10t} = 100{,}000e^{0.8\sqrt{t}}e^{-0.10t} = 100{,}000e^{0.8\sqrt{t}-0.10t}$

    $\dfrac{dA}{dt} = 100{,}000 \left(\dfrac{0.4}{\sqrt{t}} - 0.10 \right) e^{0.8\sqrt{t}-0.10t} = 0$ when $t = 16$.

    The timber should be harvested in the year 2010, (1994 + 16). **Note:** You could also use a graphing utility to graph $A(t)$ and find the maximum of $A(t)$.

48. a.
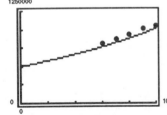

    b. The rate of growth of the receipts is 7%. The rate of growth of the interest is 8%. If these rates are not changed, receipts (i.e., taxes) continue to increase to pay the increasing interest on the national debt.

    c. $P(t) = \dfrac{I(t)}{R(t)}$

        $= \dfrac{80{,}983e^{0.08t}}{502{,}826e^{0.07t}}(100)$

        $\approx 0.1611e^{0.01t}(100)$

        $= 16.11e^{0.01t}$

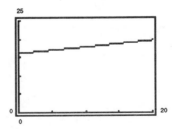

49. $\beta(I) = 10 \log_{10} \dfrac{I}{I_0}, \quad I_0 = 10^{-16}$

  a. $\beta(10^{-14}) = 10 \log_{10} \dfrac{10^{-14}}{10^{-16}} = 20$ decibels

  b. $\beta(10^{-9}) = 10 \log_{10} \dfrac{10^{-9}}{10^{-16}} = 70$ decibels

  c. $\beta(10^{-6.5}) = 10 \log_{10} \dfrac{10^{-6.5}}{10^{-16}} = 95$ decibels

  d. $\beta(10^{-4}) = 10 \log_{10} \dfrac{10^{-4}}{10^{-16}} = 120$ decibels

50. $\quad 93 = 10 \log_{10} \dfrac{I}{10^{-16}} = 10(\log_{10} I + 16)$

$-6.7 = \log_{10} I \Rightarrow I = 10^{-6.7}$

$\quad 80 = 10 \log_{10} \dfrac{I}{10^{-16}} = 10(\log_{10} I + 16)$

$-8 = \log_{10} I \Rightarrow I = 10^{-8}$

Percentage decrease: $\left(\dfrac{10^{-6.7} - 10^{-8}}{10^{-6.7}} \right)(100) \approx 95\%$

51. $R = \dfrac{\ln I - 0}{\ln 10}, \quad I = e^{R \ln 10} = 10^R$

  a. $8.3 = \dfrac{\ln I - 0}{\ln 10}$

  $I = 10^{8.3} \approx 199{,}526{,}231.5$

  b. $2R = \dfrac{\ln I - 0}{\ln 10}$

  $I = e^{2R \ln 10} = e^{2R \ln 10} = (e^{R \ln 10})^2 = (10^R)^2$

  Increases by a factor of $e^{2R \ln 10}$ or 10^R.

  c. $\dfrac{dR}{dI} = \dfrac{1}{I \ln 10}$

52. Since $dy/dt = k(y - 90)$,

$$\int \dfrac{1}{y - 90} \, dy = \int k \, dt$$

$\ln(y - 90) = kt + C.$

When $t = 0, \quad y = 1500.$ Thus, $C = \ln 1410.$

When $t = 1$,

$y = 1120$

$k(1) = \ln 1030 - \ln 1410$

$k = \ln \dfrac{103}{141}.$

Thus, $y = e^{[\ln(103/141)t + \ln 1410]} + 90$

$= 1410 e^{[\ln(103/141)]t} + 90.$

When $t = 5, \quad y = 1410 e^{5 \ln(103/141)} + 90 \approx 383.298°.$

53. Since $dy/dt = k(y - 20)$,

$$\int \dfrac{1}{y - 20} \, dy = \int k \, dt$$

$\ln(y - 20) = kt + C$

$y = Ce^{kt} + 20.$

When $t = 0, \quad y = 72.$ Therefore, $C = 52.$

When $t = 1, \quad y = 48.$ Therefore, $48 = 52e^k + 20, \quad e^k = \frac{28}{52} = \frac{7}{13},$ and $k = \ln \frac{7}{13}.$ Thus, $y = 52 e^{[\ln(7/13)]t} + 20.$

When $t = 5, \quad y = 52 e^{5 \ln(7/13)} + 20 \approx 22.35°.$

54. Let T be the outdoor temperature. Since

$$\frac{dy}{dt} = k(y - T)$$

$$\int \frac{1}{y - T} \, dy = \int k \, dt$$

$$\ln(y - T) = kt + C.$$

When $t = 0$, $y = 20 \Rightarrow \ln(20 - T) = k(0) + C$. Thus, $C = \ln(20 - T)$. When $t = \frac{1}{2}$, $y = 12 \Rightarrow \ln(12 - T) = k\left(\frac{1}{2}\right) + C$. When $t = 1$, $y = 5.5 \Rightarrow \ln(5.5 - T) = k(1) + C$. Thus, we have

$$k = \ln(5.5 - T) - \ln(20 - T) = \ln\left(\frac{5.5 - T}{20 - T}\right) \quad \text{and} \quad k = 2[\ln(12 - T) - \ln(20 - T)] = \ln\left(\frac{12 - T}{20 - T}\right)^2.$$

Solving for T, we have

$$\frac{5.5 - T}{20 - T} = \left(\frac{12 - T}{20 - T}\right)^2$$

$$(5.5 - T)(20 - T) = (12 - T)^2$$

$$T^2 - 25.5T + 110 = T^2 - 24T + 144$$

$$-1.5T = 34$$

$$T \approx -22.67° \text{ C.}$$

Section 5.7   Inverse Trigonometric Functions and Differentiation

1. $\arcsin \dfrac{1}{2} = \dfrac{\pi}{6}$

2. $\arcsin 0 = 0$

3. $\arccos \dfrac{1}{2} = \dfrac{\pi}{3}$

4. $\arccos 0 = \dfrac{\pi}{2}$

5. $\arctan \dfrac{\sqrt{3}}{3} = \dfrac{\pi}{6}$

6. $\operatorname{arccot}(-1) = \dfrac{3\pi}{4}$

7. $\operatorname{arccsc}\left(-\sqrt{2}\right) = -\dfrac{\pi}{4}$

8. $\arccos\left(-\dfrac{\sqrt{3}}{2}\right) = \dfrac{5\pi}{6}$

9. $\arccos(-0.8) \approx 2.50$

10. $\arcsin(-0.39) \approx -0.40$

11. $\operatorname{arcsec}(1.269) = \arccos\left(\dfrac{1}{1.269}\right) \approx 0.66$

12. $\arctan(-3) \approx -1.25$

13. $\arctan 0 = 0$; π is not in the range of $y = \arctan x$.

14. $y = \arcsin x$ represents the inverse of the restricted sine function where $-1 \leq x \leq 1$ and $-\pi/2 \leq y \leq \pi/2$.

15. $f(x) = \tan x$

$g(x) = \arctan x$

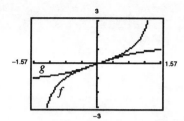

16. $f(x) = \sin x$

$g(x) = \arcsin x$

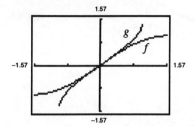

17. $\cos[\arccos(-0.1)] = -0.1$

18. $\arcsin(\sin 3\pi) = \arcsin(0) = 0$

19. a. $\sin\left(\arcsin\dfrac{1}{2}\right) = \sin\left(\dfrac{\pi}{6}\right) = \dfrac{1}{2}$

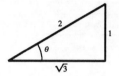

b. $\cos\left(\arcsin\dfrac{1}{2}\right) = \cos\left(\dfrac{\pi}{6}\right) = \dfrac{\sqrt{3}}{2}$

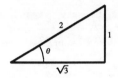

20. a. $\tan\left(\arccos\dfrac{\sqrt{2}}{2}\right) = \tan\left(\dfrac{\pi}{4}\right) = 1$

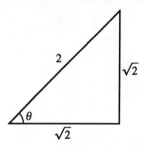

b. $\cos\left(\arcsin\dfrac{5}{13}\right) = \dfrac{12}{13}$

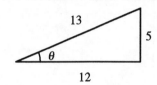

21. a. $\sin\left(\arctan\dfrac{3}{4}\right) = \dfrac{3}{5}$

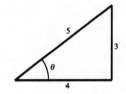

b. $\sec\left(\arcsin\dfrac{4}{5}\right) = \dfrac{5}{3}$

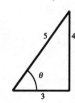

22. a. $\tan(\text{arccot } 2) = \dfrac{1}{2}$

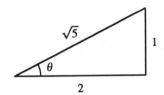

b. $\cos\left(\text{arcsec } \sqrt{5}\right) = \dfrac{\sqrt{5}}{5}$

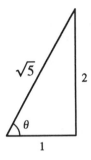

23. a. $\cot\left[\arcsin\left(-\dfrac{1}{2}\right)\right] = \cot\left(-\dfrac{\pi}{6}\right) = -\sqrt{3}$

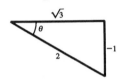

b. $\csc\left[\arctan\left(-\dfrac{5}{12}\right)\right] = -\dfrac{13}{5}$

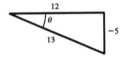

24. a. $\sec\left[\arctan\left(-\dfrac{3}{5}\right)\right] = \dfrac{\sqrt{34}}{5}$

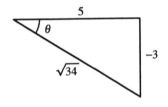

b. $\tan\left[\arcsin\left(-\dfrac{5}{6}\right)\right] = -\dfrac{5\sqrt{11}}{11}$

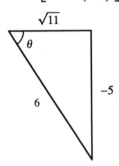

25. $y = \tan(\arctan x)$

$\theta = \arctan x$

$y = \tan\theta = x$

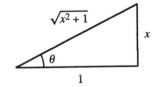

26. $y = \sin(\arccos x)$

$\theta = \arccos x$

$y = \sin\theta = \sqrt{1 - x^2}$

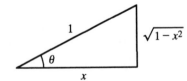

27. $y = \cos(\arcsin 2x)$

$\theta = \arcsin 2x$

$y = \cos\theta = \sqrt{1 - 4x^2}$

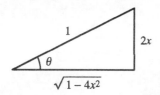

28. $y = \sec(\arctan 3x)$

$\theta = \arctan 3x$

$y = \sec\theta = \sqrt{9x^2 + 1}$

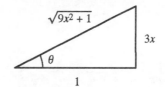

29. $y = \sin(\text{arcsec } x)$

$\theta = \text{arcsec } x, \ 0 \le \theta \le \pi, \ \theta \ne \dfrac{\pi}{2}$

$y = \sin\theta = \dfrac{\sqrt{x^2 - 1}}{|x|}$

The absolute value bars on x are necessary because of the restriction $0 \le \theta \le \pi$, $\theta \ne \pi/2$, and $\sin\theta$ for this domain must always be nonnegative.

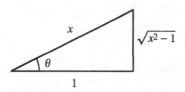

30. $y = \cos(\text{arccot } x)$

$\theta = \text{arccot } x$

$y = \cos\theta = \dfrac{x}{\sqrt{x^2 + 1}}$

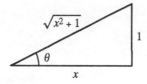

31. $y = \tan\left(\text{arcsec } \dfrac{x}{3}\right)$

$\theta = \text{arcsec } \dfrac{x}{3}$

$y = \tan\theta = \dfrac{\sqrt{x^2 - 9}}{3}$

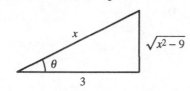

32. $y = \sec[\arcsin(x - 1)]$

$\theta = \arcsin(x - 1)$

$y = \sec\theta = \dfrac{1}{\sqrt{2x - x^2}}$

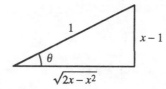

33. $y = \csc\left(\arctan \dfrac{x}{\sqrt{2}}\right)$

$\theta = \arctan \dfrac{x}{\sqrt{2}}$

$y = \csc\theta = \dfrac{\sqrt{x^2 + 2}}{x}$

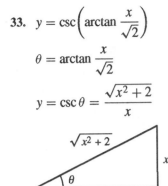

34. $y = \cos\left(\arcsin \dfrac{x - h}{r}\right)$

$\theta = \arcsin \dfrac{x - h}{r}$

$y = \cos\theta = \dfrac{\sqrt{r^2 - (x - h)^2}}{r}$

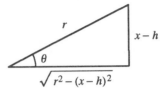

35. $\arctan \dfrac{9}{x} = \arcsin \dfrac{9}{\sqrt{x^2 + 81}}$

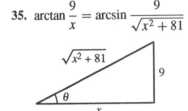

36. $\arcsin \dfrac{\sqrt{36 - x^2}}{6} = \arccos \dfrac{x}{6}$

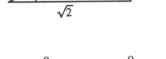

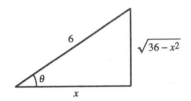

37. a. $\operatorname{arccsc} x = \arcsin \dfrac{1}{x}, \quad |x| \geq 1$

Let $y = \operatorname{arccsc} x$.

Then for $-\dfrac{\pi}{2} \leq y < 0$ and $0 < y \leq \dfrac{\pi}{2}$,

$\csc y = x \Rightarrow \sin y = \dfrac{1}{x}$.

Thus, $y = \arcsin(1/x)$. Therefore,

$\operatorname{arccsc} x = \arcsin(1/x)$.

b. $\arctan x + \arctan \dfrac{1}{x} = \dfrac{\pi}{2}, \quad x > 0$

Let $y = \arctan x + \arctan \dfrac{1}{x}$. Then:

$\tan y = \dfrac{\tan(\arctan x) + \tan[\arctan(1/x)]}{1 - \tan(\arctan x)\tan[\arctan(1/x)]}$

$= \dfrac{x + (1/x)}{1 - x(1/x)}$

$= \dfrac{x + (1/x)}{0}$ which is undefined

Thus, $y = \pi/2$. Therefore,

$\arctan x + \arctan(1/x) = \pi/2$.

38. a. $\arcsin(-x) = -\arcsin x, \quad |x| \leq 1$

Let $y = \arcsin(-x)$.

Then $-x = \sin y \Rightarrow x = -\sin y \Rightarrow x = \sin(-y)$.

Thus, $-y = \arcsin x \Rightarrow y = -\arcsin x$. Therefore, $\arcsin(-x) = -\arcsin x$.

b. $\arccos(-x) = \pi - \arccos x, \quad |x| \leq 1$

Let $y = \arccos(-x)$.

Then $-x = \cos y \Rightarrow x = -\cos y \Rightarrow x = \cos(\pi - y)$.

Thus, $\pi - y = \arccos x \Rightarrow y = \pi - \arccos x$. Therefore,

$\arccos(-x) = \pi - \arccos x$.

39. $f(x) = \arcsin(x - 1)$

$x - 1 = \sin y$

$x = 1 + \sin y$

Domain: $[0, 2]$

Range: $[-\pi/2, \pi/2]$

$f(x)$ is the graph of $\arcsin x$ shifted 1 unit to the right.

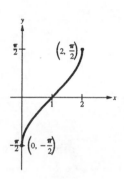

40. $f(x) = \arctan x + \dfrac{\pi}{2}$

$x = \tan\left(y - \dfrac{\pi}{2}\right)$

Domain: $(-\infty, \infty)$

Range: $(0, \pi)$

$f(x)$ is the graph of $\arctan x$ shifted $\pi/4$ unit upward.

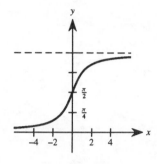

41. $f(x) = \operatorname{arcsec} 2x$

$2x = \sec y$

$x = \dfrac{1}{2} \sec y$

Domain: $(-\infty, -1/2], [1/2, \infty)$

Range: $[0, \pi/2), (\pi/2, \pi]$

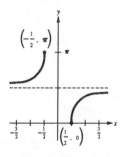

42. $f(x) = \arccos\left(\dfrac{x}{4}\right)$

$\dfrac{x}{4} = \cos y$

$x = 4 \cos y$

Domain: $[-4, 4]$

Range: $[0, \pi]$

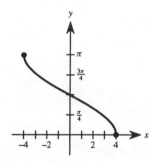

43. $\arcsin(3x - \pi) = \dfrac{1}{2}$

$3x - \pi = \sin\left(\dfrac{1}{2}\right)$

$x = \dfrac{1}{3}\left[\sin\left(\dfrac{1}{2}\right) + \pi\right] \approx 1.207$

44. $\arctan(2x) = -1$

$2x = \tan(-1)$

$x = \dfrac{1}{2}\tan(-1) \approx -0.779$

45. $\arcsin\sqrt{2x} = \arccos\sqrt{x}$

$\sqrt{2x} = \sin(\arccos\sqrt{x})$

$\sqrt{2x} = \sqrt{1-x}, \quad 0 \le x \le 1$

$2x = 1 - x$

$3x = 1$

$x = \frac{1}{3}$

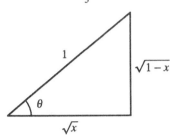

46. $\arccos x = \operatorname{arcsec} x$

$x = \cos(\operatorname{arcsec} x)$

$x = \dfrac{1}{x}$

$x^2 = 1$

$x = \pm 1$

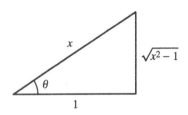

47. $f(x) = 2\arcsin(x-1)$

$f'(x) = \dfrac{2}{\sqrt{1-(x-1)^2}} = \dfrac{2}{\sqrt{2x-x^2}}$

48. $f(t) = \arcsin t^2$

$f'(t) = \dfrac{2t}{\sqrt{1-t^4}}$

49. $g(x) = 3\arccos\dfrac{x}{2}$

$g'(x) = \dfrac{-3(1/2)}{\sqrt{1-(x^2/4)}} = \dfrac{-3}{\sqrt{4-x^2}}$

50. $g(x) = \arccos\sqrt{x}$

$g'(x) = \left(\dfrac{-1}{\sqrt{1-x}}\right)\left(\dfrac{1}{2\sqrt{x}}\right) = \dfrac{-1}{2\sqrt{x}\sqrt{1-x}}$

51. $f(x) = \arctan\dfrac{x}{a}$

$f'(x) = \dfrac{1/a}{1+(x^2/a^2)} = \dfrac{a}{a^2+x^2}$

52. $f(x) = \arctan\sqrt{x}$

$f'(x) = \left(\dfrac{1}{1+x}\right)\left(\dfrac{1}{2\sqrt{x}}\right) = \dfrac{1}{2\sqrt{x}(1+x)}$

53. $g(x) = \dfrac{\arcsin 3x}{x}$

$g'(x) = \dfrac{x\left(3/\sqrt{1-9x^2}\right) - \arcsin 3x}{x^2}$

$= \dfrac{3x - \sqrt{1-9x^2}\,\arcsin 3x}{x^2\sqrt{1-9x^2}}$

54. $h(x) = x\arctan x$

$h'(x) = \dfrac{x}{1+x^2} + \arctan x$

55. $h(x) = \operatorname{arccot} 6x$

$h'(x) = \dfrac{-6}{1+36x^2}$

56. $f(x) = \operatorname{arcsec} 2x$

$f'(x) = \dfrac{2}{|2x|\sqrt{4x^2-1}} = \dfrac{1}{|x|\sqrt{4x^2-1}}$

57. $h(t) = \sin(\arccos t) = \sqrt{1-t^2}$

$h'(t) = \dfrac{1}{2}(1-t^2)^{-1/2}(-2t)$

$= \dfrac{-t}{\sqrt{1-t^2}}$

58. $g(t) = \tan(\arcsin t) = \dfrac{t}{\sqrt{1-t^2}}$

$g'(t) = \dfrac{\sqrt{1-t^2} - t\left(-t/\sqrt{1-t^2}\right)}{1-t^2}$

$= \dfrac{1}{(1-t^2)^{3/2}}$

59. $f(x) = \arcsin x + \arccos x = \dfrac{\pi}{2}$

$f'(x) = 0$

60. $f(x) = \text{arcsec } x + \text{arccsc } x = \dfrac{\pi}{2}$

$f'(x) = 0$

61. $y = \dfrac{1}{2}\left(\dfrac{1}{2}\ln\dfrac{x+1}{x-1} + \arctan x\right) = \dfrac{1}{4}[\ln(x+1) - \ln(x-1)] + \dfrac{1}{2}\arctan x$

$\dfrac{dy}{dx} = \dfrac{1}{4}\left(\dfrac{1}{x+1} - \dfrac{1}{x-1}\right) + \dfrac{1/2}{1+x^2} = \dfrac{1}{1-x^4}$

62. $y = \dfrac{1}{2}\left(x\sqrt{1-x^2} + \arcsin x\right)$

$\dfrac{dy}{dx} = \dfrac{1}{2}\left[x\left(\dfrac{-x}{\sqrt{1-x^2}}\right) + \sqrt{1-x^2} + \dfrac{1}{\sqrt{1-x^2}}\right] = \sqrt{1-x^2}$

63. $y = x\arcsin x + \sqrt{1-x^2}$

$\dfrac{dy}{dx} = x\left(\dfrac{1}{\sqrt{1-x^2}}\right) + \arcsin x - \dfrac{x}{\sqrt{1-x^2}} = \arcsin x$

64. $y = x\arctan 2x - \dfrac{1}{4}\ln(1+4x^2)$

$\dfrac{dy}{dx} = \dfrac{2x}{1+4x^2} + \arctan(2x) - \dfrac{1}{4}\left(\dfrac{8x}{1+4x^2}\right) = \arctan(2x)$

65. $f(x) = \arccos x$

$f'(x) = \dfrac{-1}{\sqrt{1-x^2}} = -2$ when $x = \pm\dfrac{\sqrt{3}}{2}$

When $x = \sqrt{3}/2,\; f(\sqrt{3}/2) = \pi/6.$ When $x = -\sqrt{3}/2,\; f(-\sqrt{3}/2) = 5\pi/6.$

Tangent lines: $y - \dfrac{\pi}{6} = -2\left(x - \dfrac{\sqrt{3}}{2}\right) \Rightarrow y = -2x + \left(\dfrac{\pi}{6} + \sqrt{3}\right)$

$\qquad\qquad\quad y - \dfrac{5\pi}{6} = -2\left(x + \dfrac{\sqrt{3}}{2}\right) \Rightarrow y = -2x + \left(\dfrac{5\pi}{6} - \sqrt{3}\right)$

66. $g(x) = \arctan x, \quad x = 2, \; dx = 0.1$

$dg = g'(x)\,dx = \dfrac{1}{1+x^2}\,dx = \left(\dfrac{1}{5}\right)(0.1) = 0.02$

67. $f(x) = \arcsin x, \quad a = \dfrac{1}{2}$

$f'(x) = \dfrac{1}{\sqrt{1-x^2}}$

$f''(x) = \dfrac{x}{(1-x^2)^{3/2}}$

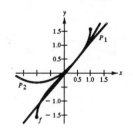

$P_1(x) = f\left(\dfrac{1}{2}\right) + f'\left(\dfrac{1}{2}\right)\left(x - \dfrac{1}{2}\right) = \dfrac{\pi}{6} + \dfrac{2\sqrt{3}}{3}\left(x - \dfrac{1}{2}\right)$

$P_2(x) = f\left(\dfrac{1}{2}\right) + f'\left(\dfrac{1}{2}\right)\left(x - \dfrac{1}{2}\right) + \dfrac{1}{2}f''\left(\dfrac{1}{2}\right)\left(x - \dfrac{1}{2}\right)^2 = \dfrac{\pi}{6} + \dfrac{2\sqrt{3}}{3}\left(x - \dfrac{1}{2}\right) + \dfrac{2\sqrt{3}}{9}\left(x - \dfrac{1}{2}\right)^2$

68. $f(x) = \arctan x, \quad a = 1$

$$f'(x) = \frac{1}{1+x^2}$$

$$f''(x) = \frac{-2x}{(1+x^2)^2}$$

$$P_1(x) = f(1) + f'(1)(x-1) = \frac{\pi}{4} + \frac{1}{2}(x-1)$$

$$P_2(x) = f(1) + f'(1)(x-1) + \frac{1}{2}f''(1)(x-1)^2$$

$$= \frac{\pi}{4} + \frac{1}{2}(x-1) - \frac{1}{4}(x-1)^2$$

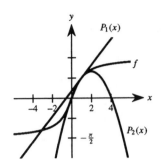

69. $f(x) = \text{arcsec } x - x$

$$f'(x) = \frac{1}{|x|\sqrt{x^2-1}} - 1$$

$$= 0 \text{ when } |x|\sqrt{x^2-1} = 1$$

$$x^2(x^2-1) = 1$$

$$x^4 - x^2 - 1 = 0 \text{ when } x^2 = \frac{1+\sqrt{5}}{2} \text{ or}$$

$$x = \pm\sqrt{\frac{1+\sqrt{5}}{2}} = \pm 1.272$$

Relative maximum: $(1.272, \ -0.606)$

Relative minimum: $(-1.272, \ 3.747)$

70. $f(x) = \arcsin x - 2x$

$$f'(x) = \frac{1}{\sqrt{1-x^2}} - 2 = 0 \text{ when } \sqrt{1-x^2} = \frac{1}{2} \text{ or}$$

$$x = \pm\frac{\sqrt{3}}{2}$$

$$f''(x) = \frac{x}{(1-x^2)^{3/2}}$$

$$f''(\sqrt{3}/2) > 0$$

Relative minimum: $(\sqrt{3}/2, \ -0.68)$

$$f''(-\sqrt{3}/2) < 0$$

Relative maximum: $(-\sqrt{3}/2, \ 0.68)$

71. $y = \arccos x$

$y = \arctan x$

The point of intersection is given by

$f(x) = \arccos x - \arctan x = 0, \quad \cos(\arccos x) = \cos(\arctan x)$.

$$x = \frac{1}{\sqrt{1+x^2}}$$

$$x^2(1+x^2) = 1$$

$$x^4 + x^2 - 1 = 0 \text{ when } x^2 = \frac{-1+\sqrt{5}}{2}$$

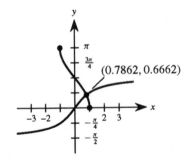

Therefore, $x = \pm\sqrt{\frac{-1+\sqrt{5}}{2}} \approx \pm 0.7862$.

Point of intersection: $(0.7862, 0.6662)$ [Since $f(-0.7862) = \pi \neq 0$.]

72. $y = \arcsin x$

$y = \arccos x$

The point of intersection is given by

$f(x) = \arcsin x - \arccos x = 0, \quad \sin(\arcsin x) = \sin(\arccos x).$

$x = \sqrt{1 - x^2}$

$x^2 = 1 - x^2$

$x = \pm \dfrac{1}{\sqrt{2}} = \pm \dfrac{\sqrt{2}}{2}$

Point of intersection: $(\sqrt{2}/2, \ \pi/4)$ [Since $f(-\sqrt{2}/2) = -\pi.$]

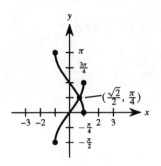

73. a. $h(t) = -16t^2 + 256$

     $-16t^2 + 256 = 0$ when $t = 4$ sec.

b. $\tan\theta = \dfrac{h}{500} = \dfrac{-16t^2 + 256}{500}$

     $\theta = \arctan\left[\dfrac{16}{500}(-t^2 + 16)\right]$

     $\dfrac{d\theta}{dt} = \dfrac{-8t/125}{1 + [(4/125)(-t^2 + 16)]^2} = \dfrac{-1000t}{15{,}625 + 16(16 - t^2)^2}$

     When $t = 1$, $d\theta/dt \approx -0.0520$ rad/sec.

     When $t = 2$, $d\theta/dt \approx -0.1116$ rad/sec.

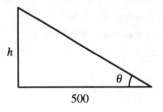

74. $\tan\theta = \dfrac{h}{300}$

     $\dfrac{dh}{dt} = 5$ ft/sec

     $\theta = \arctan\left(\dfrac{h}{300}\right)$

     $\dfrac{d\theta}{dt} = \dfrac{1/300}{1 + (h^2/300^2)}\left(\dfrac{dh}{dt}\right) = \dfrac{300}{300^2 + h^2}(5)$

     $= \dfrac{1500}{300^2 + h^2} = \dfrac{3}{200}$ rad/sec when $h = 100$.

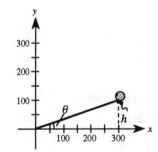

75. a. Let $y = \arctan u$. Then

$$\tan y = u$$

$$\sec^2 y \frac{dy}{dx} = u'$$

$$\frac{dy}{dx} = \frac{u'}{\sec^2 y} = \frac{u'}{1 + u^2}.$$

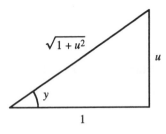

b. Let $y = \text{arcsec } u$. Then

$$\sec y = u$$

$$\sec y \tan y \frac{dy}{dx} = u'$$

$$\frac{dy}{dx} = \frac{u'}{\sec y \tan y} = \frac{u'}{|u|\sqrt{u^2 - 1}}.$$

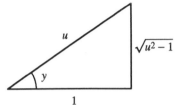

Note: The absolute value sign in the formula for the derivative of arcsec u is necessary because the inverse secant function has a positive slope at every value in its domain.

c. Let $y = \arccos u$. Then

$$\cos y = u$$

$$-\sin y \frac{dy}{dx} = u'$$

$$\frac{dy}{dx} = -\frac{u'}{\sin y} = -\frac{u'}{\sqrt{1 - u^2}}.$$

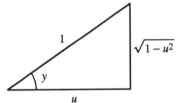

d. Let $y = \text{arccot } u$. Then

$$\cot y = u$$

$$-\csc^2 y \frac{dy}{dx} = u'$$

$$\frac{dy}{dx} = \frac{u'}{-\csc^2 y} = -\frac{u'}{1 + u^2}.$$

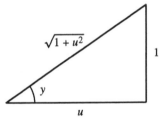

e. Let $y = \text{arccsc } u$. Then

$$\csc y = u$$

$$-\csc y \cot y \frac{dy}{dx} = u'$$

$$\frac{dy}{dx} = \frac{u'}{-\csc y \cot y} = -\frac{u'}{|u|\sqrt{u^2 - 1}}.$$

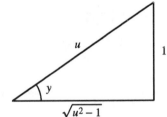

Note: The absolute value sign in the formula for the derivative of arccsc u is necessary because the inverse cosecant function has a negative slope at every value in its domain.

76. $f(x) = kx + \sin x$

$f'(x) = k + \cos x \geq 0$ for $k \geq 1$

$f'(x) = k + \cos x \leq 0$ for $k \leq -1$

Therefore, $f(x) = kx + \sin x$ is strictly monotonic and has an inverse for $k \leq -1$ or $k \geq 1$.

77. $f(x) = \sin x$

$g(x) = \arcsin(\sin x)$

a. The range of $y = \arcsin x$ is
$-\pi/2 \le y \le \pi/2$.

b. Maximum: $\pi/2$
Minimum: $-\pi/2$

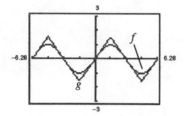

78. $f(x) = \cos x$

$g(x) = \arccos(\cos x)$

a. The range of $y = \arccos x$ is
$0 \le y \le \pi$.

b. Maximum: π
Minimum: 0

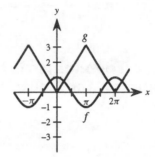

79. True

$\dfrac{d}{dx}[\arctan x] = \dfrac{1}{1 + x^2} > 0$ for all x

80. False

The range of $y = \arcsin x$ is

$\left[-\dfrac{\pi}{2}, \dfrac{\pi}{2} \right].$

81. True

$\dfrac{d}{dx}[\arctan(\tan x)] = \dfrac{\sec^2 x}{1 + \tan^2 x} = \dfrac{\sec^2 x}{\sec^2 x} = 1$

82. False

$\arcsin^2 0 + \arccos^2 0 = 0 + \left(\dfrac{\pi}{2} \right)^2 \ne 1$

83. $f(x) = \arcsin\left(\dfrac{x-2}{2} \right) - 2\arcsin\dfrac{\sqrt{x}}{2}, \quad 0 \le x \le 4$

$f'(x) = \dfrac{1/2}{\sqrt{1 - [(x-2)/2]^2}} - 2\left[\dfrac{1/(4\sqrt{x})}{\sqrt{1 - (\sqrt{x}/2)^2}} \right]$

$= \dfrac{1}{2\sqrt{1 - (1/4)(x^2 - 4x + 4)}} - \dfrac{1}{2\sqrt{x}\sqrt{1 - (x/4)}} = \dfrac{1}{2\sqrt{x - (x^2/4)}} - \dfrac{1}{2\sqrt{x - (x^2/4)}} = 0$

Since the derivative is zero, we conclude that the function is constant. (By letting $x = 0$ in $f(x)$, you can see that the constant is $-\pi/2$.)

84. $y_1 = \sin(\arcsin x) = x$
Domain: $[-1, 1]$
Range: $[-1, 1]$

$y_2 = \arcsin(\sin x)$
Domain: $(-\infty, \infty)$
Range: $\left[-\dfrac{\pi}{2}, \dfrac{\pi}{2} \right]$

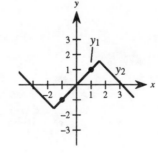

85. $\tan(\arctan x + \arctan y) = \dfrac{\tan(\arctan x) + \tan(\arctan y)}{1 - \tan(\arctan x)\tan(\arctan y)} = \dfrac{x + y}{1 - xy}, \quad xy \neq 1$

Therefore,

$$\arctan x + \arctan y = \arctan\left(\frac{x+y}{1-xy}\right), \quad xy \neq 1.$$

Let $x = \frac{1}{2}$ and $y = \frac{1}{3}$.

$$\arctan\left(\frac{1}{2}\right) + \arctan\left(\frac{1}{3}\right) = \arctan\frac{\frac{1}{2}+\frac{1}{3}}{1-\left(\frac{1}{2}\cdot\frac{1}{3}\right)} = \arctan\frac{\frac{5}{6}}{1-\frac{1}{6}} = \arctan\frac{\frac{5}{6}}{\frac{5}{6}} = \arctan 1 = \frac{\pi}{4}$$

Section 5.8   Inverse Trigonometric Functions: Integration and Completing the Square

1. Let $u = 3x, \quad du = 3\,dx$.

$$\int_0^{1/6} \frac{1}{\sqrt{1-9x^2}}\,dx = \frac{1}{3}\int_0^{1/6} \frac{1}{\sqrt{1-(3x)^2}}(3)\,dx = \left[\frac{1}{3}\arcsin(3x)\right]_0^{1/6} = \frac{\pi}{18}$$

2. $\displaystyle\int_0^1 \frac{1}{\sqrt{4-x^2}}\,dx = \left[\arcsin\frac{x}{2}\right]_0^1 = \frac{\pi}{6}$

3. Let $u = 2x, \quad du = 2\,dx$.

$$\int_0^{\sqrt{3}/2} \frac{1}{1+4x^2}\,dx = \frac{1}{2}\int_0^{\sqrt{3}/2} \frac{2}{1+(2x)^2}\,dx = \left[\frac{1}{2}\arctan(2x)\right]_0^{\sqrt{3}/2} = \frac{\pi}{6}$$

4. $\displaystyle\int_{\sqrt{3}}^3 \frac{1}{9+x^2}\,dx = \left[\frac{1}{3}\arctan\frac{x}{3}\right]_{\sqrt{3}}^3 = \frac{\pi}{36}$

5. $\displaystyle\int \frac{1}{x\sqrt{4x^2-1}}\,dx = \int \frac{2}{2x\sqrt{(2x)^2-1}}\,dx$
$$= \operatorname{arcsec}|2x| + C$$

6. $\displaystyle\int \frac{1}{4+(x-1)^2}\,dx = \frac{1}{2}\arctan\left(\frac{x-1}{2}\right) + C$

7. $\displaystyle\int \frac{x^3}{x^2+1}\,dx = \int\left[x - \frac{x}{x^2+1}\right]dx$
$$= \int x\,dx - \frac{1}{2}\int \frac{2x}{x^2+1}\,dx$$
$$= \frac{1}{2}x^2 - \frac{1}{2}\ln(x^2+1) + C$$
(Use long division.)

8. $\displaystyle\int \frac{x^4-1}{x^2+1}\,dx = \int (x^2-1)\,dx$
$$= \frac{1}{3}x^3 - x + C$$

9. $\displaystyle\int \frac{1}{\sqrt{1-(x+1)^2}}\,dx = \arcsin(x+1) + C$

10. Let $u = t^2, \quad du = 2t\,dt$.
$$\int \frac{t}{t^4+16}\,dt = \frac{1}{2}\int \frac{1}{(4)^2+(t^2)^2}(2t)\,dt$$
$$= \frac{1}{8}\arctan\frac{t^2}{4} + C$$

11. Let $u = t^2$, $du = 2t\,dt$.

$$\int \frac{t}{\sqrt{1-t^4}}\,dt = \frac{1}{2}\int \frac{1}{\sqrt{1-(t^2)^2}}(2t)\,dt$$

$$= \frac{1}{2}\arcsin(t^2) + C$$

12. Let $u = x^2$, $du = 2x\,dx$.

$$\int \frac{1}{x\sqrt{x^4-4}}\,dx = \frac{1}{2}\int \frac{1}{x^2\sqrt{(x^2)^2-2^2}}(2x)\,dx$$

$$= \frac{1}{4}\operatorname{arcsec}\frac{x^2}{2} + C$$

13. Let $u = \arctan x$, $du = \dfrac{1}{1+x^2}\,dx$.

$$\int \frac{\arctan x}{1+x^2}\,dx = \frac{1}{2}\arctan^2 x + C$$

14. $\displaystyle\int \frac{1}{(x-1)\sqrt{(x-1)^2-4}}\,dx = \frac{1}{2}\operatorname{arcsec}\frac{|x-1|}{2} + C$

15. Let $u = \arcsin x$, $du = \dfrac{1}{\sqrt{1-x^2}}\,dx$.

$$\int_0^{1/\sqrt{2}} \frac{\arcsin x}{\sqrt{1-x^2}}\,dx = \left[\frac{1}{2}\arcsin^2 x\right]_0^{1/\sqrt{2}} = \frac{\pi^2}{32} \approx 0.308$$

16. Let $u = \arccos x$, $du = -\dfrac{1}{\sqrt{1-x^2}}\,dx$.

$$\int_0^{1/\sqrt{2}} \frac{\arccos x}{\sqrt{1-x^2}}\,dx = -\int_0^{1/\sqrt{2}} \frac{-\arccos x}{\sqrt{1-x^2}}\,dx = \left[-\frac{1}{2}\arccos^2 x\right]_0^{1/\sqrt{2}} = \frac{3\pi^2}{32} \approx 0.925$$

17. Let $u = 1 - x^2$, $du = -2x\,dx$.

$$\int_{-1/2}^0 \frac{x}{\sqrt{1-x^2}}\,dx = -\frac{1}{2}\int_{-1/2}^0 (1-x^2)^{-1/2}(-2x)\,dx = \left[-\sqrt{1-x^2}\right]_{-1/2}^0 = \frac{\sqrt{3}-2}{2} \approx -0.134$$

18. Let $u = 1 + x^2$, $du = 2x\,dx$.

$$\int_{-\sqrt{3}}^0 \frac{x}{1+x^2}\,dx = \frac{1}{2}\int_{-\sqrt{3}}^0 \frac{1}{1+x^2}(2x)\,dx = \left[\frac{1}{2}\ln(1+x^2)\right]_{-\sqrt{3}}^0 = -\ln 2$$

19. Let $u = e^x$, $du = e^x\,dx$.

$$\int \frac{e^x}{\sqrt{1-e^{2x}}}\,dx = \arcsin e^x + C$$

20. Let $u = \sin x$, $du = \cos x\,dx$.

$$\int \frac{\cos x}{\sqrt{4-\sin^2 x}}\,dx = \arcsin\frac{\sin x}{2} + C$$

21. $\displaystyle\int \frac{1}{9+(x-3)^2}\,dx = \frac{1}{3}\arctan\left(\frac{x-3}{3}\right) + C$

22. $\displaystyle\int \frac{x+1}{x^2+1}\,dx = \frac{1}{2}\int \frac{2x}{x^2+1}\,dx + \int \frac{1}{1+x^2}\,dx$

$$= \frac{1}{2}\ln(x^2+1) + \arctan x + C$$

23. Let $u = e^{2x}$, $du = 2e^{2x}\,dx$.

$$\int \frac{e^{2x}}{4+e^{4x}}\,dx = \frac{1}{2}\int \frac{2e^{2x}}{4+(e^{2x})^2}\,dx = \frac{1}{4}\arctan\frac{e^{2x}}{2} + C$$

24. $\displaystyle\int_1^2 \frac{1}{3+(x-2)^2}\,dx = \int_1^2 \frac{1}{\left(\sqrt{3}\right)^2+(x-2)^2}\,dx = \left[\frac{1}{\sqrt{3}}\arctan\left(\frac{x-2}{\sqrt{3}}\right)\right]_1^2 = \frac{\sqrt{3}\,\pi}{18}$

25. Let $u = \cos x$, $du = -\sin x\,dx$.

$$\int_{\pi/2}^\pi \frac{\sin x}{1+\cos^2 x}\,dx = -\int_{\pi/2}^\pi \frac{-\sin x}{1+\cos^2 x}\,dx = \left[-\arctan(\cos x)\right]_{\pi/2}^\pi = \frac{\pi}{4}$$

26. Let $u = \sin x$, $du = \cos x \, dx$.

$$\int_0^\pi \frac{\cos x}{1 + \sin^2 x} \, dx = \left[\arctan(\sin x) \right]_0^\pi = 0$$

27. $\displaystyle\int_0^2 \frac{1}{x^2 - 2x + 2} \, dx = \int_0^2 \frac{1}{1 + (x - 1)^2} \, dx = \left[\arctan(x - 1) \right]_0^2 = \frac{\pi}{2}$

28. $\displaystyle\int_{-3}^{-1} \frac{1}{x^2 + 6x + 13} \, dx = \int_{-3}^{-1} \frac{1}{(x + 3)^2 + 4} \, dx = \left[\frac{1}{2} \arctan \frac{x + 3}{2} \right]_{-3}^{-1} = \frac{\pi}{8}$

29. $\displaystyle\int \frac{2x}{x^2 + 6x + 13} \, dx = \int \frac{2x + 6}{x^2 + 6x + 13} \, dx - 6 \int \frac{1}{x^2 + 6x + 13} \, dx$

$$= \int \frac{2x + 6}{x^2 + 6x + 13} \, dx - 6 \int \frac{1}{4 + (x + 3)^2} \, dx = \ln |x^2 + 6x + 13| - 3 \arctan \left(\frac{x + 3}{2} \right) + C$$

30. $\displaystyle\int \frac{2x - 5}{x^2 + 2x + 2} \, dx = \int \frac{2x + 2}{x^2 + 2x + 2} \, dx - 7 \int \frac{1}{1 + (x + 1)^2} \, dx = \ln |x^2 + 2x + 2| - 7 \arctan(x + 1) + C$

31. $\displaystyle\int \frac{1}{\sqrt{-x^2 - 4x}} \, dx = \int \frac{1}{\sqrt{4 - (x + 2)^2}} \, dx = \arcsin \left(\frac{x + 2}{2} \right) + C$

32. $\displaystyle\int \frac{1}{\sqrt{-x^2 + 2x}} \, dx = \int \frac{1}{\sqrt{1 - (x - 1)^2}} \, dx = \arcsin(x - 1) + C$

33. Let $u = -x^2 - 4x$, $du = (-2x - 4) \, dx$.

$$\int \frac{x + 2}{\sqrt{-x^2 - 4x}} \, dx = -\frac{1}{2} \int (-x^2 - 4x)^{-1/2}(-2x - 4) \, dx = -\sqrt{-x^2 - 4x} + C$$

34. Let $u = x^2 - 2x$, $du = (2x - 2) \, dx$.

$$\int \frac{x - 1}{\sqrt{x^2 - 2x}} \, dx = \frac{1}{2} \int (x^2 - 2x)^{-1/2}(2x - 2) \, dx = \sqrt{x^2 - 2x} + C$$

35. $\displaystyle\int_2^3 \frac{2x - 3}{\sqrt{4x - x^2}} \, dx = \int_2^3 \frac{2x - 4}{\sqrt{4x - x^2}} \, dx + \int_2^3 \frac{1}{\sqrt{4x - x^2}} \, dx$

$$= -\int_2^3 (4x - x^2)^{-1/2}(4 - 2x) \, dx + \int_2^3 \frac{1}{\sqrt{4 - (x - 2)^2}} \, dx$$

$$= \left[-2\sqrt{4x - x^2} + \arcsin \left(\frac{x - 2}{2} \right) \right]_2^3 = 4 - 2\sqrt{3} + \frac{\pi}{6} \approx 1.059$$

36. $\displaystyle\int \frac{1}{(x - 1)\sqrt{x^2 - 2x}} \, dx = \int \frac{1}{(x - 1)\sqrt{(x - 1)^2 - 1}} \, dx = \text{arcsec} \, |x - 1| + C$

37. Let $u = x^2 + 1$, $du = 2x \, dx$.

$$\int \frac{x}{x^4 + 2x^2 + 2} \, dx = \frac{1}{2} \int \frac{2x}{(x^2 + 1)^2 + 1} \, dx = \frac{1}{2} \arctan(x^2 + 1) + C$$

38. Let $u = x^2 - 4$, $du = 2x \, dx$.

$$\int \frac{x}{\sqrt{9 + 8x^2 - x^4}} \, dx = \frac{1}{2} \int \frac{2x}{\sqrt{25 - (x^2 - 4)^2}} \, dx = \frac{1}{2} \arcsin \left(\frac{x^2 - 4}{5} \right) + C$$

39. Let $u = 4\left(x - \frac{1}{2}\right)$, $du = 4\,dx$.

$$\int \frac{1}{\sqrt{-16x^2 + 16x - 3}}\,dx = \frac{1}{4}\int \frac{4}{\sqrt{1 - \left[4\left(x - \frac{1}{2}\right)\right]^2}}\,dx = \frac{1}{4}\arcsin(4x - 2) + C$$

40. Let $u = 3(x - 1)$, $du = 3\,dx$.

$$\int \frac{1}{(x-1)\sqrt{9x^2 - 18x + 5}}\,dx = \int \frac{3}{3(x-1)\sqrt{[3(x-1)]^2 - 4}}\,dx = \frac{1}{2}\operatorname{arcsec}\frac{|3(x-1)|}{2} + C$$

41. Let $u = \sqrt{x - 1}$, $u^2 + 1 = x$, $2u\,du = dx$.

$$\int \frac{\sqrt{x-1}}{x}\,dx = \int \frac{2u^2}{u^2 + 1}\,du = \int \frac{2u^2 + 2 - 2}{u^2 + 1}\,du = 2\int du - 2\int \frac{1}{u^2 + 1}\,du$$

$$= 2u - 2\arctan u + C = 2\left(\sqrt{x - 1} - \arctan\sqrt{x - 1}\right) + C$$

42. Let $u = \sqrt{x - 2}$, $u^2 + 2 = x$, $2u\,du = dx$.

$$\int \frac{\sqrt{x-2}}{x+1}\,dx = \int \frac{2u^2}{u^2 + 3}\,du = \int \frac{2u^2 + 6 - 6}{u^2 + 3}\,du = 2\int du - 6\int \frac{1}{u^2 + 3}\,du$$

$$= 2u - \frac{6}{\sqrt{3}}\arctan\frac{u}{\sqrt{3}} + C = 2\sqrt{x - 2} - 2\sqrt{3}\arctan\sqrt{\frac{x-2}{3}} + C$$

43. Let $u = \sqrt{e^t - 3}$.

$$u^2 + 3 = e^t$$

$$2u\,du = e^t\,dt$$

$$\frac{2u\,du}{u^2 + 3} = dt$$

$$\int \sqrt{e^t - 3}\,dt = \int \frac{2u^2}{u^2 + 3}\,du \quad \text{Same as Exercise 42}$$

$$= 2u - 2\sqrt{3}\arctan\frac{u}{\sqrt{3}} + C = 2\sqrt{e^t - 3} - 2\sqrt{3}\arctan\sqrt{\frac{e^t - 3}{3}} + C$$

44. Let $u = \sqrt{t}$.

$$u^2 = t$$

$$2u\,du = dt$$

$$\int \frac{1}{t\sqrt{t} + \sqrt{t}}\,dt = 2\int \frac{u}{u^3 + u}\,du = 2\int \frac{1}{u^2 + 1}\,du = 2\arctan u + C = 2\arctan\sqrt{t} + C$$

45. $(1 + x^6)y' = 3x^2$

$$y' = \frac{3x^2}{1 + x^6}$$

$$y = \int \frac{3x^2}{1 + x^6}\,dx$$

$$= \arctan(x^3) + C$$

46. $\sqrt{1 - 9x^2}\,y' = 1$

$$y' = \frac{1}{\sqrt{1 - 9x^2}}$$

$$y = \int \frac{1}{\sqrt{1 - 9x^2}}\,dx$$

$$= \frac{1}{3}\arcsin(3x) + C$$

47. $\dfrac{dy}{dx} = \dfrac{xy}{1+x^2}$

$\displaystyle\int \dfrac{dy}{y} = \int \dfrac{x}{1+x^2}\,dx$

$\ln y = \dfrac{1}{2}\ln(1+x^2) + C_1$

$\quad\quad = \ln\sqrt{1+x^2} + \ln C$

$\quad\quad = \ln C\sqrt{1+x^2}$

$\quad y = C\sqrt{1+x^2}$

48. $\dfrac{dy}{dx} = x\sec y$

$\displaystyle\int \dfrac{dy}{\sec y} = \int x\,dx$

$\displaystyle\int \cos y\,dy = \int x\,dx$

$\sin y = \dfrac{1}{2}x^2 + C$

$y = \arcsin\left(\dfrac{x^2}{2} + C\right)$

49. $A = \displaystyle\int_0^1 \dfrac{1}{1+x^2}\,dx$

$\quad = \Big[\arctan x\Big]_0^1 = \dfrac{\pi}{4}$

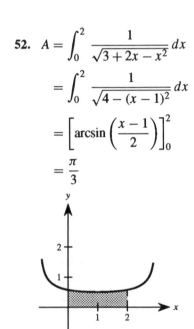

50. $A = \displaystyle\int_0^1 \dfrac{1}{\sqrt{4-x^2}}\,dx$

$\quad = \Big[\arcsin \dfrac{x}{2}\Big]_0^1 = \dfrac{\pi}{6}$

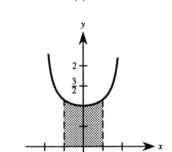

51. $A = \displaystyle\int_1^3 \dfrac{1}{x^2 - 2x + 5}\,dx$

$\quad = \displaystyle\int_1^3 \dfrac{1}{(x-1)^2 + 4}\,dx$

$\quad = \Big[\dfrac{1}{2}\arctan\left(\dfrac{x-1}{2}\right)\Big]_1^3$

$\quad = \dfrac{\pi}{8}$

52. $A = \displaystyle\int_0^2 \dfrac{1}{\sqrt{3 + 2x - x^2}}\,dx$

$\quad = \displaystyle\int_0^2 \dfrac{1}{\sqrt{4 - (x-1)^2}}\,dx$

$\quad = \Big[\arcsin\left(\dfrac{x-1}{2}\right)\Big]_0^2$

$\quad = \dfrac{\pi}{3}$

53. $F(x) = \dfrac{1}{2}\displaystyle\int_x^{x+2} \dfrac{2}{t^2+1}\,dt$

a. $F(x)$ represents the average value of $f(x)$ over the interval $[x,\ x+2]$. Maximum at $x = -1$.

b. $F(x) = \Big[\arctan t\Big]_x^{x+2}$

$\quad\quad = \arctan(x+2) - \arctan x$

$F'(x) = \dfrac{1}{1+(x+2)^2} - \dfrac{1}{1+x^2}$

$\quad = \dfrac{(1+x^2) - (x^2 + 4x + 5)}{(x^2+1)(x^2 + 4x + 5)}$

$\quad = \dfrac{-4(x+1)}{(x^2+1)(x^2 + 4x + 5)}$

$\quad = 0$ when $x = -1$

54. Area $\approx (1)(1) = 1$

Matches (c)

55. a. $v(t) = -32t + 500$

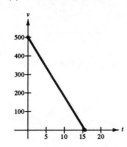

b. $s(t) = \int v(t)\, dt = \int (-32t + 500)\, dt$

$\qquad = -16t^2 + 500t + C$

$\quad s(0) = -16(0) + 500(0) + C = 0 \Rightarrow C = 0$

$\quad s(t) = -16t^2 + 500t$

When the object reaches its maximum height, $v(t) = 0$.

$\qquad v(t) = -32t + 500 = 0$

$\qquad -32t = -500$

$\qquad\quad t = 15.625$

$\quad s(15.625) = -16(15.625)^2 + 500(15.625)$

$\qquad\qquad = 3906.25$ ft (Maximum height)

c. $\displaystyle\int \frac{1}{32 + kv^2}\, dv = -\int dt$

$\quad \dfrac{1}{\sqrt{32k}} \arctan\left(\sqrt{\dfrac{k}{32}}v\right) = -t + C_1$

$\qquad \arctan\left(\sqrt{\dfrac{k}{32}}v\right) = -\sqrt{32k}\,t + C$

$\qquad\qquad \sqrt{\dfrac{k}{32}}v = \tan\left(C - \sqrt{32k}\,t\right)$

$\qquad\qquad\quad v = \sqrt{\dfrac{32}{k}}\tan\left(C - \sqrt{32k}\,t\right)$

When $t = 0$, $v = 500$,

$\quad C = \arctan\left(500\sqrt{k/32}\right)$, and we have

$\quad v(t) = \sqrt{\dfrac{32}{k}}\tan\left[\arctan\left(500\sqrt{\dfrac{k}{32}}\right) - \sqrt{32k}\,t\right].$

d. When $k = 0.001$,

$v(t) = \sqrt{32{,}000}\tan\left[\arctan\left(500\sqrt{0.00003125}\right)\right.$

$\qquad\qquad\qquad \left. - \sqrt{0.032}\,t\right]$

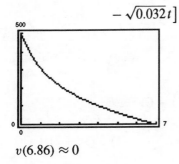

$v(6.86) \approx 0$

e. $h = \displaystyle\int_0^{6.86} \sqrt{32{,}000}\tan\left[\arctan\left(500\sqrt{0.00003125}\right)\right.$

$\qquad\qquad\qquad\qquad\left. - \sqrt{0.032}\,t\right] dt$

Simpson's Rule: $n = 10$; $h \approx 1088$ feet

f. Air resistance lowers the maximum height.

56. $\displaystyle\int \frac{1}{\sqrt{A^2 - y^2}}\, dy = \int \sqrt{\frac{k}{m}}\, dt$

$\qquad \arcsin\dfrac{y}{A} = \sqrt{\dfrac{k}{m}}\,t + C$

$\qquad\qquad y = A\sin\left(\sqrt{\dfrac{k}{m}}\,t + C\right)$

Since $y = 0$ when $t = 0$, $\ C = 0$ and we have $y = A\sin\left(\sqrt{\dfrac{k}{m}}\,t\right)$.

57. a. $\int_0^1 \frac{4}{1+x^2}\,dx = \left[4\arctan x\right]_0^1 = 4\arctan 1 - 4\arctan 0 = 4\left(\frac{\pi}{4}\right) - 4(0) = \pi$

b. Let $n = 6$.

$$4\int_0^1 \frac{1}{1+x^2}\,dx \approx 4\left(\frac{1}{3(6)}\right)\left[1 + \frac{4}{1+(1/36)} + \frac{2}{1+(1/9)} + \frac{4}{1+(1/4)} + \frac{2}{1+(4/9)} + \frac{4}{1+(25/36)} + \frac{1}{2}\right] \approx 3.14159$$

58. a. $\dfrac{d}{dx}\left[\dfrac{1}{a}\arctan\dfrac{u}{a} + C\right] = \dfrac{1}{a}\left[\dfrac{u'/a}{1+(u/a)^2}\right] = \dfrac{1}{a^2}\left[\dfrac{u'}{(a^2+u^2)/a^2}\right] = \dfrac{u'}{a^2+u^2}$

Thus, $\displaystyle\int \frac{du}{a^2+u^2} = \int \frac{u'}{a^2+u^2}\,dx = \frac{1}{a}\arctan\frac{u}{a} + C.$

b. Assume $u > 0$.

$$\frac{d}{dx}\left[\frac{1}{a}\operatorname{arcsec}\frac{u}{a} + C\right] = \frac{1}{a}\left[\frac{u'/a}{(u/a)\sqrt{(u/a)^2-1}}\right] = \frac{1}{a}\left[\frac{u'}{u\sqrt{(u^2-a^2)/a^2}}\right] = \frac{u'}{u\sqrt{u^2-a^2}}$$

Thus, $\displaystyle\int \frac{du}{u\sqrt{u^2-a^2}} = \int \frac{u'}{u\sqrt{u^2-a^2}}\,dx = \frac{1}{a}\operatorname{arcsec}\frac{|u|}{a} + C.$ The case $u < 0$ is handled in a similar manner.

59. a. $\displaystyle\int \frac{1}{\sqrt{1-x^2}}\,dx = \arcsin x + C, \quad u = x$

b. $\displaystyle\int \frac{x}{\sqrt{1-x^2}}\,dx = -\sqrt{1-x^2} + C, \quad u = 1 - x^2$

c. $\displaystyle\int \frac{1}{x\sqrt{1-x^2}}\,dx$ cannot be evaluated using the basic integration rules.

60. a. $\displaystyle\int e^{x^2}\,dx$ cannot be evaluated using the basic integration rules.

b. $\displaystyle\int xe^{x^2}\,dx = \frac{1}{2}e^{x^2} + C, \quad u = x^2$

c. $\displaystyle\int \frac{1}{x^2}e^{1/x}\,dx = -e^{1/x} + C, \quad u = \frac{1}{x}$

61. a. $\displaystyle\int \sqrt{x-1}\,dx = \frac{2}{3}(x-1)^{3/2} + C, \quad u = x - 1$

b. Let $u = \sqrt{x-1}$. Then $x = u^2 + 1$ and $dx = 2u\,du$.

$$\int x\sqrt{x-1}\,dx = \int (u^2+1)(u)(2u)\,du = 2\int (u^4+u^2)\,du = 2\left(\frac{u^5}{5} + \frac{u^3}{3}\right) + C$$

$$= \frac{2}{15}u^3(3u^2+5) + C = \frac{2}{15}(x-1)^{3/2}[3(x-1)+5] + C = \frac{2}{15}(x-1)^{3/2}(3x+2) + C$$

c. Let $u = \sqrt{x-1}$. Then $x = u^2 + 1$ and $dx = 2u\,du$.

$$\int \frac{x}{\sqrt{x-1}}\,dx = \int \frac{u^2+1}{u}(2u)\,du = 2\int (u^2+1)\,du = 2\left(\frac{u^3}{3} + u\right) + C$$

$$= \frac{2}{3}u(u^2+3) + C = \frac{2}{3}\sqrt{x-1}\,(x+2) + C$$

Note: In b and c, substitution was necessary *before* the basic integration rules could be used.

62. a. $\int \dfrac{1}{1+x^4}\,dx$ cannot be evaluated using the basic integration rules.

b. $\int \dfrac{x}{1+x^4}\,dx = \dfrac{1}{2}\int \dfrac{2x}{1+(x^2)^2}\,dx = \dfrac{1}{2}\arctan(x^2) + C, \quad u = x^2$

c. $\int \dfrac{x^3}{1+x^4}\,dx = \dfrac{1}{4}\int \dfrac{4x^3}{1+x^4}\,dx = \dfrac{1}{4}\ln(1+x^4) + C, \quad u = 1+x^4$

63. The area of the rectangle in the figure is

$A = \text{base} \times \text{height} = 1 \cdot (\pi/2) = \pi/2.$

The area of the unshaded portion is equivalent to the area under the sine curve from 0 to $\pi/2$.

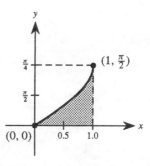

$\text{Unshaded portion} = \displaystyle\int_0^{\pi/2} \sin x\,dx = -\cos x \Big]_0^{\pi/2} = 1$

Hence, the area given by $\int_0^1 \arcsin x\,dx$ is the difference $(\pi/2) - 1 \approx 0.57$.

64.

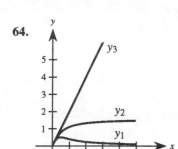

Let $f(x) = \arctan x - \dfrac{x}{1+x^2}$

$f'(x) = \dfrac{1}{1+x^2} - \dfrac{1-x^2}{(1+x^2)^2} = \dfrac{2x^2}{(1+x^2)^2} > 0$ for $x > 0.$

Since $f(0) = 0$ and f is increasing for $x > 0$,

$\arctan x - \dfrac{x}{1+x^2} > 0$ for $x > 0.$

Thus, $\arctan x > x/(1+x^2).$

Let $g(x) = x - \arctan x$

$g'(x) = 1 - \dfrac{1}{1+x^2} = \dfrac{x^2}{1+x^2} > 0$ for $x > 0.$

Since $g(0) = 0$ and g is increasing for $x > 0$, $x - \arctan x > 0$ for $x > 0$. Thus, $x > \arctan x$. Therefore,

$\dfrac{x}{1+x^2} < \arctan x < x.$

Section 5.9 Hyperbolic Functions

1. a. $\sinh 3 = \dfrac{e^3 - e^{-3}}{2} \approx 10.018$

b. $\tanh(-2) = \dfrac{\sinh(-2)}{\cosh(-2)} = \dfrac{e^{-2} - e^2}{e^{-2} + e^2} \approx -0.964$

2. a. $\cosh(0) = \dfrac{e^0 + e^0}{2} = 1$

b. $\operatorname{sech}(1) = \dfrac{2}{e + e^{-1}} \approx 0.648$

3. a. $\operatorname{csch}(\ln 2) = \dfrac{2}{e^{\ln 2} - e^{-\ln 2}} = \dfrac{2}{2 - (1/2)} = \dfrac{4}{3}$

b. $\coth(\ln 5) = \dfrac{\cosh(\ln 5)}{\sinh(\ln 5)} = \dfrac{e^{\ln 5} + e^{-\ln 5}}{e^{\ln 5} - e^{-\ln 5}}$

$= \dfrac{5 + (1/5)}{5 - (1/5)} = \dfrac{13}{12}$

4. a. $\sinh^{-1}(0) = 0$

b. $\tanh^{-1}(0) = 0$

5. a. $\cosh^{-1}(2) = \ln(2 + \sqrt{3}) \approx 1.317$

b. $\operatorname{sech}^{-1}\left(\dfrac{2}{3}\right) = \ln\left(\dfrac{1 + \sqrt{1 - (4/9)}}{2/3}\right) \approx 0.962$

6. a. $\operatorname{csch}^{-1}(2) = \ln\left(\dfrac{1 + \sqrt{5}}{2}\right) \approx 0.481$

b. $\coth^{-1}(3) = \dfrac{1}{2}\ln\left(\dfrac{4}{2}\right) \approx 0.347$

7. $\tanh^2 x + \operatorname{sech}^2 x = \left(\dfrac{e^x - e^{-x}}{e^x + e^{-x}}\right)^2 + \left(\dfrac{2}{e^x + e^{-x}}\right)^2 = \dfrac{e^{2x} - 2 + e^{-2x} + 4}{(e^x + e^{-x})^2} = \dfrac{e^{2x} + 2 + e^{-2x}}{e^{2x} + 2 + e^{-2x}} = 1$

8. $\dfrac{1 + \cosh 2x}{2} = \dfrac{1 + (e^{2x} + e^{-2x})/2}{2} = \dfrac{e^{2x} + 2 + e^{-2x}}{4} = \left(\dfrac{e^x + e^{-x}}{2}\right)^2 = \cosh^2 x$

9. $\sinh x \cosh y + \cosh x \sinh y = \left(\dfrac{e^x - e^{-x}}{2}\right)\left(\dfrac{e^y + e^{-y}}{2}\right) + \left(\dfrac{e^x + e^{-x}}{2}\right)\left(\dfrac{e^y - e^{-y}}{2}\right)$

$$= \dfrac{1}{4}[e^{x+y} - e^{-x+y} + e^{x-y} - e^{-(x+y)} + e^{x+y} + e^{-x+y} - e^{x-y} - e^{-(x+y)}]$$

$$= \dfrac{1}{4}[2(e^{x+y} - e^{-(x+y)})] = \dfrac{e^{(x+y)} - e^{-(x+y)}}{2} = \sinh(x + y)$$

10. $2 \sinh x \cosh x = 2\left(\dfrac{e^x - e^{-x}}{2}\right)\left(\dfrac{e^x + e^{-x}}{2}\right) = \dfrac{e^{2x} - e^{-2x}}{2} = \sinh 2x$

11. $3 \sinh x + 4 \sinh^3 x = \sinh x(3 + 4 \sinh^2 x) = \left(\dfrac{e^x - e^{-x}}{2}\right)\left[3 + 4\left(\dfrac{e^x - e^{-x}}{2}\right)^2\right]$

$$= \left(\dfrac{e^x - e^{-x}}{2}\right)[3 + e^{2x} - 2 + e^{-2x}]$$

$$= \dfrac{1}{2}(e^x - e^{-x})(e^{2x} + e^{-2x} + 1)$$

$$= \dfrac{1}{2}[e^{3x} + e^{-x} + e^x - e^x - e^{-3x} - e^{-x}] = \dfrac{e^{3x} - e^{-3x}}{2} = \sinh(3x)$$

12. $2 \cosh\left(\dfrac{x + y}{2}\right)\cosh\left(\dfrac{x - y}{2}\right) = 2\left[\dfrac{e^{(x+y)/2} + e^{-(x+y)/2}}{2}\right]\left[\dfrac{e^{(x-y)/2} + e^{-(x-y)/2}}{2}\right]$

$$= 2\left[\dfrac{e^x + e^y + e^{-y} + e^{-x}}{4}\right] = \dfrac{e^x + e^{-x}}{2} + \dfrac{e^y + e^{-y}}{2} = \cosh x + \cosh y$$

13. $\sinh x = \dfrac{3}{2}$

$\cosh^2 x - \left(\dfrac{3}{2}\right)^2 = 1 \Rightarrow \cosh^2 x = \dfrac{13}{4} \Rightarrow \cosh x = \dfrac{\sqrt{13}}{2}$

$\tanh x = \dfrac{3/2}{\sqrt{13}/2} = \dfrac{3\sqrt{13}}{13}$

$\operatorname{csch} x = \dfrac{1}{3/2} = \dfrac{2}{3}$

$\operatorname{sech} x = \dfrac{1}{\sqrt{13}/2} = \dfrac{2\sqrt{13}}{13}$

$\coth x = \dfrac{1}{3/\sqrt{13}} = \dfrac{\sqrt{13}}{3}$

14. $\tanh x = \dfrac{1}{2}$

$\left(\dfrac{1}{2}\right)^2 + \mathrm{sech}^2 x = 1 \Rightarrow \mathrm{sech}^2 x = \dfrac{3}{4} \Rightarrow \mathrm{sech}\, x = \dfrac{\sqrt{3}}{2}$

$\cosh x = \dfrac{1}{\sqrt{3}/2} = \dfrac{2\sqrt{3}}{3}$

$\coth x = \dfrac{1}{1/2} = 2$

$\sinh x = \tanh x \cosh x = \left(\dfrac{1}{2}\right)\left(\dfrac{2\sqrt{3}}{3}\right) = \dfrac{\sqrt{3}}{3}$

$\mathrm{csch}\, x = \dfrac{1}{\sqrt{3}/3} = \sqrt{3}$

Putting these in order:

$\sinh x = \dfrac{\sqrt{3}}{3}$ $\mathrm{csch}\, x = \sqrt{3}$

$\cosh x = \dfrac{2\sqrt{3}}{3}$ $\mathrm{sech}\, x = \dfrac{\sqrt{3}}{2}$

$\tanh x = \dfrac{1}{2}$ $\coth x = 2$

15. $y = \sinh(1 - x^2)$

$y' = -2x \cosh(1 - x^2)$

16. $y = \coth(3x)$

$y' = -3 \, \mathrm{csch}^2(3x)$

17. $f(x) = \ln(\sinh x)$

$f'(x) = \dfrac{1}{\sinh x}(\cosh x) = \coth x$

18. $g(x) = \ln(\cosh x)$

$g'(x) = \dfrac{1}{\cosh x}(\sinh x) = \tanh x$

19. $y = \ln\left(\tanh \dfrac{x}{2}\right)$

$y' = \dfrac{1/2}{\tanh(x/2)} \mathrm{sech}^2\left(\dfrac{x}{2}\right)$

$= \dfrac{1}{2 \sinh(x/2) \cosh(x/2)} = \dfrac{1}{\sinh x} = \mathrm{csch}\, x$

20. $y = x \sinh x - \cosh x$

$y' = x \cosh x + \sinh x - \sinh x = x \cosh x$

21. $h(x) = \dfrac{1}{4} \sinh(2x) - \dfrac{x}{2}$

$h'(x) = \dfrac{1}{2} \cosh(2x) - \dfrac{1}{2} = \dfrac{\cosh(2x) - 1}{2} = \sinh^2 x$

22. $h(t) = t - \coth t$

$h'(t) = 1 + \mathrm{csch}^2 t = \coth^2 t$

23. $f(t) = \arctan(\sinh t)$

$f'(t) = \dfrac{1}{1 + \sinh^2 t}(\cosh t) = \dfrac{\cosh t}{\cosh^2 t} = \mathrm{sech}\, t$

24. $f(x) = e^{\sinh x}$

$f'(x) = (\cosh x)(e^{\sinh x})$

25. Let $y = g(x)$.

$y = x^{\cosh x}$

$\ln y = \cosh x \ln x$

$\dfrac{1}{y}\left(\dfrac{dy}{dx}\right) = \dfrac{\cosh x}{x} + \sinh x \ln x$

$\dfrac{dy}{dx} = \dfrac{y}{x}[\cosh x + x(\sinh x) \ln x]$

$= \dfrac{x^{\cosh x}}{x}[\cosh x + x(\sinh x) \ln x]$

26. $g(x) = \mathrm{sech}^2\, 3x$

$g'(x) = -2 \, \mathrm{sech}(3x) \, \mathrm{sech}(3x) \tanh(3x)(3)$

$= -6 \, \mathrm{sech}^2\, 3x \tanh 3x$

27. $y = (\cosh x - \sinh x)^2$

$y' = 2(\cosh x - \sinh x)(\sinh x - \cosh x)$

$\quad = -2(\cosh x - \sinh x)^2 = -2e^{-2x}$

28. $y = \operatorname{sech}(x + 1)$

$y' = -\operatorname{sech}(x + 1)\tanh(x + 1)$

29. $f(x) = \sin x \sinh x - \cos x \cosh x, \quad -4 \le x \le 4$

$f'(x) = \sin x \cosh x + \cos x \sinh x - \cos x \sinh x + \sin x \cosh x$

$\quad = 2\sin x \cosh x = 0$ when $x = 0, \ \pm\pi$

Relative maxima: $(\pm\pi, \ \cosh\pi)$

Relative minimum: $(0, \ -1)$

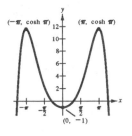

30. $f(x) = x\cosh(x - 1) - \sinh(x - 1)$

$f'(x) = x\sinh(x - 1) + \cosh(x - 1) - \cosh(x - 1)$

$\quad = x\sinh(x - 1) = 0$ when $x = 0$ or $x = 1$

Relative maximum: $(0, \ -\sinh(-1))$

Relative minimum: $(1, \ 1)$

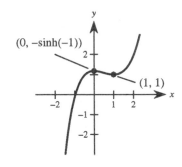

31. $g(x) = x\operatorname{sech} x = \dfrac{x}{\cosh x}$

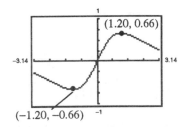

Relative maximum: $(1.20, \ 0.66)$

Relative minimum: $(-1.20, \ -0.66)$

32. $h(x) = 2\tanh x - x$

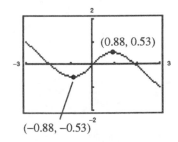

Relative maximum: $(0.88, \ 0.53)$

Relative minimum: $(-0.88, \ -0.53)$

33. $y = a\sinh x$

$y' = a\cosh x$

$y'' = a\sinh x$

$y''' = a\cosh x$

Therefore, $y''' - y' = 0.$

34. $y = a\cosh x$

$y' = a\sinh x$

$y'' = a\cosh x$

Therefore, $y'' - y = 0.$

35. Let $u = 1 - 2x, \quad du = -2\,dx.$

$$\int \sinh(1 - 2x)\,dx = -\frac{1}{2}\int \sinh(1 - 2x)(-2)\,dx = -\frac{1}{2}\cosh(1 - 2x) + C$$

36. Let $u = \sqrt{x}, \quad du = \dfrac{1}{2\sqrt{x}}\,dx.$

$$\int \frac{\cosh\sqrt{x}}{\sqrt{x}}\,dx = 2\int \cosh\sqrt{x}\left(\frac{1}{2\sqrt{x}}\right)dx = 2\sinh\sqrt{x} + C$$

37. Let $u = \cosh(x - 1)$, $du = \sinh(x - 1)\,dx$.

$$\int \cosh^2(x - 1)\sinh(x - 1)\,dx = \frac{1}{3}\cosh^3(x - 1) + C$$

38. Let $u = \cosh x$, $du = \sinh x\,dx$.

$$\int \frac{\sinh x}{1 + \sinh^2 x}\,dx = \int \frac{\sinh x}{\cosh^2 x}\,dx = \frac{-1}{\cosh x} + C = -\operatorname{sech} x + C$$

39. Let $u = \sinh x$, $du = \cosh x\,dx$.

$$\int \frac{\cosh x}{\sinh x}\,dx = \ln|\sinh x| + C$$

40. Let $u = 2x - 1$, $du = 2\,dx$.

$$\int \operatorname{sech}^2(2x - 1)\,dx = \frac{1}{2}\int \operatorname{sech}^2(2x - 1)(2)\,dx = \frac{1}{2}\tanh(2x - 1) + C$$

41. Let $u = \dfrac{x^2}{2}$, $du = x\,dx$.

$$\int x\,\operatorname{csch}^2\frac{x^2}{2}\,dx = \int \left(\operatorname{csch}^2\frac{x^2}{2}\right)x\,dx = -\coth\frac{x^2}{2} + C$$

42. Let $u = \operatorname{sech} x$, $du = -\operatorname{sech} x \tanh x\,dx$.

$$\int \operatorname{sech}^3 x \tanh x\,dx = -\int \operatorname{sech}^2 x(-\operatorname{sech} x \tanh x)\,dx = -\frac{1}{3}\operatorname{sech}^3 x + C$$

43. Let $u = \dfrac{1}{x}$, $du = -\dfrac{1}{x^2}\,dx$.

$$\int \frac{\operatorname{csch}(1/x)\coth(1/x)}{x^2}\,dx = -\int \operatorname{csch}\frac{1}{x}\coth\frac{1}{x}\left(-\frac{1}{x^2}\right)dx = \operatorname{csch}\frac{1}{x} + C$$

44. $\displaystyle \int \sinh^2 x\,dx = \frac{1}{2}\int(\cosh 2x - 1)\,dx = \frac{1}{2}\left(\frac{1}{2}\sinh 2x - x\right) + C = \frac{1}{4}(-2x + \sinh 2x) + C$

45. $\displaystyle \int_0^4 \frac{1}{25 - x^2}\,dx = \left[\frac{1}{10}\ln\left|\frac{5 + x}{5 - x}\right|\right]_0^4 = \frac{1}{10}\ln 9 = \frac{1}{5}\ln 3$

46. $\displaystyle \int_0^4 \frac{1}{\sqrt{25 - x^2}}\,dx = \left[\arcsin\frac{x}{5}\right]_0^4 = \arcsin\frac{4}{5}$

47. Let $u = 2x$, $du = 2\,dx$.

$$\int_0^{\sqrt{2}/4} \frac{1}{\sqrt{1 - (2x)^2}}(2)\,dx = \left[\arcsin(2x)\right]_0^{\sqrt{2}/4} = \frac{\pi}{4}$$

48. $\displaystyle \int \frac{2}{x\sqrt{1 + 4x^2}}\,dx = 2\int \frac{1}{(2x)\sqrt{1 + (2x)^2}}(2)\,dx = -2\ln\left(\frac{1 + \sqrt{1 + 4x^2}}{|2x|}\right) + C$

49. Let $u = x^2$, $du = 2x\,dx$.

$$\int \frac{x}{x^4 + 1}\,dx = \frac{1}{2}\int \frac{2x}{(x^2)^2 + 1}\,dx = \frac{1}{2}\arctan(x^2) + C$$

50. Let $u = \sinh x$, $du = \cosh x\,dx$.

$$\int \frac{\cosh x}{\sqrt{9 - \sinh^2 x}}\,dx = \arcsin\left(\frac{\sinh x}{3}\right) + C = \arcsin\left(\frac{e^x - e^{-x}}{6}\right) + C$$

51. $y = \cosh^{-1}(3x)$

$$y' = \frac{3}{\sqrt{9x^2 - 1}}$$

52. $y = \tanh^{-1}\left(\dfrac{x}{2}\right)$

$$y' = \frac{1}{1 - (x/2)^2}\left(\frac{1}{2}\right)$$

$$= \frac{2}{4 - x^2}$$

53. $y = \sinh^{-1}(\tan x)$

$$y' = \frac{1}{\sqrt{\tan^2 x + 1}}(\sec^2 x)$$

$$= |\sec x|$$

54. $y = \operatorname{sech}^{-1}(\cos 2x), \quad 0 < x < \dfrac{\pi}{4}$

$$y' = \frac{-1}{\cos 2x\sqrt{1 - \cos^2 2x}}(-2\sin 2x) = \frac{2\sin 2x}{\cos 2x|\sin 2x|} = \frac{2}{\cos 2x}, \quad \text{since } \sin 2x \ge 0 \text{ for } 0 < x < \frac{\pi}{4}$$

$$= 2\sec 2x$$

55. $y = \coth^{-1}(\sin 2x)$

$$y' = \frac{1}{1 - \sin^2 2x}(2\cos 2x) = 2\sec 2x$$

56. $y = (\operatorname{csch}^{-1} x)^2$

$$y' = 2\operatorname{csch}^{-1} x\left(\frac{-1}{|x|\sqrt{1 + x^2}}\right) = \frac{-2\operatorname{csch}^{-1} x}{|x|\sqrt{1 + x^2}}$$

57. $y = 2x\sinh^{-1}(2x) - \sqrt{1 + 4x^2}$

$$y' = 2x\left(\frac{2}{\sqrt{1 + 4x^2}}\right) + 2\sinh^{-1}(2x) - \frac{4x}{\sqrt{1 + 4x^2}} = 2\sinh^{-1}(2x)$$

58. $y = x\tanh^{-1} x + \ln\sqrt{1 - x^2} = x\tanh^{-1} x + \dfrac{1}{2}\ln(1 - x^2)$

$$y' = x\left(\frac{1}{1 - x^2}\right) + \tanh^{-1} x + \frac{-x}{1 - x^2} = \tanh^{-1} x$$

59. $y = a\operatorname{sech}^{-1}\left(\dfrac{x}{a}\right) - \sqrt{a^2 - x^2}$

$$\frac{dy}{dx} = \frac{-1}{(x/a)\sqrt{1 - (x^2/a^2)}} + \frac{x}{\sqrt{a^2 - x^2}} = \frac{-a^2}{x\sqrt{a^2 - x^2}} + \frac{x}{\sqrt{a^2 - x^2}} = \frac{-\sqrt{a^2 - x^2}}{x}$$

60. Equation of tangent line:

$$y - a\operatorname{sech}^{-1}\frac{x_0}{a} + \sqrt{a^2 - x_0^2} = \frac{-\sqrt{a^2 - x_0^2}}{x_0}(x - x_0)$$

When $x = 0$,

$$y = a\operatorname{sech}^{-1}\frac{x_0}{a} - \sqrt{a^2 - x_0^2} + \sqrt{a^2 - x_0^2} = a\operatorname{sech}^{-1}\frac{x_0}{a}.$$

Hence, Q is the point $\left(0, \ a\operatorname{sech}^{-1}\dfrac{x_0}{a}\right)$.

Distance from P to Q: $d = \sqrt{x_0^2 + \left(-\sqrt{a^2 - x_0^2}\right)^2} = a$

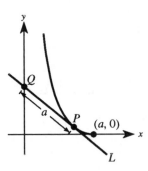

61. $\displaystyle \int \frac{1}{\sqrt{1 + e^{2x}}}\,dx = \int \frac{e^x}{e^x\sqrt{1 + (e^x)^2}}\,dx = -\operatorname{csch}^{-1}(e^x) + C = -\ln\left(\frac{1 + \sqrt{1 + e^{2x}}}{e^x}\right) + C$

62. Let $u = e^x, \quad du = e^x\,dx$.

$$\int \frac{e^x}{1 - e^{2x}}\,dx = \int \frac{1}{1 - (e^x)^2}(e^x)\,dx = \tanh^{-1}(e^x) + C = \frac{1}{2}\ln\left|\frac{1 + e^x}{1 - e^x}\right| + C$$

63. Let $u = \sqrt{x}$, $du = \dfrac{1}{2\sqrt{x}}\,dx$.

$$\int \frac{1}{\sqrt{x}\sqrt{1+x}}\,dx = 2\int \frac{1}{\sqrt{1+(\sqrt{x})^2}}\left(\frac{1}{2\sqrt{x}}\right)dx = 2\sinh^{-1}\sqrt{x} + C = 2\ln\left(\sqrt{x} + \sqrt{1+x}\right) + C$$

64. Let $u = x^{3/2}$, $du = \dfrac{3}{2}\sqrt{x}\,dx$.

$$\int \frac{\sqrt{x}}{\sqrt{1+x^3}}\,dx = \frac{2}{3}\int \frac{1}{\sqrt{1+(x^{3/2})^2}}\left(\frac{3}{2}\sqrt{x}\right)dx = \frac{2}{3}\sinh^{-1}(x^{3/2}) + C = \frac{2}{3}\ln\left(x^{3/2} + \sqrt{1+x^3}\right) + C$$

65. $\displaystyle \int \frac{1}{(x-1)\sqrt{x^2-2x+2}}\,dx = \int \frac{1}{(x-1)\sqrt{(x-1)^2+1}}\,dx = -\ln\left[\frac{1+\sqrt{(x-1)^2+1}}{|x-1|}\right] + C$

66. $\displaystyle \int \frac{-1}{4x-x^2}\,dx = \int \frac{1}{(x-2)^2-4}\,dx = \frac{1}{4}\ln\left|\frac{(x-2)-2}{(x-2)+2}\right| = \frac{1}{4}\ln\left|\frac{x-4}{x}\right| + C$

67. $\displaystyle \int \frac{1}{1-4x-2x^2}\,dx = \int \frac{1}{3-2(x+1)^2}\,dx = \frac{-1}{\sqrt{2}}\int \frac{\sqrt{2}}{[\sqrt{2}(x+1)]^2 - (\sqrt{3})^2}\,dx$

$$= \frac{-1}{2\sqrt{6}}\ln\left|\frac{\sqrt{2}(x+1)-\sqrt{3}}{\sqrt{2}(x+1)+\sqrt{3}}\right| + C = \frac{1}{2\sqrt{6}}\ln\left|\frac{\sqrt{2}(x+1)+\sqrt{3}}{\sqrt{2}(x+1)-\sqrt{3}}\right| + C$$

68. $\displaystyle \int \frac{1}{(x+1)\sqrt{2x^2+4x+8}}\,dx = \int \frac{1}{(x+1)\sqrt{2(x+1)^2+6}}\,dx = \frac{1}{\sqrt{2}}\int \frac{1}{(x+1)\sqrt{(x+1)^2+(\sqrt{3})^2}}\,dx$

$$= -\frac{1}{\sqrt{6}}\ln\left(\frac{\sqrt{3}+\sqrt{(x+1)^2+3}}{x+1}\right) + C$$

69. Let $u = 4x-1$, $du = 4\,dx$.

$$y = \int \frac{1}{\sqrt{80+8x-16x^2}}\,dx = \frac{1}{4}\int \frac{4}{\sqrt{81-(4x-1)^2}}\,dx = \frac{1}{4}\arcsin\left(\frac{4x-1}{9}\right) + C$$

70. Let $u = 2(x-1)$, $du = 2\,dx$.

$$y = \int \frac{1}{(x-1)\sqrt{-4x^2+8x-1}}\,dx$$

$$= \int \frac{2}{2(x-1)\sqrt{(\sqrt{3})^2 - [2(x-1)]^2}}\,dx = -\frac{1}{\sqrt{3}}\ln\left|\frac{\sqrt{3}+\sqrt{-4x^2+8x-1}}{2(x-1)}\right| + C$$

71. $\displaystyle y = \int \frac{x^3-21x}{5+4x-x^2}\,dx = \int\left(-x-4+\frac{20}{5+4x-x^2}\right)dx = \int(-x-4)\,dx - 20\int\frac{1}{(x-2)^2-3^2}\,dx$

$$= -\frac{x^2}{2} - 4x - \frac{20}{6}\ln\left|\frac{(x-2)-3}{(x-2)+3}\right| + C = -\frac{x^2}{2} - 4x - \frac{10}{3}\ln\left|\frac{x-5}{x+1}\right| + C$$

72. $\displaystyle y = \int \frac{1-2x}{4x-x^2}\,dx = \int \frac{4-2x}{4x-x^2}\,dx + 3\int \frac{1}{(x-2)^2-4}\,dx$

$$= \ln|4x-x^2| + \frac{3}{4}\ln\left|\frac{(x-2)-2}{(x-2)+2}\right| + C = \ln|4x-x^2| + \frac{3}{4}\ln\left|\frac{x-4}{x}\right| + C$$

73. $A = 2 \int_0^4 \text{sech} \frac{x}{2} \, dx$

$= 2 \int_0^4 \frac{2}{e^{x/2} + e^{-x/2}} \, dx$

$= 4 \int_0^4 \frac{e^{x/2}}{(e^{x/2})^2 + 1} \, dx$

$= \left[8 \arctan(e^{x/2}) \right]_0^4$

$= 8 \arctan(e^2) - 2\pi \approx 5.207$

74. $A = \int_0^2 \tanh 2x \, dx$

$= \int_0^2 \frac{e^{2x} - e^{-2x}}{e^{2x} + e^{-2x}} \, dx$

$= \frac{1}{2} \int_0^2 \frac{1}{e^{2x} + e^{-2x}} (2)(e^{2x} - e^{-2x}) \, dx$

$= \left[\frac{1}{2} \ln(e^{2x} + e^{-2x}) \right]_0^2$

$= \frac{1}{2} \ln(e^4 + e^{-4}) - \frac{1}{2} \ln 2$

$= \ln \sqrt{\frac{e^4 + e^{-4}}{2}} \approx 1.654$

75. $A = \int_0^2 \frac{5x}{\sqrt{x^4 + 1}} \, dx$

$= \frac{5}{2} \int_0^2 \frac{2x}{\sqrt{(x^2)^2 + 1}} \, dx$

$= \left[\frac{5}{2} \ln \left(x^2 + \sqrt{x^4 + 1} \right) \right]_0^2$

$= \frac{5}{2} \ln \left(4 + \sqrt{17} \right) \approx 5.237$

76. $A = \int_3^5 \frac{6}{\sqrt{x^2 - 4}} \, dx$

$= \left[6 \ln \left(x + \sqrt{x^2 - 4} \right) \right]_3^5$

$= 6 \ln \left(5 + \sqrt{21} \right) - 6 \ln \left(3 + \sqrt{5} \right)$

$= 6 \ln \left(\frac{5 + \sqrt{21}}{3 + \sqrt{5}} \right) \approx 3.626$

77. $\int \frac{3k}{16} \, dt = \int \frac{1}{x^2 - 12x + 32} \, dx$

$\frac{3kt}{16} = \int \frac{1}{(x - 6)^2 - 4} \, dx = \frac{1}{2(2)} \ln \left| \frac{(x - 6) - 2}{(x - 6) + 2} \right| + C = \frac{1}{4} \ln \left| \frac{x - 8}{x - 4} \right| + C$

When $x = 0$: $t = 0$

$$C = -\frac{1}{4} \ln(2)$$

When $x = 1$: $t = 10$

$$\frac{30k}{16} = \frac{1}{4} \ln \left| \frac{-7}{-3} \right| - \frac{1}{4} \ln(2) = \frac{1}{4} \ln \left(\frac{7}{6} \right)$$

$$k = \frac{2}{15} \ln \left(\frac{7}{6} \right)$$

When $t = 20$: $\left(\frac{3}{16} \right) \left(\frac{2}{15} \right) \ln \left(\frac{7}{6} \right) (20) = \frac{1}{4} \ln \frac{x - 8}{2x - 8}$

$$\ln \left(\frac{7}{6} \right)^2 = \ln \frac{x - 8}{2x - 8}$$

$$\frac{49}{36} = \frac{x - 8}{2x - 8}$$

$$62x = 104$$

$$x = \frac{104}{62} = \frac{52}{31} \approx 1.677 \text{ kg}$$

78. a. $v(t) = -32t$

b. $s(t) = \int v(t)\,dt = \int (-32t)\,dt = -16t^2 + C$

$$s(0) = -16(0)^2 + C = 400 \Rightarrow C = 400$$

$$s(t) = -16t^2 + 400$$

c.
$$\frac{dv}{dt} = -32 + kv^2$$

$$\int \frac{dv}{kv^2 - 32} = \int dt$$

$$\int \frac{dv}{32 - kv^2} = -\int dt$$

Let $u = \sqrt{k}\,v$, then $du = \sqrt{k}\,dv$.

$$\frac{1}{\sqrt{k}} \cdot \frac{1}{2\sqrt{32}} \ln \left| \frac{\sqrt{32} + \sqrt{k}\,v}{\sqrt{32} - \sqrt{k}\,v} \right| = -t + C$$

Since $v(0) = 0,\ C = 0$.

$$\ln \left| \frac{\sqrt{32} + \sqrt{k}\,v}{\sqrt{32} - \sqrt{k}\,v} \right| = -2\sqrt{32k}\,t$$

$$\frac{\sqrt{32} + \sqrt{k}\,v}{\sqrt{32} - \sqrt{k}\,v} = e^{-2\sqrt{32k}\,t}$$

$$\sqrt{32} + \sqrt{k}\,v = e^{-2\sqrt{32k}\,t}\left(\sqrt{32} - \sqrt{k}\,v\right)$$

$$v\left(\sqrt{k} + \sqrt{k}\,e^{-2\sqrt{32k}\,t}\right) = \sqrt{32}\left(e^{-2\sqrt{32k}\,t} - 1\right)$$

$$v = \frac{\sqrt{32}\left(e^{-2\sqrt{32k}\,t} - 1\right)}{\sqrt{k}\left(e^{-2\sqrt{32k}\,t} + 1\right)} \cdot \frac{e^{\sqrt{32k}\,t}}{e^{\sqrt{32k}\,t}}$$

$$= \frac{\sqrt{32}}{\sqrt{k}} \left[\frac{-\left(e^{\sqrt{32k}\,t} - e^{-\sqrt{32k}\,t}\right)}{e^{\sqrt{32k}\,t} + e^{-\sqrt{32k}\,t}} \right] = -\frac{\sqrt{32}}{\sqrt{k}} \tanh\left(\sqrt{32k}\,t\right)$$

d. $\displaystyle\lim_{t \to \infty} \left[-\frac{\sqrt{32}}{\sqrt{k}} \tanh\left(\sqrt{32k}\,t\right) \right] = -\frac{\sqrt{32}}{\sqrt{k}}$. The velocity is bounded by $-\sqrt{32}/\sqrt{k}$.

e. Since $\int \tanh(ct)\,dt = (1/c)\ln\cosh(ct)$ (which can be verified by differentiation), then

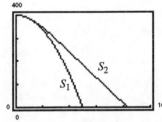

$$s(t) = \int -\frac{\sqrt{32}}{\sqrt{k}} \tanh\left(\sqrt{32k}\,t\right) dt$$

$$= -\frac{\sqrt{32}}{\sqrt{k}} \frac{1}{\sqrt{32k}} \ln\left[\cosh\left(\sqrt{32k}\,t\right)\right] + C$$

$$= -\frac{1}{k} \ln\left[\cosh\left(\sqrt{32k}\,t\right)\right] + C.$$

When $t = 0,\ s(0) = C = 400 \Rightarrow s(t) = 400 - (1/k)\ln\left[\cosh\left(\sqrt{32k}\,t\right)\right]$.

When $k = 0.01,\ s_2(t) = 400 - 100\ln\left(\cosh\sqrt{0.32}\,t\right)$

$$s_1(t) = -16t^2 + 400.$$

$$s_1(t) = 0 \text{ when } t = 5 \text{ seconds.}$$

$$s_2(t) = 0 \text{ when } t \approx 8.3 \text{ seconds.}$$

When air resistance is not neglected, it takes approximately 3.3 more seconds to reach the ground.

79. As k increases, the time required for the object to reach the ground increases.

80. a. $y = 693.8597 - 68.7672 \cosh 0.0100333x$

$y' = -68.7672(0.0100333) \sinh 0.0100333x = 0$ when $x = 0$.

Maximum height $(x = 0)$: $y = 693.8597 - 68.7672 \cosh 0 \approx 625$ feet

b. The height of the arch equals y (when $x = 0$) plus the distance from the center of mass of the triangle to the top of the highest triangle.

$A = \dfrac{1}{2}(2c)\left(\sqrt{3}\,c\right) = \sqrt{3}\,c^2$

When $x = 0$, $A \approx 125$.

$125 = \sqrt{3}\,c^2$

$c = \sqrt{\dfrac{125}{\sqrt{3}}} \approx 8.5$

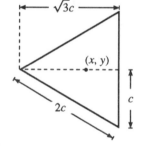

Height of the triangle: $8.5\sqrt{3}$

Distance from the center of mass of the triangle to the top of the highest triangle: $\frac{1}{3} \cdot 8.5\sqrt{3} \approx 5$ feet

Height of the arch: $625 + 5 = 630$ feet

c. The width of the arch equals $2 \cdot 299.2239 + 2$ times the distance from the center of mass of the triangle to the outer side of the triangle at the arch's base.

$A = \sqrt{3}\,c^2$

When $x \approx 299.2239$, $A \approx 1263$.

$1263 = \sqrt{3}\,c^2$

$c = \sqrt{\dfrac{1263}{\sqrt{3}}} \approx 27.0$

Height of the triangle: $27\sqrt{3}$

Distance from the center of mass of the triangle to the outer side of the triangle at the arch's base: $\frac{1}{3} \cdot 27\sqrt{3} \approx 16$ feet

Width of the arch: $598 + 32 = 630$ feet

81. $y = \cosh x = \dfrac{e^x + e^{-x}}{2}$

$y' = \dfrac{e^x - e^{-x}}{2} = \sinh x$

82. $y = \operatorname{sech} x = \dfrac{2}{e^x + e^{-x}}$

$y' = -2(e^x + e^{-x})^{-2}(e^x - e^{-x})$

$= \left(\dfrac{-2}{e^x + e^{-x}}\right)\left(\dfrac{e^x - e^{-x}}{e^x + e^{-x}}\right)$

$= -\operatorname{sech} x \tanh x$

83. $y = \cosh^{-1} x$

$\cosh y = x$

$(\sinh y)(y') = 1$

$y' = \dfrac{1}{\sinh y}$

$= \dfrac{1}{\sqrt{\cosh^2 y - 1}} = \dfrac{1}{\sqrt{x^2 - 1}}$

84. $y = \sinh^{-1} x$

$\sinh y = x$

$(\cosh y)y' = 1$

$y' = \dfrac{1}{\cosh y}$

$= \dfrac{1}{\sqrt{\sinh^2 y + 1}} = \dfrac{1}{\sqrt{x^2 + 1}}$

85.
$$y = \operatorname{sech}^{-1} x$$
$$\operatorname{sech} y = x$$
$$-(\operatorname{sech} y)(\tanh y)y' = 1$$
$$y' = \frac{-1}{(\operatorname{sech} y)(\tanh y)} = \frac{-1}{(\operatorname{sech} y)\sqrt{1 - \operatorname{sech}^2 y}} = \frac{-1}{x\sqrt{1 - x^2}}$$

86. Let $y = \arcsin(\tanh x)$. Then

$$\sin y = \tanh x = \frac{e^x - e^{-x}}{e^x + e^{-x}} \quad \text{and} \quad \tan y = \frac{e^x - e^{-x}}{2} = \sinh x.$$

Thus, $y = \arctan(\sinh x)$. Therefore, $\arctan(\sinh x) = \arcsin(\tanh x)$.

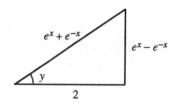

Chapter 5   Review Exercises

1. $f(x) = \ln x + 3$
Vertical shift 3 units upward
Vertical asymptote: $x = 0$

2. $f(x) = \ln(x - 3)$
Horizontal shift 3 units to the right
Vertical asymptote: $x = 3$

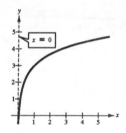

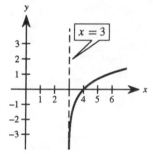

3. $\ln \sqrt[5]{\dfrac{4x^2 - 1}{4x^2 + 1}} = \dfrac{1}{5} \ln \dfrac{(2x - 1)(2x + 1)}{4x^2 + 1} = \dfrac{1}{5}[\ln(2x - 1) + \ln(2x + 1) - \ln(4x^2 + 1)]$

4. $\ln[(x^2 + 1)(x - 1)] = \ln(x^2 + 1) + \ln(x - 1)$

5. $\ln 3 + \dfrac{1}{3} \ln(4 - x^2) - \ln x = \ln 3 + \ln \sqrt[3]{4 - x^2} - \ln x = \ln\left(\dfrac{3\sqrt[3]{4 - x^2}}{x}\right)$

6. $3[\ln x - 2\ln(x^2 + 1)] + 2\ln 5 = 3\ln x - 6\ln(x^2 + 1) + \ln 5^2$

$$= \ln x^3 - \ln(x^2 + 1)^6 + \ln 25 = \ln\left[\dfrac{25x^3}{(x^2 + 1)^6}\right]$$

7. False; the domain of $f(x) = \ln x$ is the set of all **positive** real numbers.

8. False

$$\ln x + \ln y = \ln(xy) \neq \ln(x + y)$$

9. $\ln \sqrt{x+1} = 2$

$\quad\quad \sqrt{x+1} = e^2$

$\quad\quad\quad x+1 = e^4$

$\quad\quad\quad\quad x = e^4 - 1 \approx 53.598$

10. $\ln x + \ln(x-3) = 0$

$\quad\quad \ln x(x-3) = 0$

$\quad\quad\quad x(x-3) = e^0$

$\quad\quad x^2 - 3x - 1 = 0$

$$x = \frac{3 \pm \sqrt{13}}{2}$$

$$x = \frac{3 + \sqrt{13}}{2} \quad \text{only since}$$

$$\frac{3 - \sqrt{13}}{2} < 0.$$

11. $g(x) = \ln \sqrt{x} = \dfrac{1}{2} \ln x$

$\quad g'(x) = \dfrac{1}{2x}$

12. $h(x) = \ln \dfrac{x(x-1)}{x-2}$

$\quad\quad\quad = \ln x + \ln(x-1) - \ln(x-2)$

$\quad h'(x) = \dfrac{1}{x} + \dfrac{1}{x-1} - \dfrac{1}{x-2} = \dfrac{x^2 - 4x + 2}{x^3 - 3x^2 + 2x}$

13. $f(x) = x\sqrt{\ln x}$

$\quad f'(x) = \left(\dfrac{x}{2}\right)(\ln x)^{-1/2}\left(\dfrac{1}{x}\right) + \sqrt{\ln x}$

$\quad\quad = \dfrac{1}{2\sqrt{\ln x}} + \sqrt{\ln x} = \dfrac{1 + 2\ln x}{2\sqrt{\ln x}}$

14. $f(x) = \ln[x(x^2-2)^{2/3}] = \ln x + \dfrac{2}{3}\ln(x^2-2)$

$\quad f'(x) = \dfrac{1}{x} + \dfrac{2}{3}\left(\dfrac{2x}{x^2-2}\right) = \dfrac{7x^2 - 6}{3x^3 - 6x}$

15. $y = \dfrac{1}{b^2}\left[\ln(a+bx) + \dfrac{a}{a+bx}\right]$

$\quad \dfrac{dy}{dx} = \dfrac{1}{b^2}\left[\dfrac{b}{a+bx} - \dfrac{ab}{(a+bx)^2}\right] = \dfrac{x}{(a+bx)^2}$

16. $y = \dfrac{1}{b^2}[a+bx - a\ln(a+bx)]$

$\quad \dfrac{dy}{dx} = \dfrac{1}{b^2}\left(b - \dfrac{ab}{a+bx}\right) = \dfrac{x}{a+bx}$

17. $y = -\dfrac{1}{a}\ln\left(\dfrac{a+bx}{x}\right)$

$\quad\quad = -\dfrac{1}{a}[\ln(a+bx) - \ln x]$

$\quad \dfrac{dy}{dx} = -\dfrac{1}{a}\left(\dfrac{b}{a+bx} - \dfrac{1}{x}\right) = \dfrac{1}{x(a+bx)}$

18. $y = -\dfrac{1}{ax} + \dfrac{b}{a^2}\ln\dfrac{a+bx}{x}$

$\quad\quad = -\dfrac{1}{ax} + \dfrac{b}{a^2}[\ln(a+bx) - \ln x]$

$\quad \dfrac{dy}{dx} = -\dfrac{1}{a}\left(-\dfrac{1}{x^2}\right) + \dfrac{b}{a^2}\left[\dfrac{b}{a+bx} - \dfrac{1}{x}\right]$

$\quad\quad = \dfrac{1}{ax^2} + \dfrac{b}{a^2}\left[\dfrac{-a}{x(a+bx)}\right]$

$\quad\quad = \dfrac{1}{ax^2} - \dfrac{b}{ax(a+bx)}$

$\quad\quad = \dfrac{(a+bx) - bx}{ax^2(a+bx)}$

$\quad\quad = \dfrac{1}{x^2(a+bx)}$

19. $u = 7x - 2, \quad du = 7\,dx$

$$\int \frac{1}{7x-2}\,dx = \frac{1}{7} \int \frac{1}{7x-2}(7)\,dx$$

$$= \frac{1}{7} \ln |7x - 2| + C$$

20. $u = x^2 - 1, \quad du = 2x\,dx$

$$\int \frac{x}{x^2-1}\,dx = \frac{1}{2} \int \frac{2x}{x^2-1}\,dx$$

$$= \frac{1}{2} \ln |x^2 - 1| + C$$

21. $\displaystyle\int \frac{\sin x}{1+\cos x}\,dx = -\int \frac{-\sin x}{1+\cos x}\,dx$

$$= -\ln |1 + \cos x| + C$$

22. $u = \ln x, \quad du = \dfrac{1}{x}\,dx$

$$\int \frac{\ln \sqrt{x}}{x}\,dx = \frac{1}{2} \int (\ln x)\left(\frac{1}{x}\right)\,dx$$

$$= \frac{1}{4}(\ln x)^2 + C$$

23. $\displaystyle\int_1^4 \frac{x+1}{x}\,dx = \int_1^4 \left(1 + \frac{1}{x}\right)\,dx$

$$= \left[x + \ln |x| \right]_1^4 = 3 + \ln 4$$

24. $\displaystyle\int_1^e \frac{\ln x}{x}\,dx = \int_1^e (\ln x)^1 \left(\frac{1}{x}\right)\,dx$

$$= \left[\frac{1}{2}(\ln x)^2\right]_1^e = \frac{1}{2}$$

25. $\displaystyle\int_0^{\pi/3} \sec\theta\,d\theta = \left[\ln |\sec\theta + \tan\theta|\right]_0^{\pi/3}$

$$= \ln(2 + \sqrt{3})$$

26. $\displaystyle\int_0^{\pi/4} \tan\left(\frac{\pi}{4} - x\right)dx = \left[\ln \left|\cos\left(\frac{\pi}{4} - x\right)\right|\right]_0^{\pi/4}$

$$= 0 - \ln\left(\frac{1}{\sqrt{2}}\right) = \frac{1}{2}\ln 2$$

27. $y = \displaystyle\int \frac{x^2+3}{x}\,dx = \int \left(x + \frac{3}{x}\right)dx = \frac{x^2}{2} + 3\ln|x| + C$

28. $y = \displaystyle\int \frac{x+2}{2x+3}\,dx = \int \left(\frac{1}{2} + \frac{1/2}{2x+3}\right)dx = \frac{1}{2}\left(\int dx + \int \frac{1}{2x+3}\,dx\right)$

$$= \frac{1}{2}\left(x + \frac{1}{2}\ln|2x+3|\right) + C = \frac{1}{4}(2x + \ln|2x+3|) + C$$

29. a. $\quad f(x) = \frac{1}{2}x - 3$

$$y = \frac{1}{2}x - 3$$

$$2(y+3) = x$$

$$2(x+3) = y$$

$$f^{-1}(x) = 2x + 6$$

c. $f^{-1}(f(x)) = f^{-1}\left(\frac{1}{2}x - 3\right) = 2\left(\frac{1}{2}x - 3\right) + 6 = x$

$$f(f^{-1}(x)) = f(2x+6) = \frac{1}{2}(2x+6) - 3 = x$$

b.

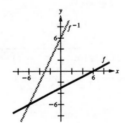

30. a. $f(x) = 5x - 7$

$$y = 5x - 7$$

$$\frac{y + 7}{5} = x$$

$$\frac{x + 7}{5} = y$$

$$f^{-1}(x) = \frac{x + 7}{5}$$

b.

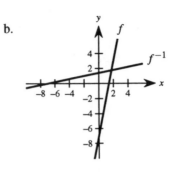

c. $f^{-1}(f(x)) = f^{-1}(5x - 7) = \dfrac{(5x - 7) + 7}{5} = x$

$f(f^{-1}(x)) = f\left(\dfrac{x + 7}{5}\right) = 5\left(\dfrac{x + 7}{5}\right) - 7 = x$

31. a. $f(x) = \sqrt{x + 1}$

$$y = \sqrt{x + 1}$$

$$y^2 - 1 = x$$

$$x^2 - 1 = y$$

$$f^{-1}(x) = x^2 - 1, \quad x \geq 0$$

b.

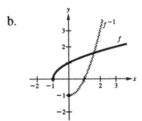

c. $f^{-1}(f(x)) = f^{-1}\left(\sqrt{x - 1}\right) = \left(\sqrt{x + 1}\right)^2 - 1 = x$

$f(f^{-1}(x)) = f(x^2 - 1) = \sqrt{(x^2 - 1) + 1}$

$$= \sqrt{x^2} = x \text{ for } x \geq 0$$

32. a. $f(x) = x^3 + 2$

$$y = x^3 + 2$$

$$\sqrt[3]{y - 2} = x$$

$$\sqrt[3]{x - 2} = y$$

$$f^{-1}(x) = \sqrt[3]{x - 2}$$

b.

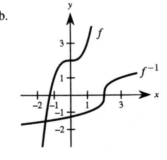

c. $f^{-1}(f(x)) = f^{-1}(x^3 + 2) = \sqrt[3]{(x^3 + 2) - 2} = x$

$f(f^{-1}(x)) = f(\sqrt[3]{x - 2}) = (\sqrt[3]{x - 2})^3 + 2 = x$

33. a. $f(x) = x^2 - 5, \quad x \geq 0$

$$y = x^2 - 5$$

$$\sqrt{y + 5} = x$$

$$\sqrt{x + 5} = y$$

$$f^{-1}(x) = \sqrt{x + 5}$$

b.

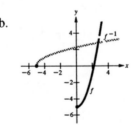

c. $f^{-1}(f(x)) = f^{-1}(x^2 - 5) = \sqrt{(x^2 - 5) + 5} = x \text{ for } x \geq 0$

$f(f^{-1}(x)) = f(\sqrt{x + 5}) = (\sqrt{x + 5})^2 - 5 = x$

34. a. $f(x) = \sqrt[3]{x+1}$

$y = \sqrt[3]{x+1}$

$y^3 - 1 = x$

$x^3 - 1 = y$

$f^{-1}(x) = x^3 - 1$

b.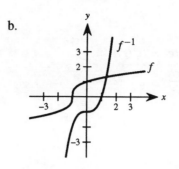

c. $f^{-1}(f(x)) = f^{-1}\left(\sqrt[3]{x+1}\right) = \left(\sqrt[3]{x+1}\right)^3 - 1 = x$

$f(f^{-1}(x)) = f(x^3 - 1) = \sqrt[3]{(x^3 - 1) + 1} = x$

35. a. $f(x) = \ln\sqrt{x}$

$y = \ln\sqrt{x}$

$e^y = \sqrt{x}$

$e^{2y} = x$

$e^{2x} = y$

$f^{-1}(x) = e^{2x}$

b.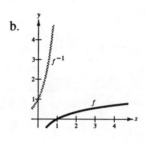

c. $f^{-1}(f(x)) = f^{-1}\left(\ln\sqrt{x}\right) = e^{2\ln\sqrt{x}} = e^{\ln x} = x$

$f(f^{-1}(x)) = f(e^{2x}) = \ln\sqrt{e^{2x}} = \ln e^x = x$

36. a. $f(x) = e^{1-x}$

$y = e^{1-x}$

$\ln y = 1 - x$

$x = 1 - \ln y$

$y = 1 - \ln x$

$f^{-1}(x) = 1 - \ln x$

b.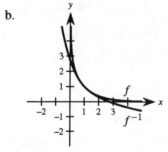

c. $f^{-1}(f(x)) = f^{-1}(e^{1-x}) = 1 - \ln e^{1-x} = 1 - (1 - x) = x$

$f(f^{-1}(x)) = f(1 - \ln x) = e^{1-(1-\ln x)} = e^{\ln x} = x$

37. $f(x) = 2(x-4)^2, \quad x \geq 4$

$y = 2(x-4)^2$

$\sqrt{y/2} + 4 = x$

$\sqrt{x/2} + 4 = y$

$f^{-1}(x) = \sqrt{x/2} + 4, \quad x \geq 0$

38. $f(x) = |x - 2| = x - 2, \quad x \geq 2$

$y = x - 2$

$y + 2 = x$

$x + 2 = y$

$f^{-1}(x) = x + 2, \quad x \geq 0$

39. $f(x) = \ln(e^{-x^2}) = -x^2$

$f'(x) = -2x$

40. $g(x) = \ln\left(\dfrac{e^x}{1 + e^x}\right)$

$= \ln e^x - \ln(1 + e^x) = x - \ln(1 + e^x)$

$g'(x) = 1 - \dfrac{e^x}{1 + e^x} = \dfrac{1}{1 + e^x}$

41. $g(t) = t^2 e^t$

$g'(t) = t^2 e^t + 2te^t = te^t(t+2)$

42. $h(z) = e^{-z^2/2}$

$h'(z) = -ze^{-z^2/2}$

43. $y = \sqrt{e^{2x} + e^{-2x}}$

$y' = \frac{1}{2}(e^{2x} + e^{-2x})^{-1/2}(2e^{2x} - 2e^{-2x})$

$= \frac{e^{2x} - e^{-2x}}{\sqrt{e^{2x} + e^{-2x}}}$

44. $y = x^{2x+1}$

$\ln y = (2x+1)\ln x$

$\frac{y'}{y} = \frac{2x+1}{x} + 2\ln x$

$y' = y\left(\frac{2x+1}{x} + 2\ln x\right)$

$= x^{2x+1}\left(\frac{2x+1}{x} + 2\ln x\right)$

45. $f(x) = 3^{x-1}$

$f'(x) = 3^{x-1}\ln 3$

46. $f(x) = 4^x e^x$

$f'(x) = 4^x e^x + (\ln 4)4^x e^x = 4^x e^x(1 + \ln 4)$

47. $g(x) = \dfrac{x^2}{e^x}$

$g'(x) = \dfrac{e^x(2x) - x^2 e^x}{e^{2x}} = \dfrac{x(2-x)}{e^x}$

48. $f(\theta) = \frac{1}{2}e^{\sin 2\theta}$

$f'(\theta) = \cos 2\theta\, e^{\sin 2\theta}$

49. $y = \tan(\arcsin x) = \dfrac{x}{\sqrt{1-x^2}}$

$y' = \dfrac{(1-x^2)^{1/2} + x^2(1-x^2)^{-1/2}}{1-x^2}$

$= (1-x^2)^{-3/2}$

50. $y = \arctan(x^2 - 1)$

$y' = \dfrac{2x}{1 + (x^2-1)^2} = \dfrac{2x}{x^4 - 2x^2 + 2}$

51. $y = x\,\text{arcsec}\,x$

$y' = \dfrac{x}{|x|\sqrt{x^2-1}} + \text{arcsec}\,x$

52. $y = \dfrac{1}{2}\arctan e^{2x}$

$y' = \dfrac{1}{2}\left(\dfrac{1}{1+e^{4x}}\right)(2e^{2x}) = \dfrac{e^{2x}}{1+e^{4x}}$

53. $y = x(\arcsin x)^2 - 2x + 2\sqrt{1-x^2}\,\arcsin x$

$y' = \dfrac{2x\arcsin x}{\sqrt{1-x^2}} + (\arcsin x)^2 - 2 + \dfrac{2\sqrt{1-x^2}}{\sqrt{1-x^2}} + \dfrac{-2x\arcsin x}{\sqrt{1-x^2}} = (\arcsin x)^2$

54. $y = \sqrt{x^2-4} - 2\,\text{arcsec}\dfrac{x}{2}, \quad 2 < x < 4$

$y' = \dfrac{x}{\sqrt{x^2-4}} - \dfrac{1}{(|x|/2)\sqrt{(x/2)^2-1}} = \dfrac{x}{\sqrt{x^2-4}} - \dfrac{4}{|x|\sqrt{x^2-4}} = \dfrac{x^2-4}{|x|\sqrt{x^2-4}} = \dfrac{\sqrt{x^2-4}}{x}$

55. $y = 2x - \cosh\sqrt{x}$

$y' = 2 - \dfrac{1}{2\sqrt{x}}(\sinh\sqrt{x})$

$= 2 - \dfrac{\sinh\sqrt{x}}{2\sqrt{x}}$

56. $y = x\tanh^{-1} 2x$

$y' = x\left(\dfrac{2}{1-4x^2}\right) + \tanh^{-1} 2x$

$= \dfrac{2x}{1-4x^2} + \tanh^{-1} 2x$

57.
$$y(\ln x) + y^2 = 0$$

$$y\left(\frac{1}{x}\right) + (\ln x)\left(\frac{dy}{dx}\right) + 2y\left(\frac{dy}{dx}\right) = 0$$

$$(2y + \ln x)\frac{dy}{dx} = \frac{-y}{x}$$

$$\frac{dy}{dx} = \frac{-y}{x(2y + \ln x)}$$

58.
$$\ln(x + y) = x$$

$$\frac{1}{x+y}\left(1 + \frac{dy}{dx}\right) = 1$$

$$\left(\frac{1}{x+y}\right)\frac{dy}{dx} = 1 - \frac{1}{x+y}$$

$$\frac{dy}{dx} = x + y - 1$$

59.
$$\cos x^2 = xe^y$$

$$-2x\sin x^2 = xe^y\frac{dy}{dx} + e^y$$

$$\frac{dy}{dx} = -\frac{2x\sin x^2 + e^y}{xe^y}$$

60.
$$ye^x + xe^y = xy$$

$$ye^x + \frac{dy}{dx}e^x + x\frac{dy}{dx}e^y + e^y = y + x\frac{dy}{dx}$$

$$\frac{dy}{dx}(e^x + xe^y - x) = y - ye^x - e^y$$

$$\frac{dy}{dx} = \frac{y - ye^x - e^y}{e^x + xe^y - x}$$

61. a. $y = x^a$

$y' = ax^{a-1}$

b. $y = a^x$

$y' = (\ln a)a^x$

c. $y = x^x$

$\ln y = x \ln x$

$\frac{1}{y}y' = x \cdot \frac{1}{x} + (1)\ln x$

$y' = y(1 + \ln x)$

$y' = x^x(1 + \ln x)$

d. $y = a^a$

$y' = 0$

62.
$$y = e^x(a\cos 3x + b\sin 3x)$$

$$y' = e^x(-3a\sin 3x + 3b\cos 3x) + e^x(a\cos 3x + b\sin 3x)$$

$$= e^x[(-3a + b)\sin 3x + (a + 3b)\cos 3x]$$

$$y'' = e^x[3(-3a + b)\cos 3x - 3(a + 3b)\sin 3x] + e^x[(-3a + b)\sin 3x + (a + 3b)\cos 3x]$$

$$= e^x[(-6a - 8b)\sin 3x + (-8a + 6b)\cos 3x]$$

$$y'' - 2y' + 10y = e^x\{[(-6a - 8b) - 2(-3a + b) + 10b]\sin 3x + [(-8a + 6b) - 2(a + 3b) + 10a]\cos 3x\} = 0$$

63. Let $u = -3x^2$, $du = -6x\,dx$.

$$\int xe^{-3x^2}\,dx = -\frac{1}{6}\int e^{-3x^2}(-6x)\,dx$$

$$= -\frac{1}{6}e^{-3x^2} + C$$

64. Let $u = \frac{1}{x}$, $du = -\frac{1}{x^2}\,dx$.

$$\int \frac{e^{1/x}}{x^2}\,dx = -\int e^{1/x}\left(-\frac{1}{x^2}\right)dx$$

$$= -e^{1/x} + C$$

65. $\int \dfrac{e^{4x} - e^{2x} + 1}{e^x}\,dx = \int (e^{3x} - e^x + e^{-x})\,dx$

$$= \frac{1}{3}e^{3x} - e^x - e^{-x} + C$$

$$= \frac{e^{4x} - 3e^{2x} - 3}{3e^x} + C$$

66. Let $u = e^{2x} + e^{-2x}$, $du = (2e^{2x} - 2e^{-2x})\,dx$.

$$\int \frac{e^{2x} - e^{-2x}}{e^{2x} + e^{-2x}}\,dx = \frac{1}{2}\int \frac{2e^{2x} - 2e^{-2x}}{e^{2x} + e^{-2x}}\,dx$$

$$= \frac{1}{2}\ln(e^{2x} + e^{-2x}) + C$$

67. Let $u = e^x - 1$, $du = e^x\,dx$.

$$\int \frac{e^x}{e^x - 1}\,dx = \ln|e^x - 1| + C$$

68. Let $u = x^3 + 1$, $du = 3x^2\,dx$.

$$\int x^2 e^{x^3+1}\,dx = \frac{1}{3}\int e^{x^3+1}(3x^2)\,dx = \frac{1}{3}e^{x^3+1} + C$$

69. Let $u = e^{2x}$, $du = 2e^{2x} dx$.

$$\int \frac{1}{e^{2x} + e^{-2x}} dx = \int \frac{e^{2x}}{1 + e^{4x}} dx = \frac{1}{2} \int \frac{1}{1 + (e^{2x})^2} (2e^{2x}) dx = \frac{1}{2} \arctan(e^{2x}) + C$$

70. Let $u = 5x$, $du = 5 dx$.

$$\int \frac{1}{3 + 25x^2} dx = \frac{1}{5} \int \frac{1}{(\sqrt{3})^2 + (5x)^2} (5) dx = \frac{1}{5\sqrt{3}} \arctan \frac{5x}{\sqrt{3}} + C$$

71. Let $u = x^2$, $du = 2x dx$.

$$\int \frac{x}{\sqrt{1 - x^4}} dx = \frac{1}{2} \int \frac{1}{\sqrt{1 - (x^2)^2}} (2x) dx = \frac{1}{2} \arcsin x^2 + C$$

72. $$\int \frac{1}{16 + x^2} dx = \frac{1}{4} \arctan \frac{x}{4} + C$$

73. Let $u = 16 + x^2$, $du = 2x dx$.

$$\int \frac{x}{16 + x^2} dx = \frac{1}{2} \int \frac{1}{16 + x^2} (2x) dx = \frac{1}{2} \ln(16 + x^2) + C$$

74. $$\int \frac{4 - x}{\sqrt{4 - x^2}} dx = 4 \int \frac{1}{\sqrt{4 - x^2}} dx + \frac{1}{2} \int (4 - x^2)^{-1/2}(-2x) dx = 4 \arcsin \frac{x}{2} + \sqrt{4 - x^2} + C$$

75. Let $u = \arctan\left(\dfrac{x}{2}\right)$, $du = \dfrac{2}{4 + x^2} dx$.

$$\int \frac{\arctan(x/2)}{4 + x^2} dx = \frac{1}{2} \int \left(\arctan \frac{x}{2}\right)\left(\frac{2}{4 + x^2}\right) dx = \frac{1}{4}\left(\arctan \frac{x}{2}\right)^2 + C$$

76. Let $u = \arcsin x$, $du = \dfrac{1}{\sqrt{1 - x^2}} dx$.

$$\int \frac{\arcsin x}{\sqrt{1 - x^2}} dx = \frac{1}{2}(\arcsin x)^2 + C$$

77. Let $u = x^2$, $du = 2x dx$.

$$\int \frac{x}{\sqrt{x^4 - 1}} dx = \frac{1}{2} \int \frac{1}{\sqrt{(x^2)^2 - 1}} (2x) dx = \frac{1}{2} \ln\left(x^2 + \sqrt{x^4 - 1}\right) + C$$

78. Let $u = x^3$, $du = 3x^2 dx$.

$$\int x^2 (\operatorname{sech} x^3)^2 dx = \frac{1}{3} \int (\operatorname{sech} x^3)^2 (3x^2) dx = \frac{1}{3} \tanh x^3 + C$$

79. $y = \displaystyle\int t e^{-t^2/2} dt$

$\quad = -\displaystyle\int e^{-t^2/2}(-t) dt$

$\quad = -e^{-t^2/2} + C$

80. $y = \displaystyle\int \frac{x - 1}{3x^2 - 6x - 1} dx$

$\quad = \dfrac{1}{6} \displaystyle\int \frac{6x - 6}{3x^2 - 6x - 1} dx$

$\quad = \dfrac{1}{6} \ln |3x^2 - 6x - 1| + C$

81. $y = \displaystyle\int \frac{e^{-2x}}{1 + e^{-2x}} dx = -\frac{1}{2} \int \frac{-2e^{-2x}}{1 + e^{-2x}} dx = -\frac{1}{2} \ln(1 + e^{-2x}) + C$

82. Let $u = \cos\left(\dfrac{1}{t}\right)$, $\quad du = -\sin\left(\dfrac{1}{t}\right)\left(-\dfrac{1}{t^2}\right) dt = \left(\dfrac{1}{t^2}\right)\sin\left(\dfrac{1}{t}\right) dt$.

$$r = \int \frac{\tan(1/t)}{t^2}\, dt = \int \frac{(1/t^2)\sin(1/t)}{\cos(1/t)}\, dt = \ln\left|\cos\left(\dfrac{1}{t}\right)\right| + C$$

83. Area $= \displaystyle\int_0^{\pi/3} \tan x\, dx$

$$= \Big[-\ln|\cos x|\Big]_0^{\pi/3} = -\ln\frac{1}{2} = \ln 2$$

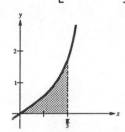

84. Area $= \displaystyle\int_0^1 \frac{1}{x+1}\, dx$

$$= \Big[\ln|x+1|\Big]_0^1 = \ln 2$$

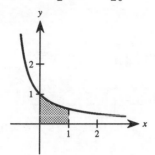

85. Area $= \displaystyle\int_0^4 xe^{-x^2}\, dx$

$$= \left[-\frac{1}{2}e^{-x^2}\right]_0^4$$

$$= -\frac{1}{2}(e^{-16} - 1) \approx 0.500$$

86. Area $= \displaystyle\int_0^4 3e^{-x/2}\, dx$

$$= \left[-6e^{-x/2}\right]_0^4$$

$$= -6(e^{-2} - 1) \approx 5.188$$

87. $\displaystyle\int_0^1 e^{x^3}\, dx \approx \frac{1}{3(4)}\Big[1 + 4e^{(0.25)^3} + 2e^{(0.5)^3} + 4e^{(0.75)^3} + e\Big] \approx 1.346$

88. $\displaystyle\int_0^1 e^{-x^2}\, dx \approx \frac{1}{3(4)}\Big[1 + 4e^{-(0.25)^2} + 2e^{-(0.5)^2} + 4e^{-(0.75)^2} + e^{-1}\Big] \approx 0.747$

89. a. $A = 500e^{(0.05)(1)} \approx \$525.64$

 b. $A = 500e^{(0.05)(10)} \approx \$824.36$

 c. $A = 500e^{(0.05)(100)} \approx \$74,206.58$

90. $2P = Pe^{10r}$

$$2 = e^{10r}$$

$$\ln 2 = 10r$$

$$r = \frac{\ln 2}{10} \approx 6.93\%$$

91. $10,000 = Pe^{(0.07)(15)}$

$$P = \frac{10,000}{e^{1.05}} \approx \$3499.38$$

92. $\dfrac{1}{5}\displaystyle\int_0^5 2500e^{0.12t}\, dt = \left[\dfrac{12,500}{3}e^{0.12t}\right]_0^5 \approx \$3425.50$

93. $A = A_0 e^{kt}$

$A_0 = 500$

When $t = 40$:

$A = 300$

$300 = 500 e^{40k}$

$\ln \frac{3}{5} = 40k$

$k \approx -0.0128$

$A \approx 500 e^{-0.0128t}$

94. $P(h) = 30 e^{kh}$

$P(18{,}000) = 30 e^{18{,}000k} = 15$

$k = \dfrac{\ln(1/2)}{18{,}000} = \dfrac{-\ln 2}{18{,}000}$

$P(h) = 30 e^{-(h \ln 2)/18{,}000}$

$P(35{,}000) = 30 e^{-(35{,}000 \ln 2)/18{,}000}$

≈ 7.79

95. $p(t) = 80 e^{-0.5t} + 20$

$p'(t) = -40 e^{-0.5t}$

$p'(1) = -40 e^{-0.5(1)}$

$\approx -24.26\%$

$p'(2) = -40 e^{-0.5(2)}$

$\approx -14.72\%$

96. a. $\dfrac{1}{2} \displaystyle\int_0^2 (80 e^{-0.5t} + 20)\, dt = \left[\dfrac{1}{2}(-160 e^{-0.5t} + 20t) \right]_0^2 = \left[10(-8 e^{-0.5t} + t) \right]_0^2 \approx 70.57\%$

b. $\dfrac{1}{2} \displaystyle\int_2^4 (80 e^{-0.5t} + 20)\, dt = \left[10(-8 e^{-0.5t} + t) \right]_2^4 \approx 38.60\%$

97. $\displaystyle\int_{t_1}^{t_2} \dfrac{1}{20} e^{-t/20}\, dt = \left[-e^{-t/20} \right]_{t_1}^{t_2}$

a. $\displaystyle\int_0^{10} \dfrac{1}{20} e^{-t/20}\, dt = \left[-e^{-t/20} \right]_0^{10} \approx 0.3935$

b. $\displaystyle\int_0^{30} \dfrac{1}{20} e^{-t/20}\, dt = \left[-e^{-t/20} \right]_0^{30} \approx 0.7769$

c. $\displaystyle\int_{15}^{30} \dfrac{1}{20} e^{-t/20}\, dt = \left[-e^{-t/20} \right]_{15}^{30} \approx 0.2492$

d. $\displaystyle\int_0^{60} \dfrac{1}{20} e^{-t/20}\, dt = \left[-e^{-t/20} \right]_0^{60} \approx 0.9502$

98. a. $P = \dfrac{1}{100} \left(25 + \displaystyle\int_{2.5}^{10} \dfrac{25}{x}\, dx \right) = \dfrac{1}{100} \left(25 + \left[25 \ln x \right]_{2.5}^{10} \right) = \dfrac{1}{4}(1 + \ln 4) \approx 0.60$

b. $P = \dfrac{1}{100} \left(50 + \displaystyle\int_{5}^{10} \dfrac{50}{x}\, dx \right) = \dfrac{1}{100} \left(50 + \left[50 \ln x \right]_{5}^{10} \right) = \dfrac{1}{2}(1 + \ln 2) \approx 0.85$

99. a. $\displaystyle \int \frac{1}{kv - 32} \, dv = \int dt$

$\displaystyle \frac{1}{k} \ln |kv - 32| = t + C_1$

$\displaystyle kv - 32 = Ce^{kt}$

$\displaystyle kv = 32 + Ce^{kt}$

$\displaystyle v(t) = \frac{1}{k}(32 + Ce^{kt})$

$\displaystyle v(0) = \frac{1}{k}(32 + C) = v_0 \Rightarrow C = kv_0 - 32$

$\displaystyle v(t) = \frac{1}{k}\left[32 + (kv_0 - 32)\, e^{kt}\right]$

Note that $k < 0$ since the object is moving downward.

b. $\displaystyle \lim_{t \to \infty} \frac{1}{k}\left[32 + (kv_0 - 32)\, e^{kt}\right] = \frac{32}{k}$

c. $\displaystyle s(t) = \frac{1}{k}\int \left[32 + (kv_0 - 32)\, e^{kt}\right]dt = \frac{1}{k}\left[32t + \frac{1}{k}(kv_0 - 32)\, e^{kt}\right] + C = \frac{32}{k}t + \frac{1}{k^2}(kv_0 - 32)\, e^{kt} + C$

$\displaystyle s(0) = \frac{1}{k^2}(kv_0 - 32) + C = s_0 \Rightarrow C = s_0 - \frac{1}{k^2}(kv_0 - 32)$

$\displaystyle s(t) = \frac{32}{k}t + \frac{1}{k^2}(kv_0 - 32)\, e^{kt} - \frac{1}{k^2}(kv_0 - 32) + s_0$

$\displaystyle = \frac{32}{k}t + \frac{1}{k^2}(kv_0 - 32)(e^{kt} - 1) + s_0 = \frac{32}{k}t - \frac{1}{k^2}(kv_0 - 32)(1 - e^{kt}) + s_0$

100. a. $\displaystyle \frac{dy}{-0.012y} = ds$

$\displaystyle -\frac{1}{0.012}\int \frac{dy}{y} = \int ds$

$\displaystyle -\frac{1}{0.012} \ln y = s + C_1 \quad \textbf{(Note: } y > 0\text{)}$

$\displaystyle \ln y = -0.012s + C_2$

$\displaystyle y = Ce^{-0.012s}$

When $s = 50$, $y = 28$.

$\displaystyle 28 = Ce^{-0.012(50)}$

$\displaystyle 28e^{0.012(50)} = C$

$\displaystyle C = 28e^{0.6}$

$\displaystyle y = 28e^{0.6 - 0.012s}$

b.

Speed (s)	50	55	60	65	70
Mileage (y)	28	26.4	24.8	23.4	22.0

101. $N = \dfrac{157}{1 + 5.4e^{-0.12t}}$

    a. $\displaystyle\lim_{t \to \infty} \dfrac{157}{1 + 5.4e^{-0.12t}} = 157$ words/minute

    b. $\dfrac{dN}{dt} = \dfrac{101.736e^{-0.12t}}{(1 + 5.4e^{-0.12t})^2}$

        When $t = 5$: $\dfrac{dN}{dt} \approx 3.55$ words/minute/week

        When $t = 14$: $\dfrac{dN}{dt} \approx 4.71$ words/minute/week

102. $t = 50 \log_{10}\left(\dfrac{18{,}000}{18{,}000 - h}\right)$

    a. Domain: $0 \le h < 18{,}000$

    c. $\qquad t = 50 \log_{10}\left(\dfrac{18{,}000}{18{,}000 - h}\right)$

        $10^{t/50} = \dfrac{18{,}000}{18{,}000 - h}$

      $18{,}000 - h = 18{,}000(10^{-t/50})$

          $h = 18{,}000(1 - 10^{-t/50})$

      As $h \to 18{,}000$, $t \to \infty$.

    d. $\quad t = 50 \log_{10} 18{,}000 - 50 \log_{10}(18{,}000 - h)$

      $\dfrac{dt}{dh} = \dfrac{50}{(\ln 10)(18{,}000 - h)}$

      $\dfrac{d^2 t}{dh^2} = \dfrac{50}{(\ln 10)(18{,}000 - h)^2}$

      No critical numbers

      As t increases, the rate of change of the altitude is increasing.

    b.

    Vertical asymptote: $h = 18{,}000$

CHAPTER 6
Applications of Integration

Section 6.1 Area of a Region Between Two Curves

1. $A = \int_0^6 [0 - (x^2 - 6x)]\, dx = -\left[\dfrac{x^3}{3} - 3x^2\right]_0^6 = 36$

2. $A = \int_{-2}^2 [(2x + 5) - (x^2 + 2x + 1)]\, dx = \int_{-2}^2 (-x^2 + 4)\, dx = \left[-\dfrac{x^3}{3} + 4x\right]_{-2}^2 = \dfrac{32}{3}$

3. $A = \int_0^3 [(-x^2 + 2x + 3) - (x^2 - 4x + 3)]\, dx = \int_0^3 (-2x^2 + 6x)\, dx = \left[-\dfrac{2x^3}{3} + 3x^2\right]_0^3 = 9$

4. $A = \int_0^1 (x^2 - x^3)\, dx = \left[\dfrac{x^3}{3} - \dfrac{x^4}{4}\right]_0^1 = \dfrac{1}{12}$

5. $A = 2\int_{-1}^0 3(x^3 - x)\, dx = 6\left[\dfrac{x^4}{4} - \dfrac{x^2}{2}\right]_{-1}^0 = 6\left(\dfrac{1}{4}\right) = \dfrac{3}{2}$

6. $A = 2\int_0^1 [(x - 1)^3 - (x - 1)]\, dx = 2\left[\dfrac{(x - 1)^4}{4} - \left(\dfrac{x^2}{2} - x\right)\right]_0^1 = \dfrac{1}{2}$

7. $\int_0^4 \left[(x + 1) - \dfrac{x}{2}\right] dx$

8. $\int_{-1}^1 [(1 - x^2) - (x^2 - 1)]\, dx$

9. $\int_0^6 \left[4(2^{-x/3}) - \dfrac{x}{6}\right] dx$

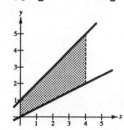

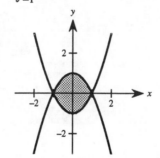

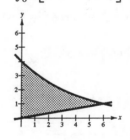

10. $\displaystyle\int_{-\pi/3}^{\pi/3} [2 - \sec x]\,dx$

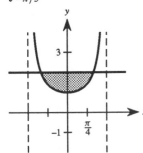

11. $f(x) = x + 1$
$g(x) = (x-1)^2$
$A \approx 4$
Matches (d)

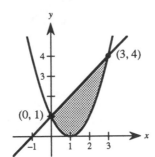

12. $f(x) = 2 - \frac{1}{2}x$
$g(x) = 2 - \sqrt{x}$
$A \approx 1$
Matches (a)

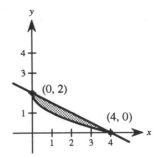

13. The points of intersection are given by:

$$x^2 - 4x = 0$$

$$x(x-4) = 0 \text{ when } x = 0,\ 4$$

$$A = \int_0^4 [g(x) - f(x)]\,dx$$

$$= -\int_0^4 (x^2 - 4x)\,dx = -\left[\frac{x^3}{3} - 2x^2\right]_0^4 = \frac{32}{3}$$

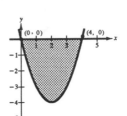

14. The points of intersection are given by:

$$3 - 2x - x^2 = 0$$

$$(3 + x)(1 - x) = 0 \text{ when } x = -3,\ 1$$

$$A = \int_{-3}^1 [f(x) - g(x)]\,dx$$

$$= \int_{-3}^1 (3 - 2x - x^2)\,dx$$

$$= \left[3x - x^2 - \frac{x^3}{3}\right]_{-3}^1 = \frac{32}{3}$$

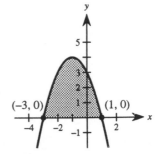

15. The points of intersection are given by:

$$x^2 + 2x + 1 = 3x + 3$$

$$(x - 2)(x + 1) = 0 \text{ when } x = -1,\ 2$$

$$A = \int_{-1}^2 [g(x) - f(x)]\,dx$$

$$= \int_{-1}^2 [(3x + 3) - (x^2 + 2x + 1)]\,dx$$

$$= \int_{-1}^2 (2 + x - x^2)\,dx$$

$$= \left[2x + \frac{x^2}{2} - \frac{x^3}{3}\right]_{-1}^2 = \frac{9}{2}$$

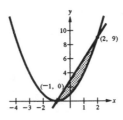

16. The points of intersection are given by:

$$-x^2 + 4x + 2 = x + 2$$

$$x(3 - x) = 0 \text{ when } x = 0, \ 3$$

$$A = \int_0^3 [f(x) - g(x)] \, dx$$

$$= \int_0^3 [(-x^2 + 4x + 2) - (x + 2)] \, dx$$

$$= \int_0^3 (-x^2 + 3x) \, dx = \left[\frac{-x^3}{3} + \frac{3}{2}x^2 \right]_0^3 = \frac{9}{2}$$

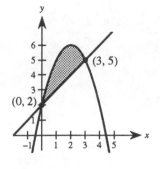

17. The points of intersection are given by:

$$x = 2 - x \quad \text{and} \quad x = 0 \quad \text{and} \quad 2 - x = 0$$

$$x = 1 \qquad\qquad\quad x = 0 \qquad\qquad x = 2$$

$$A = \int_0^1 [(2 - y) - (y)] \, dy = \left[2y - y^2 \right]_0^1 = 1$$

Note that if we integrate with respect to x, we need two integrals. Also, note that the region is a triangle.

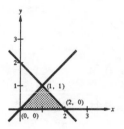

18. $A = \int_1^5 \left(\frac{1}{x^2} - 0 \right) dx = \left[-\frac{1}{x} \right]_1^5 = \frac{4}{5}$

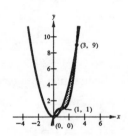

19. The points of intersection are given by:

$$x^3 - 3x^2 + 3x = x^2$$

$$x(x - 1)(x - 3) = 0 \text{ when } x = 0, \ 1, \ 3$$

$$A = \int_0^1 [f(x) - g(x)] \, dx + \int_1^3 [g(x) - f(x)] \, dx$$

$$= \int_0^1 [(x^3 - 3x^2 + 3x) - x^2] \, dx + \int_1^3 [x^2 - (x^3 - 3x^2 + 3x)] \, dx$$

$$= \int_0^1 (x^3 - 4x^2 + 3x) \, dx + \int_1^3 (-x^3 + 4x^2 - 3x) \, dx$$

$$= \left[\frac{x^4}{4} - \frac{4}{3}x^3 + \frac{3}{2}x^2 \right]_0^1 + \left[\frac{-x^4}{4} + \frac{4}{3}x^3 - \frac{3}{2}x^2 \right]_1^3 = \frac{37}{12}$$

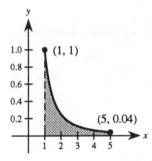

20. The point of intersection is given by:

$$x^3 - 2x + 1 = -2x$$

$$x^3 + 1 = 0 \text{ when } x = -1$$

$$A = \int_{-1}^{1} [f(x) - g(x)] \, dx$$

$$= \int_{-1}^{1} [(x^3 - 2x + 1) - (-2x)] \, dx$$

$$= \int_{-1}^{1} (x^3 + 1) \, dx = \left[\frac{x^4}{4} + x \right]_{-1}^{1} = 2$$

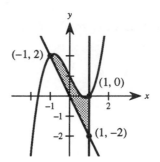

21. The points of intersection are given by:

$$x^2 - 4x + 3 = 3 + 4x - x^2$$

$$2x(x - 4) = 0 \text{ when } x = 0, \ 4$$

$$A = \int_{0}^{4} [(3 + 4x - x^2) - (x^2 - 4x + 3)] \, dx$$

$$= \int_{0}^{4} (-2x^2 + 8x) \, dx$$

$$= \left[-\frac{2x^3}{3} + 4x^2 \right]_{0}^{4} = \frac{64}{3}$$

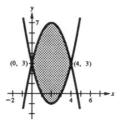

22. The points of intersection are given by:

$$x^4 - 2x^2 = 2x^2$$

$$x^2(x^2 - 4) = 0 \text{ when } x = 0, \ \pm 2$$

$$A = 2 \int_{0}^{2} [2x^2 - (x^4 - 2x^2)] \, dx$$

$$= 2 \int_{0}^{2} (4x^2 - x^4) \, dx$$

$$= 2 \left[\frac{4x^3}{3} - \frac{x^5}{5} \right]_{0}^{2} = \frac{128}{15}$$

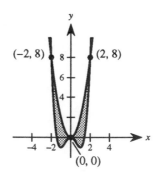

23. The points of intersection are given by:

$$\sqrt{3x} + 1 = x + 1$$

$$\sqrt{3x} = x \text{ when } x = 0, \ 3$$

$$A = \int_{0}^{3} [f(x) - g(x)] \, dx$$

$$= \int_{0}^{3} [(\sqrt{3x} + 1) - (x + 1)] \, dx$$

$$= \int_{0}^{3} [(3x)^{1/2} - x] \, dx$$

$$= \left[\frac{2}{9} (3x)^{3/2} - \frac{x^2}{2} \right]_{0}^{3} = \frac{3}{2}$$

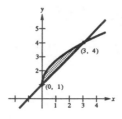

24. The points of intersection are given by:

$$\sqrt[3]{x} = x$$

$$x = -1,\ 0,\ 1$$

$$A = 2 \int_0^1 [f(x) - g(x)]\,dx$$

$$= 2 \int_0^1 (\sqrt[3]{x} - x)\,dx$$

$$= 2 \int_0^1 (x^{1/3} - x)\,dx$$

$$= 2 \left[\frac{3}{4}x^{4/3} - \frac{1}{2}x^2 \right]_0^1 = \frac{1}{2}$$

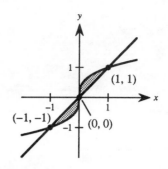

25. The points of intersection are given by:

$$\frac{1}{1+x^2} = \frac{x^2}{2}$$

$$x^4 + x^2 - 2 = 0$$

$$x^2 = \frac{-1 \pm \sqrt{1+8}}{2}$$

$$x = \pm\sqrt{\frac{-1+3}{2}} = \pm 1$$

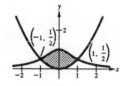

$$A = 2 \int_0^1 [f(x) - g(x)]\,dx = 2 \int_0^1 \left[\frac{1}{1+x^2} - \frac{x^2}{2} \right] dx = 2 \left[\arctan x - \frac{x^3}{6} \right]_0^1 = 2\left(\frac{\pi}{4} - \frac{1}{6} \right) = \frac{\pi}{2} - \frac{1}{3} \approx 1.237$$

26. $A = \int_0^3 \left[\frac{6x}{x^2+1} - 0 \right] dx$

$$= \left[3\ln(x^2+1) \right]_0^3$$

$$= 3\ln 10$$

$$\approx 6.908$$

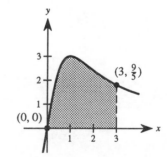

27. The points of intersection are given by:

$$y^2 = y + 2$$

$$(y-2)(y+1) = 0 \quad \text{when } y = -1,\ 2$$

$$A = \int_{-1}^2 [g(y) - f(y)]\,dy$$

$$= \int_{-1}^2 [(y+2) - y^2]\,dy$$

$$= \left[2y + \frac{y^2}{2} - \frac{y^3}{3} \right]_{-1}^2 = \frac{9}{2}$$

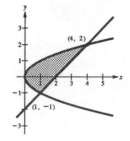

484 CHAPTER 6 Applications of Integration

28. The points of intersection are given by:

$$2y - y^2 = -y$$

$$y(y - 3) = 0 \text{ when } y = 0, \ 3$$

$$A = \int_0^3 [f(y) - g(y)]\,dy$$

$$= \int_0^3 [(2y - y^2) - (-y)]\,dy$$

$$= \int_0^3 (3y - y^2)\,dy = \left[\frac{3}{2}y^2 - \frac{1}{3}y^3\right]_0^3 = \frac{9}{2}$$

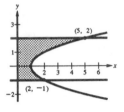

29. $$A = \int_{-1}^2 [f(y) - g(y)]\,dy$$

$$= \int_{-1}^2 [(y^2 + 1) - 0]\,dy$$

$$= \left[\frac{y^3}{3} + y\right]_{-1}^2 = 6$$

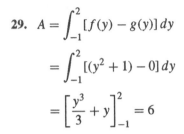

30. $$A = \int_0^3 [f(y) - g(y)]\,dy$$

$$= \int_0^3 \left[\frac{y}{\sqrt{16 - y^2}} - 0\right]\,dy$$

$$= -\frac{1}{2}\int_0^3 (16 - y^2)^{-1/2}(-2y)\,dy$$

$$= \left[-\sqrt{16 - y^2}\right]_0^3 = 4 - \sqrt{7} \approx 1.354$$

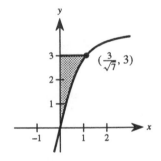

31. $$y = \frac{4}{x} \Rightarrow x = \frac{4}{y}$$

$$A = \int_1^4 \left[\frac{4}{y} - 0\right]\,dy$$

$$= \left[4\ln|y|\right]_1^4 = 4\ln 4 = 8\ln 2 \approx 5.545$$

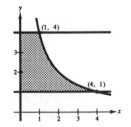

32. $$A = \int_0^1 \left(4 - \frac{4}{2 - x}\right)\,dx$$

$$= \left[4x + 4\ln|2 - x|\right]_0^1$$

$$= 4 - 4\ln 2$$

$$\approx 1.227$$

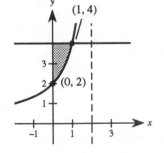

33. $A \approx \displaystyle\int_{-1}^{1.743} (3x + 4 - x^4)\,dx \approx 10.612$

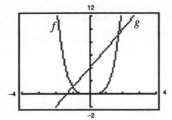

34. $A \approx \displaystyle\int_{-1}^{1.215} (x + 2 - x^6)\,dx \approx 3.967$

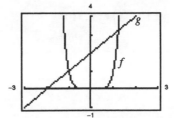

35. $A = 2\displaystyle\int_{0}^{\pi/3} [f(x) - g(x)]\,dx$

$= 2\displaystyle\int_{0}^{\pi/3} (2\sin x - \tan x)\,dx$

$= 2\Big[-2\cos x + \ln|\cos x|\Big]_{0}^{\pi/3}$

$= 2(1 - \ln 2) \approx 0.614$

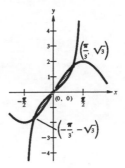

36. $A = 2\displaystyle\int_{\pi/6}^{\pi/2} [f(x) - g(x)]\,dx$

$= 2\displaystyle\int_{\pi/6}^{\pi/2} [\sin 2x - \cos x]\,dx$

$= 2\Big[-\dfrac{1}{2}\cos 2x - \sin x\Big]_{\pi/6}^{\pi/2}$

$= \dfrac{1}{2}$

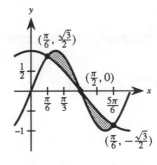

37. $A = \displaystyle\int_{0}^{\pi} [(2\sin x + \sin 2x) - 0]\,dx$

$= \Big[-2\cos x - \dfrac{1}{2}\cos 2x\Big]_{0}^{\pi} = 4$

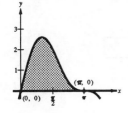

38. $A = \displaystyle\int_{0}^{\pi} [(2\sin x + \cos 2x) - 0]\,dx$

$= \Big[-2\cos x + \dfrac{1}{2}\sin 2x\Big]_{0}^{\pi} = 4$

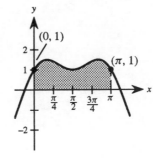

39. $A = \int_0^1 [xe^{-x^2} - 0] \, dx$

$= \left[-\frac{1}{2} e^{-x^2} \right]_0^1 = \frac{1}{2} \left(1 - \frac{1}{e} \right) \approx 0.316$

40. $A = \int_1^3 \left[\frac{1}{x^2} e^{1/x} - 0 \right] dx$

$= \left[-e^{1/x} \right]_1^3 = e - e^{1/3} \approx 1.323$

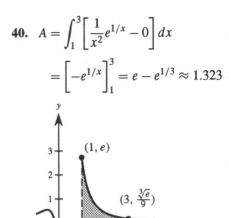

41. $A = \int_1^5 \left[\frac{4 \ln x}{x} - 0 \right] dx$

$= \left[2(\ln x)^2 \right]_1^5$

$= 2(\ln 5)^2$

≈ 5.181

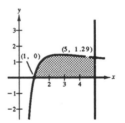

42. From the graph we see that f and g intersect twice at $x = 0$ and $x = 1$.

$A = \int_0^1 [g(x) - f(x)] \, dx$

$= \int_0^1 [(2x + 1) - 3^x] \, dx$

$= \left[x^2 + x - \frac{1}{\ln 3} (3^x) \right]_0^1$

$= 2 \left(1 - \frac{1}{\ln 3} \right) \approx 0.180$

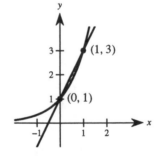

43. $A = \int_0^c \left[\left(\frac{b-a}{c} y + a \right) - \frac{b}{c} y \right] dy$

$= \int_0^c \left(-\frac{a}{c} y + a \right) dy$

$= \left[-\frac{a}{2c} y^2 + ay \right]_0^c$

$= -\frac{ac}{2} + ac = \frac{ac}{2}$

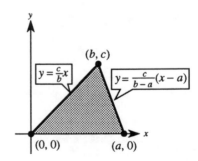

44. $A = \int_2^4 \left[\left(\frac{9}{2}x - 12 \right) - (x-5) \right] dx + \int_4^6 \left[\left(-\frac{5}{2}x + 16 \right) - (x-5) \right] dx$

$= \int_2^4 \left(\frac{7}{2}x - 7 \right) dx + \int_4^6 \left(-\frac{7}{2}x + 21 \right) dx$

$= \left[\frac{7}{4}x^2 - 7x \right]_2^4 + \left[-\frac{7}{4}x^2 + 21x \right]_4^6 = 14$

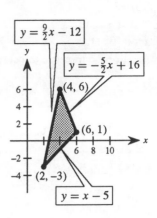

45. $A = \int_{-3}^3 (9 - x^2)\, dx = 36$

$\int_{-\sqrt{9-b}}^{\sqrt{9-b}} [(9 - x^2) - b]\, dx = 18$

$\int_0^{\sqrt{9-b}} [(9 - b) - x^2]\, dx = 9$

$\left[(9 - b)x - \frac{x^3}{3} \right]_0^{\sqrt{9-b}} = 9$

$\frac{2}{3}(9 - b)^{3/2} = 9$

$(9 - b)^{3/2} = \frac{27}{2}$

$9 - b = \frac{9}{\sqrt[3]{4}}$

$b = 9 - \frac{9}{\sqrt[3]{4}} \approx 3.330$

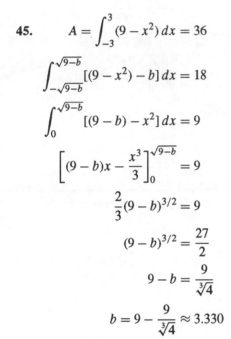

46. $A = 2 \int_0^9 (9 - x)\, dx = 2 \left[9x - \frac{x^2}{2} \right]_0^9 = 81$

$2 \int_0^{9-b} [(9 - x) - b]\, dx = \frac{81}{2}$

$2 \int_0^{9-b} [(9 - b) - x]\, dx = \frac{81}{2}$

$2 \left[(9 - b)x - \frac{x^2}{2} \right]_0^{9-b} = \frac{81}{2}$

$(9 - b)(9 - b) = \frac{81}{2}$

$9 - b = \frac{9}{\sqrt{2}}$

$b = 9 - \frac{9}{\sqrt{2}} \approx 2.636$

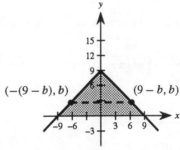

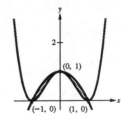

47. $x^4 - 2x^2 + 1 \leq 1 - x^2$ on $[-1,\ 1]$

$A = \int_{-1}^1 [(1 - x^2) - (x^4 - 2x^2 + 1)]\, dx$

$= \int_{-1}^1 (x^2 - x^4)\, dx$

$= \left[\frac{x^3}{3} - \frac{x^5}{5} \right]_{-1}^1 = \frac{4}{15}$

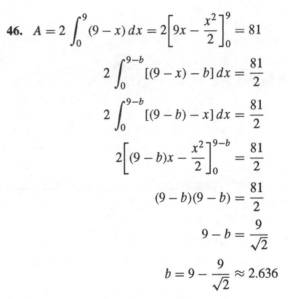

48. $x^3 \geq x$ on $[-1, \ 0]$

$x^3 \leq x$ on $[0, \ 1]$

Both functions symmetric to origin

$$\int_{-1}^{0} (x^3 - x)\, dx = -\int_{0}^{1} (x^3 - x)\, dx$$

Thus, $\int_{-1}^{1} (x^3 - x)\, dx = 0.$

$$A = 2\int_{0}^{1} (x - x^3)\, dx = 2\left[\frac{x^2}{2} - \frac{x^4}{4}\right]_{0}^{1} = \frac{1}{2}$$

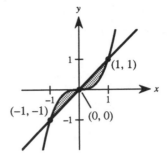

49. $f(x) = x^3$

$f'(x) = 3x^2$

At $(1, \ 1), \ \ f'(1) = 3.$

Tangent line:

$y - 1 = 3(x - 1)$ or $y = 3x - 2$

The tangent line intersects $f(x) = x^3$ at $x = -2$.

$$A = \int_{-2}^{1} [x^3 - (3x - 2)]\, dx = \left[\frac{x^4}{4} - \frac{3x^2}{2} + 2x\right]_{-2}^{1} = \frac{27}{4}$$

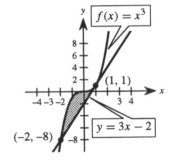

50. $f(x) = \dfrac{1}{x^2 + 1}$

$f'(x) = -\dfrac{2x}{(x^2 + 1)^2}$

At $\left(1, \ \frac{1}{2}\right), \ \ f'(1) = -\frac{1}{2}.$

Tangent line:

$$y - \frac{1}{2} = -\frac{1}{2}(x - 1) \text{ or } y = -\frac{1}{2}x + 1$$

The tangent line intersects $f(x) = \dfrac{1}{x^2 + 1}$ at $x = 0$.

$$A = \int_{0}^{1}\left[\frac{1}{x^2 + 1} - \left(-\frac{1}{2}x + 1\right)\right] dx = \left[\arctan x + \frac{x^2}{4} - x\right]_{0}^{1} = \frac{\pi - 3}{4}$$

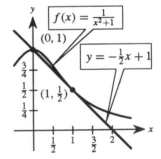

51. $\sqrt{1 + x^3} \leq \dfrac{1}{2}x + 2$ on $[0, \ 2]$

Use Simpson's Rule with $n = 10$.

$$A = \int_{0}^{2}\left[\frac{1}{2}x + 2 - \sqrt{1 + x^3}\right] dx \approx 1.7587$$

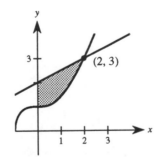

52. Use Simpson's Rule with $n = 10$.

$$A = \int_{0}^{1} \sqrt{x}\, e^x\, dx \approx 1.2531$$

53. Use Simpson's Rule with $n = 10$.

$$A = \int_{0}^{3} \sqrt{\frac{x^3}{4 - x}}\, dx \approx 4.7731$$

54. Use Simpson's Rule with $n = 10$.

$$A = \int_{0}^{4} x\sqrt{\frac{4 - x}{4 + x}}\, dx \approx 3.4042$$

55. $\lim\limits_{\|\Delta\|\to 0} \sum\limits_{i=1}^{n} (x_i - x_i^2)\Delta x$

where $x_i = \dfrac{i}{n}$ and $\Delta x = \dfrac{1}{n}$ is the same as

$$\int_0^1 (x - x^2)\,dx = \left[\frac{x^2}{2} - \frac{x^3}{3}\right]_0^1 = \frac{1}{6}.$$

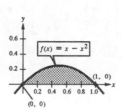

56. $\lim\limits_{\|\Delta\|\to 0} \sum\limits_{i=1}^{n} (4 - x_i^2)\Delta x$

where $x_i = -2 + \dfrac{4i}{n}$ and $\Delta x = \dfrac{4}{n}$ is the same as

$$\int_{-2}^2 (4 - x^2)\,dx = \left[4x - \frac{x^3}{3}\right]_{-2}^2 = \frac{32}{3}.$$

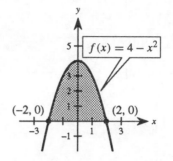

57. $\displaystyle\int_0^5 [(7.21 + 0.58t) - (7.21 + 0.45t)]\,dt = \int_0^5 0.13t\,dt = \left[\frac{0.13t^2}{2}\right]_0^5 = \$1.625 \text{ billion}$

58. $\displaystyle\int_0^5 [(7.21 + 0.26t + 0.02t^2) - (7.21 + 0.1t + 0.01t^2)]\,dt = \int_0^5 (0.01t^2 + 0.16t)\,dt = \left[\frac{0.01t^3}{3} + \frac{0.16t^2}{2}\right]_0^5$

$$= \frac{29}{12} \text{ billion} \approx \$2.417 \text{ billion}$$

59. a.

```
30

0
 0                              10
```

b. $\displaystyle\int_5^{10} [(23.613 + 0.270t + 0.019t^2) - (23.333 + 0.784t - 0.073t^2)]\,dt$

$$= \int_5^{10} (0.092t^2 - 0.514t + 0.28)\,dt$$

$$= \left[\frac{0.092t^3}{3} - 0.257t^2 + 0.28t\right]_5^{10}$$

$$\approx 8.9583 \text{ billion pounds}$$

60. $\displaystyle\int_0^{10} [(568.50 + 7.15t) - (525.60 + 6.43t)]\,dt = \int_0^{10} (42.90 + 0.72t)\,dt = \left[42.90t + 0.36t^2\right]_0^{10} = \$465 \text{ million}$

61. a. $A = 2\left[\displaystyle\int_0^5 \left(1 - \frac{1}{3}\sqrt{5 - x}\right)dx + \int_5^{5.5} (1 - 0)\,dx\right]$

$$= 2\left(\left[x + \frac{2}{9}(5 - x)^{3/2}\right]_0^5 + \left[x\right]_5^{5.5}\right) = 2\left(5 - \frac{10\sqrt{5}}{9} + 5.5 - 5\right) \approx 6.031 \text{ m}^2$$

b. $V = 2A \approx 2(6.031) \approx 12.062 \text{ m}^3$

c. $5000V \approx 5000(12.062) = 60{,}310 \text{ pounds}$

62. a. $A \approx 6.031 - 2\left[\pi\left(\frac{1}{16}\right)^2\right] - 2\left[\pi\left(\frac{1}{8}\right)^2\right] \approx 5.908$

b. $V = 2A \approx 2(5.908) \approx 11.816 \text{ m}^3$

c. $5000V \approx 5000(11.816) = 59{,}080 \text{ pounds}$

63. $50 - 0.5x = 0.125x$

$x = 80$

$P_1(80) = P_2(80) = 10$

Point of equilibrium: (80, 10)

$CS = \int_0^{80} [(50 - 0.5x) - 10]\, dx$

$= \left[-\frac{0.5x^2}{2} + 40x\right]_0^{80} = 1600$

$PS = \int_0^{80} [10 - 0.125x]\, dx$

$= \left[10x - \frac{0.125x^2}{2}\right]_0^{80} = 400$

64. $1000 - 0.4x^2 = 42x$

$x = 20$

$P_1(20) = P_2(20) = 840$

Point of equilibrium: (20, 840)

$CS = \int_0^{20} [(1000 - 0.4x^2) - 840]\, dx$

$= \left[160x - \frac{0.4x^3}{3}\right]_0^{20} \approx 2133.33$

$PS = \int_0^{20} [840 - 42x]\, dx$

$= \left[840x - 21x^2\right]_0^{20} = 8400$

65. $\frac{10{,}000}{\sqrt{x + 100}} = 100\sqrt{0.05x + 10}$

$100 = \sqrt{(x + 100)(0.05x + 10)}$

$10{,}000 = 0.05x^2 + 15x + 1000$

$0 = x^2 + 300x - 180{,}000$

$0 = (x + 600)(x - 300)$

$x = 300$

$P_1(300) = P_2(300) = 500$

Point of equilibrium: (300, 500)

$CS = \int_0^{300}\left[\frac{10{,}000}{\sqrt{x + 100}} - 500\right] dx = \left[20{,}000\sqrt{x + 100} - 500x\right]_0^{300} = 250{,}000 - 200{,}000 = 50{,}000$

$PS = \int_0^{300} (500 - 100\sqrt{0.05x + 10})\, dx$

$= \left[500x - \frac{4000}{3}(0.05x + 10)^{3/2}\right]_0^{300} = \frac{-50{,}000}{3} + \frac{40{,}000\sqrt{10}}{3} = \frac{10{,}000}{3}(4\sqrt{10} - 5) \approx 25{,}497$

66. $\sqrt{25 - 0.1x} = \sqrt{9 + 0.1x} - 2$

$x^2 - 160x = 0, \quad 0 < x < 250$

$x(x - 160) = 0$

$x = 160$

$P_1(160) = P_2(160) = 3$

Point of equilibrium: (160, 3)

$$CS = \int_0^{160} (\sqrt{25 - 0.1x} - 3)\, dx = \left[-\frac{20}{3}(25 - 0.1x)^{3/2} - 3x \right]_0^{160} = \frac{520}{3}$$

$$PS = \int_0^{160} [3 - (\sqrt{0.1x + 9} - 2)]\, dx = \left[5x - \frac{20}{3}(0.1x + 9)^{3/2} \right]_0^{160} = \frac{440}{3}$$

67. True   **68.** True

69. $D(f,\ g) = \sqrt{\dfrac{1}{b - a} \int_a^b [f(x) - g(x)]^2\, dx}$

$$= \sqrt{\frac{1}{b - a} \int_a^b [c - d]^2\, dx} = \sqrt{\frac{1}{b - a}\left[(c - d)^2 x\right]_a^b} = \sqrt{\frac{1}{b - a}(c - d)^2(b - a)} = |c - d|$$

70. $\displaystyle\int_{-2}^{2} |x^3 + x^2 - 2x|\, dx = \int_{-2}^{0} (x^3 + x^2 - 2x)\, dx + \int_0^1 -(x^3 + x^2 - 2x)\, dx + \int_1^2 (x^3 + x^2 - 2x)\, dx$

$$= \frac{8}{3} + \frac{5}{12} + \frac{37}{12} = \frac{37}{6} \approx 6.167$$

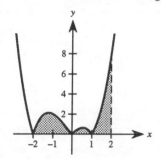

Section 6.2   Volume: The Disc Method

1. $V = \pi \displaystyle\int_0^1 (-x + 1)^2\, dx = \pi \int_0^1 (x^2 - 2x + 1)\, dx = \pi \left[\frac{x^3}{3} - x^2 + x \right]_0^1 = \frac{\pi}{3}$

2. $V = \pi \displaystyle\int_0^2 (4 - x^2)^2\, dx = \pi \int_0^2 (x^4 - 8x^2 + 16)\, dx = \pi \left[\frac{x^5}{5} - \frac{8x^3}{3} + 16x \right]_0^2 = \frac{256\pi}{15}$

3. $V = \pi \displaystyle\int_0^2 \left(\sqrt{4 - x^2} \right)^2\, dx = \pi \int_0^2 (4 - x^2)\, dx = \pi \left[4x - \frac{x^3}{3} \right]_0^2 = \frac{16\pi}{3}$

4. $V = \pi \displaystyle\int_0^1 (x^2)^2\, dx = \pi \int_0^1 x^4\, dx = \pi \left[\frac{x^5}{5} \right]_0^1 = \frac{\pi}{5}$

5. $V = \pi \displaystyle\int_1^4 (\sqrt{x})^2\, dx = \pi \displaystyle\int_1^4 x\, dx = \pi \left[\dfrac{x^2}{2}\right]_1^4 = \dfrac{15\pi}{2}$

6. $V = \pi \displaystyle\int_{-2}^2 \left(\sqrt{4-x^2}\right)^2 dx = 2\pi \displaystyle\int_0^2 (4-x^2)\, dx = 2\pi\left[4x - \dfrac{x^3}{3}\right]_0^2 = \dfrac{32\pi}{3}$

7. $V = \pi \displaystyle\int_0^1 [(x^2)^2 - (x^3)^2]\, dx = \pi \displaystyle\int_0^1 (x^4 - x^6)\, dx = \pi\left[\dfrac{x^5}{5} - \dfrac{x^7}{7}\right]_0^1 = \dfrac{2\pi}{35}$

8. $V = \pi \displaystyle\int_{-2}^2 \left[\left(4 - \dfrac{x^2}{2}\right)^2 - (2)^2\right] dx = 2\pi \displaystyle\int_0^2 \left(\dfrac{x^4}{4} - 4x^2 + 12\right) dx = 2\pi\left[\dfrac{x^5}{20} - \dfrac{4x^3}{3} + 12x\right]_0^2 = \dfrac{448\pi}{15}$

9. $y = x^2 \Rightarrow x = \sqrt{y}$

$\quad V = \pi \displaystyle\int_0^4 (\sqrt{y})^2\, dy = \pi \displaystyle\int_0^4 y\, dy = \pi\left[\dfrac{y^2}{2}\right]_0^4 = 8\pi$

10. $y = \sqrt{16 - x^2} \Rightarrow x = \sqrt{16 - y^2}$

$\quad V = \pi \displaystyle\int_0^4 \left(\sqrt{16 - y^2}\right)^2 dy = \pi \displaystyle\int_0^4 (16 - y^2)\, dy = \pi\left[16y - \dfrac{y^3}{3}\right]_0^4 = \dfrac{128\pi}{3}$

11. $y = x^{2/3} \Rightarrow x = y^{3/2}$

$\quad V = \pi \displaystyle\int_0^1 (y^{3/2})^2\, dy = \pi \displaystyle\int_0^1 y^3\, dy = \pi\left[\dfrac{y^4}{4}\right]_0^1 = \dfrac{\pi}{4}$

12. $V = \pi \displaystyle\int_1^4 (-y^2 + 4y)^2\, dy = \pi \displaystyle\int_1^4 (y^4 - 8y^3 + 16y^2)\, dy = \pi\left[\dfrac{y^5}{5} - 2y^4 + \dfrac{16y^3}{3}\right]_1^4 = \dfrac{459\pi}{15} = \dfrac{153\pi}{5}$

13. $y = \sqrt{x}, \quad y = 0, \quad x = 4$

    a. $R(x) = \sqrt{x}, \quad r(x) = 0$ b. $R(y) = 4, \quad r(y) = y^2$

$\qquad V = \pi \displaystyle\int_0^4 (\sqrt{x})^2\, dx$ $V = \pi \displaystyle\int_0^2 (16 - y^4)\, dy$

$\qquad = \pi \displaystyle\int_0^4 x\, dx = \left[\dfrac{\pi}{2}x^2\right]_0^4 = 8\pi$ $= \pi\left[16y - \dfrac{1}{5}y^5\right]_0^2 = \dfrac{128\pi}{5}$

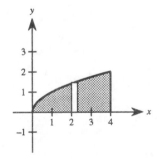
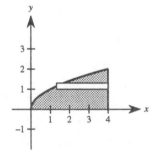

13. –CONTINUED–

c. $R(y) = 4 - y^2$, $r(y) = 0$

$$V = \pi \int_0^2 (4 - y^2)^2 \, dy$$

$$= \pi \int_0^2 (16 - 8y^2 + y^4) \, dy$$

$$= \pi \left[16y - \frac{8}{3}y^3 + \frac{1}{5}y^5 \right]_0^2 = \frac{256\pi}{15}$$

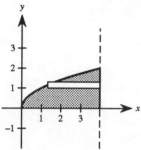

d. $R(y) = 6 - y^2$, $r(y) = 2$

$$V = \pi \int_0^2 [(6 - y^2)^2 - 4] \, dy$$

$$= \pi \int_0^2 (32 - 12y^2 + y^4) \, dy$$

$$= \pi \left[32y - 4y^3 + \frac{1}{5}y^5 \right]_0^2 = \frac{192\pi}{5}$$

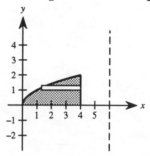

14. $y = 2x^2$, $y = 0$, $x = 2$

a. $R(y) = 2$, $r(y) = \sqrt{y/2}$

$$V = \pi \int_0^8 \left(4 - \frac{y}{2} \right) dy = \pi \left[4y - \frac{y^2}{4} \right]_0^8 = 16\pi$$

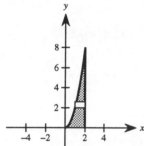

b. $R(x) = 2x^2$, $r(x) = 0$

$$V = \pi \int_0^2 4x^4 \, dx = \pi \left[\frac{4x^5}{5} \right]_0^2 = \frac{128\pi}{5}$$

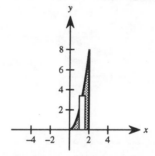

c. $R(x) = 8$, $r(x) = 8 - 2x^2$

$$V = \pi \int_0^2 [64 - (64 - 32x^2 + 4x^4)] \, dx$$

$$= \pi \int_0^2 (32x^2 - 4x^4) \, dx = 4\pi \int_0^2 (8x^2 - x^4) \, dx$$

$$= 4\pi \left[\frac{8}{3}x^3 - \frac{1}{5}x^5 \right]_0^2 = \frac{896\pi}{15}$$

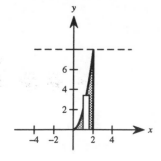

d. $R(y) = 2 - \sqrt{y/2}$, $r(y) = 0$

$$V = \pi \int_0^8 \left(2 - \sqrt{\frac{y}{2}} \right)^2 dy$$

$$= \pi \int_0^8 \left(4 - 4\sqrt{\frac{y}{2}} + \frac{y}{2} \right) dy$$

$$= \pi \left[4y - \frac{4\sqrt{2}}{3}y^{3/2} + \frac{y^2}{4} \right]_0^8 = \frac{16\pi}{3}$$

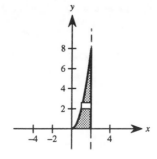

15. $y = x^2$, $y = 4x - x^2$

a. $R(x) = 4x - x^2$, $r(x) = x^2$

$$V = \pi \int_0^2 [(4x - x^2)^2 - x^4] \, dx$$

$$= \pi \int_0^2 (16x^2 - 8x^3) \, dx$$

$$= \pi \left[\frac{16}{3}x^3 - 2x^4 \right]_0^2 = \frac{32\pi}{3}$$

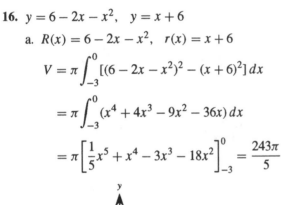

b. $R(x) = 6 - x^2$, $r(x) = 6 - (4x - x^2)$

$$V = \pi \int_0^2 [(6 - x^2)^2 - (6 - 4x + x^2)^2] \, dx$$

$$= 8\pi \int_0^2 (x^3 - 5x^2 + 6x) \, dx$$

$$= 8\pi \left[\frac{x^4}{4} - \frac{5}{3}x^3 + 3x^2 \right]_0^2 = \frac{64\pi}{3}$$

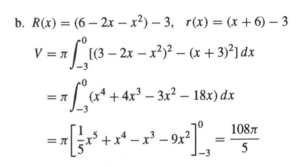

16. $y = 6 - 2x - x^2$, $y = x + 6$

a. $R(x) = 6 - 2x - x^2$, $r(x) = x + 6$

$$V = \pi \int_{-3}^0 [(6 - 2x - x^2)^2 - (x + 6)^2] \, dx$$

$$= \pi \int_{-3}^0 (x^4 + 4x^3 - 9x^2 - 36x) \, dx$$

$$= \pi \left[\frac{1}{5}x^5 + x^4 - 3x^3 - 18x^2 \right]_{-3}^0 = \frac{243\pi}{5}$$

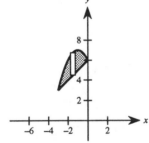

b. $R(x) = (6 - 2x - x^2) - 3$, $r(x) = (x + 6) - 3$

$$V = \pi \int_{-3}^0 [(3 - 2x - x^2)^2 - (x + 3)^2] \, dx$$

$$= \pi \int_{-3}^0 (x^4 + 4x^3 - 3x^2 - 18x) \, dx$$

$$= \pi \left[\frac{1}{5}x^5 + x^4 - x^3 - 9x^2 \right]_{-3}^0 = \frac{108\pi}{5}$$

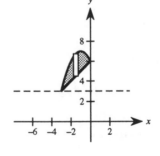

17. $R(x) = 4 - x$, $r(x) = 1$

$$V = \pi \int_0^3 [(4 - x)^2 - (1)^2] \, dx$$

$$= \pi \int_0^3 (x^2 - 8x + 15) \, dx$$

$$= \pi \left[\frac{x^3}{3} - 4x^2 + 15x \right]_0^3 = 18\pi$$

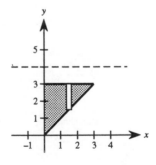

18. $R(x) = 4 - x^2, \quad r(x) = 0$

$$V = 2\pi \int_0^2 (4 - x^2)^2 \, dx$$

$$= 2\pi \int_0^2 (x^4 - 8x^2 + 16) \, dx$$

$$= 2\pi \left[\frac{x^5}{5} - \frac{8x^3}{3} + 16x \right]_0^2 = \frac{512\pi}{15}$$

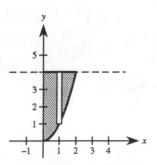

19. $R(x) = 4, \quad r(x) = 4 - \dfrac{1}{x}$

$$V = \pi \int_1^4 \left[(4)^2 - \left(4 - \frac{1}{x}\right)^2 \right] dx$$

$$= \pi \int_1^4 \left(\frac{8}{x} - \frac{1}{x^2} \right) dx$$

$$= \pi \left[8 \ln |x| + \frac{1}{x} \right]_1^4$$

$$= \pi \left(8 \ln 4 - \frac{3}{4} \right) \approx 32.49$$

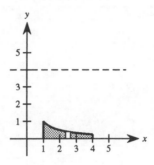

20. $R(x) = 4, \quad r(x) = 4 - \sec x$

$$V = \pi \int_0^{\pi/3} [(4)^2 - (4 - \sec x)^2] \, dx$$

$$= \pi \int_0^{\pi/3} (8 \sec x - \sec^2 x) \, dx$$

$$= \pi \left[8 \ln | \sec x + \tan x | - \tan x \right]_0^{\pi/3}$$

$$= \pi \left[\left(8 \ln |2 + \sqrt{3}| - \sqrt{3} \right) - \left(8 \ln |1 + 0| - 0 \right) \right]$$

$$= \pi \left[8 \ln \left(2 + \sqrt{3} \right) - \sqrt{3} \right] \approx 27.66$$

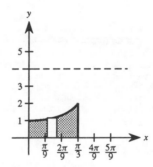

21. $R(y) = 6 - y, \quad r(y) = 0$

$$V = \pi \int_0^4 (6 - y)^2 \, dy$$

$$= \pi \int_0^4 (y^2 - 12y + 36) \, dy$$

$$= \pi \left[\frac{y^3}{3} - 6y^2 + 36y \right]_0^4 = \frac{208\pi}{3}$$

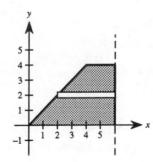

22. $R(y) = 6$, $\quad r(y) = 6 - (6 - y) = y$

$$V = \pi \int_0^4 [(6)^2 - (y)^2] \, dy$$

$$= \pi \left[36y - \frac{y^3}{3} \right]_0^4 = \frac{368\pi}{3}$$

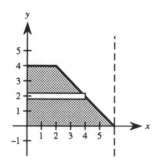

23. $R(y) = 6 - y^2$, $\quad r(y) = 2$

$$V = \pi \int_{-2}^2 [(6 - y^2)^2 - (2)^2] \, dy$$

$$= 2\pi \int_0^2 (y^4 - 12y^2 + 32) \, dy$$

$$= 2\pi \left[\frac{y^5}{5} - 4y^3 + 32y \right]_0^2 = \frac{384\pi}{5}$$

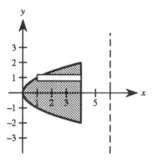

24. $R(y) = 6 - \dfrac{6}{y}$, $\quad r(y) = 0$

$$V = \pi \int_2^6 \left(6 - \frac{6}{y} \right)^2 dy$$

$$= 36\pi \int_2^6 \left(1 - \frac{2}{y} + \frac{1}{y^2} \right) dy$$

$$= 36\pi \left[y - 2 \ln |y| - \frac{1}{y} \right]_2^6$$

$$= 36\pi \left[\left(\frac{35}{6} - 2 \ln 6 \right) - \left(\frac{3}{2} - 2 \ln 2 \right) \right]$$

$$= 36\pi \left(\frac{13}{3} + 2 \ln \frac{1}{3} \right)$$

$$= 12\pi (13 - 6 \ln 3) \approx 241.59$$

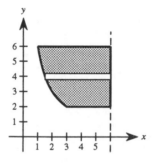

25. $R(x) = \dfrac{1}{\sqrt{x + 1}}$, $\quad r(x) = 0$

$$V = \pi \int_0^3 \left(\frac{1}{\sqrt{x + 1}} \right)^2 dx$$

$$= \pi \int_0^3 \frac{1}{x + 1} \, dx$$

$$= \left[\pi \ln |x + 1| \right]_0^3 = \pi \ln 4$$

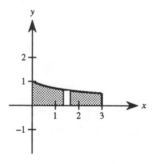

26. $R(x) = x\sqrt{4 - x^2}, \quad r(x) = 0$

$$V = 2\pi \int_0^2 \left[x\sqrt{4 - x^2} \right]^2 dx$$

$$= 2\pi \int_0^2 (4x^2 - x^4)\, dx$$

$$= 2\pi \left[\frac{4x^3}{3} - \frac{x^5}{5} \right]_0^2 = \frac{128\pi}{15}$$

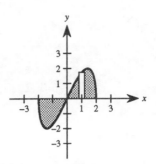

27. $R(x) = \dfrac{1}{x}, \quad r(x) = 0$

$$V = \pi \int_1^4 \left(\frac{1}{x} \right)^2 dx$$

$$= \pi \left[-\frac{1}{x} \right]_1^4$$

$$= \frac{3\pi}{4}$$

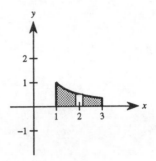

28. $R(x) = \dfrac{3}{x + 1}, \quad r(x) = 0$

$$V = \pi \int_0^8 \left(\frac{3}{x + 1} \right)^2 dx$$

$$= 9\pi \int_0^8 (x + 1)^{-2}\, dx$$

$$= 9\pi \left[-\frac{1}{x + 1} \right]_0^8 = 8\pi$$

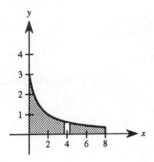

29. $R(x) = e^{-x}, \quad r(x) = 0$

$$V = \pi \int_0^1 (e^{-x})^2\, dx$$

$$= \pi \int_0^1 e^{-2x}\, dx$$

$$= \left[-\frac{\pi}{2} e^{-2x} \right]_0^1$$

$$= \frac{\pi}{2}(1 - e^{-2}) \approx 1.358$$

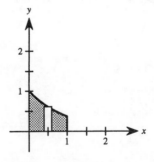

30. $R(x) = e^{x/2}, \quad r(x) = 0$

$$V = \pi \int_0^4 (e^{x/2})^2\, dx$$

$$= \pi \int_0^4 e^x\, dx$$

$$= \left[\pi e^x \right]_0^4$$

$$= \pi(e^4 - 1) \approx 168.38$$

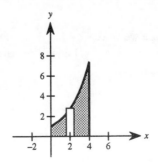

31. $R(x) = \sqrt{\sin x}, \quad r(x) = 0$

$$V = \pi \int_0^{\pi/2} (\sqrt{\sin x})^2 \, dx$$

$$= \pi \int_0^{\pi/2} \sin x \, dx$$

$$= \left[-\pi \cos x \right]_0^{\pi/2} = \pi$$

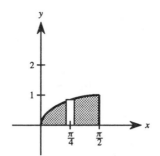

32. $R(x) = \sqrt{\cos x}, \quad r(x) = 0$

$$V = \pi \int_0^{\pi/2} (\sqrt{\cos x})^2 \, dx$$

$$= \pi \int_0^{\pi/2} \cos x \, dx$$

$$= \left[\pi \sin x \right]_0^{\pi/2} = \pi$$

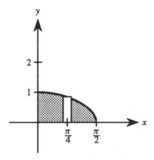

33. $A \approx 3$

Matches (a)

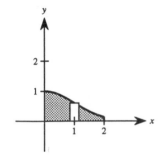

34. $A \approx \frac{3}{4}$

Matches (b)

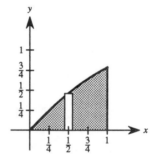

35. a. $R(x) = 4x - x^2, \quad r(x) = 0$

$$V = \pi \int_0^4 (4x - x^2)^2 \, dx$$

$$= \pi \int_0^4 (16x^2 - 8x^3 + x^4) \, dx$$

$$= \pi \left[\frac{16}{3} x^3 - 2x^4 + \frac{x^5}{5} \right]_0^4 = \frac{512\pi}{15}$$

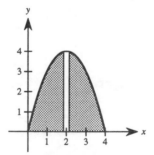

b. Completing the square we have $4x - x^2 = 4 - (x^2 - 4x + 4) = 4 - (x - 2)^2$. Thus, $y = 4 - x^2$ has the same volume as in part a since the solid has been translated only horizontally.

36. a. $R(x) = \frac{3}{5}\sqrt{25 - x^2},\quad r(x) = 0$

$$V = \frac{9\pi}{25}\int_{-5}^{5}(25 - x^2)\,dx$$

$$= \frac{18\pi}{25}\int_{0}^{5}(25 - x^2)\,dx$$

$$= \frac{18\pi}{25}\left[25x - \frac{x^3}{3}\right]_{0}^{5} = 60\pi$$

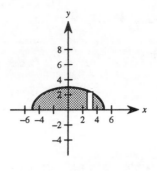

b. $R(y) = \frac{5}{3}\sqrt{9 - y^2},\quad r(y) = 0$

$$V = \frac{25\pi}{9}\int_{-3}^{3}(9 - y^2)\,dy$$

$$= \frac{50\pi}{9}\int_{0}^{3}(9 - y^2)\,dy$$

$$= \frac{50\pi}{9}\left[9y - \frac{y^3}{3}\right]_{0}^{3} = 100\pi$$

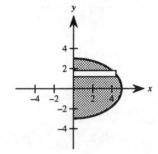

37. $R(x) = \frac{1}{2}x,\quad r(x) = 0$

$$V = \pi\int_{0}^{6}\frac{1}{4}x^2\,dx$$

$$= \left[\frac{\pi}{12}x^3\right]_{0}^{6} = 18\pi$$

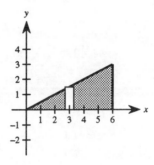

38. $R(x) = \frac{r}{h}x,\quad r(x) = 0$

$$V = \pi\int_{0}^{h}\frac{r^2}{h^2}x^2\,dx$$

$$= \left[\frac{r^2\pi}{3h^2}x^3\right]_{0}^{h}$$

$$= \frac{r^2\pi}{3h^2}h^3 = \frac{1}{3}\pi r^2 h$$

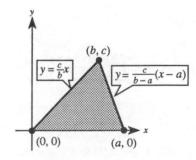

39. $R(x) = \sqrt{r^2 - x^2},\quad r(x) = 0$

$$V = \pi\int_{-r}^{r}(r^2 - x^2)\,dx$$

$$= 2\pi\int_{0}^{r}(r^2 - x^2)\,dx$$

$$= 2\pi\left[r^2 x - \frac{1}{3}x^3\right]_{0}^{r}$$

$$= 2\pi\left(r^3 - \frac{1}{3}r^3\right) = \frac{4}{3}\pi r^3$$

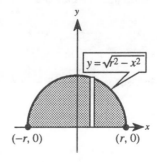

40. $x = \sqrt{r^2 - y^2}$, $R(y) = \sqrt{r^2 - y^2}$, $r(y) = 0$

$$V = \pi \int_h^r \left(\sqrt{r^2 - y^2}\right)^2 dy$$

$$= \pi \int_h^r (r^2 - y^2)\, dy$$

$$= \pi \left[r^2 y - \frac{y^3}{3} \right]_h^r$$

$$= \pi \left[\left(r^3 - \frac{r^3}{3} \right) - \left(r^2 h - \frac{h^3}{3} \right) \right]$$

$$= \pi \left(\frac{2r^3}{3} - r^2 h + \frac{h^3}{3} \right)$$

$$= \frac{\pi}{3}(2r^3 - 3r^2 h + h^3)$$

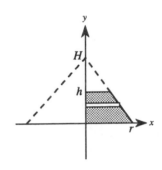

41. $x = r - \frac{r}{H} y = r\left(1 - \frac{y}{H}\right)$, $R(y) = r\left(1 - \frac{y}{H}\right)$, $r(y) = 0$

$$V = \pi \int_0^h \left[r\left(1 - \frac{y}{H}\right) \right]^2 dy = \pi r^2 \int_0^h \left(1 - \frac{2}{H} y + \frac{1}{H^2} y^2 \right) dy$$

$$= \pi r^2 \left[y - \frac{1}{H} y^2 + \frac{1}{3H^2} y^3 \right]_0^h$$

$$= \pi r^2 \left(h - \frac{h^2}{H} + \frac{h^3}{3H^2} \right)$$

$$= \pi r^2 h \left(1 - \frac{h}{H} + \frac{h^2}{3H^2} \right)$$

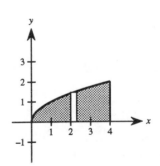

42. a. $V = \pi \int_0^4 (\sqrt{x})^2 dx = \pi \int_0^4 x\, dx = \left[\frac{\pi x^2}{2} \right]_0^4 = 8\pi$

Let $0 < c < 4$ and set $\pi \int_0^c x\, dx = \left[\frac{\pi x^2}{2} \right]_0^c = \frac{\pi c^2}{2} = 4\pi$.

$c^2 = 8$

$c = \sqrt{8} = 2\sqrt{2}$

Thus, when $x = 2\sqrt{2}$, the solid is divided into two parts of equal volume.

b. Set $\pi \int_0^c x\, dx = \frac{8\pi}{3}$ (one third of the volume). Then

$$\frac{\pi c^2}{2} = \frac{8\pi}{3}, \quad c^2 = \frac{16}{3}, \quad c = \frac{4}{\sqrt{3}} = \frac{4\sqrt{3}}{3}.$$

To find the other value, set $\pi \int_0^d x\, dx = \frac{16\pi}{3}$ (two thirds of the volume). Then

$$\frac{\pi d^2}{2} = \frac{16\pi}{3}, \quad d^2 = \frac{32}{3}, \quad d = \frac{\sqrt{32}}{\sqrt{3}} = \frac{4\sqrt{6}}{3}.$$

The x-values that divide the solid into three parts of equal volume are
$x = (4\sqrt{3})/3$ and $x = (4\sqrt{6})/3$.

---

43. $V = \pi \int_0^2 \left(\frac{1}{8}x^2\sqrt{2-x}\right)^2 dx = \frac{\pi}{64}\int_0^2 x^4(2-x)\,dx = \frac{\pi}{64}\left[\frac{2x^5}{5} - \frac{x^6}{6}\right]_0^2 = \frac{\pi}{30}$

44. $y = \begin{cases} \sqrt{0.1x^3 - 2.2x^2 + 10.9x + 22.2}, & 0 \le x \le 11.5 \\ 2.95, & 11.5 < x \le 15 \end{cases}$

$V = \pi \int_0^{11.5} \left(\sqrt{0.1x^3 - 2.2x^2 + 10.9x + 22.2}\right)^2 dx + \pi \int_{11.5}^{15} 2.95^2\,dx$

$= \pi\left[\frac{0.1x^4}{4} - \frac{2.2x^3}{3} + \frac{10.9x^2}{2} + 22.2x\right]_0^{11.5} + \pi\left[2.95^2 x\right]_{11.5}^{15}$

≈ 1031.9016 cubic centimeters

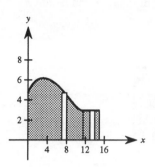

45. Total volume: $V = \frac{4\pi(50)^3}{3} = \frac{500{,}000\pi}{3}$ ft³

Volume of water in the tank:

$\pi \int_{-50}^{y_0} \left(\sqrt{2500 - y^2}\right)^2 dy = \pi \int_{-50}^{y_0} (2500 - y^2)\,dy$

$= \pi\left[2500y - \frac{y^3}{3}\right]_{-50}^{y_0}$

$= \pi\left(2500y_0 - \frac{y_0^3}{3} + \frac{250{,}000}{3}\right)$

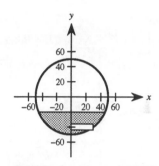

When the tank is one-fourth of its capacity:

$\frac{1}{4}\left(\frac{500{,}000\pi}{3}\right) = \pi\left(2500y_0 - \frac{y_0^3}{3} + \frac{250{,}000}{3}\right)$

$125{,}000 = 7500y_0 - y_0^3 + 250{,}000$

$y_0^3 - 7500y_0 - 125{,}000 = 0$

$y_0 \approx -17.36$

Depth: $-17.36 - (-50) = 32.64$ feet

When the tank is three-fourths of its capacity the depth is $100 - 17.36 = 82.64$ feet.

46. $V \approx \left(\frac{10-0}{3(10)}\right)\pi\left[2^2 + 4(2.05)^2 + 2(2.1)^2 + 4(2.35)^2 + 2(2.6)^2 + 4(2.85)^2 + 2(3)^2 + 4(2.95)^2\right.$

$\left. + 2(2.45)^2 + 4(2.2)^2 + (2.3)^2\right] \approx 196.030$ cubic centimeters

47. $V = \pi \int_0^2 (e^{-x^2})^2\,dx \approx 1.969$

48. $V = \pi \int_1^3 (\ln x)^2\,dx \approx 3.233$

49. $V = \pi \int_{-1}^2 (e^{x/2} + e^{-x/2})^2\,dx = \pi \int_{-1}^2 (e^x + 2 + e^{-x})\,dx \approx 49.023$

50. a. $\pi \int_0^h r^2 \, dx$ is the volume of a right circular cylinder with radius r and height h.

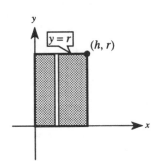

b. $\pi \int_{-b}^b \left(a \sqrt{1 - \dfrac{x^2}{b^2}} \right)^2 dx$

is the volume of an ellipsoid with axes $2a$ and $2b$.

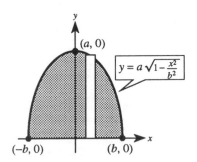

c. $\pi \int_{-r}^r \left(\sqrt{r^2 - x^2} \right)^2 dx$ is the volume of a sphere with radius r.

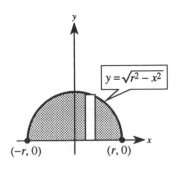

d. $\pi \int_0^h \left(\dfrac{rx}{h} \right)^2 dx$ is the volume of a right circular cone with the radius of the base as r and height h.

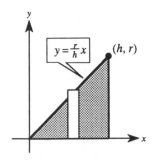

e. $\pi \int_{-r}^r \left[\left(R + \sqrt{r^2 - x^2} \right)^2 - \left(R - \sqrt{r^2 - x^2} \right)^2 \right] dx$

is the volume of a torus with the radius of its circular cross section as r and the distance from the axis of the torus to the center of its cross section as R.

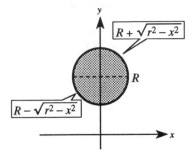

51.

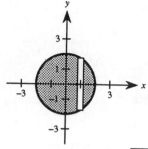

Base of Cross Section $= 2\sqrt{4-x^2}$

a. $A(x) = b^2 = \left(2\sqrt{4-x^2}\right)^2$

$V = \displaystyle\int_{-2}^{2} 4(4-x^2)\,dx$

$\quad = 4\left[4x - \dfrac{x^3}{3}\right]_{-2}^{2} = \dfrac{128}{3}$

$2\sqrt{4-x^2}$

$\longleftarrow 2\sqrt{4-x^2} \longrightarrow$

b. $A(x) = \dfrac{1}{2}bh = \dfrac{1}{2}\left(2\sqrt{4-x^2}\right)\left(\sqrt{3}\sqrt{4-x^2}\right)$

$\qquad = \sqrt{3}(4-x^2)$

$V = \sqrt{3}\displaystyle\int_{-2}^{2}(4-x^2)\,dx$

$\quad = \sqrt{3}\left[4x - \dfrac{x^3}{3}\right]_{-2}^{2} = \dfrac{32\sqrt{3}}{3}$

$2\sqrt{4-x^2}$ $2\sqrt{4-x^2}$

$\longleftarrow 2\sqrt{4-x^2} \longrightarrow$

c. $A(x) = \dfrac{1}{2}\pi r^2 = \dfrac{\pi}{2}\left(\sqrt{4-x^2}\right)^2 = \dfrac{\pi}{2}(4-x^2)$

$V = \dfrac{\pi}{2}\displaystyle\int_{-2}^{2}(4-x^2)\,dx = \dfrac{\pi}{2}\left[4x - \dfrac{x^3}{3}\right]_{-2}^{2} = \dfrac{16\pi}{3}$

$\longleftarrow 2\sqrt{4-x^2} \longrightarrow$

d. $A(x) = \dfrac{1}{2}bh$

$\qquad = \dfrac{1}{2}\left(2\sqrt{4-x^2}\right)\left(\sqrt{4-x^2}\right) = 4-x^2$

$V = \displaystyle\int_{-2}^{2}(4-x^2)\,dx = \left[4x - \dfrac{x^3}{3}\right]_{-2}^{2} = \dfrac{32}{3}$

$\sqrt{4-x^2}$

$\longleftarrow 2\sqrt{4-x^2} \longrightarrow$

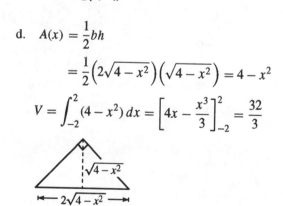

52.

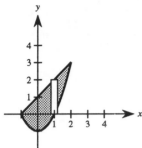

Base of Cross Section
$$= (x+1) - (x^2 - 1) = 2 + x - x^2$$

a. $A(x) = b^2 = (2 + x - x^2)^2$

$$= 4 + 4x - 3x^2 - 2x^3 + x^4$$

$$V = \int_{-1}^{2} (4 + 4x - 3x^2 - 2x^3 + x^4)\,dx$$

$$= \left[4x + 2x^2 - x^3 - \frac{1}{2}x^4 + \frac{1}{5}x^5 \right]_{-1}^{2} = \frac{81}{10}$$

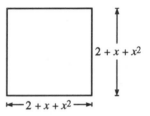

b. $A(x) = bh = 2 + x - x^2$

$$V = \int_{-1}^{2} (2 + x - x^2)\,dx = \left[2x + \frac{x^2}{2} - \frac{x^3}{3} \right]_{-1}^{2} = \frac{9}{2}$$

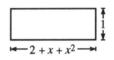

c. $A(x) = \frac{1}{2}\pi ab = \left(\frac{1}{2} \right)\pi(2)\left(\frac{2 + x - x^2}{2} \right) = \frac{\pi}{2}(2 + x - x^2)$

$$V = \frac{\pi}{2}\int_{-1}^{2} (2 + x - x^2)\,dx = \frac{\pi}{2}\left[2x + \frac{x^2}{2} - \frac{x^3}{3} \right]_{-1}^{2}$$

$$= \frac{\pi}{2}\left(\frac{9}{2} \right) = \frac{9\pi}{4}$$

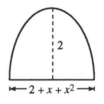

d. $A(x) = \frac{1}{2}bh = \frac{1}{2}(2 + x - x^2)\dfrac{\sqrt{3}(2 + x - x^2)}{2} = \frac{\sqrt{3}}{4}(2 + x - x^2)^2$

$$V = \frac{\sqrt{3}}{4}\int_{-1}^{2} (2 + x - x^2)^2\,dx$$

$$= \frac{\sqrt{3}}{4}\left[4x + 2x^2 - x^3 - \frac{1}{2}x^4 + \frac{1}{5}x^5 \right]_{-1}^{2}$$

$$= \frac{\sqrt{3}}{4}\left(\frac{81}{10} \right) = \frac{81\sqrt{3}}{40}$$

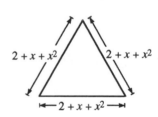

53.

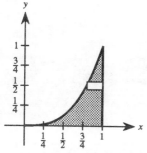

Base of Cross Section $= 1 - \sqrt[3]{y}$

a. $A(y) = b^2 = (1 - \sqrt[3]{y})^2$

$$V = \int_0^1 (1 - \sqrt[3]{y})^2 \, dy$$

$$= \int_0^1 (1 - 2y^{1/3} + y^{2/3}) \, dy$$

$$= \left[y - \frac{3}{2}y^{4/3} + \frac{3}{5}y^{5/3} \right]_0^1 = \frac{1}{10}$$

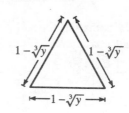

b. $A(y) = \frac{1}{2}\pi r^2 = \frac{1}{2}\pi \left(\frac{1 - \sqrt[3]{y}}{2} \right)^2 = \frac{1}{8}\pi(1 - \sqrt[3]{y})^2$

$$V = \frac{1}{8}\pi \int_0^1 (1 - \sqrt[3]{y})^2 \, dy = \frac{\pi}{8}\left(\frac{1}{10} \right) = \frac{\pi}{80}$$

c. $A(y) = \frac{1}{2}bh = \frac{1}{2}(1 - \sqrt[3]{y})\left(\frac{\sqrt{3}}{2} \right)(1 - \sqrt[3]{y})$

$$= \frac{\sqrt{3}}{4}(1 - \sqrt[3]{y})^2$$

$$V = \frac{\sqrt{3}}{4} \int_0^1 (1 - \sqrt[3]{y})^2 \, dy = \frac{\sqrt{3}}{4}\left(\frac{1}{10} \right) = \frac{\sqrt{3}}{40}$$

d. $A(y) = \frac{h}{2}(b_1 + b_2) = \frac{1}{2}\left[\frac{1 - \sqrt[3]{y}}{2} + (1 - \sqrt[3]{y}) \right]\left(\frac{1 - \sqrt[3]{y}}{2} \right)$

$$= \frac{3}{8}(1 - \sqrt[3]{y})^2$$

$$V = \frac{3}{8} \int_0^1 (1 - \sqrt[3]{y})^2 \, dy = \frac{3}{8}\left(\frac{1}{10} \right) = \frac{3}{80}$$

e. $A(y) = \frac{1}{2}\pi ab = \frac{\pi}{2}(2)(1 - \sqrt[3]{y})\frac{1 - \sqrt[3]{y}}{2}$

$$= \frac{\pi}{2}(1 - \sqrt[3]{y})^2$$

$$V = \frac{\pi}{2} \int_0^1 (1 - \sqrt[3]{y})^2 \, dy = \frac{\pi}{2}\left(\frac{1}{10} \right) = \frac{\pi}{20}$$

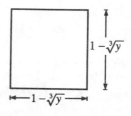

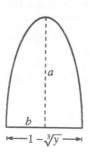

54. Since the cross sections are square:

$$A(y) = b^2 = \left(\sqrt{r^2 - y^2}\right)^2$$

By symmetry, we can set up an integral for an eighth of the volume and multiply by 8.

$$V = 8\int_0^r (r^2 - y^2)\,dy$$

$$= 8\left[r^2 y - \frac{1}{3}y^3\right]_0^r$$

$$= \frac{16}{3}r^3$$

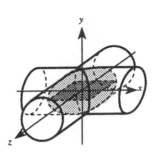

55. a. Since the cross sections are isosceles right triangles:

$$A(x) = \frac{1}{2}bh = \frac{1}{2}\left(\sqrt{r^2 - y^2}\right)\left(\sqrt{r^2 - y^2}\right) = \frac{1}{2}(r^2 - y^2)$$

$$V = \frac{1}{2}\int_{-r}^r (r^2 - y^2)\,dy = \int_0^r (r^2 - y^2)\,dy = \left[r^2 y - \frac{y^3}{3}\right]_0^r = \frac{2}{3}r^3$$

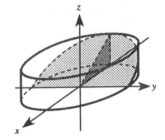

b. $A(x) = \frac{1}{2}bh = \frac{1}{2}\sqrt{r^2 - y^2}\left(\sqrt{r^2 - y^2}\,\tan\theta_0\right) = \frac{\tan\theta_0}{2}(r^2 - y^2)$

$$V = \frac{\tan\theta_0}{2}\int_{-r}^r (r^2 - y^2)\,dy = \tan\theta_0\int_0^r (r^2 - y^2)\,dy = \tan\theta_0\left[r^2 y - \frac{y^3}{3}\right]_0^r = \frac{2}{3}r^3\tan\theta_0$$

As $\theta_0 \to 90°$, $v \to \infty$.

56. Using the facts that the top surface of the oil is parallel to the ground and that the volume is one half of the volume of the cylinder, we have:

$$\sin 20° = \frac{d\sin 70°}{h}$$

$$h = \frac{d\sin 70°}{\sin 20°}$$

$$\text{Volume} = \frac{1}{2}(\pi r^2 h)$$

$$= \frac{\pi}{2}\left(\frac{d}{2}\right)^2\left(\frac{d\sin 70°}{\sin 20°}\right)$$

$$= \frac{\pi d^3 \sin 70°}{8\sin 20°}$$

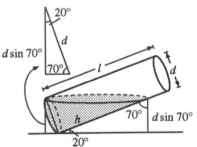

57. Let $A_1(x)$ and $A_2(x)$ equal the areas of the cross sections of the two solids for $a \le x \le b$. Since $A_1(x) = A_2(x)$, we have

$$V_1 = \int_a^b A_1(x)\,dx = \int_a^b A_2(x)\,dx = V_2.$$

Thus, the volumes are the same.

58. $V = \pi \int_{-\sqrt{R^2-r^2}}^{\sqrt{R^2-r^2}} \left[\left(\sqrt{R^2 - x^2} \right)^2 - r^2 \right] dx$

$\quad = 2\pi \int_0^{\sqrt{R^2-r^2}} (R^2 - r^2 - x^2) \, dx$

$\quad = 2\pi \left[(R^2 - r^2)x - \frac{x^3}{3} \right]_0^{\sqrt{R^2-r^2}}$

$\quad = 2\pi \left[(R^2 - r^2)^{3/2} - \frac{(R^2 - r^2)^{3/2}}{3} \right]$

$\quad = \frac{4}{3}\pi (R^2 - r^2)^{3/2}$

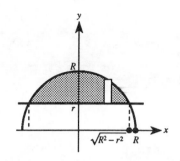

59. $\frac{4}{3}\pi (25 - r^2)^{3/2} = \frac{1}{2} \left(\frac{4}{3} \right) \pi (125)$

$\quad (25 - r^2)^{3/2} = \frac{125}{2}$

$\quad 25 - r^2 = \left(\frac{125}{2} \right)^{2/3}$

$\quad 25 - \frac{25}{(2^{2/3})} = r^2$

$\quad 25(1 - 2^{-2/3}) = r^2$

$\quad r = 5\sqrt{1 - 2^{-2/3}} \approx 3.0415$

60. a. When $a = 1$: $|x| + |y| = 1$ represents a square.
   When $a = 2$: $|x|^2 + |y|^2 = 1$ represents a circle.

   b. $|y| = (1 - |x|^a)^{1/a}$

   $A = 2 \int_{-1}^1 (1 - |x|^a)^{1/a} \, dx = 4 \int_0^1 (1 - x^a)^{1/a} \, dx$

   To approximate the volume of the solid, form n slices, each of whose area is approximated by the integral above. Then sum the volumes of these n slices.

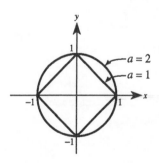

Section 6.3   Volume: The Shell Method

1. $p(x) = x$

$\quad h(x) = x$

$\quad V = 2\pi \int_0^2 x(x) \, dx = \left[\frac{2\pi x^3}{3} \right]_0^2 = \frac{16\pi}{3}$

2. $p(x) = x$

$\quad h(x) = 1 - x$

$\quad V = 2\pi \int_0^1 x(1 - x) \, dx$

$\quad = 2\pi \int_0^1 (x - x^2) \, dx = 2\pi \left[\frac{x^2}{2} - \frac{x^3}{3} \right]_0^1 = \frac{\pi}{3}$

3. $p(x) = x$

$h(x) = \sqrt{x}$

$V = 2\pi \displaystyle\int_0^4 x\sqrt{x}\, dx$

$\quad = 2\pi \displaystyle\int_0^4 x^{3/2}\, dx$

$\quad = \left[\dfrac{4\pi}{5} x^{5/2}\right]_0^4 = \dfrac{128\pi}{5}$

4. $p(x) = x$

$h(x) = 8 - (x^2 + 4) = 4 - x^2$

$V = 2\pi \displaystyle\int_0^2 x(4 - x^2)\, dx$

$\quad = 2\pi \displaystyle\int_0^2 (4x - x^3)\, dx$

$\quad = 2\pi \left[2x^2 - \dfrac{x^4}{4}\right]_0^2 = 8\pi$

5. $p(x) = x$

$h(x) = x^2$

$V = 2\pi \displaystyle\int_0^2 x^3\, dx = \left[\dfrac{\pi}{2} x^4\right]_0^2 = 8\pi$

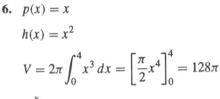

6. $p(x) = x$

$h(x) = x^2$

$V = 2\pi \displaystyle\int_0^4 x^3\, dx = \left[\dfrac{\pi}{2} x^4\right]_0^4 = 128\pi$

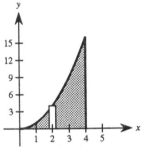

7. $p(x) = x$

$h(x) = (4x - x^2) - x^2 = 4x - 2x^2$

$V = 2\pi \displaystyle\int_0^2 x(4x - 2x^2)\, dx$

$\quad = 4\pi \displaystyle\int_0^2 (2x^2 - x^3)\, dx$

$\quad = 4\pi \left[\dfrac{2}{3} x^3 - \dfrac{1}{4} x^4\right]_0^2 = \dfrac{16\pi}{3}$

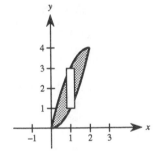

8. $p(x) = x$

$h(x) = 4 - x^2$

$V = 2\pi \displaystyle\int_0^2 (4x - x^3)\, dx$

$\quad = 2\pi \left[2x^2 - \dfrac{1}{4} x^4\right]_0^2 = 8\pi$

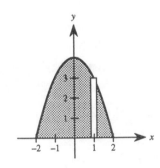

9. $p(x) = x$

$h(x) = 4 - (4x - x^2) = x^2 - 4x + 4$

$V = 2\pi \int_0^2 (x^3 - 4x^2 + 4x)\, dx$

$\quad = 2\pi \left[\dfrac{x^4}{4} - \dfrac{4}{3}x^3 + 2x^2 \right]_0^2 = \dfrac{8\pi}{3}$

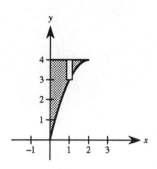

10. $p(x) = x$

$h(x) = 4 - 2x$

$V = 2\pi \int_0^2 x(4 - 2x)\, dx$

$\quad = 2\pi \int_0^2 (4x - 2x^2)\, dx$

$\quad = 2\pi \left[2x^2 - \dfrac{2}{3}x^3 \right]_0^2 = \dfrac{16\pi}{3}$

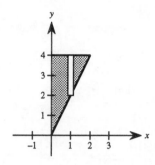

11. $p(x) = x$

$h(x) = \dfrac{1}{\sqrt{2\pi}} e^{-x^2/2}$

$V = 2\pi \int_0^1 x \left(\dfrac{1}{\sqrt{2\pi}} e^{-x^2/2} \right) dx$

$\quad = \sqrt{2\pi} \int_0^1 e^{-x^2/2} x\, dx$

$\quad = \left[-\sqrt{2\pi}\, e^{-x^2/2} \right]_0^1 = \sqrt{2\pi} \left(1 - \dfrac{1}{\sqrt{e}} \right)$

$\quad \approx 0.986$

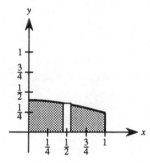

12. $p(x) = x$

$h(x) = \dfrac{\sin x}{x}$

$V = 2\pi \int_0^\pi x \left[\dfrac{\sin x}{x} \right] dx$

$\quad = 2\pi \int_0^\pi \sin x\, dx = \left[-2\pi \cos x \right]_0^\pi = 4\pi$

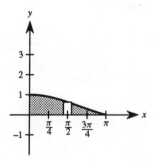

13. $p(y) = y$

$h(y) = 2 - y$

$V = 2\pi \displaystyle\int_0^2 y(2 - y)\,dy$

$= 2\pi \displaystyle\int_0^2 (2y - y^2)\,dy$

$= 2\pi \left[y^2 - \dfrac{y^3}{3} \right]_0^2 = \dfrac{8\pi}{3}$

14. $p(y) = -y$

$h(y) = 4 - (2 - y) = 2 + y$

$V = 2\pi \displaystyle\int_{-2}^0 (-y)(2 + y)\,dy$

$= 2\pi \displaystyle\int_{-2}^0 (-2y - y^2)\,dy$

$= 2\pi \left[-y^2 - \dfrac{y^3}{3} \right]_{-2}^0 = \dfrac{8\pi}{3}$

15. $p(y) = y$ and $h(y) = 1$ if $0 \le y < \dfrac{1}{2}$.

$p(y) = y$ and $h(y) = \dfrac{1}{y} - 1$ if $\dfrac{1}{2} \le y \le 1$.

$V = 2\pi \displaystyle\int_0^{1/2} y\,dy + 2\pi \displaystyle\int_{1/2}^1 (1 - y)\,dy$

$= 2\pi \left[\dfrac{y^2}{2} \right]_0^{1/2} + 2\pi \left[y - \dfrac{y^2}{2} \right]_{1/2}^1 = \dfrac{\pi}{2}$

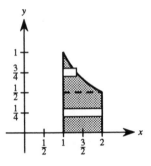

16. $p(y) = y$

$h(y) = 9 - y^2$

$V = 2\pi \displaystyle\int_0^3 y(9 - y^2)\,dy$

$= 2\pi \displaystyle\int_0^3 (9y - y^3)\,dy$

$= 2\pi \left[\dfrac{9}{2}y^2 - \dfrac{1}{4}y^4 \right]_0^3 = \dfrac{81\pi}{2}$

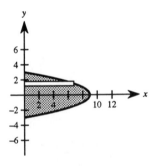

17. $p(x) = 4 - x$

$h(x) = 4x - x^2 - x^2 = 4x - 2x^2$

$V = 2\pi \displaystyle\int_0^2 (4 - x)(4x - 2x^2)\,dx$

$= 2\pi(2) \displaystyle\int_0^2 (x^3 - 6x^2 + 8x)\,dx$

$= 4\pi \left[\dfrac{x^4}{4} - 2x^3 + 4x^2 \right]_0^2 = 16\pi$

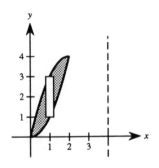

18. $p(x) = 2 - x$

$h(x) = 4x - x^2 - x^2 = 4x - 2x^2$

$V = 2\pi \displaystyle\int_0^2 (2 - x)(4x - 2x^2)\,dx$

$= 2\pi \displaystyle\int_0^2 (8x - 8x^2 + 2x^3)\,dx$

$= 2\pi \left[4x^2 - \dfrac{8}{3}x^3 + \dfrac{1}{2}x^4 \right]_0^2 = \dfrac{16\pi}{3}$

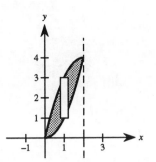

19. $p(x) = 5 - x$

$h(x) = 4x - x^2$

$V = 2\pi \displaystyle\int_0^4 (5 - x)(4x - x^2)\,dx$

$= 2\pi \displaystyle\int_0^4 (x^3 - 9x^2 + 20x)\,dx$

$= 2\pi \left[\dfrac{x^4}{4} - 3x^3 + 10x^2 \right]_0^4 = 64\pi$

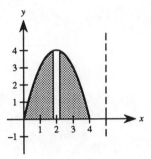

20. $p(x) = 6 - x$

$h(x) = \sqrt{x}$

$V = 2\pi \displaystyle\int_0^4 (6 - x)\sqrt{x}\,dx$

$= 2\pi \displaystyle\int_0^4 (6x^{1/2} - x^{3/2})\,dx$

$= 2\pi \left[4x^{3/2} - \dfrac{2}{5}x^{5/2} \right]_0^4 = \dfrac{192\pi}{5}$

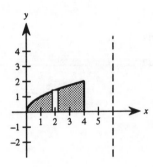

21. a. Disc

$R(x) = x^3$

$r(x) = 0$

$V = \pi \displaystyle\int_0^2 x^6\,dx = \pi \left[\dfrac{x^7}{7} \right]_0^2 = \dfrac{128\pi}{7}$

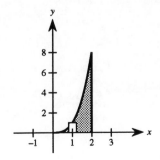

b. Shell

$p(x) = x$

$h(x) = x^3$

$V = 2\pi \displaystyle\int_0^2 x^4\,dx = 2\pi \left[\dfrac{x^5}{5} \right]_0^2 = \dfrac{64\pi}{5}$

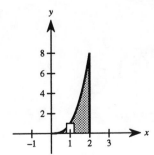

21. –CONTINUED–

c. Shell

$$p(x) = 4 - x$$

$$h(x) = x^3$$

$$V = 2\pi \int_0^2 (4 - x)x^3 \, dx$$

$$= 2\pi \int_0^2 (4x^3 - x^4) \, dx$$

$$= 2\pi \left[x^4 - \frac{1}{5}x^5 \right]_0^2 = \frac{96\pi}{5}$$

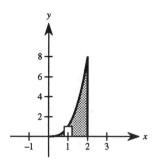

d. Disc

$$R(x) = 8$$

$$r(x) = 8 - x^3$$

$$V = \pi \int_0^2 [64 - (8 - x^3)^2] \, dx$$

$$= \pi \int_0^2 (16x^3 - x^6) \, dx$$

$$= \pi \left[4x^4 - \frac{x^7}{7} \right]_0^2 = \frac{320\pi}{7}$$

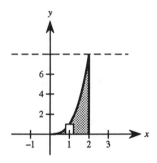

22. a. Disc

$$R(x) = \frac{1}{x^2}$$

$$r(x) = 0$$

$$V = \pi \int_1^4 \left(\frac{1}{x^2} \right)^2 \, dx$$

$$= \pi \left[-\frac{1}{3x^3} \right]_1^4 = \frac{21\pi}{64}$$

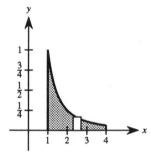

b. Shell

$$p(x) = x$$

$$h(x) = \frac{1}{x^2}$$

$$V = 2\pi \int_1^4 x \left(\frac{1}{x^2} \right) \, dx$$

$$= 2\pi \int_1^4 \frac{1}{x} \, dx$$

$$= 2\pi \left[\ln |x| \right]_1^4 = 2\pi \ln 4$$

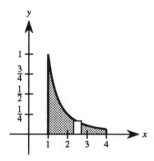

22. –CONTINUED–

c. **Shell**

$p(x) = 4 - x$

$h(x) = \dfrac{1}{x^2}$

$V = 2\pi \displaystyle\int_1^4 (4 - x)\left(\dfrac{1}{x^2}\right) dx$

$\quad = 2\pi \displaystyle\int_1^4 \left(\dfrac{4}{x^2} - \dfrac{1}{x}\right) dx$

$\quad = 2\pi \left[-\dfrac{4}{x} - \ln|x|\right]_1^4$

$\quad = 2\pi(3 - \ln 4) = \dfrac{75\pi}{64}$

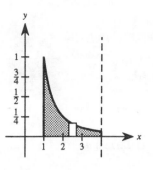

d. **Disc**

$R(x) = 1$

$r(x) = 1 - \dfrac{1}{x^2}$

$V = \pi \displaystyle\int_1^4 \left[1 - \left(1 - \dfrac{1}{x^2}\right)^2\right] dx$

$\quad = \pi \displaystyle\int_1^4 \left(\dfrac{2}{x^2} - \dfrac{1}{x^4}\right) dx$

$\quad = \pi \left[-\dfrac{2}{x} + \dfrac{1}{3x^3}\right]_1^4 = \dfrac{225\pi}{192}$

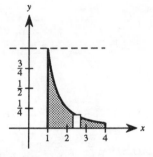

23. a. Shell

$p(y) = y$

$h(y) = (a^{1/2} - y^{1/2})^2$

$V = 2\pi \displaystyle\int_0^a y(a - 2a^{1/2}y^{1/2} + y)\, dy$

$\quad = 2\pi \displaystyle\int_0^a (ay - 2a^{1/2}y^{3/2} + y^2)\, dy$

$\quad = 2\pi \left[\dfrac{a}{2}y^2 - \dfrac{4a^{1/2}}{5}y^{5/2} + \dfrac{y^3}{3}\right]_0^a$

$\quad = 2\pi \left[\dfrac{a^3}{2} - \dfrac{4a^3}{5} + \dfrac{a^3}{3}\right] = \dfrac{\pi a^3}{15}$

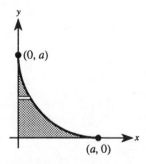

b. Same as part a by symmetry

23. –CONTINUED–

  c. **Shell**

$$p(x) = a - x$$

$$h(x) = (a^{1/2} - x^{1/2})^2$$

$$V = 2\pi \int_0^a (a - x)(a^{1/2} - x^{1/2})^2 \, dx$$

$$= 2\pi \int_0^a (a^2 - 2a^{3/2}x^{1/2} + 2a^{1/2}x^{3/2} - x^2) \, dx$$

$$= 2\pi \left[a^2 x - \frac{4}{3}a^{3/2}x^{3/2} + \frac{4}{5}a^{1/2}x^{5/2} - \frac{1}{3}x^3 \right]_0^a = \frac{4\pi a^3}{15}$$

  d. Same as part c by symmetry

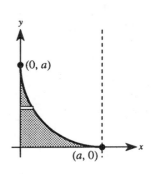

24. a. Disc

$$R(x) = (a^{2/3} - x^{2/3})^{3/2}$$

$$r(x) = 0$$

$$V = \pi \int_{-a}^a (a^{2/3} - x^{2/3})^3 \, dx$$

$$= 2\pi \int_0^a (a^2 - 3a^{4/3}x^{2/3} + 3a^{2/3}x^{4/3} - x^2) \, dx$$

$$= 2\pi \left[a^2 x - \frac{9}{5}a^{4/3}x^{5/3} + \frac{9}{7}a^{2/3}x^{7/3} - \frac{1}{3}x^3 \right]_0^a$$

$$= 2\pi \left(a^3 - \frac{9}{5}a^3 + \frac{9}{7}a^3 - \frac{1}{3}a^3 \right) = \frac{32\pi a^3}{105}$$

  b. Same as part a by symmetry

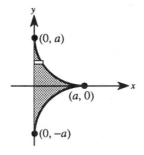

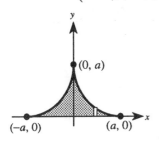

25. $y = 2e^{-x}$, $y = 0$, $x = 0$, $x = 2$

  Volume ≈ 7.5

  Matches (d)

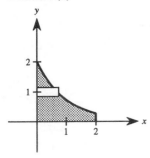

26. $y = \tan x$, $y = 0$, $x = 0$, $x = \dfrac{\pi}{4}$

  Volume ≈ 1

  Matches (e)

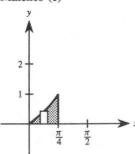

27. $p(x) = x$

$h(x) = 2 - \dfrac{1}{2}x^2$

$V = 2\pi \displaystyle\int_0^2 x\left(2 - \dfrac{1}{2}x^2\right) dx = 2\pi \int_0^2 \left(2x - \dfrac{1}{2}x^3\right) dx = 2\pi\left[x^2 - \dfrac{1}{8}x^4\right]_0^2 = 4\pi$ (total volume)

Now find x_0 such that

$$\pi = 2\pi \int_0^{x_0}\left(2x - \dfrac{1}{2}x^3\right) dx$$

$$1 = 2\left[x^2 - \dfrac{1}{8}x^4\right]_0^{x_0}$$

$$1 = 2x_0^2 - \dfrac{1}{4}x_0^4$$

$$x_0^4 - 8x_0^2 + 4 = 0$$

$$x_0^2 = 4 \pm 2\sqrt{3}.$$

Take $x_0 = \sqrt{4 - 2\sqrt{3}}$ since the other root is too large.

Diameter: $2\sqrt{4 - 2\sqrt{3}} \approx 1.464$

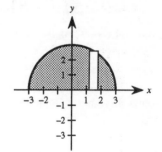

28. Total volume of the hemisphere is $\dfrac{1}{2}\left(\dfrac{4}{3}\right)\pi r^3 = \dfrac{2}{3}\pi(3)^3 = 18\pi$.

By the Shell Method, $p(x) = x$, $h(x) = \sqrt{9 - x^2}$. Find x_0 such that

$$6\pi = 2\pi \int_0^{x_0} x\sqrt{9 - x^2}\, dx$$

$$6 = -\int_0^{x_0} (9 - x^2)^{1/2}(-2x)\, dx$$

$$= \left[-\dfrac{2}{3}(9 - x^2)^{3/2}\right]_0^{x_0} = 18 - \dfrac{2}{3}(9 - x_0^2)^{3/2}$$

$$(9 - x_0^2)^{3/2} = 18$$

$$x_0 = \sqrt{9 - 18^{2/3}}.$$

Diameter: $2\sqrt{9 - 18^{2/3}} \approx 2.920$

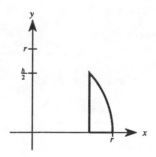

29. $x = \sqrt{r^2 - \left(\dfrac{h}{2}\right)^2} = \dfrac{\sqrt{4r^2 - h^2}}{2}$

$V = 4\pi \displaystyle\int_{\sqrt{4r^2-h^2}/2}^{r} x\sqrt{r^2 - x^2}\, dx$

$= \left[-2\pi\left(\dfrac{2}{3}\right)(r^2 - x^2)^{3/2}\right]_{\sqrt{4r^2-h^2}/2}^{r}$

$= 0 + \dfrac{4\pi}{3}\left[r^2 - \dfrac{4r^2 - h^2}{4}\right]^{3/2}$

$= \dfrac{4\pi}{3}\left(\dfrac{h^2}{4}\right)^{3/2}$

$= \dfrac{4\pi}{3}\cdot\dfrac{h^3}{8} = \dfrac{\pi h^3}{6}$

30. $V = 4\pi \int_{-1}^{1} (2-x)\sqrt{1-x^2}\,dx$

$$= 8\pi \int_{-1}^{1} \sqrt{1-x^2}\,dx - 4\pi \int_{-1}^{1} x\sqrt{1-x^2}\,dx$$

$$= 8\pi \left(\frac{\pi}{2}\right) + 2\pi \int_{-1}^{1} x(1-x^2)^{1/2}(-2)\,dx = 4\pi^2 + \left[2\pi \left(\frac{2}{3}\right)(1-x^2)^{3/2}\right]_{-1}^{1} = 4\pi^2$$

31. $V = 4\pi \int_{-r}^{r} (R-x)\sqrt{r^2-x^2}\,dx$

$$= 4\pi R \int_{-r}^{r} \sqrt{r^2-x^2}\,dx - 4\pi \int_{-r}^{r} x\sqrt{r^2-x^2}\,dx = 4\pi R\left(\frac{\pi r^2}{2}\right) + \left[2\pi \left(\frac{2}{3}\right)(r^2-x^2)^{3/2}\right]_{-r}^{r} = 2\pi^2 r^2 R$$

32. Disc

$R(y) = \sqrt{r^2 - y^2}$

$r(y) = 0$

$V = \pi \int_{r-h}^{r} (r^2 - y^2)\,dy$

$= \pi \left[r^2 y - \dfrac{y^3}{3} \right]_{r-h}^{r} = \dfrac{1}{3}\pi h^2 (3r - h)$

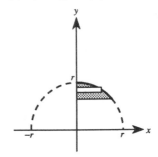

33. $\pi \int_{1}^{5} (x-1)\,dx = \pi \int_{1}^{5} (\sqrt{x-1})^2\,dx$

This integral represents the volume of the solid generated by revolving the region bounded by $y = \sqrt{x-1}$, $y = 0$, and $x = 5$ about the x-axis by using the Disc Method. $2\pi \int_{0}^{2} y[5 - (y^2+1)]\,dy$ represents this same volume by using the Shell Method.

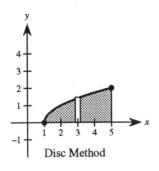

Disc Method

34. $2\pi \int_{0}^{4} x\left(\dfrac{x}{2}\right)\,dx$

represents the volume of the solid generated by revolving the region bounded by $y = x/2$, $y = 0$, and $x = 4$ about the y-axis by using the Shell Method.

$$\pi \int_{0}^{2} [16 - (2y)^2]\,dy = \pi \int_{0}^{2} [(4)^2 - (2y)^2]\,dy$$

represents this same volume by using the Disc Method.

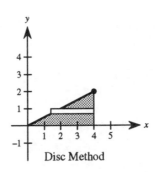

Disc Method

35. $x^{4/3} + y^{4/3} = 1$, $x = 0$, $y = 0$

$$y = \pm(1 - x^{4/3})^{3/4}$$

$$V = 2\pi \int_0^1 x(1 - x^{4/3})^{3/4}\, dx \approx 1.5016$$

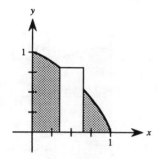

36. $V = 2\pi \int_0^1 x\sqrt{1 - x^3}\, dx \approx 2.2935$

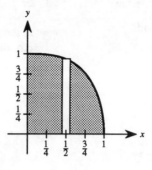

37. $V = 2\pi \int_2^6 x\sqrt[3]{(x-2)^2(x-6)^2}\, dx$

≈ 186.0552

38. $V = 2\pi \int_1^3 \dfrac{2x}{1 + e^{1/x}}\, dx \approx 19.0162$

39. $V = 2\pi \int_0^{40} x\, f(x)\, dx \approx \dfrac{2\pi(40)}{3(4)}\big[0 + 4(10)(46) + 2(20)(40) + 4(30)(20) + 0\big] \approx 122{,}313$ cubic feet

40. $V = 2\pi \int_0^{200} x\, f(x)\, dx$

$\approx \dfrac{2\pi(200)}{3(8)}\big[0 + 4(25)(19) + 2(50)(19) + 4(75)(17) + 2(100)(15) + 4(125)(14) + 2(150)(10) + 4(175)(6) + 0\big]$

$\approx 1{,}366{,}593$ cubic feet

41. a. $2\pi \int_0^r hx\left(1 - \dfrac{x}{r}\right) dx$

is the volume of a right circular cone with the radius of the base as r and height h.

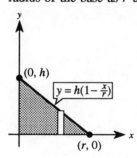

b. $2\pi \int_{-r}^r (R - x)\left(2\sqrt{r^2 - x^2}\right) dx$

is the volume of a torus with the radius of its circular cross section as r and the distance from the axis of the torus to the center of its cross section as R.

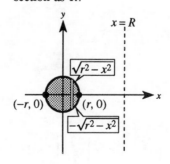

41. –CONTINUED–

c. $2\pi \displaystyle\int_0^r 2x\sqrt{r^2 - x^2}\, dx$ is the volume of a sphere with radius r.

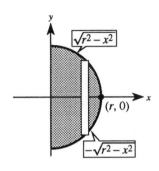

d. $2\pi \displaystyle\int_0^r hx\, dx$ is the volume of a right circular cylinder with a radius of r and a height of h.

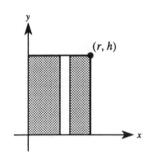

e. $2\pi \displaystyle\int_0^b 2ax\sqrt{1 - (x^2/b^2)}\, dx$ is the volume of an ellipsoid with axes $2a$ and $2b$.

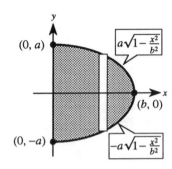

42. Volume of sphere:

$$V = \frac{4}{3}\pi(60{,}268)^3 \approx 9.16957 \times 10^{14}$$

Volume of oblate ellipsoid:

$$V = 2\pi \int_0^{60{,}268} 2(54{,}364)x\sqrt{1 - \frac{x^2}{60{,}268^2}}\, dx$$

$$= 217{,}456\pi \int_0^{60{,}268} \left(1 - \frac{x^2}{60{,}268^2}\right)^{1/2} x\, dx$$

$$= \left[(217{,}456\pi)\left(-\frac{60{,}268^2}{2}\right)\left(\frac{2}{3}\right)\left(1 - \frac{x^2}{60{,}268^2}\right)^{3/2}\right]_0^{60{,}268} \approx 8.27130 \times 10^{14}$$

Ratio: $\dfrac{8.27130 \times 10^{14}}{9.16957 \times 10^{14}} \approx 0.902$

Section 6.4   Arc Length and Surfaces of Revolution

1. (0, 0), (5, 12)

a. $d = \sqrt{(5 - 0)^2 + (12 - 0)^2} = 13$

b. $y = \dfrac{12}{5}x$

$y' = \dfrac{12}{5}$

$s = \displaystyle\int_0^5 \sqrt{1 + \left(\frac{12}{5}\right)^2}\, dx = \left[\frac{13}{5}x\right]_0^5 = 13$

2. (1, 2), (7, 10)

a. $d = \sqrt{(7 - 1)^2 + (10 - 2)^2} = 10$

b. $y = \dfrac{4}{3}x + \dfrac{2}{3}$

$y' = \dfrac{4}{3}$

$s = \displaystyle\int_1^7 \sqrt{1 + \left(\frac{4}{3}\right)^2}\, dx = \left[\frac{5}{3}x\right]_1^7 = 10$

3. $y = \frac{2}{3}x^{3/2} + 1$

$y' = x^{1/2}, \quad [0, 1]$

$s = \int_0^1 \sqrt{1 + x}\, dx$

$= \left[\frac{2}{3}(1 + x)^{3/2} \right]_0^1$

$= \frac{2}{3}(\sqrt{8} - 1) \approx 1.219$

4. $y = x^{3/2} - 1$

$y' = \frac{3}{2}x^{1/2}, \quad [0, 4]$

$s = \int_0^4 \sqrt{1 + \frac{9}{4}x}\, dx$

$= \frac{4}{9}\int_0^4 \left(1 + \frac{9}{4}x\right)^{1/2}\left(\frac{9}{4}\right) dx$

$= \left[\frac{8}{27}\left(1 + \frac{9}{4}x\right)^{3/2} \right]_0^4$

$= \frac{8}{27}(10^{3/2} - 1) \approx 9.073$

5. $y = \frac{3}{2}x^{2/3}$

$y' = \frac{1}{x^{1/3}}, \quad [1, 8]$

$s = \int_1^8 \sqrt{1 + \left(\frac{1}{x^{1/3}}\right)^2}\, dx$

$= \int_1^8 \sqrt{\frac{x^{2/3} + 1}{x^{2/3}}}\, dx$

$= \frac{3}{2}\int_1^8 \sqrt{x^{2/3} + 1}\left(\frac{2}{3x^{1/3}}\right) dx$

$= \frac{3}{2}\left[\frac{2}{3}(x^{2/3} + 1)^{3/2} \right]_1^8$

$= 5\sqrt{5} - 2\sqrt{2} \approx 8.352$

6. $y = \frac{x^5}{10} + \frac{1}{6x^3}$

$y' = \frac{1}{2}x^4 - \frac{1}{2x^4}$

$1 + (y')^2 = \left(\frac{1}{2}x^4 + \frac{1}{2x^4}\right)^2, \quad [1, 2]$

$s = \int_a^b \sqrt{1 + (y')^2}\, dx$

$= \int_1^2 \sqrt{\left(\frac{1}{2}x^4 + \frac{1}{2x^4}\right)^2}\, dx$

$= \int_1^2 \left(\frac{1}{2}x^4 + \frac{1}{2x^4}\right) dx$

$= \left[\frac{1}{10}x^5 - \frac{1}{6x^3} \right]_1^2 = \frac{779}{240} \approx 3.246$

7. $y = \frac{x^4}{8} + \frac{1}{4x^2}$

$y' = \frac{1}{2}x^3 - \frac{1}{2x^3}$

$1 + (y')^2 = \left(\frac{1}{2}x^3 + \frac{1}{2x^3}\right)^2, \quad [1, 2]$

$s = \int_a^b \sqrt{1 + (y')^2}\, dx$

$= \int_1^2 \left(\frac{1}{2}x^3 + \frac{1}{2x^3}\right) dx$

$= \left[\frac{1}{8}x^4 - \frac{1}{4x^2} \right]_1^2 = \frac{33}{16} \approx 2.063$

8. $y = \frac{1}{2}(e^x + e^{-x})$

$y' = \frac{1}{2}(e^x - e^{-x})$

$1 + (y')^2 = \left[\frac{1}{2}(e^x + e^{-x})\right]^2, \quad [0, 2]$

$s = \int_0^2 \sqrt{\left[\frac{1}{2}(e^x + e^{-x})\right]^2}\, dx$

$= \frac{1}{2}\int_0^2 (e^x + e^{-x})\, dx$

$= \frac{1}{2}\left[e^x - e^{-x} \right]_0^2 = \frac{1}{2}\left(e^2 - \frac{1}{e^2}\right) \approx 3.627$

9. $y = x^2 + x - 2$

$y' = 2x + 1, \quad [-2, \ 1]$

$s = \displaystyle\int_{-2}^{1} \sqrt{1 + (2x + 1)^2} \, dx$

$ = \displaystyle\int_{-2}^{1} \sqrt{2 + 4x + 4x^2} \, dx$

10. $y = \dfrac{1}{x + 1}$

$y' = -\dfrac{1}{(x + 1)^2}, \quad [0, \ 1]$

$s = \displaystyle\int_{0}^{1} \sqrt{1 + \dfrac{1}{(x + 1)^4}} \, dx$

$ = \displaystyle\int_{0}^{1} \dfrac{\sqrt{(x + 1)^4 + 1}}{(x + 1)^2} \, dx$

11. $y = 4 - x^2$

$y' = -2x, \quad [0, \ 2]$

$s = \displaystyle\int_{0}^{2} \sqrt{1 + (-2x)^2} \, dx$

$ = \displaystyle\int_{0}^{2} \sqrt{1 + 4x^2} \, dx$

12. $y = \cos x$

$y' = -\sin x, \quad \left[-\dfrac{\pi}{2}, \ \dfrac{\pi}{2}\right]$

$s = \displaystyle\int_{-\pi/2}^{\pi/2} \sqrt{1 + (-\sin x)^2} \, dx$

$ = \displaystyle\int_{-\pi/2}^{\pi/2} \sqrt{1 + \sin^2 x} \, dx$

13. $x = e^{-y}$

$x' = -e^{-y}, \quad [0, \ 2]$

$s = \displaystyle\int_{0}^{2} \sqrt{1 + e^{-2y}} \, dy$

$ = \displaystyle\int_{0}^{2} \dfrac{\sqrt{e^{2y} + 1}}{e^y} \, dy$

14. $x = \sqrt{a^2 - y^2}$

$x' = \dfrac{-y}{\sqrt{a^2 - y^2}}, \quad \left[0, \ \dfrac{a}{2}\right]$

$s = \displaystyle\int_{0}^{a/2} \sqrt{1 + \dfrac{y^2}{a^2 - y^2}} \, dy$

$ = \displaystyle\int_{0}^{a/2} \dfrac{a}{\sqrt{a^2 - y^2}} \, dy$

15. $\displaystyle\int_{0}^{2} \sqrt{1 + \left[\dfrac{d}{dx}\left(\dfrac{5}{x^2 + 1}\right)\right]^2} \, dx$

$s \approx 5$

Matches (b)

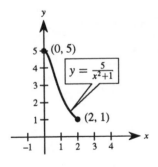

16. $\displaystyle\int_{0}^{\pi/4} \sqrt{1 + \left[\dfrac{d}{dx}(\tan x)\right]^2} \, dx$

$s \approx 1$

Matches (e)

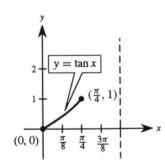

17. $y = x^3$, $[0, 4]$

   a. $d = \sqrt{(4-0)^2 + (64-0)^2} \approx 64.125$

   b. $d = \sqrt{(1-0)^2 + (1-0)^2} + \sqrt{(2-1)^2 + (8-1)^2} + \sqrt{(3-2)^2 + (27-8)^2} + \sqrt{(4-3)^2 + (64-27)^2}$

   ≈ 64.525

   c. $s = \displaystyle\int_0^4 \sqrt{1 + (3x^2)^2}\,dx = \int_0^4 \sqrt{1 + 9x^4}\,dx \approx 64.666$

18. $f(x) = (x^2 - 4)^2$, $[0, 4]$

   a. $d = \sqrt{(4-0)^2 + (144-16)^2} \approx 128.062$

   b. $d = \sqrt{(1-0)^2 + (9-16)^2} + \sqrt{(2-1)^2 + (0-9)^2} + \sqrt{(3-2)^2 + (25-0)^2} + \sqrt{(4-3)^2 + (144-25)^2}$

   ≈ 160.151

   c. $s = \displaystyle\int_0^4 \sqrt{1 + [4x(x^2-4)]^2}\,dx \approx 159.087$

19. $y = \dfrac{1}{x}$, $y' = -\dfrac{1}{x^2}$, $[1, 3]$

   $s = \displaystyle\int_1^3 \sqrt{1 + \dfrac{1}{x^4}}\,dx \approx 2.1468$

20. $y = x^2$, $y' = 2x$, $[0, 1]$

   $s = \displaystyle\int_0^1 \sqrt{1 + 4x^2}\,dx \approx 1.4789$

21. $y = \sin x$, $y' = \cos x$, $[0, \pi]$

   $s = \displaystyle\int_0^\pi \sqrt{1 + \cos^2 x}\,dx \approx 3.8202$

22. $y = \ln x$, $y' = \dfrac{1}{x}$, $[1, 5]$

   $s = \displaystyle\int_1^5 \sqrt{1 + \dfrac{1}{x^2}}\,dx \approx 4.3681$

23. $y = \dfrac{1}{3}[x^{3/2} - 3x^{1/2} + 2]$

   When $x = 0$, $y = \frac{2}{3}$. Thus, the fleeing object has traveled $\frac{2}{3}$ units when it is caught.

   $$y' = \frac{1}{3}\left[\frac{3}{2}x^{1/2} - \frac{3}{2}x^{-1/2}\right] = \left(\frac{1}{2}\right)\frac{x-1}{x^{1/2}}$$

   $$1 + (y')^2 = 1 + \frac{(x-1)^2}{4x} = \frac{(x+1)^2}{4x}$$

   $$s = \int_0^1 \frac{x+1}{2x^{1/2}}\,dx = \frac{1}{2}\int_0^1 (x^{1/2} + x^{-1/2})\,dx = \frac{1}{2}\left[\frac{2}{3}x^{3/2} + 2x^{1/2}\right]_0^1 = \frac{4}{3} = 2\left(\frac{2}{3}\right)$$

   The pursuer has traveled twice the distance that the fleeing object has traveled when it is caught.

24. $y = 31 - 10(e^{x/20} + e^{-x/20})$

$y' = -\dfrac{1}{2}(e^{x/20} - e^{-x/20})$

$1 + (y')^2 = 1 + \dfrac{1}{4}(e^{x/10} - 2 + e^{-x/10}) = \left[\dfrac{1}{2}(e^{x/20} + e^{-x/20})\right]^2$

$s = \displaystyle\int_{-20}^{20} \sqrt{\left[\dfrac{1}{2}(e^{x/20} + e^{-x/20})\right]^2} \, dx$

$= \dfrac{1}{2}\displaystyle\int_{-20}^{20}(e^{x/20} + e^{-x/20}) \, dx = \left[10(e^{x/20} - e^{-x/20})\right]_{-20}^{20} = 20\left(e - \dfrac{1}{e}\right) \approx 47 \text{ ft}$

Thus, there are $100(47) = 4700$ square feet of roofing on the barn.

25. $y = 60 \cosh \dfrac{x}{60}$

$y' = \sinh \dfrac{x}{60}$

$s = \displaystyle\int_{-60}^{60} \sqrt{1 + \sin^2 \dfrac{x}{60}} \, dx$

$= \displaystyle\int_{-60}^{60} \sqrt{\cosh^2 \dfrac{x}{60}} \, dx = \int_{-60}^{60} \cosh \dfrac{x}{60} \, dx = \left[60 \sinh \dfrac{x}{60}\right]_{-60}^{60} = 60[\sinh 1 - \sinh(-1)] \approx 141 \text{ feet}$

26. $y = 693.8597 - 68.7672 \cosh 0.0100333x$

$y' = -0.6899619478 \sinh 0.0100333x$

$s = \displaystyle\int_{-299.2239}^{299.2239} \sqrt{1 + (-0.6899619478 \sinh 0.0100333x)^2} \, dx \approx 1480$

(Use Simpson's Rule with $n = 100$.)

27. $y = \sqrt{9 - x^2}$

$y' = \dfrac{-x}{\sqrt{9 - x^2}}$

$1 + (y')^2 = \dfrac{9}{9 - x^2}$

$s = \displaystyle\int_0^2 \sqrt{\dfrac{9}{9 - x^2}} \, dx$

$= \displaystyle\int_0^2 \dfrac{3}{\sqrt{9 - x^2}} \, dx$

$= \left[3 \arcsin \dfrac{x}{3}\right]_0^2$

$= 3\left(\arcsin \dfrac{2}{3} - \arcsin 0\right)$

$= 3 \arcsin \dfrac{2}{3} \approx 2.1892$

28. $y = \sqrt{25 - x^2}$

$y' = \dfrac{-x}{\sqrt{25 - x^2}}$

$1 + (y')^2 = \dfrac{25}{25 - x^2}$

$s = \displaystyle\int_{-3}^4 \sqrt{\dfrac{25}{25 - x^2}} \, dx$

$= \displaystyle\int_{-3}^4 \dfrac{5}{\sqrt{25 - x^2}} \, dx$

$= \left[5 \arcsin \dfrac{x}{5}\right]_{-3}^4$

$= 5\left[\arcsin \dfrac{4}{5} - \arcsin\left(-\dfrac{3}{5}\right)\right] \approx 7.8540$

$\dfrac{1}{4}[2\pi(5)] \approx 7.8540 = s$

29. $y = \dfrac{x^3}{3}$

$y' = x^2, \quad [0, 3]$

$S = 2\pi \displaystyle\int_0^3 \dfrac{x^3}{3}\sqrt{1+x^4}\,dx$

$= \dfrac{\pi}{6} \displaystyle\int_0^3 (1+x^4)^{1/2}(4x^3)\,dx$

$= \left[\dfrac{\pi}{9}(1+x^4)^{3/2}\right]_0^3$

$= \dfrac{\pi}{9}(82\sqrt{82}-1) \approx 258.85$

30. $y = \sqrt{x}$

$y' = \dfrac{1}{2\sqrt{x}}, \quad [1, 4]$

$S = 2\pi \displaystyle\int_1^4 \sqrt{x}\sqrt{1+\dfrac{1}{4x}}\,dx$

$= 2\pi \displaystyle\int_1^4 \dfrac{\sqrt{x}}{2\sqrt{x}}\sqrt{4x+1}\,dx$

$= \pi \displaystyle\int_1^4 \sqrt{4x+1}\,dx$

$= \dfrac{\pi}{4} \displaystyle\int_1^4 (4x+1)^{1/2}(4)\,dx$

$= \left[\dfrac{\pi}{6}(4x+1)^{3/2}\right]_1^4$

$= \dfrac{\pi}{6}(17\sqrt{17}-5\sqrt{5}) \approx 30.85$

31. $y = \dfrac{x^3}{6} + \dfrac{1}{2x}$

$y' = \dfrac{x^2}{2} - \dfrac{1}{2x^2}$

$1+(y')^2 = \left(\dfrac{x^2}{2} + \dfrac{1}{2x^2}\right)^2, \quad [1, 2]$

$S = 2\pi \displaystyle\int_1^2 \left(\dfrac{x^3}{6} + \dfrac{1}{2x}\right)\left(\dfrac{x^2}{2} + \dfrac{1}{2x^2}\right)dx$

$= 2\pi \displaystyle\int_1^2 \left(\dfrac{x^5}{12} + \dfrac{x}{3} + \dfrac{1}{4x^3}\right)dx$

$= 2\pi \left[\dfrac{x^6}{72} + \dfrac{x^2}{6} - \dfrac{1}{8x^2}\right]_1^2 = \dfrac{47\pi}{16}$

32. $y = \dfrac{x}{2}$

$y' = \dfrac{1}{2}$

$1+(y')^2 = \dfrac{5}{4}, \quad [0, 6]$

$S = 2\pi \displaystyle\int_0^6 \dfrac{x}{2}\sqrt{\dfrac{5}{4}}\,dx$

$= \left[\dfrac{2\pi\sqrt{5}}{8}x^2\right]_0^6 = 9\sqrt{5}\,\pi$

33. $y = \sqrt[3]{x} + 2$

$y' = \dfrac{1}{3x^{2/3}}, \quad [1, 8]$

$S = 2\pi \displaystyle\int_1^8 x\sqrt{1+\dfrac{1}{9x^{4/3}}}\,dx$

$= \dfrac{2\pi}{3} \displaystyle\int_1^8 x^{1/3}\sqrt{9x^{4/3}+1}\,dx$

$= \dfrac{\pi}{18} \displaystyle\int_1^8 (9x^{4/3}+1)^{1/2}(12x^{1/3})\,dx$

$= \left[\dfrac{\pi}{27}(9x^{4/3}+1)^{3/2}\right]_1^8$

$= \dfrac{\pi}{27}(145\sqrt{145}-10\sqrt{10}) \approx 199.48$

34. $y = 4 - x^2$

$y' = -2x, \quad [0, 2]$

$S = 2\pi \displaystyle\int_0^2 x\sqrt{1+4x^2}\,dx$

$= \dfrac{\pi}{4} \displaystyle\int_0^2 (1+4x^2)^{1/2}(8x)\,dx$

$= \left[\dfrac{\pi}{6}(1+4x^2)^{3/2}\right]_0^2$

$= \dfrac{\pi}{6}(17\sqrt{17}-1) \approx 36.18$

35. $y = \sin x$

$y' = \cos x$, $[0,\ \pi]$

$S = 2\pi \int_0^\pi \sin x \sqrt{1 + \cos^2 x}\, dx$

≈ 14.4260

36. $y = \ln x$

$y' = \dfrac{1}{x}$

$1 + (y')^2 = \dfrac{x^2 + 1}{x^2}$, $[1,\ e]$

$S = 2\pi \int_1^e x \sqrt{\dfrac{x^2 + 1}{x^2}}\, dx$

$= 2\pi \int_1^e \sqrt{x^2 + 1}\, dx \approx 22.9430$

37. $y = \dfrac{hx}{r}$

$y' = \dfrac{h}{r}$

$1 + (y')^2 = \dfrac{r^2 + h^2}{r^2}$

$S = 2\pi \int_0^r x \sqrt{\dfrac{r^2 + h^2}{r^2}}\, dx$

$= \left[\dfrac{2\pi \sqrt{r^2 + h^2}}{r} \left(\dfrac{x^2}{2} \right) \right]_0^r = \pi r \sqrt{r^2 + h^2}$

38. $y = \sqrt{r^2 - x^2}$

$y' = \dfrac{-x}{\sqrt{r^2 - x^2}}$

$1 + (y')^2 = \dfrac{r^2}{r^2 - x^2}$

$S = 2\pi \int_{-r}^r \sqrt{r^2 - x^2} \sqrt{\dfrac{r^2}{r^2 - x^2}}\, dx$

$= 2\pi \int_{-r}^r r\, dx = \Big[2\pi r x \Big]_{-r}^r = 4\pi r^2$

39. $y = \sqrt{9 - x^2}$

$y' = \dfrac{-x}{\sqrt{9 - x^2}}$

$\sqrt{1 + (y')^2} = \dfrac{3}{\sqrt{9 - x^2}}$

$S = 2\pi \int_0^2 \dfrac{3x}{\sqrt{9 - x^2}}\, dx = -3\pi \int_0^2 \dfrac{-2x}{\sqrt{9 - x^2}}\, dx = \Big[-6\pi \sqrt{9 - x^2} \Big]_0^2 = 6\pi(3 - \sqrt{5}) \approx 14.40$

See figure in Exercise 40.

40. From Exercise 39 we have:

$S = 2\pi \int_0^a \dfrac{rx}{\sqrt{r^2 - x^2}}\, dx$

$= -r\pi \int_0^a \dfrac{-2x\, dx}{\sqrt{r^2 - x^2}}$

$= \Big[-2r\pi \sqrt{r^2 - x^2} \Big]_0^a$

$= 2r^2\pi - 2r\pi \sqrt{r^2 - a^2}$

$= 2r\pi \left(r - \sqrt{r^2 - a^2} \right)$

$= 2\pi r h$ (where h is the height of the zone)

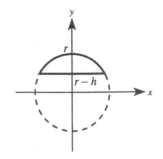

41.
$$y = \frac{1}{3}x^{1/2} - x^{3/2}$$

$$y' = \frac{1}{6}x^{-1/2} - \frac{3}{2}x^{1/2} = \frac{1}{6}(x^{-1/2} - 9x^{1/2})$$

$$1 + (y')^2 = 1 + \frac{1}{36}(x^{-1} - 18 + 81x) = \frac{1}{36}(x^{-1/2} + 9x^{1/2})^2$$

$$S = 2\pi \int_0^{1/3} \left(\frac{1}{3}x^{1/2} - x^{3/2}\right)\sqrt{\frac{1}{36}(x^{-1/2} + 9x^{1/2})^2}\, dx$$

$$= \frac{2\pi}{6}\int_0^{1/3} \left(\frac{1}{3}x^{1/2} - x^{3/2}\right)(x^{-1/2} + 9x^{1/2})\, dx$$

$$= \frac{\pi}{3}\int_0^{1/3} \left(\frac{1}{3} + 2x - 9x^2\right) dx = \frac{\pi}{3}\left[\frac{1}{3}x + x^2 - 3x^3\right]_0^{1/3} = \frac{\pi}{27} \text{ ft}^2 \approx 16.8 \text{ in}^2$$

Amount of glass needed: $V = \frac{\pi}{27}\left(\frac{0.015}{12}\right) \approx 0.00015 \text{ ft}^3 \approx 0.25 \text{ in}^3$

42. a. $V \approx \sum \pi r_i{}^2(3) = \sum 3\pi \left(\frac{C_i}{2\pi}\right)^2 = \frac{3}{4\pi}\sum C_i{}^2$

$$= \frac{3}{4\pi}\left[\left(\frac{50+65.5}{2}\right)^2 + \left(\frac{65.5+67}{2}\right)^2 + \left(\frac{67+68}{2}\right)^2 + \left(\frac{68+65}{2}\right)^2 + \left(\frac{65+51}{2}\right)^2 + \left(\frac{51+48}{2}\right)^2\right]$$

$$\approx 5375.5 \text{ cubic inches}$$

b. $S = \sum 2\pi r_i(3) = \sum C_i(3)$

$$\approx \left(\frac{50+65.5}{2}\right)(3) + \left(\frac{65.5+67}{2}\right)(3) + \left(\frac{67+68}{2}\right)(3) + \left(\frac{68+65}{2}\right)(3) + \left(\frac{65+51}{2}\right)(3) + \left(\frac{51+48}{2}\right)(3)$$

$$= 3[25 + 65.5 + 67 + 68 + 65 + 51 + 24] = 1096.5 \text{ square inches}$$

43. Individual project, see Exercise 42.

44. Area of circle with radius L: $A = \pi L^2$

Area of sector with central angle θ (in radians): $S = \frac{\theta}{2\pi}A = \frac{\theta}{2\pi}(\pi L^2) = \frac{1}{2}L^2\theta$

45. $S = \frac{1}{2}L^2\theta$ (from Exercise 44)

Let s be the arc length of the sector. Then $s = L\theta$. Therefore,

$$S = \frac{1}{2}L^2\left(\frac{s}{L}\right) = \frac{1}{2}Ls = \frac{1}{2}L(2\pi r) = \pi r L .$$

46. The lateral surface area of the frustrum is the difference of the large cone and the small cone.

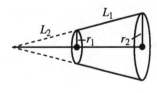

$$S = \pi r_2(L + L_1) - \pi r_1 L_1 \quad \text{(from Exercise 45)}$$

$$= \pi r_2 L + \pi L_1(r_2 - r_1).$$

By similar triangles,

$$\frac{L+L_1}{r_2} = \frac{L_1}{r_1} \Rightarrow Lr_1 = L_1(r_2 - r_1).$$

Hence, $S = \pi r_2 L + \pi(Lr_1) = \pi L(r_1 + r_2).$

47. Essay

48. a. $V = \pi \int_1^b \frac{1}{x^2}\,dx = \left[-\frac{\pi}{x}\right]_1^b = \pi\left(1 - \frac{1}{b}\right)$

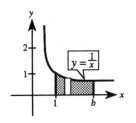

b. $S = 2\pi \int_1^b \frac{1}{x}\sqrt{1 + \left(-\frac{1}{x^2}\right)^2}\,dx$

$= 2\pi \int_1^b \frac{1}{x}\sqrt{1 + \frac{1}{x^4}}\,dx$

$= 2\pi \int_1^b \frac{\sqrt{x^4 + 1}}{x^3}\,dx$

c. $\displaystyle\lim_{b\to\infty} V = \lim_{b\to\infty} \pi\left(1 - \frac{1}{b}\right) = \pi$

d. Since

$$\frac{\sqrt{x^4 + 1}}{x^3} > \frac{\sqrt{x^4}}{x^3} = \frac{1}{x} > 0 \text{ on } [1,\ b]$$

we have

$$\int_1^b \frac{\sqrt{x^4 + 1}}{x^3}\,dx > \int_1^b \frac{1}{x}\,dx = \left[\ln x\right]_1^b = \ln b \quad \text{and} \quad \lim_{b\to\infty} \ln b \to \infty.$$

Thus, $\displaystyle\lim_{b\to\infty} 2\pi \int_1^b \frac{\sqrt{x^4 + 1}}{x^3}\,dx \to \infty.$

49. a. $\dfrac{ds}{dx} = \sqrt{1 + [f'(x)]^2}$

b. $ds = \sqrt{1 + [f'(x)]^2}\,dx$

$(ds)^2 = \left[1 + [f'(x)]^2\right](dx)^2 = \left[1 + \left(\frac{dy}{dx}\right)^2\right](dx)^2 = (dx)^2 + (dy)^2$

c. $s(x) = \displaystyle\int_1^x \sqrt{1 + \left(\frac{3}{2}t^{1/2}\right)^2}\,dt = \int_1^x \sqrt{1 + \frac{9}{4}t}\,dt = \left[\frac{4}{9}\left(\frac{2}{3}\right)\left(1 + \frac{9}{4}t\right)^{3/2}\right]_1^x$

$= \dfrac{8}{27}\left[\left(1 + \frac{9}{4}x\right)^{3/2} - \left(\frac{13}{4}\right)^{3/2}\right] = \dfrac{8}{27}\left[\dfrac{(4 + 9x)^{3/2}}{8} - \dfrac{13\sqrt{13}}{8}\right] = \dfrac{1}{27}\left[(4 + 9x)^{3/2} - 13\sqrt{13}\right]$

$s(2) = \dfrac{1}{27}\left[22\sqrt{22} - 13\sqrt{13}\right] = \dfrac{22}{27}\sqrt{22} - \dfrac{13}{27}\sqrt{13}$

50.

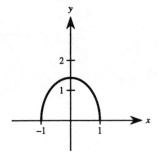

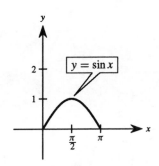

For the ellipse:

$$s = \int_{-1}^{1} \sqrt{1 + \left(-\frac{2x}{\sqrt{2(1-x^2)}}\right)^2}\, dx = 2\int_{0}^{1} \sqrt{1 + \frac{4x^2}{2(1-x^2)}}\, dx = 2\int_{0}^{1} \sqrt{\frac{1+x^2}{1-x^2}}\, dx$$

Let $x = \cos\theta$, then $\sqrt{1-x^2} = \sin\theta$ and $dx = -\sin\theta\, d\theta$. When $x = 0$, $\theta = \pi/2$ and when $x = 1$, $\theta = 0$.

$$s = 2\int_{\pi/2}^{0} \frac{\sqrt{1 + \cos^2\theta}}{\sin\theta}(-\sin\theta)\, d\theta = 2\int_{0}^{\pi/2} \sqrt{1 + \cos^2\theta}\, d\theta$$

For the sine curve:

$$s = \int_{0}^{\pi} \sqrt{1 + (\cos x)^2}\, dx = 2\int_{0}^{\pi/2} \sqrt{1 + \cos^2 x}\, dx$$

These arc lengths are equal.

Section 6.5   Work

1. $W = Fd = (100)(10) = 1000\ \text{ft} \cdot \text{lb}$

2. $W = Fd = (2400)(6) = 14{,}400\ \text{ft} \cdot \text{lb}$

3. $W = Fd = (25)(12) = 300\ \text{ft} \cdot \text{lb}$

4. $W = Fd = [9(2000)]\left[\frac{1}{2}(5280)\right] = 47{,}520{,}000\ \text{ft} \cdot \text{lb}$

5. $F(x) = kx$

$\quad 5 = k(4)$

$\quad k = \dfrac{5}{4}$

$\quad W = \int_{0}^{7} \frac{5}{4}x\, dx = \left[\frac{5}{8}x^2\right]_{0}^{7}$

$\quad\quad = \dfrac{245}{8}\ \text{in} \cdot \text{lb}$

$\quad\quad = 30.625\ \text{in} \cdot \text{lb} \approx 2.55\ \text{ft} \cdot \text{lb}$

6. $W = \int_{5}^{9} \frac{5}{4}x\, dx = \left[\frac{5}{8}x^2\right]_{5}^{9}$

$\quad\quad = 35\ \text{in} \cdot \text{lb} \approx 2.92\ \text{ft} \cdot \text{lb}$

7. $F(x) = kx$

$\quad 60 = 12k$

$\quad k = 5$

$\quad W = \int_{9}^{15} 5x\, dx = \left[\frac{5}{2}x^2\right]_{9}^{15}$

$\quad\quad = 360\ \text{in} \cdot \text{lb} = 30\ \text{ft} \cdot \text{lb}$

8. $F(x) = kx$

$\quad 200 = 2k$

$\quad k = 100$

$\quad W = \int_{0}^{2} 100x\, dx = \left[50x^2\right]_{0}^{2}$

$\quad\quad = 200\ \text{ft} \cdot \text{lb}$

9. $F(x) = kx$

$15 = 6k$

$k = \dfrac{5}{2}$

$W = \displaystyle\int_0^{12} \dfrac{5}{2} x \, dx = \left[\dfrac{5}{4} x^2 \right]_0^{12}$

$= 180 \text{ in} \cdot \text{lb} = 15 \text{ ft} \cdot \text{lb}$

10. $F(x) = kx$

$15 = k$

$W = 2 \displaystyle\int_0^4 15x \, dx = \left[15x^2 \right]_0^4$

$= 240 \text{ ft} \cdot \text{lb}$

11. Assume that the earth has a radius of 4000 miles.

$F(x) = \dfrac{k}{x^2}$

$4 = \dfrac{k}{(4000)^2}$

$k = 64{,}000{,}000$

$F(x) = \dfrac{64{,}000{,}000}{x^2}$

a. $W = \displaystyle\int_{4000}^{4200} \dfrac{64{,}000{,}000}{x^2} \, dx = \left[-\dfrac{64{,}000{,}000}{x} \right]_{4000}^{4200} \approx -15{,}238.095 + 16{,}000$

$= 761.905 \text{ mi} \cdot \text{ton} \approx 8.05 \times 10^9 \text{ ft} \cdot \text{lb}$

b. $W = \displaystyle\int_{4000}^{4400} \dfrac{64{,}000{,}000}{x^2} \, dx = \left[-\dfrac{64{,}000{,}000}{x} \right]_{4000}^{4400} \approx -14{,}545.455 + 16{,}000$

$= 1454.545 \text{ mi} \cdot \text{ton} \approx 1.54 \times 10^{10} \text{ ft} \cdot \text{lb}$

12. $\quad W = \displaystyle\int_{4000}^{h} \dfrac{64{,}000{,}000}{x^2} \, dx = \left[-\dfrac{64{,}000{,}000}{x} \right]_{4000}^{h} = -\dfrac{64{,}000{,}000}{h} + 16{,}000$

$\displaystyle\lim_{h \to \infty} W = 16{,}000 \text{ mi} \cdot \text{ton} \approx 1.690 \times 10^{11} \text{ ft} \cdot \text{lb}$

13. Assume that the earth has a radius of 4000 miles.

$F(x) = \dfrac{k}{x^2}$

$10 = \dfrac{k}{(4000)^2}$

$k = 160{,}000{,}000$

$F(x) = \dfrac{160{,}000{,}000}{x^2}$

a. $W = \displaystyle\int_{4000}^{15{,}000} \dfrac{160{,}000{,}000}{x^2} \, dx = \left[-\dfrac{160{,}000{,}000}{x} \right]_{4000}^{15{,}000} \approx -10{,}666.667 + 40{,}000$

$= 29{,}333.333 \text{ mi} \cdot \text{ton}$

$\approx 2.93 \times 10^4 \text{ mi} \cdot \text{ton}$

$\approx 3.10 \times 10^{11} \text{ ft} \cdot \text{lb}$

b. $W = \displaystyle\int_{4000}^{26{,}000} \dfrac{160{,}000{,}000}{x^2} \, dx = \left[-\dfrac{160{,}000{,}000}{x} \right]_{4000}^{26{,}000} \approx -6{,}153.846 + 40{,}000$

$= 33{,}846.154 \text{ mi} \cdot \text{ton}$

$\approx 3.38 \times 10^4 \text{ mi} \cdot \text{ton}$

$\approx 3.57 \times 10^{11} \text{ ft} \cdot \text{lb}$

14. Weight on surface of moon: $\frac{1}{6}(12) = 2$ tons

Weight varies inversely as the square of distance from the center of the moon. Therefore,

$$F(x) = \frac{k}{x^2}$$

$$2 = \frac{k}{(1100)^2}$$

$$k = 2.42 \times 10^6$$

$$W = \int_{1100}^{1150} \frac{2.42 \times 10^6}{x^2}\, dx = \left[\frac{-2.42 \times 10^6}{x}\right]_{1100}^{1150} = 2.42 \times 10^6 \left(\frac{1}{1100} - \frac{1}{1150}\right)$$

$$\approx 95.652 \text{ mi} \cdot \text{ton} \approx 1.01 \times 10^9 \text{ ft} \cdot \text{lb}$$

15. Weight of each layer: $62.4(20)\,\Delta y$

Distance: $4 - y$

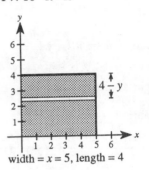

a. $W = \displaystyle\int_2^4 62.4(20)(4 - y)\, dy = \left[4992y - 624y^2\right]_2^4 = 2496 \text{ ft} \cdot \text{lb}$

b. $W = \displaystyle\int_0^4 62.4(20)(4 - y)\, dy = \left[4992y - 624y^2\right]_0^4 = 9984 \text{ ft} \cdot \text{lb}$

width $= x = 5$, length $= 4$

16. The bottom half had to be pumped a greater distance than the top half.

17. Volume of disc of water: $\pi(64)\,\Delta y$

Weight of disc of water: $62.4(64\pi\,\Delta y)$

Distance the disc of water is moved: $15 - y$

$$W = \int_0^{12} (15 - y)(62.4)(64\pi)\, dy = 3993.6\pi \int_0^{12} (15 - y)\, dy = 3993.6\pi\left[15y - \frac{y^2}{2}\right]_0^{12} = 431,308.8\pi \text{ ft} \cdot \text{lb}$$

18. $W = \displaystyle\int_{20}^{26} y(62.4)\pi(64)\, dy = 3993.6\pi \int_{20}^{26} y\, dy = \left[1996.8\pi y^2\right]_{20}^{26} = 551,116.8\pi \text{ ft} \cdot \text{lb}$

19. Volume of disc: $\pi\left(\sqrt{36 - y^2}\right)^2 \Delta y$

Weight of disc: $62.4\pi(36 - y^2)\,\Delta y$

Distance: y

$$W = 62.4\pi \int_0^6 y(36 - y^2)\, dy$$

$$= 62.4\pi \int_0^6 (36y - y^3)\, dy = 62.4\pi\left[18y^2 - \frac{1}{4}y^4\right]_0^6 = 20,217.6\pi \text{ ft} \cdot \text{lb}$$

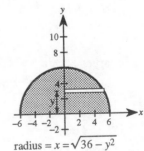

radius $= x = \sqrt{36 - y^2}$

C H A P T E R   6   Applications of Integration

20. Volume of each layer: $\left(\dfrac{y+3}{3}\right)(3)\Delta y = (y+3)\,dy$

Weight of each layer: $55.6(y+3)\,dy$

Distance: $5-y$

$$W = 55.6\int_0^3 (5-y)(y+3)\,dy$$

$$= 55.6\int_0^3 (15+2y-y^2)\,dy$$

$$= 55.6\left[15y+y^2-\frac{y^3}{3}\right]_0^3 = 2502 \text{ ft} \cdot \text{lb}$$

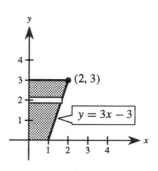

21. Volume of disc: $\pi\left(\dfrac{2}{3}y\right)^2 \Delta y$

Weight of disc: $62.4\pi\left(\dfrac{2}{3}y\right)^2 \Delta y$

Distance: $6-y$

$$W = \frac{4(62.4)\pi}{9}\int_0^6 (6-y)y^2\,dy = \frac{4}{9}(62.4)\pi\left[2y^3-\frac{1}{4}y^4\right]_0^6 = 2995.2\pi \text{ ft} \cdot \text{lb}$$

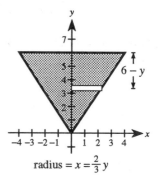

radius $= x = \frac{2}{3}y$

22. Volume of disc: $\pi\left(\dfrac{2}{3}y\right)^2 \Delta y$

Weight of disc: $62.4\pi\left(\dfrac{2}{3}y\right)^2 \Delta y$

Distance: y

a. $W = \dfrac{4}{9}(62.4)\pi\displaystyle\int_0^2 y^3\,dy = \left[\dfrac{4}{9}(62.4)\pi\left(\dfrac{1}{4}y^4\right)\right]_0^2 \approx 110.9\pi \text{ ft} \cdot \text{lb}$

b. $W = \dfrac{4}{9}(62.4)\pi\displaystyle\int_4^6 y^3\,dy = \left[\dfrac{4}{9}(62.4)\pi\left(\dfrac{1}{4}y^4\right)\right]_4^6 \approx 7210.7\pi \text{ ft} \cdot \text{lb}$

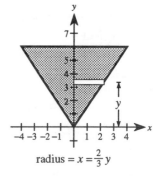

radius $= x = \frac{2}{3}y$

23. Volume of layer: $V = lwh = 4(2)\sqrt{(9/4)-y^2}\,\Delta y$

Weight of layer: $W = 42(8)\sqrt{(9/4)-y^2}\,\Delta y$

Distance: $\dfrac{13}{2}-y$

$$W = \int_{-1.5}^{1.5} 42(8)\sqrt{(9/4)-y^2}\left(\frac{13}{2}-y\right)dy$$

$$= 336\left[\frac{13}{2}\int_{-1.5}^{1.5}\sqrt{(9/4)-y^2}\,dy - \int_{-1.5}^{1.5}\sqrt{(9/4)-y^2}\,y\,dy\right]$$

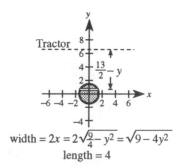

width $= 2x = 2\sqrt{\frac{9}{4}-y^2} = \sqrt{9-4y^2}$

length $= 4$

The second integral is zero since the integrand is odd and the limits of integration are symmetric to the origin. The first integral represents the area of a semicircle of radius $\frac{3}{2}$. Thus, the work is

$$W = 336\left(\frac{13}{2}\right)\pi\left(\frac{3}{2}\right)^2\left(\frac{1}{2}\right) = 2457\pi \text{ ft} \cdot \text{lb}.$$

24. Volume of layer: $V = 12(2)\sqrt{(25/4) - y^2}\, \Delta y$

Weight of layer: $W = 42(24)\sqrt{(25/4) - y^2}\, \Delta y$

Distance: $\dfrac{19}{2} - y$

$$W = \int_{-2.5}^{2.5} 42(24)\sqrt{(25/4) - y^2}\left(\frac{19}{2} - y\right) dy$$

$$= 1008\left[\frac{19}{2}\int_{-2.5}^{2.5}\sqrt{(25/4) - y^2}\, dy + \int_{-2.5}^{2.5}\sqrt{(25/4) - y^2}(-y)\, dy\right]$$

The second integral is zero since the integrand is odd and the limits of integration are symmetric to the origin. The first integral represents the area of a semicircle of radius $\frac{5}{2}$. Thus, the work is

$$W = 1008\left(\frac{19}{2}\right)\pi\left(\frac{5}{2}\right)^2\left(\frac{1}{2}\right) = 29{,}925\pi \ \text{ft} \cdot \text{lb} \approx 94{,}012.16 \ \text{ft} \cdot \text{lb}.$$

width $= 2x = 2\sqrt{\frac{25}{4} - y^2} = \sqrt{25 - 4y^2}$

length $= 12$

25. Weight of section of chain: $3\,\Delta y$

Distance: $15 - y$

$$W = 3\int_0^{15} (15 - y)\, dy$$

$$= \left[-\frac{3}{2}(15 - y)^2\right]_0^{15}$$

$$= 337.5 \ \text{ft} \cdot \text{lb}$$

26. The lower 10 feet of chain are raised 5 feet with a constant force.

$$W_1 = 3(10)5 = 150 \ \text{ft} \cdot \text{lb}$$

The top 5 feet will be raised with variable force.

Weight of section: $3\,\Delta y$

Distance: $5 - y$

$$W_2 = 3\int_0^5 (5 - y)\, dy = \left[-\frac{3}{2}(5 - y)^2\right]_0^5 = \frac{75}{2} \ \text{ft} \cdot \text{lb}$$

$$W = W_1 + W_2 = 150 + \frac{75}{2} = \frac{375}{2} \ \text{ft} \cdot \text{lb}$$

27. The lower 5 feet of chain are raised 10 feet with a constant force.

$$W_1 = 3(5)(10) = 150 \ \text{ft} \cdot \text{lb}$$

The top 10 feet of chain are raised with a variable force.

Weight per section: $3\,\Delta y$

Distance: $10 - y$

$$W_2 = 3\int_0^{10} (10 - y)\, dy$$

$$= \left[-\frac{3}{2}(10 - y)^2\right]_0^{10} = 150 \ \text{ft} \cdot \text{lb}$$

$$W = W_1 + W_2 = 300 \ \text{ft} \cdot \text{lb}$$

28. The work required to lift the chain is 337.5 ft · lb (from Exercise 25). The work required to lift the 100-pound load is $W = (100)(15) = 1500$. The work required to lift the chain with a 100-pound load attached is $W = 337.5 + 1500 = 1837.5 \ \text{ft} \cdot \text{lb}$.

29. Weight of section of chain: $3\,\Delta y$

Distance: $15 - 2y$

$$W = 3\int_0^{7.5} (15 - 2y)\, dy = \left[-\frac{3}{4}(15 - 2y)^2\right]_0^{7.5} = \frac{3}{4}(15)^2 = 168.75 \ \text{ft} \cdot \text{lb}$$

30. $W = 3\displaystyle\int_0^6 (12 - 2y)\, dy = \left[-\dfrac{3}{4}(12 - 2y)^2\right]_0^6 = \dfrac{3}{4}(12)^2 = 108 \ \text{ft} \cdot \text{lb}$

31. Work to pull up the ball: $W_1 = 500(15) = 7500$ ft · lb

Work to wind up the top 15 feet of cable: force is variable

Weight per section: $1\,\Delta x$

Distance: $15 - x$

$$W_2 = \int_0^{15} (15 - x)\,dx = \left[-\frac{1}{2}(15 - x)^2\right]_0^{15} = 112.5 \text{ ft · lb}$$

Work to lift the lower 25 feet of cable with a constant force:

$$W_3 = (1)(25)(15) = 375 \text{ ft · lb}$$

$$W = W_1 + W_2 + W_3 = 7500 + 112.5 + 375 = 7987.5 \text{ ft · lb}$$

32. Work to pull up the ball: $W_1 = 500(40) = 20{,}000$ ft · lb

Work to pull up the cable: force is variable

Weight per section: $1\,\Delta x$

Distance: $40 - x$

$$W_2 = \int_0^{40} (40 - x)\,dx = \left[-\frac{1}{2}(40 - x)^2\right]_0^{40} = 800 \text{ ft · lb}$$

$$W = W_1 + W_2 = 20{,}000 + 800 = 20{,}800 \text{ ft · lb}$$

33. $p = \dfrac{k}{V}$

$1000 = \dfrac{k}{2}$

$k = 2000$

$$W = \int_2^3 \frac{2000}{V}\,dV = \Big[2000 \ln|V|\Big]_2^3$$

$$= 2000 \ln\left(\frac{3}{2}\right) \approx 810.93 \text{ ft · lb}$$

34. $p = \dfrac{k}{V}$

$2000 = \dfrac{k}{1}$

$k = 2000$

$$W = \int_1^4 \frac{2000}{V}\,dV = \Big[2000 \ln|V|\Big]_1^4$$

$$= 2000 \ln 4 \approx 2772.59 \text{ ft · lb}$$

35. $F(x) = \dfrac{k}{(2 - x)^2}$

$$W = \int_{-2}^1 \frac{k}{(2 - x)^2}\,dx = \left[\frac{k}{2 - x}\right]_{-2}^1 = k\left(1 - \frac{1}{4}\right) = \frac{3k}{4} \text{ (units of work)}$$

36. a. $W = FD = (8000\pi)(2) = 16,000\pi$ ft · lb

b. $W \approx \dfrac{2-0}{3(6)}[0 + 4(20,000) + 2(22,000) + 4(15,000) + 2(10,000) + 4(5000) + 0]$

$\approx 24,888.889$ ft · lb

c. $F(x) = 258 - 41,957x + 58,188\sqrt{x}$

$F'(x) = -41,957 + \dfrac{29,094}{\sqrt{x}} = 0$ when

$x = \left(\dfrac{29,094}{41,957}\right)^2 \approx 0.481$ ft.

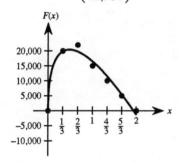

d. $W = \displaystyle\int_0^2 (258 - 41,957x + 58,188\sqrt{x})\, dx$

$= \left[258x - 20,978.5x^2 + 38,792x^{3/2}\right]_0^2$

$\approx 26,322.345$ ft · lb

37. $W = \displaystyle\int_0^5 1000[1.8 - \ln(x+1)]\, dx \approx 3250$ ft · lb

38. $W = \displaystyle\int_0^4 \left(\dfrac{e^{x^2} - 1}{100}\right) dx \approx 14,205$ ft · lb

39. $W = \displaystyle\int_0^5 100x\sqrt{125 - x^3}\, dx \approx 10,203$ ft · lb

40. $W = \displaystyle\int_0^2 1000 \sinh x\, dx \approx 2762$ ft · lb

41. $W = \displaystyle\int_{x_1}^{x_2} F(x)\, dx$

From Newton's Second Law of Motion,

$F = m\dfrac{dv}{dt}$

and we have

$W = \displaystyle\int_{x_1}^{x_2} m\dfrac{dv}{dt}\, dx = \int_{x_1}^{x_2} m\dfrac{dv}{dx}\dfrac{dx}{dt}\, dx$

$= \displaystyle\int_{x_1}^{x_2} mv\dfrac{dv}{dx}\, dx = \int_{v_1}^{v_2} mv\, dv$

$= \left[\dfrac{mv^2}{2}\right]_{v_1}^{v_2} = \dfrac{mv_2^2}{2} - \dfrac{mv_1^2}{2}$

which is the change in kinetic energy.

42. a. $V = 500 \displaystyle\int_0^{1000} \left(25 - \dfrac{x^2}{40,000}\right) dx$

$= 500\left[25x - \dfrac{x^3}{120,000}\right]_0^{1000}$

$\approx 8.33 \times 10^6$ ft^3

b. $W = 500(64) \displaystyle\int_0^{25} y(200\sqrt{y})\, dy$

$= \left[6,400,000\left(\dfrac{2}{5}y^{5/2}\right)\right]_0^{25}$

$= 8.00 \times 10^9$ ft · lb

Section 6.6   Fluid Pressure and Fluid Force

1. $F = PA = [62.4(5)](3) = 936$ lb

2. $F = PA = [62.4(5)](18) = 5616$ lb

3. $F = 62.4(h + 2)(6) - (62.4)(h)(6)$

   $\quad = 62.4(2)(6) = 748.8$ lb

4. $F = 62.4(h + 4)(48) - (62.4)(h)(48)$

   $\quad = 62.4(4)(48) = 11{,}980.8$ lb

5. $h(y) = 3 - y$

   $L(y) = 4$

   $F = 62.4 \displaystyle\int_0^3 (3 - y)(4)\, dy$

   $\quad = 249.6 \displaystyle\int_0^3 (3 - y)\, dy$

   $\quad = 249.6 \left[3y - \dfrac{y^2}{2} \right]_0^3 = 1123.2$ lb

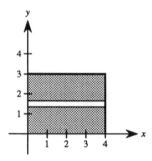

6. $h(y) = 3 - y$

   $L(y) = \dfrac{4}{3} y$

   $F = 62.4 \displaystyle\int_0^3 (3 - y)\left(\dfrac{4}{3} y \right) dy$

   $\quad = \dfrac{4}{3}(62.4) \displaystyle\int_0^3 (3y - y^2)\, dy$

   $\quad = \dfrac{4}{3}(62.4) \left[\dfrac{3y^2}{2} - \dfrac{y^3}{3} \right]_0^3 = 374.4$ lb

   Force is one-third that of Exercise 5.

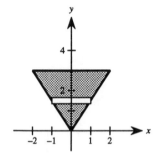

7. $h(y) = 3 - y$

   $L(y) = 2\left(\dfrac{y}{3} + 1 \right)$

   $F = 2(62.4) \displaystyle\int_0^3 (3 - y)\left(\dfrac{y}{3} + 1 \right) dy$

   $\quad = 124.8 \displaystyle\int_0^3 \left(3 - \dfrac{y^2}{3} \right) dy$

   $\quad = 124.8 \left[3y - \dfrac{y^3}{9} \right]_0^3 = 748.8$ lb

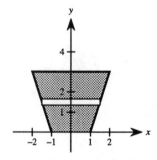

8. $h(y) = -y$

   $L(y) = 2\sqrt{4 - y^2}$

   $F = 62.4 \displaystyle\int_{-2}^0 (-y)(2)\sqrt{4 - y^2}\, dy$

   $\quad = \left[62.4 \left(\dfrac{2}{3} \right)(4 - y^2)^{3/2} \right]_{-2}^0 = 332.8$ lb

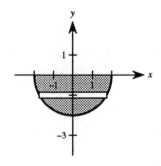

9. $h(y) = 4 - y$

$L(y) = 2\sqrt{y}$

$F = 2(62.4)\displaystyle\int_0^4 (4 - y)\sqrt{y}\,dy$

$\quad = 124.8 \displaystyle\int_0^4 (4y^{1/2} - y^{3/2})\,dy$

$\quad = 124.8\left[\dfrac{8y^{3/2}}{3} - \dfrac{2y^{5/2}}{5}\right]_0^4 = 1064.96 \text{ lb}$

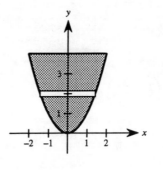

10. $h(y) = -y$

$L(y) = \dfrac{4}{3}\sqrt{9 - y^2}$

$F = 62.4\displaystyle\int_{-3}^0 (-y)\dfrac{4}{3}\sqrt{9 - y^2}\,dy$

$\quad = 62.4\left(\dfrac{2}{3}\right)\displaystyle\int_{-3}^0 (9 - y^2)^{1/2}(-2y)\,dy$

$\quad = \left[62.4\left(\dfrac{4}{9}\right)(9 - y^2)^{3/2}\right]_{-3}^0 = 748.8 \text{ lb}$

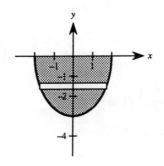

11. $h(y) = 4 - y$

$L(y) = 2$

$F = 62.4\displaystyle\int_0^2 2(4 - y)\,dy$

$\quad = 62.4\left[8y - y^2\right]_0^2 = 748.8 \text{ lb}$

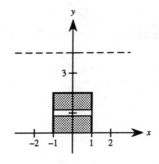

12. $h(y) = (1 + 2\sqrt{2}) - y$

$L_1(y) = 2y$ [lower part]

$L_2(y) = 2(2\sqrt{2} - y)$ [upper part]

$F = 124.8\left[\displaystyle\int_0^{\sqrt{2}} (1 + 2\sqrt{2} - y)y\,dy + \displaystyle\int_{\sqrt{2}}^{2\sqrt{2}} (1 + 2\sqrt{2} - y)(2\sqrt{2} - y)\,dy\right]$

$\quad = 124.8\left(\left[\dfrac{y^2}{2} + \sqrt{2}\,y^2 - \dfrac{y^3}{3}\right]_0^{\sqrt{2}} + \left[2\sqrt{2}\,y + 8y - 2\sqrt{2}\,y^2 - \dfrac{y^2}{2} + \dfrac{y^3}{3}\right]_{\sqrt{2}}^{2\sqrt{2}}\right)$

$\quad = 249.6(1 + \sqrt{2}) \approx 602.6 \text{ lb}$

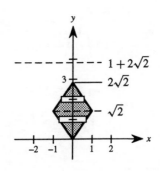

13. $h(y) = 12 - y$

$L(y) = 6 - \dfrac{2y}{3}$

$F = 62.4 \displaystyle\int_0^9 (12 - y)\left(6 - \dfrac{2y}{3}\right) dy$

$ = 62.4\left[72y - 7y^2 + \dfrac{2y^3}{9}\right]_0^9 = 15{,}163.2 \text{ lb}$

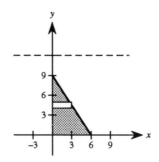

14. $h(y) = 6 - y$

$L(y) = 1$

$F = 62.4 \displaystyle\int_0^5 1(6 - y)\, dy$

$ = 62.4\left[6y - \dfrac{y^2}{2}\right]_0^5 = 1092 \text{ lb}$

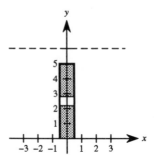

15. $h(y) = 2 - y$

$L(y) = 10$

$F = 140.7 \displaystyle\int_0^2 (2 - y)(10)\, dy$

$ = 1407 \displaystyle\int_0^2 (2 - y)\, dy$

$ = 1407\left[2y - \dfrac{y^2}{2}\right]_0^2 = 2814 \text{ lb}$

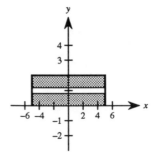

16. $h(y) = 4 - y$

$L(y) = 6$

$F = 140.7 \displaystyle\int_0^4 (4 - y)(6)\, dy$

$ = 844.2 \displaystyle\int_0^4 (4 - y)\, dy$

$ = 844.2\left[4y - \dfrac{y^2}{2}\right]_0^4 = 6753.6 \text{ lb}$

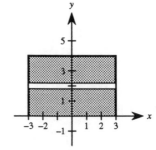

17. $h(y) = -y$

$L(y) = 2\left(\dfrac{4}{3}\sqrt{9 - y^2}\right)$

$F = 140.7 \displaystyle\int_{-3}^0 (-y)(2)\left(\dfrac{4}{3}\sqrt{9 - y^2}\right) dy$

$ = \dfrac{(140.7)(4)}{3} \displaystyle\int_{-3}^0 \sqrt{9 - y^2}(-2y)\, dy$

$ = \left[\dfrac{(140.7)(4)}{3}\left(\dfrac{2}{3}\right)(9 - y^2)^{3/2}\right]_{-3}^0 = 3376.8 \text{ lb}$

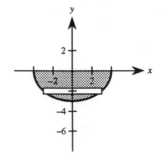

18. $h(y) = -y$

$L(y) = 6 + \dfrac{3}{2}y$

$F = 140.7 \displaystyle\int_{-4}^{0} (-y)\left(6 + \dfrac{3}{2}y\right) dy$

$= \left[-140.7\left(3y^2 + \dfrac{y^3}{2}\right)\right]_{-4}^{0} = 2251.2$ lb

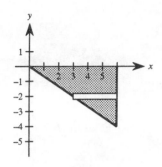

19. $h(y) = -y$

$L(y) = 2\left(\dfrac{1}{2}\right)\sqrt{9 - 4y^2}$

$F = 42 \displaystyle\int_{-3/2}^{0} (-y)\sqrt{9 - 4y^2}\, dy$

$= \dfrac{42}{8} \displaystyle\int_{-3/2}^{0} (9 - 4y^2)^{1/2}(-8y)\, dy$

$= \left[\left(\dfrac{21}{4}\right)\left(\dfrac{2}{3}\right)(9 - 4y^2)^{3/2}\right]_{-3/2}^{0} = 94.5$ lb

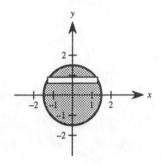

20. $h(y) = \dfrac{3}{2} - y$

$L(y) = 2\left(\dfrac{1}{2}\right)\sqrt{9 - 4y^2}$

$F = 42 \displaystyle\int_{-3/2}^{3/2} \left(\dfrac{3}{2} - y\right)\sqrt{9 - 4y^2}\, dy = 63 \displaystyle\int_{-3/2}^{3/2} \sqrt{9 - 4y^2}\, dy + \dfrac{21}{4} \displaystyle\int_{-3/2}^{3/2} \sqrt{9 - 4y^2}(-8y)\, dy$

The second integral is zero since it is an odd function and the limits of integration are symmetric to the origin. The first integral is twice the area of a semicircle of radius $\dfrac{3}{2}$.

$\left(\sqrt{9 - 4y^2} = 2\sqrt{(9/4) - y^2}\right)$

Thus, the force is $63\left(\dfrac{9}{4}\pi\right) = 141.75\pi \approx 445.32$ lb.

21. $h(y) = k - y$

$L(y) = 2\sqrt{r^2 - y^2}$

$F = w \displaystyle\int_{-r}^{r} (k - y)\sqrt{r^2 - y^2}(2)\, dy$

$= w\left[2k \displaystyle\int_{-r}^{r} \sqrt{r^2 - y^2}\, dy + \displaystyle\int_{-r}^{r} \sqrt{r^2 - y^2}(-2y)\, dy\right]$

The second integral is zero since its integrand is odd and the limits of integration are symmetric to the origin. The first integral is the area of a semicircle with radius r.

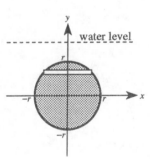

water level

$F = w\left[(2k)\dfrac{\pi r^2}{2} + 0\right] = wk\pi r^2$

22. $h(y) = k - y$

$L(y) = b$

$F = w \int_{-h/2}^{h/2} (k - y)b \, dy$

$= wb \left[ky - \dfrac{y^2}{2} \right]_{-h/2}^{h/2} = wb(hk) = wkhb$

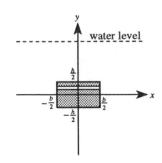

23. From Exercise 22:

$F = 64(15)(1)(1) = 960$ lb

24. From Exercise 21:

$F = 64(15)\pi \left(\dfrac{1}{2} \right)^2 \approx 753.98$ lb

25. a. Wall at shallow end

From Exercise 22: $F = 62.4(2)(4)(20) = 9984$ lb

b. Wall at deep end

From Exercise 22: $F = 62.4(4)(8)(20) = 39{,}936$ lb

c. Side wall

From Exercise 22: $F_1 = 62.4(2)(4)(40) = 19{,}968$ lb

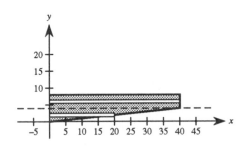

$F_2 = 62.4 \int_0^4 (8 - y)(10y) \, dy$

$= 624 \int_0^4 (8y - y^2) \, dy = 624 \left[4y^2 - \dfrac{y^3}{3} \right]_0^4 = 26{,}624$ lb

Total force: $F_1 + F_2 = 46{,}592$ lb

26. $h(y) = 4 - y$

$F = 62.4 \int_0^4 (4 - y)L(y) \, dy$

Using Simpson's Rule with $n = 8$ we have:

$F \approx 62.4 \left(\dfrac{4 - 0}{3(8)} \right) [0 + 4(3.5)(3) + 2(3)(5) + 4(2.5)(8) + 2(2)(9) + 4(1.5)(10)$

$+ 2(1)(10.25) + 4(0.5)(10.5) + 0] = 3010.8$ lb

27. $h(y) = 12 - y$

$L(y) = 2(4^{2/3} - y^{2/3})^{3/2}$

$F = 62.4 \int_0^4 2(12 - y)(4^{2/3} - y^{2/3})^{3/2} \, dy$

≈ 6483 lb

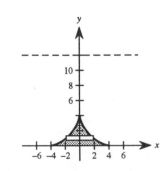

28. $h(y) = 12 - y$

$$L(y) = 2\frac{\sqrt{7(16 - y^2)}}{2} = \sqrt{7(16 - y^2)}$$

$$F = 62.4 \int_0^4 (12 - y)\sqrt{7(16 - y^2)}\, dy$$

$$= 62.4\sqrt{7} \int_0^4 (12 - y)\sqrt{16 - y^2}\, dy \approx 21,297 \text{ lb}$$

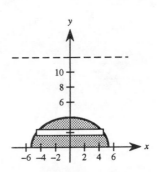

29. $h(y) = 3 - y$

Solving $y = 5x^2/(x^2 + 4)$ for x, you obtain $x = \sqrt{4y/(5 - y)}$.

$$L(y) = 2\sqrt{\frac{4y}{5 - y}}$$

$$F = 62.4(2) \int_0^3 (3 - y)\sqrt{\frac{4y}{5 - y}}\, dy = 2(124.8) \int_0^3 (3 - y)\sqrt{\frac{y}{5 - y}}\, dy \approx 536.4 \text{ lb}$$

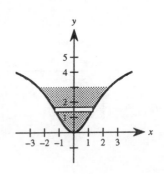

Section 6.7 Moments, Centers of Mass, and Centroids

1. $\bar{x} = \dfrac{6(-5) + 3(1) + 5(3)}{6 + 3 + 5} = -\dfrac{6}{7}$

2. $\bar{x} = \dfrac{7(-3) + 4(-2) + 3(5) + 8(6)}{7 + 4 + 3 + 8} = \dfrac{17}{11}$

3. $\bar{x} = \dfrac{1(7) + 1(8) + 1(12) + 1(15) + 1(18)}{1 + 1 + 1 + 1 + 1} = 12$

4. $\bar{x} = \dfrac{12(-3) + 1(-2) + 6(-1) + 3(0) + 11(4)}{12 + 1 + 6 + 3 + 11} = 0$

5. $\bar{x} = \dfrac{(7 + 5) + (8 + 5) + (12 + 5) + (15 + 5) + (18 + 5)}{5} = 17 = 12 + 5$

6. $\bar{x} = \dfrac{12(-3 - 3) + 1(-2 - 3) + 6(-1 - 3) + 3(0 - 3) + 11(4 - 3)}{12 + 1 + 6 + 3 + 11} = -3 = 0 - 3$

7.
$$\bar{x} = \frac{5(2) + 1(-3) + 3(1)}{5 + 1 + 3} = \frac{10}{9}$$

$$\bar{y} = \frac{5(2) + 1(1) + 3(-4)}{5 + 1 + 3} = -\frac{1}{9}$$

$$(\bar{x}, \bar{y}) = \left(\frac{10}{9}, -\frac{1}{9}\right)$$

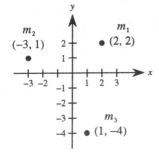

8.
$$\bar{x} = \frac{10(1) + 2(5) + 5(-4)}{10 + 2 + 5} = 0$$

$$\bar{y} = \frac{10(-1) + 2(5) + 5(0)}{10 + 2 + 5} = 0$$

$$(\bar{x}, \bar{y}) = (0, \ 0)$$

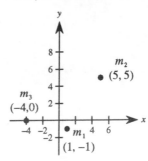

9.
$$\bar{x} = \frac{3(-2) + 4(-1) + 2(7) + 1(0) + 6(-3)}{3 + 4 + 2 + 1 + 6} = -\frac{7}{8}$$

$$\bar{y} = \frac{3(-3) + 4(0) + 2(1) + 1(0) + 6(0)}{3 + 4 + 2 + 1 + 6} = -\frac{7}{16}$$

$$(\bar{x}, \bar{y}) = \left(-\frac{7}{8}, -\frac{7}{16}\right)$$

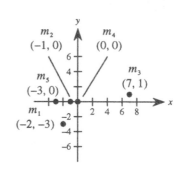

10.
$$\bar{x} = \frac{4(2) + 2(-1) + 2.5(6) + 5(2)}{4 + 2 + 2.5 + 5} = \frac{31}{13.5} = \frac{62}{27}$$

$$\bar{y} = \frac{4(3) + 2(5) + 2.5(8) + 5(-2)}{4 + 2 + 2.5 + 5} = \frac{32}{13.5} = \frac{64}{27}$$

$$(\bar{x}, \bar{y}) = \left(\frac{62}{27}, \frac{64}{27}\right)$$

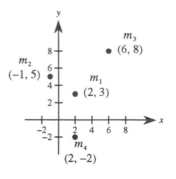

11.
$$m = \rho \int_0^4 \sqrt{x}\, dx = \left[\frac{2\rho}{3} x^{3/2}\right]_0^4 = \frac{16\rho}{3}$$

$$M_x = \rho \int_0^4 \frac{\sqrt{x}}{2}(\sqrt{x})\, dx = \left[\rho \frac{x^2}{4}\right]_0^4 = 4\rho$$

$$\bar{y} = \frac{M_x}{m} = 4\rho \left(\frac{3}{16\rho}\right) = \frac{3}{4}$$

$$M_y = \rho \int_0^4 x\sqrt{x}\, dx = \left[\rho \frac{2}{5} x^{5/2}\right]_0^4 = \frac{64\rho}{5}$$

$$\bar{x} = \frac{M_y}{m} = \frac{64\rho}{5}\left(\frac{3}{16\rho}\right) = \frac{12}{5}$$

$$(\bar{x}, \bar{y}) = \left(\frac{12}{5}, \frac{3}{4}\right)$$

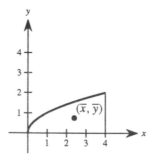

12.
$$m = \rho \int_0^4 x^2\, dx = \left[\rho \frac{x^3}{3}\right]_0^4 = \frac{64}{3}\rho$$

$$M_x = \rho \int_0^4 \frac{x^2}{2}(x^2)\, dx = \left[\rho \frac{x^5}{10}\right]_0^4 = \frac{512\rho}{5}$$

$$\bar{y} = \frac{M_x}{m} = \frac{512\rho}{5}\left(\frac{3}{64\rho}\right) = \frac{24}{5}$$

$$M_y = \rho \int_0^4 x(x^2)\, dx = \left[\rho \frac{x^4}{4}\right]_0^4 = 64\rho$$

$$\bar{x} = \frac{M_y}{m} = 64\rho \left(\frac{3}{64\rho}\right) = 3$$

$$(\bar{x}, \bar{y}) = \left(3, \frac{24}{5}\right)$$

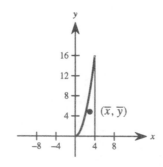

13.
$$m = \rho \int_0^1 (\sqrt{x} - x)\, dx = \rho \left[\frac{2}{3}x^{3/2} - \frac{x^2}{2} \right]_0^1 = \frac{\rho}{6}$$

$$M_x = \rho \int_0^1 \frac{(\sqrt{x} + x)}{2}(\sqrt{x} - x)\, dx = \frac{\rho}{2}\int_0^1 (x - x^2)\, dx = \frac{\rho}{2}\left[\frac{x^2}{2} - \frac{x^3}{3} \right]_0^1 = \frac{\rho}{12}$$

$$\bar{y} = \frac{M_x}{m} = \frac{\rho}{12}\left(\frac{6}{\rho} \right) = \frac{1}{2}$$

$$M_y = \rho \int_0^1 x(\sqrt{x} - x)\, dx = \rho \int_0^1 (x^{3/2} - x^2)\, dx = \rho \left[\frac{2}{5}x^{5/2} - \frac{x^3}{3} \right]_0^1 = \frac{\rho}{15}$$

$$\bar{x} = \frac{M_y}{m} = \frac{\rho}{15}\left(\frac{6}{\rho} \right) = \frac{2}{5}$$

$$(\bar{x},\ \bar{y}) = \left(\frac{2}{5},\ \frac{1}{2} \right)$$

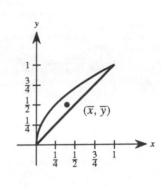

14.
$$m = \rho \int_0^1 (x^2 - x^3)\, dx = \rho \left[\frac{x^3}{3} - \frac{x^4}{4} \right]_0^1 = \frac{\rho}{12}$$

$$M_x = \rho \int_0^1 \frac{(x^2 + x^3)}{2}(x^2 - x^3)\, dx = \frac{\rho}{2}\int_0^1 (x^4 - x^6)\, dx = \frac{\rho}{2}\left[\frac{x^5}{5} - \frac{x^7}{7} \right]_0^1 = \frac{\rho}{35}$$

$$\bar{y} = \frac{M_x}{m} = \frac{\rho}{35}\left(\frac{12}{\rho} \right) = \frac{12}{35}$$

$$M_y = \rho \int_0^1 x(x^2 - x^3)\, dx = \rho \int_0^1 (x^3 - x^4)\, dx = \rho \left[\frac{x^4}{4} - \frac{x^5}{5} \right]_0^1 = \frac{\rho}{20}$$

$$\bar{x} = \frac{M_y}{m} = \frac{\rho}{20}\left(\frac{12}{\rho} \right) = \frac{3}{5}$$

$$(\bar{x},\ \bar{y}) = \left(\frac{3}{5},\ \frac{12}{35} \right)$$

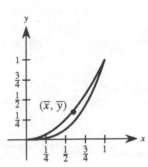

15.
$$m = \rho \int_0^3 [(-x^2 + 4x + 2) - (x + 2)]\, dx = -\rho \left[\frac{x^3}{3} + \frac{3x^2}{2} \right]_0^3 = \frac{9\rho}{2}$$

$$M_x = \rho \int_0^3 \left[\frac{(-x^2 + 4x + 2) + (x + 2)}{2} \right][(-x^2 + 4x + 2) - (x + 2)]\, dx$$

$$= \frac{\rho}{2}\int_0^3 (-x^2 + 5x + 4)(-x^2 + 3x)\, dx = \frac{\rho}{2}\int_0^3 (x^4 - 8x^3 + 11x^2 + 12x)\, dx$$

$$= \frac{\rho}{2}\left[\frac{x^5}{5} - 2x^4 + \frac{11x^3}{3} + 6x^2 \right]_0^3 = \frac{99\rho}{5}$$

$$\bar{y} = \frac{M_x}{m} = \frac{99\rho}{5}\left(\frac{2}{9\rho} \right) = \frac{22}{5}$$

$$M_y = \rho \int_0^3 x[(-x^2 + 4x - 2) - (x + 2)]\, dx = \rho \int_0^3 (-x^3 + 3x^2)\, dx = \rho \left[-\frac{x^4}{4} + x^3 \right]_0^3 = \frac{27\rho}{4}$$

$$\bar{x} = \frac{M_y}{m} = \frac{27\rho}{4}\left(\frac{2}{9\rho} \right) = \frac{3}{2}$$

$$(\bar{x},\ \bar{y}) = \left(\frac{3}{2},\ \frac{22}{5} \right)$$

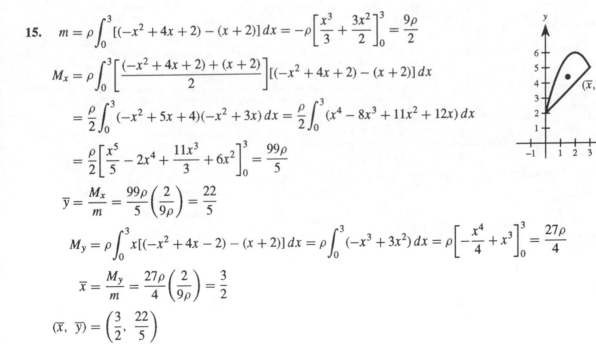

16. $m = \rho \displaystyle\int_0^3 [(\sqrt{3x} + 1) - (x + 1)]\,dx = \rho \left[\dfrac{2}{9}(3x)^{3/2} - \dfrac{x^2}{2} \right]_0^3 = \dfrac{3\rho}{2}$

$M_x = \rho \displaystyle\int_0^3 \dfrac{[(\sqrt{3x} + 1) + (x + 1)]}{2} [(\sqrt{3x} + 1) - (x + 1)]\,dx$

$\quad = \dfrac{\rho}{2} \displaystyle\int_0^3 (\sqrt{3x} + x + 2)(\sqrt{3x} - x)\,dx$

$\quad = \dfrac{\rho}{2} \displaystyle\int_0^3 (2\sqrt{3x} + x - x^2)\,dx$

$\quad = \dfrac{\rho}{2} \left[\dfrac{4}{9}(3x)^{3/2} + \dfrac{x^2}{2} - \dfrac{x^3}{3} \right]_0^3 = \dfrac{15\rho}{4}$

$\bar{y} = \dfrac{M_x}{m} = \dfrac{15\rho}{4}\left(\dfrac{2}{3\rho} \right) = \dfrac{5}{2}$

$M_y = \rho \displaystyle\int_0^3 x[(\sqrt{3x} + 1) - (x + 1)]\,dx = \rho \displaystyle\int_0^3 (x\sqrt{3x} - x^2)\,dx = \rho \left[\dfrac{2\sqrt{3}}{5}x^{5/2} - \dfrac{x^3}{3} \right]_0^3 = \dfrac{9\rho}{5}$

$\bar{x} = \dfrac{M_y}{m} = \dfrac{9\rho}{5}\left(\dfrac{2}{3\rho} \right) = \dfrac{6}{5}$

$(\bar{x},\ \bar{y}) = \left(\dfrac{6}{5},\ \dfrac{5}{2} \right)$

17. $m = \rho \displaystyle\int_0^8 x^{2/3}\,dx = \rho \left[\dfrac{3}{5}x^{5/3} \right]_0^8 = \dfrac{96\rho}{5}$

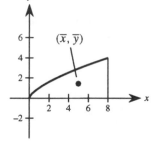

$M_x = \rho \displaystyle\int_0^8 \dfrac{x^{2/3}}{2}(x^{2/3})\,dx = \dfrac{\rho}{2}\left[\dfrac{3}{7}x^{7/3} \right]_0^8 = \dfrac{192\rho}{7}$

$\bar{y} = \dfrac{M_x}{m} = \dfrac{192\rho}{7}\left(\dfrac{5}{96\rho} \right) = \dfrac{10}{7}$

$M_y = \rho \displaystyle\int_0^8 x(x^{2/3})\,dx = \rho \left[\dfrac{3}{8}x^{8/3} \right]_0^8 = 96\rho$

$\bar{x} = \dfrac{M_y}{m} = 96\rho \left(\dfrac{5}{96\rho} \right) = 5$

$(\bar{x},\ \bar{y}) = \left(5,\ \dfrac{10}{7} \right)$

18. $m = 2\rho \displaystyle\int_0^8 (4 - x^{2/3})\,dx = 2\rho \left[4x - \dfrac{3}{5}x^{5/3} \right]_0^8 = \dfrac{128\rho}{5}$

By symmetry, M_y and $\bar{x} = 0$.

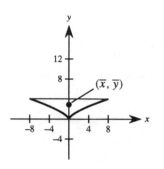

$M_x = 2\rho \displaystyle\int_0^8 \left(\dfrac{4 + x^{2/3}}{2} \right)(4 - x^{2/3})\,dx = \rho \left[16x - \dfrac{3}{7}x^{7/3} \right]_0^8 = \dfrac{512\rho}{7}$

$\bar{y} = \dfrac{512\rho}{7}\left(\dfrac{5}{128\rho} \right) = \dfrac{20}{7}$

$(\bar{x},\ \bar{y}) = \left(0,\ \dfrac{20}{7} \right)$

19. $\quad m = 2\rho \displaystyle\int_0^2 (4-y^2)\,dy = 2\rho \left[4y - \dfrac{y^3}{3}\right]_0^2 = \dfrac{32\rho}{3}$

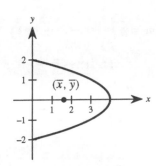

$\quad M_y = 2\rho \displaystyle\int_0^2 \left(\dfrac{4-y^2}{2}\right)(4-y^2)\,dy = \rho \left[16y - \dfrac{8}{3}y^3 + \dfrac{y^5}{5}\right]_0^2 = \dfrac{256\rho}{15}$

$\quad \overline{x} = \dfrac{M_y}{m} = \dfrac{256\rho}{15}\left(\dfrac{3}{32\rho}\right) = \dfrac{8}{5}$

By symmetry, M_x and $\overline{y} = 0$.

$(\overline{x},\ \overline{y}) = \left(\dfrac{8}{5},\ 0\right)$

20. $\quad m = \rho \displaystyle\int_0^2 (2y - y^2)\,dy = \rho\left[y^2 - \dfrac{y^3}{3}\right]_0^2 = \dfrac{4\rho}{3}$

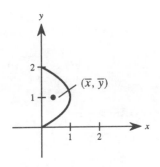

$\quad M_y = \rho \displaystyle\int_0^2 \left(\dfrac{2y-y^2}{2}\right)(2y-y^2)\,dy = \dfrac{\rho}{2}\left[\dfrac{4y^3}{3} - y^4 + \dfrac{y^5}{5}\right]_0^2 = \dfrac{8\rho}{15}$

$\quad \overline{x} = \dfrac{M_y}{m} = \dfrac{8\rho}{15}\left(\dfrac{3}{4\rho}\right) = \dfrac{2}{5}$

$\quad M_x = \rho \displaystyle\int_0^2 y(2y - y^2)\,dy = \rho\left[\dfrac{2y^3}{3} - \dfrac{y^4}{4}\right]_0^2 = \dfrac{4\rho}{3}$

$\quad \overline{y} = \dfrac{M_x}{m} = \dfrac{4\rho}{3}\left(\dfrac{3}{4\rho}\right) = 1$

$(\overline{x},\ \overline{y}) = \left(\dfrac{2}{5},\ 1\right)$

21. $\quad m = \rho \displaystyle\int_0^3 [(2y - y^2) - (-y)]\,dy = \rho\left[\dfrac{3y^2}{2} - \dfrac{y^3}{3}\right]_0^3 = \dfrac{9\rho}{2}$

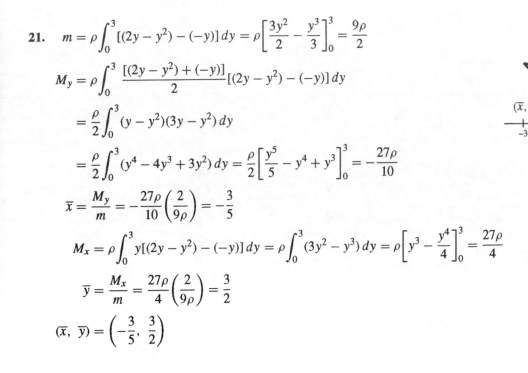

$\quad M_y = \rho \displaystyle\int_0^3 \dfrac{[(2y - y^2) + (-y)]}{2}[(2y - y^2) - (-y)]\,dy$

$\quad\quad = \dfrac{\rho}{2}\displaystyle\int_0^3 (y - y^2)(3y - y^2)\,dy$

$\quad\quad = \dfrac{\rho}{2}\displaystyle\int_0^3 (y^4 - 4y^3 + 3y^2)\,dy = \dfrac{\rho}{2}\left[\dfrac{y^5}{5} - y^4 + y^3\right]_0^3 = -\dfrac{27\rho}{10}$

$\quad \overline{x} = \dfrac{M_y}{m} = -\dfrac{27\rho}{10}\left(\dfrac{2}{9\rho}\right) = -\dfrac{3}{5}$

$\quad M_x = \rho \displaystyle\int_0^3 y[(2y - y^2) - (-y)]\,dy = \rho\displaystyle\int_0^3 (3y^2 - y^3)\,dy = \rho\left[y^3 - \dfrac{y^4}{4}\right]_0^3 = \dfrac{27\rho}{4}$

$\quad \overline{y} = \dfrac{M_x}{m} = \dfrac{27\rho}{4}\left(\dfrac{2}{9\rho}\right) = \dfrac{3}{2}$

$(\overline{x},\ \overline{y}) = \left(-\dfrac{3}{5},\ \dfrac{3}{2}\right)$

22.
$$m = \rho \int_{-1}^{2} [(y+2) - y^2]\,dy = \rho \left[\frac{y^2}{2} + 2y - \frac{y^3}{3} \right]_{-1}^{2} = \frac{9\rho}{2}$$

$$M_y = \rho \int_{-1}^{2} \frac{[(y+2) + y^2]}{2}[(y+2) - y^2]\,dy$$

$$= \frac{\rho}{2} \int_{-1}^{2} [(y+2)^2 - y^4]\,dy = \frac{\rho}{2}\left[\frac{(y+2)^3}{3} - \frac{y^5}{5} \right]_{-1}^{2} = \frac{36\rho}{5}$$

$$\bar{x} = \frac{M_y}{m} = \frac{36\rho}{5}\left(\frac{2}{9\rho} \right) = \frac{8}{5}$$

$$M_x = \rho \int_{-1}^{2} y[(y+2) - y^2]\,dy$$

$$= \rho \int_{-1}^{2} (2y + y^2 - y^3)\,dy = \rho \left[y^2 + \frac{y^3}{3} - \frac{y^4}{4} \right]_{-1}^{2} = \frac{9\rho}{4}$$

$$\bar{y} = \frac{M_x}{m} = \frac{9\rho}{4}\left(\frac{2}{9\rho} \right) = \frac{1}{2}$$

$$(\bar{x}, \ \bar{y}) = \left(\frac{8}{5}, \ \frac{1}{2} \right)$$

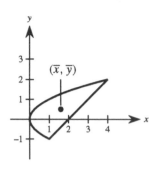

23.
$$A = \int_{0}^{1} (x - x^2)\,dx = \left[\frac{1}{2}x^2 - \frac{1}{3}x^3 \right]_{0}^{1} = \frac{1}{6}$$

$$\frac{1}{A} = 6$$

$$\bar{x} = 6\int_{0}^{1} (x^2 - x^3)\,dx = 6\left[\frac{1}{3}x^3 - \frac{1}{4}x^4 \right]_{0}^{1} = \frac{1}{2}$$

$$\bar{y} = (6)\frac{1}{2}\int_{0}^{1} (x^2 - x^4)\,dx = 3\left[\frac{1}{3}x^3 - \frac{1}{5}x^5 \right]_{0}^{1} = \frac{2}{5}$$

$$(\bar{x}, \ \bar{y}) = \left(\frac{1}{2}, \ \frac{2}{5} \right)$$

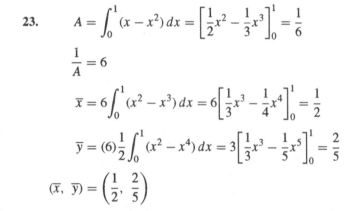

24.
$$A = \int_{1}^{4} \frac{1}{x}\,dx = \left[\ln|x| \right]_{1}^{4} = \ln 4$$

$$\frac{1}{A} = \frac{1}{\ln 4}$$

$$\bar{x} = \frac{1}{\ln 4}\int_{1}^{4} x\left(\frac{1}{x} \right)\,dx = \left[\frac{1}{\ln 4}x \right]_{1}^{4} = \frac{3}{\ln 4} \approx 2.1640$$

$$\bar{y} = \left(\frac{1}{\ln 4} \right)\frac{1}{2}\int_{1}^{4} \frac{1}{x^2}\,dx = \left[\frac{1}{2\ln 4}\left(-\frac{1}{x} \right) \right]_{1}^{4} = \frac{3}{8\ln 4} \approx 0.2705$$

$$(\bar{x}, \ \bar{y}) = \left(\frac{3}{\ln 4}, \ \frac{3}{8\ln 4} \right)$$

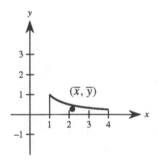

25.
$$A = \int_0^3 (2x + 4)\, dx = \left[x^2 + 4x \right]_0^3 = 21$$

$$\frac{1}{A} = \frac{1}{21}$$

$$\bar{x} = \frac{1}{21}\int_0^3 (2x^2 + 4x)\, dx = \frac{1}{21}\left[\frac{2}{3}x^3 + 2x^2 \right]_0^3 = \frac{12}{7}$$

$$\bar{y} = \left(\frac{1}{21}\right)\frac{1}{2}\int_0^3 (2x + 4)^2\, dx = \frac{1}{42}\left[\frac{4}{3}x^3 + 8x^2 + 16x \right]_0^3 = \frac{26}{7}$$

$$(\bar{x},\ \bar{y}) = \left(\frac{12}{7},\ \frac{26}{7} \right)$$

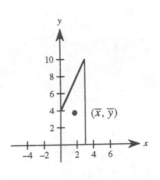

26.
$$A = \int_{-2}^2 (4 - x^2)\, dx = \left[4x - \frac{x^3}{3} \right]_{-2}^2 = \frac{32}{3}$$

$$\frac{1}{A} = \frac{3}{32}$$

By symmetry, $\bar{x} = 0$.

$$\bar{y} = \left(\frac{3}{32}\right)\frac{1}{2}\int_{-2}^2 (x^2 - 4)(4 - x^2)\, dx = -\frac{3}{64}\int_{-2}^2 (x^4 - 8x^2 + 16)\, dx$$

$$= -\frac{3}{64}\left[\frac{x^5}{5} - \frac{8}{3}x^3 + 16x \right]_{-2}^2 = -\frac{3}{64}\left(\frac{512}{15} \right) = -\frac{8}{5}$$

$$(\bar{x},\ \bar{y}) = \left(0,\ -\frac{8}{5} \right)$$

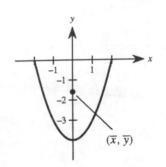

27.
$$A = \frac{1}{2}(2a)c = ac$$

$$\frac{1}{A} = \frac{1}{ac}$$

$$\bar{x} = \left(\frac{1}{ac}\right)\frac{1}{2}\int_0^c \left[\left(\frac{b-a}{c}y + a \right)^2 - \left(\frac{b+a}{c}y - a \right)^2 \right] dy$$

$$= \frac{1}{2ac}\int_0^c \left[\frac{4ab}{c}y - \frac{4ab}{c^2}y^2 \right] dy$$

$$= \frac{1}{2ac}\left[\frac{2ab}{c}y^2 - \frac{4ab}{3c^2}y^3 \right]_0^c = \frac{1}{2ac}\left(\frac{2}{3}abc \right) = \frac{b}{3}$$

$$\bar{y} = \frac{1}{ac}\int_0^c y\left[\left(\frac{b-a}{c}y + a \right) - \left(\frac{b+a}{c}y - a \right) \right] dy$$

$$= \frac{1}{ac}\int_0^c y\left(-\frac{2a}{c}y + 2a \right) dy = \frac{2}{c}\int_0^c \left(y - \frac{y^2}{c} \right) dy = \frac{2}{c}\left[\frac{y^2}{2} - \frac{y^3}{3c} \right]_0^c = \frac{c}{3}$$

$$(\bar{x},\ \bar{y}) = \left(\frac{b}{3},\ \frac{c}{3} \right)$$

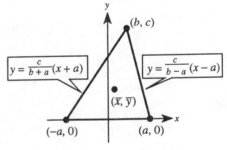

In Exercise 58 of Section P.4, you found that $(b/3,\ c/3)$ is the point of intersection of the medians.

28. $A = bh = ac$

$$\frac{1}{A} = \frac{1}{ac}$$

$$\bar{x} = \frac{1}{ac} \frac{1}{2} \int_0^c \left[\left(\frac{b}{c}y + a \right)^2 - \left(\frac{b}{c}y \right)^2 \right] dy$$

$$= \frac{1}{2ac} \int_0^c \left(\frac{2ab}{c}y + a^2 \right) dy$$

$$= \frac{1}{2ac} \left[\frac{ab}{c}y^2 + a^2 y \right]_0^c = \frac{1}{2ac}[abc + a^2 c] = \frac{1}{2}(b + a)$$

$$\bar{y} = \frac{1}{ac} \int_0^c y \left[\left(\frac{b}{c}y + a \right) - \left(\frac{b}{c}y \right) \right] dy = \left[\frac{1}{c} \frac{y^2}{2} \right]_0^c = \frac{c}{2}$$

$$(\bar{x}, \bar{y}) = \left(\frac{b+a}{2}, \frac{c}{2} \right)$$

This is the point of intersection of the diagonals.

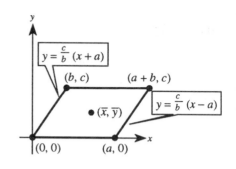

29. $A = \frac{c}{2}(a + b)$

$$\frac{1}{A} = \frac{2}{c(a + b)}$$

$$\bar{x} = \frac{2}{c(a+b)} \int_0^c x \left(\frac{b-a}{c}x + a \right) dx = \frac{2}{c(a+b)} \int_0^c \left(\frac{b-a}{c}x^2 + ax \right) dx = \frac{2}{c(a+b)} \left[\frac{b-a}{c} \frac{x^3}{3} + \frac{ax^2}{2} \right]_0^c$$

$$= \frac{2}{c(a+b)} \left[\frac{(b-a)c^2}{3} + \frac{ac^2}{2} \right] = \frac{2}{c(a+b)} \left[\frac{2bc^2 - 2ac^2 + 3ac^2}{6} \right] = \frac{c(2b+a)}{3(a+b)} = \frac{(a+2b)c}{3(a+b)}$$

$$\bar{y} = \frac{2}{c(a+b)} \frac{1}{2} \int_0^c \left(\frac{b-a}{c}x + a \right)^2 dx = \frac{1}{c(a+b)} \int_0^c \left[\left(\frac{b-a}{c} \right)^2 x^2 + \frac{2a(b-a)}{c}x + a^2 \right] dx$$

$$= \frac{1}{c(a+b)} \left[\left(\frac{b-a}{c} \right)^2 \frac{x^3}{3} + \frac{2a(b-a)}{c} \frac{x^2}{2} + a^2 x \right]_0^c = \frac{1}{c(a+b)} \left[\frac{(b-a)^2 c}{3} + ac(b-a) + a^2 c \right]$$

$$= \frac{1}{3c(a+b)} [(b^2 - 2ab + a^2)c + 3ac(b-a) + 3a^2 c]$$

$$= \frac{1}{3(a+b)} [b^2 - 2ab + a^2 + 3ab - 3a^2 + 3a^2] = \frac{a^2 + ab + b^2}{3(a+b)}$$

Thus,

$$(\bar{x}, \bar{y}) = \left(\frac{(a+2b)c}{3(a+b)}, \frac{a^2 + ab + b^2}{3(a+b)} \right).$$

The one line passes through $(0, a/2)$ and $(c, b/2)$. It's equation is

$$y = \frac{b-a}{2c}x + \frac{a}{2}.$$

The other line passes through $(0, -b)$ and $(c, a+b)$. It's equation is

$$y = \frac{a+2b}{c}x - b.$$

$(\bar{x}, \bar{y})$ is the point of intersection of these two lines.

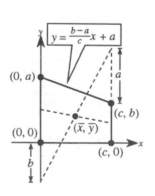

30. $\bar{x} = 0$ by symmetry

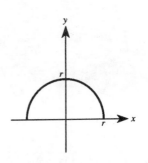

$$A = \frac{1}{2}\pi r^2$$

$$\frac{1}{A} = \frac{2}{\pi r^2}$$

$$\bar{y} = \frac{2}{\pi r^2}\frac{1}{2}\int_{-r}^{r}\left(\sqrt{r^2 - x^2}\right)^2 dx$$

$$= \frac{1}{\pi r^2}\left[r^2 x - \frac{x^3}{3}\right]_{-r}^{r} = \frac{1}{\pi r^2}\left[\frac{4r^3}{3}\right] = \frac{4r}{3\pi}$$

$$(\bar{x},\ \bar{y}) = \left(0,\ \frac{4r}{3\pi}\right)$$

31. $\bar{x} = 0$ by symmetry

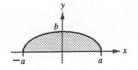

$$A = \frac{1}{2}\pi ab$$

$$\frac{1}{A} = \frac{2}{\pi ab}$$

$$\bar{y} = \frac{2}{\pi ab}\frac{1}{2}\int_{-a}^{a}\left(\frac{b}{a}\sqrt{a^2 - x^2}\right)^2 dx$$

$$= \frac{1}{\pi ab}\left(\frac{b^2}{a^2}\right)\left[a^2 x - \frac{x^3}{3}\right]_{-a}^{a} = \frac{b}{\pi a^3}\left[\frac{4a^3}{3}\right] = \frac{4b}{3\pi}$$

$$(\bar{x},\ \bar{y}) = \left(0,\ \frac{4b}{3\pi}\right)$$

32.

$$A = \int_{0}^{1}[1 - (2x - x^2)]\,dx = \frac{1}{3}$$

$$\frac{1}{A} = 3$$

$$\bar{x} = 3\int_{0}^{1}x[1 - (2x - x^2)]\,dx = 3\int_{0}^{1}[x - 2x^2 + x^3]\,dx = 3\left[\frac{x^2}{2} - \frac{2}{3}x^3 + \frac{x^4}{4}\right]_{0}^{1} = \frac{1}{4}$$

$$\bar{y} = 3\int_{0}^{1}\frac{[1 + (2x - x^2)]}{2}[1 - (2x - x^2)]\,dx$$

$$= \frac{3}{2}\int_{0}^{1}[1 - (2x - x^2)^2]\,dx = \frac{3}{2}\int_{0}^{1}[1 - 4x^2 + 4x^3 - x^4]\,dx = \frac{3}{2}\left[x - \frac{4}{3}x^3 + x^4 - \frac{x^5}{5}\right]_{0}^{1} = \frac{7}{10}$$

$$(\bar{x},\ \bar{y}) = \left(\frac{1}{4},\ \frac{7}{10}\right)$$

33. $m = \rho \int_0^5 10x\sqrt{125 - x^3}\, dx \approx 1020.3\rho$

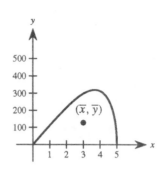

$M_x = \rho \int_0^5 \left(\frac{10x\sqrt{125 - x^3}}{2} \right) \left(10x\sqrt{125 - x^3} \right) dx$

$= 50\rho \int_0^5 x^2(125 - x^3)\, dx = \frac{3,124,375\rho}{24} \approx 130,182\rho$

$M_y = \rho \int_0^5 10x^2\sqrt{125 - x^3}\, dx$

$= -\frac{10\rho}{3} \int_0^5 \sqrt{125 - x^3}(-3x^2)\, dx = \frac{12,500\sqrt{5}\,\rho}{9} \approx 3040.5\rho$

$\bar{x} = \frac{M_y}{m} \approx 3.0$

$\bar{y} = \frac{M_x}{m} \approx 127.6$

Therefore, the centroid is (3.0, 127.6).

34. $m = \rho \int_0^4 xe^{-x/2}\, dx \approx 2.3759\rho$

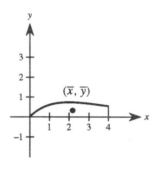

$M_x = \rho \int_0^4 \left(\frac{xe^{-x/2}}{2} \right)(xe^{-x/2})\, dx = \frac{\rho}{2} \int_0^4 x^2 e^{-x}\, dx \approx 0.7623\rho$

$M_y = \rho \int_0^4 x^2 e^{-x/2}\, dx \approx 5.1736\rho$

$\bar{x} = \frac{M_y}{m} \approx 2.2$

$\bar{y} = \frac{M_x}{m} \approx 0.3$

Therefore, the centroid is (2.2, 0.3).

35. $m = \rho \int_{-20}^{20} 5\sqrt[3]{400 - x^2}\, dx \approx 1210.0\rho$

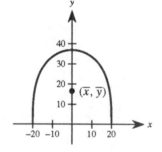

$M_x = \rho \int_{-20}^{20} \frac{5\sqrt[3]{400 - x^2}}{2} \left(5\sqrt[3]{400 - x^2} \right) dx$

$= \frac{25\rho}{2} \int_{-20}^{20} (400 - x^2)^{2/3}\, dx \approx 19,936\rho$

$\bar{y} = \frac{M_x}{m} \approx 16.5$

$\bar{x} = 0$ by symmetry. Therefore, the centroid is (0, 16.5).

36. $m = \rho \int_{-2}^{2} \frac{8}{x^2+4}\, dx \approx 6.2832\rho$

$M_x = \rho \int_{-2}^{2} \frac{1}{2}\left(\frac{8}{x^2+4}\right)\left(\frac{8}{x^2+4}\right) dx = 32\rho \int_{-2}^{2} \frac{1}{(x^2+4)^2}\, dx \approx 5.14149\rho$

$\bar{y} = \dfrac{M_x}{m} \approx 0.8$

$\bar{x} = 0$ by symmetry. Therefore, the centroid is $(0, 0.8)$.

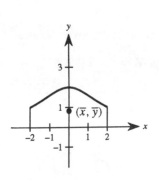

37. $\bar{x} = 0$ by symmetry. Using Simpson's Rule to approximate the area and M_x, we have:

$A \approx \dfrac{40-(-40)}{3(8)}(2)[30 + 4(29) + 2(26) + 4(20) + 0] = \dfrac{5560}{3}$

$m = \rho A = \dfrac{5560\rho}{3}$

$M_x = \rho \int_{-40}^{40} \frac{1}{2}[f(x)]^2$

$\approx \rho \left(\dfrac{40-(-40)}{3(8)}\right)(2)\left(\dfrac{1}{2}\right)[(30)^2 + 4(29)^2 + 2(26)^2 + 4(20)^2 + 0] = \dfrac{72,160\rho}{3}$

$\bar{y} = \dfrac{M_x}{m} = \dfrac{72,160\rho}{3} \cdot \dfrac{3}{5560\rho} \approx 12.98$

Center of mass: $(\bar{x},\ \bar{y}) = (0,\ 12.98)$

38. Assume that the top curve is the semicircle $y = \sqrt{4-x^2}$. By symmetry, $\bar{x} = 0$.
Using Simpson's Rule we have

$A \approx 2\left[\dfrac{2-0}{3(4)}\right][1.50 + 4(1.45) + 2(1.30) + 4.(0.99) + 0] = \dfrac{231}{50} = 4.62$

$m = \rho A = 4.62\rho$

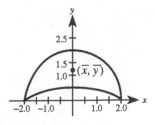

$M_x \approx \rho 2 \left(\dfrac{1}{2}\right)\left[\dfrac{2-0}{3(4)}\right][1.50(2.0) + 4(1.45)(1.93) + 2(1.3)(1.73) + 4(0.99)(1.32) + 0] \approx 3.9865\rho$

$\bar{y} = \dfrac{M_x}{m} \approx 0.86$

Center of mass: $(\bar{x},\ \bar{y}) \approx (0,\ 0.86)$

Note: To approximate $\dfrac{f(x)+g(x)}{2}$ at x_0, evaluate $\sqrt{4-x_0{}^2} - \dfrac{l}{2}$.

39. Centroids of the given regions: $(1, 0)$ and $(3, 0)$

Area: $A = 4 + \pi$

$\bar{x} = \dfrac{4(1) + \pi(3)}{4+\pi} = \dfrac{4+3\pi}{4+\pi}$

$\bar{y} = \dfrac{4(0) + \pi(0)}{4+\pi} = 0$

$(\bar{x},\ \bar{y}) = \left(\dfrac{4+3\pi}{4+\pi},\ 0\right)$

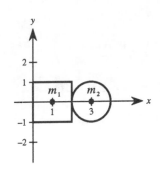

40. Centroids of the given regions: $\left(\frac{1}{2}, \frac{3}{2}\right)$, $\left(2, \frac{1}{2}\right)$, and $\left(\frac{7}{2}, 1\right)$

Area: $A = 3 + 2 + 2 = 7$

$$\bar{x} = \frac{3(1/2) + 2(2) + 2(7/2)}{7} = \frac{25/2}{7} = \frac{25}{14}$$

$$\bar{y} = \frac{3(3/2) + 2(1/2) + 2(1)}{7} = \frac{15/2}{7} = \frac{15}{14}$$

$$(\bar{x}, \bar{y}) = \left(\frac{25}{14}, \frac{15}{14}\right)$$

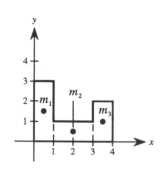

41. Centroids of the given regions: $\left(0, \frac{3}{2}\right)$, $(0, 5)$, and $\left(0, \frac{15}{2}\right)$

Area: $A = 15 + 12 + 7 = 34$

$$\bar{x} = \frac{15(0) + 12(0) + 7(0)}{34} = 0$$

$$\bar{y} = \frac{15(3/2) + 12(5) + 7(15/2)}{34} = \frac{135}{34}$$

$$(\bar{x}, \bar{y}) = \left(0, \frac{135}{34}\right)$$

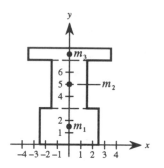

42. $m_1 = \frac{7}{8}(2) = \frac{7}{4}, \quad P_1 = \left(0, \frac{7}{16}\right)$

$m_2 = \frac{7}{8}\left(6 - \frac{7}{8}\right) = \frac{287}{64}, \quad P_2 = \left(0, \frac{55}{16}\right)$

By symmetry, $\bar{x} = 0$.

$$\bar{y} = \frac{(7/4)(7/16) + (287/64)(55/16)}{(7/4) + (287/64)} = \frac{16{,}569}{6384} = \frac{5523}{2128}$$

$$(\bar{x}, \bar{y}) = \left(0, \frac{5523}{2128}\right) \approx (0, 2.595)$$

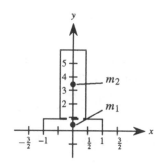

43. Centroids of the given regions: $(1, 0)$ and $(3, 0)$

Mass: $4 + 2\pi$

$$\bar{x} = \frac{4(1) + 2\pi(3)}{4 + 2\pi} = \frac{2 + 3\pi}{2 + \pi}$$

$$\bar{y} = 0$$

$$(\bar{x}, \bar{y}) = \left(\frac{2 + 3\pi}{2 + \pi}, 0\right)$$

44. Centroids of the given regions: $(3, 0)$ and $(1, 0)$

Mass: $8 + \pi$

$$\bar{y} = 0$$

$$\bar{x} = \frac{8(1) + \pi(3)}{8 + \pi} = \frac{8 + 3\pi}{8 + \pi}$$

$$(\bar{x}, \bar{y}) = \left(\frac{8 + 3\pi}{8 + \pi}, 0\right)$$

45. $V = 2\pi r A = 2\pi(5)(16\pi) = 160\pi^2 \approx 1579.14$

46. $V = 2\pi r A = 2\pi(a)(\pi b^2) = 2ab^2\pi^2 = 2a(b\pi)^2$

47. $A = \dfrac{1}{2}(4)(4) = 8$

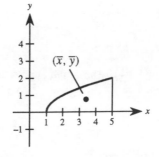

$$\bar{y} = \left(\frac{1}{8}\right)\frac{1}{2}\int_0^4 (4+x)(4-x)\,dx$$

$$= \frac{1}{16}\left[16x - \frac{x^3}{3}\right]_0^4 = \frac{8}{3}$$

$$r = \bar{y} = \frac{8}{3}$$

$$V = 2\pi r A = 2\pi\left(\frac{8}{3}\right)(8) = \frac{128\pi}{3} \approx 134.04$$

48. $A = \displaystyle\int_1^5 \sqrt{x-1}\,dx = \left[\frac{2}{3}(x-1)^{3/2}\right]_1^5 = \frac{16}{3}$

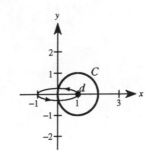

$$\bar{x} = \frac{3}{16}\int_1^5 x\sqrt{x-1}\,dx$$

$$= \frac{3}{16}\left[\frac{2}{5}(x-1)^{5/2} + \frac{2}{3}(x-1)^{3/2}\right]_1^5 = \frac{17}{5}$$

$$r = \bar{x} = \frac{17}{5}$$

$$V = 2\pi r A = 2\pi\left(\frac{17}{5}\right)\left(\frac{16}{3}\right) = \frac{544\pi}{15} \approx 113.94$$

49. The surface area of the sphere is $S = 4\pi r^2$.
The arc length of C is $s = \pi r$.
The distance traveled by the centroid is

$$d = \frac{S}{s} = \frac{4\pi r^2}{\pi r} = 4r.$$

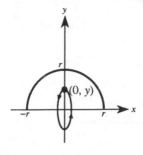

This distance is also the circumference of the circle of radius y.

$$d = 2\pi y$$

Thus, $2\pi y = 4r$ and we have $y = 2r/\pi$. Therefore, the centroid of the semicircle $y = \sqrt{r^2 - x^2}$ is $(0,\ 2r/\pi)$.

50. The centroid is $(1, 0)$.
The distance traveled by the centroid is 2π.
The arc length of the circle is also 2π.
Therefore, $S = (2\pi)(2\pi) = 4\pi^2$.

51. $A = \int_0^1 x^n \, dx = \left[\dfrac{x^{n+1}}{n+1} \right]_0^1 = \dfrac{1}{n+1}$

$m = \rho A = \dfrac{\rho}{n+1}$

$M_x = \dfrac{\rho}{2} \int_0^1 (x^n)^2 \, dx = \left[\dfrac{\rho}{2} \cdot \dfrac{x^{2n+1}}{2n+1} \right]_0^1 = \dfrac{\rho}{2(2n+1)}$

$M_y = \rho \int_0^1 x(x^n) \, dx = \left[\rho \cdot \dfrac{x^{n+2}}{n+2} \right]_0^1 = \dfrac{\rho}{n+2}$

$\bar{x} = \dfrac{M_y}{m} = \dfrac{n+1}{n+2}$

$\bar{y} = \dfrac{M_x}{m} = \dfrac{n+1}{2(2n+1)} = \dfrac{n+1}{4n+2}$

Centroid: $\left(\dfrac{n+1}{n+2}, \dfrac{n+1}{4n+2} \right)$

As $n \to \infty$, $(\bar{x}, \bar{y}) \to \left(1, \frac{1}{4}\right)$. The graph approaches the x-axis and the line $x = 1$ as $n \to \infty$.

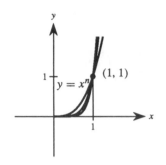

Chapter 6 Review Exercises

1. $A = \int_1^5 \dfrac{1}{x^2} \, dx = \left[-\dfrac{1}{x} \right]_1^5 = \dfrac{4}{5}$

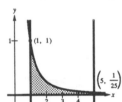

2. $A = \int_{1/2}^5 \left(4 - \dfrac{1}{x^2} \right) dx$

$= \left[4x + \dfrac{1}{x} \right]_{1/2}^5 = \dfrac{81}{5}$

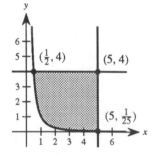

3. $A = \int_{-1}^1 \dfrac{1}{x^2+1} \, dx$

$= \left[\arctan x \right]_{-1}^1$

$= \dfrac{\pi}{4} - \left(-\dfrac{\pi}{4} \right) = \dfrac{\pi}{2}$

4. $A = \int_0^1 [(y^2 - 2y) - (-1)] \, dy$

$= \int_0^1 (y^2 - 2y + 1) \, dy$

$= \int_0^1 (y - 1)^2 \, dy$

$= \left[\dfrac{(y-1)^3}{3} \right]_0^1 = \dfrac{1}{3}$

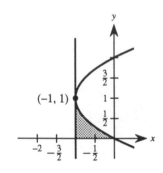

5. $A = 2 \int_0^1 (x - x^3)\, dx$

$\quad = 2 \left[\dfrac{1}{2}x^2 - \dfrac{1}{4}x^4 \right]_0^1$

$\quad = \dfrac{1}{2}$

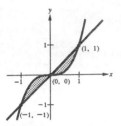

6. $A = \int_{-1}^2 [(y + 3) - (y^2 + 1)]\, dy$

$\quad = \int_{-1}^2 (2 + y - y^2)\, dy$

$\quad = \left[2y + \dfrac{1}{2}y^2 - \dfrac{1}{3}y^3 \right]_{-1}^2 = \dfrac{9}{2}$

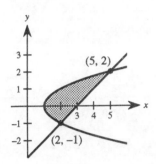

7. $A = \int_0^8 [(3 + 8x - x^2) - (x^2 - 8x + 3)]\, dx$

$\quad = \int_0^8 (16x - 2x^2)\, dx$

$\quad = \left[8x^2 - \dfrac{2}{3}x^3 \right]_0^8 = \dfrac{512}{3}$

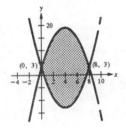

8. Point of intersection is given by:

$\quad x^3 - x^2 + 4x - 3 = 0 \Rightarrow x \approx 0.783.$

$\quad A \approx \int_0^{0.783} (3 - 4x + x^2 - x^3)\, dx$

$\quad = \left[3x - 2x^2 + \dfrac{1}{3}x^3 - \dfrac{1}{4}x^4 \right]_0^{0.783}$

$\quad \approx 1.189$

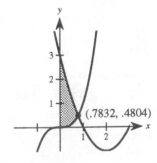

9. $y = (1 - \sqrt{x})^2$

$\quad A = \int_0^1 (1 - \sqrt{x})^2\, dx$

$\quad = \int_0^1 (1 - 2x^{1/2} + x)\, dx$

$\quad = \left[x - \dfrac{4}{3}x^{3/2} + \dfrac{1}{2}x^2 \right]_0^1 = \dfrac{1}{6}$

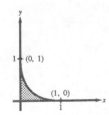

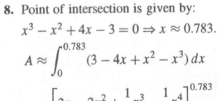

10. $A = 2 \int_0^2 [2x^2 - (x^4 - 2x^2)] \, dx$

$= 2 \int_0^2 (4x^2 - x^4) \, dx$

$= 2 \left[\frac{4}{3}x^3 - \frac{1}{5}x^5 \right]_0^2 = \frac{128}{15}$

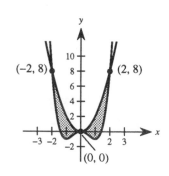

11. $A = \int_0^2 (e^2 - e^x) \, dx$

$= \left[xe^2 - e^x \right]_0^2$

$= e^2 + 1$

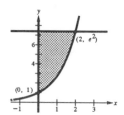

12. $A = 2 \int_{\pi/6}^{\pi/2} (2 - \csc x) \, dx$

$= 2 \left[2x - \ln|\csc x - \cot x| \right]_{\pi/6}^{\pi/2}$

$= 2 \left([\pi - 0] - \left[\frac{\pi}{3} - \ln(2 - \sqrt{3}) \right] \right)$

$= 2 \left[\frac{2\pi}{3} + \ln(2 - \sqrt{3}) \right] \approx 1.555$

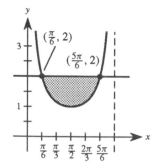

13. $A = \int_{\pi/4}^{5\pi/4} (\sin x - \cos x) \, dx$

$= \left[-\cos x - \sin x \right]_{\pi/4}^{5\pi/4}$

$= \left(\frac{1}{\sqrt{2}} + \frac{1}{\sqrt{2}} \right) - \left(-\frac{1}{\sqrt{2}} - \frac{1}{\sqrt{2}} \right)$

$= \frac{4}{\sqrt{2}} = 2\sqrt{2}$

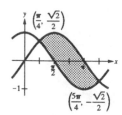

14. $A = \int_{\pi/3}^{5\pi/3} \left(\frac{1}{2} - \cos y \right) dy + \int_{5\pi/3}^{7\pi/3} \left(\cos y - \frac{1}{2} \right) dy$

$= \left[\frac{y}{2} - \sin y \right]_{\pi/3}^{5\pi/3} + \left[\sin y - \frac{y}{2} \right]_{5\pi/3}^{7\pi/3}$

$= \frac{\pi}{3} + 2\sqrt{3}$

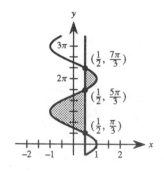

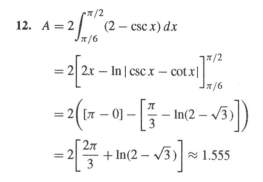

15. $x = y^2 - 2y \Rightarrow x + 1 = (y-1)^2 \Rightarrow y = 1 \pm \sqrt{x+1}$

$$A = \int_{-1}^{0} \left[\left(1 + \sqrt{x+1}\right) - \left(1 - \sqrt{x+1}\right) \right] dx = \int_{-1}^{0} 2\sqrt{x+1}\, dx$$

$$A = \int_{0}^{2} [0 - (y^2 - 2y)]\, dy = \int_{0}^{2} (2y - y^2)\, dy = \left[y^2 - \frac{1}{3}y^3 \right]_{0}^{2} = \frac{4}{3}$$

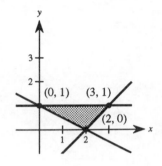

16. $\displaystyle A = \int_{0}^{2} \left[1 - \left(1 - \frac{x}{2}\right) \right] dx + \int_{2}^{3} [1 - (x-2)]\, dx$

$$= \int_{0}^{2} \frac{x}{2}\, dx + \int_{2}^{3} (3 - x)\, dx$$

$$y = 1 - \frac{x}{2} \Rightarrow x = 2 - 2y$$

$$y = x - 2 \Rightarrow x = y + 2, \quad y = 1$$

$$A = \int_{0}^{1} [(y+2) - (2 - 2y)]\, dy$$

$$= \int_{0}^{1} 3y\, dy = \left[\frac{3}{2} y^2 \right]_{0}^{1} = \frac{3}{2}$$

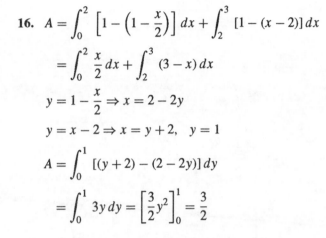

17. $\displaystyle A = \int_{0}^{1} 2\, dx + \int_{1}^{5} \left[2 - \sqrt{x-1} \right] dx$

$$x = y^2 + 1$$

$$A = \int_{0}^{2} (y^2 + 1)\, dy$$

$$= \left[\frac{1}{3} y^3 + y \right]_{0}^{2} = \frac{14}{3}$$

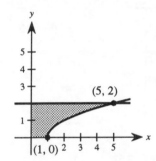

18. $y = \sqrt{x-1} \Rightarrow x = y^2 + 1$

$$y = \frac{x-1}{2} \Rightarrow x = 2y + 1$$

$$A = \int_{0}^{2} [(2y+1) - (y^2 + 1)]\, dy$$

$$= \int_{1}^{5} \left[\sqrt{x-1} - \frac{x-1}{2} \right] dx$$

$$= \left[\frac{2}{3}(x-1)^{3/2} - \frac{1}{4}(x-1)^2 \right]_{1}^{5} = \frac{4}{3}$$

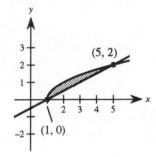

19. a. Disc

$$V = \pi \int_0^4 x^2 \, dx = \left[\frac{\pi x^3}{3} \right]_0^4 = \frac{64\pi}{3}$$

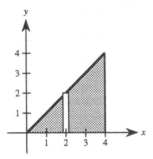

b. Shell

$$V = 2\pi \int_0^4 x^2 \, dx = \left[\frac{2\pi}{3} x^3 \right]_0^4 = \frac{128\pi}{3}$$

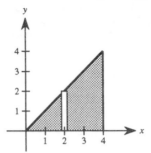

c. Shell

$$V = 2\pi \int_0^4 (4 - x)x \, dx$$

$$= 2\pi \int_0^4 (4x - x^2) \, dx$$

$$= 2\pi \left[2x^2 - \frac{x^3}{3} \right]_0^4 = \frac{64\pi}{3}$$

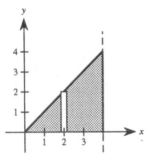

d. Shell

$$V = 2\pi \int_0^4 (6 - x)x \, dx$$

$$= 2\pi \int_0^4 (6x - x^2) \, dx$$

$$= 2\pi \left[3x^2 - \frac{1}{3}x^3 \right]_0^4 = \frac{160\pi}{3}$$

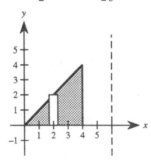

20. a. Shell

$$V = 2\pi \int_0^2 y^3 \, dy = \left[\frac{\pi}{2} y^4 \right]_0^2 = 8\pi$$

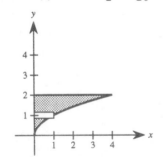

b. Shell

$$V = 2\pi \int_0^2 (2 - y)y^2 \, dy$$

$$= 2\pi \int_0^2 (2y^2 - y^3) \, dy$$

$$= 2\pi \left[\frac{2}{3}y^3 - \frac{1}{4}y^4 \right]_0^2 = \frac{8\pi}{3}$$

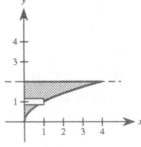

20. –CONTINUED–

c. Disc

$$V = \pi \int_0^2 y^4 \, dy$$

$$= \left[\frac{\pi}{5} y^5 \right]_0^2 = \frac{32\pi}{5}$$

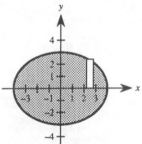

d. Disc

$$V = \pi \int_0^2 [(y^2 + 1)^2 - 1^2] \, dy$$

$$= \pi \int_0^2 (y^4 + 2y^2) \, dy$$

$$= \pi \left[\frac{1}{5} y^5 + \frac{2}{3} y^3 \right]_0^2 = \frac{176\pi}{15}$$

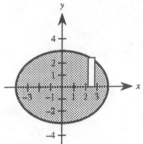

21. a. Shell

$$V = 4\pi \int_0^4 x \left(\frac{3}{4} \right) \sqrt{16 - x^2} \, dx$$

$$= \left[3\pi \left(-\frac{1}{2} \right) \left(\frac{2}{3} \right) (16 - x^2)^{3/2} \right]_0^4 = 64\pi$$

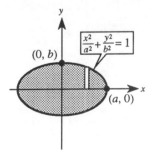

b. Disc

$$V = 2\pi \int_0^4 \left[\frac{3}{4} \sqrt{16 - x^2} \right]^2 \, dx$$

$$= \frac{9\pi}{8} \left[16x - \frac{x^3}{3} \right]_0^4 = 48\pi$$

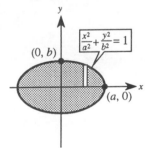

22. a. Shell

$$V = 4\pi \int_0^a (x) \frac{b}{a} \sqrt{a^2 - x^2} \, dx$$

$$= \frac{-2\pi b}{a} \int_0^a (a^2 - x^2)^{1/2} (-2x) \, dx$$

$$= \left[\frac{-4\pi b}{3a} (a^2 - x^2)^{3/2} \right]_0^a = \frac{4}{3} \pi a^2 b$$

b. Disc

$$V = 2\pi \int_0^a \frac{b^2}{a^2} (a^2 - x^2) \, dx$$

$$= \frac{2\pi b^2}{a^2} \left[a^2 x - \frac{1}{3} x^3 \right]_0^a$$

$$= \frac{4}{3} \pi a b^2$$

23. Shell

$$V = 2\pi \int_0^1 \frac{x}{x^4 + 1} \, dx$$

$$= \pi \int_0^1 \frac{(2x)}{(x^2)^2 + 1} \, dx$$

$$= \left[\pi \arctan(x^2) \right]_0^1$$

$$= \pi \left[\frac{\pi}{4} - 0 \right] = \frac{\pi^2}{4}$$

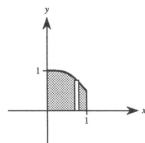

24. Disc

$$V = 2\pi \int_0^1 \left[\frac{1}{\sqrt{1 + x^2}} \right]^2 dx$$

$$= \left[2\pi \arctan x \right]_0^1$$

$$= 2\pi \left(\frac{\pi}{4} - 0 \right)$$

$$= \frac{\pi^2}{2}$$

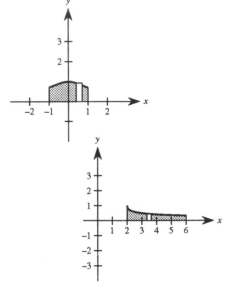

25. Shell

$$u = \sqrt{x - 2}$$

$$x = u^2 + 2$$

$$dx = 2u \, du$$

$$V = 2\pi \int_2^6 \frac{x}{1 + \sqrt{x - 2}} \, dx = 4\pi \int_0^2 \frac{(u^2 + 2)u}{1 + u} \, du$$

$$= 4\pi \int_0^2 \frac{u^3 + 2u}{1 + u} \, du = 4\pi \int_0^2 \left(u^2 - u + 3 - \frac{3}{1 + u} \right) du$$

$$= 4\pi \left[\frac{1}{3}u^3 - \frac{1}{2}u^2 + 3u - 3\ln(1 + u) \right]_0^2 = \frac{4\pi}{3}(20 - 9\ln 3) \approx 42.359$$

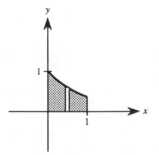

26. Disc

$$V = \pi \int_0^1 (e^{-x})^2 \, dx$$

$$= \pi \int_0^1 e^{-2x} \, dx = \left[-\frac{\pi}{2}e^{-2x} \right]_0^1$$

$$= \left(\frac{-\pi}{2e^2} + \frac{\pi}{2} \right) = \frac{\pi}{2}\left(1 - \frac{1}{e^2} \right)$$

27. Since $y \le 0$, $A = -\displaystyle\int_{-1}^{0} x\sqrt{x+1}\,dx$.

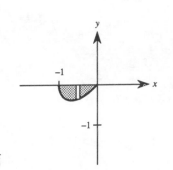

$u = x + 1$

$x = u - 1$

$dx = du$

$A = -\displaystyle\int_{0}^{1} (u-1)\sqrt{u}\,du = -\int_{0}^{1} (u^{3/2} - u^{1/2})\,du = -\left[\frac{2}{5}u^{5/2} - \frac{2}{3}u^{3/2}\right]_{0}^{1} = \frac{4}{15}$

28. a. Disc

$V = \pi\displaystyle\int_{-1}^{0} x^2(x+1)\,dx$

$= \pi\displaystyle\int_{-1}^{0} (x^3 + x^2)\,dx$

$= \pi\left[\dfrac{x^4}{4} + \dfrac{x^3}{3}\right]_{-1}^{0} = \dfrac{\pi}{12}$

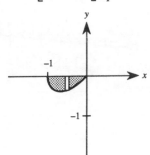

b. Shell

$u = \sqrt{x+1}$

$x = u^2 - 1$

$dx = 2u\,du$

$V = 2\pi\displaystyle\int_{-1}^{0} x^2\sqrt{x+1}\,dx$

$= 4\pi\displaystyle\int_{0}^{1} (u^2 - 1)^2 u^2\,du$

$= 4\pi\displaystyle\int_{0}^{1} (u^6 - 2u^4 + u^2)\,du$

$= 4\pi\left[\dfrac{1}{7}u^7 - \dfrac{2}{5}u^5 + \dfrac{1}{3}u^3\right]_{0}^{1} = \dfrac{32\pi}{105}$

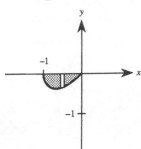

29. From Exercise 21a we have: $\qquad V = 64\pi \ \text{ft}^3$

$\dfrac{1}{4}V = 16\pi$

Disc: $\qquad \pi\displaystyle\int_{-3}^{y_0} \frac{16}{9}(9 - y^2)\,dy = 16\pi$

$\dfrac{1}{9}\displaystyle\int_{-3}^{y_0} (9 - y^2)\,dy = 1$

$\left[9y - \dfrac{1}{3}y^3\right]_{-3}^{y_0} = 9$

$\left(9y_0 - \dfrac{1}{3}y_0^3\right) - (-27 + 9) = 9$

$y_0^3 - 27y_0 - 27 = 0$

By Newton's Method, $y_0 \approx -1.042$ and the depth of the gasoline is $3 - 1.042 = 1.958$ ft.

30. $A(x) = \frac{1}{2}bh = \frac{1}{2}\left(2\sqrt{a^2 - x^2}\right)\left(\sqrt{3}\sqrt{a^2 - x^2}\right) = \sqrt{3}(a^2 - x^2)$

$$V = \sqrt{3}\int_{-a}^{a}(a^2 - x^2)\,dx = \sqrt{3}\left[a^2x - \frac{x^3}{3}\right]_{-a}^{a} = \sqrt{3}\left(\frac{4a^3}{3}\right)$$

Since $(4\sqrt{3}\,a^3)/3 = 10$, we have $a^3 = (5\sqrt{3})/2$. Thus,

$$a = \sqrt[3]{\frac{5\sqrt{3}}{2}} \approx 1.630 \text{ inches.}$$

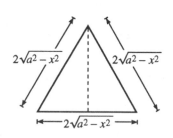

31. $f(x) = \frac{4}{5}x^{5/4}$

$f'(x) = x^{1/4}$

$1 + [f'(x)]^2 = 1 + \sqrt{x}$

$u = 1 + \sqrt{x}$

$x = (u - 1)^2$

$dx = 2(u - 1)\,du$

$$s = \int_{0}^{4}\sqrt{1 + \sqrt{x}}\,dx = 2\int_{1}^{3}\sqrt{u}(u - 1)\,du = 2\int_{1}^{3}(u^{3/2} - u^{1/2})\,du$$

$$= 2\left[\frac{2}{5}u^{5/2} - \frac{2}{3}u^{3/2}\right]_{1}^{3} = \frac{4}{15}\left[u^{3/2}(3u - 5)\right]_{1}^{3} = \frac{8}{15}(1 + 6\sqrt{3}) \approx 6.076$$

32. $y = \frac{x^3}{6} + \frac{1}{2x}$

$y' = \frac{1}{2}x^2 - \frac{1}{2x^2}$

$1 + (y')^2 = \left(\frac{1}{2}x^2 + \frac{1}{2x^2}\right)^2$

$$s = \int_{1}^{3}\left(\frac{1}{2}x^2 + \frac{1}{2x^2}\right)dx = \left[\frac{1}{6}x^3 - \frac{1}{2x}\right]_{1}^{3} = \frac{14}{3}$$

33. $y = \frac{3}{4}x$

$y' = \frac{3}{4}$

$1 + (y')^2 = \frac{25}{16}$

$$S = 2\pi\int_{0}^{4}\left(\frac{3}{4}x\right)\sqrt{\frac{25}{16}}\,dx = \left[\left(\frac{15\pi}{8}\right)\frac{x^2}{2}\right]_{0}^{4} = 15\pi$$

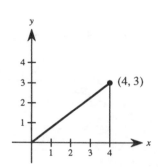

34.
$$y = 2\sqrt{x}$$

$$y' = \frac{1}{\sqrt{x}}$$

$$1 + (y')^2 = 1 + \frac{1}{x} = \frac{x+1}{x}$$

$$S = 2\pi \int_0^3 2\sqrt{x}\sqrt{\frac{x+1}{x}}\, dx = 4\pi \int_0^3 \sqrt{x+1}\, dx = 4\pi\left[\left(\frac{2}{3}\right)(x+1)^{3/2}\right]_0^3 = \frac{56\pi}{3}$$

35. $F = kx$

$4 = k(1)$

$F = 4x$

$$W = \int_0^5 4x\, dx = \left[2x^2\right]_0^5$$

$$= 50 \text{ in} \cdot \text{lb} \approx 4.167 \text{ ft} \cdot \text{lb}$$

36. $F = kx$

$$50 = k(9) \Rightarrow k = \frac{50}{9}$$

$$F = \frac{50}{9}x$$

$$W = \int_0^9 \frac{50}{9}x\, dx = \left[\frac{25}{9}x^2\right]_0^9$$

$$= 225 \text{ in} \cdot \text{lb} = 18.75 \text{ ft} \cdot \text{lb}$$

37. Volume of disc: $\pi\left(\dfrac{1}{3}\right)^2 \Delta y$

Weight of disc: $62.4\pi\left(\dfrac{1}{3}\right)^2 \Delta y$

Distance: $175 - y$

$$W = \frac{62.4\pi}{9}\int_0^{150}(175 - y)\, dy = \frac{62.4\pi}{9}\left[175y - \frac{y^2}{2}\right]_0^{150} = 104{,}000\pi \text{ ft} \cdot \text{lb} \approx 163.4 \text{ ft} \cdot \text{ton}$$

38. We know that
$$\frac{dV}{dt} = \frac{4 \text{ gal/min} - 12 \text{ gal/min}}{7.481 \text{ gal/ft}^3} = -\frac{8}{7.481} \text{ ft}^3/\text{min}$$

$$V = \pi r^2 h = \pi\left(\frac{1}{9}\right)h$$

$$\frac{dV}{dt} = \frac{\pi}{9}\left(\frac{dh}{dt}\right)$$

$$\frac{dh}{dt} = \frac{9}{\pi}\left(\frac{dV}{dt}\right) = \frac{9}{\pi}\left(-\frac{8}{7.481}\right) \approx -3.064 \text{ ft/min}$$

Depth of water: $-3.064t + 150$

Time to drain well: $t = \dfrac{150}{3.064} \approx 49$ minutes

$(49)(12) = 588$ gallons pumped

Volume of water pumped in Exercise 37: 391.7 gallons

$$\frac{391.7}{52\pi} = \frac{588}{x\pi}$$

$$x = \frac{588(52)}{391.7} \approx 78$$

Work $\approx 78\pi$ ft $\cdot$ ton

39. Weight of section of chain: $5\,\Delta x$

Distance moved: $10 - x$

$$W = 5\int_0^{10} (10 - x)\,dx = \left[-\frac{5}{2}(10 - x)^2\right]_0^{10} = 250 \text{ ft} \cdot \text{lb}$$

40. a. Weight of section of cable: $4\,\Delta x$

Distance: $200 - x$

$$W = 4\int_0^{200} (200 - x)\,dx = \left[-2(200 - x)^2\right]_0^{200} = 80,000 \text{ ft} \cdot \text{lb} = 40 \text{ ft} \cdot \text{ton}$$

b. Work to move 300 pounds 200 feet vertically: $200(300) = 60,000$ ft $\cdot$ lb $= 30$ ft $\cdot$ ton

Total work = work for drawing up the cable + work of lifting the load

$$= 40 \text{ ft} \cdot \text{ton} + 30 \text{ ft} \cdot \text{ton} = 70 \text{ ft} \cdot \text{ton}$$

41. Wall at shallow end:

$$F = 62.4\int_0^5 y(20)\,dy = \left[(1248)\frac{y^2}{2}\right]_0^5 = 15,600 \text{ lb}$$

Wall at deep end:

$$F = 62.4\int_0^{10} y(20)\,dy = \left[(624)y^2\right]_0^{10} = 62,400 \text{ lb}$$

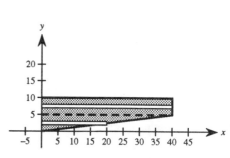

Side wall:

$$F_1 = 62.4\int_0^5 y(40)\,dy = \left[(1248)y^2\right]_0^5 = 31,200 \text{ lb}$$

$$F_2 = 62.4\int_0^5 (10 - y)8y\,dy = 62.4\int_0^5 (80y - 8y^2)\,dy = 62.4\left[40y^2 - \frac{8}{3}y^3\right]_0^5 = 41,600 \text{ lb}$$

$$F = F_1 + F_2 = 72,800 \text{ lb}$$

42. Let $D =$ surface of liquid; $\rho =$ weight per cubic volume.

$$F = \rho\int_c^d (D - y)[f(y) - g(y)]\,dy$$

$$= \rho\left[\int_c^d D[f(y) - g(y)]\,dy - \int_c^d y[f(y) - g(y)]\,dy\right]$$

$$= \rho\left[\int_c^d [f(y) - g(y)]\,dy\right]\left[D - \frac{\int_c^d y[f(y) - g(y)]\,dy}{\int_c^d [f(y) - g(y)]\,dy}\right]$$

$$= \rho(\text{Area})(D - \bar{y})$$

$$= \rho(\text{Area})(\text{depth of centroid})$$

43. $F = 62.4(16\pi)5 = 4992\pi$ lb

44. Raise water level 5 more feet.

45.

$$A = \int_0^a (\sqrt{a} - \sqrt{x})^2 \, dx = \int_0^a (a - 2\sqrt{a}\, x^{1/2} + x) \, dx = \left[ax - \frac{4}{3}\sqrt{a}\, x^{3/2} + \frac{1}{2}x^2 \right]_0^a = \frac{a^2}{6}$$

$$\frac{1}{A} = \frac{6}{a^2}$$

$$\bar{x} = \frac{6}{a^2} \int_0^a x(\sqrt{a} - \sqrt{x})^2 \, dx = \frac{6}{a^2} \int_0^a (ax - 2\sqrt{a}\, x^{3/2} + x^2) \, dx = \frac{6}{a^2} \left[\frac{ax^2}{2} - \frac{4}{5}\sqrt{a}\, x^{5/2} + \frac{1}{3}x^3 \right]_0^a = \frac{a}{5}$$

$$\bar{y} = \left(\frac{6}{a^2} \right) \frac{1}{2} \int_0^a (\sqrt{a} - \sqrt{x})^4 \, dx$$

$$= \frac{3}{a^2} \int_0^a (a^2 - 4a^{3/2}x^{1/2} + 6ax - 4a^{1/2}x^{3/2} + x^2) \, dx$$

$$= \frac{3}{a^2} \left[a^2 x - \frac{8}{3}a^{3/2}x^{3/2} + 3ax^2 - \frac{8}{5}a^{1/2}x^{5/2} + \frac{1}{3}x^3 \right]_0^a = \frac{a}{5}$$

$$(\bar{x},\ \bar{y}) = \left(\frac{a}{5},\ \frac{a}{5} \right)$$

46.

$$A = \int_{-1}^3 [(2x+3) - x^2] \, dx = \left[x^2 + 3x - \frac{1}{3}x^3 \right]_{-1}^3 = \frac{32}{3}$$

$$\frac{1}{A} = \frac{3}{32}$$

$$\bar{x} = \frac{3}{32} \int_{-1}^3 x(2x + 3 - x^2) \, dx = \frac{3}{32} \int_{-1}^3 (3x + 2x^2 - x^3) \, dx = \frac{3}{32} \left[\frac{3}{2}x^2 + \frac{2}{3}x^3 - \frac{1}{4}x^4 \right]_{-1}^3 = 1$$

$$\bar{y} = \left(\frac{3}{32} \right) \frac{1}{2} \int_{-1}^3 [(2x+3)^2 - x^4] \, dx$$

$$= \frac{3}{64} \int_{-1}^3 (9 + 12x + 4x^2 - x^4) \, dx$$

$$= \frac{3}{64} \left[9x + 6x^2 + \frac{4}{3}x^3 - \frac{1}{5}x^5 \right]_{-1}^3 = \frac{17}{5}$$

$$(\bar{x},\ \bar{y}) = \left(1,\ \frac{17}{5} \right)$$

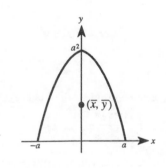

47. By symmetry, $x = 0$.

$$A = 2 \int_0^1 (a^2 - x^2) \, dx = 2 \left[a^2 x - \frac{x^3}{3} \right]_0^a = \frac{4a^3}{3}$$

$$\frac{1}{A} = \frac{3}{4a^3}$$

$$\bar{y} = \left(\frac{3}{4a^3} \right) \frac{1}{2} \int_{-a}^a (a^2 - x^2)^2 \, dx = \frac{6}{8a^3} \int_0^a (a^4 - 2a^2 x^2 + x^4) \, dx$$

$$= \frac{6}{8a^3} \left[a^4 x - \frac{2a^2}{3}x^3 + \frac{1}{5}x^5 \right]_0^a = \frac{6}{8a^3} \left(a^5 - \frac{2}{3}a^5 + \frac{1}{5}a^5 \right) = \frac{2a^2}{5}$$

$$(\bar{x},\ \bar{y}) = \left(0,\ \frac{2a^2}{5} \right)$$

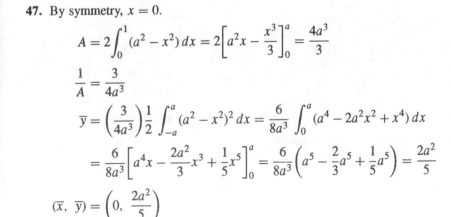

48.
$$A = \int_0^8 \left(x^{2/3} - \frac{1}{2}x\right) dx = \left[\frac{3}{5}x^{5/3} - \frac{1}{4}x^2\right]_0^8 = \frac{16}{5}$$

$$\frac{1}{A} = \frac{5}{16}$$

$$\bar{x} = \frac{5}{16}\int_0^8 x\left(x^{2/3} - \frac{1}{2}x\right) dx = \frac{5}{16}\left[\frac{3}{8}x^{8/3} - \frac{1}{6}x^3\right]_0^8 = \frac{10}{3}$$

$$\bar{y} = \left(\frac{5}{16}\right)\frac{1}{2}\int_0^8 \left(x^{4/3} - \frac{1}{4}x^2\right) dx = \frac{1}{2}\left(\frac{5}{16}\right)\left[\frac{3}{7}x^{7/3} - \frac{1}{12}x^3\right]_0^8 = \frac{40}{21}$$

$$(\bar{x}, \bar{y}) = \left(\frac{10}{3}, \frac{40}{21}\right)$$

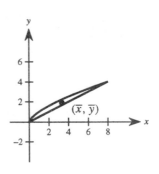

49. $\bar{y} = 0$ by symmetry

For the trapezoid:
$$m = [(4)(6) - (1)(6)]\rho = 18\rho$$

$$M_y = \rho\int_0^6 x\left[\left(\frac{1}{6}x + 1\right) - \left(-\frac{1}{6}x - 1\right)\right] dx$$

$$= \rho\int_0^6 \left(\frac{1}{3}x^2 + 2x\right) dx = \rho\left[\frac{x^3}{9} + x^2\right]_0^6 = 60\rho$$

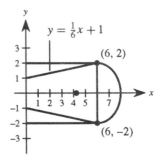

For the semicircle:
$$m = \left(\frac{1}{2}\right)(\pi)(2)^2\rho = 2\pi\rho$$

$$M_y = \rho\int_6^8 x\left[\sqrt{4 - (x - 6)^2} - \left(-\sqrt{4 - (x - 6)^2}\right)\right] dx = 2\rho\int_6^8 x\sqrt{4 - (x - 6)^2}\, dx$$

Let $u = x - 6$, then $x = u + 6$ and $dx = du$. When $x = 6$, $u = 0$. When $x = 8$, $u = 2$.

$$M_y = 2\rho\int_0^2 (u + 6)\sqrt{4 - u^2}\, du = 2\rho\int_0^2 u\sqrt{4 - u^2}\, du + 12\rho\int_0^2 \sqrt{4 - u^2}\, du$$

$$= 2\rho\left[\left(-\frac{1}{2}\right)\left(\frac{2}{3}\right)(4 - u^2)^{3/2}\right]_0^2 + 12\rho\left[\frac{\pi(2)^2}{4}\right] = \frac{16\rho}{3} + 12\pi\rho = \frac{4\rho(4 + 9\pi)}{3}$$

Thus, we have:
$$\bar{x}(18\rho + 2\pi\rho) = 60\rho + \frac{4\rho(4 + 9\pi)}{3}$$

$$\bar{x} = \frac{180\rho + 4\rho(4 + 9\pi)}{3} \cdot \frac{1}{2\rho(9 + \pi)} = \frac{2(9\pi + 49)}{3(\pi + 9)}$$

The centroid of the blade is
$$\left(\frac{2(9\pi + 49)}{3(\pi + 9)}, 0\right).$$

50. The numerical centroid should agree with the centroid located in part a.

51. $y = 300\cosh\left(\dfrac{x}{2000}\right) - 280, \quad -2000 \le x \le 2000$

$$y' = \frac{3}{20}\sinh\left(\frac{x}{2000}\right)$$

$$s = \int_{-2000}^{2000} \sqrt{1 + \left[\frac{3}{20}\sinh\left(\frac{x}{2000}\right)\right]^2}\, dx = \frac{1}{20}\int_{-2000}^{2000} \sqrt{400 + 9\sinh^2\left(\frac{x}{2000}\right)}\, dx \approx 4018.2 \text{ ft (by Simpson's Rule)}$$